CHILTON'S

Covers all U.S. and Canadian models of
Dodge Ram 1500, 2500 and 3500 Pick-Ups, Dakota and Durango;
2 and 4 wheel drive; gasoline and Cummins diesel engines

by Joe Pellicciotti

PUBLISHED BY **HAYNES NORTH AMERICA**, Inc.

Manufactured in USA
© 2001 Haynes North America, Inc.
ISBN 1 56392 416 1
Library of Congress Control No. 2001092095

Haynes Publishing Group
Sparkford Nr Yeovil
Somerset BA22 7JJ England

Haynes North America, Inc
861 Lawrence Drive
Newbury Park
California 91320 USA

ABCDE
FGHIJ
KLMNO
PQR

Chilton is a registered trademark of W.G. Nichols, Inc., and has been licensed to Haynes North America, Inc.

Contents

1 GENERAL INFORMATION AND MAINTENANCE

1-2	HOW TO USE THIS BOOK	1-36	FLUIDS AND LUBRICANTS
1-4	SERVICING YOUR VEHICLE	1-49	TRAILER TOWING
1-5	FASTENERS, MEASUREMENTS AND CONVERSIONS	1-50	JUMP STARTING A DEAD BATTERY
1-10	SERIAL NUMBER IDENTIFICATION	1-52	MAINTENANCE INTERVAL CHARTS
1-12	ROUTINE MAINTENANCE AND TUNE-UP	1-54	CAPACITIES

2 ENGINE ELECTRICAL

2-9	CHARGING SYSTEM	2-8	FIRING ORDERS
2-2	DISTRIBUTOR IGNITION SYSTEM	2-2	PDC & PCM LOCATIONS
2-7	DISTRIBUTORLESS IGNITION - 8.0L ENGINE	2-15	SENDING UNITS AND SENSORS
2-6	DISTRIBUTORLESS IGNITION SYSTEM - 4.7L ENGINE	2-12	STARTING SYSTEM

3 ENGINE AND ENGINE OVERHAUL

3-2	ENGINE MECHANICAL	3-49	ENGINE RECONDITIONING
3-47	EXHAUST SYSTEM		

4 DRIVEABILITY AND EMISSION CONTROLS

4-2	AIR POLLUTION	4-21	TROUBLE CODES
4-3	AUTOMOTIVE EMISSIONS	4-26	CLEARING CODES
4-4	EMISSION CONTROLS	4-27	VACUUM DIAGRAMS
4-12	ELECTRONIC ENGINE CONTROLS		

5 FUEL SYSTEM

5-2	BASIC FUEL SYSTEM DIAGNOSIS	5-10	DIESEL FUEL SYSTEM
5-2	FUEL LINES AND FITTINGS	5-17	FUEL TANK
5-4	GASOLINE FUEL INJECTION SYSTEM		

6 CHASSIS ELECTRICAL

6-7	AIR BAG (SUPPLEMENTAL RESTRAINT SYSTEM)	6-17	INSTRUMENTS AND SWITCHES
6-7	BATTERY CABLES	6-18	LIGHTING
6-26	CIRCUIT PROTECTION	6-25	TRAILER WIRING
6-11	CRUISE CONTROL	6-2	UNDERSTANDING AND TROUBLESHOOTING ELECTRICAL SYSTEMS
6-12	ENTERTAINMENT SYSTEMS		
6-27	FUSES	6-14	WINDSHIELD WIPER SYSTEM
6-9	HEATING AND AIR CONDITIONING	6-29	WIRING DIAGRAMS

Contents

7 DRIVE TRAIN

- **7-2** MANUAL TRANSMISSION
- **7-4** CLUTCH
- **7-8** AUTOMATIC TRANSMISSION
- **7-12** TRANSFER CASE
- **7-14** DRIVELINE
- **7-19** FRONT DRIVE AXLE
- **7-25** REAR AXLE

8 SUSPENSION AND STEERING

- **8-2** WHEELS
- **8-4** FRONT SUSPENSION
- **8-16** REAR SUSPENSION
- **8-18** STEERING

9 BRAKES

- **9-2** BRAKE OPERATING SYSTEM
- **9-12** DISC BRAKES
- **9-17** DRUM BRAKES
- **9-23** PARKING BRAKE
- **9-27** REAR WHEEL ANTILOCK BRAKE SYSTEM
- **9-29** FOUR WHEEL ANTILOCK BRAKE SYSTEM

10 BODY AND TRIM

- **10-2** EXTERIOR
- **10-7** INTERIOR

GLOSSARY

- **GL-1** GLOSSARY

MASTER INDEX

- **IND-1** MASTER INDEX

SAFETY NOTICE

Proper service and repair procedures are vital to the safe, reliable operation of all motor vehicles, as well as the personal safety of those performing repairs. This manual outlines procedures for servicing and repairing vehicles using safe, effective methods. The procedures contain many NOTES, CAUTIONS and WARNINGS which should be followed, along with standard procedures to eliminate the possibility of personal injury or improper service which could damage the vehicle or compromise its safety.

It is important to note that repair procedures and techniques, tools and parts for servicing motor vehicles, as well as the skill and experience of the individual performing the work vary widely. It is not possible to anticipate all of the conceivable ways or conditions under which vehicles may be serviced, or to provide cautions as to all possible hazards that may result. Standard and accepted safety precautions and equipment should be used when handling toxic or flammable fluids, and safety goggles or other protection should be used during cutting, grinding, chiseling, prying, or any other process that can cause material removal or projectiles.

Some procedures require the use of tools specially designed for a specific purpose. Before substituting another tool or procedure, you must be completely satisfied that neither your personal safety, nor the performance of the vehicle will be endangered.

Although information in this manual is based on industry sources and is complete as possible at the time of publication, the possibility exists that some car manufacturers made later changes which could not be included here. While striving for total accuracy, the authors or publishers cannot assume responsibility for any errors, changes or omissions that may occur in the compilation of this data.

PART NUMBERS

Part numbers listed in this reference are not recommendations by Haynes North America, Inc. for any product brand name. They are references that can be used with interchange manuals and aftermarket supplier catalogs to locate each brand supplier's discrete part number.

SPECIAL TOOLS

Special tools are recommended by the vehicle manufacturer to perform their specific job. Use has been kept to a minimum, but where absolutely necessary, they are referred to in the text by the part number of the tool manufacturer. These tools can be purchased, under the appropriate part number, from your local dealer or regional distributor, or an equivalent tool can be purchased locally from a tool supplier or parts outlet. Before substituting any tool for the one recommended, read the SAFETY NOTICE at the top of this page.

ACKNOWLEDGMENTS

We are grateful to the Chrysler Corporation for assistance with technical information and certain illustrations. Technical authors who contributed to this project include Bob Doughten and Kevin M. G. Maher.

All rights reserved. No part of this book may be reproduced or transmitted in any form or by any means, electronic or mechanical, including photocopying, recording or by any information storage or retrieval system, without permission in writing from the copyright holder.

While every attempt is made to ensure that the information in this manual is correct, no liability can be accepted by the authors or publishers for loss, damage or injury caused by any errors in, or omissions from, the information given.

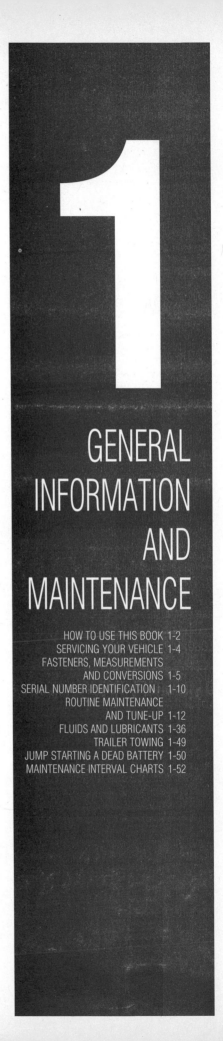

1

GENERAL INFORMATION AND MAINTENANCE

HOW TO USE THIS BOOK 1-2
SERVICING YOUR VEHICLE 1-4
FASTENERS, MEASUREMENTS AND CONVERSIONS 1-5
SERIAL NUMBER IDENTIFICATION 1-10
ROUTINE MAINTENANCE AND TUNE-UP 1-12
FLUIDS AND LUBRICANTS 1-36
TRAILER TOWING 1-49
JUMP STARTING A DEAD BATTERY 1-50
MAINTENANCE INTERVAL CHARTS 1-52

CAPACITIES 1-54

FASTENERS MEASUREMENTS AND CONVERSIONS
BOLTS NUTS AND OTHER THREADED RETAINERS 1-5
STANDARD AND METRIC MEASUREMENTS 1-9
TORQUE 1-6

FLUIDS AND LUBRICANTS
AUTOMATIC TRANSMISSION 1-40
BODY LUBRICATION AND MAINTENANCE 1-47
BRAKE MASTER CYLINDER 1-45
CHASSIS GREASING 1-46
CLUTCH MASTER CYLINDER 1-45
COOLING SYSTEM 1-42
DRIVE AXLE (REAR AND/OR FRONT) 1-41
ENGINE 1-37
FLUID DISPOSAL 1-36
FUEL AND ENGINE OIL RECOMMENDATIONS 1-36
MANUAL TRANSMISSION 1-38
POWER STEERING PUMP 1-46
TRANSFER CASE 1-41
WHEEL BEARINGS 1-48
WINDSHIELD WASHER RESERVOIR 1-44

HOW TO USE THIS BOOK
AVOIDING THE MOST COMMON MISTAKES 1-2
AVOIDING TROUBLE 1-2
MAINTENANCE OR REPAIR? 1-2
SPECIAL TOOLS 1-4
TOOLS AND EQUIPMENT 1-2
WHERE TO BEGIN 1-2

JUMP STARTING A DEAD BATTERY
BATTERY 1-50
JACKING 1-52
JACKING PRECAUTIONS 1-52
JUMP STARTING PRECAUTIONS 1-50
JUMP STARTING PROCEDURE 1-50

MAINTENANCE INTERVAL CHARTS 1-52

ROUTINE MAINTENANCE AND TUNE-UP 1-12
AIR CLEANER (ELEMENT) 1-13
AIR CONDITIONING SYSTEM 1-31
BATTERY 1-18
BELTS 1-20
CV-BOOTS 1-23
DISTRIBUTOR CAP AND ROTOR 1-26
EVAPORATIVE CANISTER 1-17
FUEL FILTER (GASOLINE ENGINES) 1-15
FUEL FILTER/WATER SEPARATOR (DIESEL ENGINES) 1-15
HOSES 1-23
IDLE SPEED AND MIXTURE ADJUSTMENTS 1-30
IGNITION TIMING (GASOLINE ENGINES) 1-27
PCV VALVE 1-16
REPLACEMENT 1-20
SPARK PLUG WIRES 1-26
SPARK PLUGS 1-24
SYSTEM INSPECTION 1-31
TIRES AND WHEELS 1-33
TIRES AND WHEELS CARE OF SPECIAL WHEELS 1-35
TIRES AND WHEELS INFLATION & INSPECTION 1-35
TIRES AND WHEELS TIRE DESIGN 1-34
TIRES AND WHEELS TIRE ROTATION 1-33
TIRES AND WHEELS TIRE STORAGE 1-35
VALVE LASH 1-29
WINDSHIELD WIPER (ELEMENTS) 1-32

SERIAL NUMBER IDENTIFICATION
DRIVE AXLE 1-11
ENGINE 1-10
EQUIPMENT IDENTIFICATION 1-10
TRANSFER CASE 1-12
TRANSMISSION 1-10
VEHICLE 1-10
VEHICLE SAFETY CERTIFICATION LABEL 1-10

SERVICING YOUR VEHICLE
DO'S 1-4
DON'TS 1-5
SAFELY 1-4

TRAILER TOWING
COOLING 1-49
GENERAL RECOMMENDATIONS 1-49
HANDLING A TRAILER 1-50
HITCH (TONGUE) WEIGHT 1-49
MANUFACTURER'S RECOMMENDATIONS 1-50
TOWING THE VEHICLE 1-50
TRAILER WEIGHT 1-49

1-2 GENERAL INFORMATION AND MAINTENANCE

HOW TO USE THIS BOOK

Chilton's Total Car Care manual for Dodge Dakota/Durango/Ram trucks is intended to help you learn more about the inner workings of your vehicle while saving money on its upkeep and operation.

The beginning of the book will likely be referred to the most, since that is where you will find information for maintenance and tune-up. The other sections deal with the more complex systems of your vehicle. Operating systems from engine through brakes are covered to the extent that the average do-it-yourselfer becomes mechanically involved. This book will not explain such things as rebuilding a differential for the simple reason that the expertise required and the investment in special tools make this task uneconomical. It will, however, give you detailed instructions to help you change your own brake pads and shoes, replace spark plugs, and perform many more jobs that can save you money, give you personal satisfaction and help you avoid expensive problems.

A secondary purpose of this book is a reference for owners who want to understand their vehicle and/or their mechanics. In this case, no tools at all are required.

Where to Begin

Before beginning any job, read through the entire procedure. This will give you the overall view of what tools and supplies will be required. There is nothing more frustrating than having to walk to the bus stop on Monday morning because you were short one bolt on Sunday afternoon. So read ahead and plan ahead. Each operation should be approached logically and all procedures thoroughly understood before attempting any work.

All sections contain adjustments, maintenance, removal and installation procedures, and in some cases, repair or overhaul procedures. When repair is not considered practical, we tell you how to remove the part and then how to install the new or rebuilt replacement. In this way, you at least save labor costs. "Backyard" repair of some components is just not practical.

Avoiding Trouble

Many procedures in this book require you to "label and disconnect... " a group of lines, hoses or wires. Don't be lulled into thinking you can remember where everything goes - you won't. If you hook up vacuum or fuel lines incorrectly, the vehicle may run poorly, if at all. If you hook up electrical wiring incorrectly, you may instantly learn a very expensive lesson.

You don't need to know the official or engineering name for each hose or line. A piece of masking tape on the hose and a piece on its fitting will allow you to assign your own label such as the letter A or a short name. As long as you remember your own code, the lines can be reconnected by matching similar letters or names. Do remember that tape will dissolve in gasoline or other fluids; if a component is to be washed or cleaned, use another method of identification. A permanent felt-tipped marker or a metal scribe can be very handy for marking metal parts. Remove any tape or paper labels after assembly.

Maintenance or Repair?

It's necessary to mention the difference between maintenance and repair. Maintenance includes routine inspections, adjustments, and replacement of parts which show signs of normal wear. Maintenance compensates for wear or deterioration. Repair implies that something has broken or is not working. A need for repair is often caused by lack of maintenance. Example: draining and refilling the automatic transmission fluid is maintenance recommended by the manufacturer at specific mileage intervals. Failure to do this can shorten the life of the transmission, requiring very expensive repairs. While no maintenance program can prevent items from breaking or wearing out, a general rule can be stated: MAINTENANCE IS CHEAPER THAN REPAIR.

Two basic mechanic's rules should be mentioned here. First, whenever the left side of the vehicle or engine is referred to, it is meant to specify the driver's side. Conversely, the right side of the vehicle means the passenger's side. Second, screws and bolts are loosened by turning counterclockwise and tightened by turning clockwise unless specifically noted.

Safety is always the most important rule. Constantly be aware of the dangers involved in working on an automobile and take the proper precautions. See the information in this section regarding SERVICING YOUR VEHICLE SAFELY and the SAFETY NOTICE on the acknowledgment page.

Avoiding the Most Common Mistakes

Pay attention to the instructions provided. There are three common mistakes in mechanical work:

1. Incorrect order of assembly, disassembly or adjustment. When taking something apart or putting it together, performing steps in the wrong order usually just costs you extra time; however, it CAN break something. Read the entire procedure before beginning disassembly. Perform everything in the order in which the instructions say you should, even if you can't immediately see a reason for it. When you're taking apart something that is very intricate, you might want to draw a picture of how it looks when assembled at one point in order to make sure you get everything back in its proper position. We will supply exploded views whenever possible. When making adjustments, perform them in the proper order. One adjustment possibly will affect another.

2. Overtorquing (or undertorquing). While it is more common for overtorquing to cause damage, undertorquing may allow a fastener to vibrate loose causing serious damage. Especially when dealing with aluminum parts, pay attention to torque specifications and utilize a torque wrench in assembly. If a torque figure is not available, remember that if you are using the right tool to perform the job, you will probably not have to strain yourself to get a fastener tight enough. The pitch of most threads is so slight that the tension you put on the wrench will be multiplied many times in actual force on what you are tightening. A good example of how critical torque is can be seen in the case of spark plug. Too little torque can fail to crush the gasket, causing leakage of combustion gases and consequent overheating of the plug and engine parts. Too much torque can damage the threads or distort the plug, changing the spark gap.

There are many commercial products available for ensuring that fasteners won't come loose, even if they are not torqued just right (a popular product family is made by Loctite,). If you're worried about getting something together tight enough to hold, but loose enough to avoid mechanical damage during assembly, one of these products might offer substantial insurance. Before choosing a threadlocking compound, read the label on the package and make sure the product is compatible with the materials, fluids, etc. involved.

3. Crossthreading. This occurs when a part such as a bolt is screwed into a nut or casting at the wrong angle and forced. Crossthreading is more likely to occur if access is difficult. It helps to clean and lubricate fasteners, then to start threading the bolt, spark plug, etc. with your fingers. If you encounter resistance, unscrew the part and start over again at a different angle until it can be inserted and turned several times without much effort. Keep in mind that many parts, especially spark plugs, have tapered threads, so that gentle turning will automatically bring the part you're threading to the proper angle. Don't put a wrench on the part until it's been tightened a couple of turns by hand. If you suddenly encounter resistance, and the part has not seated fully, don't force it. Pull it back out to make sure it's clean and threading properly.

Be sure to take your time and be patient, and always plan ahead. Allow yourself ample time to perform repairs and maintenance. You may find maintaining your car a satisfying and enjoyable experience.

TOOLS AND EQUIPMENT

See Figures 1, 2, 3, 4, 5, 6, 7, 8, 9, 10, 11, 12, 13, 14 and 15

Naturally, without the proper tools and equipment it is impossible to properly service your vehicle. It would also be virtually impossible to catalog every tool that you would need to perform all of the operations in this book. Of course, it would be unwise for the amateur to rush out and buy an expensive set of tools on the theory that he/she may need one or more of them at some time.

The best approach is to proceed slowly, gathering a good quality set of those tools that are used most frequently. Don't be misled by the low cost of bargain tools. It is far better to spend a little more for better quality. Forged wrenches, 6 - point sockets and fine tooth or radian ratchets are by far preferable to their less expensive counterparts. As any good mechanic can tell you, there are few worse experiences than trying to work on a vehicle with bad tools. Your monetary savings will be far outweighed by frustration, rounded fasteners and mangled knuckles.

Begin accumulating those tools that are used most frequently: those associated with routine maintenance and tune-up. Engine and transmission type will certainly have an impact on helping you to determine which tools are most essential. In addition to the normal assortment of screwdrivers and pliers, you should have the following tools:

• Wrenches/sockets and combination open end/box end wrenches in sizes from 1/8 - 3/4 in. or 3 - 19mm. Most socket sets incorporate one or more

GENERAL INFORMATION AND MAINTENANCE 1-3

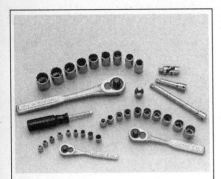

Fig. 1 All but the most basic procedures will require an assortment of ratchets and sockets

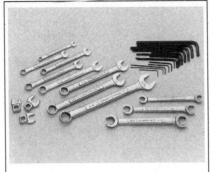

Fig. 2 In addition to sockets, a good set of wrenches and hex keys may be necessary

Fig. 3 A hydraulic floor jack and a set of jackstands are essential for lifting and supporting the vehicle

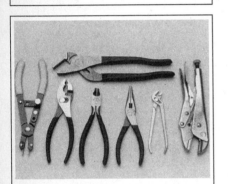

Fig. 4 An assortment of pliers, grippers and cutters will be handy for old rusted parts and stripped bolt heads

special spark plug sockets, but plugs come in a number of sizes so check to ensure you have a socket that will fit.
- Jackstands for support.
- Oil filter wrench.
- Spout or funnel for pouring fluids.
- Grease gun for chassis lubrication.
- Hydrometer for checking the battery (unless equipped with a sealed, maintenance-free battery).
- Containers for draining oil and other fluids.
- Absorbent rags for wiping up the occasional mess.

Note: If possible, buy various length socket drive extensions. Universal-joint and wobble extensions can be extremely useful, but be careful when using them, as they can change the amount of torque applied to the socket.

In addition to the above items there are several others that are not absolutely necessary, but handy to have around. These include Oil Dry®, (or an equivalent oil absorbent gravel - such as cat litter) and the usual supply of lubricants, antifreeze and fluids, although these can be purchased as needed. This is a basic list for routine maintenance, but only your personal needs and desire can accurately determine your list of tools.

After performing a few projects on the vehicle, you'll be amazed at the other tools and non-tools on your workbench. Some useful household items are: a large turkey baster or siphon, empty coffee cans and ice trays (to store parts), a ball of twine, electrical tape for wiring, small rolls of colored tape for tagging lines or hoses, markers and pens, a note pad, golf tees (for plugging vacuum lines), metal coat hangers

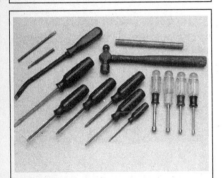

Fig. 5 Various drivers, chisels and prybars are great tools to have in your toolbox

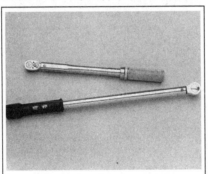

Fig. 6 Many repairs will require the use of a torque wrench to assure the components are properly fastened

Fig. 7 Although not always necessary, using specialized brake tools will save time

Fig. 8 A few inexpensive lubrication tools will make maintenance easier

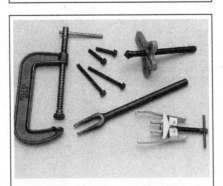

Fig. 9 Various pullers, clamps and separator tools are needed for many larger, more complicated repairs

Fig. 10 A variety of tools and gauges should be used for spark plug gapping and installation

1-4 GENERAL INFORMATION AND MAINTENANCE

Fig. 11 Inductive type timing light

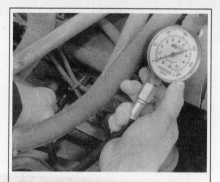

Fig. 12 A screw-in type compression gauge should come with a range of fittings and hose adapters

Fig. 13 A vacuum/pressure tester is necessary for many testing procedures

or a roll of mechanic's wire (to hold things out of the way), dental pick or similar long, pointed probe, a strong magnet, and a small mirror (to see into recesses and under manifolds).

A more advanced set of tools, suitable for tune-up work, can be drawn up easily. While the tools are slightly more sophisticated, they need not be outrageously expensive. The key to these purchases is to make them with an eye towards adaptability and wide range. A basic list of tune-up tools could include:

• A torque wrench. Necessary for all but the most basic work, since proper torque on fasteners is critical to safe operation of the vehicle and often to avoid damage to complex assemblies. Beam type models are inexpensive and normally adequate, although the click types (breakaway) are easier to use. They are usually more expensive. Also keep in mind that all types of torque wrenches should be periodically checked and/or recalibrated. You will have to decide for yourself which better fits your pocketbook, and purpose.

• Spark plug gauge/gapping tool. Even new plugs should be checked before installation.

• Feeler gauges for various clearance adjustments.

• Compression gauge. Be sure that fittings and hose extensions are suitable for your engine.

• Manifold vacuum gauge.

• Tach/dwell meter suitable for your engine (4,6,8,10 cyl.)

• 12V test light.

• A multimeter or volt/ohmmeter (VOM).

• Induction Ammeter. This is used for determining whether or not there is current in a wire. These are handy for use if a wire is broken somewhere in a wiring harness.

• Timing light. Although ignition timing on these vehicles is controlled by the on-board computer, a timing light has some uses in maintenance operations. The choice of a timing light should be made carefully. A light which works on the DC current supplied by the vehicle's battery is the best choice. It should have a xenon tube for brightness. Inductive timing lights are preferred.

• An impact driver, 1/2 " drive. Very handy for breaking free large and/or rusted fasteners.

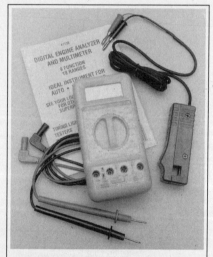

Fig. 14 Most modern automotive multimeters incorporate many helpful features

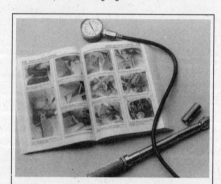

Fig. 15 Proper information is vital, so always have a Chilton Total Car Care manual handy

Special Tools

Normally, the use of special factory tools is avoided for repair procedures, since these are not readily available for the do-it-yourself mechanic. When it is possible to perform the job with more commonly available tools, it will be pointed out, but occasionally, a special tool was designed to perform a specific function and should be used. Before substituting another tool, you should be convinced that neither your safety nor the performance of the vehicle will be compromised.

Special tools can usually be purchased from an automotive parts store or from your dealer. In some cases special tools may be available directly from the tool manufacturer.

SERVICING YOUR VEHICLE

SAFELY

See Figures 16, 17, 18 and 19

It is virtually impossible to anticipate all of the hazards involved with automotive maintenance and service, but care and common sense will prevent most accidents.

The rules of safety for mechanics range from "don't smoke around gasoline," to "use the proper tool(s) for the job." The trick to avoiding injuries is to develop safe work habits and to take every possible precaution.

Do's

• Do keep a fire extinguisher and first aid kit handy.

• Do wear safety glasses or goggles when cutting, drilling, grinding or prying. If you wear glasses for the sake of vision, wear safety goggles over your regular glasses.

• Do shield your eyes whenever you work around the battery. Batteries contain sulfuric acid. In case of contact with the eyes or skin, flush the area with water or a mixture of water and baking soda, then seek immediate medical attention.

• Do use safety stands (jackstands) for any undervehicle service. Jacks are for raising vehicles; jackstands are for making sure the vehicle stays raised until you want it to come down. Whenever the vehicle is raised, block the wheels remaining on the ground and set the parking brake.

• Do use adequate ventilation when working with any chemicals, hazardous materials or when performing operations that generate dust.

• Do disconnect the negative battery cable when working on the electrical system. The secondary ignition system contains EXTREMELY HIGH VOLTAGE. In some cases it can even exceed 50,000 volts.

GENERAL INFORMATION AND MAINTENANCE 1-5

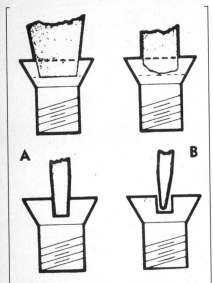

Fig. 16 Screwdrivers should be kept in good condition to prevent injury or damage which could result if the blade slips from the screw

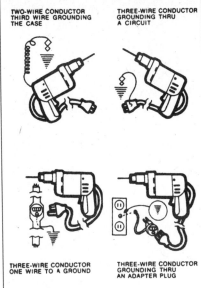

Fig. 17 Power tools should always be properly grounded

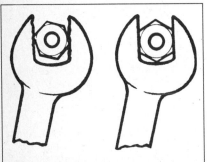

Fig. 18 Using the correct size and standard (metric or SAE) wrench will help prevent the possibility of rounding off a nut

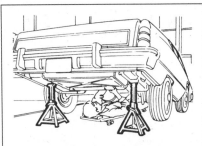

Fig. 19 NEVER work under a vehicle unless it is supported using safety stands (jackstands)

- Do follow manufacturer's directions whenever working with potentially hazardous materials. Most chemicals and fluids are poisonous if taken internally.
- Do properly maintain your tools. Loose hammerheads, mushroomed punches and chisels, frayed or poorly grounded electrical cords, excessively worn screwdrivers, spread wrenches (open end), cracked sockets, slipping ratchets, or faulty droplight sockets can cause accidents.
- Likewise, keep your tools clean; a greasy wrench can slip off a bolt head, ruining the bolt and often harming your knuckles in the process.
- Do use the proper size and type of tool for the job at hand. Do select a wrench or socket that fits the nut or bolt. The wrench or socket should sit straight, not cocked.
- Do, when possible, pull on a wrench handle rather than push on it, and adjust your stance to prevent a fall.
- Do be sure that adjustable wrenches are tightly closed on the nut or bolt and pulled so that the force is on the side of the fixed jaw.
- Do strike squarely with a hammer; avoid glancing blows.
- Do set the parking brake and block the drive wheels if the work requires a running engine.

Don'ts

- Don't run the engine in a garage or anywhere else without proper ventilation - EVER! Carbon monoxide is poisonous; it takes a long time to leave the human body and you can build up a deadly supply of it in your system by simply breathing in a little every day. You may not realize you are slowly poisoning yourself. Always use power vents, windows, fans and/or open the garage door.
- Don't work around moving parts while wearing loose clothing. Short sleeves are much safer than long, loose sleeves. Hard-toed shoes with neoprene soles protect your toes and give a better grip on slippery surfaces. Jewelry such as watches, fancy belt buckles, beads or body adornment of any kind is not safe working around a vehicle. Long hair should be tied back under a hat or cap.
- Don't use pockets for toolboxes. A fall or bump can drive a screwdriver deep into your body. Even a rag hanging from your back pocket can wrap around a spinning shaft or fan.
- Don't smoke when working around gasoline, cleaning solvent or other flammable material.
- Don't smoke when working around the battery. When the battery is being charged, it gives off explosive hydrogen gas.
- Don't use gasoline to wash your hands; there are excellent soaps available. Gasoline contains dangerous additives which can enter the body through a cut or through your pores. Gasoline also removes all the natural oils from the skin so that bone dry hands will suck up oil and grease.
- Don't service the air conditioning system unless you are equipped with the necessary tools and training. When liquid or compressed gas refrigerant is released to atmospheric pressure it will absorb heat from whatever it contacts. This will chill or freeze anything it touches.
- Don't use screwdrivers for anything other than driving screws! A screwdriver used as a prying tool can snap when you least expect it, causing injuries. At the very least, you'll ruin a good screwdriver.
- Don't use an emergency jack (that little ratchet, scissors, or pantograph jack supplied with the vehicle) for anything other than changing a flat! These jacks are only intended for emergency use out on the road; they are NOT designed as a maintenance tool. If you are serious about maintaining your vehicle yourself, invest in a hydraulic floor jack of at least a 1-1/2 ton capacity, and at least two sturdy jackstands.

FASTENERS, MEASUREMENTS AND CONVERSIONS

Bolts, Nuts and Other THREADED RETAINERS

See Figures 20, 21, 22, 23 and 24

Although there are a great variety of fasteners found in the modern car or truck, the most commonly used retainer is the threaded fastener (nuts, bolts, screws, studs, etc.). Most threaded retainers may be reused, provided that they are not damaged in use or during the repair. Some retainers (such as stretch bolts or torque prevailing nuts) are designed to deform when tightened or in use and should not be reinstalled.

Whenever possible, we will note any special retainers which should be replaced during a procedure. But you should always inspect the condition of a retainer when it is removed and replace any that show signs of damage. Check all threads for rust or corrosion which can increase the torque necessary to achieve the desired clamp load for which that fastener was originally selected. Additionally, be sure that the driver surface of the fastener has not been compromised by rounding or other damage. In some cases a driver surface may become only partially rounded, allowing the driver to catch in only one direction. In many of these occurrences, a fastener may be installed and tightened, but the driver would not be able to grip and loosen the fastener again. (This could lead to frustration down the line should that component ever need to be disassembled again.)

If you must replace a fastener, whether due to de-

1-6 GENERAL INFORMATION AND MAINTENANCE

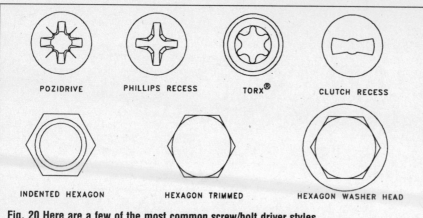

Fig. 20 Here are a few of the most common screw/bolt driver styles

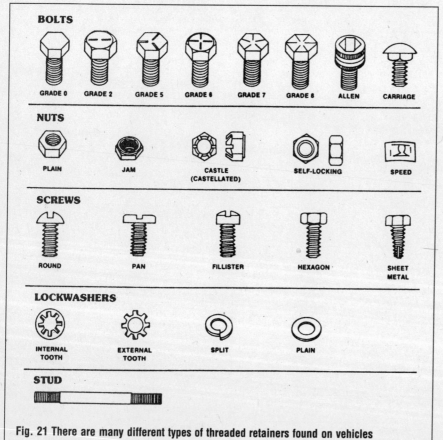

Fig. 21 There are many different types of threaded retainers found on vehicles

sign or damage, you must ALWAYS be sure to use the proper replacement. In all cases, a retainer of the same design, material and strength should be used. Markings on the heads of most bolts will help determine the proper strength of the fastener. The same material, thread and pitch must be selected to assure proper installation and safe operation of the vehicle afterwards.

Thread gauges are available to help measure a bolt or stud's thread. Most automotive and hardware stores keep gauges available to help you select the proper size. In a pinch, you can use another nut or bolt for a thread gauge. If the bolt you are replacing is not too badly damaged, you can select a match by finding another bolt which will thread in its place. If you find a nut which threads properly onto the damaged bolt, then use that nut to help select the replacement bolt. If however, the bolt you are replacing is so badly damaged (broken or drilled out) that its threads cannot be used as a gauge, you might start by looking for another bolt (from the same assembly or a similar location on your vehicle) which will thread into the damaged bolt's mounting. If so, the other bolt can be used to select a nut; the nut can then be used to select the replacement bolt.

In all cases, be absolutely sure you have selected the proper replacement. Don't be shy, you can always ask the store clerk for help.

WARNING:

Be aware that when you find a bolt with damaged threads, you may also find the nut or hole it was threaded into has also been damaged. If this is the case, you may have to drill and tap the hole, replace the nut or otherwise repair the threads. NEVER try to force a replacement bolt to fit into the damaged threads.

There are a number of thread repair compounds available at auto parts stores which can be very useful under certain circumstance. However, they will not work if threads are badly damaged and should not be used if the fastener requires a high torque loading, such as cylinder head or suspension use.

Torque

Torque is defined as the measurement of resistance to turning or rotating. It tends to twist a body about an axis of rotation. A common example of this would be tightening a threaded retainer such as a nut, bolt or screw. Measuring torque is one of the most common ways to help assure that a threaded retainer has been properly fastened.

When tightening a threaded fastener, torque is applied in three distinct areas, the head, the bearing surface and the clamp load. About 50 percent of the measured torque is used in overcoming bearing friction. This is the friction between the bearing surface of the bolt head, screw head or nut face and the base material or washer (the surface on which the fastener is rotating). Approximately 40 percent of the applied torque is used in overcoming thread friction. This leaves only about 10 percent of the applied torque to develop a useful clamp load (the force which holds a joint together). This means that friction can account for as much as 90 percent of the applied torque on a fastener.

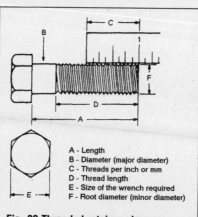

A - Length
B - Diameter (major diameter)
C - Threads per inch or mm
D - Thread length
E - Size of the wrench required
F - Root diameter (minor diameter)

Fig. 22 Threaded retainer sizes are determined using these measurements

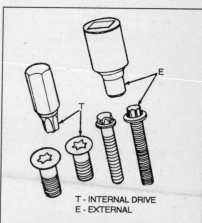

T - INTERNAL DRIVE
E - EXTERNAL

Fig. 23 Special fasteners such as these Torx, head bolts may discourage people from working on vehicles without the proper tools

GENERAL INFORMATION AND MAINTENANCE

Standard Torque Specifications and Fastener Markings

In the absence of specific torques, the following chart can be used as a guide to the maximum safe torque of a particular size/grade of fastener.
- There is no torque difference for fine or coarse threads.
- Torque values are based on clean, dry threads. Reduce the value by 10% if threads are oiled prior to assembly.
- The torque required for aluminum components or fasteners is considerably less.

U.S. Bolts

SAE Grade Number	1 or 2			5			6 or 7		
Number of lines always 2 less than the grade number.									
	Maximum Torque			Maximum Torque			Maximum Torque		
Bolt Size (Inches)—(Thread)	Ft./Lbs.	Kgm	Nm	Ft./Lbs.	Kgm	Nm	Ft./Lbs.	Kgm	Nm
¼—20	5	0.7	6.8	8	1.1	10.8	10	1.4	13.5
—28	6	0.8	8.1	10	1.4	13.6			
5/16—18	11	1.5	14.9	17	2.3	23.0	19	2.6	25.8
—24	13	1.8	17.6	19	2.6	25.7			
⅜—16	18	2.5	24.4	31	4.3	42.0	34	4.7	46.0
—24	20	2.75	27.1	35	4.8	47.5			
7/16—14	28	3.8	37.0	49	6.8	66.4	55	7.6	74.5
—20	30	4.2	40.7	55	7.6	74.5			
½—13	39	5.4	52.8	75	10.4	101.7	85	11.75	115.2
—20	41	5.7	55.6	85	11.7	115.2			
9/16—12	51	7.0	69.2	110	15.2	149.1	120	16.6	162.7
—18	55	7.6	74.5	120	16.6	162.7			
⅝—11	83	11.5	112.5	150	20.7	203.3	167	23.0	226.5
—18	95	13.1	128.8	170	23.5	230.5			
¾—10	105	14.5	142.3	270	37.3	366.0	280	38.7	379.6
—16	115	15.9	155.9	295	40.8	400.0			
⅞—9	160	22.1	216.9	395	54.6	535.5	440	60.9	596.5
—14	175	24.2	237.2	435	60.1	589.7			
1—8	236	32.5	318.6	590	81.6	799.9	660	91.3	894.8
—14	250	34.6	338.9	660	91.3	849.8			

Metric Bolts

Relative Strength Marking	4.6, 4.8			8.8		
Bolt Markings						
	Maximum Torque			Maximum Torque		
Bolt Size Thread Size x Pitch (mm)	Ft./Lbs.	Kgm	Nm	Ft./Lbs.	Kgm	Nm
6 x 1.0	2–3	.2–.4	3–4	3–6	4–.8	5–8
8 x 1.25	6–8	.8–1	8–12	9–14	1.2–1.9	13–19
10 x 1.25	12–17	1.5–2.3	16–23	20–29	2.7–4.0	27–39
12 x 1.25	21–32	2.9–4.4	29–43	35–53	4.8–7.3	47–72
14 x 1.5	35–52	4.8–7.1	48–70	57–85	7.8–11.7	77–110
16 x 1.5	51–77	7.0–10.6	67–100	90–120	12.4–16.5	130–160
18 x 1.5	74–110	10.2–15.1	100–150	130–170	17.9–23.4	180–230
20 x 1.5	110–140	15.1–19.3	150–190	190–240	26.2–46.9	160–320
22 x 1.5	150–190	22.0–26.2	200–260	250–320	34.5–44.1	340–430
24 x 1.5	190–240	26.2–46.9	260–320	310–410	42.7–56.5	420–550

Fig. 24 Torque specifications based on fastener size and grade. Use only as a guide in the event specific torque is not supplied

1-8 GENERAL INFORMATION AND MAINTENANCE

TORQUE WRENCHES

See Figure 25

In most applications, a torque wrench can be used to assure proper installation of a fastener. Torque wrenches come in various designs and most automotive supply stores will carry a variety to suit your needs. A torque wrench should be used any time we supply a specific torque value for a fastener. A torque wrench can also be used if you are following the general guidelines in the accompanying chart. Keep in mind that because there is no worldwide standardization of fasteners, the charts are a general guideline and should be used with caution. Again, the general rule of "if you are using the right tool for the job, you should not have to strain to tighten a fastener" applies here.

Beam Type

See Figure 26

The beam type torque wrench is one of the most popular types. It consists of a pointer attached to the head that runs the length of the flexible beam (shaft) to a scale located near the handle. As the wrench is pulled, the beam bends and the pointer indicates the torque using the scale.

Click (Breakaway) Type

See Figure 27

Another popular design of torque wrench is the click type. To use the click type wrench you pre-adjust it to a torque setting. Once the torque is reached, the wrench has a reflex signaling feature that causes a momentary breakaway of the torque wrench body, sending an impulse to the operator's hand.

Pivot Head Type

See Figures 27 and 28

Some torque wrenches (usually of the click type) may be equipped with a pivot head which can allow it to be used in areas of limited access. BUT, it must be used properly. To hold a pivot head wrench, grasp the handle lightly, and as you pull on the handle, it should be floated on the pivot point. If the handle comes in contact with the yoke extension during the process of pulling, there is a very good chance the torque readings will be inaccurate because this could alter the wrench loading point. The design of the handle is usually such as to make it inconvenient to deliberately misuse the wrench.

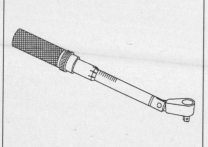

Fig. 27 A click type or breakaway torque wrench - note that this one has a pivoting head

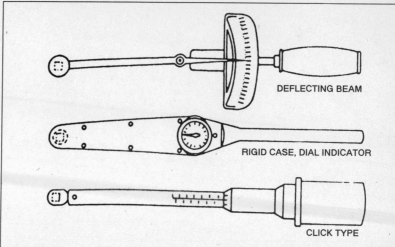

Fig. 25 Various styles of torque wrenches are usually available at your local automotive supply store

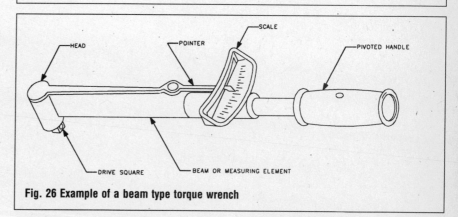

Fig. 26 Example of a beam type torque wrench

Note: It should be mentioned that the use of any U-joint, wobble or extension will have an effect on the torque readings, no matter what type of wrench you are using. For the most accurate readings, install the socket directly on the wrench driver. If necessary, straight extensions (which hold a socket directly under the wrench driver) will have the least effect on the torque reading. Avoid any extension that alters the length of the wrench from the handle to the head/driving point (such as a crow's foot). U-joint or wobble extensions can greatly affect the readings; avoid their use at all times.

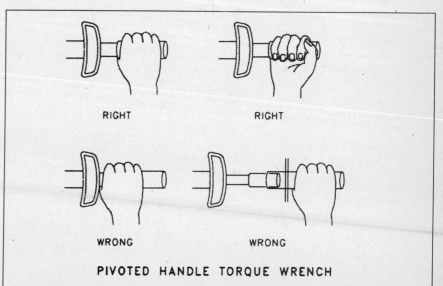

Fig. 28 Torque wrenches with pivoting heads must be grasped and used properly to prevent an incorrect reading

GENERAL INFORMATION AND MAINTENANCE

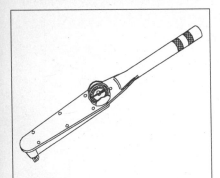

Fig. 29 The rigid case (direct reading) torque wrench uses a dial indicator to show torque

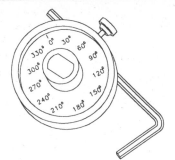

Fig. 30 Some specifications require the use of a torque angle meter (mechanical protractor)

Rigid Case (Direct Reading)

See Figure 29

A rigid case or direct reading torque wrench is equipped with a dial indicator to show torque values. One advantage of these wrenches is that they can be held at any position on the wrench without affecting accuracy. These wrenches are often preferred because they tend to be compact, easy to read and have a great degree of accuracy.

TORQUE ANGLE METERS

See Figure 30

Because the frictional characteristics of each fastener or threaded hole will vary, clamp loads which are based strictly on torque will vary as well. In most applications, this variance is not significant enough to cause worry. But, in certain applications, a manufacturer's engineers may determine that more precise clamp loads are necessary (such is the case with many aluminum cylinder heads). In these cases, a torque angle method of installation would be specified. When installing fasteners which are torque angle tightened, a predetermined seating torque and standard torque wrench are usually used first to remove any compliance from the joint. The fastener is then tightened the specified additional portion of a turn measured in degrees. A torque angle gauge (mechanical protractor) is used for these applications.

Standard and Metric MEASUREMENTS

See Figure 31

Throughout this manual, specifications are given to help you determine the condition of various components on your vehicle, or to assist you in their installation. Some of the most common measurements include length (in. or cm/mm), torque (ft. lbs., inch lbs. or Nm) and pressure (psi, in. Hg, kPa or mm Hg). In most cases, we strive to provide the proper measurement as determined by the manufacturer's engineers.

Though, in some cases, that value may not be conveniently measured with what is available in your toolbox. Luckily, many of the measuring devices which are available today will have two scales so the SAE or metric measurements may easily be taken. If any of the various measuring tools which are available to you do not contain the same scale as listed in the specifications, use the accompanying conversion factors to determine the proper value.

The conversion factor chart is used by taking the given specification and multiplying it by the necessary conversion factor. For instance, looking at the first line, if you have a measurement in inches such as "free-play should be 2 in." but your ruler reads only in millimeters, multiply 2 in. by the conversion factor of 25.4 to get the metric equivalent of 50.8mm. Likewise, if the specification was given only in a metric measurement, for example in Newton Meters (Nm), then look at the center column first. If the measurement is 100 Nm, multiply it by the conversion factor of 0.738 to get 73.8 ft. lbs.

CONVERSION FACTORS

LENGTH–DISTANCE

Inches (in.)	x 25.4	= Millimeters (mm)	x .0394	= Inches
Feet (ft.)	x .305	= Meters (m)	x 3.281	= Feet
Miles	x 1.609	= Kilometers (km)	x .0621	= Miles

VOLUME

Cubic Inches (in3)	x 16.387	= Cubic Centimeters	x .061	= in3
IMP Pints (IMP pt.)	x .568	= Liters (L)	x 1.76	= IMP pt.
IMP Quarts (IMP qt.)	x 1.137	= Liters (L)	x .88	= IMP qt.
IMP Gallons (IMP gal.)	x 4.546	= Liters (L)	x .22	= IMP gal.
IMP Quarts (IMP qt.)	x 1.201	= US Quarts (US qt.)	x .833	= IMP qt.
IMP Gallons (IMP gal.)	x 1.201	= US Gallons (US gal.)	x .833	= IMP gal.
Fl. Ounces	x 29.573	= Milliliters	x .034	= Ounces
US Pints (US pt.)	x .473	= Liters (L)	x 2.113	= Pints
US Quarts (US qt.)	x .946	= Liters (L)	x 1.057	= Quarts
US Gallons (US gal.)	x 3.785	= Liters (L)	x .264	= Gallons

MASS–WEIGHT

Ounces (oz.)	x 28.35	= Grams (g)	x .035	= Ounces
Pounds (lb.)	x .454	= Kilograms (kg)	x 2.205	= Pounds

PRESSURE

Pounds Per Sq. In. (psi)	x 6.895	= Kilopascals (kPa)	x .145	= psi
Inches of Mercury (Hg)	x .4912	= psi	x 2.036	= Hg
Inches of Mercury (Hg)	x 3.377	= Kilopascals (kPa)	x .2961	= Hg
Inches of Water (H$_2$O)	x .07355	= Inches of Mercury	x 13.783	= H$_2$O
Inches of Water (H$_2$O)	x .03613	= psi	x 27.684	= H$_2$O
Inches of Water (H$_2$O)	x .248	= Kilopascals (kPa)	x 4.026	= H$_2$O

TORQUE

Pounds–Force Inches (in-lb)	x .113	= Newton Meters (N·m)	x 8.85	= in–lb
Pounds–Force Feet (ft-lb)	x 1.356	= Newton Meters (N·m)	x .738	= ft–lb

VELOCITY

Miles Per Hour (MPH)	x 1.609	= Kilometers Per Hour (KPH)	x .621	= MPH

POWER

Horsepower (Hp)	x .745	= Kilowatts	x 1.34	= Horsepower

FUEL CONSUMPTION*

Miles Per Gallon IMP (MPG)	x .354	= Kilometers Per Liter (Km/L)		
Kilometers Per Liter (Km/L)	x 2.352	= IMP MPG		
Miles Per Gallon US (MPG)	x .425	= Kilometers Per Liter (Km/L)		
Kilometers Per Liter (Km/L)	x 2.352	= US MPG		

*It is common to covert from miles per gallon (mpg) to liters/100 kilometers (l/100 km), where mpg (IMP) x l/100 km = 282 and mpg (US) x l/100 km = 235.

TEMPERATURE

Degree Fahrenheit (°F)	= (°C x 1.8) + 32
Degree Celsius (°C)	= (°F – 32) x .56

Fig. 31 Standard and metric conversion factors chart

1-10 GENERAL INFORMATION AND MAINTENANCE

SERIAL NUMBER IDENTIFICATION

Vehicle

See Figure 32

The Vehicle Information Number (VIN) plate is located on the left (driver's side) upper panel of the dashboard and is visible through the lower left of the windshield. The VIN consists of 17 characters which represent codes supplying important information about your vehicle. This information may be useful when ordering replacement parts. Refer to the illustration of an example of VIN interpretation.

The eighth digit of the VIN indicates the engine type fitted.

The tenth digit of the VIN indicates vehicle model year.

Refer to the Engine and Vehicle Identification Chart for code interpretation.

Equipment Identification

See Figure 33

The Equipment Identification plate provides build information that may be very useful in carrying out repairs. The plate is usually located on the underside of the hood. Your dealer will be able to provide interpretations of these codes if they are needed.

Vehicle Safety Certification Label

See Figure 34

The Vehicle Safety Certification Label is located on the driver's side door jamb. It certifies that the vehicle conforms to all applicable Federal Motor Vehicle Safety Standards. It also lists month and year of man-

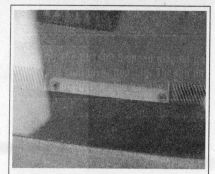

Fig. 32 The VIN plate (2) is permanently affixed to the instrument panel/dashboard (1) and is visible through the lower left (driver's) side of the windshield

ufacture, VIN, Gross Vehicle Weight Rating (GVWR), and GVWR for each axle based on minimum rim size and maximum cold tire inflation pressure.

Engine

See Figure 35

The engine code is represented by the eighth character in the VIN.

The engine itself has an identification number normally indicating date and place of manufacture, displacement and serial number. ID location will depend on engine type and year. It may be stamped into a machined pad on the right side of the cylinder block between cylinders #3 and #4 (2.5L), in the left front

Fig. 33 The Equipment Identification plate provides build information.

corner of the cylinder block (3.9L, 5.2L, 5.9L gasoline, 8.0L), on the right side of the block near the oil pan rail (older V-8s), or on the left side of the engine on the gear housing (diesel). The 4.7L has the VIN stamped on the right front side of the block. Build date code is normally included on the yellow bar code sticker on the oil fill housing. When engine components need to be replaced, refer to the engine type and serial number.

Transmission

See Figures 36, 37, 38, 39 and 40

The transmission code is located on the identification nameplate. There are many different transmissions options: be sure to identify the one with which your vehicle is equipped when ordering replacement parts.

Fig. 34 Vehicle Safety Certification label

```
        X M AAA YYYY 0000

X = Last Digit of Model Year
M = Plant-M Mound Road
        S Saltillo
        T Trenton
        K Toluca
AAA = Engine Displacement (CID)
YYYY = Month/Day
0000 = Engine Serial Code
```

Fig. 35 Engine Identification (serial) number key - typical

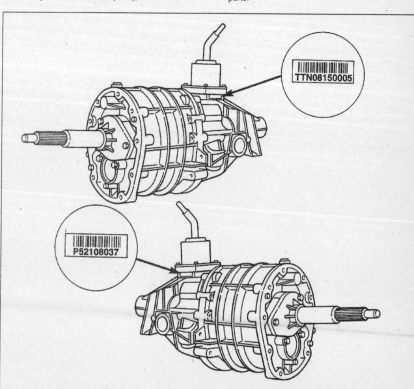

Fig. 36 The NV1500 5-speed manual transmission has two bar code tags located near the gear shift lever housing. The tag on the right side contains the part number, the tag on the left side has the build sequence and date information

GENERAL INFORMATION AND MAINTENANCE 1-11

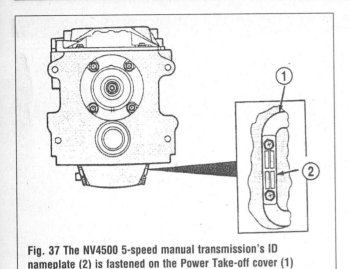

Fig. 37 The NV4500 5-speed manual transmission's ID nameplate (2) is fastened on the Power Take-off cover (1)

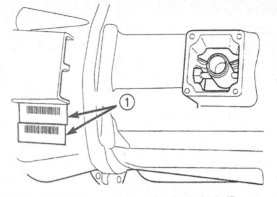

Fig. 38 The NV5600 6-speed manual transmission's ID nameplate (1) consists of two tags. One tag is the transmission part number, the other is the manufacturer's build date information. Both should be used when ordering replacement parts

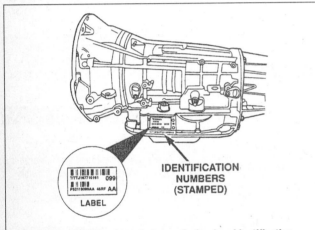

Fig. 39 The 45RFE automatic transmission has identification numbers stamped above the oil pan gasket surface, and a bar code label containing additional information which may be useful

Fig. 40 The 42RE automatic transmission ID numbers are stamped onto the left side of the case, just above the oil pan gasket surface. The part number (1) build date (2) and serial numbers (3) are provided

The manual transmissions use a tag, nameplate, or barcode located on the driver's side, or near the gear shift lever housing, depending on model. The automatic transmissions have a three part code, or a two part bar code tag, depending on model, stamped on the left side of the case just above the oil pan gasket surface.

Drive Axle

See Figures 41, 42, 43 and 44

The tag attached to most differential covers may be used in identifying the axle and/or the gear ratio. However, the tag may have been removed during service and not replaced. Then the axle differential covers may be used for axle identification. The number of bolts securing the differential cover, the size and shape of the differential cover, the gear ratio as given on the ratio tag, and the location and type of fill plug are all useful tools in axle identification. Be sure to note all this information before beginning any maintenance or repairs to the axle.

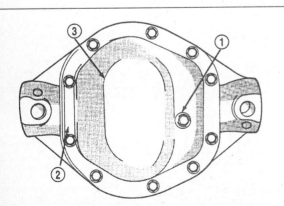

Fig. 41 The fill plug (1) on the 216 FBI axle's differential cover (3) is a hex-head, and the identification tag (3) is affixed by two cover bolts

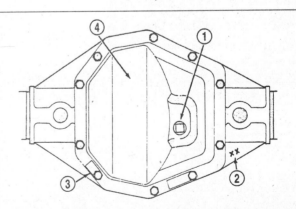

Fig. 42 The fill plug (1) on the 248 FBI axle's differential cover (4) is a square drive plug, and the ratio tag (3) and model numbers (2) are in separate locations

1-12 GENERAL INFORMATION AND MAINTENANCE

Transfer Case

See Figure 45

All Four Wheel Drive (4WD) models use a transfer case. The transfer case delivers power to both the front and rear axles when 4WD is selected.

An identification tag is attached to the rear case of every transfer case. The tag information is useful when identifying the transfer case in your 4WD vehicle.

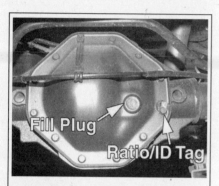

Fig. 43 The fill plug on the 9 1/4 inch axle's differential cover is a rubber push-in type, and the ratio tag is fastened by a cover bolt

Fig. 44 The drive axle is originally equipped with a gear ratio identification tag attached by one of the cover bolts

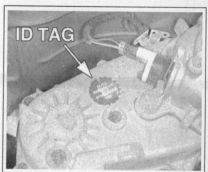

Fig. 45 Transfer cases have identification tags (1) located on the rear side of the case. The fill plugs (2) and drain plugs (3) may be different than the ones shown here, but their location should be correct.

ROUTINE MAINTENANCE AND TUNE-UP

Fig. 46 UNDERHOOD MAINTENANCE COMPONENT LOCATIONS

1. Radiator cap
2. Coolant reservoir
3. Engine oil dipstick
4. Automatic transmission dipstick
5. Air filter housing
6. Engine oil filler cap
7. Brake master cylinder
8. Battery
9. Windshield washer reservoir

See Figure 46

Proper maintenance and tune-up is the key to long and trouble-free vehicle life, and the work can yield its own rewards. Studies have shown that a properly tuned and maintained vehicle can achieve better gas mileage than an out-of-tune vehicle. As a conscientious owner and driver, set aside a Saturday morning, say once a month, to check or replace items which could cause major problems later. Keep your own personal log to jot down which services you performed, how much the parts cost you, the date, and the exact odometer reading at the time. Keep all receipts for such items as engine oil and filters, so that they may be referred to in case of related problems or to determine operating expenses. As a do-it-yourselfer, these receipts are the only proof you have that the required maintenance was performed. In the event

GENERAL INFORMATION AND MAINTENANCE 1-13

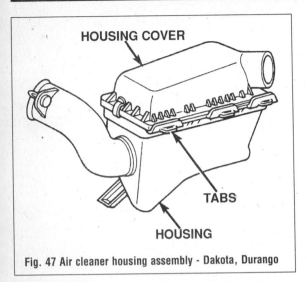

Fig. 47 Air cleaner housing assembly - Dakota, Durango

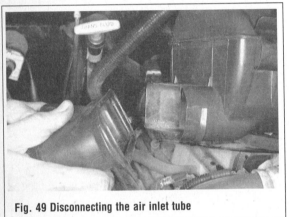

Fig. 49 Disconnecting the air inlet tube

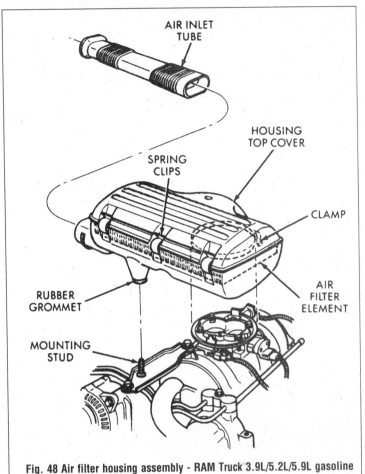

Fig. 48 Air filter housing assembly - RAM Truck 3.9L/5.2L/5.9L gasoline engines

Fig. 50 Air cleaner housing mounting stud

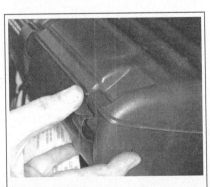

Fig. 51 Unlatching the air cleaner housing spring clips

Fig. 52 Remove the air cleaner element and check condition

of a warranty problem, these receipts will be invaluable.

The literature provided with your vehicle when it was originally delivered includes the factory recommended maintenance schedule. If you no longer have this literature, replacement copies are usually available from the dealer. A maintenance schedule is provided later in this section, in case you do not have the factory literature.

Air Cleaner (Element)

REMOVAL & INSTALLATION

Dakota, Durango

See Figure 47

1. Housing removing is not necessary for filter replacement.
2. Pry up the spring clips from the housing cover.
3. Release the housing cover from the locating the locating tabs on the housing and remove the cover.
4. Remove the old element.

To install:

5. Clean the inside of the housing removing any debris or foreign matter.
6. Install the new filter element. Be sure it is correctly positioned in the housing.
7. Place the housing cover into position in the locating tabs. Be sure the cover is properly seated.
8. Pry up the spring clips to secure the cover.

RAM Truck 3.9L/5.2L/5.9L Gasoline Engines

See Figures 48, 49, 50, 51, 52 and 53

WARNING:

Do not attempt to remove the filter element by removing the housing top cover only. The entire assembly must be removed from the engine for filter replacement.

1-14 GENERAL INFORMATION AND MAINTENANCE

Fig. 53 Align tabs and slots when refitting the housing cover

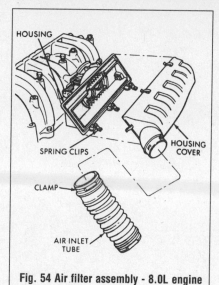

Fig. 54 Air filter assembly - 8.0L engine

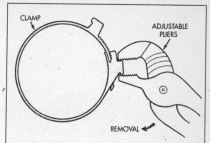

Fig. 55 Use adjustable pliers as shown to loosen the air inlet tube clamp

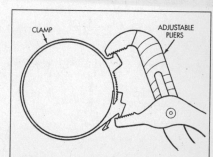

Fig. 56 To tighten the air inlet tube clamp, use adjustable pliers as shown

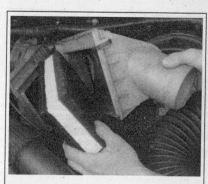

Fig. 57 Remove the air cleaner element from the housing - 8.0L engine

1. Remove the air inlet tube.
2. Loosen, but do not remove, the screw clamp that secures the air cleaner housing to the throttle body. (Note clamp positioning tabs on air cleaner housing.)
3. Disconnect the breather hose at the rear of the air cleaner housing.
4. On the 5.9L HDC V-8 engine, disconnect the air pump hose at the air cleaner housing.
5. Lift the assembly from the throttle body while slipping it from the mounting stud.
6. Unlatch the spring clips and tilt the top cover up and to the rear to remove.
7. Remove the filter element.

To install:

8. Clean the inside of the housing, removing any debris or foreign matter.
9. Check condition of the throttle body seal and replace it if damaged.
10. Install a new filter element and refit the housing top cover, inserting tabs into the slots. Latch the three clips.
11. Check position of the mounting stud rubber grommet, then install the housing on the throttle body. Push down to ensure it is seated.
12. Tighten the throttle body clamp to 15 inch lbs (2 Nm). Check that the housing has been secured.
13. Reconnect all hoses.

RAM Truck 8.0L Engine

See Figures 54, 55, 56 and 57

1. Loosen the clamp and remove the air inlet tube at the front of the air cleaner housing.
2. Unlatch the spring clips and remove the housing cover.
3. Remove the old filter element.
4. Clean the inside of the housing. A small amount of oil wetting is normal.
5. Install the new element, replace the cover and latch the spring clips.
6. Connect the air inlet tube and tighten the clamp.

Diesel engines

See Figures 58 and 59

Note: Diesel engines are equipped with a Filter Minder gauge. This is an air flow restriction gauge which will indicate when it is time

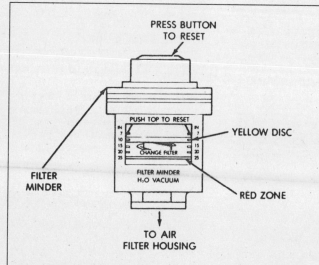

Fig. 58 The Filter Minder gauge found on the air cleaner will help determine when filter replacement is needed - diesel engines

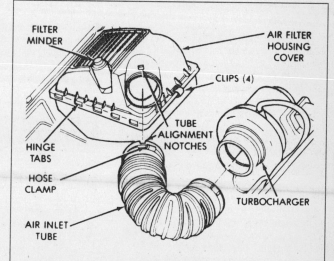

Fig. 59 Exploded view of the air cleaner assembly - diesel engines

GENERAL INFORMATION AND MAINTENANCE 1-15

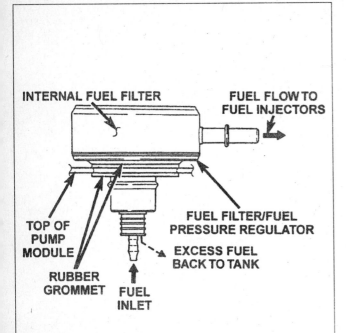

Fig. 60 The fuel filter/fuel regulator for the gasoline engines does not need to be replaced periodically. Only replace it if it is clogged or if directed to after a diagnostic test - gasoline engines

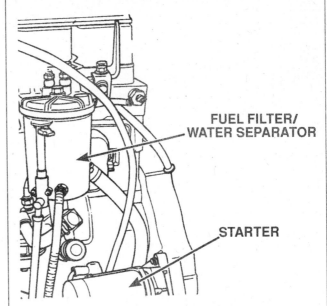

Fig. 61 The fuel filter/water separator is located directly above the starter motor on the left hand side of the engine. The fuel filter element is removed from the housing after all water (if present) has been drained from the assembly - Diesel engines

to replace the air filter. The yellow disc inside will move along a graduated scale. Once the engine has been shut off, the disc will be held at the highest restriction the air filter has experienced. Check with the engine OFF. If the yellow disk has reached the red zone on the graduated scale, the filter should be replaced. Do not remove the top of the filter housing unless the scale on the unit indicates it is time to do so.

1. Remove the air inlet tube.
2. Unlatch the four spring clips located around the outer edges of the air cleaner seal.
3. Tilt the upper half of the air cleaner upward, pull it forward, and remove the cover.
4. Remove the air cleaner element.

To install:

5. Clean the inside of the air cleaner of any leaves, debris, or excessive dirt before installing a new air filter.
6. Place the filter element inside the lower half of the air cleaner.
7. Align the tabs on the top half to the slots on the bottom half of the air cleaner while tilting the air inlet end upward.
8. Push the tabs into the slots and push the top half downward to seat on the air filter seal.

WARNING:

Do not use excessive force when aligning the two halves of the air cleaner assembly. A broken air cleaner assembly could cause damage to internal engine parts. The air filter seal should be seated within the lower housing with no excess material outside or above the air cleaner seam. Make sure the proper air filter has been selected, that the air filter is properly placed in the housing, and that the tabs on the top half are properly

aligned with the slots on the lower half of the air cleaner.

9. Noting the alignment notches, install the air inlet tube and tighten the clamp to 25 inch lbs. (3 Nm).
10. After replacing the filter element, press the button on top of the Filter Minder, to reset. The yellow disk should spring back to the UP position.

WARNING:

Some commercial engine cleaners may discolor or damage the plastic housing of the Filter Minder,. Cover the unit when using these compounds.

Note: Sometimes a temporary blockage of the air passages - as might be caused by snow or heavy rain - will cause the disk to move into the red zone. If the filter is found to be clean, allow it to dry and retest under normal condition.

Fuel Filter (Gasoline Engines)

See Figure 60

The gasoline-powered vehicles covered by this manual all use a fuel filter/fuel pressure regulator mounted on top of the fuel pump module, located on the top of the fuel tank. No periodic service is required. The fuel filter/fuel regulator should only be replaced if it is clogged or if a diagnostic procedure indicates it is time to do so. No separate frame mounted filter is used on any of these vehicles.

See "Fuel System" for fuel tank service procedures.

Fuel Filter/Water Separator (Diesel Engines)

See Figure 61

The fuel filter/water separator protects the fuel injection pump on the diesel powered vehicles covered in this manual. The fuel filter/water separator is located on the left hand side of the engine, above the starter motor. The fuel filter is located within the fuel filter/water separator housing.

Water should be removed from the system whenever the water-in-fuel warning lamp remains lit.

The procedure for simple maintenance of the fuel filter/water separator is given here. For more detailed procedures for the replacement or service of the assembly, refer to "Fuel System".

REMOVAL & INSTALLATION

See Figures 62, 63, 64 and 65

Note: Remove all water present in the housing assembly before attempting to remove the fuel filter. Follow the procedures for draining the water separator even if the water-in-fuel light is not illuminated. A drain hose is located on the bottom of the housing for convenience.

CAUTION:

Fuel lines on diesel equipped vehicles carry diesel fuel under extremely high pressure. This pressure may be as high as 18,000 psi. Use extreme caution when inspecting or performing service on these systems. High-pressure diesel fuel may cause personal injury if contact is made with the skin or eyes.

1-16 GENERAL INFORMATION AND MAINTENANCE

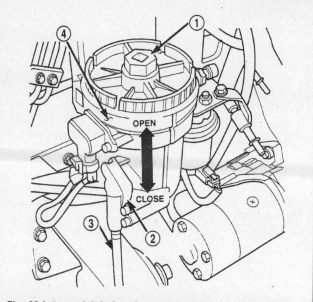

Fig. 62 Late model drain valve system showing handle (2) and drain hose (3) on fuel filter/water separator (4)

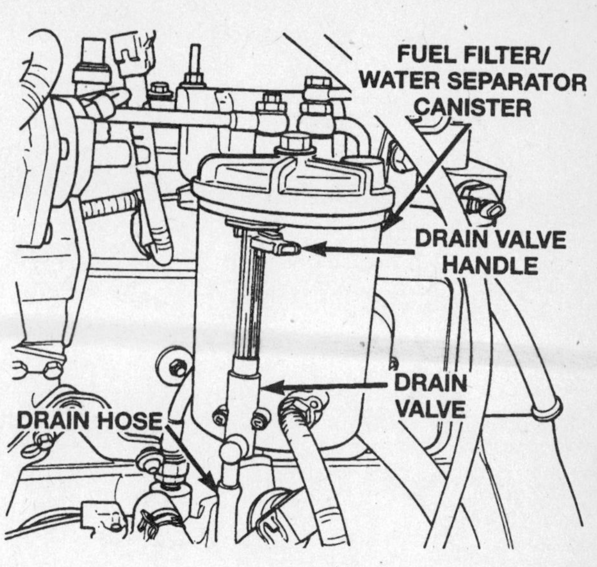

Fig. 63 Earlier model drain valve system components

1. Drain all the water and diesel fuel from the housing assembly as follows;
   a. Place a drain pan under the drain hose.
   b. With the engine off, either rotate the valve to "DRAIN" (earlier models) or lift the valve upward (later models) until all water and contaminants have been removed and the fluid stream is clear fuel.
   c. If the filter is being replaced, drain the housing completely until all water and fuel is removed.
   d. After the filter/separator is empty, close the valve handle.
2. Remove the fuel filter cap by removing the nut on top of the canister. On late models the "nut" is part of the cap.
3. The fuel filter will be removed with the cap.
4. Inspect the o-rings for cracks, distortion, or any other signs of deterioration. Replace as needed.
5. Clean and inspect the inside of the canister.
6. Install new o-ring(s) as needed.
7. Install proper filter element in the canister.
8. Fill the canister with clean diesel fuel.

Note: If the canister is not filled with clean diesel fuel prior to installation, manual air bleeding of the system will be necessary, follow the procedure outlined in "Driveability And Emission Controls".

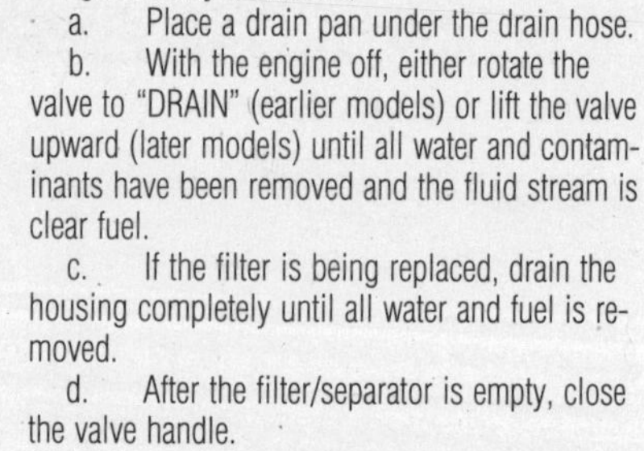

9. Apply a coating of diesel fuel to all seals.
10. Position cap to canister. On 1997-99 models, be sure locating tabs are properly lined up to avoid breaking the canister or having the drain valve handle contact the throttle linkage.
11. Install canister cap. On 1997-99 models, install and tighten nut to 10 ft. lbs. (14 Nm). On later models, tighten cap to 25 ft. lbs. (34 Nm).
12. Start engine and check for leaks by holding a piece of cardboard around seal area. Check cardboard for signs of diesel fuel.

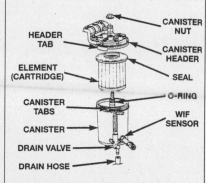

Fig. 64 On 1997-99 diesel powered models the canister nut and cap are separate parts...

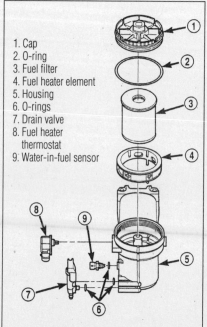

1. Cap
2. O-ring
3. Fuel filter
4. Fuel heater element
5. Housing
6. O-rings
7. Drain valve
8. Fuel heater thermostat
9. Water-in-fuel sensor

Fig. 65 ...on later models, cap and nut are one piece

PCV Valve

Note: All 3.9L, 4.7L, 5.2L, and 5.9L gasoline engines are equipped with a PCV valve and a closed crankcase ventilation system (CCV). The 2.5L and 8.0L engines are not fitted with PCV valves and do not require any maintenance.

When the engine is running, a small portion of gases which are formed in the combustion chamber leak by the piston rings entering the crankcase. Since these gases are under pressure they tend to escape the crankcase and enter into the atmosphere. If the gases are allowed to remain in the crankcase for any length of time, they would contaminate the engine oil and cause sludge to build. If the gases are allowed to escape into the atmosphere, they would pollute the air, as they contain unburned hydrocarbons. The crankcase ventilation system recycles these gases back into the combustion chambers, where they are burned.

Crankcase gases are recycled in the following manner. While the engine is running, fresh air is drawn into the engine and mixes with crankcase vapors. Manifold vacuum draws the crankcase vapors up into the intake and they are burned during the engine's normal combustion.

The Positive Crankcase Ventilation (PCV) system utilizes a vacuum-controlled PCV valve located in the valve cover or the oil filler housing (depending on model). This valve regulates the amount of gases which are recycled into the combustion chamber. At low engine speeds the valve is partially closed, limiting the flow of gases into the intake manifold. As engine speed increases, the valve opens to admit greater quantities of the gases into the intake manifold.

If the valve should become blocked or plugged, the gases will be prevented from escaping the crankcase by the normal route. Since these gases are under pressure, they will find their own way out of the crankcase. This alternate route is usually a weak oil seal or gasket in the engine. As the gas escapes by

GENERAL INFORMATION AND MAINTENANCE 1-17

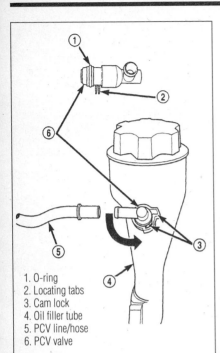

1. O-ring
2. Locating tabs
3. Cam lock
4. Oil filler tube
5. PCV line/hose
6. PCV valve

Fig. 66 Some models have the PCV valve located in the oil fill tube. This PCV valve must be rotated counterclockwise before pulling it out - 4.7L engine shown

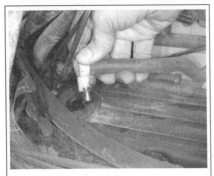

Fig. 67 Pull the PCV valve out from the rubber grommet

WARNING:

The 8.0L engine has a fixed orifice located in the right valve cover and is GREY in color. DO NOT interchange it or confuse it with the black fitting on the left valve cover.

REMOVAL & INSTALLATION

See Figures 66, 67, 68 and 69

Note: Procedures for basic maintenance are given here. For more detailed troubleshooting and diagnostic procedures, refer to "Driveability And Emission Controls".

1. The PCV valve can be found on the valve cover, except on 4.7L engines which have it on the oil filler tube.
2. To remove a valve cover PCV valve, simply pull it out of the rubber grommet. Disconnect the hose, if required.
3. To remove oil filler PCV valve, first disconnect the hose. Rotate the valve CCW until the locating tabs have been freed at cam lock. Then pull the valve straight out.
4. See "Driveability And Emission Controls" for PCV valve tests.

Fig. 68 Separate the PCV valve from the hose

5. Inspect the inside of the hose. If it is dirty, disconnect it from the intake manifold and clean it with a safe solvent.

Note: Do not attempt to clean a PCV valve. If you suspect it is not working, replace it.

To install:

6. If the PCV valve hose was removed, connect it to the intake manifold.
7. Connect the PCV valve to its hose.
8. Install the PCV valve into the rubber grommet.

Evaporative Canister

See Figures 70, 71 and 72

Gasoline-engined vehicles are fitted with an EVAP control system.

The EVAP control system prevents gasoline vapors from escaping into the atmosphere. When fuel evaporates from the fuel tank, the vapors pass through vent hoses or tubes to the carbon-filled EVAP canister. They are temporarily held in the canister until they can be drawn into the intake manifold when the engine is running.

Location and configuration of EVAP canister varies with model, year and engine. Most are mounted on a frame rail (left or right) forward of the fuel tank. 1997 Ram 1500-3500 trucks and all Dakota and Durango models use one EVAP canister,

the gasket, it also creates an oil leak. Besides causing oil leaks, a clogged PCV valve also allows these gases to remain in the crankcase for an extended period of time, promoting the formation of sludge in the engine.

The Closed Crankcase Ventilation (CCV) system operates the same way as a PCV system, however it does not utilize a vacuum controlled valve. A fitting of a calibrated size, referred to as a fixed orifice, meters the amount of crankcase vapors that the engine burns. This fitting can be found in the valve covers of engines which employ this system. No maintenance is required.

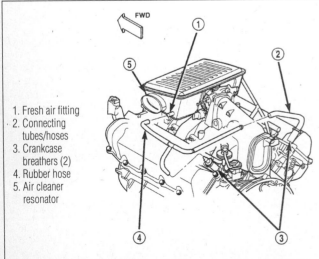

1. Fresh air fitting
2. Connecting tubes/hoses
3. Crankcase breathers (2)
4. Rubber hose
5. Air cleaner resonator

Fig. 69 The hoses used in the crankcase ventilation systems are just as important as the valves. Be sure to inspect the entire system whenever a problem is evident - 4.7L engine shown

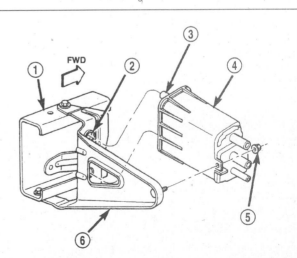

Fig. 70 Typical EVAP canister mounting - Dakota/Durango shown

1. Left frame rail
2. Rubber grommets (2)
3. Locating pins (2)
4. EVAP canister
5. Mounting nut
6. Mounting bracket

1-18 GENERAL INFORMATION AND MAINTENANCE

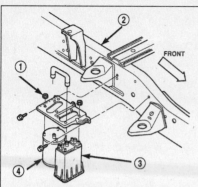

Fig. 71 Typical EVAP canister mounting - Ram 1500-3500 (1998 and later)

1. Mounting nuts
2. Frame rail (right)
3. Front EVAP canister
4. Rear EVAP canister

mounted to the left frame rail, in front of the fuel tank. 1998 and later Ram 1500-3500 trucks utilize two EVAP canisters, located side-by-side on the outside of the right frame rail, in front of the fuel tank.

The canister is purged by the EVAP canister purge solenoid at predetermined times and certain engine operating conditions.

The EVAP canister purge solenoid controls the vacuum that draws the vapors from the canister. The solenoid is controlled by the Powertrain Control Module (PCM). The PCM operates the solenoid by switching the ground circuit on or off. When energized the solenoid prevents vacuum from reaching the EVAP canister. When not energized, the solenoid allows vacuum to flow to the EVAP canister. The purge solenoid is mounted to a bracket located near the left-rear side of the intake manifold.

The solenoid opens when the engine reaches normal operating temperature and the vehicle is moving.

SERVICING

Note: The hoses used in this system are specifically manufactured for this application. If replacement becomes necessary, it is important to use only fuel resistant hose.

The EVAP canister is sealed and maintenance-free. However, a periodic inspection of the unit and vent hoses is advisable. Although the factory does not give recommendations here, you should inspect the unit at least every six months or 6000 miles (9600 km). During your inspection, make sure the unit is mounted firmly in its bracket, that it is not cracked and that the vent hoses are connected and not cracked. Correct any problems you may discover, or refer to the removal and installation procedure under "Driveability And Emission Controls" if you suspect the unit itself is malfunctioning.

Battery

Note: Diesel trucks utilize a two-battery system. Two batteries are required by these vehicles due to the immense amperage needed by the starter to turn the high-compression engine fast enough to facilitate combustion. Attention to proper battery maintenance is even more important for owners of these vehicles.

PRECAUTIONS

Always use caution when working on or near the battery. Never allow a tool to bridge the gap between the negative and positive battery terminals. Also, be careful not to allow a tool to provide a ground between the positive cable/terminal and any metal component on the vehicle. Either of these conditions will cause a short circuit, leading to sparks and possible personal injury.

Do not smoke, have an open flame or create sparks near a battery; the gases contained in the battery are very explosive and, if ignited, could cause severe injury or death.

All batteries, regardless of type, should be carefully secured by a battery hold-down device. If this is not done, the battery terminals or casing may crack from stress applied to the battery during vehicle operation. A battery which is not secured may allow acid to leak out, making it discharge faster; such leaking corrosive acid can also eat away at components under the hood.

Always visually inspect the battery case for cracks, leakage and corrosion. A white corrosive substance on the battery case or on nearby components would indicate a leaking or cracked battery. If the battery is cracked, it should be replaced immediately.

GENERAL MAINTENANCE

See Figure 73

A battery that is not sealed must be checked periodically for electrolyte level. You cannot add water to a sealed maintenance-free battery (though not all maintenance-free batteries are sealed); however, a sealed battery must also be checked for proper electrolyte level, as indicated by the color of the built-in hydrometer "eye."

Always keep the battery cables and terminals free of corrosion. Check these components about once a year. Refer to the removal, installation and cleaning procedures outlined in this section.

Keep the top of the battery clean, as a film of dirt can help completely discharge a battery that is not used for long periods. A solution of baking soda and water may be used for cleaning, but be careful to flush this off with clear water. DO NOT let any of the solution into the filler holes. Baking soda neutralizes battery acid and will de-activate a battery cell.

Batteries in vehicles which are not operated on a regular basis can fall victim to parasitic loads (small current drains which are constantly drawing current from the battery). Normal parasitic loads may drain a battery on a vehicle that is in storage and not used for 6 to 8 weeks. Vehicles that have additional accessories such as a cellular phone, an alarm system or other devices that increase parasitic load may discharge a battery sooner. If the vehicle is to be stored for longer than 6 weeks in a secure area and the alarm system, if present, is not necessary, the negative battery cable should be disconnected at the onset of storage to protect the battery charge.

Remember that constantly discharging and recharging will shorten battery life. Take care not to allow a battery to be needlessly discharged.

BATTERY FLUID

Check the battery electrolyte level at least once a month, or more often in hot weather or during periods of extended vehicle operation. On non-sealed batteries, the level can be checked either through the case on translucent batteries or by removing the cell caps on opaque-cased types. The electrolyte level in each cell should be kept filled to the split ring inside each cell, or the line marked on the outside of the case.

If the level is low, add only distilled water through the opening until the level is correct. Each cell is separate from the others, so each must be checked and filled individually. Distilled water should be used, because the chemicals and minerals found in most drinking water are harmful to the battery and could significantly shorten its life.

If water is added in freezing weather, the vehicle should be driven several miles to allow the water to mix with the electrolyte. Otherwise, the battery could freeze.

Although some maintenance-free batteries have removable cell caps for access to the electrolyte, the electrolyte condition and level on all sealed maintenance-free batteries must be checked using the built-in hydrometer "eye." The exact type of eye varies between battery manufacturers, but most apply a sticker to the battery itself explaining the possible readings. When in doubt, refer to the battery manufacturer's instructions to interpret battery condition using the built-in hydrometer.

Note: Although the readings from built-in hydrometers found in sealed batteries may vary, a green eye usually indicates a properly charged battery with sufficient fluid level. A dark eye is normally an indicator of a battery with sufficient fluid, but one which may be low in charge. And a light or yellow eye is usually an indication that electrolyte supply has dropped below the necessary

Fig. 72 EVAP canister mounted on the left frame rail of a 97 Ram 1500

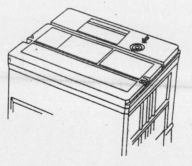

Fig. 73 A typical location for the built-in hydrometer on maintenance-free batteries

GENERAL INFORMATION AND MAINTENANCE 1-19

Fig. 75 On non-maintenance-free batteries, the fluid level can be checked through the case on translucent models; the cell caps must be removed on other models

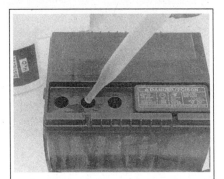

Fig. 76 If the fluid level is low, add only distilled water through the opening until the level is correct

level for battery (and hydrometer) operation. In this last case, sealed batteries with an insufficient electrolyte level must usually be discarded.

Checking the Specific Gravity

See Figures 75, 76 and 77

A hydrometer is required to check the specific gravity on all batteries that are not maintenance-free. On batteries that are maintenance-free, the specific gravity is checked by observing the built-in hydrometer "eye" on the top of the battery case. Check with your battery's manufacturer for proper interpretation of its built-in hydrometer readings.

CAUTION:

Battery electrolyte contains sulfuric acid. If you should splash any on your skin or in your eyes, flush the affected area with plenty of clear water. If it lands in your eyes, get medical help immediately.

The fluid (sulfuric acid solution) contained in the battery cells will tell you many things about the condition of the battery. Because the cell plates must be kept submerged below the fluid level in order to operate, maintaining the fluid level is extremely important. And, because the specific gravity of the acid is an indication of electrical charge, testing the fluid can be an aid in determining if the battery must be replaced. A battery in a vehicle with a properly operating charging system should require little maintenance, but careful, periodic inspection should reveal problems before they leave you stranded.

As stated earlier, the specific gravity of a battery's electrolyte level can be used as an indication of battery charge. At least once a year, check the specific gravity of the battery. It should be between 1.20 and 1.26 on the gravity scale. Most auto supply stores carry a variety of inexpensive battery testing hydrometers. These can be used on any non-sealed battery to test the specific gravity in each cell.

The battery testing hydrometer has a squeeze bulb at one end and a nozzle at the other. Battery electrolyte is sucked into the hydrometer until the float is lifted from its seat. The specific gravity is then read by noting the position of the float. If gravity is low in one or more cells, the battery should be slowly charged and checked again to see if the gravity has come up. Generally, if after charging, the specific gravity between any two cells varies more than 50 points (0.50), the battery should be replaced, as it can no longer produce sufficient voltage to guarantee proper operation.

CABLES

See Figures 78, 79, 80, 81 and 82

Once a year (or as necessary), the battery terminals and the cable clamps should be cleaned. Loosen the clamps and remove the cables, negative cable first. On batteries with posts on top, the use of a puller specially made for this purpose is recommended. These are inexpensive and available in most auto parts stores. Side terminal battery cables are secured with a small bolt.

Clean the cable clamps and the battery terminal with a wire brush, until all corrosion, grease, etc., is removed and the metal is shiny. It is especially important to clean the inside of the clamp thoroughly (an old knife is useful here), since a small deposit of foreign material or oxidation there will prevent a sound electrical connection and inhibit either starting or charging. Special tools are available for cleaning these parts, one type for conventional top post batteries and another type for side terminal batteries. It is also a good idea to apply some dielectric grease or petroleum jelly to the terminal, as this will aid in the prevention of corrosion.

After the clamps and terminals are clean, reinstall the cables, negative cable last; DO NOT hammer the clamps onto battery posts. Tighten the clamps securely, but do not distort them. Give the clamps and terminals a thin external coating of grease after installation, to retard corrosion.

WARNING:

The positive and negative terminals on top post batteries are not the same size. Keep this in mind when performing maintenance or installing replacement clamps.

Check the cables at the same time that the terminals are cleaned. If the cable insulation is cracked or broken, or if the ends are frayed, the cable should be replaced with a new cable of the same length and gauge. Note that battery cables can corrode beneath the insulation which may be hard to detect. The result may be an open circuit, poor starter operation, as well as other battery problems. Replacement of the cable(s) is the only solution.

CHARGING

CAUTION:

The chemical reaction which takes place in all batteries generates explosive hydrogen gas. A spark can cause the battery to explode and splash acid. To avoid serious personal injury, be sure there is proper ventilation and take appropriate fire safety precautions when connecting, disconnecting, or charging a battery and when using jumper cables.

A battery should be charged at a slow rate to keep the plates inside from getting too hot. However, if some maintenance-free batteries are allowed to discharge until they are almost "dead," they may have to be charged at a high rate to bring them back to "life." Always follow the charger manufacturer's instructions on charging the battery.

Fig. 77 Check the specific gravity of the battery's electrolyte with a hydrometer

Fig. 78 Maintenance is performed with household items and with special tools like this post cleaner

Fig. 79 The underside of this special battery tool has a wire brush to clean post terminals

1-20 GENERAL INFORMATION AND MAINTENANCE

Fig. 80 Place the tool over the battery posts and twist to clean until the metal is shiny

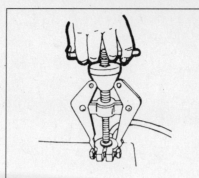

Fig. 81 A special tool is available to pull the clamp from the post

Fig. 82 The cable ends should be cleaned as well

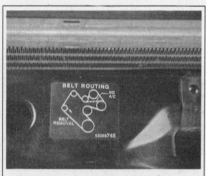

Fig. 83 Vehicle-specific belt routing is provided on a plate in the engine compartment

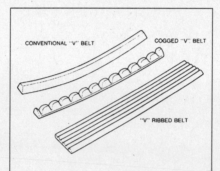

Fig. 84 There are typically 3 types of accessory drive belts found on vehicles today

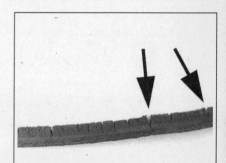

Fig. 85 Deep cracks in this belt will cause flex, building up heat that will eventually lead to belt failure

REPLACEMENT

When it becomes necessary to replace the battery, select one with an amperage rating equal to or greater than the battery originally installed. Deterioration and just plain aging of the battery cables, starter motor, and associated wires makes the battery's job harder in successive years. The slow increase in electrical resistance over time makes it prudent to install a new battery with a greater capacity than the old.

Belts

See Figure 83

WARNING:

Belt routing schematics are based on the latest information available at the time of publication. The vehicle's Belt Routing Label, located in the engine compartment, should be consulted when replacing belt.

INSPECTION

See Figures 84, 85, 86, 87, 88 and 89

CAUTION:

If equipped, the electrically operated cooling fan may come on under certain circumstances, even though the ignition is OFF. Be sure to disconnect the negative battery cable before servicing your vehicle.

Inspect the belts for signs of glazing or cracking. A glazed belt will be perfectly smooth from slippage, while a good belt will have a slight texture of fabric visible. Cracks will usually start at the inner edge of the belt and run outward. All worn or damaged drive belts should be replaced immediately. It is best to replace all drive belts at one time, as a preventive maintenance measure, during this service operation.

Check belt tension. For 2.5L engines, tension is checked with a belt tension gauge. For models with automatic tensioners, alignment marks are used to when belt replacement is required:

For 4.7L engines:

Check the belt if the distance between index marks exceeds 0.94 in. (24 mm).

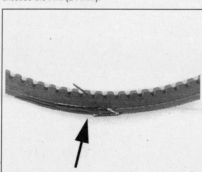

Fig. 86 The cover of this belt is worn, exposing the critical reinforcing cords to excessive wear.

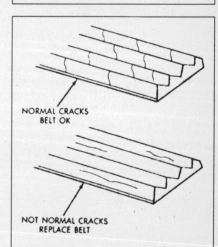

Fig. 87 Typical wear patterns for a serpentine drive belt

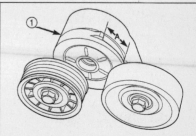

Fig. 88 Accessory drive belt wear indicator - 4.7L engine

GENERAL INFORMATION AND MAINTENANCE 1-21

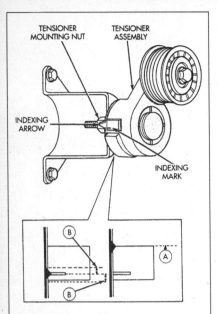

Fig. 89 The belt wear limit occurs when the indexing arrow reaches the travel limit indicator (A) - typical

For other engines

Check the belt if indexing arrow has moved to the tensioner travel limit as shown in the illustration. The arrow should be within 1/8 inch (3mm) of the indexing mark if a new belt is installed.

If wear indicators show that belt replacement is needed, check that belt wear is the cause. False readings can be caused by loose or worn pulleys, damaged bearings, loose or misaligned accessories or incorrectly fitted belts.

ADJUSTMENT

The 2.5L engine requires periodic adjustment. All other models are fitted with automatic belt tensioners. Be sure, however, to inspect/replace the belts at the proper intervals.

Belt Adjustment (2.5L)

See Figure 90

Belt tension is adjusted at the power steering pump bracket and idler pulley assembly. This procedure requires the use of a belt tension gauge.
1. Disconnect the battery negative cable.
2. Loosen the idler pulley center bolt.

Fig. 93 Rotating tensioner and removing serpentine belt

Fig. 90 Belt tension adjusting bolt (1) and idler pulley bolt (2) - 2.5L engine

3. Using the adjusting bolt, tension the belt to 140 - 160 lbs. for in-service belts. Tension for new belts (used 15 mins. or less) should be 180 - 200 lbs.)
4. Tighten the idler pulley center bolt to 42 ft. lbs. (57 Nm)
5. Recheck belt tension.

REMOVAL & INSTALLATION

See Figures 91, 92, 93, 94, 95, 96, 97, 98, 99, 100, 101, 102 and 103

2.5L Engine

1. Loosen the idler pulley bolt, then the adjusting bolt to remove tension from the belt.
2. Remove the belt.
3. Check pulleys for damage.
4. Install belt per diagram.
5. Adjust tension as described in the "Belt Adjustment (2.5L)" section, above.

All Other Models

See Figure 93

Use a socket, wrench or ratchet on the tensioner pulley to rotate the tensioner arm, loosen the belt and disengage it from one of the fixed pulleys. To install, place the belt over all pulleys but the tensioner pulley, then rotate the tensioner arm in the same manner and slip the belt over the pulley. When you release it, the spring-loaded tensioner will automatically apply the correct tension to the belt.

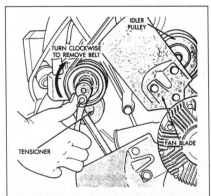

Fig. 94 Some tensioners may be rotated clockwise to remove tension...

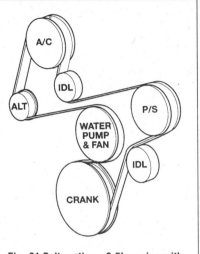

Fig. 91 Belt routing - 2.5L engine with A/C

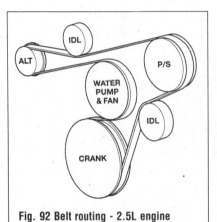

Fig. 92 Belt routing - 2.5L engine without A/C

Note: Depending on model, engine and belt driven accessories, the tensioner arm may be rotated clockwise or counter clockwise to relieve tension. Be sure to check this by attempting to move the tensioner in either direction. Also, either a wrench, a socket and ratchet, a plain ratchet, or a special belt tensioner tool may be used.

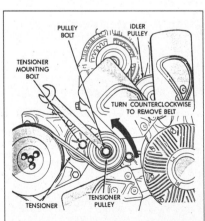

Fig. 95 ... while others must be rotated counterclockwise. Check before applying excessive force

1-22 GENERAL INFORMATION AND MAINTENANCE

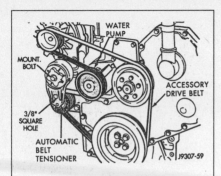

Fig. 96 Diesel engines may use a square hole instead of a bolt or nut to rotate the tensioner arm. Use a ratchet without any socket for these vehicles

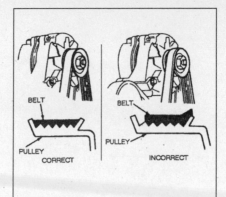

Fig. 97 Verify serpentine belt alignment in the pulley

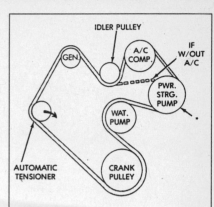

Fig. 98 Serpentine belt routing - 3.9L, V6 5.2L/5.9L V-8 LDC-gas engines

CAUTION:
Be sure the tool used to rotate the tensioner is the proper size. Bodily harm may occur from tool slippage due to the pressure applied by the tensioner.

1. Disconnect the negative battery cable.
2. Attach the proper size socket, wrench, or ratchet to the automatic belt tensioner pulley bolt or square hole.
3. Rotate the tensioner arm until tension has been removed from the belt.
4. Remove the belt from the tensioner pulley, then slowly allow the tensioner arm to rotate back until it stops.
5. Remove the belt from the remaining pulleys.

To install:

6. Position the belt over all of the pulleys except the tensioner.

WARNING:
The drive belt must be installed exactly as the manufacturer states. Refer to the Belt Routing Label in the engine compartment to ensure proper installation. Unless the belt is correctly routed, engine or accessories may be damaged due components rotating in the wrong direction.

7. Attach the proper size socket, wrench, or ratchet to the automatic belt tensioner pulley bolt.
8. Rotate the tensioner until the stop is reached. Carefully position the belt over the tensioner pulley

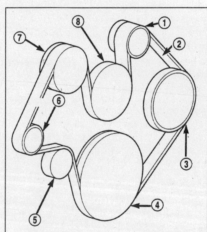

Fig. 99 Serpentine belt routing - 4.7L engine

1. Generator
2. Accessory drive belt
3. Power steering pump pulley
4. Crankshaft pulley
5. Idler pulley
6. Tensioner pulley
7. A/C compressor pulley
8. Water pump pulley

and release it slowly until it contacts the belt.

9. Check the tensioner indexing marks to make sure the proper new belt is installed and the tensioner is not worn beyond its limits.
10. Check to make sure the belt is seated properly on all the pulleys. Check for proper belt routing.
11. Connect the negative battery cable.
12. Start the engine and observe the belt to ensure proper operation.

CAUTION:
Do not attempt to disassemble the automatic belt tensioner. High spring pressure may cause flying parts and bodily injury.

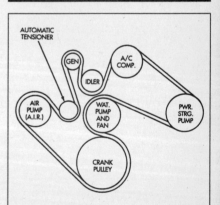

Fig. 100 Serpentine belt routing - 5.9L HDC-gas engines and 8.0L V-10 engines with A/C

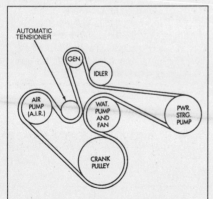

Fig. 101 Serpentine belt routing - 5.9L HDC-gas engines and 8.0L V-10 engines without A/C

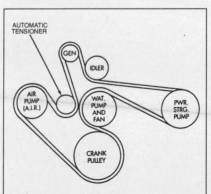

Fig. 102 Serpentine belt routing - 5.9L diesel engines without air conditioning

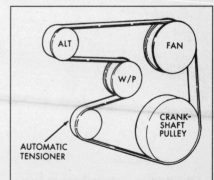

Fig. 103 Serpentine belt routing - 5.9L diesel engines with air conditioning

GENERAL INFORMATION AND MAINTENANCE 1-23

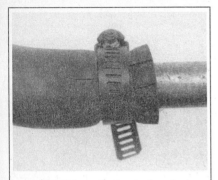

Fig. 104 The cracks developing along this hose are a result of age and hardening

Fig. 105 A hose clamp that is too tight can cause older hoses to separate and tear

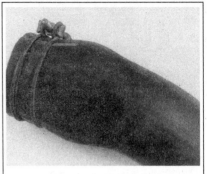

Fig. 106 A soft, spongy hose like this is identifiable by the swollen section near the clamp.

Hoses

INSPECTION

See Figures 104, 105, 106 and 107

CAUTION:

On models equipped with an electric cooling fan, disconnect the negative battery cable or fan motor wiring harness connector, before replacing any radiator/heater hose. The fan may come on, under certain circumstances, even though the ignition is OFF.

Upper and lower radiator hoses along with the heater hoses should be checked for deterioration, leaks and loose hose clamps at least every 15,000 miles (24,000 km). Early spring and at the beginning of the fall or winter, when you are performing other maintenance, are good times to perform the inspection. Make sure the engine and cooling system are cold. Visually inspect for cracking, rotting or collapsed hoses, replace as necessary. Run your hand along the length of the hose. If a weak or swollen spot is noted when squeezing the hose wall, replace the hose.

REMOVAL & INSTALLATION

CAUTION:

Never remove the pressure cap while the engine is running or personal injury from scalding hot coolant or steam may result. If possi-

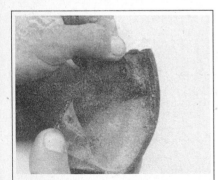

Fig. 107 Debris in the coolant can weaken the hose from the inside

ble, wait until the engine has cooled to remove the pressure cap. If this is not possible, wrap a thick cloth around the pressure cap and turn it slowly to the stop. Step back while the pressure is released from the cooling system. When you are sure all the pressure has been released, still using the cloth, turn and remove the cap.

Note: The vehicles covered in this manual come from the manufacturer with both the worm-drive hose clamp and the spring type hose clamp. The worm-drive is the clamp which uses a screw or bolt held against the band to hold tension. The spring type maintains constant tension by being constructed of a spring steel band. It may be necessary to replace the clamps with the exact replacement due to space limitations and accessibility.

CAUTION:

Use care when removing or installing the spring type clamps. Special constant-tension clamp pliers are available. If these pliers are not used, the tool may slip and cause bodily harm.

1. Drain the cooling system into a suitable container (if the coolant is to be reused, make sure the container is clean) by loosening the draincock on the bottom of the radiator.

CAUTION:

When draining the coolant, keep in mind that cats and dogs are attracted by ethylene glycol antifreeze, and are quite likely to drink any that is left in an uncovered container or in puddles on the ground. This will prove fatal in sufficient quantity. Always drain the coolant into a sealable container. Coolant should be reused unless it is contaminated or several years old.

2. Loosen the hose clamps at each end of the hose that requires replacement.
   a. Rotate the screw or bolt on the worm-drive clamps counterclockwise to loosen.
   b. Squeeze the tabs on the spring type together to remove tension.
3. Twist, pull and slide the hose off the radiator, water pump, thermostat or heater connection.

Note: If the hose is stuck at the connection, do not try to insert a screwdriver or other sharp tool under the hose end in an effort to free it, as the connection and/or hose may become damaged. Heater connections especially may be easily damaged by such a procedure. If the hose is to be replaced, use a single-edged razor blade to make a slice along the portion of the hose which is stuck on the connection, perpendicular to the end of the hose. Do not cut deep so as to prevent damaging the connection. The hose can then be peeled from the connection and discarded.

4. Clean both hose mounting connections. Inspect the condition of the hose clamps and replace them, if necessary.

To install:

5. Dip the ends of the new hose into clean engine coolant to ease installation.
6. Position the clamps on the new hose.
7. Coat the connection surfaces with a water resistant sealer and slide the hose into position. Make sure the hose clamps are located beyond the raised bead of the connector (if equipped) and centered in the clamping area of the connection.
8. Tighten the worm-drive clamps to 20 - 30 inch lbs. (2 - 3 Nm). Do not overtighten.
9. Be sure the radiator draincock is closed, then fill the cooling system.

CAUTION:

If you are checking for leaks with the system at normal operating temperature, BE EXTREMELY CAREFUL not to touch any moving or hot engine parts. Once temperature has been reached, shut the engine OFF, and check for leaks around the hose fittings and connections which were removed earlier.

10. Start the engine and allow it to reach normal operating temperature. Check for leaks.

CV-Boots

INSPECTION

See Figures 108 and 109

The CV (Constant Velocity) boots should be checked for damage each time the oil is changed and

1-24 GENERAL INFORMATION AND MAINTENANCE

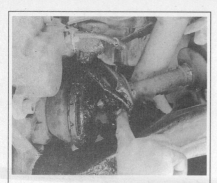

Fig. 108 A torn boot should be replaced immediately

Fig. 109 CV-boots must be inspected periodically for damage

any other time the vehicle is raised for service. These boots keep water, grime, dirt and other damaging matter from entering the CV-joints. Any of these could cause early CV-joint failure which can be expensive to repair. Heavy grease thrown around the inside of the front wheel(s) and on the brake caliper/drum can be an indication of a torn boot. Thoroughly check the boots for missing clamps and tears. If the boot is damaged, it should be replaced immediately. Please refer to "Suspension And Steering" for procedures.

Spark Plugs

SPARK PLUG HEAT RANGE

See Figure 110

Spark plug heat range is the ability of the plug to dissipate heat. The longer the insulator (or the farther it extends into the engine), the hotter the plug will operate; the shorter the insulator (the closer the electrode is to the block's cooling passages) the cooler it will operate. A plug that absorbs little heat and remains too cool will quickly accumulate deposits of oil and carbon since it is not hot enough to burn them off. This leads to plug fouling and consequently to misfiring. A plug that absorbs too much heat will have no deposits but, due to the excessive heat, the electrodes will burn away quickly and might possibly lead to preignition or other ignition problems. Preignition takes place when plug tips get so hot that they glow sufficiently to ignite the air/fuel mixture before the actual spark occurs. This early ignition will usually cause a pinging during low speeds and heavy loads.

The general rule of thumb for choosing the correct heat range when picking a spark plug is: if most of your driving is long distance, high speed travel, use a colder plug; if most of your driving is stop and go, use a hotter plug. Original equipment plugs are generally a good compromise between the two styles and most people never have the need to change their plugs from the factory-recommended heat range.

RECOMMENDED SPARK PLUGS

Manufacturer's spark plug recommendations may change due to field experience. Refer to your owner's manual for the correct plug for your vehicle. Typical recommendations for the vehicles covered in this manuals are:
- 2.5L: RC12ECC
- 3.9L: RC12LC4
- 4.7L: RC12MCC4
- 5.2L: RC12LC4
- 5.9L ('97): RC12YC
- 5.9L (later): RC12LC4
- 8.0L: QC9MC4

Refer to the Gasoline Engine Tune-Up Specifications chart for the proper gap setting.

WARNING:

The 4.7L V-8 engine is equipped with copper core ground electrode spark plugs. They must be replaced with the same type/number. If another type of plug is used, pre-ignition will result.

REMOVAL & INSTALLATION

See Figure 111

A set of spark plugs usually requires replacement after about 20,000 - 30,000 miles (32,000 - 48,000 km), depending on your style of driving. In normal operation plug gap increases about 0.001 in. (0.025mm) for every 2500 miles (4000 km). As the gap increases, the plug's voltage requirement also increases. It requires a greater voltage to jump the wider gap and about two to three times as much voltage to fire the plug at high speeds than at idle. The improved air/fuel ratio control of modern fuel injection combined with the higher voltage output of modern ignition systems will often allow an engine to run significantly longer on a set of standard spark plugs, but keep in mind that efficiency will drop as the gap widens (along with fuel economy and power).

When you're removing spark plugs, work on one at a time. Don't start by removing the plug wires all at once, because, unless you number them, they may become mixed up. Take a minute before you begin and number the wires with tape.

1. Disconnect the negative battery cable, and if the vehicle has been run recently, allow the engine to thoroughly cool.
2. Except 4.7L: Carefully twist the spark plug wire boot to loosen it, then pull upward and remove the boot from the plug. Be sure to pull on the boot and not on the wire, otherwise the connector located inside the boot may become separated. With modern plug boots, the use of the special pliers is recommended. This inexpensive tool is widely available.
3. 4.7L: Plugs are located beneath the ignition coils. Disconnect the coil wiring, remove the mounting nut and lift the coil up with a slight twisting motion to remove.
4. Using compressed air, blow any water or debris from the spark plug well to assure that no harmful contaminants are allowed to enter the combustion

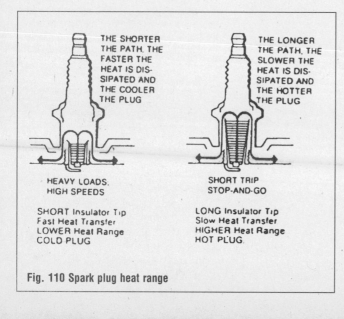

Fig. 110 Spark plug heat range

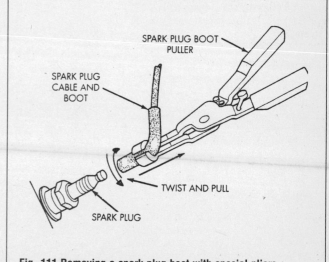

Fig. 111 Removing a spark plug boot with special pliers

GENERAL INFORMATION AND MAINTENANCE 1-25

A **normally worn** spark plug should have light tan or gray deposits on the firing tip.

A **carbon fouled** plug, identified by soft, sooty, black deposits, may indicate an improperly tuned vehicle. Check the air cleaner, ignition components and engine control system.

A **physically damaged** spark plug may be evidence of severe detonation in that cylinder. Watch that cylinder carefully between services, as a continued detonation will not only damage the plug, but could also damage the engine.

This spark plug has been **left in the engine too long**, as evidenced by the extreme gap- Plugs with such an extreme gap can cause misfiring and stumbling accompanied by a noticeable lack of power.

An **oil fouled** spark plug indicates an engine with worn piston rings and/or bad valve seals allowing excessive oil to enter the chamber.

A **bridged or almost bridged** spark plug, identified by a build-up between the electrodes caused by excessive carbon or oil build-up on the plug.

Fig. 112 Inspect the spark plug to determine engine running conditions

chamber when the spark plug is removed. If compressed air is not available, use a rag or a brush to clean the area.

Note: Remove the spark plugs when the engine is cold, if possible, to prevent damage to the threads. If removal of the plugs is difficult, apply a few drops of penetrating oil or silicone spray to the area around the base of the plug, and allow it a few minutes to work.

5. Using a spark plug socket that is equipped with a rubber insert to properly hold the plug, turn the spark plug counterclockwise to loosen and remove the spark plug from the bore.

WARNING:

Be sure not to use a flexible extension on the socket. Use of a flexible extension may allow a shear force to be applied to the plug. A shear force could break the plug off in the cylinder head, leading to costly and frustrating repairs.

To install:

6. Inspect the spark plug boot for tears or damage. If a damaged boot is found, the spark plug wire must be replaced.
7. Using a wire feeler gauge, check and adjust the spark plug gap (refer to the **Gasoline Engine Tune-Up Specifications** chart for the proper gap setting). When using a gauge, the proper size should pass between the electrodes with a slight drag. The next larger size should not be able to pass while the next smaller size should pass freely.
8. Carefully thread the plug into the bore by hand. If resistance is felt before the plug is almost completely threaded, back the plug out and begin threading again. In small, hard to reach areas, an old spark plug wire and boot could be used as a threading tool. The boot will hold the plug while you twist the end of the wire and the wire is supple enough to twist before it would allow the plug to crossthread.

WARNING:

Do not use the spark plug socket to thread the plugs. Always carefully thread the plug by hand or using an old plug wire to prevent the possibility of crossthreading and damaging the cylinder head bore.

9. Carefully tighten the spark plug. If the plug you are installing is equipped with a crush washer, seat the plug, then tighten about 1/4 turn to crush the washer.
10. Tighten the plugs to proper torque: 30 ft. lbs. on all engines except the 4.7L V8, which are tightened to 20 ft. lbs.
11. Apply a small amount of silicone dielectric compound to the end of the spark plug lead or inside the spark plug boot to prevent sticking, then install the boot to the spark plug and push until it clicks into place. The click may be felt or heard, then gently pull back on the boot to assure proper contact.

INSPECTION & GAPPING

See Figures 112, 113, 114, 115, 116 and 117

Check the plugs for deposits and wear. If they are not going to be replaced, clean the plugs thoroughly. Remember that any kind of deposit will decrease the efficiency of the plug. Plugs can be cleaned on a spark plug cleaning machine, which can sometimes be found in service stations, or you can do an acceptable job of cleaning with a small, hand-held wire brush. Do not use a powered wire wheel to clean plugs.

If the plugs are cleaned, the electrodes must be filed flat. Use an ignition points file, not an emery board or the like, which will leave deposits. The electrodes must be filed perfectly flat with sharp edges; rounded edges reduce the spark plug voltage by as much as 50%.

Check spark plug gap before installation. The ground (side) must be parallel to the center electrode and the specified size wire gauge (please refer to the Tune-Up Specifications chart for details) must pass between the electrodes with a slight drag.

Fig. 113 A variety of tools and gauges are needed for spark plug service

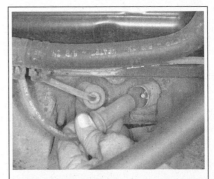
Fig. 114 Removing the boot from the heat shield

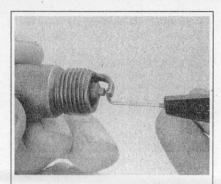

Fig. 115 Checking the spark plug gap with a feeler gauge

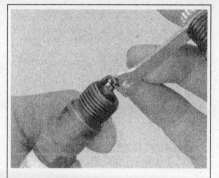

Fig. 116 Adjusting the spark plug gap

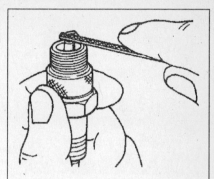

Fig. 117 If a standard plug is in good condition, the electrode may be filed flat - WARNING: do not file platinum plugs

Note: NEVER adjust the gap on a used platinum type spark plug.

Always check the gap on new plugs as they are not always set correctly at the factory. Do not use a flat feeler gauge when measuring the gap on a used plug, because the reading may be inaccurate. A round-wire type gapping tool is the best way to check the gap. The correct gauge should pass through the electrode gap with a slight drag. If you're in doubt, try one size smaller and one larger. The smaller gauge should go through easily, while the larger one shouldn't go through at all. Wire gapping tools usually have a bending tool attached. Use that to adjust the side electrode until the proper distance is obtained. Absolutely never attempt to bend the center electrode. Also, be careful not to bend the side electrode too far or too often as it may weaken and break off within the engine, requiring removal of the cylinder head to retrieve it.

Spark Plug Wires

TESTING

See Figure 118

At every tune-up/inspection, visually check the spark plug wires for burns, cuts, or breaks in the insulation. Check the boots and the nipples on the distributor cap and/or coil. Replace any damaged wiring.

Every 50,000 miles (80,000 Km) or 60 months, the resistance of the wires should be checked with an ohmmeter. Wires with excessive resistance will cause misfiring, and may make the engine difficult to start in damp weather.

Minimum spark plug wire resistance should be 250 ohms per inch (63.5 ohms per centimeter) and maximum resistance should be 1000 ohms per inch (250 ohms per centimeter).

REMOVAL & INSTALLATION

See Figure 119

Note: The vehicles covered in this manual may have heat shields, which were installed by the manufacturer, located around the spark plugs. These are designed to protect the spark plug boots from the intense heat of the exhaust manifold. If they are still in place, do not remove them. If they are not in place, be sure to inspect the boots more frequently.

1. Mark all the spark plug wires near the cap with their cylinder number for identification purposes. On some models factory wires have the cylinder number already on them.
2. Grasp and twist the spark plug boot until the boot comes free of the plug. Do not pull on the wire.
3. Repeat for the other spark plug wires.

To install:

4. When installing a new set of spark plug cables, replace the cables one at a time so there will be no mix-up. Start by replacing the longest cable first.

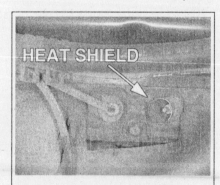

Fig. 119 These heat shields, which were installed by the manufacturer, are designed to protect the spark plug boots from the intense heat of the exhaust manifold. If they are still in place, do not remove them. If they are not in place, be sure to inspect the boots more frequently

Install the boot firmly over the spark plug. Route the wire exactly the same as the original. Insert the nipple firmly into the tower on the distributor cap.

5. Repeat the process for each cable.

Distributor Cap and Rotor

Note: Not all vehicles covered by this manual use a distributor type ignition system. On vehicles with a distributorless ignition system, the coil towers should be inspected for wear or signs of misfire. Refer to the procedure for distributor cap inspection.

REMOVAL & INSTALLATION

See Figures 120, 121, 122 and 123

1. Disconnect the negative battery cable.
2. Remove the distributor cap retaining screws and spark plug wires from the cap. If there is no wire retainer, be sure to identify all wires before disconnecting them from the distributor cap. This will help preserve the proper firing order and greatly ease cap installation.

Note: An alternate way to change caps is to situate the new cap next to the old (with the wires still attached) and simply swap the wires one-by-one from the old cap to the new cap in the exact same order.

3. On 2.5L engines, remove the distributor splash shield. Remove the cap screws.
4. On V6 and V8 engines, unfasten the retaining clips or remove the screws and lift off the cap assembly.

Fig. 118 Checking individual plug wire resistance with a digital ohmmeter

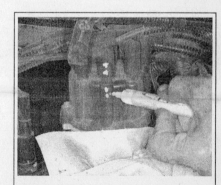

Fig. 120 Mark distributor towers and spark plug wires to ease installation

GENERAL INFORMATION AND MAINTENANCE 1-27

Fig. 121 Removing the distributor cap screws

Fig. 122 Lift off the cap

Fig. 123 Removing the rotor from the shaft

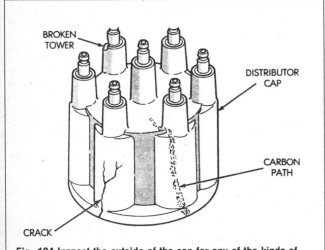

Fig. 124 Inspect the outside of the cap for any of the kinds of damage or wear that would warrant replacement

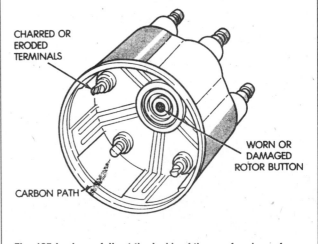

Fig. 125 Look carefully at the inside of the cap for signs of significant damage or wear

5. Scribe or paint a mark on the edge of the distributor housing to indicate the position of the rotor. This will serve as a reference upon reinstallation, if needed.

Note: Although most rotors can only be installed in one direction, it is still wise to note the position of the rotor before removal to assure proper installation and ignition timing.

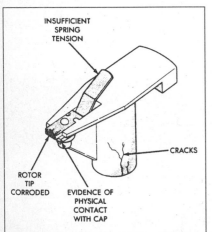

Fig. 126 Carefully inspect the rotor after removal, replacement is not necessary unless wear is evident

6. Remove the rotor attaching screw(s), if equipped, or simply pull the rotor from the distributor.

To install:

7. Install the rotor facing in the direction noted earlier, then if applicable, secure the rotor using the attaching screw(s).
8. Install the cap to the distributor assembly and secure using the cap retaining clips or screws.
9. Install the spark plug wires and retainer to the cap. If no wire retainer is used, carefully connect the wires to the cap as tagged during removal.
10. Connect the negative battery cable.

INSPECTION

See Figures 124, 125 and 126

Remove the distributor cap and wipe it clean with a dry lint-free cloth. Visually inspect the cap for cracks, carbon paths, broken towers, or damaged rotor buttons. Also check for white deposits on the inside (caused by condensation entering through cracks). Replace any cap that has charred or eroded terminals. The inside flat surface of a terminal end (faces toward the rotor) will indicate some evidence of erosion from normal operation. Examine the terminal ends for evidence of mechanical interference with the rotor tip.

Ignition Timing (Gasoline Engines)

GENERAL INFORMATION

See Figures 127, 128, 129, 130 and 131

Ignition timing is the measurement, in degrees of crankshaft rotation, of the point at which the spark plugs fire in each of the cylinders. It is measured in degrees before or after Top Dead Center (TDC) of the compression stroke.

Ideally, the air/fuel mixture in the cylinder will be ignited by the spark plug just as the piston passes TDC of the compression stroke. If this happens, the piston will be beginning the power stroke just as the compressed and ignited air/fuel mixture starts to expand. The expansion of the air/fuel mixture then forces the piston down on the power stroke and turns the crankshaft.

Because it takes a fraction of a second for the spark plug to ignite the mixture in the cylinder, the spark plug must fire a little before the piston reaches TDC. Otherwise, the mixture will not be completely ignited as the piston passes TDC, and the full power of the explosion will not be used by the engine.

As the engine speed increases, the pistons go faster. The spark plugs therefore have to ignite the fuel even sooner if it is to be completely ignited when the piston reaches TDC.

1-28 GENERAL INFORMATION AND MAINTENANCE

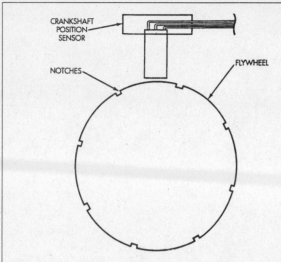

Fig. 127 The notches on the reluctor wheel trigger the Crankshaft Position (CKP) sensor to signal the Powertrain Control Module (PCM) on crankshaft position and speed - 5.2L and 5.9L gasoline engines shown

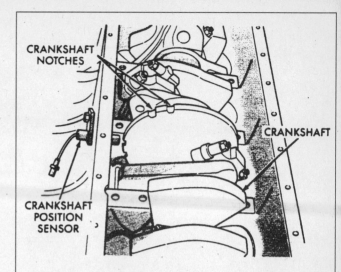

Fig. 128 The reluctor wheel is located on the crankshaft - 8.0L engine shown

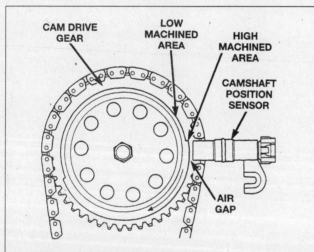

Fig. 129 The Camshaft Position (CMP) sensor is used along with the CKP sensor to determine spark timing and fuel injection synchronization - 8.0L engine shown

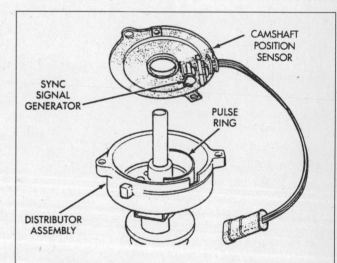

Fig. 130 On some engines, the CMP is located within the distributor assembly - 3.9L and 5.2L engines shown

The Powertrain Control Module (PCM) receives inputs from Crankshaft Position (CKP) sensors and Camshaft Position (CMP) sensors along with various other sensors to control spark timing. Refer "Engine Electrical" for more information.

INSPECTION & ADJUSTMENT

Note: Base ignition timing is not adjustable on any engine covered in this manual. The distributors do not have centrifugal or vacuum advance. Instead, base ignition timing and advance on each of these engines is controlled by its Powertrain Control Module (PCM).

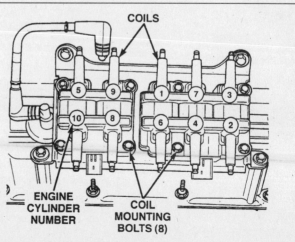

Fig. 131 Distributorless ignition systems fire two plugs at the same time. Refer to "Engine Electrical" for more information - 8.0L engine

GENERAL INFORMATION AND MAINTENANCE 1-29

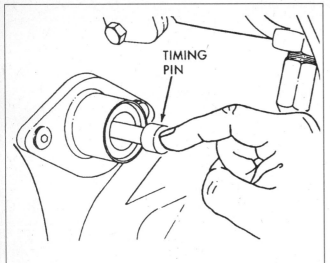

Fig. 132 Use the timing pin to locate TDC for diesel engines - 1997 - 98 models

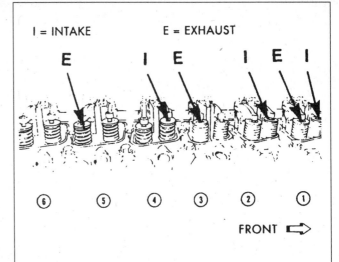

Fig. 133 Adjust the intake and exhaust valves shown in this first phase of the diesel valve adjustment procedure - 1997 - 98 models

Valve Lash

Gasoline Engines

No valve lash adjustment is necessary or possible on any of the gasoline engines covered in this manual. Hydraulic valve lifters automatically maintain zero clearance. After engine reassembly, these lifters adjust themselves as soon as engine oil pressure builds up.

Diesel Engine (1997 - 98)

See Figures 132, 133 and 134

Valve adjustment is required on the 12 - valve diesel engine every 24,000 miles (38,000 km).

Note: The timing pin is used in this procedure to locate Top Dead Center (TDC). It is found at the back of the gear housing and below the injection pump. Be sure to disengage the timing pin after locating TDC.

WARNING:

Do not set the valve lash closer than specified in an attempt to quiet the lifters. This will only result in burned valves.

1. The engine must be cold for this adjustment (below 140-degrees F/60-degrees C).
2. Disconnect the negative battery cable.
3. Remove the valve cover.
4. Manually turn the engine and use the timing pin to locate Top Dead Center (TDC) for cylinder No. 1. Disengage the timing pin after locating TDC.
5. Perform the following two-step procedure to adjust the valves. Refer to the accompanying illustrations to determine which valves to adjust in each of the steps:

   a. Valve lash is measured between the rocker arm and the end of the valve. Check the lash by inserting a feeler gauge between the rocker arm and the valve. Adjust the clearance, if necessary. The clearance for the intake valves is 0.010 in. (0.254mm). The clearance for the exhaust valves is 0.20 in. (0.508mm). Adjust the lash by loosening the locknut on the rocker arm and turning the adjusting screw. After the adjustment is made, tighten the locknut to 18 ft. lbs. (24 Nm) and recheck the lash to be sure it did not change as the locknut was being tightened.

   b. Double-check that the timing pin is disengaged, then mark the pulley and turn the engine crankshaft 360-degrees in the normal direction of rotation (clockwise). Check and adjust the clearance of the indicated valves following the same specifications as in Step 4a.

6. Install the rocker cover with a new gasket, then attach the fuel line to the injector. Start the engine and check for leaks.

Diesel Engine (1999 - 2000)

See Figures 135 and 136

Valve adjustment is not a routine maintenance item. Measurements should be taken while troubleshooting an operating problem or after valve train components have been removed/replaced.

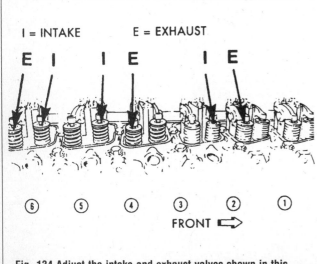

Fig. 134 Adjust the intake and exhaust valves shown in this second stage of the diesel valve adjustment - 1997 - 98 models

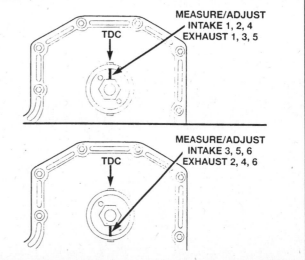

Fig. 135 Align the timing marks and check clearances at valves shown for each position.

GENERAL INFORMATION AND MAINTENANCE

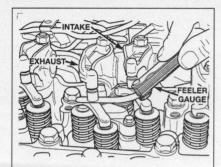

Fig. 136 Checking clearance with a feeler gauge

1. Measurements must be taken on a COLD engine (coolant temperature below 140-degrees F/60-degrees C).
2. Refer to the accompanying illustration. Rotate the crankshaft to align the timing marks as shown and measure the clearance with a feeler gauge between rocker arm socket and crosshead at the appropriate valves as shown in the illustration.
3. If clearance is not within specification, loosen the locknut and turn the adjusting screw as required to achieve the proper clearance. Tighten the locknut and recheck.

Idle Speed and Mixture Adjustments

GASOLINE ENGINES

Unlike carbureted vehicles, the fuel injected engines covered in this manual do not require idle speed adjustment checks as a part of a regular tune-up. In fact, normal idle speed is NOT adjustable and should not be attempted. Base idle speed is controlled by the engine controller (computer).

Note: This adjustment is the minimum idle speed with the Automatic Idle Speed (AIS) closed.

1. Before adjusting the idle on an electronic fuel injected vehicle, the following items must be checked (and corrected, if necessary):
- AIS motor operation
- Engine vacuum and/or EGR leaks
- Engine timing
- Coolant temperature sensor operation

2. Connect a tachometer and timing light to the engine, according to the manufacturer's instructions.
3. Close the AIS by using ATM tester C-4805 or equivalent, and ATM test code #03.
4. Connect a jumper to the radiator fan, so that it will run continuously.
5. Start and run the engine for 3 minutes to allow the idle speed to stabilize.
6. Check engine rpm and compare the result with the specifications listed on the underhood emission control sticker.
7. If idle rpm is not within specifications, use tool C-4804 or equivalent to turn the idle speed adjusting screw to obtain 790-810 rpm. If the underhood emission sticker specifications are different, use those values for adjustment.

Note: If idle will not adjust down check for binding linkage, speed control servo cable adjustment or throttle shaft binding.

8. Turn the engine OFF, then disconnect the tachometer, reattach the AIS wire, and remove the jumper wire from the fan motor.

DIESEL ENGINE

The low speed idle is the only adjustment which can be made to the injection pump. All other injection pump service requires a qualified diesel engine technician.

Use an optical diesel engine tachometer to read engine rpm. If necessary, adjust low idle speed at the low idle speed screw. If the vehicle is equipped with an automatic transmission, with the A/C on, set the low speed idle to 700 rpm. If equipped with a manual transmission, with the A/C on, set the idle to 750 rpm. A special Cummins tool No. 3823750 is required to loosen the idle speed locknut for adjustment.

GASOLINE ENGINE TUNE-UP SPECIFICATIONS

Year	Engine Displacement Liters (cc)	Engine ID/VIN	Spark Plug Gap (in.)	Ignition Timing (deg.)	Fuel Pump (psi)	Idle Speed (rpm)	Valve Clearance Intake	Valve Clearance Exhaust
1997	2.5 (2458)	P	0.035	①	49.2	②	HYD	HYD
	3.9 (3916)	X	0.035	①	49.2	②	HYD	HYD
	5.2 (5211)	Y	0.035	①	49.2	②	HYD	HYD
	5.9 (5899)	5	0.035	①	49.2	②	HYD	HYD
	5.9 (5899)	Z	0.035	①	49.2	②	HYD	HYD
	8.0 (7997)	W	0.045	①	49.2	②	HYD	HYD
1998	2.5 (2464)	P	0.035	①	44.2 - 54.2	②	HYD	HYD
	3.9 (3916)	X	0.040	①	44.2 - 54.2	②	HYD	HYD
	5.2 (5208)	Y	0.040	①	44.2 - 54.2	②	HYD	HYD
	5.9 (5825)	5	0.040	①	44.2 - 54.2	②	HYD	HYD
	5.9 (5899)	Z	0.040	①	44.2 - 54.2	②	HYD	HYD
	8.0 (7994)	W	0.045	①	44.2 - 54.2	②	HYD	HYD
1999	2.5 (2464)	P	0.035	①	44.2 - 54.2	②	HYD	HYD
	3.9 (3916)	X	0.040	①	44.2 - 54.2	②	HYD	HYD
	5.2 (5208)	Y	0.040	①	44.2 - 54.2	②	HYD	HYD
	5.9 (5825)	5	0.040	①	44.2 - 54.2	②	HYD	HYD
	5.9 (5899)	Z	0.040	①	44.2 - 54.2	②	HYD	HYD
	8.0 (7994)	W	0.045	①	44.2 - 54.2	②	HYD	HYD
2000	2.5 (2464)	P	0.035	①	44.2 - 54.2	②	HYD	HYD
	3.9 (3916)	X	0.040	①	44.2 - 54.2	②	HYD	HYD
	4.7 (4698)	N	0.040	①	44.2 - 54.2	②	HYD	HYD
	5.2 (5208)	Y	0.040	①	44.2 - 54.2	②	HYD	HYD
	5.9 (5825)	5	0.040	①	44.2 - 54.2	②	HYD	HYD
	5.9 (5899)	Z	0.040	①	44.2 - 54.2	②	HYD	HYD
	8.0 (7994)	W	0.045	①	44.2 - 54.2	②	HYD	HYD

NOTE: The Vehicle Emission Control Information (VECI) label often reflects specification changes made during production. The label figures must be used if they differ from those in this chart.

HYD - Hydraulic

① Ignition timing cannot be adjusted. Base engine timing is set during assembly, and controlled by the PCM.
② Idle speed is not adjustable

GENERAL INFORMATION AND MAINTENANCE 1-31

DIESEL ENGINE TUNE-UP SPECIFICATIONS

Year	Engine Displacement cu. in. (cc)	Engine ID/VIN	Valve Clearance		Intake Valve Opens (deg.)	Injection Pump Setting (deg.)	Injection Nozzle Pressure (psi)		Idle Speed (rpm)	Cranking Compression Pressure (psi)
			Intake (in.)	Exhaust (in.)			New	Used		
1997	5.9 (5882)	D	0.010	0.020	NA	①	3822	NA	④	NA
1998	5.9 (5882)	D	0.010	0.020	NA	②	3394-3887	NA	⑤	NA
1999	5.9 (5882)	6	0.006-0.015	0.0015-0.030	NA	③	4250-4750	NA	⑥	NA
2000	5.9 (5882)	7	0.006-0.015	0.0015-0.030	NA	③	4250-4750	NA	⑥	NA

NOTE: The Vehicle Emission Control Information (VECI) label often reflects specification changes made during production. The label figures must be used if they differ from those in this chart
NA - Not Available

① Align marks on pump flange and gear housing
② Align the marks on the crankshaft, camshaft and pump sprockets.
③ Federal models with manual transmissions: 13.5 degrees BTDC
   Except Federal models with manual transmissions: 14.0 degrees BTDC
④ Automatic transmission with A/C: 750-800 rpm
   Manual transmission with A/C: 780 rpm
⑤ The idle speed is computer-controlled and cannot be adjusted.
⑥ Automatic transmission: 750-800 rpm with trans. in drive and A/C on.
   Manual transmission: 780 rpm with trans. in Neutral and A/C on.

Air Conditioning System

SYSTEM SERVICE & REPAIR

See Figure 137

Note: It is recommended that the A/C system be serviced by an EPA Section 609 certified automotive technician utilizing a refrigerant recovery/recycling machine.

The do-it-yourselfer should not service his/her own vehicle's A/C system for many reasons, including legal concerns, personal injury, environmental damage and cost. The following are some of the reasons why you may decide not to service your own vehicle's A/C system.

According to the U.S. Clean Air Act, it is a federal crime to service or repair (involving the refrigerant) a Motor Vehicle Air Conditioning (MVAC) system for money, without being EPA certified. It is also illegal to vent R-134a refrigerant into the atmosphere.

State and/or local laws may be more strict than the federal regulations, so be sure to check with your state and/or local authorities for further information. For further federal information on the legality of servicing your A/C system, call the EPA Stratospheric Ozone Hotline.

Note: Federal law dictates that a fine of up to $25,000 may be levied on people convicted of venting refrigerant into the atmosphere. Additionally, the EPA may pay up to $10,000 for information or services leading to a criminal conviction of the violation of these laws.

When servicing an A/C system you run the risk of handling or coming in contact with refrigerant, which may result in skin or eye irritation or frostbite. Although low in toxicity (due to chemical stability), inhalation of concentrated refrigerant fumes is dangerous and can result in death; cases of fatal cardiac arrhythmia have been reported in people accidentally subjected to high levels of refrigerant. Some early symptoms include loss of concentration and drowsiness.

Also, refrigerants can decompose at high temperatures (near gas heaters or open flame), which may result in hydrofluoric acid, hydrochloric acid and phosgene (a fatal nerve gas).

It is usually more economically feasible to have a certified MVAC automotive technician perform A/C system service to your vehicle. While it is illegal to service an A/C system without the proper equipment, the home mechanic would have to purchase an expensive refrigerant recovery/recycling machine to service his/her own vehicle.

PREVENTIVE MAINTENANCE

Although the A/C system should not be serviced by the do-it-yourselfer, preventive maintenance can be practiced and A/C system inspections can be performed to help maintain the efficiency of the vehicle's A/C system. For preventive maintenance, perform the following:

• The easiest and most important preventive maintenance for your A/C system is to be sure that it is used on a regular basis. Running the system for five minutes each month (no matter what the season) will help ensure that the seals and all internal components remain lubricated.

Note: Some newer vehicles automatically operate the A/C system compressor whenever the windshield defroster is activated. When running, the compressor lubricates the A/C system components; therefore, the A/C system would not need to be operated each month.

• In order to prevent heater core freeze-up during A/C operation, it is necessary to maintain a proper antifreeze protection. Use a hand-held coolant tester (hydrometer) to periodically check the condition of the antifreeze in your engine's cooling system.

Note: Antifreeze should not be used longer than the manufacturer specifies.

• For efficient operation of an air conditioned vehicle's cooling system, the radiator cap should have a holding pressure which meets manufacturer's specifications. A cap which fails to hold these pressures should be replaced.

• Any obstruction of or damage to the condenser configuration will restrict air flow which is essential to its efficient operation. It is, therefore, a good rule to keep this unit clean and in proper physical shape.

Note: Bug screens which are mounted in front of the condenser (unless they are original equipment) are regarded as obstructions.

• The condensation drain tube expels any water, which accumulates on the bottom of the evaporator housing, into the engine compartment. If this tube is obstructed, the air conditioning performance can be restricted and condensation buildup can spill over onto the vehicle's floor.

SYSTEM INSPECTION

See Figure 138

Although the A/C system should not be serviced by the do-it-yourselfer, preventive maintenance can be practiced and A/C system inspections can be performed to help maintain the efficiency of the vehicle's A/C system. For A/C system inspection, perform the following:

The easiest and often most important check for the air conditioning system consists of a visual inspection of the system components. Visually inspect the air conditioning system for refrigerant leaks, damaged compressor clutch, abnormal compressor drive belt tension and/or condition, plugged evaporator drain tube, blocked condenser fins, disconnected or broken wires, blown fuses, corroded connections and poor insulation.

A refrigerant leak will usually appear as an oily residue at the leakage point in the system. The oily

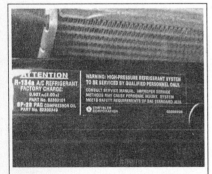

Fig. 137 The A/C certification label states the refrigerant capacity as well as the proper oil

1-32 GENERAL INFORMATION AND MAINTENANCE

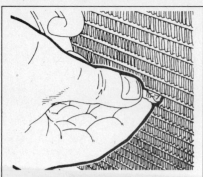

Fig. 138 Periodically remove any debris from the condenser and radiator fins

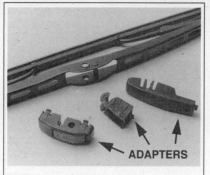

Fig. 139 Bosch® wiper blade and fit kit

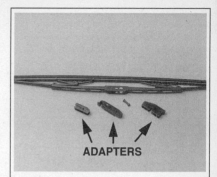

Fig. 140 Lexor® wiper blade and fit kit

residue soon picks up dust or dirt particles from the surrounding air and appears greasy. Through time, this will build up and appear to be a heavy dirt impregnated grease.

For a thorough visual and operational inspection, check the following:

- Check the surface of the radiator and condenser for dirt, leaves or other material which might block air flow.
- Check for kinks in hoses and lines. Check the system for leaks.
- Make sure the drive belt is properly tensioned. When the air conditioning is operating, make sure the drive belt is free of noise or slippage.
- Make sure the blower motor operates at all appropriate positions, then check for distribution of the air from all outlets with the blower on HIGH or MAX.

Note: Keep in mind that under conditions of high humidity, air discharged from the A/C vents may not feel as cold as expected, even if the system is working properly. This is because vaporized moisture in humid air retains heat more effectively than dry air, thereby making humid air more difficult to cool.

- Make sure the air passage selection lever is operating correctly. Start the engine and warm it to normal operating temperature, then make sure the temperature selection lever is operating correctly.

Windshield Wiper (Elements)

ELEMENT (REFILL) CARE & REPLACEMENT

See Figures 139, 140, 141, 142, 143, 144, 145, 146, 147 and 148

For maximum effectiveness and longest element life, the windshield and wiper blades should be kept clean. Dirt, tree sap, road tar and so on will cause streaking, smearing and blade deterioration if left on the glass. It is advisable to wash the windshield carefully with a commercial glass cleaner at least once a month. Wipe off the rubber blades with the wet rag afterwards. Do not attempt to move wipers across the windshield by hand; damage to the motor and drive mechanism will result.

To inspect and/or replace the wiper blade elements, place the wiper switch in the LOW speed position and the ignition switch in the ACC position. When the wiper blades are approximately vertical on the windshield, turn the ignition switch to OFF.

Examine the wiper blade elements. If they are found to be cracked, broken or torn, they should be replaced immediately. Replacement intervals will vary with usage, although ozone deterioration usually limits its element life to about one year. If the wiper pattern is smeared or streaked, or if the blade chatters across the glass, the elements should be replaced. It is easiest and most sensible to replace the elements in pairs.

If your vehicle is equipped with aftermarket blades, there are several different types of refills and your vehicle might have any kind. Aftermarket blades and arms rarely use the exact same type blade or refill as the original equipment. Here are some typical aftermarket blades; not all may be available for your vehicle:

The Anco® type uses a release button that is pushed down to allow the refill to slide out of the

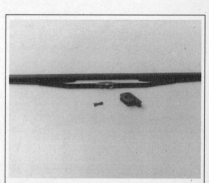

Fig. 141 Pylon® wiper blade and adapter

Fig. 142 Trico® wiper blade and fit kit

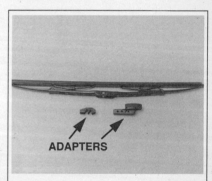

Fig. 143 Tripledge® wiper blade and fit kit

Fig. 144 To remove and install a Lexor® wiper blade refill, slip out the old insert and slide in a new one

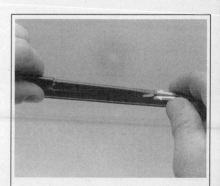

Fig. 145 On Pylon® inserts, the clip at the end has to be removed prior to sliding the insert off

GENERAL INFORMATION AND MAINTENANCE 1-33

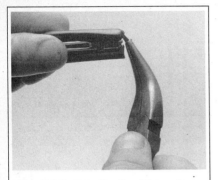

Fig. 146 On Trico® wiper blades, the tab at the end of the blade must be turned up...

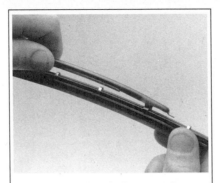

Fig. 147 ... then the insert can be removed. After installing the replacement insert, bend the tab back

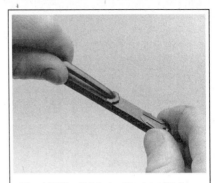

Fig. 148 The Tripledge® wiper blade insert is removed and installed using a securing clip

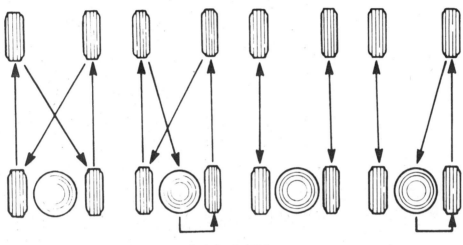

Fig. 149 Recommended method of tire rotation - except dual wheel vehicles

yoke jaws. The new refill slides back into the frame and locks in place.

Some Trico® refills are removed by locating where the metal backing strip or the refill is wider. Insert a small screwdriver blade between the frame and metal backing strip. Press down to release the refill from the retaining tab.

Other types of Trico® refills have two metal tabs which are unlocked by squeezing them together. The rubber filler can then be withdrawn from the frame jaws. A new refill is installed by inserting the refill into the front frame jaws and sliding it rearward to engage the remaining frame jaws. There are usually four jaws; be certain when installing that the refill is engaged in all of them. At the end of its travel, the tabs will lock into place on the front jaws of the wiper blade frame.

Another type of refill is made from polycarbonate. The refill has a simple locking device at one end which flexes downward out of the groove into which the jaws of the holder fit, allowing easy release. By sliding the new refill through all the jaws and pushing through the slight resistance when it reaches the end of its travel, the refill will lock into position.

To replace the Tridon® refill, it is necessary to remove the wiper blade. This refill has a plastic backing strip with a notch about 1 in. (25mm) from the end. Hold the blade (frame) on a hard surface so that the frame is tightly bowed. Grip the tip of the backing strip and pull up while twisting counterclockwise. The backing strip will snap out of the retaining tab.

Do this for the remaining tabs until the refill is free of the blade. The length of these refills is molded into the end and they should be replaced with identical types.

Regardless of the type of refill used, be sure to follow the part manufacturer's instructions closely. Make sure that all of the frame jaws are engaged as the refill is pushed into place and locked. If the metal blade holder and frame are allowed to touch the glass during wiper operation, the glass will be scratched.

Tires and Wheels

Common sense and good driving habits will afford maximum tire life. Fast starts, sudden stops and hard cornering are hard on tires and will shorten their useful life span. Make sure that you don't overload the vehicle or run with incorrect pressure in the tires. Both of these practices will increase tread wear.

Note: For optimum tire life, keep the tires properly inflated, rotate them often and have the wheel alignment checked periodically.

Inspect your tires frequently. Be especially careful to watch for bubbles in the tread or sidewall, deep cuts or underinflation. Replace any tires with bubbles in the sidewall. If cuts are so deep that they penetrate to the cords, discard the tire. Any cut in the sidewall of a radial tire renders it unsafe. Also look for uneven tread wear patterns that may indicate the front end is out of alignment or that the tires are out of balance.

TIRE ROTATION

See Figures 149 and 150

Tires must be rotated periodically to equalize wear patterns that vary with a tire's position on the vehicle. Tires will also wear in an uneven way as the front steering/suspension system wears to the point where the alignment should be reset.

Rotating the tires will ensure maximum life for the tires as a set, so you will not have to discard a tire early due to wear on only part of the tread. Regular rotation is required to equalize wear.

Note: The compact or space-saver spare, if so equipped, is strictly for emergency use. It must never be included in the tire rotation or placed on the vehicle for everyday use.

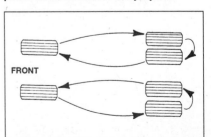

Fig. 150 Tire rotation pattern for all dual wheel vehicles

1-34 GENERAL INFORMATION AND MAINTENANCE

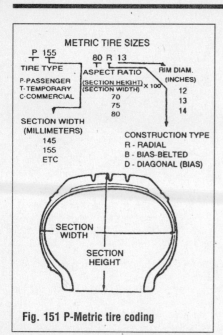

Fig. 151 P-Metric tire coding

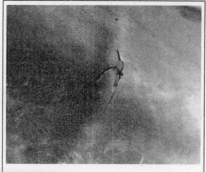

Fig. 152 Tires should be checked frequently for any sign of puncture or damage

Fig. 153 Tires with deep cuts, or cuts which bulge, should be replaced immediately

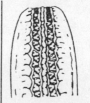

Fig. 154 Examples of inflation-related tire wear patterns

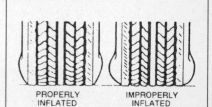

Fig. 155 Radial tires have a characteristic sidewall bulge; don't try to measure pressure by looking at the tire. Use a quality air pressure gauge

TIRE DESIGN

See Figure 151

For maximum satisfaction, tires should be used in sets of four. Mixing of different types (radial, bias belted, fiberglass belted) must be avoided. In most cases, the vehicle manufacturer has designated a type of tire on which the vehicle will perform best. Your first choice when replacing tires should be to use the same type of tire that the manufacturer recommends.

When radial tires are used, tire sizes and wheel diameters should be selected to maintain ground clearance and tire load capacity equivalent to the original specified tire.

CAUTION:

Radial tires should always be used in sets of four.

When selecting tires, pay attention to the original size as marked on the tire. Most tires are described using an industry size code sometimes referred to as P-Metric. This allows the exact identification of the tire specifications, regardless of the manufacturer. If selecting a different tire size or brand, remember to check the installed tire for any sign of interference with the body or suspension while the vehicle is stopping, turning sharply or heavily loaded.

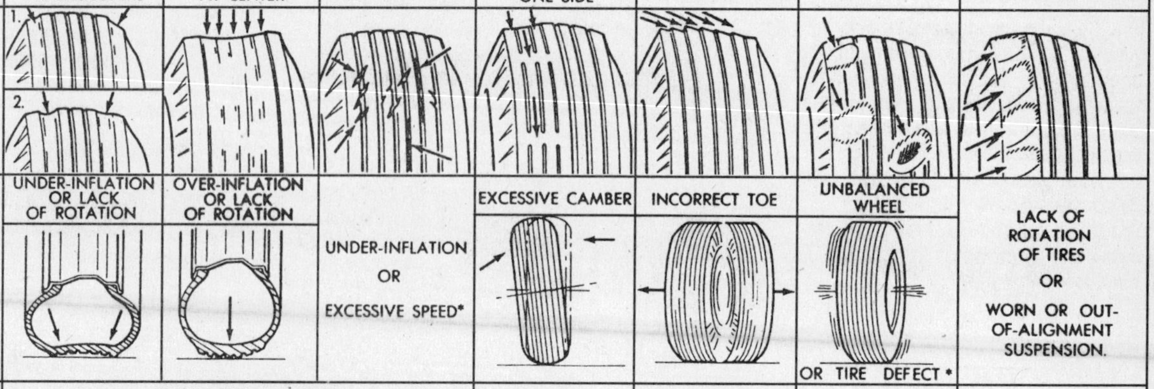

CONDITION	RAPID WEAR AT SHOULDERS	RAPID WEAR AT CENTER	CRACKED TREADS	WEAR ON ONE SIDE	FEATHERED EDGE	BALD SPOTS	SCALLOPED WEAR
EFFECT							
CAUSE	UNDER-INFLATION OR LACK OF ROTATION	OVER-INFLATION OR LACK OF ROTATION	UNDER-INFLATION OR EXCESSIVE SPEED*	EXCESSIVE CAMBER	INCORRECT TOE	UNBALANCED WHEEL OR TIRE DEFECT*	LACK OF ROTATION OF TIRES OR WORN OR OUT-OF-ALIGNMENT SUSPENSION.
CORRECTION	ADJUST PRESSURE TO SPECIFICATIONS WHEN TIRES ARE COOL ROTATE TIRES			ADJUST CAMBER TO SPECIFICATIONS	ADJUST TOE-IN TO SPECIFICATIONS	DYNAMIC OR STATIC BALANCE WHEELS	ROTATE TIRES AND INSPECT SUSPENSION

*HAVE TIRE INSPECTED FOR FURTHER USE.

Fig. 156 Common tire wear patterns and causes

GENERAL INFORMATION AND MAINTENANCE 1-35

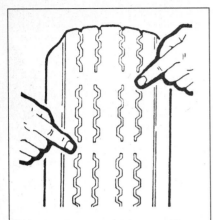

Fig. 157 Tread wear indicators will appear when the tire is worn

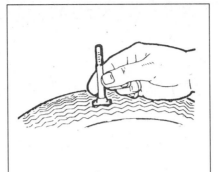

Fig. 158 Accurate tread depth indicators are inexpensive and handy

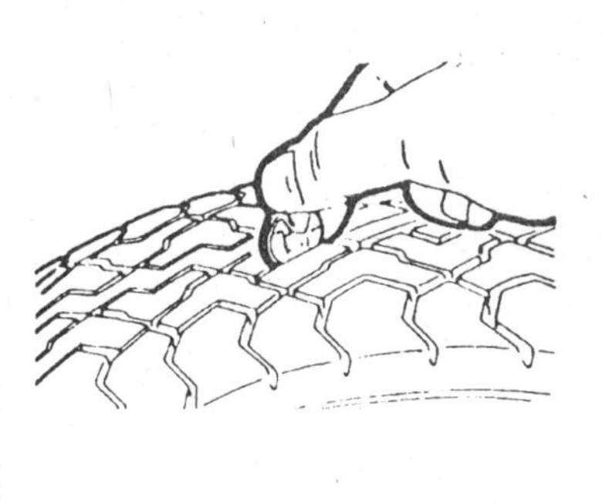

Fig. 159 A penny works well for a quick check of tread depth

Snow Tires

Good radial tires can produce a big advantage in slippery weather, but in snow, a street radial tire does not have sufficient tread to provide traction and control. The small grooves of a street tire quickly pack with snow and the tire behaves like a billiard ball on a marble floor. The more open, chunky tread of a snow tire will self-clean as the tire turns, providing much better grip on snowy surfaces.

To satisfy municipalities requiring snow tires during weather emergencies, most snow tires carry either an M + S designation after the tire size stamped on the sidewall, or the designation "all-season." In general, no change in tire size is necessary when buying snow tires.

Most manufacturers strongly recommend the use of 4 snow tires on their vehicles for reasons of stability. If snow tires are fitted only to the drive wheels, the opposite end of the vehicle may become very unstable when braking or turning on slippery surfaces. This instability can lead to unpleasant endings if the driver can't counteract the slide in time.

Note that snow tires, whether 2 or 4, will affect vehicle handling in all non-snow situations. The stiffer, heavier snow tires will noticeably change the turning and braking characteristics of the vehicle. Once the snow tires are installed, you must re-learn the behavior of the vehicle and drive accordingly.

Note: Consider buying extra wheels on which to mount the snow tires. Once done, the "snow wheels" can be installed and removed as needed. This eliminates the potential damage to tires or wheels from seasonal removal and installation. Even if your vehicle has styled wheels, see if inexpensive steel wheels are available. Although the look of the vehicle will change, the expensive wheels will be protected from salt, curb hits and pothole damage.

TIRE STORAGE

If they are mounted on wheels, store the tires at proper inflation pressure. All tires should be kept in a cool, dry place. If they are stored in the garage or basement, do not let them stand on a concrete floor; set them on strips of wood, a mat or a large stack of newspaper. Keeping them away from direct moisture is of paramount importance. Tires should not be stored upright, but in a flat position.

INFLATION & INSPECTION

See Figures 152, 153, 154, 155, 156, 157, 158 and 159

The importance of proper tire inflation cannot be overemphasized. A tire employs air as part of its structure. It is designed around the supporting strength of the air at a specified pressure. For this reason, improper inflation drastically reduces the tire's ability to perform as intended. A tire will lose some air in day-to-day use; having to add a few pounds of air periodically is not necessarily a sign of a leaking tire.

Two items should be a permanent fixture in every glove compartment: an accurate tire pressure gauge and a tread depth gauge. Check the tire pressure (including the spare) regularly with a pocket type gauge. Too often, the gauge on the end of the air hose at your corner garage is not accurate because it suffers too much abuse. Always check tire pressure when the tires are cold, as pressure increases with temperature. If you must move the vehicle to check the tire inflation, do not drive more than a mile before checking. A cold tire is generally one that has not been driven for at least three hours prior to checking.

A plate or sticker is normally provided somewhere in the vehicle (door post, hood, tailgate or trunk lid) which shows the proper pressure for the tires. Never counteract excessive pressure build-up by bleeding off air pressure (letting some air out). This will cause the tire to run hotter and wear quicker.

CAUTION:

Never exceed the maximum tire pressure embossed on the tire! This is the pressure to be used when the tire is at maximum loading, but it is rarely the correct pressure for everyday driving. Consult the owner's manual or the tire pressure sticker for the correct tire pressure.

Once you've maintained the correct tire pressures for several weeks, you'll be familiar with the vehicle's braking and handling personality. Slight adjustments in tire pressures can fine-tune these characteristics, but never change the cold pressure specification by more than 2 psi. A slightly softer tire pressure will give a softer ride but also yield lower fuel mileage. A slightly harder tire will give crisper dry road handling but can cause skidding on wet surfaces. Unless you're fully attuned to the vehicle, stick to the recommended inflation pressures.

All tires made since 1968 have built-in tread wear indicator bars that show up as 1/2 in. (13mm) wide smooth bands across the tire when 1/16 in. (1.5mm) of tread remains. The appearance of tread wear indicators means that the tires should be replaced. In fact, many states have laws prohibiting the use of tires with less than this amount of tread.

You can check your own tread depth with an inexpensive gauge or by using a Lincoln head penny. Slip the Lincoln penny (with Lincoln's head upside-down) into several tread grooves. If you can see the top of Lincoln's head in 2 adjacent grooves, the tire has less than 1/16 in. (1.5mm) tread left and should be replaced. You can measure snow tires in the same manner by using the "tails" side of the Lincoln penny. If you can see the top of the Lincoln memorial, it's time to replace the snow tire(s).

CARE OF SPECIAL WHEELS

If you have invested money in magnesium, aluminum alloy or sport wheels, special precautions should be taken to make sure your investment is not wasted and that your special wheels look good for the life of the vehicle.

Special wheels are easily damaged and/or scratched. Occasionally check the rims for cracking, impact damage or air leaks. If any of these are found, replace the wheel. But in order to prevent this type of damage and the costly replacement of a special wheel, observe the following precautions:

- Use extra care not to damage the wheels during removal, installation, balancing, etc. After removal of the wheels from the vehicle, place them on a mat or other protective surface. If they are to be stored for any length of time, support them on strips of wood. Never store tires and wheels upright; the tread may develop flat spots.
- When driving, watch for hazards; it doesn't take much to crack a wheel.
- When washing, use a mild soap or non-abrasive dish detergent (keeping in mind that detergent tends to remove wax). Avoid cleansers with abrasives or the use of hard brushes. There are many cleaners and polishes for special wheels.
- If possible, remove the wheels during the winter. Salt and sand used for snow removal can severely damage the finish of a wheel.
- Make certain the recommended lug nut torque is never exceeded or the wheel may crack. Never use snow chains on special wheels; severe scratching will occur.

1-36 GENERAL INFORMATION AND MAINTENANCE

FLUIDS AND LUBRICANTS

Fluid Disposal

Used fluids such as engine oil, transmission fluid, antifreeze and brake fluid are hazardous wastes and must be disposed of properly. Before draining any fluids, consult with your local authorities; in many areas, waste oil, antifreeze, etc. is being accepted as a part of recycling programs. A number of service stations and auto parts stores are also accepting waste fluids for recycling.

Be sure of the recycling center's policies before draining any fluids, as many will not accept different fluids that have been mixed together.

Fuel and Engine Oil Recommendations

See Figure 160

GASOLINE ENGINES

See Figure 161

All late model trucks are designed to run on unleaded gasoline. Any vehicle originally equipped with a catalytic converter MUST use unleaded gasoline.

The recommended oil viscosities for sustained temperatures ranging from below - 10-degrees F (- 23-degrees C) to above 100-degrees F (38-degrees C) are listed in this section. They are broken down into multi-viscosities and single viscosities. Multi-viscosity oils are recommended because of their wider range of acceptable temperatures and driving conditions.

When adding oil to the crankcase or when changing the oil and filter, it is important that oil of an equal quality to original equipment be used in your truck. The use of inferior oils may void the warranty, damage your engine, or both.

The Society of Automotive Engineers (SAE) grade number indicates the oil's viscosity (its ability to lubricate at a given temperature). The lower the SAE number, the lighter the oil; the lower the viscosity, the easier it is to crank the engine in cold weather but the less the oil will lubricate and protect the engine in high temperatures. This number is marked on every oil container.

Oil viscosities should be chosen from those oils recommended for the lowest anticipated temperatures during the oil change interval. Due to the need for an oil that embodies both good lubrication at high temperatures and easy cranking in cold weather, multi-grade oils have been developed. Basically, a multi-grade oil is thinner at low temperatures and thicker at high temperatures. For example, a 10W - 40 oil exhibits the characteristics of a 10 weight (SAE 10) oil when the truck is first started and the oil is cold. Its lighter weight allows it to travel to the lubricating surfaces quicker and offer less resistance to starter motor cranking than, say, a straight 30 weight (SAE 30) oil. But after the engine reaches operating temperature, the 10W - 40 oil has about the same viscosity that straight 40 weight (SAE 40) oil would have at that temperature.

The American Petroleum Institute (API) designations, also found on the oil container, indicates the classification of engine oil used under certain given operating conditions. Only oils designated for use Service SJ (or a later superceding designation) heavy duty detergent should be used in your truck. Oils of the SJ type perform many functions inside the engine besides their basic lubrication. Through a balanced system of metallic detergents and polymeric dispersants, the oil prevents high and low temperature deposits, while keeping sludge and dirt particles in suspension. Acids, particularly sulfuric acid, as well as other by products of engine combustion, are neutralized by the oil. If these acids are allowed to concentrate, they can cause corrosion and rapid wear of the internal engine parts.

WARNING:

Non-detergent motor oils or straight mineral oils should not be used in your gasoline engine.

DIESEL ENGINES

See Figure 162

Diesel engines require different engine oil from those used in gasoline engines. Besides providing the protection that gasoline engine oil does, diesel oil must also deal with increased engine heat and the diesel blow-by gases, which create sulfuric acid, a highly corrosive substance.

Under the American Petroleum Institute (API) classifications, gasoline engine oil codes begin with an S, and diesel engine oil codes begin with a C. This first letter designation is followed by a second letter code which explains what type of service (heavy, moderate, light) the oil is meant for. For example, the label of a typical oil bottle may well include: API SERVICES SF, CD. This means the oil is a superior, heavy duty engine oil and can be used in a diesel engine.

Many diesel manufacturers recommend an oil with both gasoline and diesel engine API classifications.

Note: The manufacturer specifies the use of an engine oil conforming to API quality CE, or CE/SG. Oils with a high ash content are not recommended as they may cause deposits on the valves. A maximum of 1.85 % sulfated ash content is recommended for the diesel engines covered in this manual.

Fuel makers commonly produce two grades of diesel fuel for use in automotive diesel engines No. 1 and No. 2. Generally speaking, No. 2 fuel is recommended over No. 1 for driving in temperatures above 20-degrees F (7-degrees C). In fact, in many areas, No. 2 diesel is the only fuel available. By comparison, No. 2 diesel fuel is less volatile than No. 1 fuel, and gives better fuel economy. Also, No. 2 fuel is a better injection pump lubricant.

The cetane number of a diesel fuel refers to the ease with which a diesel fuel ignites. High cetane numbers mean that the fuel will ignite with relative ease or that it ignites well at low temperatures. Natu-

Fig. 160 Look for the API oil identification label when choosing your engine oil

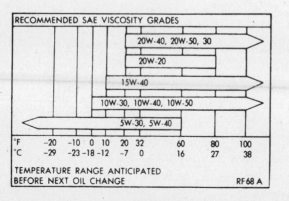

Fig. 161 Gasoline engine oil viscosity chart

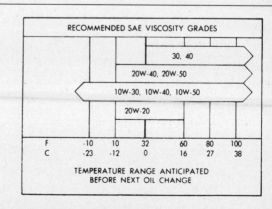

Fig. 162 Diesel engine viscosity chart

GENERAL INFORMATION AND MAINTENANCE 1-37

rally, the lower the cetane number, the higher the temperature must be to ignite the fuel. Most commercial fuels have cetane numbers that range from 35 - 65. No. 1 diesel fuel generally has a higher cetane rating than No. 2 fuel.

As the temperature goes down, diesel fuel tends to thicken. Diesel fuel contains paraffins (wax) and at low ambient temperatures, wax crystals begin forming in the fuel. The temperature at which this occurs is known as the cloud point. The cloud point for diesel fuel varies due to its composition and that information should be available from your fuel supplier or gas station. A typical cloud point temperature is 10-degrees F (- 12-degrees C). This is an important piece of information as in extremely cold weather, diesel fuel can stop flowing altogether. This can result in no start condition or poor engine performance.

Depending on local climate, most fuel manufacturers make winterized No. 2 fuel available seasonally. The manufacturers often winterize No. 2 diesel fuel using various fuel additives and blends (No. 1 diesel fuel, kerosene, etc.) to lower its wintertime viscosity. Generally speaking, though, No. 1 diesel fuel is more satisfactory in extremely cold weather.

Note: No. 1 and No. 2 diesel fuels will mix and burn with no ill effects, although the engine manufacturer will undoubtedly recommend one or the other. Consult the owner's manual for information.

Many automobile manufacturers publish pamphlets giving the locations of diesel fuel stations nationwide. Contact a local dealer for information.

When planning a trip with a diesel powered vehicle, take into account the temperature of your destination. While your local temperature may be high enough for good running, lower temperatures at the destination may cause clouding and plugging.

Do not substitute home heating oil for automotive diesel fuel. While in some cases, home heating oil refinement levels equal those of diesel fuel, many times they are far below diesel engine requirements. The result of using dirty home heating oil will be a clogged fuel system, in which case the entire system may have to be dismantled and cleaned.

One more word on diesel fuels. Don't thin diesel fuel with gasoline in cold weather. The lighter gasoline, which is more explosive, will cause rough running at the very least, and may cause extensive damage to the engine if enough is used.

Engine

OIL LEVEL CHECK

See Figures 163, 164, 165 and 166

Check the engine oil level every time you fill the gas tank. The oil level should be above the ADD mark and not above the FULL mark on the dipstick. Make sure that the dipstick is inserted into the crankcase as far as possible and that the vehicle is resting on level ground. Also, allow a few minutes after turning off the engine for the oil to drain into the pan, or an inaccurate reading will result.

1. Open the hood and remove the engine oil dipstick.
2. Wipe the dipstick with a clean, lint-free rag and reinsert it. Be sure to insert it all the way.
3. Pull out the dipstick and note the oil level. It should be between the SAFE/FULL (MAX) mark and the ADD (MIN) mark.

Fig. 163 Dipstick handles are marked for application

4. If the level is below the lower mark, insert the dipstick and add fresh oil to bring the level within the proper range. Do not overfill.
5. Recheck the oil level and close the hood.

OIL & FILTER CHANGE

See Figures 167, 168, 169 and 170

Note: You will need a container which is capable of holding a minimum of 7 quarts of oil for gasoline engines or 11 quarts for the diesel. (Check the "Capacities" chart for your engine.) A container which is larger than the oil capacity is recommended so that it can be easily slid out from underneath the truck without the danger of spillage.

CAUTION:

The EPA warns that prolonged contact with used engine oil may cause a number of skin disorders, including cancer! You should make every effort to minimize your exposure to used engine oil. Protective gloves should be worn when changing the oil. Wash your hands and any other exposed skin areas as soon as possible after exposure to used engine oil. Soap and water, or waterless hand cleaner should be used.

The oil should be changed more frequently if the vehicle is being operated in very dusty areas. Before draining the oil, make sure that the engine is at operating temperature. Hot oil will hold more impurities in suspension and will flow better, allowing the removal of more oil and dirt.

Fig. 166 Even with "easy pour" bottles, it is still wise to use a funnel to keep oil from spilling

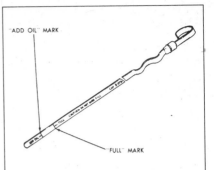

Fig. 164 Engine oil dipsticks are marked with both ADD and FULL marks

Fig. 165 On some models, remove the oil filler cap from the rocker arm cover to add oil

Note: Though some manufacturers have at times recommended replacement of the filter at every other oil change, Chilton recommends the filter be replaced with each engine oil service. The small amount saved by reusing an oil filter rarely justifies the risk. A clogged or dirty filter may fail to protect the expensive internal parts of your engine.

1. Loosen the drain plug with a wrench.
2. Unscrew the plug using a rag to shield your fingers from the heat. Push in on the plug as you unscrew it (this should prevent oil from escaping past the threads until the plug is removed).
3. Once the plug is unthreaded, quickly pull it and your arm back, away from the hot oil. Watch the oil drain and, if necessary, move the pan to keep it underneath the stream of oil. Be careful of the oil. If it

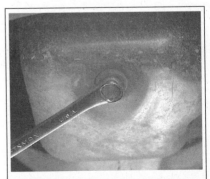

Fig. 167 Loosen the oil drain plug with a wrench, then carefully remove it with your fingers

1-38 GENERAL INFORMATION AND MAINTENANCE

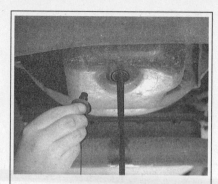

Fig. 168 Unscrew the bolt the rest of the way - be cautious with the stream of heated oil that will follow

Fig. 169 Be sure to clean and inspect the drain plug threads. Do not use self-tapping bolts

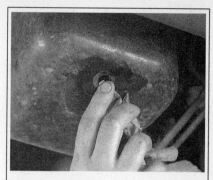

Fig. 170 Clean the area around the drain hole

Fig. 171 Spin-on oil filters are easily removed using a filter wrench. For installation, most filters should be tightened 3/4 to 1 turn after contact. Proper torque is on the filter body

is at operating temperatures, it is hot enough to burn you.

4. Give the oil sufficient time to drain in order to assure you have removed the most oil and dirt possible, then carefully install the drain plug. Make sure the plug is properly tightened, but DO NOT overtighten the plug, as the threads are easily stripped.

WARNING:

Do not reuse self-tapping drain bolts!

5. Change the filter as described below.
6. Add the correct quantity of oil, ensuring the correct grade and viscosity.
7. Start and run the engine for a few minutes. The oil pressure light may stay on for a few seconds while the filter fills up. Shut off the engine and let the oil settle for a few minutes. Check level and top up if necessary. The vehicle should be on a level surface for an accurate dipstick reading.
8. Check for leaks. Be sure to check the drain plug for looseness or seepage after the engine has been fully warmed.

Spin-On Filter Replacement

See Figures 171, 172, 173 and 174

To remove the filter you will need a quality oil filter wrench. The filter may be fitted tightly and the heat from the engine may make it even tighter. A filter wrench can be obtained at an auto parts store and is well worth the investment, since it will save you a lot of grief. Two types are available: the strap wrench and the cup type. The one which is best for you may depend on filter location and accessibility. Also note that strap wrenches may not work on smaller oil filters, while some cup wrenches are often designed specifically to fit only the filters made by a single manufacturer.

Note: Oil filter wrenches are available in different size ranges. Be sure to select a wrench of the proper circumference.

1. Drain the crankcase oil as outlined above.
2. Position a drain pan under the filter before you start to remove it from the engine; should some of the hot oil happen to get on you, you will have a place to dump the filter in a hurry.
3. Loosen the filter with the filter wrench. With a rag wrapped around the filter, unscrew the filter from the boss on the side of the engine. Be careful of hot oil that will run down the side of the filter.
4. Wipe the base of the mounting boss with a clean, dry cloth.

Note: Make sure the old filter gasket was removed with the filter and is not left on the engine adapter. If the old gasket is left in place, you are almost assured to have an oil leak.

5. When you install the new filter, smear a small amount of new engine oil on the gasket with your finger, just enough to coat the entire surface, where it comes in contact with the mounting plate. Partially filling the new filter with oil will reduce the time the engine must run dry when starting.
6. When you tighten the filter follow the part manufacturer's instructions. If none are provided, tighten it about 3/4 to 1 turn after it comes in contact with the mounting boss.
7. Add crankcase oil as outlined above.

Manual Transmission

FLUID RECOMMENDATIONS

Mopar AX 15:
- 75W90, API grade GL-5 gear oil

NV1500
- Mopar manual transmission lubricant P/N 4761526 (ONLY)

NV3500:
- Mopar manual transmission lubricant P/N 4761526 (ONLY)

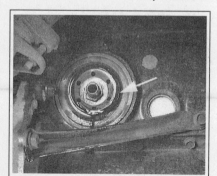

Fig. 172 When the filter is removed, make sure the gasket does not stick to the mounting boss, as in this photo

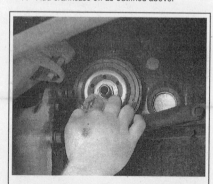

Fig. 173 Be sure to clean the mounting boss' gasket surface before installing the new filter

Fig. 174 Coat the gasket on the new filter with fresh engine oil. Partially filling the new filter with oil will reduce the time the engine must run dry on start up

GENERAL INFORMATION AND MAINTENANCE 1-39

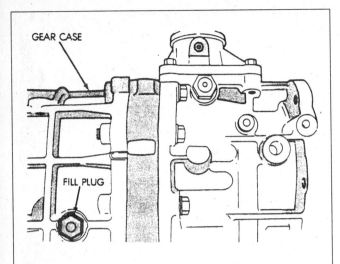

Fig. 175 The fill plug for the AX15 manual transmission is located on the driver's side ...

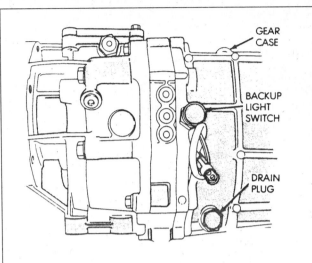

Fig. 176 ... while the drain plug is located on the opposite side - AX15 manual transmission

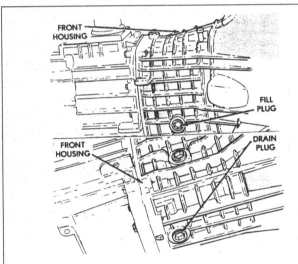

Fig. 177 The fill and drain plugs for the NV1500 and NV3500 manual transmissions require an Allen wrench for removal and installation

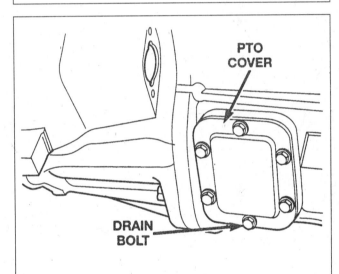

Fig. 178 The drain bolt on the NV4500 and NV5600 is located on the Power Take-Off cover

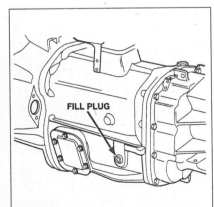

Fig. 179 The fill plug requires a 3/8 inch drive ratchet without a socket on it - NV4500 and NV5600

NV4500:
- Mopar manual transmission lubricant P/N 4761526 (ONLY)

NV5600
- Mopar manual transmission lubricant P/N 4874464 (ONLY)

LEVEL CHECK

See Figures 175, 176, 177, 178 and 179

The fluid level should be checked every 6 months/6,000 miles (9500 km), whichever comes first.

1. Park the truck on a level surface, turn the engine off, apply the parking brake and block the wheels.

Note: Some models covered will need Allen wrenches or a 3/8 inch drive ratchet without a socket to remove and install the drain and fill plugs. Be sure to use the correct tool for the job.

2. Remove the filler plug from the side of the transmission case with a proper size wrench. The fluid level should be even with the bottom of the filler hole.
3. If additional fluid is necessary, add it through the filler hole using a siphon pump or squeeze bottle.
4. Install the filler plug; do not overtighten.

DRAIN AND REFILL

Note: Refer to the capacities chart for more information.

1. Oil should be drained hot to ensure thorough purge of old oil and foreign matter.
2. Position the truck on a level surface.
3. Place a pan of sufficient capacity under the transmission and remove the upper (fill) plug to provide a vent opening.

1-40 GENERAL INFORMATION AND MAINTENANCE

Fig. 180 With the engine running, remove the automatic transmission fluid dipstick to check the level

Fig. 181 Automatic transmission dipstick markings

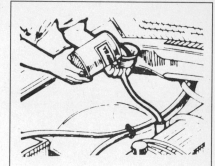

Fig. 182 Use a long-mouth funnel to add fluid to the automatic transmission dipstick tube. Do not overfill

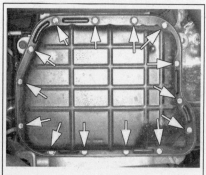

Fig. 183 Automatic transmission pan bolts

Fig. 184 Leave a bolt at each corner of the pan to keep it from bending and spilling the fluid

Fig. 185 As the four corner bolts are slowly loosened, the fluid will begin to drain; it may be necessary to pry gently on the pan to break the seal

4. Remove the lower (drain) plug and allow all of the fluid to drain from the transmission.
5. Reinstall the drain plug.
6. Pump in sufficient lubricant to bring the level to the bottom of the filler plug opening.

Automatic Transmission

FLUID RECOMMENDATIONS

Mopar, ATF Plus 3, Type 7176 automatic transmission fluid is the recommended fluid for all Chrysler automatic transmissions covered in this manual.

LEVEL CHECK

See Figures 180, 181 and 182

It is very important to maintain the proper fluid level in an automatic transmission. If the level is either too high or too low, poor shifting operation and internal damage are likely to occur. For this reason a regular check of the fluid level is essential.

1. Drive the vehicle for 15 - 20 minutes to allow the transmission to reach operating temperature. If the fluid temperature gets very hot, the results of the check may be inaccurate. In this case, give the fluid some time to cool off before checking.
2. Park the truck on a level surface, apply the parking brake and leave the engine idling. Shift the transmission and engage each gear, then place the gear selector in P (PARK).
3. Open the hood and locate the transmission dipstick. Most models will be marked "TRANS FLUID".
4. Wipe away any dirt in the area of the transmission dipstick to prevent it from falling into the filler tube. Withdraw the dipstick, wipe it with a clean, lint-free rag and reinsert it until it seats.
5. Withdraw the dipstick and note the fluid level. It should be between the upper (FULL) mark and the lower (ADD) mark.
6. If the level is below the lower mark, use a funnel and add fluid in small quantities through the dipstick filler neck. Keep the engine running while adding fluid and check the level after each small amount. Do not overfill.
7. The fluid on the dipstick should be a bright red color. If it is discolored (brown or black), or smells burnt, serious transmission troubles, probably due to overheating, should be suspected. The transmission should be inspected to locate the cause of the burnt fluid.
8. Replace the dipstick and make sure it is fully seated.

DRAIN AND REFILL

See Figures 183, 184, 185, 186, 187, 188, 189 and 190

It is a good idea to warm the transmission fluid first so it will flow better. This can be accomplished by 15 - 20 miles (24 - 32 km) of highway driving. Fluid which is warmed to normal operating temperature will flow faster, drain more completely and remove more contaminants from the engine.

CAUTION:

The EPA warns that prolonged contact with used transmission fluid may cause a number of skin disorders, including cancer. You should make every effort to minimize your exposure to used transmission fluid. Protective gloves should be worn when changing the fluid. Wash your hands and any other exposed skin areas as soon as possible after exposure to used transmission fluid. Soap and water, or waterless hand cleaner, should be used.

1. Raise the front of the truck and support it safely on jackstands. Place a large drain pan under the transmission.
2. Loosen, but do not remove, the transmission pan bolts.
3. Begin removing the bolts, but leave one in each of the four corners of the pan.
4. Back off the remaining bolts slowly. If fluid begins to come out, skip the next step and proceed to the following one.
5. If the pan is stuck, break the seal by prying gently around the lip or striking it with a plastic mallet.
6. After the seal is broken, support the pan with one hand and remove the two remaining bolts on one side, allowing it to tilt. Fluid will begin to pour out. When it stops, support the pan and remove the two remaining bolts, lowering the pan and draining off the fluid.
7. Remove the filter attaching screws and remove the filter. These may be Torx® style.

GENERAL INFORMATION AND MAINTENANCE 1-41

Fig. 186 Remove the transmission filter attaching screws

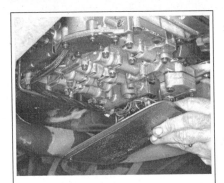

Fig. 187 Removing the filter

Fig. 188 Remove all old gasket material, but avoid damage to the mating surfaces

Fig. 189 Clean the pan with a safe solvent and a clean lint-free cloth

Fig. 190 Make sure the magnet(s) are clean and that you place them back inside the pan before installation

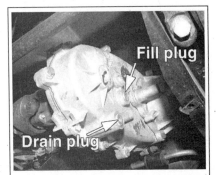

Fig. 191 Depending on model, some transfer case fill and drain plugs may require Allen wrenches, standard wrenches, or even a 3/8 inch drive ratchet for removal or installation - Durango shown

To install:

8. Installing a new filter is recommended.
9. Clean all the sediment out of the transmission pan with solvent. Remove the magnet(s) from the pan and clean them thoroughly, then return them to the pan. The magnets are used to trap any metallic particles floating in the transmission fluid.

Note: Some transmission pans and cases are made of aluminum and can be easily damaged. When removing old gasket material, take care not to gouge the aluminum.

10. Clean the transmission pan and transmission gasket mating surfaces of all old gasket material. After scraping all the old material off, wipe the area with a lint-free rag moistened with solvent. Allow it to dry completely.
11. Lay the new gasket on the transmission pan and check for proper alignment. Remove the gasket, place a few dabs of adhesive on the pan and place the gasket down again. Make sure the gasket is aligned before the adhesive is dry.
12. Position the pan on the transmission and install the pan bolts. Tighten the bolts in a crisscross pattern to 150 inch lbs. (17 Nm).

WARNING:

Never operate the transmission without fluid, otherwise severe damage will result.

13. Remove the jackstands and carefully lower the vehicle, then immediately refill the transmission with the proper amount and type of fluid.

14. Refill the transmission slowly, checking the level often. You may notice that it usually takes less than the amount of fluid listed in the Capacities chart to refill the transmission. This is due to the torque converter being partially filled with fluid. To make sure the proper level is obtained, start the engine and shift the transmission, engaging each gear. Then place the selector in P. Adjust the fluid to the proper level.

Note: If the vehicle is not resting on level ground, the fluid level reading on the dipstick may be slightly off. Be sure to check the level only when the vehicle is sitting level.

15. Empty the used fluid into a suitable container for recycling.

Transfer Case

FLUID RECOMMENDATIONS

The recommended fluid for all transfer cases covered in this manual is Mopar, Dexron III, or ATF Plus 3.

LEVEL CHECK

See Figure 191

Note: On some vehicles, the skid plate must be removed to access the transfer case.

Position the vehicle on level ground. Remove the transfer case fill plug located on the left side of the transfer case. The fluid level should be up to the fill hole. If lubricant doesn't flow from the hole when the plug is removed, add lubricant until it does run out, then install the plug.

DRAIN AND REFILL

The transfer case is serviced at the same time and in the same manner as the transmission. Draining while hot is recommended. Clean the area around the fill and drain plugs and remove both. Allow the lubricant to drain completely.

Clean and install the drain plug. Add the proper lubricant. See the section on level checks.

Drive Axle (Rear and/or Front)

FLUID RECOMMENDATIONS

WARNING:

The axle and differential fluid must be changed whenever the axle has been submerged in water. Immediately replace fluid to avoid axle and differential damage.

Multipurpose hypoid gear lubricant meeting API GL - 5 requirements and having a viscosity rating of 80W-90 can be used in conventional differential

1-42 GENERAL INFORMATION AND MAINTENANCE

Fig. 192 The fill plug on some axles may be the rubber plug type. Be careful not to damage the plug when removing or installing it - 9 1/4 inch rear

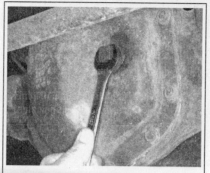

Fig. 193 Check the front differential also whenever the rear differential is being serviced. The skid plate on some vehicles must be removed

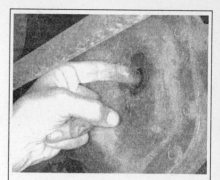

Fig. 194 Lubricant level should be just below the bottom of the filler plug hole

axles. Trak-Lok®, and Power-Lok® limited slip units must use a friction modifier. A MOPAR friction modifier additive is available to cure chatter and noise in these units. For more information on rear axle identification, please refer to the serial number identification information earlier in this section.

Note: Friction modifiers need to be added whenever the differentials are serviced. Add the friction modifiers first, then continue to fill the differential to the proper level with multipurpose hypoid gear lubricant

LEVEL CHECK

See Figures 192, 193 and 194

Note: The skid plate on some vehicles must be removed to service the front differential.

To check the axle lubricant level, remove the axle filler plug with the truck parked on a level surface. This may be a rubber plug or threaded plug depending on model.

You can use a finger for a dipstick, being careful of sharp threads. If necessary, add lubricant. Gear oil can be purchased in plastic quart bottles with a handy spout. Just squeeze to add lube. The lubricant level should be just below the bottom of the filler plug hole.

DRAIN AND REFILL

See Figures 195, 196 and 197

Note: Axles on vehicles covered by this manual are not equipped with drain plugs. The

old lubricant should be drained by removing the differential housing cover, although a suction device can be used effectively. Use a bead of RTV silicone sealant as a gasket when replacing the cover.

Sealant should be applied as follows:
1. Scrape away any remains of the paper gasket or old traces of sealant, as applicable.
2. Clean the cover surface with mineral spirits. Any axle lubricant on the cover or axle housing will prevent the sealant from taking.
3. Apply a 1/16 - 3/32 in. (1.6 - 2.4mm) bead of sealant to the clean and dry cover flange. Apply the bead continuously along the bolt circle of the cover, looping inside the bolt holes as shown.
4. Allow the sealant to air dry.
5. Clean the carrier gasket flange (if solvent is used, allow it to air dry). Install the cover. If, for any reason, the cover is not installed within 20 minutes of applying the sealant, remove the sealant and start over.

Cooling System

CAUTION:

Never remove the radiator cap under any conditions while the engine is running! Failure to follow these instructions could result in damage to the cooling system or engine and/or in personal injury. To avoid having scalding hot coolant or steam blow out of the radiator, DO NOT remove the cap from a hot radiator. Wait until the engine has cooled

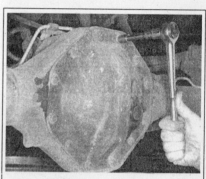

Fig. 195 Removing the differential cover bolts

sufficiently, then wrap a thick cloth around the radiator cap and turn it SLOWLY to the first stop. Step back while the pressure is released from the cooling system. When you are sure the pressure has been released, press down on the radiator cap (with the cloth still in position), then turn and remove the cap.

LEVEL CHECK

See Figure 198

1. Check the fluid level in the reservoir tank. Marks are provided. Check when the engine is at operating temperature. Add a 50/50 mix to maintain performance.

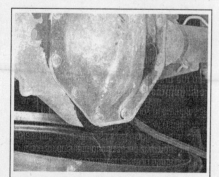

Fig. 196 Use a pry bar to break the cover loose

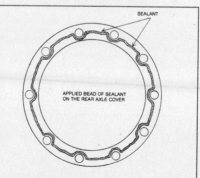

Fig. 197 Apply sealant as shown when replacing the axle cover

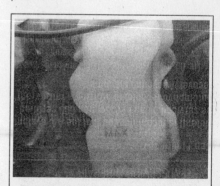

Fig. 198 The coolant level is checked at the overflow tank. Check when hot. Level should be between the lines marked "MAX" and "MIN" or "FULL" and "ADD".

GENERAL INFORMATION AND MAINTENANCE 1-43

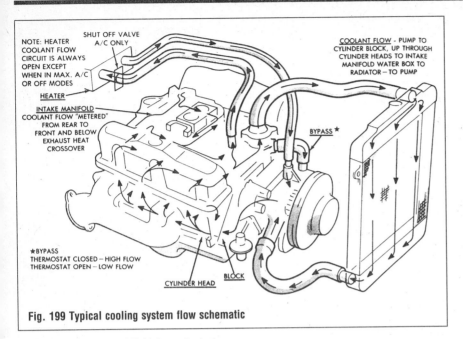

Fig. 199 Typical cooling system flow schematic

2. If it is necessary to add fluid, be sure to check level again soon.

3. A coolant level that drops consistently is usually a sign of a small, hard-to-detect leak. In most cases this is caused by a loose or damaged hose. However, check the heater core. Check the coolant for droplets of engine oil and the engine oil for milky white contamination (emulsified oil). This would indicate an internal leak (blown head gasket or worse), which must be addressed.

INSPECTION

See Figures 199 and 200

At least once every 2 years, the engine cooling system should be inspected, flushed, and refilled with fresh coolant. If the coolant is left in the system too long, it loses its ability to prevent rust and corrosion. If the coolant has been diluted with too much water, it won't protect against freezing.

The radiator cap should be examined for signs of age or deterioration. Fan belts should be inspected and, if necessary, adjusted to the proper tension (please refer to Belt Tension Adjustment in this section).

Hose clamps should be tightened, and soft or cracked hoses replaced. Damp spots, or accumulations of rust or dye near hoses, the water pump or other areas, indicate possible leakage. This must be corrected before filling the system with fresh coolant.

Check the overflow tank cap for a worn or cracked gasket. If the cap doesn't seal properly, fluid will be lost and the engine will overheat. A worn cap should be replaced with a new one. The coolant should be free of rust and oil. If oil is found in the coolant, there may be a major mechanical problem.

Checking The Radiator Cap

See Fig. 201

While you are checking the coolant level, check the radiator cap for a worn or cracked gasket. If the cap doesn't seal properly, fluid will be lost in the form of steam and the engine will overheat. If necessary, replace the cap with a new one.

Radiator Debris

Periodically clean any debris (leaves, paper, insects, etc.) from the radiator fins. Pick the large pieces off by hand. The smaller pieces can be washed away with water pressure from a hose.

Carefully straighten any bent radiator fins with a pair of needle nose pliers. Be careful; the fins are very soft! Don't wiggle the fins back and forth too much. Straighten them once and try not to move them again.

FLUID RECOMMENDATIONS

Coolant found in late model trucks is normally a 50/50 mixture of ethylene glycol and water which can be used year round. Always use a good quality antifreeze with water pump lubricants, rust and other corrosion inhibitors, and acid neutralizers.

Also available is another type of antifreeze, propylene glycol, which is non-toxic.

Keep in mind that should you decide to use a propylene glycol antifreeze, you should follow the antifreeze manufacturer's instructions closely. Do not mix ethylene and propylene glycol together, as the benefits of the non-toxic propylene glycol would be lost. In the event you decide to change to propylene glycol, make sure to completely flush the cooling system of all ethylene glycol traces.

CAUTION:

When draining coolant, keep in mind that cats and dogs are attracted to ethylene glycol antifreeze, and are likely to drink any that is left in an uncovered container or in puddles on the ground. This will prove fatal in sufficient quantity. Always drain coolant into a sealable container. Coolant may be reused unless it is contaminated or several years old.

FLUSHING & CLEANING THE SYSTEM

See Figures 202, 203, 204 and 205

The system should be completely drained and refilled at least every two years in order to remove accumulated rust, scale and other deposits.

Note: Ensure that the engine is completely cool prior to starting this service.

1. Remove the radiator and overflow tank caps.
2. Place a drain pan of sufficient capacity under the radiator and open the petcock (drain).

Fig. 200 Cooling systems should be pressure tested for leaks periodically

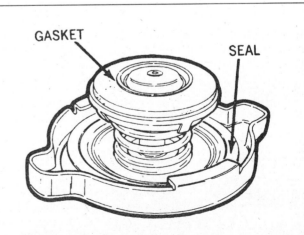

Fig. 201 Be sure the rubber gasket on the radiator cap has a tight seal

1-44 GENERAL INFORMATION AND MAINTENANCE

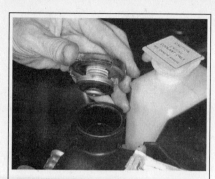

Fig. 202 Only remove the radiator cap when the engine is cool

Note: The petcock may be plastic and should be handled carefully. Before opening the radiator petcock, spray it with some penetrating lubricant.

3. Drain the coolant from the engine block either by removing the drain plug or disconnecting the lower radiator hose.
4. When the system is completely drained, close the petcock and fill the system with a radiator flush. Clean water may also be used, but is not as efficient.
5. Idle the engine until the upper radiator hose gets hot. Be sure to put the heater on to circulate the water or cleaning fluid through the entire system.
6. Allow the engine to cool and drain the system again.
7. Repeat this process until the drained water is clear and free of scale.
8. Flush the overflow tank with water and leave it empty.

Note: If you decide to add the antifreeze and water separately (instead of pre-mixing them), be sure that you add a sufficient amount of antifreeze, before topping off with water. Add a half gallon of each at a time so that you won't find yourself with a filled system with too much of one or the other.

9. Determine the capacity of the coolant system, then properly refill the cooling system with a 50/50 mixture of fresh coolant (antifreeze and water), as follows:
   a. Fill the radiator with coolant mixture until it reaches the radiator filler neck seat.
   b. Start the engine and allow it to idle (heater on) until the thermostat opens (the upper radiator hose will become hot).
   c. Turn the engine OFF and refill the radiator until the coolant level is at the filler neck seat.
   d. Fill the overflow tank with the coolant mixture until the level is midway between the upper and lower level marks, then install the radiator cap.
10. If available, install a pressure tester and check for leaks. If a pressure tester is not available, run the engine until normal operating temperature is reached (allowing the system to naturally pressurize), then check for leaks.
11. Check the level of protection with an antifreeze/coolant hydrometer.
12. After the system has cycled between operating temperature and cold a couple of times, check the overflow tank level and top up if needed.

Windshield Washer Reservoir

See Figure 206

The windshield washer solvent bottle is located adjacent to the radiator on most models. Remove the cap on the bottle and check the fluid by looking down into the bottle; if it is not full, pour the proper mixture of washer solvent into the bottle. Washer solvent comes in a concentrated formula. Check the directions on the bottle you are using to determine the proper solvent and water mixture.

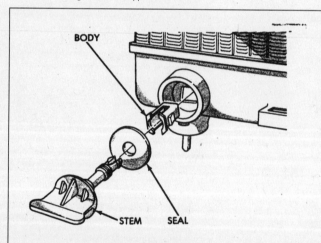

Fig. 203 Radiator petcock components

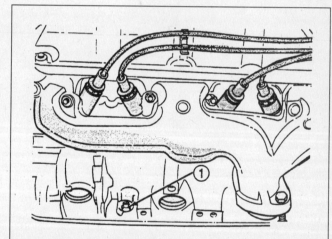

Fig. 204 Cylinder block drain plug location (1) - typical

Fig. 205 Fill the system, run the engine, then check and top off as necessary

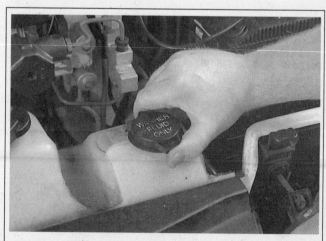

Fig. 206 On this model, the washer solvent reservoir is located in the front driver's side of the engine compartment, adjacent to the radiator

GENERAL INFORMATION AND MAINTENANCE 1-45

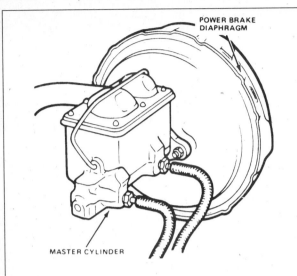

Fig. 207 Brake master cylinder with one-piece metal cap secured by wire bail

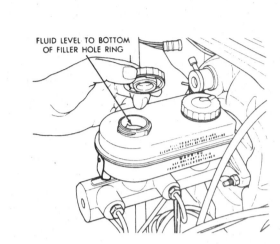

Fig. 208 Brake master cylinder with screw-on caps. Fluid compartments are independent; be sure to check both

Fig. 209 Brake master cylinder with level inspection window

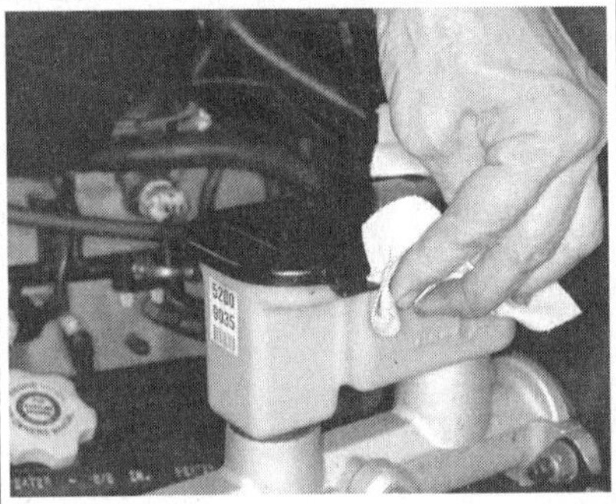

Fig. 210 Clean the area around the cap before removal

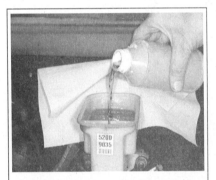

Fig. 211 Fill the brake master cylinder reservoir to the correct level with FRESH brake fluid

Brake Master Cylinder

LEVEL CHECK

See Figures 207, 208, 209 and 210

The master cylinder reservoir is located on the left side of the firewall. Several different types are used. Older styles have a metal cap secured by a wire bail. Other versions use screw-on caps or a one-piece snap-on plastic cap.

If an inspection window is provided, check fluid level relative to the mark(s). Add fluid if level is at or near the "MIN" mark.

On master cylinders with screw-on caps, the level should reach the bottom of the filler hole ring.

On older style master cylinders (one-piece metal caps), the level should be about 1/4" below the reservoir lip.

Before removing a reservoir cap, clean all dirt away from the top of the master cylinder

Note: Any sudden drop in fluid level probably indicates a leak in the system which must be investigated immediately.

There is a rubber diaphragm in the top of the master cylinder cap. As the fluid level falls in the reservoir due to normal brake wear, the diaphragm takes up the space. This is to keep air out of the system due to sloshing as well as to prevent leaks. Make sure to inspect this cap for any tears or cracks, it will require replacement if any damage has occurred.

After refilling the master cylinder to the proper level (using the proper brake fluid), but before installing the cap, be sure to fold the rubber diaphragm up into the cap. Then install the cap in the reservoir and tighten the retaining bolt or snap the retaining clip into place.

If the level of the brake fluid is less than half the volume of the reservoir, it is advised that you check the brake system for leaks. Leaks in the hydraulic brake system most commonly occur at the wheel.

FLUID RECOMMENDATIONS

See Figure 211

Note: When making additions of fluid, use only fresh, uncontaminated brake fluid meeting or exceeding DOT 3 standards. Be careful not to spill any brake fluid on painted surfaces, because it will damage the paint. Do not allow the brake fluid to contact any plastic parts like grill or trim. It may damage some types of plastic. Do not allow the fluid container or brake fluid reservoir to remain open any longer than necessary.

WARNING:

Do not use old brake fluid that has been sitting on your shelf for any length of time, even if you think the bottle cap was tight. Brake fluid is hygroscopic, meaning it acts like a sponge to soak up water vapor from the air. Old brake fluid is likely to be water-contaminated which will have a detrimental affect on your braking system, reducing effectiveness and causing corrosion.

Clutch Master Cylinder

See Figure 212

Some models are equipped with a hydraulic clutch.

The clutch master cylinder is very similar to the brake master cylinder. The hydraulic fluid reservoir on these systems is mounted on the firewall. After removing the cap and diaphragm, check fluid against the level ring provided on the reservoir.

The system should be inspected periodically when other underhood services are performed. To prevent contamination of the system with dirt, ALWAYS clean the top and sides of the reservoir before opening.

1-46 GENERAL INFORMATION AND MAINTENANCE

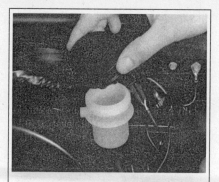

Fig. 212 Remove the cap to check the clutch master cylinder reservoir's fluid level

Keep the reservoir topped up with DOT 3 fluid; do not overfill. Note that the fluid level may actually rise as the clutch wears. A drop in level may indicate a problem.

WARNING:

Refer to the "Brake Master Cylinder" section preceding for information on use and handling of brake fluid. The same information applies for the hydraulic clutch.

WARNING:

Do not allow any petroleum-based fluids to enter the clutch hydraulic system, this could cause seal damage.

Fig. 214 To check level, insert the dipstick and screw it all the way in

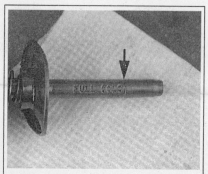

Fig. 215 The arrow indicates the "FULL COLD" level; the other side has the "FULL HOT" mark

Fig. 213 Remove the power steering cap (which in many cases is clearly marked as on this V10 engine)

Power Steering Pump

FLUID RECOMMENDATIONS

If it is necessary to add fluid, use MOPAR Power Steering Fluid or its equivalent.

WARNING:

Use power steering fluid only. Do not overfill.

LEVEL CHECK

See Figures 213, 214 and 215

Note: Always clean the outside of the reservoir cover before removal.

1. Level is checked with the engine OFF. Check fluid level with reference to the dipstick marks. Most are marked "FULL HOT" and "FULL COLD". Take system temperature into account.
2. Clean the area around the power steering pump cap with a clean rag.
3. Unscrew the cap and wipe off the dipstick.
4. Reinsert the dipstick, screwing the cap all the way in. Remove and check level.
5. Top up if needed, adding a little fluid at a time. Use fresh, clean power steering fluid only.

WARNING:

Never add gear oil or automatic transmission fluid!

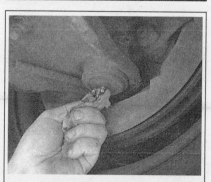

Fig. 216 Clean grease nipples with a rag and solvent before and after applying grease

Chassis Greasing

See Figures 216 and 217

The vehicle should be greased according to the intervals in the Preventive Maintenance Schedule at the end of this section.

Water resistant EP chassis lubricant (grease) conforming to Mopar specification NLGI Grade 2 should be used for all chassis grease points.

Every year or 7,500 miles (12,000 km) the front suspension ball joints, both upper and lower on each side of the truck, must be greased. Most trucks covered in this guide should be equipped with grease nipples on the ball joints, although some may have plugs which must be removed and nipples fitted. These nipples can be obtained through your local parts distributor.

WARNING:

Do not pump so much grease into the ball joint that excess grease squeezes out of the rubber boot. This destroys the watertight seal.

Jack up the front end of the truck and safely support it with jackstands. Block the rear wheels and firmly apply the parking brake. If the truck has been parked in temperatures below 20-degrees F for any length of time, park it in a heated garage for an hour or so until the ball joints loosen up enough to accept the grease.

Depending on which front wheel you work on first, turn the wheel and tire outward, either full-lock right or full-lock left. You now have the ends of the upper and lower suspension control arms in front of you; the grease nipples are visible pointing up (top ball joint) and down (lower ball joint) through the end of each control arm. If the nipples are not accessible, remove the wheel and tire. Wipe all dirt and crud from the nipples or from around the plugs (if installed). If plugs are on the truck, remove them and install grease nipples in the holes (nipples are available in various thread sizes at most auto parts stores). Using a hand operated, low pressure grease gun loaded with a quality chassis grease, lubricate the ball joint only until the rubber joint boot begins to swell out. DO NOT OVER GREASE!

STEERING LINKAGE

See Figures 218 and 219

The steering linkage should be greased at the

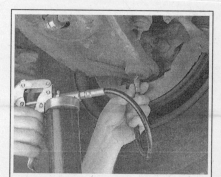

Fig. 217 Be sure the gun fitting is completely seated before pressurizing

GENERAL INFORMATION AND MAINTENANCE 1-47

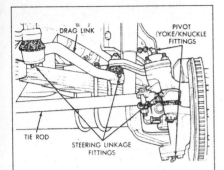

Fig. 218 Steering linkage for a 4WD system

same interval as the ball joints. Grease nipples are installed on the steering tie rod ends on most models. Wipe all dirt and crud from around the nipples at each tie rod end. Using a hand operated, low pressure grease gun loaded with a suitable chassis grease, lubricate the linkage until the old grease begins to squeeze out around the tie rod ends. Wipe off the nipples and any excess grease. Also lubricate the nipples on the steering idler arms.

PARKING BRAKE LINKAGE

Use chassis grease on the parking brake cable where it contacts the cable guides, levers and linkage.

AUTOMATIC TRANSMISSION LINKAGE

Apply a small amount of clean engine oil to the kickdown and shift linkage points at 7,500 mile (12,000 km) intervals.

Body Lubrication and Maintenance

CARE OF YOUR TRUCK

Glass Surfaces

All glass surfaces should be kept clean at all times for safe driving. You should use a window cleaner, the same cleaner you use on windows in your home. Use caution and never use abrasive cleaners; this will cause scratching to the window surfaces.

Exterior Care

Your truck is exposed to all kinds of corrosive effects from nature and chemicals. Some of these are road salt, oils, rain, hail and sleet just to name a few. To protect not only the paint and trim, but also the many exposed mounts and fixtures, it is important to wash your vehicle often and thoroughly. After washing, allow all surfaces to drain and dry before parking in a closed garage. Washing may not clean all deposits off your truck, so you may need additional cleaners. When using professional cleaners, make sure they are suitable for acrylic painted surfaces, chrome, tires etc. These supplies can be purchased in your local auto parts store. You also should wax and polish your vehicle every few months to keep the paint in good shape.

HOOD LATCH AND HINGES

See Figure 220

Clean the latch surfaces and apply light lithium grease (the white stuff) or clean engine oil to the latch pilot bolts and the spring anchor. Also lubricate the hood hinges. Use a lithium or chassis grease to lubricate all the pivot points in the latch release mechanism.

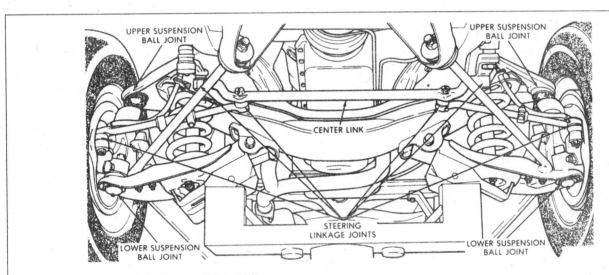

Fig. 219 Steering linkage lube fittings - 2WD vehicles

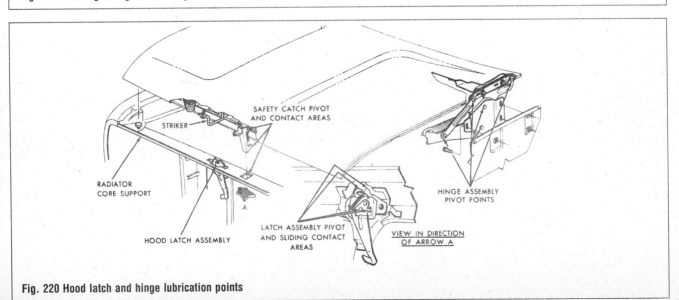

Fig. 220 Hood latch and hinge lubrication points

1-48 GENERAL INFORMATION AND MAINTENANCE

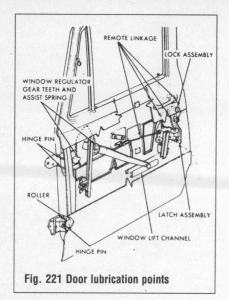

Fig. 221 Door lubrication points

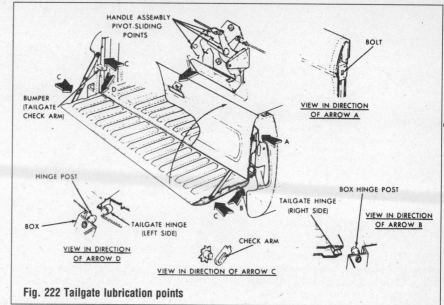

Fig. 222 Tailgate lubrication points

TAILGATE HINGES

See Figures 221 and 222

The gas tank filler door and truck doors should be wiped clean and lubricated with clean engine oil once a year. The door lock cylinders and latch mechanisms should be lubricated periodically with a few drops of graphite lock lubricant or a few shots of silicone spray.

Wheel Bearings

See Figures 223, 224 and 225

REMOVAL, REPACKING, & INSTALLATION

Before handling the bearings, there are a few things that you should remember to do and NOT to do.
Remember to DO the following:
• Remove all outside dirt from the housing before exposing the bearing.
• Treat a used bearing as gently as you would a new one.
• Work with clean tools in clean surroundings.
• Use clean, dry canvas gloves, or at least clean, dry hands.
• Clean solvents and flushing fluids are a must.
• Use clean paper when laying out the bearings to dry.
• Protect disassembled bearings from rust and dirt. Cover them up.
• Use clean rags to wipe bearings.
• Keep the bearings in oil-proof paper when they are to be stored or are not in use.
• Clean the inside of the housing before replacing the bearing.
Do NOT do the following:
• Don't work in dirty surroundings.
• Don't use dirty, chipped or damaged tools.
• Try not to work on wooden benches or use wooden mallets.
• Don't handle bearings with dirty or moist hands.
• Do not use gasoline for cleaning; use a safe solvent.
• Do not spin-dry bearings with compressed air. They will likely be damaged and could even fly apart.

• Do not spin dirty bearings.
• Avoid using cotton, waste, or dirty cloths to wipe bearings.
• Try not to scratch or nick bearing surfaces.
• Do not allow the bearing to come in contact with dirt or rust at any time.

Note: For information on wheel bearing removal and installation, refer to the "Drive Train" section.

1. Remove the wheel bearing.
2. Clean all parts in a non-flammable solvent and let them air dry.

Note: Only use lint-free rags to dry the bearings. Never spin-dry a bearing with compressed air, as this will damage the rollers.

3. Check for excessive wear and damage. Replace the bearing as necessary.

Note: Packing wheel bearings with grease is best accomplished by using a wheel bearing packer (available at most automotive parts stores).

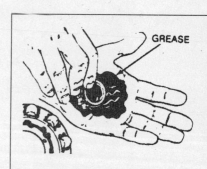

Fig. 223 Packing the wheel bearing by hand

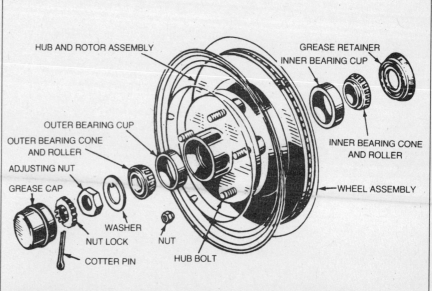

Fig. 224 Exploded view of a 2WD front wheel and bearing assembly

GENERAL INFORMATION AND MAINTENANCE 1-49

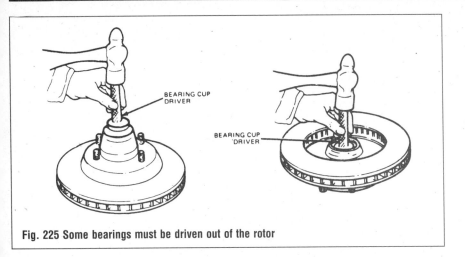

Fig. 225 Some bearings must be driven out of the rotor

4. If a wheel bearing packer is not available, the bearings may be packed by hand.
   a. Place a "healthy" glob of grease in the palm of one hand.
   b. Force the edge of the bearing into the grease so that the grease fills the space between the rollers and the bearing cage.
   c. Keep rotating the bearing while continuing to push the grease through.
   d. Continue until the grease is forced out the other side of the bearing.
5. Place the packed bearing on a clean surface and cover it until it is time for installation.
6. Install the wheel bearing.

TRAILER TOWING

See Figure 226

General Recommendations

Your vehicle was primarily designed to carry passengers and cargo. It is important to remember that towing a trailer will place additional loads on your vehicle's engine, drive train, steering, braking and other systems. However, if you decide to tow a trailer, using the proper equipment is a must.

Local laws may require specific equipment such as trailer brakes or fender mounted mirrors. Check your local laws.

Trailer Weight

The weight of the trailer is the most important factor. A good weight-to-horsepower ratio is about 35:1, 35 lbs. of Gross Combined Weight (GCW) for every horsepower your engine develops. Multiply the engine's rated horsepower by 35 and subtract the weight of the vehicle passengers and luggage. The result is the approximate ideal maximum weight you should tow, although a numerically higher axle ratio can help compensate for heavier weight.

Hitch (Tongue) Weight

Figure the hitch weight to select a proper hitch. Hitch weight is usually 9 - 11% of the trailer gross weight and should be measured with the trailer loaded. Hitches fall into various categories: those that mount on the frame and rear bumper, the bolt-on or weld-on distribution type used for larger trailers. Axle mounted or clamp-on bumper hitches should never be used.

Check the gross weight rating of your trailer. Tongue weight is usually figured as 10% of gross trailer weight. Therefore, a trailer with a maximum gross weight of 2000 lbs. will have a maximum tongue weight of 200 lbs. Class I trailers fall into this category. Class II trailers are those with a gross weight rating of 2000 - 3000 lbs., while Class III trailers fall into the 3500 - 6000 lbs. category. Class IV trailers are those over 6000 lbs. and are for use with fifth wheel trucks, only.

When you've determined the hitch that you'll need, follow the manufacturer's installation instructions, exactly, especially when it comes to fastener torque. The hitch will be subjected to a lot of stress and good hitches come with hardened bolts. Never substitute an inferior bolt for a hardened bolt.

Cooling

ENGINE

One of the most common, if not THE most common, problems associated with trailer towing is engine overheating. If you have a standard cooling system, without an expansion tank, you'll definitely need to get an aftermarket expansion tank kit, preferably one with at least a 2 quart capacity. These kits are easily installed on the radiator's overflow hose, and come with a pressure cap designed for expansion tanks.

Another helpful accessory for vehicles using a belt-driven radiator fan is a flex fan. These fans are large diameter units and are designed to provide more airflow at low speeds, with blades that have deeply cupped surfaces. The blades then flex, or flatten out, at high speed, when less cooling air is needed. These fans are far lighter in weight than stock fans, requiring less horsepower to drive them. Also, they are far quieter than stock fans. If you do decide to replace your stock fan with a flex fan, note that if your vehicle has a fan clutch, a spacer will be needed between the flex fan and water pump hub.

Aftermarket engine oil coolers are helpful for prolonging engine oil life and reducing overall engine temperatures. Both of these factors increase engine life. While not absolutely necessary in towing Class I and some Class II trailers, they are recommended for heavier Class II and all Class III towing. Engine oil cooler systems consists of an adapter, screwed on in place of the oil filter, a remote filter mounting and a multi-tube, finned heat exchanger, which is mounted in front of the radiator or air conditioning condenser.

TRANSMISSION

An automatic transmission is usually recommended for trailer towing. Modern automatics have proven reliable and, of course, easy to operate, in trailer towing. The increased load of a trailer, however, causes an increase in the temperature of the automatic transmission fluid. Heat is the worst enemy of an automatic transmission. As the temperature of the fluid increases, the life of the fluid decreases.

It is essential, therefore, that you install an automatic transmission cooler. The cooler, which consists of a multi-tube, finned heat exchanger, is usually installed in front of the radiator or air conditioning compressor, and hooked in-line with the transmission cooler tank inlet line. Follow the cooler manufacturer's installation instructions.

Select a cooler of at least adequate capacity, based upon the combined gross weights of the vehicle and trailer.

Cooler manufacturers recommend that you use an aftermarket cooler in addition to, and not instead of, the present cooling tank in your radiator. If you do want to use it in place of the radiator cooling tank, get a cooler at least two sizes larger than normally necessary.

Note: A transmission cooler can, sometimes, cause slow or harsh shifting in the transmission during cold weather, until the fluid has a chance to come up to normal operating temperature. Some coolers can be purchased with or retrofitted with a temperature bypass valve which will allow fluid flow through the cooler only when the fluid has reached above a certain operating temperature.

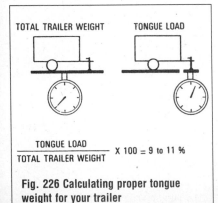

Fig. 226 Calculating proper tongue weight for your trailer

1-50 GENERAL INFORMATION AND MAINTENANCE

Handling A Trailer

Towing a trailer with ease and safety requires a certain amount of experience. It's a good idea to learn the feel of a trailer by practicing turning, stopping and backing in an open area such as an empty parking lot.

TOWING THE VEHICLE

See Figure 227

Manufacturer's Recommendations

A vehicle equipped with SAE approved sling-type towing equipment can be used to tow all short-bed 2WD vehicles covered by this manual. When towing 4WD vehicles and long-bed vehicles, Chrysler Corporation recommends the use of a flatbed-type transportation vehicle. If one is not available, a wheel-lift, sling-type device can be used provided ALL the wheels are lifted off the ground using tow dollies. The flatbed, if used, should have an approach ramp angle not exceeding 15 degrees.

CAUTION:

Do not use the steering column lock to secure the steering wheel during a towing operation.

TOWING WITH REAR END LIFTED

This is the preferable method for towing a 2WD vehicle, but not required. Towing 2WD vehicles with the rear end lifted is permissible. To do so, unlock the steering column and secure the steering wheel in a straight-ahead position with a clamp device designed for towing. Also verify that the steering components are in good condition.

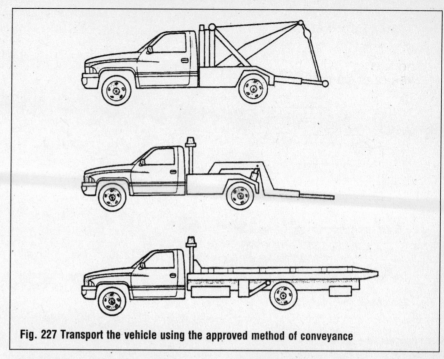

Fig. 227 Transport the vehicle using the approved method of conveyance

TOWING WITH FRONT END LIFTED

If a 2WD vehicle cannot be towed with the rear wheels lifted, it can be towed with the front wheels lifted with the transmission in NEUTRAL.

SAFETY PRECAUTIONS

- Always observe state and local laws regulating towing.
- Do not tow a heavily loaded vehicle. Damage to the cab, cargo box or frame may result. Use a flatbed to transport a loaded vehicle.
- Secure any loose or protruding parts.
- Always use a safety chain that is independent of the lifting/towing equipment.
- Do not allow any towing equipment to contact the vehicle's fuel tank.
- Do not allow anyone under the disabled vehicle while it is lifted by the tow vehicle.
- Do not allow passengers to ride in the vehicle being towed.
- Tow the vehicle in a safe manner that does not jeopardize the operator, pedestrians, motorists, or property.
- Do not attach tow chains, T-hooks, J-hooks, or a tow sling to a bumper, steering linkage, driveshafts, halfshafts, or a non-reinforced frame hole.

JUMP STARTING A DEAD BATTERY

Whenever a vehicle is jump started, precautions must be followed in order to prevent the possibility of personal injury. Remember that batteries contain a small amount of explosive hydrogen gas which is a by-product of battery charging. Sparks should always be avoided when working around batteries, especially when attaching jumper cables. To minimize the possibility of accidental sparks, follow the procedure carefully.

CAUTION:

NEVER hook up the batteries in a series circuit or the entire electrical system, including the starter, will go up in smoke!

Trucks equipped with a diesel engine may utilize two 12 volt batteries. If so, the batteries are connected in a parallel circuit (positive terminal to positive terminal, negative terminal to negative terminal). Hooking the batteries up in parallel circuit increases battery cranking power (amperage) without increasing total battery voltage output. Output remains at 12 volts. On the other hand, hooking two 12 volt batteries up in a series circuit (positive terminal to negative terminal, positive terminal to negative terminal) increases total battery output to 24 volts (12 volts plus 12 volts).

Jump Starting Precautions

- Be sure that both batteries are of the same voltage. Vehicles covered by this manual and most vehicles on the road today utilize a 12 volt charging system.
- Be sure that both batteries are of the same polarity (have the same terminal, in most cases NEGATIVE grounded).
- Be sure that the vehicles are not touching or a short could occur.
- On serviceable batteries, be sure the vent cap holes are not obstructed.
- Do not smoke or allow sparks anywhere near the batteries.
- In cold weather, make sure the battery electrolyte is not frozen. This can occur more readily in a battery that has been in a state of discharge.
- Do not allow electrolyte to contact your skin or clothing.

Jump Starting Procedure

SINGLE BATTERY VEHICLES

See Figure 228

1. Make sure that the voltages of the two batteries are the same. Most batteries and charging systems are of the 12 volt variety.
2. Pull the jumping vehicle (with the good battery) into a position so the jumper cables can reach the dead battery and that vehicle's engine. Make sure that the vehicles do NOT touch.
3. Place the transmissions of both vehicles in NEUTRAL or PARK, as applicable, then firmly set their parking brakes.

Note: If necessary for safety reasons, the hazard lights on both vehicles may be oper-

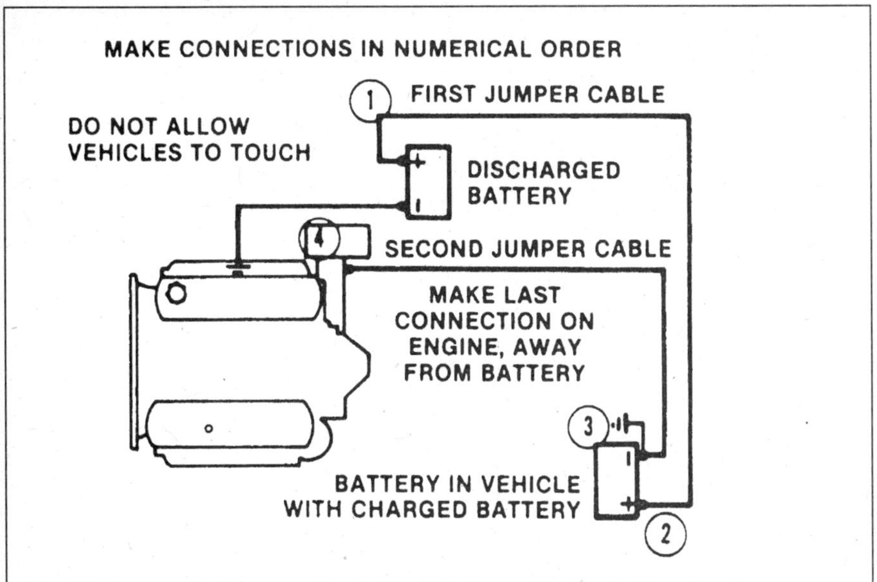

GENERAL INFORMATION AND MAINTENANCE 1-51

Fig. 228 Connect the jumper cables to the batteries and engine in the order shown

ated throughout the entire procedure without significantly increasing the difficulty of jumping the dead battery.

4. Turn all lights and accessories off on both vehicles. Make sure the ignition switches on both vehicles are turned to the OFF position.
5. Cover the battery cell caps with a rag, but do not cover the terminals.
6. Make sure the terminals on both batteries are clean and free of corrosion or proper electrical connection will be impeded. If necessary, clean the battery terminals before proceeding.
7. Identify the positive (+) and negative (−) terminals on both battery posts.
8. Connect the first jumper cable to the positive (+) terminal of the dead battery, then connect the other end of that cable to the positive (+) terminal of the booster (good) battery.
9. Connect one end of the other jumper cable to the negative terminal on the booster battery and the other cable clamp to an engine bolt head, alternator bracket or other solid, metallic point on the engine with the dead battery. Try to pick a ground on the engine that is positioned away from the battery in order to minimize the possibility of the two clamps touching should one loosen during the procedure. DO NOT connect this clamp to the negative (-) terminal of the bad battery.

CAUTION:

Be very careful to keep the jumper cables away from moving parts (cooling fan, belts, etc.) on both engines.

10. Check to make sure that the cables are routed away from any moving parts, then start the donor vehicle's engine. Run the engine at moderate speed for several minutes to allow the dead battery a chance to receive some initial charge.
11. With the donor vehicle's engine still running slightly above idle, try to start the vehicle with the dead battery. Crank the engine for no more than 10 seconds at a time and let the starter cool for at least 20 seconds between tries. If the vehicle does not start in 3 tries, it is likely that something else is also wrong or that the battery needs additional time to charge.
12. Once the vehicle is started, allow it to run at idle for a few seconds to make sure that it is operating properly operating.
13. Turn on the headlights, heater blower and, if equipped, the rear defroster of both vehicles in order to reduce the severity of voltage spikes and subsequent risk of damage to the vehicles' electrical systems when the cables are disconnected. This step is especially important to late model vehicles equipped with computer control modules.
14. Carefully disconnect the cables in the reverse order of connection. Start with the negative cable that is attached to the engine ground, then the negative cable on the donor battery. Disconnect the positive cable from the donor battery and finally, disconnect the positive cable from the formerly dead battery. Be careful when disconnecting the cables from the positive terminals not to allow the alligator clips to touch any metal on either vehicle or a short and sparks will occur.

DUAL BATTERY DIESEL MODELS

See Figure 229

Diesel model vehicles utilize two 12 volt batteries, one on either side of the engine compartment. The batteries are connected in a parallel circuit (positive terminal to positive terminal and negative terminal to negative terminal). Hooking the batteries up in a parallel circuit increases battery cranking power without increasing total battery voltage output. The output will remain at 12 volts. On the other hand, hooking two 12 volt batteries in a series circuit (positive terminal to negative terminal and negative terminal to positive terminal) increases the total battery output to 24 volts (12 volts plus 12 volts).

WARNING:

Never hook the batteries up in a series circuit or the entire electrical system will be damaged, including the starter motor.

In the event that a dual battery vehicle needs to be jump started, use the following procedure:
1. Turn the heater blower motor ON to help protect the electrical system from voltage surges when the jumper cables are connected and disconnected.
2. Turn all lights and other switches OFF.

Note: The battery cables connected to one of the diesel vehicle's batteries may be thicker than those connected to its other battery. (The passenger side battery often has thicker cables.) This set-up allows relatively high jump starting current to pass without damage. If so, be sure to connect the positive jumper cable to the appropriate battery in the disabled vehicle. If there is no difference in cable thickness, connect the jumper cable to either battery's positive terminal. Similarly, if the donor vehicle also utilizes two batteries, the jumper cable connections should be made to the battery with the thicker cables; if there is no difference in thickness, the connections can be made to either donor battery.

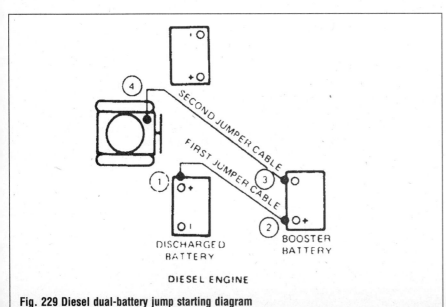

Fig. 229 Diesel dual-battery jump starting diagram

1-52 GENERAL INFORMATION AND MAINTENANCE

3. Connect the end of a jumper cable to one of the disabled diesel's positive (+) battery terminals, then connect the clamp at the other end of the same cable to the positive terminal (+) on the jumper battery.

4. Connect one end of the other jumper cable to the negative battery terminal (−) on the jumper battery, then connect the other cable clamp to an engine bolt head, alternator bracket or other solid, metallic point on the disabled vehicle's engine. DO NOT connect this clamp to the negative terminal (−) of the disabled vehicle's battery.

CAUTION:

Be careful to keep the jumper cables away from moving parts (cooling fan, belts, etc.) on both engines.

5. Start the engine on the vehicle with the good battery and run it at a moderate speed.
6. Start the engine of the vehicle with the discharged battery.
7. When the engine starts on the vehicle with the discharged battery, remove the cable from the engine block before disconnecting the cable from the positive terminal.

JACKING

See Figure 230

Your vehicle was supplied with a jack for emergency road repairs. This jack is fine for changing a flat tire or other short term procedures not requiring you to go beneath the vehicle. If it is used in an emergency situation, carefully follow the instructions provided either with the jack or in your owner's manual. Do not attempt to use the jack on any portions of the vehicle other than specified by the vehicle manufacturer. Always block the diagonally opposite wheel when using a jack.

A more convenient way of jacking is the use of a garage or floor jack. You may use the floor jack only at approved vehicle jacking/lifting points.

Never place the jack under the radiator, engine or transmission components. Severe and expensive damage will result when the jack is raised. Additionally, never jack under the floorpan or bodywork; the metal will deform.

Whenever you plan to work under the vehicle, you must support it on jackstands or ramps. Never use cinder blocks or stacks of wood to support the vehicle, even if you're only going to be under it for a few minutes. Never crawl under the vehicle when it is supported only by the tire-changing jack or other floor jack.

Note: Always position a block of wood or small rubber pad on top of the jack or jackstand to protect the lifting point's finish when lifting or supporting the vehicle.

Small hydraulic, screw, or scissors jacks are satisfactory for raising the vehicle. Drive-on trestles or ramps are also a handy and safe way to both raise and support the vehicle. Be careful though, some ramps may be too steep to drive your vehicle onto without scraping the front bottom panels. Never support the vehicle on any suspension member (unless specifically instructed to do so by a repair manual) or by an underbody panel.

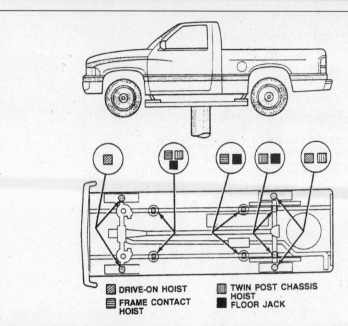

Fig. 230 Vehicle lifting locations. When raising and supporting the vehicle, be sure to use only the approved jacking points

Jacking Precautions

The following safety points cannot be overemphasized:
- Always block the opposite wheel or wheels to keep the vehicle from rolling off the jack.
- When raising the front of the vehicle, firmly apply the parking brake.
- When the drive wheels are to remain on the ground, leave the vehicle in gear to help prevent it from rolling.
- Always use jackstands to support the vehicle when you are working underneath. Place the stands beneath the vehicle's jacking brackets. Before climbing underneath, rock the vehicle a bit to make sure it is firmly supported.

MAINTENANCE INTERVAL CHARTS

The manufacturer provides maintenance schedules for each vehicle family based on "normal" operating conditions and schedules based on "severe service" conditions.

"Severe service" is considered to apply if the vehicle is operated under one or more of the following conditions:
- Frequent short trip driving of less than 5 miles (8 km).
- Frequent driving in dusty conditions.
- Trailer towing or heavy load hauling.
- Frequent long periods of engine idling.
- Sustained high speed operation.
- Desert operation.
- Frequent starting or stopping.
- Cold climate operation.
- Off-road driving.
- Commercial service.
- Snow plow operation.
- More than half of the vehicle operation occurs in heavy city traffic during hot weather (above 90-degrees F/32-degrees C).

Maintenance items for vehicles in "severe service" should be performed approximately twice as often as those in normal use.

In addition to the items listed in the charts, the following procedures should be carried out:

Note: At each stop for fuel:
- Check engine oil level.
- Check windshield washer solvent.
- Clean windshield and wiper blades.
- After completion of off-road (4WD) use, the underside of the vehicle should be thoroughly inspected. Examine threaded fasteners for looseness.

Note: Once a month:
- Check tire pressure and look for unusual wear or damage.
- Inspect battery and clean and tighten terminals as required.
- Check fluid levels of coolant reservoir, brake master cylinder, power steering, and transmission.
- Check lights and other electrical devices for proper operation.

Note: At each oil change:
- Inspect exhaust system.
- Inspect brake hoses.
- Adjust rear brakes.
- Rotate the tires. If using "severe service" maintenance schedule, do this at every other oil change.
- Check engine coolant level, hoses, clamps.
- Lubricate steering linkage.
- Lubricate the driveshaft universal joints and slip spline, if applicable.

MANUFACTURER RECOMMENDED MAINTENANCE INTERVALS (LIGHT DUTY)

TO BE SERVICED		TYPE OF SERVICE	VEHICLE MILEAGE INTERVAL (x1000)												
			7.5	15	22.5	30	37.5	45	52.5	60	67.5	75	82.5	90	97.5	miles
			12	24	36	48	60	72	84	96	108	120	132	144	156	km
			6	12	18	24	30	36	42	48	54	60	66	72	78	months
Engine oil & oil filter		D/R	✓	✓	✓	✓	✓	✓	✓	✓	✓	✓	✓	✓	✓	
Front wheel bearings	①	S/I			✓			✓			✓			✓		
Brake linings		S/I			✓			✓			✓			✓		
Air cleaner element	②	R				✓				✓				✓		
Spark plugs	②	R				✓				✓				✓		
ATF / Replace filter, adjust bands		D/R						✓				✓				
Transfer case fluid		D/R						✓				✓				
Engine Coolant	③	R/F					✓									
Distributor cap and rotor	②	R								✓						
Ignition wires	②	R								✓						
PCV valve. Replace if nec.	②	S/I								✓						
Accessory belt(s)		S/I								✓		✓		✓		
Crankcase inlet air filter		C/L								✓						

R - Replace
S/I - Service or Inspect
R/F - Replace and flush system
D/R - Drain and refill
C/L - Clean and lubricate

① Clean and repack if required (4x2).
② Emission Control System Maintenance is listed in bold type. Service must be performed on-schedule to assure continued proper functioning of the system.
③ Flush and refill at 36 mos. regardless of miles, then every 24 mos./30,000 mi. (48,000 km) thereafter.

MANUFACTURER RECOMMENDED MAINTENANCE INTERVALS (MEDIUM & HEAVY DUTY)

TO BE SERVICED		TYPE OF SERVICE	VEHICLE MILEAGE INTERVAL (x1000)												
			6	12	18	24	30	36	42	48	54	60	66	72	78	miles
			10	19	29	38	48	58	67	77	86	96	106	115	125	km
			6	12	18	24	30	36	42	48	54	60	66	72	78	months
Engine oil		D/R	✓	✓	✓	✓	✓	✓	✓	✓	✓	✓	✓	✓	✓	
Engine oil filter		R	✓	✓	✓	✓	✓	✓	✓	✓	✓	✓	✓			
Front wheel bearings	①	S/I			✓			✓			✓			✓		
Brake linings		S/I			✓			✓			✓			✓		
Air cleaner (8.0L)	②	S/I		✓		✓		✓		✓		✓		✓		
Air cleaner and air pump elements	②	R				✓				✓				✓		
Spark plugs	②	R					✓					✓				
ATF / Replace filter, adjust bands		D/R				✓ ④				✓ ④				✓ ④		
Transfer case fluid		D/R						✓				✓				
Engine Coolant	③	F/R						✓								
Distributor cap and rotor (5.9L)	②	R										✓				
Ignition wires	②	R										✓				
PCV valve (5.9L)	②	R										✓				
Crankcase inlet air filter (5.9L)		C/L				✓				✓				✓		

R - Replace
S/I - Service or Inspect
F/R - Flush and refill
D/R - Drain and refill
C/L - Clean and lubricate

① Clean and repack if required (4x2).
② Emission Control System Maintenance is listed in bold type. Service must be performed on-schedule to assure continued proper functioning of the system.
③ Flush and refill at 36 mos. regardless of miles, then every 24 mos./30,000 mi. (48,000 km) thereafter.
④ Off-the-highway operation, trailer towing, snow plowing, prolonged operation with heavy loading, especially in hot weather require more frequent transmission service.

MANUFACTURER RECOMMENDED MAINTENANCE INTERVALS (DIESEL)

TO BE SERVICED		TYPE OF SERVICE	VEHICLE MILEAGE INTERVAL (x1000)												
			7.5	15	22.5	30	37.5	45	52.5	60	67.5	75	82.5	90	97.5	miles
			12	24	36	48	60	72	84	96	108	120	132	144	156	km
			6	12	18	24	30	36	42	48	54	60	66	72	78	months
Engine oil & oil filter		D/R	✓	✓	✓	✓	✓	✓	✓	✓	✓	✓	✓	✓	✓	
Water pump weep hole clear		S/I		✓		✓		✓		✓		✓		✓		
Replace fuel filter		R		✓		✓		✓		✓		✓		✓		
Clean fuel strainer		S/I			✓			✓			✓			✓		
Front wheel bearings	①	S/I			✓			✓			✓			✓		
Brake linings		S/I			✓			✓			✓			✓		
Fan hub		S/I				✓				✓				✓		
Damper		S/I				✓				✓				✓		
Air cleaner element ④	②	R														
ATF / Replace filter, adjust bands		D/R				✓				✓				✓		
Transfer case fluid		D/R						✓				✓				
Engine Coolant	③	R/F						✓								
Accessory belt(s)		S/I			✓			✓			✓			✓		

R - Replace
S/I - Service or Inspect
R/F - Replace and flush system
D/R - Drain and refill
C/L - Clean and lubricate

① Clean and repack if required (4x2).
② Emission Control System Maintenance is listed in bold type. Service must be performed on-schedule to assure continued proper functioning of the system.
③ Flush and refill at 36 mos. regardless of miles, then every 24 mos./30,000 mi. (48,000 km) thereafter.
④ Check Filter Minder monthly; replace element if necessary.

1-54 GENERAL INFORMATION AND MAINTENANCE

CAPACITIES

CAPACITIES

Year	Model	Engine Displacement Liters (cc)	Engine ID/VIN	Oil with Filter (qts.)	Engine Transmission (pts.) Manual	Engine Transmission (pts.) Auto	Transfer Case (pts.)	Drive Axle Front (pts.)	Drive Axle Rear (pts.)	Fuel Tank (gal.)	Cooling System (qts.)
1997	Dakota	2.5 (2458)	P	4.5	①	②	③	④	⑤	⑥	9.8
		3.9 (3916)	X	4.5	①	②	③	④	⑤	⑥	14.0
		5.2 (5211)	Y	5.0	①	②	③	④	⑤	⑥	14.3
	Ram 1500-3500	3.9 (3916)	X	4.0	①	②	—	—	⑤	⑦	20.0
		5.2 (5211)	Y	5.0	①	②	③	④	⑤	⑦	20.0
		5.9 (5899)	Z	5.0	①	②	③	④	⑤	⑦	20.0
		5.9 (5882)	D	11.0	①	②	③	④	⑤	⑦	26.0
		8.0 (7997)	W	7.0	①	②	③	④	⑤	⑦	24.0
		5.9 (5899)	5	5.0	①	②	③	④	⑤	⑦	20.0
1998	Dakota	2.5 (2458)	P	4.5	①	—	—	④	⑤	⑥	9.8
		3.9 (3916)	X	4.0	①	②	③	④	⑤	⑥	14.0
		5.2 (5211)	Y	5.0	①	②	③	④	⑤	⑥	14.3
		5.9 (5899)	Z	5.0	①	②	③	④	⑤	⑥	14.3
	Durango	3.9 (3916)	X	4.0	—	②	③	④	⑤	25.0	14.0
		5.2 (5211)	Y	5.0	—	②	③	④	⑤	25.0	14.3
		5.9 (5899)	Z	5.0	—	②	③	④	⑤	25.0	14.3
	Ram 1500-3500	3.9 (3916)	X	4.0	①	②	③	④	⑤	⑦	20.0
		5.2 (5211)	Y	5.0	①	②	③	④	⑤	⑦	20.0
		5.9 (5899)	Z	5.0	—	②	③	④	⑤	⑦	20.0
		5.9 (5899)	5	5.0	①	②	③	④	⑤	⑦	20.0
		8.0 (7997)	W	7.0	①	②	③	④	⑤	⑦	24.0
		5.9 (5882)	D	11.0	①	②	③	④	⑤	⑦	26.0
1999	Dakota	2.5 (2458)	P	4.5	①	—	—	④	⑤	⑥	9.8
		3.9 (3916)	X	4.0	①	②	③	④	⑤	⑥	14.0
		5.2 (5211)	Y	5.0	①	②	③	④	⑤	⑥	14.3
		5.9 (5899)	Z	5.0	—	②	③	④	⑤	⑥	14.3
	Durango	3.9 (3916)	X	4.0	—	②	③	④	⑤	25.0	14.0
		5.2 (5211)	Y	5.0	—	②	③	④	⑤	25.0	14.3
		5.9 (5899)	Z	5.0	—	②	③	④	⑤	25.0	14.3
	Ram 1500-3500	3.9 (3916)	X	4.0	①	②	③	④	⑤	⑦	20.0
		5.2 (5211)	Y	5.0	①	②	③	④	⑤	⑦	20.0
		5.9 (5899)	Z	5.0	—	②	③	④	⑤	⑦	20.0
		5.9 (5899)	5	5.0	①	②	③	④	⑤	⑦	20.0
		8.0 (7997)	W	7.0	①	②	③	④	⑤	⑦	24.0
		5.9 (5882)	6	11.0	①	②	③	④	⑤	⑦	26.0
2000	Dakota	2.5 (2458)	P	4.5	①	—	—	④	⑤	⑥	9.8
		3.9 (3916)	X	4.0	①	②	③	④	⑤	⑥	14.0
		4.7 (4701)	N	6.0	①	②	③	④	⑤	⑥	17.0
		5.2 (5211)	Y	5.0	①	②	③	④	⑤	⑥	14.3
		5.9 (5899)	Z	5.0	—	②	③	④	⑤	⑥	14.3
	Durango	3.9 (3916)	X	4.0	—	②	③	④	⑤	25.0	14.0
		4.7 (4701)	N	6.0	—	②	③	④	⑤	25.0	17.0
		5.2 (5211)	Y	5.0	—	②	③	④	⑤	25.0	14.3
		5.9 (5899)	Z	5.0	—	②	③	④	⑤	25.0	14.3
	Ram 1500-3500	3.9 (3916)	X	4.0	①	②	③	④	⑤	⑦	20.0
		5.2 (5211)	Y	5.0	①	②	③	④	⑤	⑦	20.0
		5.9 (5899)	Z	5.0	—	②	③	④	⑤	⑦	20.0
		5.9 (5899)	5	5.0	①	②	③	④	⑤	⑦	20.0
		8.0 (7997)	W	7.0	①	②	③	④	⑤	⑦	24.0
		5.9 (5882)	6	11.0	①	②	③	④	⑤	⑦	26.0

NOTE: All capacities are approximate. Always use the dipstick or level plug or level marks for final determination. Add fluid gradually and check to be sure a proper fluid level is obtained.

① NV3500: 4.2 pts.
  NV4500/HD: 8.0 pts.
  NV5600: 9.5 pts.
  AX15: 6.6 pts.
  Getrag: 7.0 pts.

② Routine oil/filter change: approx. 3 pts.;
  after rebuild: approx. 20 pts. Use dipstick for final level.

③ NV231: 2.5 pts.
  NV231-HD: 2.5 pts.
  NV241: 5.0 pts.
  NV241HD: 6.5 pts.
  NV241HD w/PTO: 10.3 pts.
  NV242: 3.0 pts.

④ 194-FIA: 3.0 pts
  216-FBI: 4.8 pts. pts.
  248-FBI: 8.5 pts.

⑤ 8.25 in.: 4.4 pts.
  9.25: 4.9 pts.
  248-RBI (2WD): 6.3 pts.
  248-RBI (4WD): 7.0 pts.
  267-RBI (2WD): 7.0
  267-RBI (4WD): 7.5 pts.
  286-RBI (2WD): 6.8 pts.
  286-RBI (4WD): 10.1 pts.

⑥ Standard fuel tank: 15 gal.
  Optional fuel tank: 22 gal.

⑦ 22, 34 or 35 gal.

CHARGING SYSTEM
ALTERNATOR 2-10
ALTERNATOR SYSTEM
ELECTRONIC VOLTAGE REGULATOR 2-12
GENERAL INFORMATION 2-9
PRECAUTIONS 2-9

DISTRIBUTOR IGNITION SYSTEM
CRANKSHAFT AND CAMSHAFT POSITION
 SENSORS 2-6
DIAGNOSIS AND TESTING 2-2
DISTRIBUTOR 2-4
DISTRIBUTOR CAP/ROTOR 2-3
GENERAL INFORMATION 2-2
IGNITION COIL 2-3
IGNITION COIL REMOVAL &
 INSTALLATION 2-4

DISTRIBUTORLESS IGNITION - 8.0L ENGINE
CRANKSHAFT AND CAMSHAFT POSITION
 SENSORS 2-8
GENERAL INFORMATION 2-7
IGNITION COIL PACK 2-7

DISTRIBUTORLESS IGNITION SYSTEM - 4.7L ENGINE
CRANKSHAFT AND CAMSHAFT POSITION
 SENSORS 2-7
DIAGNOSIS AND TESTING 2-7
GENERAL INFORMATION 2-6
IGNITION COIL 2-7

FIRING ORDERS 2-8

PDC & PCM LOCATIONS 2-2

SENDING UNITS AND SENSORS
BATTERY TEMPERATURE SENSOR 2-17
COOLANT TEMPERATURE GAUGE
 SENDING UNIT 2-15
ELECTRIC FAN 2-17
OIL PRESSURE SENSOR 2-16

STARTING SYSTEM
GENERAL INFORMATION 2-12
STARTER 2-13

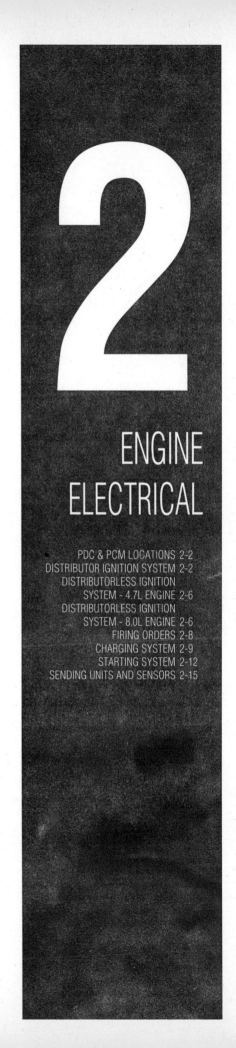

2
ENGINE ELECTRICAL

PDC & PCM LOCATIONS 2-2
DISTRIBUTOR IGNITION SYSTEM 2-2
DISTRIBUTORLESS IGNITION
 SYSTEM - 4.7L ENGINE 2-6
DISTRIBUTORLESS IGNITION
 SYSTEM - 8.0L ENGINE 2-6
FIRING ORDERS 2-8
CHARGING SYSTEM 2-9
STARTING SYSTEM 2-12
SENDING UNITS AND SENSORS 2-15

2-2 ENGINE ELECTRICAL

PDC & PCM LOCATIONS

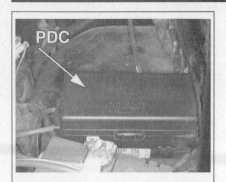

Fig. 1 Power Distribution Center (PDC) location on the inner fender - typical

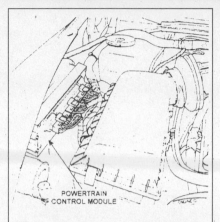

Fig. 2 Powertrain Control Module (PCM) location - right inner fender shield

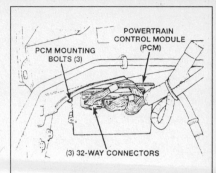

Fig. 3 Powertrain Control Module (PCM) location - firewall

See Figures 1, 2 and 3

The systems described in this section usually use circuits/components located in the Power Distribution Center (PDC) and/or the Powertrain Control Module (PCM).

1. The Powertrain Control Module is the on-board computer which manages the ignition and fuel systems, among many other functions.
2. The Power Distribution Center incorporates relays, fuses, and internal circuitry to centralize and distribute electrical power to all other systems on the vehicle.
3. These units are located as follows:
The Power Distribution Center is located on the left side of the engine compartment on the fender inner shield.
The Powertrain Control Module is located on the right side of the engine compartment, either on the fender inner shield or on the firewall.

DISTRIBUTOR IGNITION SYSTEM

Note: For information on understanding electricity and troubleshooting electrical circuits, please refer to the "Chassis Electrical" section.

General Information

This section deals with the following distributor-equipped gasoline engines: 2.5L, 3.9L, 5.2L, and 5.9L engines.

The main components of the ignition system are:
- Powertrain control module (PCM)
- Ignition coil
- Distributor
- Spark plugs
- Crankshaft and camshaft position sensors.

The PCM automatically regulates the spark advance to fire the spark plugs according to input from various engine sensors. The input signals are then used to compute the optimum ignition timing for the lowest exhaust emissions and best driveability. During the crank-start period, ignition timing advance is set to ensure quick and efficient starting.

The amount of electronic spark advance provided is determined by four input factors: coolant temperature, engine rpm, throttle position and available manifold vacuum. On some late model systems, intake manifold temperature is also used. Gear selection may also be a factor.

Engines described in the following paragraphs are equipped with a camshaft-driven mechanical distributor containing a shaft-driven rotor. The distributor is equipped with a camshaft position sensor which provides fuel injection synchronization and cylinder identification. The distributor does not have a built-in advance mechanism. Base ignition timing and all advance functions are controlled by the PCM and are not adjustable.

Battery voltage is supplied to the ignition coil from the Automatic Shut Down (ASD) relay. The PCM opens and closes the ignition coil ground circuit to operate the coil.

Diagnosis and Testing

HELPFUL TOOLS

See Figure 4

Some simple but very useful tools will go far when troubleshooting an ignition problem.

An inductive-type spark tester will allow you to check for spark at each plug without disconnecting the wires. This inexpensive tool uses a probe which is applied to a plug wire with the engine running or cranking. The current flowing in the wire (if present) will cause the probe's light to flash. This will allow you to check all cylinders in a short time.

A test light will permit checks of circuit integrity. Even more useful (but more expensive) is a VOM (volt-ohmmeter), which can be very helpful for finding short or open circuits as well as performing component tests.

A spark plug boot tool is the best way to disconnect plug leads on modern engines. These are usually difficult to remove and the tool will enable removal without damaging the wire or boot and may help to protect the user against electrical shock.

SYSTEM INSPECTION

See Figure 5

1. Check the spark plug cable connections for good contact at the coil, distributor cap and spark plugs. Push on them to ensure they are fully seated. Check that the boots are in good condition and fit tightly on their terminals. Boots that are cracked or torn should be replaced. Distributor terminals and plug boots should be smeared with dielectric grease to prevent corrosion.
2. Clean plug wires with a non-flammable solvent to remove oil, grease or dirt. Wipe dry. Check for cracks, brittle condition, melted insulation or other obvious damage. Replace any damaged wire(s).
3. Some engines are fitted with heat shields pressed into the cylinder head to protect the boots from damage. They should not be removed. When fully pressed in, the boot should show a small air gap between the boot lip and the top of the shield.
4. Check that plug wires are correctly routed, secure in wire holders (if fitted) and clear of danger areas such as exhaust manifolds or moving parts.

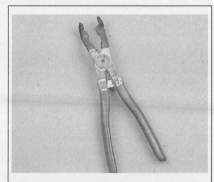

Fig. 4 An insulated spark plug boot tool helps disconnect leads quickly, safely and without damage

ENGINE ELECTRICAL 2-3

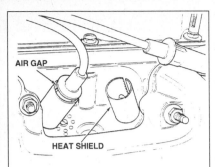

Fig. 5 Heat shield and proper boot installation

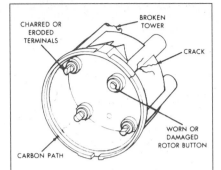

Fig. 6 Inspect the cap for any of the kinds of damage or wear that would cause problems

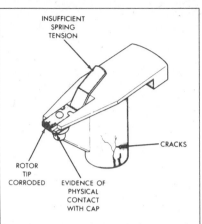

Fig. 7 Carefully inspect the rotor for damage of this sort

5. A rough inspection of the ignition system can be carried out by running the engine in the dark. Look for arcing from wires, cap, coil, etc. to ground.

WARNING:

When performing ignition system tests, note the following:

- Never run the engine in an enclosed garage.
- Engines should not be run for more than a minute or so with one or more spark plug wires disconnected. To do so risks damage to the catalytic converter.
- Disconnected plug wires must be grounded to the engine when cranking or running the engine. Failure to do so may damage the electronic ignition components.
- Make sure the area is free of any flammable materials and that the vehicle and garage floor are dry. When the engine is running, do not touch ignition system components without proper insulation on tools or hands.

SPARK DROP TEST

1. In the event of a misfire or poor running condition in which no spark at one or more plugs is suspected, check for spark at each plug.

Note: If any service was performed on the engine recently, be sure that the spark plug wires are connected to the correct cylinders before proceeding. It is possible for an engine to run (poorly) with wires switched. Refer to the firing order illustrations in this section.

2. An inductive tester, as mentioned above, is the easiest way to check. If not available, proceed as follows.
3. With the engine running, disconnect the wires from the distributor cap terminals one at a time. Be sure to properly insulate yourself before handling the wires. Do not leave the wire(s) disconnected for more than a few seconds.
4. As each one is disconnected, engine rpm should decrease or the engine should run roughly. If no difference is noted when a plug wire is disconnected, check the spark plug, wire, and distributor cap terminal for that cylinder.

SECONDARY SPARK TEST

1. Remove the plug for the cylinder in question and check condition. "General Information And Maintenance" provides illustrations of spark plug condition and the various causes.
2. Connect the boot and ground the plug on the engine. Start the engine and check for spark at the plug.
3. Remove the plug and insert a screwdriver or metal rod into the cap and hold it about 1/4-inch from the engine. If a spark jumps, the problem was the spark plug.
4. If there is still no spark, check condition of the plug wire. Replace it if burned, cracked, or shows insulation damage. Resistance should not exceed 12,000 ohms/foot.
5. If no test equipment is available, try fitting the wire from another cylinder and repeating the spark test.
6. If plug and wire are in working condition, check the distributor cap for cracks, terminals for corrosion, internal contacts for erosion, carbon paths, or other obvious faults.

NO-SPARK CONDITION

1. If the engine stops running or will not start, the problem is probably in the distributor cap, ignition coil or other upstream components.
2. Check for spark at each wire. Use an inductive tester while cranking the engine. If one is not available, disconnect and ground the plug caps (one at a time) and check for spark while cranking the engine.
3. If there is no spark at any plug, check the distributor cap and rotor.

Distributor Cap/Rotor

REMOVAL & INSTALLATION

See Figures 6 and 7

WARNING:

Before removing the distributor cap, mark the plug wires for reconnection to the correct terminals.

1. Remove the splash shield, air cleaner and hoses only if necessary for access to the distributor.
2. Mark and disconnect the plug wires. Remove the distributor cap.
3. Wipe the cap clean with a dry lint-free cloth. Visually inspect the cap for cracks, carbon paths, broken towers, or damaged rotor button. Also check for white deposits on the inside (caused by condensation entering through cracks). Replace any cap that has charred or eroded terminals. The inside flat surface of a terminal end (faces toward the rotor) will show some evidence of erosion from normal operation. Examine the terminal ends for evidence of mechanical interference with the rotor tip.
4. The rotor can be checked in place. Check for corrosion or damage to the tip, a broken spring or housing cracks.
5. Distributor contacts can be cleaned up with a small file. The rotor tip is covered with a varnish for radio noise suppression. It will appear charred. This is normal and should not be removed.
6. If the no-spark condition persists, check the ignition coil.

Ignition Coil

TESTING IN PLACE

See Figure 8

1. Disconnect the coil lead from the center terminal of the distributor cap and hold it about 1/2-inch from a good ground on the engine.
2. Crank the engine. If spark occurs, the problem is probably in the distributor. If there is no spark, the problem is the coil or an upstream component.

CAUTION:

Coil output is very high. Be sure to insulate yourself from the coil wire during the test.

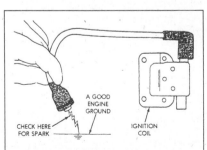

Fig. 8 Checking coil for spark with component in place. For safety, insulate yourself from the lead

ENGINE ELECTRICAL

LOCATION

On 2.5L engines, the coil is mounted on a bracket and located behind the distributor.

On 3.9L, 5.2L, 5.9L LDC engines the coil is mounted to a bracket bolted to the front of the right cylinder head. The bracket is secured to the automatic belt tensioner bracket.

On 5.9L HDC engines, the coil is mounted to a bracket bolted to the air injection pump mounting bracket.

TESTING

See Figure 9

Useful tests can be carried out with an ohmmeter.
1. Disconnect the primary wires and the lead to the distributor.
2. Check resistance across the primary wire connectors. It should be about 1 ohm.
3. Check resistance between one of the primary wire connectors and the high tension lead. It should be about 12 - 15K ohms.
4. If resistance is not within spec, replace the coil.

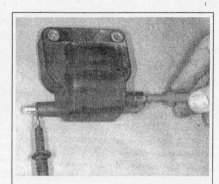

Fig. 9 Checking secondary winding resistance

REMOVAL & INSTALLATION

See Figures 10, 11, 12 and 13

1. Disconnect the battery negative terminal.
2. Disconnect the primary wires from the coils.
3. Disconnect the high tension lead.
4. Remove the securing bolts to remove the coil.
5. Installation is the reverse of removal.

CAUTION:

On 3.9L, 5.2L and 5.9L LDC engines the coil bracket is under tension. Do not attempt to remove it unless accessory drive belt tension is relieved.

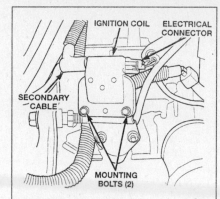

Fig. 10 Ignition coil location to the rear of the distributor - 2.5L

Fig. 11 Disconnecting the primary wires

Fig. 12 Ignition coil location is on the right side of the block - 3.9L, 5.2L, 5.9L LDC engines

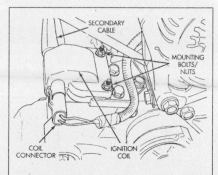

Fig. 13 Ignition coil location on the AIR pump bracket - 5.9L HDC engine

Distributor

REMOVAL

See Figures 14, 15, 16, 17, 18, 19 and 20

WARNING:

Before removing the distributor, be sure you understand the purpose and use of matchmarking the distributor for installation. Otherwise, special equipment may be necessary to reset the ignition timing. Conventional timing lights cannot be used to set ignition timing.

1. Remove the splash shield, if fitted as on 2.5L engines.
2. Remove the air cleaner assembly and connecting tubes if necessary for access.
3. On V-8 engines, remove the throttle body and plenum.
4. Disconnect the distributor low tension wires at the connector.
5. Disconnect the high tension lead from the coil.
6. Some setups may require disconnecting the spark plug wires from the cap to allow removal. If this is the case, mark each wire before disconnecting.
7. Remove the screws or unfasten the clips which secure the cap and remove it.
8. The engine need not be specially positioned to remove the distributor, but if there is any chance at all that the crankshaft might be moved after the distributor is removed, it will be best to remove it after the No. 1 cylinder is set to Top Dead Center (TDC) on the compression stroke.
9. Proceed as follows:
   a. The No. 1 cylinder is the FRONT cylinder on 2.5L engines and the FRONT LEFT cylinder on V-6 and V-8 engines.
   b. TDC can be achieved by slowly rotating the engine and noting when the index mark on the crankshaft harmonic balancer lines up with the "0" or "TDC" mark on the timing plate, but you must get the cylinder on the compression stroke. Refer to the next step.
   c. With the TDC marks aligned, check the distributor rotor. It should be pointing at the No. 1 cylinder terminal of the cap. If not, turn the crank 360-degrees and try again.
   d. The compression stroke can also be determined by feeling for pressure at the spark plug hole as the engine is rotated. If you can't feel any pressure, turn the crankshaft 360-degrees and try again.

Note: Another way to find TDC on the compression stroke is to use a clever little tool known as a "TDC whistle" available at many auto parts stores. Threaded into the plug hole, the whistle will sound as the compression rises and the piston reaches TDC.

10. At this point, check that the rotor is pointing to the "CYL NO. 1" mark on the camshaft position sensor mounting plate (if fitted).
11. After the engine is correctly positioned, proceed with distributor removal as follows.
12. Matchmark the position of the distributor housing relative to the engine block.

13. Matchmark the position of the distributor shaft or rotor relative to the distributor housing. Remember that both alignments will have to be restored when the distributor is installed.
14. Remove the distributor hold-down bolt and clamp.
15. Carefully lift the distributor assembly out of the engine. Note that the rotor will turn as the driven gear unmeshes with the drive gear. When clear of the drive gear, matchmark the position of the rotor. This will be the starting position when reinstalling the distributor.

INSTALLATION

Crankshaft Undisturbed

If the engine was not rotated after the distributor was removed, successful installation can be achieved by aligning the matchmarks. Proceed as follows:
1. Check the condition of the driven gear.
2. Check that the distributor shaft rotates without binding.
3. Check the condition of the O-ring. Replace if damage. Smear the O-ring with a bit of grease to ease assembly.
4. Position the rotor to align with the last matchmark made as the assembly was removed.
5. Carefully lower the assembly into the mounting hole. As the gears mesh, the rotor will begin to turn. Remember that the tongue at the end of the distributor shaft must engage the slot in the oil pump shaft.
6. When fully seated, check alignments: line up the engine block and distributor housing marks. The rotor and distributor housing mark should also align.
7. Install and tighten the hold-down bolt to 17 ft. lbs. (22 Nm).
8. The remainder of the installation procedure is the reverse of removal.

Crankshaft Disturbed

Use this procedure if the crankshaft was moved after the distributor assembly was removed.
1. Position the No. 1 cylinder at TDC on the compression stroke. Refer to the "Removal" procedure, above, for instructions on doing this.
2. Check the condition of the driven gear.
3. Check that the distributor shaft rotates without binding.
4. Check the condition of the O-ring. Replace if damaged. Smear the O-ring with a bit of grease to ease assembly.
5. You will be installing the distributor so that the rotor points at the No. 1 cylinder terminal on the distributor cap (see Firing Orders). Remember also that the tongue at the base of the distributor shaft must engage the slot in the oil pump shaft. This shaft can be turned with a screwdriver, if needed, to position it to accept the distributor shaft. On the 2.5L engine, the oil pump shaft slot should be just slightly CCW of the 10 o'clock position. On other engines, "eyeball" the pump shaft before attempting to install the distributor.
6. On 2.5L engines the rotor should be locked in position by using a 3/16-inch pin inserted into the plastic ring through the "2.5" alignment hole. New distributors have a plastic pin already in place.
7. Carefully lower the assembly into the mounting hole. As the gears mesh, the rotor will begin to turn. Remember that the tongue at the end of the dis-

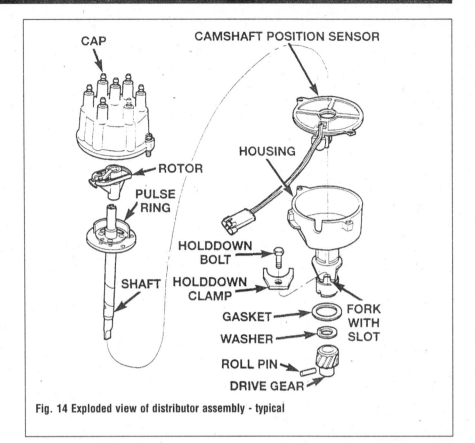

Fig. 14 Exploded view of distributor assembly - typical

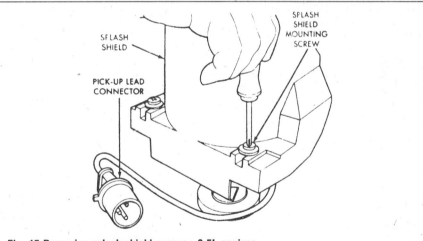

Fig. 15 Removing splash shield screws - 2.5L engines

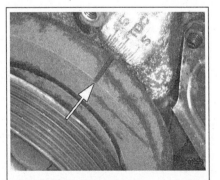

Fig. 16 TDC alignment marks - typical. Some engines may use "0" instead of "TDC"

Fig. 17 Distributor clamp bolt - V-8 shown

2-6 ENGINE ELECTRICAL

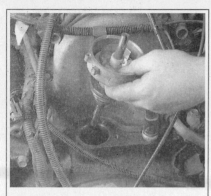

Fig. 18 Removing the distributor assembly

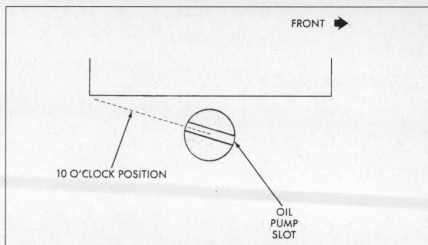

Fig. 19 Oil pump shaft slot alignment for distributor installation - 2.5L shown. Slot alignment must be taken into account on all engines

tributor shaft must engage the slot in the oil pump shaft.

8. When the assembly is seated, align the distributor housing and engine block matchmarks. The rotor should point to the No. 1 cylinder terminal on the distributor cap. On some engines the camshaft position sensor "CYL NO. 1" mark will aid in alignment: the rotor should align with the notch.

9. If no alignment marks were made, align the centerline of the distributor base slot with the hold-down bolt hole.

10. Don't forget to remove the locating pin on 2.5L engines.

11. The remainder of the procedure is the reverse of removal. Install and tighten the hold-down bolt to 17 ft. lbs. (22 Nm).

Crankshaft and Camshaft Position Sensors

Refer to "Driveability and Emission Controls" for information on these components.

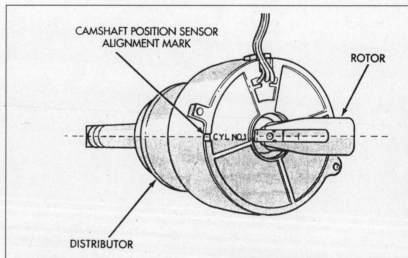

Fig. 20 Rotor pointing to "CYL NO 1" alignment mark - V-6, v-8

DISTRIBUTORLESS IGNITION SYSTEM - 4.7L ENGINE

Note: For information on understanding electricity and troubleshooting electrical circuits, please refer to the "Chassis Electrical" section.

General Information

See Figure 21

The 4.7L V-8 engine uses an advanced distributorless ignition system. A dedicated ignition coil is mounted above each spark plug and bolted to the engine. No plug wires are employed.

The main components of the ignition system are:
- Powertrain control module (PCM)
- Ignition coils
- Spark plugs
- Crankshaft and camshaft position sensors

The PCM automatically regulates the spark advance to fire the spark plugs according to input from various engine sensors. The input signals are then used to compute the optimum ignition timing for the lowest exhaust emissions and best driveability. During the crank-start period, ignition timing advance is set to ensure quick and efficient starting.

The amount of electronic spark advance provided is determined by four input factors: coolant temperature, engine rpm, throttle position and available manifold vacuum. On some late model systems, intake manifold temperature is also used. Gear selection may also be a factor.

Battery voltage is supplied to each coil by the Automatic Shutdown (ASD) relay. The PCM opens and closes each coil's ground circuit at a specific time to fire the plug.

By controlling the coil ground circuit, the PCM is able to set the base timing and adjust the ignition timing advance. This is done to meet changing engine conditions.

Base ignition timing is set by the PCM and is not adjustable.

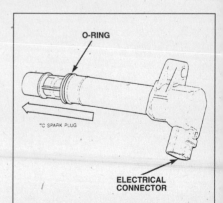

Fig. 21 Ignition coil components on the 4.7L engine. Primary winding resistance is measured at the electrical connector terminals

ENGINE ELECTRICAL 2-7

Diagnosis and Testing

Component tests are limited to checks of the spark plugs and coils. Specialized analysis equipment is needed for more in-depth troubleshooting.

CAUTION:

Do not run the engine with a coil disconnected. Do not attempt to ground a coil to the engine to check for spark. Sensitive components could be ruined or the extremely high voltage could cause serious injury or death.

Ignition Coil

TESTING

The coil(s) can be checked with an ohmmeter.
1. Disconnect the negative battery cable.
2. Remove the coil from the engine.
3. Primary winding resistance (across the low tension terminals) should be 0.6 - 0.9 ohms.
4. Secondary winding resistance (across the spark plug lead and one of the low tension terminals) should be 6 - 9K ohms.
5. Replace any coil not within specification.

REMOVAL & INSTALLATION

1. Certain coils may require removal of the throttle body air intake tube or intake box for access.
2. Disconnect the negative battery cable.
3. Detach the electrical connector from the coil by pushing downward on the release lock on top of the connector and pulling the connector from the coil.
4. Clean the area at the base of each coil with compressed air.
5. Remove the coil mounting nut. Pull the coil up with a slight twisting action and remove it from the vehicle.
6. Installation is the reverse of removal. Smear the coil O-ring with silicone grease. Tighten the mounting nut to 70 inch lbs. (8 Nm). Connect the wiring.

Crankshaft and Camshaft Position Sensors

Refer to "Driveability and Emission Controls" for information on these components.

DISTRIBUTORLESS IGNITION - 8.0L ENGINE

General Information

The 8.0L engine uses a distributorless ignition system. The main components of this ignition system are:
- Powertrain Control Module (PCM)
- Ignition coil packs
- Spark plugs
- Crankshaft and camshaft position sensors

The PCM automatically regulates the spark advance to fire the spark plugs according to input from various engine sensors. The input signals are then used to compute the optimum ignition timing for the lowest exhaust emissions and best driveability. During the crank-start period, ignition timing advance is set to ensure quick and efficient starting.

The amount of electronic spark advance provided is determined by four input factors: coolant temperature, engine rpm, throttle position and available manifold vacuum. On some systems, intake manifold temperature is also used. Gear selection may also be a factor.

Two separate coil packs, containing a total of five (5) independent ignition coils, are located on a common mounting bracket above the right side engine valve cover.

Each coil fires two cylinders simultaneously. When this occurs, one of the cylinders will be on the power stroke, the other will be on the exhaust stroke, the resulting spark being a waste spark. It is important to note that isolating one lead of the pair will eliminate the spark at both plugs.

Cylinders are paired as follows: 5/10, 9/8 for the rearmost four plug coil pack; 1/6, 7/4, 3/2 for the front six plug coil pack.

Battery voltage is supplied to each coil's positive terminal by the Automatic Shutdown (ASD) relay. If the PCM does not see a signal from the crankshaft and camshaft sensors (indicating the ignition key is ON but the engine is not running), it will shut down the ASD circuit.

By controlling the coil ground circuit, the PCM is able to set the base timing and adjust the ignition timing advance. This is done to meet changing engine conditions.

Base ignition timing is set by the PCM and is not adjustable.

CAUTION:

Do not run the engine with a spark plug lead disconnected. Do not attempt to ground a lead to the engine to check for spark. Sensitive components could be ruined or the extremely high voltage could cause serious injury or death.

Ignition Coil Pack

TESTING

See Figures 22, 23, 24 and 25

Coil packs are mounted above the right engine valve cover.
1. Disconnect the negative battery cable.
2. Mark and disconnect the spark plug leads at the coil packs. Disconnect the primary wiring at the connectors.
3. Check resistance across each pair of spark plug (high tension) terminals: 3/2, 7/4, 1/6, 9/8, 5/10. Resistance should be 11 - 15K ohms.
4. Check resistance across the (low tension) primary windings by connecting the ohmmeter between the B+ terminal and each of the coil terminals in turn. In each case, resistance should be 0.53 - 0.65 ohms.
5. If resistance is not within specification, the coil pack will have to be replaced.

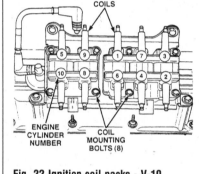

Fig. 22 Ignition coil packs - V-10

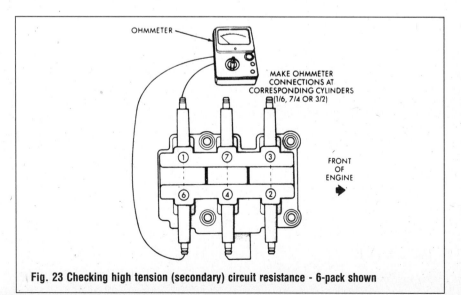

Fig. 23 Checking high tension (secondary) circuit resistance - 6-pack shown

2-8 ENGINE ELECTRICAL

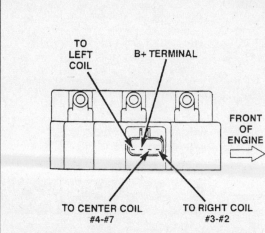

Fig. 24 Checking primary winding (low tension) circuit resistance: one probe to "B+" and the other to "Left", "Right" and "Center" terminals in turn - 6-pack shown

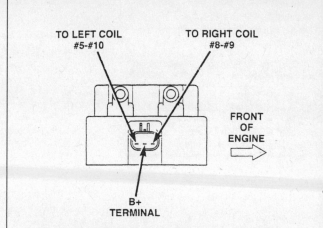

Fig. 25 Checking primary winding (low tension) circuit resistance: one probe to "B+" and the other to "Left" and "Right" terminals in turn - 4-pack shown

REMOVAL & INSTALLATION

See Figure 22

1. Disconnect the battery negative cable.
2. Tag each plug wire for location and disconnect the wires from the coil packs.
3. Disconnect the primary wire harness connectors.
4. Remove the four mounting bolts for each coil pack and remove the coil pack(s) from the engine.

To install:

5. Note that primary wire connectors face downward.
6. Mounting bolt torque is 90 inch lbs. (10 Nm).
7. When connecting primary wires, the four-wire connector goes to the front coil pack and the three-wire connector goes to the rear pack.
8. Reconnect plug wires to the proper terminals as marked.

Crankshaft and Camshaft Position Sensors

Refer to "Driveability and Emission Controls" for information on these components.

FIRING ORDERS

See Figures 26, 27, 28, 29 and 30

Before carrying out extensive tests to isolate poor running problems, check connections of spark plug wires. This is especially important if plugs, wires, distributor, etc., have been disturbed recently for service. Obviously, this is not applicable to the wireless 4.7L engine.

Note: To avoid problems, remove and tag the spark plug wires one at a time to ease reinstallation.

For vehicles with an extensive service history, keep in mind that the distributor could have been removed previously and rewired. The resultant wiring would hold the correct firing order, but could change the relative placement of the plug towers in relation to the engine. For this reason, it is very important that you label all wires before disconnecting any of them. Also, before removal, compare the wiring on your vehicle with the accompanying illustrations. If the wiring on your vehicle does not match, make notes in your book to reflect how your engine is wired.

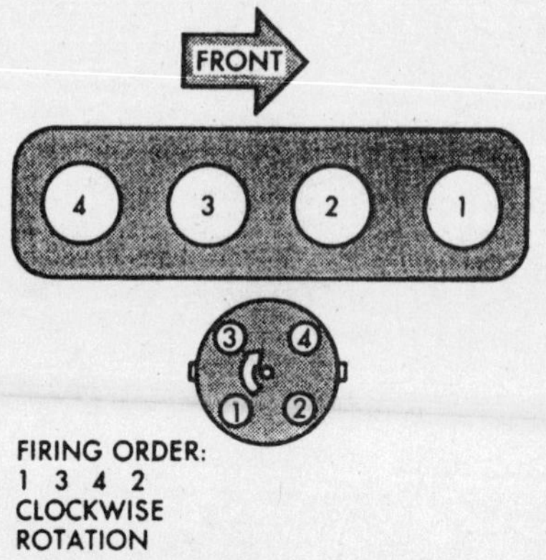

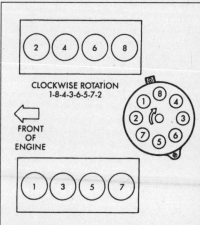

Fig. 26 2.5L Engine

Firing Order: 1 - 3 - 4 - 2

Distributor Rotation: Clockwise

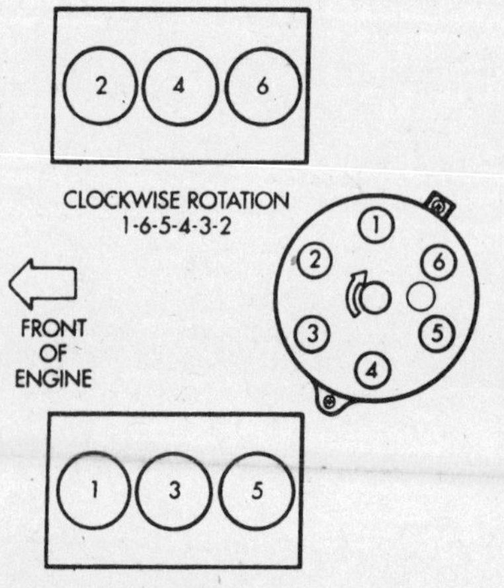

Fig. 27 3.9L Engine

Firing Order: 1 - 6 - 5 - 4 - 3 - 2

Distributor Rotation: Clockwise

Fig. 28 5.2L and 5.9L Engines

Firing Order: 1 - 8 - 4 - 3 - 6 - 5 - 7 - 2

Distributor Rotation: Clockwise

ENGINE ELECTRICAL 2-9

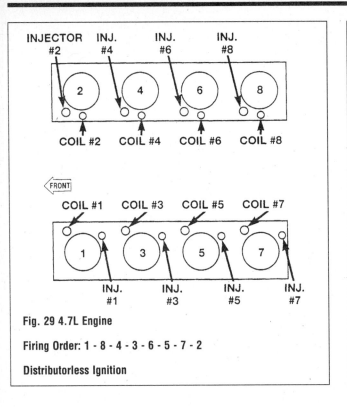

Fig. 29 4.7L Engine

Firing Order: 1 - 8 - 4 - 3 - 6 - 5 - 7 - 2

Distributorless Ignition

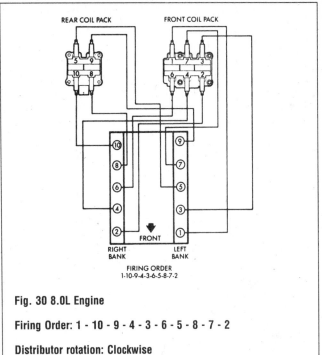

Fig. 30 8.0L Engine

Firing Order: 1 - 10 - 9 - 4 - 3 - 6 - 5 - 8 - 7 - 2

Distributor rotation: Clockwise

CHARGING SYSTEM

General Information

See Figures 31 and 32

The charging system incorporates the following major components:
- Alternator
- Alternator drive belt
- Electronic Voltage Regulator (EVR) circuitry in the PCM
- Battery
- Battery temperature sensor
- Ignition switch
- "Check Gauges" lamp
- Voltmeter (if fitted)
- Wiring harness and connections

The charging system is basically a series circuit with the battery wired in parallel. After the engine is started and running, the alternator takes over as the source of power and the battery then becomes part of the load on the charging system.

The alternator, which is driven by the belt, consists of a rotating coil of laminated wire called the rotor. Surrounding the rotor are more coils of laminated wire that remain stationary just inside the alternator case, called the stator. When current is passed through the rotor via the slip rings and brushes, the rotor becomes a rotating magnet with, of course, a magnetic field. When a magnetic field passes through a conductor (the stator), alternating current (A/C) is generated. This A/C current is rectified, or turned into direct current (D/C), by the diodes located within the alternator.

The voltage regulator circuitry controls the alternator's field voltage by grounding one end of the field windings very rapidly. The frequency varies according to current demand. The more the field is grounded, the more voltage and current the alternator produces. Voltage regulation on these vehicles is located in the Powertrain Control Module (PCM).

Voltage is maintained at about 13.5 - 15 volts. During high engine speeds and low current demands, the regulator will adjust the voltage of the alternator field to lower the alternator output voltage. Conversely, when the vehicle is idling and the current demands may be high, the regulator will increase the field voltage, increasing the output of the alternator. A major factor in determining alternator output is the battery temperature sensor whose input to the PCM is used to adjust field voltage.

The charging system is turned on and off with the ignition switch. The system is on when the engine is running and the ASD relay is energized. When the ASD relay is on, the PCM supplies voltage to the alternator field coil.

The amount of current produced by the alternator is controlled by the EVR field control circuitry in the PCM. A battery temperature sensor, located in the battery tray housing, monitors battery condition. This information, along with data from monitored line voltage, is used by the PCM to vary charging rate.

Precautions

To prevent damage to the on-board computer, alternator and regulator, the following precautionary measures must be taken when working with the electrical system.
- Wear safety glasses when working on or near the battery.

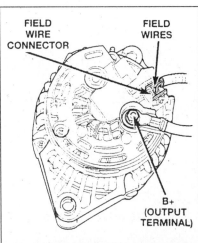

Fig. 31 Back of typical alternator illustrating wiring

Fig. 32 Another style of alternator wiring

2-10 ENGINE ELECTRICAL

- Don't wear a watch with a metal band when servicing the battery. Serious burns can result if the band completes the circuit between the positive battery terminal and ground.
- When installing a battery, make sure that the positive and negative cables are not reversed.
- Be absolutely sure of the polarity of a booster battery before making connections. Connect the cables positive to positive, and negative to negative. Connect positive cables first and then make the last connection to ground on the body of the booster vehicle, so that arcing cannot ignite hydrogen gas that may have accumulated near the battery.
- Even momentary connection of a booster battery with the polarity reversed will destroy the alternator.
- Disconnect both battery cables - negative first - before attempting to charge a battery.
- Never ground the alternator output cable.
- Be cautious when using metal tools around a battery to avoid creating a short circuit between the terminals.
- Never ground the field circuit between the alternator and regulator.
- Never attempt to polarize an alternator.
- Never operate the alternator with the battery disconnected.
- Do not short across or ground any alternator terminals.
- Do not apply full battery voltage to the field connector(s).
- Always disconnect the battery ground cable before disconnecting the alternator lead.
- Never subject the alternator to excessive heat or dampness. If you are steam cleaning the engine, cover the alternator.
- If arc welding is to be done on the car, disconnect the battery and alternator. Disconnect the wiring from the engine controller (PCM). This consists of 14 - and 60 - pin connectors on most models.

Alternator

SYSTEM TESTING

The charging system should be inspected if:
- A Diagnostic Trouble Code (DTC) is set relating to the charging system
- The charging system warning light (if equipped) is illuminated when the engine is running
- The "Check Gauges" lamp (if equipped) is illuminated when the engine is running
- The dashboard voltmeter (if equipped) indicates improper charging (either high or low)
- The battery is overcharged (electrolyte level is low and/or boiling out)
- The battery is undercharged (insufficient power to crank the starter)

The starting point for all charging system problems begins with the inspection of the battery, related wiring and the alternator drive belt. The battery must be in good condition and fully charged before system testing. If a Diagnostic Trouble Code (DTC) is set, diagnose and repair the cause of the trouble code first.

If equipped, the charging system warning light will illuminate if the charging voltage is either too high or too low. The warning light should illuminate when the key is turned to the ON position as a bulb check. When the alternator starts producing voltage due to the engine starting, the light should go out. A good sign of overly high voltage is lights that burn out and/or burn very brightly. Overcharging can also cause damage to the battery and electronic circuits.

Drive belts are often overlooked when diagnosing a charging system failure. Check the belt tension, check travel limit marks on the tensioner, if fitted. Check belt condition and replace the belt if needed. A loose belt will result in an undercharged battery and a no-start condition. This is especially true in wet weather conditions when moisture causes the belt to become more slippery.

If the vehicle is equipped with a voltmeter, this can sometimes give a clue to belt problems. If the indicated voltage stays low after starting, especially in wet weather, it may indicate that the belt is too loose. Other indications are voltage fluctuations when loads are added or removed from the system as when turning on the air conditioner.

DIAGNOSTIC TROUBLE CODES

Most charging system problems will set a Diagnostic Trouble Code and light the Malfunction Indicator Lamp (MIL) or similar warning indicator will be lit. On some models it may be possible to access the codes without the special scan tool. Try the following to see if this applies to your vehicle:

1. Turn the ignition switch ON - OFF - ON - OFF - ON within five seconds.
2. Check the odometer display or note the activity of the MIL. Any stored error codes will be displayed. If the vehicle uses the MIL, it will flash the codes (i.e.: four flashes - pause - seven flashes=47). If the odometer is used, two-digit error codes will be displayed one digit at a time with a short pause between digits.
3. The last display will be "55", indicating that all resident error codes have been displayed.
4. Check any error codes against the following:

Error Code	Problem
12	Battery disconnected
41	Open or shorted field control circuit
44	Shorted battery temp sensor
46	Charging voltage too high
47	Charging voltage too low

TESTING

Note 1: Before testing, make sure all connections and mounting bolts are clean and tight. Many charging system problems are related to loose and corroded terminals or bad grounds. Don't overlook the engine ground connection to the body. On some vehicles, it is beneficial to add an additional ground between the engine and the chassis. This may solve many intermittent problems.

Note 2: Alternators must be replaced if defective in any way. No internal service is possible.

Voltage Drop Test

Note: Before proceeding, make sure the battery is in good condition and fully charged.

Perform a voltage drop test of the positive side of the circuit as follows:
1. Start the engine and allow it to reach normal operating temperature.
2. Turn the headlamps, heater blower motor and interior lights ON.
3. Bring the engine to about 2,500 rpm and hold it there.
4. Connect the negative voltmeter lead directly to the battery positive terminal.
5. Touch the positive voltmeter lead directly to the alternator B+ output stud, not the nut. The meter should read no higher than about 0.5 volts. If it does, there is higher than normal resistance between the positive side of the battery and the B+ output at the alternator.
6. Move the positive meter lead to the nut and see if the voltage reading drops substantially. If it does, there is resistance between the stud and the nut. The theory is to keep moving closer to the battery terminal, one connection at a time, in order to find the area of high resistance (bad connection).

Perform a voltage drop test of the negative side of the circuit as follows:
1. Start the engine and allow it to reach normal operating temperature.
2. Turn the headlamps, heater blower motor and interior lights ON.
3. Bring the engine to about 2,500 rpm and hold it there.
4. Connect the negative voltmeter lead directly to the negative battery terminal.
5. Touch the positive voltmeter lead directly to the alternator case or ground connection. The meter should read no higher than about 0.3 volts. If it does, there is higher than normal resistance between the battery ground terminal and the alternator ground.
6. Move the positive meter lead to the alternator mounting bracket. If the voltage reading drops substantially, you know that there is a bad electrical connection between the alternator and mounting bracket. The theory is to keep moving closer to the battery terminal, one connection at a time, in order to find the area of high resistance (bad connection).

Alternator Isolation Test

On some models, it is possible to isolate the alternator from the regulator by grounding the field terminal. Grounding the field terminal removes the EVR from the circuit and forces full alternator output. This may help determine whether the problem is the alternator or EVR.

Note: Most alternators have two field terminals, one positive and one negative. With the engine running, the positive terminal will have battery voltage present and the negative terminal will have 3 - 5 volts less. Use a voltmeter to identify the negative terminal before carrying out the following test.

WARNING:

Do not let the voltage get higher than 18 volts. Damage to electrical circuits may occur.

1. Connect a voltmeter across the battery terminals so the voltage can be monitored.
2. Start the engine and let it reach normal operating temperature.
3. Connect a jumper lead to a good ground.

ENGINE ELECTRICAL 2-11

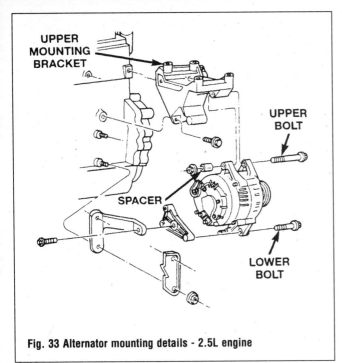

Fig. 33 Alternator mounting details - 2.5L engine

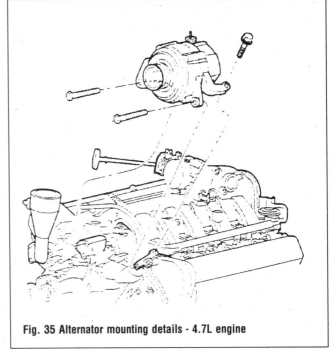

Fig. 35 Alternator mounting details - 4.7L engine

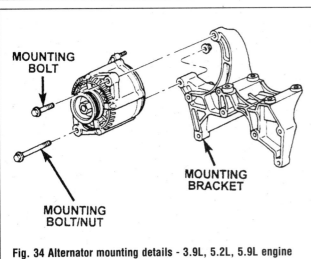

Fig. 34 Alternator mounting details - 3.9L, 5.2L, 5.9L engine

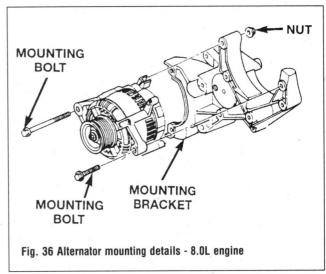

Fig. 36 Alternator mounting details - 8.0L engine

4. Locate the field terminal (negative) on the back of the alternator.

5. Momentarily connect the grounded jumper to the field terminal. If the alternator is okay, the voltage will climb rapidly. Disconnect the jumper before the output reaches 18 volts. If the voltage does not rise, replace the alternator. If the voltage rises, the regulator circuits are bad.

REMOVAL & INSTALLATION

See Figures 33, 34, 35, 36, 37, 38, 39, 40 and 41

1. Disconnect the negative battery cable (both on diesels).
2. Release tension and remove the drive belt.
3. Disconnect the wiring harness at the plastic connector.
4. Remove the output terminal nut(s) and disconnect the output cable.

5. Diesels: Loosen, but do not remove, the alternator bracket-to-engine bolt.
6. Remove alternator mounting bolts. Note any shims or spacers in the assembly.
7. Carefully remove the alternator from the engine.

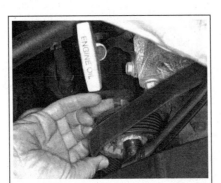

Fig. 37 Use a socket and breaker bar to rotate the tensioner and remove the serpentine belt

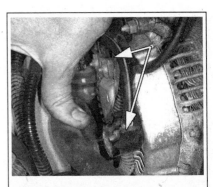

Fig. 38 Wiring arrangements may vary depending on model

2-12 ENGINE ELECTRICAL

Fig. 39 Removing one of the bracket mounting bolts

Fig. 40 Upper alternator mounting bolt

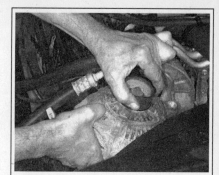

Fig. 41 Removing the alternator from the engine

To install:

8. Position the alternator on the engine. Be sure to install any shims or spacers.
9. Install the alternator mounting bolts and tighten to 40 ft. lbs. (54 Nm) for 2.5L and diesel engines, 30 ft. lbs. (41 Nm) for others.
10. Connect the electrical harness. Connect the output cable and tighten the nut.
11. Install the accessory drive belt and check tension.
12. Connect the negative battery cable.
13. Check the alternator for proper operation.

CAUTION:

Check routing of belt against the plate in the engine compartment before starting the engine.

Electronic Voltage Regulator

The Electronic Voltage Regulator (EVR) is a circuit within the Powertrain Control Module (PCM). It cannot be serviced separately. If EVR problems arise, the entire PCM must be replaced.

STARTING SYSTEM

General Information

Operation of the starting system is basically similar on all models, although starter motor specifications may differ for each engine family.

The starting system includes the battery, starter motor, solenoid, ignition switch and, in some cases, a starter relay. An inhibitor switch (neutral safety) is included in the starting system circuit to prevent the vehicle from being started while in gear.

When the ignition key is turned to the START position, current flows and energizes the starter's solenoid coil. The energized coil becomes a magnet which pulls the plunger into the coil, and the plunger closes a set of contacts which allow high current to reach the starter motor. The plunger also serves to push the starter pinion into the teeth on the flywheel/flexplate.

To prevent damage to the starter motor when the engine starts, the pinion gear incorporates an over-running (one-way) clutch which is splined to the starter armature shaft. The rotation of the running engine may speed the rotation of the pinion, but not the starter motor itself.

Some starting systems employ a starter relay in addition to the solenoid. This relay may be located under the dashboard, in the kick panel, or in the fuse/relay center under the hood. This relay is used to reduce the amount of current which the ignition switch must carry.

PRECAUTIONS

Always disconnect the negative battery cable before servicing the starter. Battery voltage is always present at the large (B) terminal on the solenoid. When removing the starter motor, be prepared to support its weight after the last bolt is removed, because the starter motor is a very heavy component.

Never operate the starter for more than 15 seconds at a time. Too much cranking will cause the starter motor to overheat, causing permanent damage. Allow the starter to cool for at least two minutes between starting attempts.

TESTING

See Figures 42, 43, 44, 45 and 46

Any troubleshooting of the starting system should start with checks of the following components:
- Battery condition
- Wiring and connections at the battery, starter motor and solenoid
- Starter relay
- Wiring at switches: clutch pedal (if equipped), Park/Neutral position switch (if auto), ignition switch

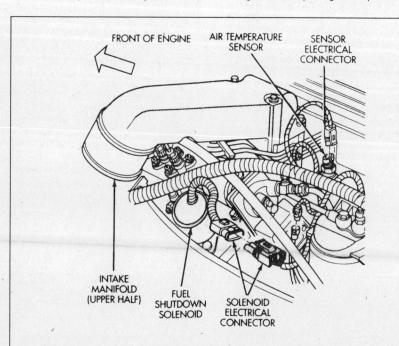

Fig. 42 On diesels, disconnect the fuel shutdown solenoid before doing any work on the starter motor

ENGINE ELECTRICAL 2-13

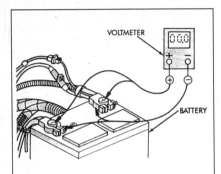

Fig. 43 Checking for voltage drop across the battery clamps

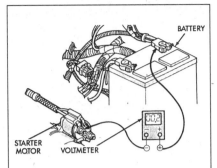

Fig. 44 Testing the battery positive cable voltage drop

CAUTION:

Before performing any of the tests below, make sure that the engine will not start accidentally by disconnecting and isolating the ignition coil wire (gasoline engines) or unplugging the fuel shutdown solenoid wire harness (diesels). Place the transmission in Neutral (manual trans.) or Park (auto trans.), parking brake applied.

1. Ensure that the battery is in good condition and fully charged. Simply checking for voltage across the terminals will not indicate battery condition. A load test is required. This requires special equipment, but a local garage or parts store may be able to help.
Check the battery and clean the connections as follows:
 a. If the battery cells have removable caps, check the water level. Add distilled water if low. Load test the battery and charge if necessary.
 b. Remove the cables and clean them with a wire brush. Reconnect the cables.
 c. Check for voltage across the battery posts and across the clamps to ensure that clamp connections are making good contact.
2. On manual trans. models, disconnect the clutch position switch and connect a jumper across the harness connector to simulate depressing the clutch pedal.
3. Connect the voltmeter leads to the battery positive terminal and the starter solenoid battery cable stud. Turn the ignition switch to START and hold it there. If the meter reads above 0.2v, suspect bad contact at the solenoid battery cable stud.
4. Connect the voltmeter leads to the battery negative terminal and a good ground on the engine. Turn the ignition switch to START and hold it there. If the meter reads above 0.2v, suspect bad connections of the battery negative cable or internal corrosion of the cable.
5. Connect the positive lead of the voltmeter to the starter motor housing and the negative meter lead to the battery's negative terminal. Turn the ignition switch to START and hold it there. If the meter reads above 0.2v, suspect poor starter-to-engine ground.
6. If the above tests do not reveal the cause of the problem, proceed to component tests, following.

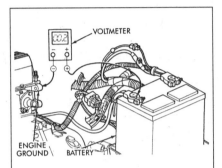

Fig. 45 Testing for ground circuit voltage drop

Starter

PRECAUTIONS

CAUTION:

Starters are removed from beneath the vehicle. This means the vehicle will have to be raised a sufficient distance for the work to be performed, and safely supported during the procedure. Block all wheels and use suitable jackstands to support the vehicle. Remember - you will be working underneath it for a considerable length of time.

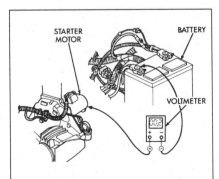

Fig. 46 Checking starter ground

On most vehicles, the starter wiring will be disconnected after the unit is unbolted from the engine. Be prepared to support the starter as this is done. Do not allow the starter to hang by the wiring harness.
 Always disconnect the negative battery cable before servicing the starter. Battery voltage is always present at the large (B) terminal on the solenoid. When removing the starter motor, be prepared to support its weight after the last bolt is removed, because the starter motor is a very heavy component.
 Never operate the starter for more than 15 seconds at a time. Too much cranking will cause the starter motor to overheat, causing permanent damage. Allow the starter to cool for at least two minutes between starting attempts.

SOLENOID TESTING

See Figures 47 and 48

The solenoid can be checked with an ohmmeter or test light. Solenoids may differ in style, but all those covered here are similar in operation. Solenoid circuitry consists of a low tension winding wired between the solenoid terminal and solenoid case and two high tension terminals. When battery voltage is applied to the solenoid terminal, the contact is closed and the two high tension terminals are electrically connected.
 1. Remove the starter motor.
 2. Remove the wire from the solenoid field coil terminal.
 3. Check for continuity between one of the field coil (high tension) terminals and the solenoid (low tension) terminal. Continuity should exist.
 4. Check for continuity between the solenoid (low tension) terminal and the solenoid case. Continuity should exist.
 5. If both tests are not met successfully, the assembly must be replaced.

Note: Solenoids are not individually replaceable. If defective, the entire starter motor assembly must be replaced.

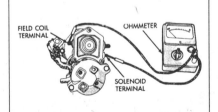

Fig. 47 Checking for continuity between the field coil and solenoid terminals -

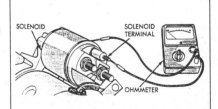

Fig. 48 Checking for continuity between the solenoid terminal and solenoid case - typical

2-14 ENGINE ELECTRICAL

RELAY

See Figures 49, 50 and 51

The relay is a micro-switch which energizes the starter solenoid when the ignition key is turned. The relay is located in the Power Distribution Center (PDC) in the engine compartment. Relay location can be found on the underside of the PDC cover.

More than one style of relay is in use, but they are electrically identical and testing is the same.

Basic tests can be carried out with an ohmmeter or test light as follows:

1. Continuity should exist between terminals 30 and 87A
2. There should be no continuity between terminals 30 and 87.
3. Resistance between terminals 85 and 86 should be 70 - 80 ohms.
4. Connect a 12v battery to terminals 85 and 86. Now, there should be continuity between terminals 30 and 87 and no continuity between terminals 30 and 87A.

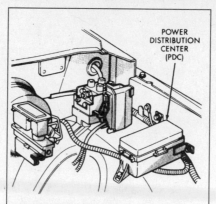

Fig. 49 Location of the Power Distribution Center (PDC) - typical

5. If the unit fails any of the above tests it must be replaced.
6. The common feed terminal (30) should have voltage at all times. Check the circuit to fuse F12 in the PDC if it does not.

STARTER MOTOR REMOVAL & INSTALLATION

See Figures 52, 53, 54, 55, 56 and 57

1. Disconnect and isolate the battery's negative cable.
2. Raise the vehicle and support it safely on jackstands. See "Precautions", above.
3. Check undercarriage for starter accessibility. On some models, it may be necessary to remove skid plates, etc., for access.
4. Disconnect any accessible wiring at this point. All starters have a battery positive cable and one or two leads to the solenoid.
5. Unbolt the securing hardware while supporting the starter motor with one hand. Diesels have three fasteners, others have two.

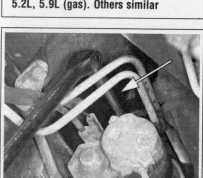

Fig. 52 Starter motor mounting - 3.9L, 5.2L, 5.9L (gas). Others similar

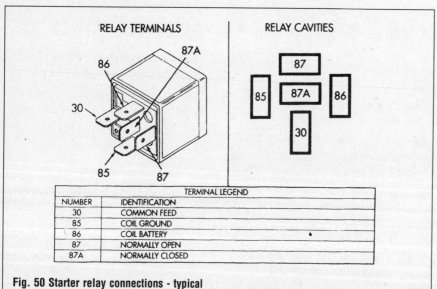

Fig. 50 Starter relay connections - typical

Fig. 53 Extensions must be rigged for access to the upper starter hardware

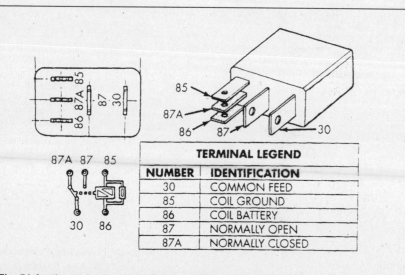

Fig. 51 Another style of starter relay connections

Fig. 54 Most units are secured at two points

ENGINE ELECTRICAL **2-15**

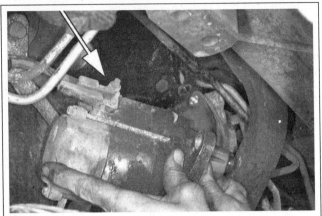

Fig. 55 Access to wiring connections occurs after removal on most starters

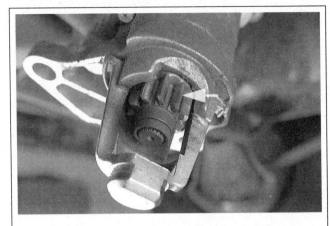

Fig. 56 Inspect the starter pinion for chipped or broken teeth

6. On some automatic transmission models, the cooler tube bracket may have to be removed from the stud. Note any shims between starter motor and mounting flange.
7. Pull the starter until the pinion nose clears the housing, then lower it until the wiring terminals are accessible. Disconnect the battery cable and solenoid wiring if not done before. Do not allow the starter motor to hang by the wiring harness.
8. Remove the starter motor from the vehicle.
9. Check the starter pinion and flywheel gear teeth for damage.

To install:
10. Installation is the reverse of removal. Be sure the motor mounting surfaces, mounting hardware and terminals are free of grease and grit. Remember to reinstall any shims which may have been used.
Tighten the mounting hardware to the following values:
- 2.5L: 33 ft. lbs.
- Diesel: 32 ft. lbs.
- Others: 50 ft lbs.

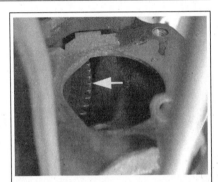

Fig. 57 Check flywheel teeth for condition

SENDING UNITS AND SENSORS

Note: This section covers sending units and sensors which are designed to provide the driver with information on vehicle condition.

Instrument panels contain a number of indicating devices (gauges and warning lights). These devices are composed of two separate components. One is the sending unit, mounted on the engine or other remote part of the vehicle, and the other is the actual gauge or light in the instrument panel.

Several types of sending units exist. However, most can be characterized as being either a pressure type or a resistance type. Pressure-type sending units are normally on-off in operation, completing or breaking a circuit to operate a warning light. Resistance-type sending units use variable resistance to control the current flow back to the indicating device like a gauge to indicate variations of pressure or temperature. Both types of sending units are connected in series by a wire to the battery (through the ignition switch). When the ignition is turned ON, current flows from the battery through the indicating device and on to the sending unit.

Coolant Temperature Gauge Sending Unit

See Figure 58

The temperature gauge sending unit controls the gauge or warning light. Some models use two temperature sensors (one for the gauge or warning light and one as a PCM input) which may be located adjacent to each other. Refer to the illustrations.

REMOVAL & INSTALLATION

See Figures 59, 60, 61, 62 and 63

1. The engine must be cold when performing this procedure.
2. Disconnect the battery negative cable.
3. Drain engine coolant to below sending unit level. A quart or so should be sufficient.
4. Disconnect the wiring from the temperature gauge sending unit.
5. Unscrew the old sending unit.

To install:
6. Coat the threads of the new unit with Teflon tape or a temperature-resistant sealer.
7. Tighten the sensor to 11 - 15 ft lbs. (15 - 20 Nm).
8. Reconnect wiring and add coolant to the radiator. Start the engine and check operation. Check for leaks at operating temperature.

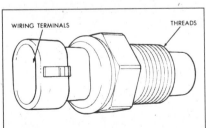

Fig. 58 Coolant temperature sensor - typical

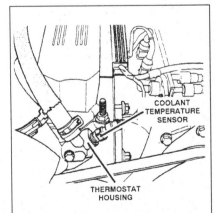

Fig. 59 Coolant temperature sensor on the 2.5L engine is in the thermostat housing

ENGINE ELECTRICAL

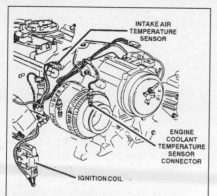

Fig. 60 Coolant temperature sensor location on 3.9L, 5.2L and 5.9L gas engines

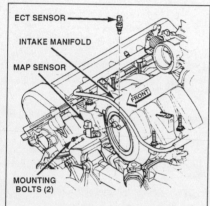

Fig. 61 Coolant temperature sensor location on the 4.7L engine

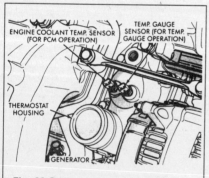

Fig. 62 Coolant temperature sensor location - 8.0L engine

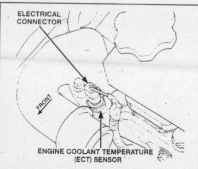

Fig. 63 Coolant temperature sensor location - 5.9L diesel engine

TESTING

1. The sending unit varies the resistance of the circuit in response to changes in coolant temperature. With a cold engine, resistance is high and gauge reading is low or the warning light is OFF. When the engine is hot, resistance is low and the gauge reading is high or the warning light is ON.
2. Disconnect the wiring to the sending unit and connect a jumper across the two leads. When the ignition is turned on, the gauge needle should move to the maximum or the warning light will go ON. If this happens, the gauge or light is functioning properly and the problem is likely to be the sensor.
3. If an ohmmeter is available, connect it across the sending unit terminals and start the engine. Resistance should decrease as the coolant temperature increases. If it does not, replace the sending unit.

Oil Pressure Sensor

TESTING

1. The oil pressure sensor is threaded into the engine block (see location illustrations). Low oil pressure (less than 10 psi) will close the switch, creating a circuit which includes the warning light. The light will go on when the switch is closed.
2. Disconnect the harness wire(s) from the sensor. Connect a jumper wire from the harness wire to ground on the engine (if single wire) or across the two harness wires.

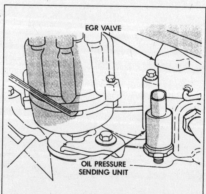

Fig. 64 Oil pressure sensor location near distributor - 3.9L, 5.2L, 5.9L (gas engine)

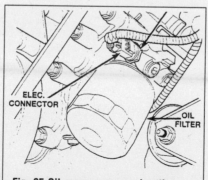

Fig. 65 Oil pressure sensor location - 4.7L engine

3. Turn the ignition switch ON. The low oil pressure warning light should go ON. Disconnect the jumper. The light should go OFF. If the lamp does not light when the circuit is completed, suspect a burned out bulb or wiring problem.
4. With the engine off and oil pressure "zero", check for continuity across the switch body. On single-terminal switches, check continuity between the switch terminal and ground on the switch body or engine block. On two-terminal switches, check for continuity between the two terminals. In either case, there should be continuity. If not, replace the switch.

REMOVAL & INSTALLATION

See Figures 64, 65, 66 and 67

1. Disconnect the battery negative cable.
2. Disconnect the wiring from the oil pressure sensor.
3. Unscrew the old sensor.
4. Coat the threads of the new sensor with Teflon® tape or an oil-resistant sealer.
5. Tighten the sensor to 11 - 15 ft lbs. (15 - 20 Nm).
6. Reconnect wiring. Start the engine and check operation. Check for leaks.

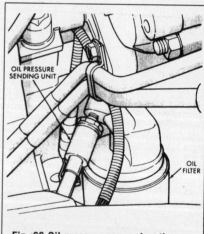

Fig. 66 Oil pressure sensor location - 8.0L

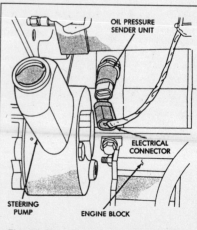

Fig. 67 Oil pressure sensor location - diesel

ENGINE ELECTRICAL 2-17

Battery Temperature Sensor

This sensor is located beneath the battery attached to a mounting hole on the battery tray. Models which have dual batteries have one sensor, on the driver's side battery.

Input from this sensor to the PCM is used to determine the charging rate.

REMOVAL AND INSTALLATION

See Figure 68

1. Remove the battery.
2. Disconnect the sensor pigtail harness from the engine wiring harness.
3. Pry the sensor straight up from the battery tray mounting hole.
4. To install, feed the harness through the mounting hole and snap in the sensor.

Electric Fan

Vehicles equipped with the 2.5L engine use an electric fan for engine cooling. Models with the 4.7L engine incorporate an auxiliary electric fan to aid low-speed cooling. Fan operation is controlled by the PCM through a relay located in the Power Distribution Center (PDC). The PDC is in the engine compartment on the left fender inner shield. The location of the fan relay is given on the inside of the PDC cover.

The fan relies upon several inputs to determine operation, notably the coolant temperature sensor.

- The fan will run whenever the engine is running and the air conditioner compressor clutch is engaged.
- When the air conditioner compressor clutch is disengaged, the fan will run at vehicle speeds above 40 mph if coolant temperature is above 230-degrees F (110-degrees C). The same is true of vehicles without air conditioning.
- The fan will turn off when coolant temperature drops to 220-degrees F (104-degrees C).
- At speeds below 40 mph, the fan turns on when coolant temperature reaches 210-degrees F (99-degrees C) and turns off when coolant temperature drops to 200-degrees F (93-degrees C).
- The fan will not run during engine cranking.

TESTING

See Figure 69

1. If the fan will not run, the easiest test is to disconnect the fan motor lead and apply 12v across the connector. Note polarity as shown in the illustration. If the fan runs, the problem is in the circuit or upstream components.
2. With the ignition and air conditioner ON, check for voltage across the wiring harness side of the fan motor connector. Note polarity as shown in the illustration. There should be 12v at the connector. If there is, the problem is the fan motor. If not, suspect the relay, circuitry, or other upstream component.
3. The PCM will set a diagnostic trouble code in the memory if it detects a problem in the fan relay or circuit, although a DRB scan tool is required for this analysis. If not available, proceed as follows:
4. Check fuses: there is a 10A fuse in the junction block and a 40A maxi fuse in the PDC.
5. Locate the fan relay in the PDC (location is shown on the cover). Apply 12v (with 14-gauge wire) to relay terminal 87. If the fan will not run, check for an open circuit in C25 or Z1. If the circuits are okay, replace the fan.
6. With the ignition key OFF, check for battery voltage at circuit C28 (relay terminal 30). If no battery voltage is present, check for open/short circuit C28 between the PDC and relay.
7. With the ignition key in the RUN position, check for battery voltage at circuit F18 (relay terminal 86). If no battery voltage is present, check for an open/short in circuit F18 between the junction block and the relay.

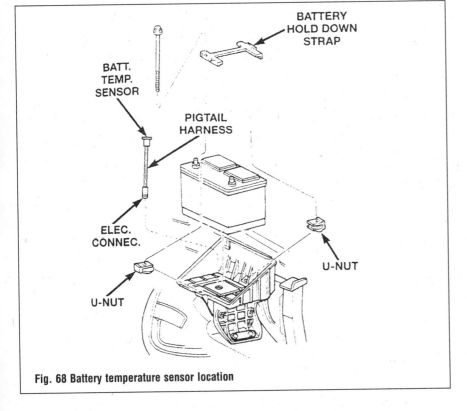

Fig. 68 Battery temperature sensor location

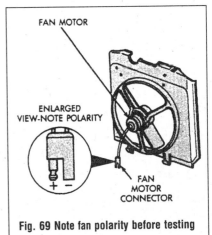

Fig. 69 Note fan polarity before testing

Notes

ENGINE MECHANICAL 3-2
BELT-DRIVEN ENGINE FAN 3-29
CAMSHAFT BEARINGS AND LIFTERS 3-43
CRANKSHAFT DAMPER 3-38
CYLINDER HEAD 3-32
ELECTRIC ENGINE FAN (2.5L) 3-29
ENGINE 3-18
EXHAUST MANIFOLD 3-27
FLYWHEEL/DRIVEPLATE 3-47
INTAKE MANIFOLD 3-23
OIL PAN 3-35
OIL PUMP 3-36
RADIATOR 3-28
REAR MAIN SEAL 3-45
ROCKER ARMS 3-20
THERMOSTAT 3-22
TIMING CHAIN AND SPROCKETS 3-40
TIMING COVER AND SEAL (GASOLINE
    ENGINES) 3-39
TIMING GEAR COVER AND SEAL
    (DIESELS) 3-40
TURBOCHARGER 3-28
VALVE COVER 3-19
WATER PUMP 3-30
ENGINE RECONDITIONING
BUY OR REBUILD? 3-50
CYLINDER HEAD 3-52
DETERMINING ENGINE CONDITION 3-49
ENGINE BLOCK 3-58
ENGINE OVERHAUL TIPS 3-51
ENGINE PREPARATION 3-52
ENGINE START-UP AND BREAK-IN 3-63
EXHAUST SYSTEM 3-47
COMPONENT REPLACEMENT 3-48
SAFETY PRECAUTIONS 3-48

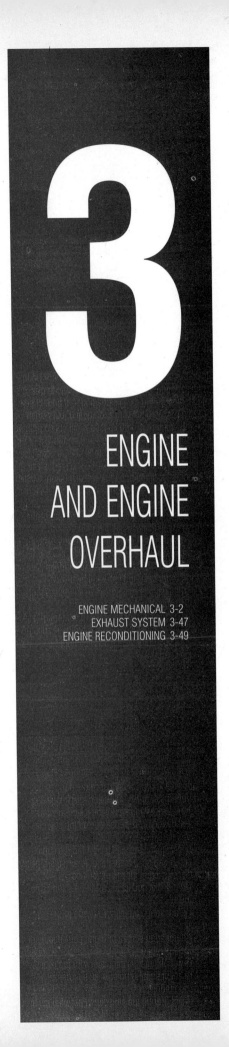

3
ENGINE AND ENGINE OVERHAUL

ENGINE MECHANICAL 3-2
EXHAUST SYSTEM 3-47
ENGINE RECONDITIONING 3-49

3-2 ENGINE AND ENGINE OVERHAUL

ENGINE MECHANICAL

2.5L (150 cu.in.) ENGINE MECHANICAL SPECIFICATIONS

Description	English Specifications	Metric Specifications
General Information		
Engine Type	In-line 4 Cylinder OHV	
Bore x Stroke	3.88 x 3.19 in.	98.4 x 81.0mm
Displacement	150 c.i.	2.5 L
Compression Ratio	9.1:1	
Firing Order	1-3-4-2	
Lubrication	Pressure Feed - Full Flow Filtration	
Cooling System	Liquid Cooled - Forced Circulation	
Cylinder Block	Cast Iron	
Crankshaft	Cast Nodular Iron	
Cylinder Head	Cast Iron	
Combustion Chambers	Double Quench	
Camshaft	Cast Iron	
Pistons	Aluminum Alloy	
Connecting Rods	Cast Malleable Iron	
Cylinder Compression Pressure	120-150 psi	827-1034 kPa
Max. Variation Between Cylinders	30 psi	206 kPa
Camshaft		
Hydraulic Tappet clearance	Zero Lash	
Cam Bearing Diameter		
No. 1	2.029 - 2.030 in.	51.54 - 51.56mm
No. 2	2.019 - 2.020 in.	51.28 - 51.31mm
No. 3	2.009 - 2.010 in.	51.03 - 51.05mm
No. 4	1.999 - 2.000 in.	50.78 - 50.80mm
Bearing-to-Journal Clearance		
Standard	0.001 - 0.003 in.	0.0254 - 0.0762mm
Max. Allowable	0.005 in.	0.127mm
Camshaft Lobe Lift		
Exhaust	0.259 in.	6.579mm
Intake	0.255 in.	6.477mm
Valve Lift		
Exhaust	0.415 in.	10.53mm
Intake	0.408 in.	10.35mm
Connecting Rods		
Piston Pin Bore Diameter	0.9288 - 0.9298 in.	23.59 - 23.62mm
Side Clearance	0.010 - 0.019 in.	0.25 - 0.48mm
Twist (max)	0.001 in./in.	0.001mm/mm
Bend (max)	0.001 in./in.	0.001mm/mm
Bearing Clearance	0.001 - 0.003 in.	0.025 - 0.076mm
Crankshaft		
Rod Journals		
Rod Journal Dia.	2.0934 - 2.0955 in.	53.17 - 53.23mm
Out Of Round (Max.)	0.0005 in.	0.013mm
Taper (Max.)	0.0005 in.	0.013mm
Rod Journal Width	1.070 - 1.076 in.	27.18 - 27.33mm
Main Bearings		
Main Bearing Clearance	0.0010 - 0.0025 in.	0.03 - 0.06mm
Main Bearing Journal Dia.	2.4996 - 2.5001 in.	63.489 - 63.502mm
Out Of Round (Max.)	0.0005 in.	0.013mm
Taper (Max.)	0.0005 in.	0.013mm
Journal Width		
No. 1	1.086 - 1.098 in.	27.58 - 27.89mm
No. 2	1.271 - 1.273 in.	32.28 - 32.33mm
Nos. 3, 4, 5	1.182 - 1.188 in.	30.02 - 30.18mm
Crankshaft End-Play	0.0015 - 0.0065 in.	0.038 - 0.165mm

ENGINE AND ENGINE OVERHAUL 3-3

Cylinder Block
   Cylinder Bore Dia. (Standard)   3.8759 - 3.8775 in.   98.45 - 98.48mm
     Out Of Round (Max.)   0.001 in.   0.025mm
     Taper (Max.)   0.001 in.   0.025mm
   Tappet Bore Dia.   0.9055 - 0.9065 in.   23.000 - 23.025mm
   Flatness (Min. Acceptable)   0.008 in./total length   0.20 mm/total length

Cylinder Head And Valves
   Valve Guide ID   0.313 - 0.314 in.   7.95 - 7.97mm
   Stem-Guide Clearance   0.001 - 0.003 in.   0.025 - 0.076mm
   Valve Seat Width   0.040 - 0.060 in.   1.01 - 1.52mm
   Face Angle   45.0 degrees
   Seat Angle   44.5 degrees
   Stem Diameter   0.311 - 0.312 in.   7.899 - 7.925mm
   Tip Refinishing (Max. Allowable)   0.010 in.   0.25mm
   Valve Springs
     Free Length   1.876 in.   47.65mm
     Spring Tension (Valve Closed)   71 - 79 lbf @ 1.64 in.   316 - 351 N @ 41.66mm
     Spring Tension (Valve Open)   202 - 218 lbf. @ 1.216 in.   899 - 970 N @ 30.89mm
     Installed Height   1.64 in.   41.66mm

Hydraulic Tappets
   Body Diameter   0.9035 - 0.9040 in.   22.949 - 22.962mm
        0.904 - 0.905 in.   22.962 - 22.974mm
   Clearance in Block   0.0010 - 0.0025 in.   0.025 - 0.063mm
   Pushrod Length   9.500 - 9.520 in.   241.3 - 241.8mm

Oil Pressure
   At Idle Speed (800 rpm)   25 - 35 psi   172 - 241 kPa
   Min. Pressure (600 rpm)   13 psi   90 kPa
   At 1600 rom & Higher   37 - 75 psi   255 - 517 kPa
   Oil Pressure Relief Valve Setting   75 psi   517 kPa

Oil Pump
        0.004 in.   0.1016mm
   Gear-to-Body Clearance (Radial)   0.002 - 0.004 in   0.051 - 0.102mm
   Gear End Clearance (Feeler Gauge)   0.004 - 0.008 in.   0.102 - 0.203mm

Pistons
        0.0005 - 0.0015 in.   0.013 - 0.038mm
   Cylinder Bore Clearance   0.008 - 0.0015 in.   0.018 - 0.038mm
   Land Clearance (Diam.)   3.19 in.   81.03mm
   Wrist Pin Bore Dia.   0.9312 - 0.9315 in.   23.650 - 23.658mm
   Ring Groove Height (Compression)   0.0602 - 0.0612 in.   1.530 - 1.555mm
   Ring Groove Height (Oil)   0.1589 - 0.1598 in.   4.035 - 4.060mm

Piston Pins
   Clearance to Piston   0.0005 - 0.0009 in.   0.0102 - 0.0208mm
   Diameter   0.9306 - 0.9307 in.   23.637 - 23.640mm

Piston Rings
   Ring Gap
     Compression Ring (Top)   0.009 - 0.024 in.   0.229 - 0.610mm
     Compression Rings (Second)   0.019 - 0.038 in.   0.483 - 0.965mm
     Oil Steel Rails   0.010 - 0.060 in.   0.254 - 1.500mm
   Ring Side Clearance
     Compression Rings   0.0017 - 0.0033 in.   0.042 - 0.084mm
     Oil Control (Steel Rails)   0.0024 - 0.0083 in.   0.06 - 0.21mm

ENGINE AND ENGINE OVERHAUL

3.9L (3916cc) ENGINE MECHANICAL SPECIFICATIONS

Description	English Specifications	Metric Specifications
General information		
Engine Type	90 degree V-6 OHV	
Bore x Stroke	3.91 x 3.31 in	99.3 x 84.0 mm
Displacement	238 c.i.	3.9 L
Compression Ratio	9.1:1	
Firing Order	1-6-5-4-3-2	
Lubrication	Pressure Feed - Full Flow Filtration	
Cooling System	Liquid Cooled - Forced Circulation	
Cylinder Block	Cast Iron	
Cylinder Head	Cast Iron	
Crankshaft	Nodular Iron	
Camshaft	Nodular Cast Iron	
Combustion Chambers	"Fast Burn" Design	
Pistons	Aluminum Alloy w/strut	
Connecting Rods	Forged Steel	
Cylinder Compression Pressure (Min.)	100 psi	689.5 kPa
Camshaft		
Bearing Diameter (Inside)		
No. 1	2.000 - 2.001 in.	50.800 - 50.825 mm
No. 2	1.984 - 1.985 in.	50.394 - 50.419 mm
No. 3	1.953 - 1.954 in.	49.606 - 49.632 mm
No. 4	1.5265 - 1.5635 in.	39.688 - 39.713 mm
Journal Diameter		
No. 1	1.998 - 1.999 in.	50.749 - 50.775 mm
No. 2	1.982 - 1.983 in.	50.343 - 50.368 mm
No. 3	1.951 - 1.952 in.	49.555 - 49.581 mm
No. 4	1.5605 - 1.5615 in.	39.637 - 39.662 mm
Bearing To Journal Clearance		
Standard	0.001 - 0.003 in.	0.0254 - 0.0762 mm
Max. Allowable	0.005 in.	0.127 mm
Camshaft End Play		
End Play	0.002 - 0.010 in.	0.051 - 0.254 mm
Connecting Rods		
Piston Pin Bore Diameter	0.9819 - 0.9834 in.	24.940 - 24.978 mm
Side Clearance (Two Rods)	0.006 - 0.014 in.	0.152 - 0.356 mm
Bearing Clearance	0.0005 - 0.0022 in.	0.013 - 0.056 mm
Max. Allowable	0.003 in.	0.08 mm
Crankshaft		
Rod Journal		
Diameter	2.124 - 2.125 in.	53.950 - 53.975 mm
Out Of Round (Max.)	0.001 in.	0.0254 mm
Taper (Max.)	0.001 in.	0.0254 mm
Bearing Clearance	0.0005 - 0.0022 in.	0.013 - 0.056 mm
Service Limit	0.003 in.	0.08 mm
Main Journal		
Diameter	2.4995 - 2.5005 in.	63.487 - 63.513 mm
Out Of Round (Max.)	0.001 in.	0.0254 mm
Taper (Max.)	0.001 in.	0.0254 mm
Bearing Clearance (#1)	0.0005 - 0.0015 in.	0.013 - 0.038 mm
Service Limit	0.0025 in.	0.064 mm
Bearing Clearance (#2 - 4)	0.0005 - 0.0020 in.	0.013 - 0.051 mm
Service Limit	0.0025 in.	0.064 mm
Crankshaft End Play		
End Play	0.002 - 0.007 in.	0.051 - 0.178 mm
Service Limit	0.010 in.	0.254 mm
Cylinder Block		
Cylinder Bore		
Diameter	3.910 - 3.912 in	99.314 - 99.365 mm
Out Of Round (Max.)	0.005 in.	0.127 mm
Taper (Max.)	0.010 in.	0.254 mm
Oversize Limit	0.040 in.	1.016 mm
Lifter Bore Diameter	0.9501 - 0.9059 in.	22.99 - 23.01 mm
Distributor Drive Bushing (Press Fit)		
Bushing to Bore Interference	0.0005 - 0.0140 in.	0.0127 - 0.3556 mm
Shaft to Bushing Clearance	0.0007 - 0.0027 in.	0.0178 - 0.0686 mm

ENGINE AND ENGINE OVERHAUL

3.9L (3916cc) ENGINE MECHANICAL SPECIFICATIONS

Description	English Specifications	Metric Specifications
Cylinder Head And Valves		
Compression Pressure (Min.)	100 psi	689 kPa
Valve Seat		
Angle	44.25 - 44.75 degrees	
Runout (Max.)	0.003 in.	0.0762 mm
Width (Finish)		
Intake	0.040 - 0.060 in.	1.016 - 1.542 mm
Exhaust	0.060 - 0.080 in.	1.524 - 2.032 mm
Valves		
Face Angle	43.25 - 43.75 degrees	
Stem Diameter - Intake/Exhaust	0.311 - 0.312 in.	7.899 - 7.925 mm
Guide Bore Diameter	0.313 - 0.314 in.	7.950 - 7.976 mm
Stem to Guide Clearance - Intake/Exhaust	0.001 - 0.003 in.	0.0254 - 0.0762 mm
Service Limit (Rocking Method)	0.017 in.	0.4318 mm
Valve Spring		
Free Length - Intake/Exhaust	1.967 in.	49.962 mm
Spring Tension (valve closed) - Intake/Exhaust	85 lbs. @ 1.64 in.	378 N @ 41.66 mm
Spring Tension (valve open) - Intake/Exhaust	200 lbs. @ 1.212 in.	890 N @ 30.98 mm
Number of Coils - Intake/Exhaust	6.8	
Installed Height - Intake/Exhaust	1.64 in.	41.66 mm
Hydraulic Tappets		
Body Diameter	0.9035 - 0.9040 in.	22.949 - 22.962 mm
Clearance in Block	0.0011 - 0.0024 in.	0.0279 - 0.0610 mm
Dry Lash	0.060 - 0.210 in.	1.524 - 5.334 mm
Push rod Length	6.915 - 6.935 in.	175.64 - 176.15 mm
Oil Pressure		
@ Curb Idle (min.)*	6 psi	41.4 kPa
@ 3000 rpm	30 - 80 psi	207 - 552 kPa
Bypass Valve Setting	9 - 15 psi	62 - 103 kPa
Switch Actuating Pressure	5 - 7 psi	34.5 - 48.3 kPa
*CAUTION: if oil pressure is ZERO at curb idle, DO NOT run engine @ 3000 rpm.		
Oil Pump		
Clearance over Rotors (Max.)	0.004 in.	0.1016 mm
Cover Out of Flat (Max.)	0.0015 in.	0.0381 mm
Inner Rotor Thickness (Min.)	0.825 in.	20.955 mm
Outer Rotor Clearance (Max.)	0.014 in.	0.3556 mm
Outer Rotor Diameter (Min.)	2.469 in.	62.7126 mm
Outer Rotor Thickness (Min.)	0.825 in.	20.955 mm
Tip Clearance between Rotors (Max.)	0.008 in.	0.2032 mm
Pistons		
Clearance at Top of Skirt	0.0005 - 0.0015 in.	0.0127 - 0.0381 mm
Land Clearance (Diam.)	0.025 - 0.040 in.	0.635 - 1.016 mm
Ring Groove Depth (#1 & 2)	0.180 - 0.190 in.	4.572 - 4.826 mm
Ring Groove Depth (#3)	0.150 - 0.160 in.	3.810 - 4.054 mm
Piston Pins		
Clearance in Piston	0.00025 - 0.00075 in.	0.0064 - 0.0191 mm
Clearance in Rod (Interference)	0.0007 - 0.0014 in.	0.0178 - 0.0356 mm
Diameter	0.9841 - 0.9843 in.	24.996 - 25.001 mm
End Play	NONE	
Piston Rings		
Ring Gap		
Compression Rings	0.010 - 0.020 in.	0.254 - 0.508 mm
Oil Control (Steel Rails)	0.010 - 0.050 in.	0.254 - 1.270 mm
Ring Side Clearance		
Compression Rings	0.0015 - 0.0030 in.	0.038 - 0.076 mm
Oil Control (Steel Rails)	0.002 - 0.008 in.	0.05 - 0.20 mm
Ring Width		
Compression Rings	0.0776 - 0.0783 in.	1.971 - 1.989 mm
Oil Control (Steel Rails)	0.1515 - 0.1565 in.	3.848 - 3.975 mm

3-6 ENGINE AND ENGINE OVERHAUL

4.7L (4701cc) ENGINE MECHANICAL SPECIFICATIONS

Description	English Specifications	Metric Specifications
General information		
Engine Type	90 degree V8 SOHC	
Bore x Stroke	3.66 x 3.40 in	93.0 x 86.5 mm
Displacement	287 c.i.	4.7 L
Compression Ratio	9.0:1	
Firing Order	1-8-4-3-6-5-7-2	
Lubrication	Pressure Feed - Full Flow Filtration	
Cooling System	Liquid Cooled - Forced Circulation	
Cylinder Block	Cast Iron	
Crankshaft	Nodular Iron	
Cylinder Head	Aluminum	
Camshaft	Powdered Metal Steel Lobes/Steel Tube	
Pistons	Aluminum Alloy	
Connecting Rods	Forged Powdered Metal	
Camshaft		
Bearing Diameter (Inside)	1.0245 - 1.0252 in.	26.02 - 26.04 mm
Bearing Journal Diameter	1.0227 - 1.0235 in.	25.975 - 25.995 mm
Bearing To Journal Clearance		
Standard	0.0010 - 0.0026 in.	0.025 - 0.065 mm
Max. Allowable	0.0026 in.	0.065 mm
Camshaft End Play	0.0030 - 0.0079 in.	0.075 - 0.200 mm
Connecting Rods		
Piston Pin Bore Clearance	0.0009 - -0.0018 in.	0.022 - 0.045 mm
Side Clearance (Two Rods)	0.004 - 0.0138 in.	0.10 - 0.35 mm
Crankshaft		
Rod Journal		
Diameter	2.0076 - 2.0082 in.	50.992 - 51.008 mm
Out Of Round (Max.)	0.002 in.	0.005 mm
Taper (Max.)	0.0004 in.	0.008 mm
Bearing Clearance	0.0004 - 0.0019 in.	0.010 - 0.048 mm
Service Limit	0.003 in.	0.0762 mm
Main Bearing Journal		
Diameter	2.4996 - 2.5005 in.	63.488 - 63.512 mm
Out Of Round (Max.)	0.0002 in.	0.005 mm
Taper (Max.)	0.0004 in.	0.008 mm
Bearing Clearance	0.0002 - 0.0013 in.	0.004 - 0.032 mm
Crankshaft End Play		
End Play	0.0021 - 0.0112 in.	0.052 - 0.282 mm
Service Limit	0.0112 in.	0.282 mm
Cylinder Block		
Cylinder Bore		
Diameter	3.6616 - 3.6622 in.	93.005 - 93.020 mm
Out Of Round (Max.)	0.003 in.	0.076 mm
Taper (Max.)	0.002 in.	0.051 mm
Cylinder Head And Valves		
Valve Seat		
Angle	44.5 - 45.0 degrees	
Runout (Max.)	0.002 in.	0.051 mm
Width (Finish) - Intake	0.0698 - 0.0928 in.	1.75 - 2.36 mm
Width (Finish) - Exhaust	0.0673 - 0.0911 in.	1.71 - 2.32 mm
Valves		
Face Angle	45.0 - 45.5 degrees	
Stem Diameter		
Intake	0.2729 - 0.2739 in.	6.931 - 6.957 mm
Exhaust	0.2717 - 0.2728 in.	6.902 - 6.928 mm
Guide Bore Diameter	0.2747 - 0.2756 mm	6.975 - 7.000 mm
Stem to Guide Clearance	0.001 - 0.003 in.	0.0254 - 0.0762 mm
Intake	0.0008 - 0.0028 in.	0.018 - 0.069 mm
Exhaust	0.0019 - 0.0039 in.	0.047 - 0.098 mm
Service Limit (Rocking Method)		
Intake	0.0028 in.	0.069 mm
Exhaust	0.0039 in.	0.098 mm

ENGINE AND ENGINE OVERHAUL 3-7

4.7L (4701cc) ENGINE MECHANICAL SPECIFICATIONS

Description	English Specifications	Metric Specifications
Valve Spring		
Free Length	1.91 in.	48.6 mm
Spring Tension (valve closed)(min.)	71 lbs. @ 1.61 in.	316 N @ 41.0 mm
Spring Tension (valve open)(min.)	177 lbs. @ 1.17 in.	786 N @ 29.6 mm
Installed Height		
Intake	1.613 in.	40.97 mm
Exhaust	1.606 in.	40.81 mm
Oil Pressure		
@ Curb Idle (min.)*	4 psi	25 kPa
@ 3000 rpm	25 - 80 psi	170 - 552 kPa
Oil Pressure Bypass Valve Setting	9 - 15 psi	62 - 103 kPa
*CAUTION: if oil pressure is ZERO at curb idle, DO NOT run engine @ 3000 rpm.		
Oil Pump		
Clearance over Rotors (Max.)	0.0014 - 0.0038 in.	0.035 - 0.095 mm
Cover Out of Flat (Max.)	0.001 in.	0.025 mm
Rotor Thickness	0.4756 in.	12.08 mm
Outer Rotor Clearance (Max.)	0.0186 in.	0.47 mm
Outer Rotor Diameter (Min.)	3.3829 in.	85.925 mm
Tip Clearance between Rotors (Max.)	0.006 in.	0.150 mm
Pistons		
Ring Groove Dia. (#1)	3.296 - 3.306 in.	83.73 - 83.97 mm
Ring Groove Dia. (#2)	3.261 - 3.269 in.	82.833 - 83.033 mm
Ring Groove Dia. (#3)	3.302 - 3.310 in.	83.88 - 84.08 mm
Piston Pins		
Clearance in Piston	0.0004 - 0.0008 in.	0.010 - 0.019 mm
Diameter	0.9454 - 0.9456 in.	24.013 - 24.016 mm
Piston Rings		
Ring Gap		
Compression Rings (Nos. 1 & 2)	0.015 - 0.025 in.	0.37 - 0.63 mm
Oil Control (Steel Rails)	0.010 - 0.030 in.	0.25 - 0.76 mm
Ring Side Clearance		
Compression Ring (Top)	0.0020 - 0.0037 in.	0.051 - 0.094 mm
Compression Ring (2nd)	0.0016 - 0.0031 in.	0.040 - 0.080 mm
Oil Control (Steel Rails)	0.007 - 0.0091 in.	0.445 - 0.470 mm
Ring Width		
Compression Rings	0.057 - 0.058 in.	1.472 - 1.490 mm
Oil Control (Steel Rails)	0.017 - 0.018 in.	0.445 - 0.470 mm

3-8 ENGINE AND ENGINE OVERHAUL

5.2L (5211cc) ENGINE MECHANICAL SPECIFICATIONS

Description	English Specifications	Metric Specifications
General information		
Engine Type	90 degree V8 OHV	
Bore x Stroke	3.91 x 3.31 in.	99.3 x 84.0 mm
Displacement	318 c.i.	5.2 L
Compression Ratio	9.1:1	
Firing Order	1-8-4-3-6-5-7-2	
Lubrication	Pressure Feed - Full Flow Filtration	
Cooling System	Liquid Cooled - Forced Circulation	
Cylinder Block	Cast Iron	
Crankshaft	Nodular Iron	
Cylinder Head	Cast Iron	
Combustion Chambers	Wedge-High Swirl Valve Shrouding	
Camshaft	Nodular Cast Iron	
Pistons	Aluminum Alloy w/Strut	
Connecting Rods	Forged Steel	
Cylinder Compression Pressure (Min.)	100 psi	689.5 kPa
Camshaft		
Bearing Diameter (Inside)		
No. 1	2.000 - 2.001 in.	50.800 - 50.825 mm
No. 2	1.984 - 1.985 in.	50.394 - 50.419 mm
No. 3	1.969 - 1.970 in.	50.013 - 50.038 mm
No. 4	1.953 - 1.954 in.	49.606 - 49.632 mm
No. 5	1.5265 - 1.5635 in.	39.688 - 39.713 mm
Journal Diameter		
No. 1	1.998 - 1.999 in.	50.749 - 50.775 mm
No. 2	1.982 - 1.983 in.	50.343 - 50.368 mm
No. 3	1.967 - 1.968 in.	49.962 - 49.987 mm
No. 4	1.951 - 1.952 in.	49.555 - 49.581 mm
No. 5	1.5605 - 1.5615 in.	39.637 - 39.662 mm
Bearing To Journal Clearance		
Standard	0.001 - 0.003 in.	0.0254 - 0.0762 mm
Max. Allowable	0.005 in.	0.127 mm
Camshaft End Play	0.002 - 0.010 in.	0.051 - 0.254 mm
Connecting Rods		
Piston Pin Bore Diameter	0.9829 - 0.9834 in.	24.966 - 24.978 mm
Side Clearance (Two Rods)	0.006 - 0.014 in.	0.152 - 0.356 mm
Crankshaft		
Rod Journal		
Diameter	2.124 - 2.125 in.	53.950 - 53.975 mm
Out Of Round (Max.)	0.001 in.	0.0254 mm
Taper (Max.)	0.001 in.	0.0254 mm
Bearing Clearance	0.0005 - 0.0022 in.	0.013 - 0.056 mm
Service Limit	0.003 in.	0.0762 mm
Main Bearing Journal		
Diameter	2.4995 - 2.5005 in.	63.487 - 63.513 mm
Out Of Round (Max.)	0.001 in.	0.127 mm
Taper (Max.)	0.001 in.	0.0254 mm
Bearing Clearance (#1 Journal)	0.0005 - 0.0015 in.	0.013 - 0.038 mm
Service Limit (#1 Journal)	0.0015 in	0.0381 mm
Bearing Clearance (#2 - 5 Journals)	0.0005 - 0.0020 in.	0.013 - 0.051 mm
Service Limit (#2 - 5 Journals)	0.0025 in.	0.064 mm
Crankshaft End Play		
End Play	0.002 - 0.007 in.	0.051 - 0.178 mm
Service Limit	0.010 in.	0.254 mm
Cylinder Block		
Cylinder Bore		
Diameter	3.910 - 3.912 in.	99.314 - 99.365 mm
Out Of Round (Max.)	0.005 in.	0.127 mm
Taper (Max.)	0.010 in.	0.254 mm
Oversize Limit	0.040 in.	1.016 mm
Lifter Bore Dia.	0.9501 - 0.9059 in.	22.99 - 23.01 mm

ENGINE AND ENGINE OVERHAUL 3-9

5.2L (5211cc) ENGINE MECHANICAL SPECIFICATIONS

Description	English Specifications	Metric Specifications
Distributor Drive Bushing (Press Fit)		
Bushing to Bore Interference	0.0005 - 0.0140 in.	0.0127 - 0.3556 mm
Shaft to Bushing Clearance	0.0007 - 0.0027 in.	0.0178 - 0.0686 mm
Cylinder Head And Valves		
Compression Pressure	100 psi	689 Kpa
Valve Seat		
Angle	44.25 - 44.75 degrees	
Runout (Max.)	0.003 in.	0.0762 mm
Width (Finish) - Intake	0.040 - 0.060 in.	1.016 - 1.542 mm
Width (Finish) - Exhaust	0.060 - 0.080 in.	1.524 - 2.032 mm
Valves		
Face Angle	43.25 - 43.75 degrees	
Stem Diameter - Intake/Exhaust	0.311 - 0.312 in.	7.899 - 7.925 mm
Guide Bore Diameter	0.313 - 0.314 in.	7.950 - 7.976 mm
Stem to Guide Clearance - Intake/Exhaust	0.001 - 0.003 in.	0.0254 - 0.0762 mm
Service Limit (Rocking Method)	0.017 in.	0.4318 mm
Valve Spring		
Free Length	1.967 in.	49.962 mm
Spring Tension (valve closed)	85 lbs. @ 1.64 in.	378 N @ 41.66 mm
Spring Tension (valve open)	200 lbs. @ 1.212 in.	890 N @ 30.89 mm
Installed Height	1.64 in.	41.66 mm
Hydraulic Tappets		
Body Diameter	0.9035 - 0.9040 in.	22.949 - 22.962 mm
Clearance in Block	0.0011 - 0.0024 in.	0.0279 - 0.0610 mm
Dry Lash	0.060 - 0.210 in.	1.524 - 5.334 mm
Push rod Length	6.915 - 6.935 in.	175.64 - 176.15 mm
Oil Pressure		
@ Curb Idle (min.)*	6 psi	41.4 kPa
@ 3000 rpm	30 - 80 psi	207 - 552 kPa
Oil Pressure Bypass Valve Setting	9 - 15 psi	62 - 103 kPa
Switch Actuating Pressure	5 - 7 psi	34.5 - 48.3 kPa
*CAUTION: if oil pressure is ZERO at curb idle, DO NOT run engine @ 3000 rpm.		
Oil Pump		
Clearance over Rotors (Max.)	0.004 in.	0.1016 mm
Cover Out of Flat (Max.)	0.0015 in.	0.0381 mm
Inner Rotor Thickness (Min.)	0.825 in.	20.955 mm
Outer Rotor Clearance (Max.)	0.014 in.	0.3556 mm
Outer Rotor Diameter (Min.)	2.469 in.	62.7126 mm
Outer Rotor Thickness (Min.)	0.825 in.	20.955 mm
Tip Clearance between Rotors (Max.)	0.008 in.	0.2032 mm
Pistons		
Clearance at Top of Skirt	0.0005 - 0.0015 in.	0.013 - 0.038 mm
Land Clearance (Diam.)	0.025 - 0.040 in.	0.635 - 1.016 mm
Ring Groove Depth (#1 & 2)	0.180 - 0.190 in.	4.572 - 4.826 mm
Ring Groove Depth (#3)	0.150 - 0.160 in.	3.810 - 4.064 mm
Piston Pins		
Clearance in Piston	0.00025 - 0.00075 in.	0.00635 - 0.01905 mm
Diameter	0.9841 - 0.9843 in.	24.996 - 25.001 mm
End Play	NONE	
Piston Rings		
Ring Gap		
Compression Rings (Nos. 1 & 2)	0.010 - 0.020 in.	0.254 - 0.508 mm
Oil Control (Steel Rails)	0.010 - 0.050 in.	0.254 - 1.270 mm
Ring Side Clearance		
Compression Rings	0.0015 - 0.0030 in.	0.038 - 0.076 mm
Oil Control (Steel Rails)	0.002 - 0.008 in.	0.05 - 0.20 mm
Ring Width		
Compression Rings	0.0776 - 0.0783 in.	1.971 - 1.989 mm
Oil Control (Steel Rails)	0.1515 - 0.1565 in.	3.848 - 3.975 mm

3-10 ENGINE AND ENGINE OVERHAUL

5.9L (5899cc) ENGINE MECHANICAL SPECIFICATIONS

Description	English Specifications	Metric Specifications
General information		
Engine Type	90 degree V8 OHV	
Bore x Stroke	4.00 x 3.58 in	101.6 x 90.09mm
Displacement	360 c.i.	5.9 L
Compression Ratio	9.1:1	
Firing Order	1-8-4-3-6-5-7-2	
Lubrication	Pressure Feed - Full Flow Filtration	
Cooling System	Liquid Cooled - Forced Circulation	
Cylinder Block	Cast Iron	
Crankshaft	Nodular Iron	
Cylinder Head	Cast Iron	
Combustion Chambers	Wedge-High Swirl Valve Shrouding	
Camshaft	Nodular Cast Iron	
Pistons	Aluminum Alloy w/Strut	
Connecting Rods	Forged Steel	
Cylinder Compression Pressure (Min.)	100 psi	689.5 kPa
Camshaft		
Bearing Diameter (Inside)		
No. 1	2.000 - 2.001 in.	50.800 - 50.825 mm
No. 2	1.984 - 1.985 in.	50.394 - 50.419 mm
No. 3	1.969 - 1.970 in.	50.013 - 50.038 mm
No. 4	1.953 - 1.954 in.	49.606 - 49.632 mm
No. 5	1.5265 - 1.5635 in.	39.688 - 39.713 mm
Journal Diameter		
No. 1	1.998 - 1.999 in.	50.749 - 50.775 mm
No. 2	1.982 - 1.983 in.	50.343 - 50.368 mm
No. 3	1.967 - 1.968 in.	49.962 - 49.987 mm
No. 4	1.951 - 1.952 in.	49.555 - 49.581 mm
No. 5	1.5605 - 1.5615 in.	39.637 - 39.662 mm
Bearing To Journal Clearance		
Standard	0.001 - 0.003 in.	0.0254 - 0.0762 mm
Max. Allowable	0.005 in.	0.127 mm
Camshaft End Play	0.002 - 0.010 in.	0.051 - 0.254 mm
Connecting Rods		
Piston Pin Bore Diameter	0.9829 - 0.9834 in.	24.966 - 24.978 mm
Side Clearance (Two Rods)	0.006 - 0.014 in.	0.152 - 0.356 mm
Crankshaft		
Rod Journal		
Diameter	2.124 - 2.125 in.	53.950 - 53.975 mm
Out Of Round (Max.)	0.001 in.	0.0254 mm
Taper (Max.)	0.001 in.	0.0254 mm
Bearing Clearance	0.0005 - 0.0022 in.	0.013 - 0.056 mm
Service Limit	0.003 in.	0.0762 mm
Main Bearing Journal		
Diameter	2.8095 - 2.8105 in.	71.361 - 71.387 mm
Out Of Round (Max.)	0.001 in.	0.127 mm
Taper (Max.)	0.001 in.	0.0254 mm
Bearing Clearance (#1 Journal)	0.0005 - 0.0015 in.	0.013 - 0.038 mm
Service Limit (#1 Journal)	0.0015 in	0.0381 mm
Bearing Clearance (#2 - 5 Journals)	0.0005 - 0.0020 in.	0.013 - 0.051 mm
Service Limit (#2 - 5 Journals)	0.0025 in.	0.064 mm
Crankshaft End Play		
End Play	0.002 - 0.007 in.	0.051 - 0.178 mm
Service Limit	0.010 in.	0.254 mm
Cylinder Block		
Cylinder Bore		
Diameter	4.000 - 4.002 in.	101.60 - 101.65 mm
Out Of Round (Max.)	0.005 in.	0.127 mm
Taper (Max.)	0.010 in.	0.254 mm
Lifter Bore Dia.	0.9501 - 0.9059 in.	22.99 - 23.01 mm
Distributor Drive Bushing (Press Fit)		

5.9L (5899cc) ENGINE MECHANICAL SPECIFICATIONS

Description	English Specifications	Metric Specifications
Bushing to Bore Interference	0.0005 - 0.0140 in.	0.0127 - 0.3556 mm
Shaft to Bushing Clearance	0.0007 - 0.0027 in.	0.0178 - 0.0686 mm
Cylinder Head And Valves		
Compression Pressure	100 psi	689 Kpa
Valve Seat		
Angle	44.25 - 44.75 degrees	
Runout (Max.)	0.003 in.	0.0762 mm
Width (Finish) - Intake	0.040 - 0.060 in.	1.016 - 1.542 mm
Width (Finish) - Exhaust	0.060 - 0.080 in.	1.524 - 2.032 mm
Valves		
Face Angle	43.25 - 43.75 degrees	
Stem Diameter		
Intake	0.372 - 0.373 in.	9.449 - 9.474 mm
Exhaust	0.371 - 0.372 in.	9.423 - 9.449 mm
Guide Bore Diameter	0.374 - 0.375 in.	9.500 - 9.525 mm
Stem to Guide Clearance - Intake/Exhaust		
Intake	0.001 - 0.003 in.	0.0254 - 0.0762 mm
Exhaust	0.002 - 0.004 in.	0.0508 - 0.1016 mm
Service Limit (Rocking Method)	0.017 in.	0.4318 mm
Valve Spring		
Free Length	1.967 in.	49.962 mm
Spring Tension (valve closed)	85 lbs. @ 1.64 in.	378 N @ 41.66 mm
Spring Tension (valve open)	200 lbs. @ 1.212 in.	890 N @ 30.89 mm
Installed Height	1.64 in.	41.66 mm
Hydraulic Tappets		
Body Diameter	0.9035 - 0.9040 in.	22.949 - 22.962 mm
Clearance in Block	0.0011 - 0.0024 in.	0.0279 - 0.0610 mm
Dry Lash	0.060 - 0.210 in.	1.524 - 5.334 mm
Push rod Length	6.915 - 6.935 in.	175.64 - 176.15 mm
Oil Pressure		
@ Curb Idle (min.)*	6 psi	41.4 kPa
@ 3000 rpm	30 - 80 psi	207 - 552 kPa
Oil Pressure Bypass Valve Setting	9 - 15 psi	62 - 103 kPa
Switch Actuating Pressure	5 - 7 psi	34.5 - 48.3 kPa
*CAUTION: if oil pressure is ZERO at curb idle, DO NOT run engine @ 3000 rpm.		
Oil Pump		
Clearance over Rotors (Max.)	0.004 in.	0.1016 mm
Cover Out of Flat (Max.)	0.0015 in.	0.0381 mm
Inner Rotor Thickness (Min.)	0.825 in.	20.955 mm
Outer Rotor Clearance (Max.)	0.014 in.	0.3556 mm
Outer Rotor Diameter (Min.)	2.469 in.	62.7126 mm
Outer Rotor Thickness (Min.)	0.825 in.	20.955 mm
Tip Clearance between Rotors (Max.)	0.008 in.	0.2032 mm
Pistons		
Clearance at Top of Skirt	0.0005 - 0.0015 in.	0.013 - 0.038 mm
Land Clearance (Diam.)	0.020 - 0.026 in.	0.508 - 0.660 mm
Ring Groove Depth (#1 & 2)	0.187 - 0.193 in.	4.761 - 4.912 mm
Ring Groove Depth (#3)	0.157 - 0.164 in.	3.996 - 4.177 mm
Piston Pins		
Clearance in Piston	0.00023 - 0.00074 in.	0.006 - 0.019 mm
Diameter	0.9845 - 0.9848 in.	25.007 - 25.015 mm
End Play	NONE	
Piston Rings		
Ring Gap		
Compression Ring (Top)	0.012 - 0.022 in.	0.30 - 0.55 mm
Compression Ring (2nd)	0.022 - 0.031 in.	0.55 - 0.80 mm
Oil Control (Steel Rails)	0.015 - 0.055 in.	0.381 - 1.397 mm
Ring Side Clearance		
Compression Rings	0.0016 - 0.0033 in.	0.040 - 0.085 mm
Oil Control (Steel Rails)	0.002 - 0.008 in.	0.05 - 0.20 mm
Ring Width		
Compression Rings	0.060 - 0.061 in.	1.530 - 1.555 mm
Oil Control (Steel Rails)	0.018 - 0.019 in.	0.447 - 0.473 mm

3-12 ENGINE AND ENGINE OVERHAUL

8.0 (7997cc) ENGINE MECHANICAL SPECIFICATIONS

Description	English Specifications	Metric Specifications
General Information		
Engine Type	90 degree V10 OHV	
Bore & Stroke	4.00 x 3.88 in.	101.6 x 98.6 mm
Displacement	488 c.i.	8.0 L
Compression Ratio	8.4:1	
Firing Order	1-10-9-4-3-6-5-8-7-2	
Lubrication	Pressure Feed - Full Flow Filtration	
Cooling System	Liquid Cooled - Forced Circulation	
Cylinder Block	Cast Iron	
Crankshaft	Nodular Iron	
Cylinder Head	Cast Iron	
Combustion Chambers	Wedge-High Swirl Valve Shrouding	
Camshaft	Nodular Cast Iron	
Pistons	Aluminum Alloy	
Connecting Rods	Forged Steel	
Cylinder Compression Pressure (Min.)	100 psi	689.5 kPa
Camshaft		
Bearing Diameter (Inside)		
No. 1	2.093 - 2.094 in.	53.16 - 53.19mm
No. 2	2.077 - 2.078 in.	52.76 - 52.78mm
No. 3	2.061 - 2.062 in.	52.35 - 52.37mm
No. 4	2.045 - 2.046 in.	51.94 - 51.97mm
No. 5	2.029 - 2.030 in.	51.54 - 51.56mm
No. 6	1.919 - 1.920 in.	48.74 - 48.77mm
Journal Diameter		
No. 1	2.091 - 2.092 in.	53.11 - 53.14mm
No. 2	2.075 - 2.076 in.	52.69 - 52.72mm
No. 3	2.059 - 2.060 in.	52.30 - 52.32mm
No. 4	2.043 - 2.044 in.	51.89 - 51.92mm
No. 5	2.027 - 2.028 in.	51.49 - 51.51mm
No. 6	1.917 - 1.918 in.	48.69 - 48.72mm
Bearing-to-Journal Clearance		
1, 3, 4, 5, 6	0.001 - 0.003 in.	0.0254 - 0.0762mm
2	0.0005 - 0.0035 in.	0.0381 - 0.0889mm
Max. Allowable	0.005 in.	0.127mm
Camshaft End-Play	0.005 - 0.015 in.	0.127 - 0.381mm
Connecting Rods		
Piston Pin Bore Diameter	0.9819 - 0.9834 in.	24.940 - 24.978mm
Side Clearance	0.010 - 0.018 in.	0.25 - 0.46mm
Crankshaft		
Rod Journal		
Diameter	2.124 - 2.125 in.	53.950 - 53.975mm
Out Of Round (Max.)	0.001 in.	0.0254mm
Taper (Max.)	0.001 in.	0.0254mm
Bearing Clearance	0.0002 - 0.0029 in.	0.005 - 0.074mm
Service Limit	0.003 in.	0.0762mm
Main Bearing Journal		
Diameter	2.8995 - 3.005 in.	76.187 - 76.213mm
Out Of Round (Max.)	0.001 in.	0.0254mm
Taper (Max.)	0.001 in.	0.0254mm
Bearing Clearance	0.0002 - 0.0023 in.	0.0051 - 0.058mm
Service Limit	0.0028 in.	0.071mm
Crankshaft End-Play		
Standard	0.003 - 0.012 in.	0.076 - 0.305mm
Service Limit	0.015 in.	0.381mm
Cylinder Block		
Cylinder Bore		
Diameter	4.0003 - 4.0008 in.	101.60 - 101.65mm
Out Of Round (Max.)	0.003 in.	0.0762mm
Taper (Max.)	0.005 in.	0.127mm
Lifter Bore Dia.	0.9048 - 0.9059 in.	22.982 - 23.010mm
Cylinder Head And Valves		
Valve Seat Angle	44.5 degrees	
Seat Run-out (Max.)	0.003 in.	0.0762mm
Seat Width (Finish)	0.040 - 0.060 in.	1.016 - 1.542mm

ENGINE AND ENGINE OVERHAUL 3-13

8.0 (7997cc) ENGINE MECHANICAL SPECIFICATIONS

Description	English Specifications	Metric Specifications
Valves		
Face Angle	45 degrees	
Lift (@ zero lash) - Intake	0.390 in.	9.91mm
Lift (@ zero lash) - Exhaust	0.407 in.	10.34mm
Stem Diameter	0.372 - 0.373 in.	9.449 - 9.474mm
Guide Bore Diameter	0.374 - 0.375 in.	9.500 - 9.525mm
Stem to Guide Clearance	0.001 - 0.003 in.	0.025 - 0.076mm
Service Limit	0.017 in.	0.432mm
Valve Spring		
Free Length	1.967 in.	49.962mm
Spring Tension (valve closed)	85 lbs.. @ 1.64 in.	378 N @ 41.66mm
Spring Tension (valve open)	200 lbs.. @ 1.212 in.	890 N @ 30.89mm
Installed Height	1.64 in.	41.66mm
Hydraulic Tappets		
Body Diameter	0.9035 - 0.9040 in.	22.949 - 22.962mm
Clearance in Block	0.0008 - 0.0024 in.	0.0203 - 0.0610mm
Dry Lash	0.060 - 0.210 in.	1.524 - 5.334mm
Pushrod Length	7.698 - 7.717 in.	195.52 - 196.02mm
Oil Pressure		
@ Curb Idle (min.)*	12 psi	83 kPa
*CAUTION: if oil pressure is ZERO at curb idle, DO NOT run engine @ 3000 rpm.		
@ 3000 rpm	50 - 60 psi	345 - 414 kPa
Oil Pressure Bypass Valve Setting	9 - 15 psi	62 - 103 kPa
Switch Actuating Pressure	5 - 7 psi	34.5 - 48.3 kPa
Oil Pump		
Clearance over Rotors (Max.)	0.0075 in.	0.1906mm
Cover Out of Flat (Max.)	0.002 in.	0.051mm
Inner Rotor Thickness (Min.)	0.5876 - 0.5886 in.	14.925 - 14.950mm
Outer Rotor Clearance (Max.)	0.006 in.	0.1626mm
Outer Rotor Diameter (Min.)	3.246 in.	82.461mm
Outer Rotor Thickness (Min.)	0.5876	14.925mm
Tip Clearance between Rotors (Max.)	0.0230 in.	0.584mm
Pistons		
Piston To Bore Clearance	0.0005 - 0.0015 in.	0.013 - 0.038mm
Service Limit	0.003 in.	0.0762mm
Piston Pins		
Clearance in Piston	0.0004 - 0.0008 in.	0.010 - 0.020mm
Diameter	0.9841 - 0.9843 in.	24.996 - 25.001mm
End-Play	NONE	
Length	2.67 - 2.69 in.	67.8 - 68.3mm
Piston Rings		
Ring Gap		
Compression Rings	0.010 - 0.020 in.	0.254 - 0.508mm
Oil Control (Steel Rails)	0.015 - 0.055 in.	0.381 - 1.397mm
Ring Side Clearance		
Compression Rings	0.0029 - 0.0038 in.	0.074 - 0.097mm
Oil Control (Steel Rails)	0.0073 - 0.0097 in.	0.185 - 0.246mm
Ring Width		
Compression Rings	0.162 - 0.172 in.	4.115 - 4.369mm
Oil Control (Steel Rails)	0.102 - 0.108 in.	2.591 - 2.743mm

ENGINE AND ENGINE OVERHAUL

5.9L DIESEL (12-VALVE) ENGINE MECHANICAL SPECIFICATIONS

Description	English Specifications	Metric Specifications
General Information		
Engine Type	In-Line 6 cyl turbo diesel	
Bore & Stroke	4.02 x 4.72 in.	102.0 x 120.0mm
Displacement	359 c.i.	5.9L
Compression Ratio	17.5:1	
Firing Order	1-5-3-6-2-4	
Lubrication	Pressure Feed - Full Flow Filtration	
Cooling System	Liquid Cooled - Forced Circulation	
Cylinder Block	Cast Iron	
Crankshaft	Induction-Hardened Forged Steel	
Cylinder Head	Cast Iron	
Combustion Chambers	High-Swirl Bowl	
Camshaft	Chilled Ductile Iron	
Pistons	Cast Aluminum Alloy	
Connecting Rods	Forged Steel	
Camshaft		
Journal Diameter (Min.)	2.1245 in.	53.962mm
Intake Lobe Min. Dia. @ Peak	1.852 in.	47.040mm
Exhaust Lobe Min. Dia. @ Peak	1.841 in.	46.770mm
Lift Pump Lobe Min. Dia. @ Peak	1.398 in.	35.500mm
Camshaft End Clearance	0.006 - 0.010 in.	0.152 - 0.254mm
Gear Backlash	0.003 - 0.013 in.	0.080 - 0.330mm
Connecting Rods		
Piston Pin Bore Diameter (Max.)	1.5764 in.	40.042mm
Side Clearance	0.004 - 0.012 in.	0.100 - 0.300mm
Crankshaft		
Rod Journal		
Diameter (Std.)	2.7150 in.	68.962mm
Out Of Round (Max.)	0.002 in.	0.050mm
Taper (Max.)	0.0005 in.	0.013mm
Bearing Clearance Service Limit	0.0035 in.	0.089mm
Main Bearing Journal		
Diameter	3.2662 in.	82.962mm
Out Of Round (Max.)	0.002 in.	0.050mm
Taper (Max.)	0.0005 in.	0.013mm
Bearing Clearance Service Limit	0.0047 in.	0.119mm
Crankshaft End-Play	0.004 - 0.017 in.	0.100 - 0.430mm
Gear Backlash	0.003 - 0.030 in.	0.080 - 0.330mm
Cylinder Block		
Cylinder Bore		
Diameter (Max. Std.)	4.0203 in.	102.116mm
Out Of Round (Max.)	0.0015 in.	0.038mm
Taper (Max.)	0.003 in.	0.076mm
Lifter Bore Dia. (Max.)	0.632 in.	16.055mm
Deck Flatness (Max.)	0.003 in.	0.075mm
Main Bearing Inside Dia.	3.2719 in.	83.106mm
Cam Bore Dia. (Max.)	2.1314 in.	54.139mm

ENGINE AND ENGINE OVERHAUL 3-15

Cylinder Head And Valves
  Head Overall Flatness (Max.)   0.012 in.   0.030mm
  Valve Seat Angle
    Intake   30 degrees
    Exhaust   45 degrees
  Seat Width
    Min.   0.060 in.   1.52mm
    Max.   0.080 in.   2.03mm
  Tappet Stem Dia.   0.627 in.   15.925mm
  Valves
    Intake Clearance   0.010 in.   0.25mm
    Exhaust Clearance   0.020 in.   0.51mm
    Stem Diameter   0.3126 - 0.3134 in.   7.935 - 7.960mm
    Guide Bore Diameter   0.3157 - 0.3185 in.   8.019 - 8.089mm
    Depth (Installed)   0.039 - 0.060 in.   0.99 - 1.52 mm
  Valve Spring
    Free Length   2.36 in.   60mm
    Inclination (Max.)   0.039 in.   1.00mm
    Spring Tension (valve closed)   81 lbs. @ 1.94 in.   359 N @ 49.25mm

Oil Pressure
  @ Curb Idle (min.)*   10 psi   69 kPa
  *CAUTION: if oil pressure is ZERO at curb idle, DO NOT run engine @ 3000 rpm.
  @ 2500 rpm   30 psi   207 kPa
  Regulating Valve Opening Pressure   65 psi   448 kPa
  Filter Bypass Opening   25 psi   172.3 kPa

Oil Pump
  Gerotor Drive/Planetary
    To Port Plate Clearance   0.005 in.   0.127mm
  Gerotor Planetary To
    Body Clearance (Max.)   0.015 in.   0.381mm
  Tip Clearance (Max.)   0.007 in.   0.1778mm
  Gear Backlash (Used Pump)   0.003 - 0.015 in.   0.080 - 0.380mm

Pistons
  Skirt Dia.   4.0110 - 4.0088 in.   101.880 - 101.823mm
  Ring Groove Depth
    Intermediate   0.006 in.   0.150mm
    Oil   0.005 in.   0.130mm

Piston Pins
  Pin Diameter (Min.)   1.5744 in.   39.990mm
  Bore Diameter (Max.)   1.5758 in.   40.025mm

Piston Rings
  Ring End Gap
    Top   0.016 - 0.0275 in.   0.400 - 0.700mm
    Intermediate   0.010 - 0.0215 in.   0.250 - 0.550mm
    Oil   0.010 - 0.0215 in.   0.250 - 0.550mm

ENGINE AND ENGINE OVERHAUL

5.9L DIESEL (24-VALVE) ENGINE MECHANICAL SPECIFICATIONS

Description	English Specifications	Metric Specifications
General Information		
Engine Type	In-Line 6 cyl turbo diesel	
Bore & Stroke	4.02 x 4.72 in.	102.0 x 120.0mm
Displacement	359 c.i.	5.9L
Compression Ratio	16.5:1	
Firing Order	1-5-3-6-2-4	
Lubrication	Pressure Feed - Full Flow Filtration	
Cooling System	Liquid Cooled - Forced Circulation	
Cylinder Block	Cast Iron	
Crankshaft	Induction-Hardened Forged Steel	
Cylinder Head	Cast Iron	
Combustion Chambers	High-Swirl Bowl	
Camshaft	Chilled Ductile Iron	
Pistons	Cast Aluminum Alloy	
Connecting Rods	Forged Steel	
Camshaft		
Journal Diameter		
#1	2.127 - 2.128 in.	54.028 - 54.048mm
#2 - 7	2.1245 - 2.1265 in.	53.987 - 54.013mm
Intake Lobe Min. Dia. @ Peak	1.857 in.	47.173mm
Exhaust Lobe Min. Dia. @ Peak	1.796 in.	45.636mm
Camshaft End Clearance	0.005 - 0.018 in.	0.100 - 0.46mm
Gear Backlash	0.003 - 0.013 in.	0.080 - 0.330mm
Connecting Rods		
Piston Pin Bore Diameter (Max.)	1.5764 in.	40.042mm
Side Clearance	0.004 - 0.012 in.	0.100 - 0.300mm
Crankshaft		
Rod Journal		
Diameter (Std.)	2.7155 in.	68.9745mm
Out Of Round (Max.)	0.002 in.	0.050mm
Taper (Max.)	0.0005 in.	0.013mm
Bearing Clearance Service Limit	0.0035 in.	0.089mm
Main Bearing Journal		
Diameter Std. Min.	3.2662 in.	82.962mm
Out Of Round (Max.)	0.002 in.	0.050mm
Taper (Max.)	0.0005 in.	0.013mm
Bearing Clearance Service Limit	0.0047 in.	0.119mm
Crankshaft End-Play	0.004 - 0.017 in.	0.100 - 0.430mm
Gear Backlash	0.003 - 0.030 in.	0.080 - 0.330mm
Cylinder Block		
Cylinder Bore		
Diameter (Max. Std.)	4.0203 in.	102.116mm
Out Of Round (Max.)	0.0015 in.	0.038mm
Taper (Max.)	0.003 in.	0.076mm
Lifter Bore Dia. (Max.)	0.632 in.	16.055mm
Deck Flatness (Max.)	0.003 in.	0.075mm
Main Bearing Inside Dia.	3.2719 in.	83.106mm
Cam Bore Dia. #1 (Max.)	2.1312 in.	54.133mm
Cam Bore Dia. #2 - 7 (Max.)	2.1314 in.	54.139mm

ENGINE AND ENGINE OVERHAUL 3-17

Cylinder Head And Valves
  Head Overall Flatness (Max.)   0.012 in.   0.030mm
  Side-To-Side Flatness (Max.)   0.003 in.   0.076 in.
  Valve Seat Angle
    Intake   30 degrees
    Exhaust   45 degrees
  Seat Width
    Min.   0.059 in.   1.49mm
    Max.   0.071 in.   1.80mm
  Tappet Stem Dia.   0.627 in.   15.925mm
  Valves
    Intake Clearance   0.006 - 0.015 in.   0.152 - 0.381mm
    Exhaust Clearance   0.015 - 0.030 in.   0.381 - 0.762mm
    Stem Diameter   0.2752 - 0.2760 in.   6.990 - 7.010mm
    Guide Bore Diameter   0.2772 - 0.2780 in.   7.042 - 7.062mm
    Depth (Installed)
      Intake   0.023 - 0.044 in.   0.59 - 1.11mm
      Exhaust   0.038 - 0.058 in.   0.96 - 1.48mm
  Valve Spring
    Free Length   2.36 in.   60mm
    Inclination (Max.)   0.059 in.   1.5mm
    Installed Height   1.39 in.   35.33mm
    Minimum Load   76.4 lbs. @ 1.39 in.   339.8 N @ 35.33mm

Oil Pressure
  @ Curb Idle (min.)*   10 psi   69 kPa
  *CAUTION: if oil pressure is ZERO at curb idle, DO NOT run engine @ 3000 rpm.
  @ 2500 rpm   30 psi   207 kPa
  Regulating Valve Opening Pressure   65 psi   448 kPa
  Filter Bypass Opening   25 psi   172.3 kPa

Oil Pump
  Gerotor Drive/Planetary
    To Port Plate Clearance   0.005 in.   0.127mm
  Gerotor Planetary To
    Body Clearance (Max.)   0.015 in.   0.381mm
  Tip Clearance (Max.)   0.007 in.   0.1778mm
  Gear Backlash (Used Pump)   0.003 - 0.015 in.   0.080 - 0.380mm

Pistons
  Skirt Dia.   4.0104 - 4.0117 in.   101.864 - 101.896mm
  Ring Groove Depth
    Intermediate (Max.)   0.0037 in.   0.095mm
    Oil (Max.)   0.0033 in.   0.085mm

Piston Pins
  Pin Diameter (Min.)   1.5744 in.   39.990mm
  Bore Diameter (Max.)   1.5758 in.   40.025mm

Piston Rings
  Ring End Gap
    Top   0.016 - 0.0275 in.   0.400 - 0.700mm
    Intermediate   0.010 - 0.0215 in.   0.250 - 0.550mm
    Oil   0.010 - 0.0215 in.   0.250 - 0.550mm

3-18 ENGINE AND ENGINE OVERHAUL

Engine

REMOVAL & INSTALLATION

In the process of removing the engine, you will come across a number of steps which call for the removal of a separate component or system, such as "disconnect the exhaust system" or "remove the radiator." In most instances, a detailed removal procedure can be found elsewhere in this manual.

It is virtually impossible to list each individual wire and hose which must be disconnected, simply because so many different model and engine combinations have been manufactured. Careful observation and common sense are the best possible approaches to any repair procedure.

Removal and installation of the engine can be made easier if you follow these basic points:

- If you have to drain any of the fluids, use a suitable container.
- Always tag any wires or hoses and, if possible, the components they came from before disconnecting them.
- Because there are so many bolts and fasteners involved, store and label the hardware from components separately in muffin pans, jars or coffee cans. This will prevent confusion during installation.
- After unbolting the transmission or transaxle, always make sure it is properly supported.
- If it is necessary to disconnect the air conditioning system, have this service performed by a qualified technician using a recovery/recycling station. If the system does not have to be disconnected, unbolt the compressor and set it aside.
- When unbolting the engine mounts, always make sure the engine is properly supported. When removing the engine, make sure that any lifting devices are properly attached to the engine. It is recommended that if your engine is supplied with lifting hooks, your lifting apparatus be attached to them.
- Lift the engine from its compartment slowly, checking that no hoses, wires or other components are still connected.
- After the engine is clear of the compartment, place it on an engine stand or workbench.
- After the engine has been removed, you can perform a partial or full teardown of the engine using the procedures outlined here.

GASOLINE ENGINES (EXCEPT 4.7L)

Note: Procedures will vary depending on vehicle style and equipment fitted.

1. Depressurize the fuel system (see Chapter 5), then disconnect the cable from the negative terminal of the battery.
2. Remove the engine compartment lamp assembly, if fitted.
3. Scribe hood hinge outlines around the brackets. Remove the hood.
4. Disconnect the battery. Remove the battery from the engine compartment.
5. Drain the coolant from the radiator and engine block.

CAUTION:
Drain the coolant into a suitable container. Reuse it unless it is more than a year old or contaminated, in which case it should be disposed of in accordance with applicable regulations.

6. Drain the engine oil. Remove the filter.

CAUTION:
Minimize contact with engine oil which may be connected with skin disorders. Wear gloves or thoroughly wash exposed areas as soon as possible.

7. RAM trucks: Remove the upper crossmember.
8. Remove the air cleaner assembly.
9. Disconnect and remove emission system vacuum hoses from throttle body and valve covers and power brake vacuum hoses, if fitted.
10. Disconnect the fuel line at the fuel rail.
11. Disconnect throttle, cruise control, transmission and any other cables attached to the throttle body assembly.
12. Depressurize the air conditioning system (if equipped), recovering the refrigerant in accordance with all applicable federal laws. This can only be done by an authorized facility.
13. Remove the air conditioner compressor.
14. Disconnect upper and lower radiator hoses and the heater hoses.
15. Remove the coolant recovery bottle and the windshield washer bottle if attached to the fan shroud.
16. Remove the fan and fan shroud. The fan nut has a RIGHTHAND thread.
17. 2.5L engine: after removing the fan assembly, install a 5/16 x 1/2 inch SAE capscrew through the fan pulley into the water pump flange. This will maintain the pulley and water pump in alignment when the crankshaft is rotated.
18. Remove the transmission oil cooler, if fitted.
19. Remove the radiator.
20. Remove the serpentine belt.
21. Remove the alternator.
22. Remove the power steering pump.
23. Tag and disconnect all vacuum lines.
24. Disconnect engine ground strap. Disconnect electrical wiring to the fuel injection system. Disconnect electrical connectors to the throttle body, manifold, and exhaust pipe(s).
25. Disconnect all emission system sensors at the manifold and throttle body.
26. Disconnect oil and coolant sensor wiring from the engine block.
27. Remove the throttle body and the intake manifold.
28. Remove the distributor cap and plug wires.
29. On manual transmission models, remove the shift lever.
30. Remove the exhaust manifold(s).
31. Remove the starter motor.
32. Using a boom hoist attached to the engine with the shortest hookup possible, take up all the tension and support the engine.

WARNING:
Do not attempt to lift the engine by the intake manifold.

33. Remove the engine front mounts and insulators.

    a. On 4WD vehicles, on the left side, remove the two bolts attaching bracket to the transmission bell housing. Remove the two bracket to pinion nose adapter bolts. Separate the engine from insulator by removing the upper nut and washer assembly and bolt from the engine support bracket. On the right side, remove the two bracket to axle bolts and one bracket to bell housing bolt. Separate the engine from the insulator by removing the upper nut washer assembly and bolt from the engine support bracket.

34. On automatic transmission models, remove the bell housing bolts and inspection plate. Attach a C-clamp on the front bottom of the transmission torque converter housing to prevent the torque converter from coming out.
35. Automatic transmission models: remove the retaining plate bolts from the torque converter driveplate. Mark the converter and driveplate to aid in reassembly.
36. Remove the driveshaft and engine rear support.
37. Support the transmission with a suitable transmission jack.
38. Remove the transmission.
39. Remove the driveplate or the clutch and flywheel.
40. Carefully remove the engine from the vehicle. Watch for any snagged or missed connections. Mount the engine on a suitable engine stand for further disassembly or service work.
41. The installation is the reverse of removal. Tighten all fasteners to specification. Do not forget to add oil, coolant and power steering fluid, etc.
42. On the V-10 engine, prime the oil pump by filling the J-trap of the front timing cover with oil and quickly installing an oil-filled filter in place just as the oil is coming out of the J-trap.

4.7L ENGINE

1. Remove the engine compartment lamp assembly, if fitted.
2. Scribe hood hinge outlines around the brackets. Remove the hood.
3. Depressurize the fuel system (see Chapter 5).
4. Disconnect and remove the battery.
5. Remove the exhaust crossover pipe.
6. On 4WD vehicles, disconnect the axle vent tube from the left side engine mount.
7. Remove the through bolt retaining nut and bolt from both left and right side engine mounts.
8. On 4WD vehicles, remove the locknut from left and right side engine mount brackets.
9. Disconnect the three ground straps from the engine.
10. Disconnect the crankshaft position sensor.
11. On 4WD vehicles with automatic transmissions, remove the axle isolator bracket from the engine, transmission and axle.
12. Remove the structural cover.
13. Remove the starter.
14. Drain the cooling system.

ENGINE AND ENGINE OVERHAUL   3-19

> **CAUTION:**
>
> Drain the coolant into a suitable container. Reuse it unless it is more than a year old or contaminated, in which case it should be disposed of in accordance to applicable regulations.

15. Drain the engine oil. Remove the filter.
16. Remove the torque converter bolts.
17. Disconnect the engine block heater power cable from the block heater, if fitted.
18. Remove the throttle body resonator assembly and air inlet hose.
19. Disconnect the throttle and cruise control cables.
20. Disconnect the tube from left and right side crankcase breathers. Remove the breathers.
21. Depressurize the air conditioning system (if equipped), recovering the refrigerant in accordance with all applicable federal laws. This can only be done by an authorized facility.
22. Remove the air conditioner compressor.
23. Remove the cooling fan and shroud assembly.
24. Remove the upper and lower radiator hoses, the radiator, the A/C condenser and transmission oil cooler.
25. Remove the alternator.
26. Remove the two heater hoses from the timing chain cover and heater core.
27. Disconnect engine electrical connectors:
    - IAT sensor
    - Fuel injectors
    - TPS switch
    - IAC motor
    - Oil pressure switch
    - ECT sensor
    - MAP sensor
    - Camshaft position sensor
    - Ignition coils
28. Disconnect throttle body and intake manifold vacuum lines.
29. Disconnect the fuel line at the fuel rail.
30. Remove the power steering pump.
31. Check for and disconnect remaining ground straps, vacuum lines, etc.
32. Support the transmission with a suitable jack.
33. Install a lifting device into the cylinder head. Using a boom hoist attached to the engine with the shortest hookup possible, take up all the tension and support the engine.
34. Remove the transmission-to-engine fasteners. Carefully lift the engine from the engine compartment.
35. Remove the driveplate or clutch and flywheel and mount the engine on an engine stand.
36. The installation is the reverse of removal. Tighten all fasteners to specification. Do not forget to add oil, coolant and power steering fluid, etc.
37. Tighten the engine mount brackets to 70 ft. lbs. on 2WD vehicles and 75 ft. lbs. on 4WD vehicles.

DIESEL ENGINE

1. Remove the engine compartment lamp assembly, if fitted.
2. Scribe hood hinge outlines around the brackets. Remove the hood.
3. Disconnect the batteries. Remove the batteries from the engine compartment.
4. Drain the coolant from the radiator and engine block.
5. Disconnect upper and lower radiator hoses and the heater hoses.

> **CAUTION:**
>
> Drain the coolant into a suitable container. Reuse it unless it is more than a year old or contaminated, in which case it should be disposed of in accordance to applicable regulations.

6. Drain the engine oil. Remove the filter.

> **CAUTION:**
>
> Minimize contact with engine oil which may be connected with skin disorders. Wear gloves or thoroughly wash exposed areas as soon as possible.

7. Remove the coolant recovery bottle and the windshield washer bottle if fitted to the fan shroud.
8. Remove the serpentine drivebelt.
9. Remove the fan and fan shroud.

Note: The fan nut is a LEFT HAND thread.

10. Remove the air cleaner assembly and all related ductwork.
11. Remove the transmission and transfer case, if equipped.
12. Disconnect the exhaust pipe from the turbocharger extension pipe.
13. Remove the starter motor.
14. Disconnect the fuel supply and return lines (see Chapter 5).
15. Depressurize the air conditioning system (if equipped), recovering the refrigerant in accordance with all applicable federal laws. This can only be done by an authorized facility.
16. Disconnect all A/C hoses and remove the compressor.
17. Remove the transmission oil cooler, if fitted.
18. Remove the radiator and upper radiator support panel.
19. Remove the front bumper assembly.
20. Disconnect the charge air cooler piping. Remove the bolts and the charge air cooler and A/C condenser, if fitted.
21. Disconnect the engine block heater connector.
22. Disconnect the A/C electrical wires.
23. Tag and disconnect all vacuum lines.
24. Disconnect engine ground strap. Disconnect electrical wiring to the fuel injection system. Disconnect electrical connectors to the throttle body, manifold, and exhaust pipe(s).
25. Remove the alternator.
26. Remove the power steering pump.
27. Remove the throttle linkage cover. Remove the Accelerator Pedal Position Sensor (APPS), leaving all cables connected and place it out of the way. Disconnect the APPS connector.
28. Disconnect the vacuum pump supply hose.
29. Remove the cylinder head cover.
30. Remove the rocker arms from the two rearmost cylinders.
31. Loosen, but do not remove, the engine mount through bolts and nuts.
32. Attach a chain and use a hoist designed to lift an engine of this size and weight.

33. Carefully remove the engine from the vehicle. Watch for any snagged or missed connections. Mount the engine on a suitable engine stand for further disassembly or service work.
34. The installation is the reverse of removal. Tighten all fasteners to specification. Do not forget to add oil, coolant and power steering fluid, etc.
35. Engine mount through bolts and nuts are tightened to 65 ft. lbs.

Valve Cover

REMOVAL & INSTALLATION

See Figures 1, 2, 3, 4, 5, 6 and 7

1. Disconnect the battery negative cable(s).
2. Remove air cleaner housing and duct work if needed for access to the valve cover(s).
3. Disconnect crankcase ventilation hose from the cover, if fitted.
4. Remove the PCV valve from the valve cover, if equipped.
5. Remove any breathers, vacuum or emission system hoses fitted to the cover(s).
6. On 4.7L engines, disconnect the injector connectors and unclip the injector harness.
7. Check engine wiring relative to the valve cover(s). On some models, wiring is run along the edge of the cover(s) and secured by clips on the valve cover studs. Remove any wiring attached to the valve cover studs.

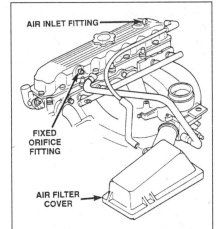

Fig. 1 Remove any emission system hoses fitted to valve cover - 2.5L engine shown

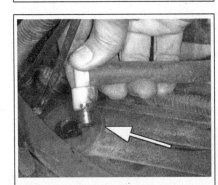

Fig. 2 Removing a PCV valve

3-20 ENGINE AND ENGINE OVERHAUL

Fig. 3 Removing the valve cover - V-8 engine shown

Fig. 4 Most valve cover gaskets are reusable

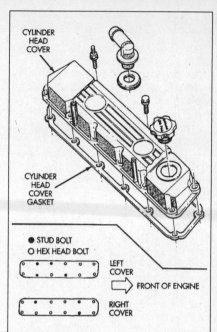

Fig. 5 V-10 valve cover components and fastener locations

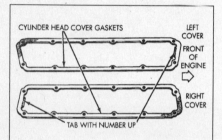

Fig. 6 V-10 cover gasket installation

8. On diesels, remove the air crossover tube.
9. On V-10 engines, the upper intake manifold must be removed to remove the right side valve cover.
10. Remove the valve cover bolts and/or studs. Note locations to ease installation.
11. Tap the valve cover with a plastic mallet to break it loose, then carefully lift the cover off.

WARNING:

Do not attempt to pry the cover off. Damage to cover or gasket may result.

To install:

12. Clean the mating surfaces.
13. Check covers for warpage.
14. Most models are equipped with reusable gaskets, normally steel-backed silicone. Replace compressed or damaged gaskets, or fill in gaps with a quality silicone sealant.
15. On the V-10, for the left side the number tab is at the front of the engine with the number up. For the right side the number tab is at the rear of the engine with the number up. Cylinder head cover fasteners have special plating. Do not use substitutes. Tighten to 12 ft. lbs. (16 Nm).
16. On the 24-valve diesel, locate the gasket with "TOP FRONT" in the proper orientation. Torque bolts to 18 ft. lbs. (24 Nm).
17. On 2.5L engines, tighten cover fasteners to 113 inch lbs. (13 Nm).
18. On 3.9L, 5.2L, and 5.7L gasoline engines, tighten fasteners to 95 inch lbs. (95 Nm).
19. On 4.7L engines, tighten fasteners to 105 inch lbs. (12 Nm).
20. Tighten cover bolts and studs evenly and in a cross pattern.
21. Check for leaks after several minutes of operation.

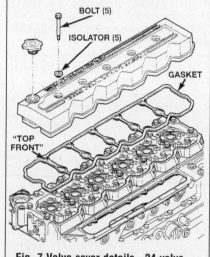

Fig. 7 Valve cover details - 24-valve diesel

Rocker Arms

REMOVAL & INSTALLATION

Note: Valve train components must be reinstalled in their original locations. Lay out and tag all components to ease assembly.

2.5L, 3.9L, 5.2L, 5.9L, 8.0L Gasoline Engines

See Figures 8, 9, 10, 11 and 12

1. Remove the valve cover(s).
2. On the 2.5L engine, adjacent rocker arms are linked by a bridge. Loosen paired bolts evenly, then remove bolts, bridge, pivot assembly and rocker arms.

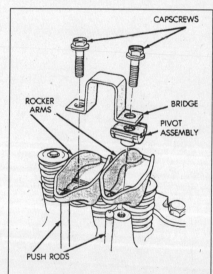

Fig. 8 2.5L engine rocker arm assembly

3. On other models, loosen the rocker arm bolt and remove the bolt, pivot assembly and rocker arm. V-10 engines have a retainer beneath the rocker arms.
4. Remove the pushrods, if required. Note locations of each for reassembly.

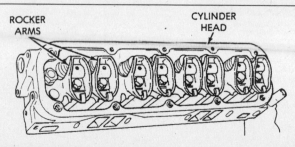

Fig. 9 5.2L engine rocker arm assembly

ENGINE AND ENGINE OVERHAUL   3-21

Fig. 10 Remove a rocker arm bolt - 5.2L engine shown

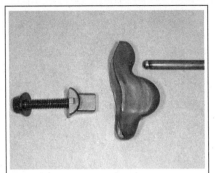

Fig. 11 Rocker arm assembly - 5.2L engine shown

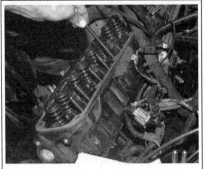

Fig. 12 Removing a pushrod. Be sure to note location for reassembly. A cardboard box with a hole for each pushrod will do the trick

To install:

5. On the V-6 and V-8 engines, rotate the crankshaft until the "V6" or "V8" mark lines up with the TDC mark on the timing chain case cover.
6. Clean and lubricate all parts.
7. Be sure the pushrods are seated in the tappets.
8. Apply clean engine oil to the pushrod tip and to the rocker arm pivot.
9. Tighten rocker arm bolts to 21 ft. lbs. (28 Nm). On the 2.5L engine, tighten the adjacent bolts gradually until the proper torque is reached.

WARNING:
Do not rotate or crank the engine during or immediately after rocker arm installation. Allow the hydraulic tappets about 5 minutes to bleed down.

4.7L Engine
See Figure 13

Note: Rocker arm removal requires the use of special tool, mfg. P/N 8516.

1. Disconnect the battery negative cable.
2. Remove the cylinder head cover(s).
3. Rotate the crankshaft so that the piston of the cylinder to be serviced is at Bottom Dead Center (BDC) and both valves are closed.
4. Use special tool 8516 to depress the valve and remove the rocker arm.
5. Repeat for each rocker arm to be serviced.

Note: Keep valvetrain components in order for reassembly.

To install:

6. Rotate the crankshaft so that the piston of the cylinder to be serviced is at BDC.
7. Compress the valve spring and install each rocker arm in its original position.
8. Repeat for each rocker arm to be installed.

WARNING:
Be sure the rocker arms are installed with the concave pocket over the lash adjusters.

12-Valve Diesel
See Figures 14 and 15

1. Remove the valve cover(s).
2. Loosen the valve lash adjusting screw locknuts. Loosen the adjusting screws until they stop.
3. Remove the bolts from the rocker arm pedestals. Remove the pedestals and rocker arms.
4. Remove the retaining ring and push out the rocker arm shaft if further disassembly is desired.

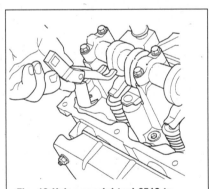

Fig. 13 Using special tool 8516 to depress valve for rocker arm removal - 4.7L engine

5. Remove the pushrods. The rearmost pushrods are removed through the cowl panel access holes.

To install:

6. Make sure the dowel rings in the pedestals are installed in the dowel bores in the cylinder head.
7. If the pushrod is holding a pedestal off its seat, turn the engine until the pedestal reaches the seat.
8. Use clean engine oil to lubricate rocker arm components, cylinder head bolt threads and under bolt heads.
9. Install bolts and tighten 12mm bolts as follows: tighten in the order shown, starting in the center of the head and working outward. Tighten to 66 ft. lbs. (90 Nm) on the first step, then 89 ft. lbs. (120 Nm), then an additional 90-degrees.

24-Valve Diesel
See Figure 16

1. Remove the cylinder head cover.

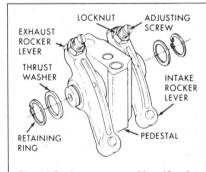

Fig. 14 Rocker arm assembly - 12-valve diesel

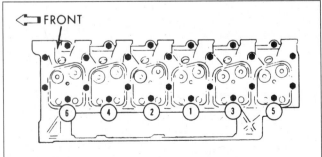

Fig. 15 Rocker arm bolt tightening sequence

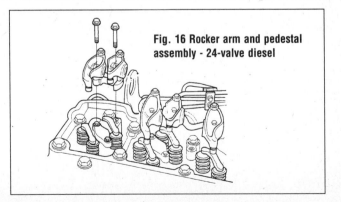

Fig. 16 Rocker arm and pedestal assembly - 24-valve diesel

3-22 ENGINE AND ENGINE OVERHAUL

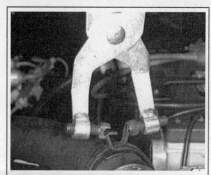

Fig. 17 Safe removal of constant tension clamps requires the proper tool

2. Unscrew and remove the rocker arm pedestal bolts.
3. Remove the pushrods. The rearmost pushrods are removed through the cowl panel access holes.

To install:
4. Be sure the pushrods are seated in the tappets.
5. Lubricate pushrod ends and other parts.
6. Tighten pedestal bolts to 27 ft. lbs. (36 Nm).
7. Adjust valve lash.

Thermostat

CAUTION:
Never open, service or drain the radiator or cooling system when hot; serious burns can occur from the steam and hot coolant. Avoid physical contact with the coolant. Wear protective clothing and eye protection. Always drain coolant into a sealable container. Clean up spills as soon as possible. Coolant should be reused unless it is contaminated or is several years old.

REMOVAL & INSTALLATION

See Figures 17, 18, 19, 20, 21, 22, 23, 24 and 25

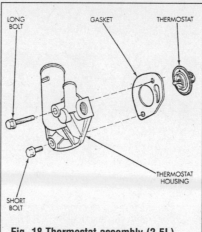

Fig. 18 Thermostat assembly (2.5L). Note bolt lengths

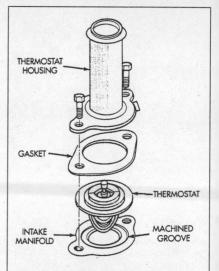

Fig. 19 On 3.9L, 5.2L, and 5.9L (gas) engines, the gasket goes above the thermostat

WARNING:
Most cooling system hoses use "constant tension" hose clamps. When removing or installing, use a tool designed for servicing this type of clamp. The clamps are stamped with a letter or number on the tongue. If replacement is necessary, use only an OEM clamp with a matching ID.

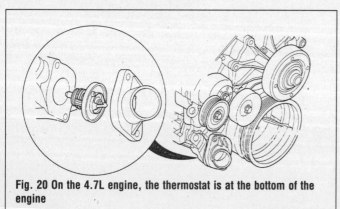

Fig. 20 On the 4.7L engine, the thermostat is at the bottom of the engine

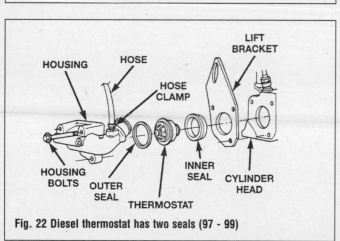

Fig. 22 Diesel thermostat has two seals (97 - 99)

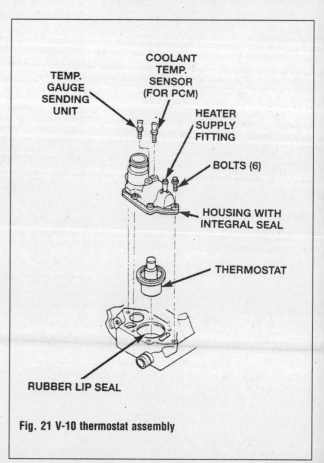

Fig. 21 V-10 thermostat assembly

ENGINE AND ENGINE OVERHAUL 3-23

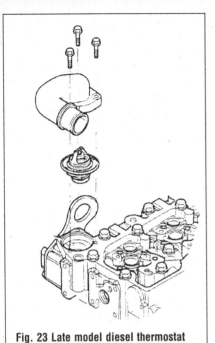

Fig. 23 Late model diesel thermostat assembly (00)

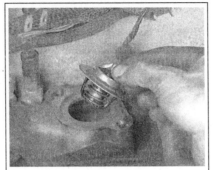

Fig. 24 Removing the thermostat from the manifold - typical

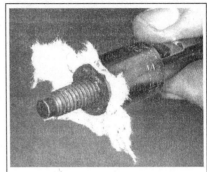

Fig. 25 A bit of paper towel will keep the bolt in the socket until the bolt hole is reached

CAUTION:

Always wear safety glasses when servicing constant tension clamps.

1. Locate the thermostat. This is located in a housing on the engine side of the upper radiator hose for all engines except the 4.7L. On the 4.7L engine, the thermostat is located on the engine side of the lower radiator hose.
2. Disconnect the negative battery cable (both batteries on diesels).
3. Drain the engine coolant from the block until the level is below the thermostat.
4. V-10 engine: remove the support rod.
5. On most V-6 and V-8 models the alternator must be removed or repositioned for access to the thermostat housing.
6. Remove the radiator hose from the thermostat housing.
7. Disconnect any sensors fitted to the thermostat housing.
8. Remove the retaining bolts from the thermostat housing. Note lengths of each for ease of installation.
9. Remove the thermostat housing, thermostat, and gasket, if fitted.
10. Note the relative positions of all components, especially gaskets and seals. Note the orientation of the thermostat in the housing.

To install:

11. Be sure the new thermostat is the correct one for your engine.
12. On V-10 engines, note that there is a rubber lip-type seal with a metal shoulder pressed into the intake manifold beneath the thermostat. The thermostat should fit snugly in the seal.
13. Clean the gasket or seal mating surfaces.
14. Paper gaskets must be replaced.
15. Install the thermostat, gasket or seals.
16. Install the thermostat housing on the engine.

Note: The thermostat housing may have the word FRONT on it (3.9L, 5.2L, 5.9L); this goes towards the front of the vehicle.

17. Be sure all components are properly seated before tightening.
18. Tighten the housing bolts to 18 ft. lbs. (24 Nm). Fasteners should be tightened evenly to avoid leaks or damage.
19. Reinstall the radiator hose onto the housing.

Note: Ensure that you have secured the system drain plug(s) before refilling with coolant.

20. Refill the radiator with a proper coolant mixture.
21. Connect the negative battery cable(s).
22. Start the engine and bleed the cooling system. Refer to "General Information And Maintenance".
23. Ensure that the thermostat is operational (by checking the upper radiator hose for warmth), and that there are no leaks.

Intake Manifold

REMOVAL AND INSTALLATION

2.5L Engine

See Figure 26

1. Depressurize the fuel system (see Chapter 5) and disconnect the battery negative cable.
2. Remove the air cleaner hose and resonator from the throttle body and air cleaner.
3. Remove the accessory drive belt from the power steering pump.
4. Remove the power steering pump and brackets from the water pump and secure it out of the way.
5. Disconnect the fuel line from the fuel rail. Note that some lines may require a special tool for removal/installation.
6. Disconnect the throttle, cruise control and transmission line pressure cable (if fitted) from the throttle body and cable bracket.

WARNING:

Do not pry the cruise control cable off with pliers or screwdrivers. Use fingers only.

7. Disconnect electrical wiring. Secure wiring out of the way.
8. Disconnect the vacuum, brake booster, CCV and MAP hoses.
9. Remove the vacuum harness.
10. Referring to the illustration of the 2.5L manifold, remove bolts 2 through 5 securing the manifold to the cylinder head. Loosen bolt 1 and nuts 6 and 7.
11. Remove the intake manifold and gaskets.

To install:

12. Use new O-rings at the quick-connect fuel line coupling.
13. Fit a new intake manifold gasket.
14. Replace the intake manifold and finger tighten the fasteners.
15. Tighten Bolt #1 to 30 ft. lbs. (41 Nm).
16. Install the intake manifold. Tighten bolts 2 through 5 to 23 ft. lbs. (31 Nm).
17. Install the exhaust manifold spacers over the mounting studs.

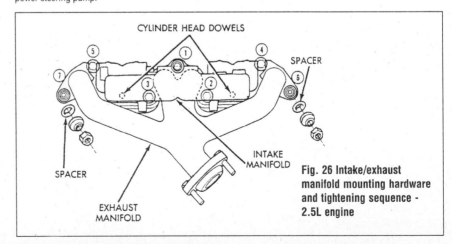

Fig. 26 Intake/exhaust manifold mounting hardware and tightening sequence - 2.5L engine

3-24 ENGINE AND ENGINE OVERHAUL

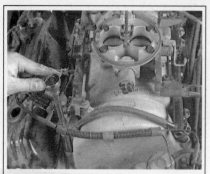

Fig. 27 Removing the intake manifold bolts

Fig. 28 Intake manifold air temperature sensor

Fig. 29 Removing the intake manifold assembly

Fig. 30 Intake manifold, throttle body and fuel rail assemblies

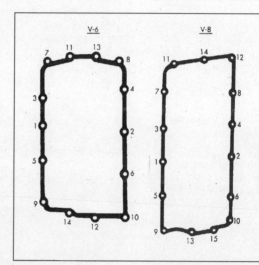

Fig. 31 The plenum pan bolt tightening sequence

18. Tighten nuts 6 and 7 to 23 ft. lbs. (31 Nm).
19. The remainder of the procedure is the reverse of removal. Tighten power steering pump bolts and tensioner bracket to 21ft. lbs. (28 Nm).

3.9L, 5.2L, 5.9L Gasoline Engines

See Figures 27, 28, 29, 30, 31, 32, 33, 34, 35 and 36

1. Depressurize the fuel system (see Chapter 5) and disconnect the battery negative cable.
2. Drain the cooling system.
3. Remove the alternator.
4. Remove the air cleaner assembly.
5. Disconnect the supply line from the fuel rail.
6. Disconnect the throttle, cruise control and transmission kick down cables (if fitted) from the throttle body bracket.
7. Remove the distributor cap and wires.
8. Disconnect the fuel system and throttle body wiring connectors.
9. Disconnect the coil wires and sensors: MAP, IAT, CTS, IAC (if equipped).
10. Disconnect the heater hoses and the bypass hose.
11. Remove any vacuum hoses fitted to the manifold.
12. Remove the intake manifold bolts.
13. Remove the manifold and throttle body from the engine as an assembly.
14. Remove the side and crossover gaskets.
15. Remove the plenum pan, if desired, by removing the bolts.

To install:

16. If the plenum pan was removed, use a new gasket. Tighten the bolts evenly and in a cross pattern beginning in the center of the plenum and working outward. Tighten bolts in three steps to 24 inch lbs. (2.7 Nm), 48 inch lbs. (5.4 Nm) and 84 inch lbs.
17. Place the plastic locator dowels (if supplied) in the engine block.

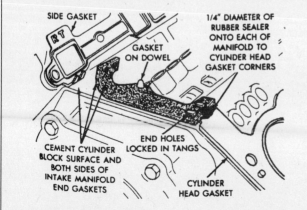

Fig. 32 When applying the intake manifold end seals, line up their holes with the dowels and tangs

Fig. 33 Apply a small bead of sealant to the corners of the end seals

ENGINE AND ENGINE OVERHAUL   3-25

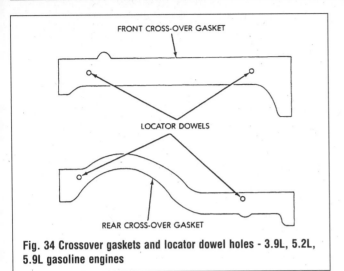

Fig. 34 Crossover gaskets and locator dowel holes - 3.9L, 5.2L, 5.9L gasoline engines

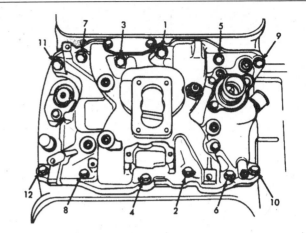

Fig. 35 Proper tightening sequence for the V6 intake manifold

18. Apply RTV sealant to the four corner joints. Bead height should be slightly higher than the crossover gaskets (about 0.2 in./5 mm).
19. Install the new front and rear crossover gaskets over the dowels.
20. Install the new flange gaskets on the cylinder heads.

Note: Ensure that the vertical port alignment tab is resting on the deck face of the block and that the horizontal alignment tabs are in position with the mating cylinder head gasket tabs. The words MANIFOLD SIDE should be visible on the center of each flange gasket.

21. Place the manifold into position on the engine block and cylinder heads.
22. Ensure that the gaskets have not shifted after the manifold is in place.
23. Install the intake manifold bolts and tighten as follows:

V6 Engine
• Tighten bolts 1 and 2 to 72 inch lbs. (8 Nm), in alternating steps, 12 inch lbs. (1.4 Nm) at a time.
• Tighten bolts 3 through 12, in sequence, to 72 inch lbs. (8 Nm).
• Tighten all bolts, in sequence, to 12 ft. lbs. (16 Nm).

V8 Engines
• Tighten bolts 1 through 4, in sequence, to 72 inch lbs. (8 Nm), in alternating steps, 12 inch lbs. (1.4 Nm) at a time.
• Tighten bolts 5 through 12, in sequence, to 72 inch lbs. (8 Nm).
• Tighten all bolts, in sequence to 12 ft. lbs. (16 Nm).

All engines
24. Install the vacuum hoses in their proper locations.
25. Install the heater and bypass hoses.
26. Install the injector harness.
27. Install the fuel line(s).
28. Reconnect the emissions and throttle body sensor wiring.
29. Reconnect the throttle linkage, speed control cable (if equipped), and transmission kickdown cable (if equipped).
30. Reconnect the oil pressure sending unit.
31. Install the distributor cap.
32. Install the alternator.
33. Install the air cleaner assembly.
34. Connect the negative battery cable.
35. Fill and bleed the cooling system.
36. Check for leaks.

4.7L Engine
See Figure 37

1. Depressurize the fuel system (see Chapter 5), then disconnect the battery negative cable.
2. Remove the air cleaner assembly.
3. Disconnect the throttle and cruise control cables.
4. Disconnect MAP, IAT, TPS, CTS and IAC sensors.
5. Disconnect the vapor purge, brake booster, speed control servo and PCV hoses.
6. Disconnect the alternator.
7. Disconnect the A/C electrical connectors.
8. Disconnect the left and right radio suppressor straps.
9. Disconnect and remove the ignition coil towers.
10. Remove the top oil dipstick tube retaining bolt and ground strap.
11. Remove the fuel rail.
12. Remove the throttle body assembly and mounting bracket.
13. Drain the cooling system below manifold level.
14. Remove the heater hoses.
15. Remove the coolant temperature sensor.
16. Remove the manifold bolts and take off the manifold.
17. Clean the manifolds and mounting surfaces with solvent and let dry or dry completely with compressed air.
18. Inspect the surfaces for cracks and for

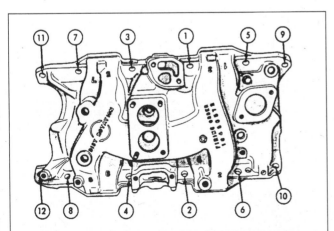

Fig. 36 Proper tightening sequence for the V8 intake manifold

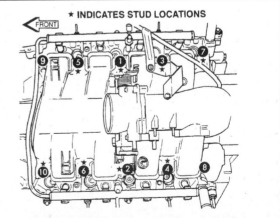

Fig. 37 Intake manifold tightening sequence - 4.7L engine

3-26 ENGINE AND ENGINE OVERHAUL

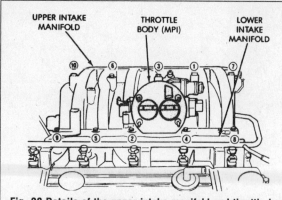

Fig. 38 Details of the upper intake manifold and throttle body assembly - V-10 engine

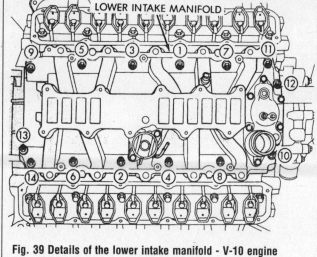

Fig. 39 Details of the lower intake manifold - V-10 engine

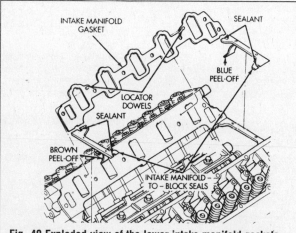

Fig. 40 Exploded view of the lower intake manifold gaskets - V-10 engine

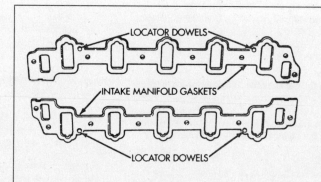

Fig. 41 Align the intake manifold flange gasket - V-10 engine

warpage using a straightedge.

19. Inspect the gaskets on the manifold for tears, cracks or hardening, replacing them if necessary. The gaskets can be reused if not damaged.

20. Installation is the reverse of removal. Tighten the manifold bolts gradually and in a cross pattern beginning in the center and working outwards. Proper torque is 105 inch lbs. (12 Nm).

8.0L Engine
See Figures 38, 39, 40 and 41

1. Drain the cooling system
2. Depressurize the fuel system (see Chapter 5) and disconnect the negative battery cable.
3. Remove alternator and air cleaner.
4. Remove the serpentine belt.
5. Remove the alternator and its brace.
6. Remove the air conditioning compressor brace, remove the compressor and set it aside without disconnecting the lines.
7. Remove the air cleaner cover and filter. Remove the filter. Discard the gasket.
8. Disconnect the fuel line from the fuel rail.
9. Disconnect the accelerator linkage, and if so equipped, the speed control and transmission kickdown cables.
10. Remove the coil assemblies. Disconnect the vacuum lines.
11. Disconnect the heater hoses and bypass hose.
12. Remove the closed crankcase ventilation and evaporation control systems.

13. Remove throttle body and lift it off the upper intake manifold. Discard the gasket.

14. Remove the front upper intake manifold bolts. Retain the three rear bolts in the up position with tape or rubber bands.

15. Lift the upper intake manifold out of the engine bay. Discard the gasket.

16. Remove the lower intake manifold bolts and remove the manifold. Discard the gasket.

To install:

17. Clean the manifolds and mounting surfaces with solvent and let dry or dry completely with compressed air.

18. Inspect the surfaces for cracks and for warpage using a straightedge.

19. With the locator dowels in place on the head, install the intake manifold side gaskets.

20. When sure that the block is OIL-FREE, peel off the paper (blue-rear, brown-front) and press firmly onto the block. Align the slots in the end seals with the notches in the intake manifold gaskets.

21. Into each of the four corner pockets, insert Mopar, Silicon Rubber Adhesive Sealant, or equivalent. DO NOT overfill.

22. The lower intake manifold must be installed within three minutes of sealant application. After installation, inspect to be sure all the gaskets and the seals are in their proper places. Finger-start all lower intake bolts.

23. Tighten in sequence to 40 ft. lbs. (54 Nm).
24. Install a new gasket and the upper intake manifold and finger-tighten all bolts. Alternate from one side to the other.
25. Tighten the bolts to 16 ft. lbs. (22 Nm).
26. Install a new gasket and the throttle body onto the upper intake manifold. Tighten the bolts to 17 ft. lbs. (23 Nm).
27. Install the closed crankcase ventilation and evaporation control systems.
28. Connect the heater hoses and bypass hose.
29. Connect the vacuum lines.
30. Install the oil assemblies and the ignition cables.
31. Connect the accelerator linkage, and if so equipped, the speed control and transmission kickdown cables.
32. Install the fuel lines.
33. Using a new gasket, install the air cleaner housing. Tighten the nuts to 96 inch lbs. (11 Nm). Install the air cleaner assembly and its cover.
34. Install the air conditioning compressor. Install the brace, tightening the bolts to 30 ft. lbs. (41 Nm).
35. Install the serpentine belt.
36. Fill the cooling system and connect the negative battery cable.

Diesel Engines
See Figure 42

1. Disconnect the negative battery cables.

ENGINE AND ENGINE OVERHAUL 3-27

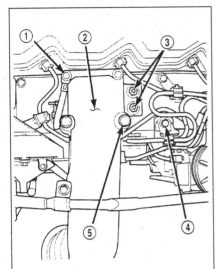

Fig. 42 Air inlet housing components - ground strap (1), air intake housing (2), heater power cable mounting studs (3), fuel bracket bolt (4), housing bolts (5)

2. Remove the charge air cooler outlet tube from the air inlet housing.
3. Remove the engine oil dipstick mounting bolt and move the dipstick aside.
4. Disconnect the air grid heater power cables at the cable mounting studs.
5. Remove the four air inlet housing mounting bolts and remove the housing from top of the heater elements.
6. Remove the intake air grid heater from the manifold.
7. Remove the high pressure fuel lines.
8. Remove the rest of the manifold bolts and remove the manifold.
9. Installation is the reverse of removal. Tighten manifold bolts to 18 ft. lbs. (24 Nm).

Exhaust Manifold

REMOVAL & INSTALLATION

See Figure 43

Common Steps

Note: The following steps are common to all models. Read this information before following the specific procedures for your vehicle.

1. New gaskets are required on installation. Purchase these before beginning.
2. Access to exhaust manifold components is easier if the vehicle is raised.

CAUTION:

Obey all applicable safety procedures for jacking, supporting and working beneath the vehicle.

3. Disconnect the battery negative cable(s).
4. Check manifolds and header pipes for any sensor(s) which may be fitted to them and disconnect the wiring if it interferes with removal.

Fig. 43 Spray hardware with penetrating fluid. Brittle metal may snap if the fastener is forced

5. Remove the heat shield(s), if fitted.
6. Disconnecting the header(s) from the exhaust pipe(s) may be easier from beneath the vehicle.
7. Wire brush threads of any accessible fasteners. Apply penetrating fluid to fasteners to ease disassembly.
8. Do not strike manifold or exhaust pipe with a hammer. Use a plastic mallet to prevent damage.
9. Clean mating surfaces before installation.
10. Anti-seize compound applied to the threads of nuts and bolts will make future removal easier.
11. Tighten all hardware evenly and to the proper torque. Manifold nuts and bolts should be tightened from the center outwards. Do not attempt to drive the manifold home by running down the nuts or bolts.
12. After installation is complete, start the engine and check for leaks.

Note: If the studs come out of the block when the nuts are loosened, the use of new studs is recommended. Coat the threads of the new studs with an adhesive/sealer. This will help prevent oil leaks.

2.5L Engine
See Figure 26

1. Perform "common steps", above.
2. Remove the exhaust pipe nuts and disconnect the exhaust pipe from the exhaust manifold.
3. Remove the intake manifold.
4. Remove the exhaust manifold.
5. Install a new intake manifold gasket over the alignment dowels.
6. Install the exhaust manifold assembly. Be sure it is centrally located over the end studs and spacer.
7. Tighten Bolt #1 to 30 ft. lbs. (41 Nm).
8. Install the intake manifold. Tighten bolts 2 through 5 to 23 ft. lbs. (31 Nm).
9. Install the exhaust manifold spacers over the mounting studs.
10. Tighten nuts 6 and 7 to 23 ft. lbs. (31 Nm).
11. The remainder of the procedure is the reverse of removal. Tighten the exhaust pipe nuts to 23 ft. lbs. (31 Nm).

4.7L Engine
See Figure 44

1. Perform "common steps", above.
2. Remove the left side manifold by undoing the exhaust pipe nuts and the manifold bolts on the cylinder head.
3. For the right side manifold:

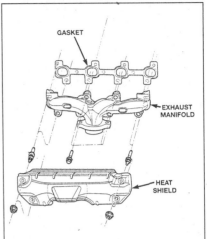

Fig. 44 Exhaust manifold and heat shield assembly - 4.7L

 a. Remove the air cleaner and resonator assemblies and the air inlet hose.
 b. Remove the accessory drive belt.
 c. Remove the A/C compressor and set it to one side.
 d. Remove the A/C accumulator support bracket fastener.
 e. Drain the radiator and remove the heater hoses at the engine.
 f. Remove the heat shield.
 g. Remove upper exhaust manifold fasteners.
 h. Disconnect the exhaust pipe.
 i. Remove the starter.
 j. Remove the lower exhaust manifold fasteners. Remove the manifold.
4. Installation is the reverse of removal. Tighten manifold hardware to 18 ft. lbs. (25 Nm).
5. Tighten heat shield fasteners to 72 inch lbs. (8 Nm), then loosen 45-degrees.

3.9L, 5.2L, 5.9L Gasoline engines
See Figures 45 and 46

1. Perform "common steps", above.
2. Remove the exhaust pipe to manifold nuts.
3. Remove the nuts and bolts securing the manifold to the head.
4. Installation is the reverse of removal. Tighten cylinder head hardware to 18 ft. lbs. (24 Nm). Tighten exhaust pipe nuts to 20 ft. lbs. (27 Nm).

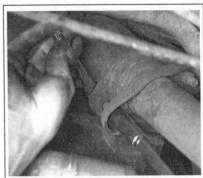

Fig. 45 Disconnecting the manifold-to-exhaust pipe nuts - typical

3-28 ENGINE AND ENGINE OVERHAUL

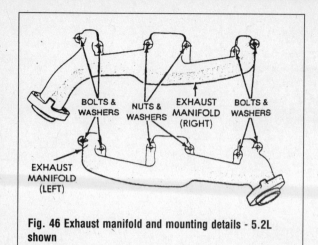

Fig. 46 Exhaust manifold and mounting details - 5.2L shown

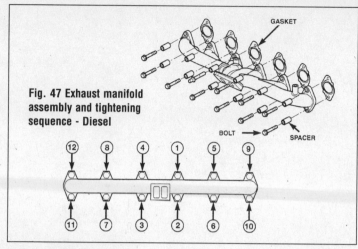

Fig. 47 Exhaust manifold assembly and tightening sequence - Diesel

8.0L Engine

1. Perform "common steps", above.
2. Remove the exhaust pipe to manifold nuts and bolts.
3. Remove the dipstick bracket (right side).
4. Remove the bolts securing the manifold to the head.
5. Installation is the reverse of removal. Tighten hardware to 16 ft. lbs. (22 Nm).

Diesel Engine
See Figure 47

1. Perform "common steps", above.
2. Disconnect the exhaust pipe from the turbocharger.
3. Remove the turbocharger. See "Turbocharger" for procedures.
4. Remove the cab heater return pipe nut from the exhaust manifold stud. Position tube out of the way.
5. Remove the fasteners and separate the exhaust manifold from the head.
6. Installation is the reverse of removal. Tighten the manifold hardware evenly and in a cross pattern starting in the center and working outwards. Proper torque is 32 ft. lbs. (43 Nm).
7. Be sure to pre-lubricate the turbocharger (see the next section).

Turbocharger

Note: This procedure applies only to diesel-engined models.

REMOVAL & INSTALLATION

See Figure 48

1. Disconnect the battery negative cables.
2. Disconnect the exhaust pipe from the turbocharger elbow.
3. Disconnect the air inlet hose.
4. Disconnect the oil supply line and oil drain tube.
5. Disconnect the charge air cooler inlet pipe.
6. Remove the nuts and remove the turbocharger from the exhaust manifold.
7. Cover the opening to prevent entry of foreign matter.

To install:

8. Apply anti-seize compound to the mounting studs. Tighten the nuts to 24 ft. lbs. (32 Nm).
9. Tighten the oil drain tube to 18 ft. lbs. (24 Nm).
10. Pour 2 - 3 oz (50 - 60 cc) of clean engine oil into the oil supply line fitting. Carefully rotate the impeller by hand to distribute the oil.
11. Tighten the oil supply line fitting nut to 11 ft. lbs. (20 Nm).
12. Tighten hose clamps to 95 inch lbs. (11 Nm). Tighten exhaust pipe bolts to 25 ft. lbs. (34 Nm).

Radiator

Procedures will vary depending on model and equipment fitted.

CAUTION:

Never open, service or drain the radiator or cooling system when hot; serious burns can occur from the steam and hot coolant. Avoid physical contact with the coolant. Wear protective clothing and eye protection. Always drain coolant into a sealable container. Clean up spills as soon as possible. Coolant should be reused unless it is contaminated or is several years old.

REMOVAL & INSTALLATION

See Figures 49, 50, 51, 52, 53 and 54

1. Disconnect the battery negative cable(s).
2. Drain the radiator.

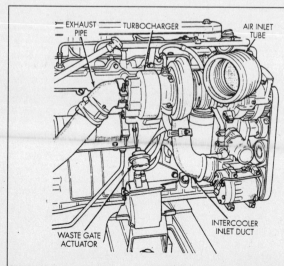

Fig. 48 Turbocharger assembly - diesels

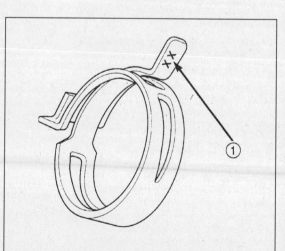

Fig. 49 Size marking on constant tension clamps (1)

ENGINE AND ENGINE OVERHAUL 3-29

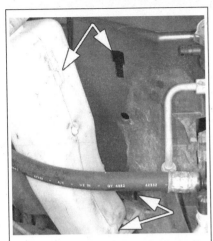

Fig. 50 Coolant bottle T-slots and studs

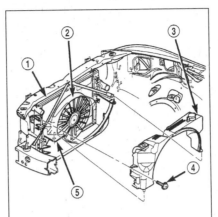

Fig. 51 Shroud assembly (4.7L): Radiator (1), Fan (2), Upper Shroud (3), Bolts (4), Lower Shroud (5)

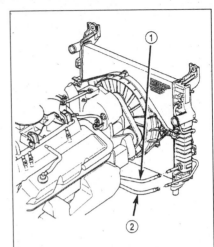

Fig. 52 Mark automatic transmission oil cooler lines before disconnecting. Supply (1) and return (2) lines in this illustration for the 4.7L engine

Fig. 53 Plug the transmission lines to prevent leakage

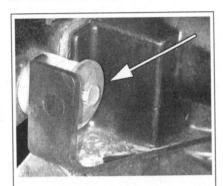

Fig. 54 Radiator mounting bolt

3. Remove any cables or lines clipped to the radiator or shrouds.
4. Remove the upper and lower heater hoses.

WARNING:

Most cooling system hoses use "constant tension" hose clamps. When removing or installing, use a tool designed for servicing this type of clamp. The clamps are stamped with a letter or number on the tongue. If replacement is necessary, use only an OEM clamp with a matching ID.

CAUTION:

Always wear safety glasses when servicing constant tension clamps.

5. If coolant overflow and windshield washer fluid bottles are fitted to the fan shroud, disconnect the lines and remove them. Pull straight up to remove the bottles from shroud.
6. On 4.7 models, remove the upper fan shroud. Disconnect the electric fan wiring.
7. On 2.5L models, disconnect the fan wiring, remove the fan and shroud hardware and remove the assembly.
8. On other models, remove the fan shroud hardware and reposition the shroud towards the engine for access to the radiator.
9. Label, then disconnect the automatic transmission lines if the fluid cooler is incorporated into the radiator. Plug the lines to prevent spillage. Be sure to note which line goes to which fitting.
10. Durango: remove the grill.
11. Check for rubber shields on the sides of the radiator. Remove them if fitted. These are normally secured with non-reusable plastic pins.
12. Remove the radiator bolts (usually two at the top). Lift the radiator up and out of the vehicle, being careful that the cooling fins do not bang against anything. They are easily damaged.
13. Some vehicles may have an auxiliary automatic transmission fluid cooler which will come away with the radiator.

To install:

14. Lower the radiator into place. There are two alignment pins at the bottom which fit into holes in the lower support.
15. Tighten radiator mounting bolts to 17 ft. lbs. (23 Nm).
16. The remainder of the procedure is the reverse of removal. Double-check the connection of all hoses and lines before adding fluids or operating the vehicle.

Electric Engine Fan (2.5L)

Vehicles equipped with the 2.5L engine utilize an electric fan for engine cooling. Models equipped with the 4.7L engine have an auxiliary electric fan to aid low speed cooling. These units are not serviceable. If defective, they must be replaced. See "Engine Electrical" for testing procedures.

The following procedure refers to 2.5L models only. On the 4.7L, the electric fan is removed with the radiator. Refer to that procedure.

REMOVAL & INSTALLATION

1. Disconnect the battery negative cable.
2. Disconnect the fan motor wire connector.
3. Remove the fan shroud mounting clips connecting the shroud to the radiator.
4. Remove the fan and shroud from the radiator as an assembly.
5. Installation is the reverse of removal.

Belt-Driven Engine Fan

The fan is threaded onto the water pump shaft with a 36mm nut. To loosen the nut, the pump must be prevented from turning. This can be accomplished with a special tool or with a prybar.

Fan and fan shroud are removed together.

REMOVAL & INSTALLATION

See Figures 55, 56, 57, 58 and 59

Note: Gasoline engine fans have a standard RIGHT-HAND thread. Turn CCW to loosen. Diesel engine fans have a LEFT-HAND thread. Turn CW to loosen.

1. Disconnect the battery negative cable(s).
2. Disconnect the throttle cable is attached to the fan shroud.
3. Remove the windshield washer bottle and coolant overflow tank from the fan shroud.
4. Stop the water pump from turning with a special tool or prybar on the nuts. Loosen the fan nut (36mm). Turn CCW to loosen on gasoline engines; turn CW to loosen on diesel engines.
5. Remove the fan shroud mounting hardware. These may be bolts, clips or a combination of the two, depending on model.
6. Remove the fan and shroud together.

3-30 ENGINE AND ENGINE OVERHAUL

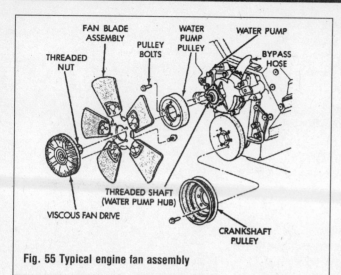

Fig. 55 Typical engine fan assembly

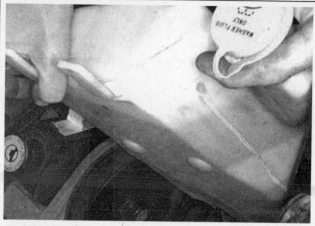

Fig. 56 Remove the windshield washer bottle from the fan shroud

Fig. 57 Loosening the fan nut

Fig. 58 To loosen the fan nut, stop the pump shaft from turning. This tool will do it; you could also use a prybar. In this photo, the fan has been removed for clarity

Fig. 59 Removing fan and shroud together

WARNING:

Store the fan in the vertical (operating) position. If placed on its side, the silicone fluid in the viscous drive could drain into the bearing assembly and contaminate the lubricant.

7. To replace the fan or viscous drive, remove the four bolts and separate the components.

To install:

8. If drive and fan were separated, tighten the bolts to 17 ft. lbs. (23 Nm).
9. The remainder of the installation procedure is the reverse of removal. Recommended fan nut torque on diesels is 42 ft. lbs. (57 Nm).
10. If a new viscous drive has been fitted, start and run the engine at 2,000 rpm for about 2 minutes to distribute fluid within the drive.

Water Pump

REMOVAL & INSTALLATION

CAUTION:

Never open, service or drain the radiator or cooling system when hot; serious burns can occur from the steam and hot coolant. Avoid physical contact with the coolant. Wear protective clothing and eye protection. Always drain coolant into a sealable container. Clean up spills as soon as possible. Coolant should be reused unless it is contaminated or is several years old.

Common Steps
See Figure 60

1. Disconnect the battery negative cable(s).

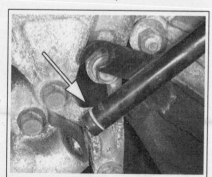

Fig. 60 Heater hose pipe O-ring (arrow): twist gently when removing or installing; check condition; lubricate with coolant before installing

2. Drain the cooling system.
3. Remove the accessory drive belt.
4. Remove the lower radiator hose (except Diesel).
5. Disconnect the heater hose from the pump, if fitted.

WARNING:

Most cooling system hoses use "constant tension" hose clamps. When removing or installing, use a tool designed for servicing this type of clamp. The clamps are stamped with a letter or number on the tongue. If replacement is necessary, use only an OEM clamp with a matching ID.

CAUTION:

Always wear safety glasses when servicing contact tension clamps.

6. Always use a new gasket when installing the water pump. Clean mating surfaces thoroughly before installation.
7. This would be a good time to replace radiator hoses. At least inspect them carefully for aging or other damage.
8. If the heater hose pipe was removed, use thread sealant or equivalent on the pipe threads (if threaded), or lubricate any O-rings with coolant before installing.

ENGINE AND ENGINE OVERHAUL 3-31

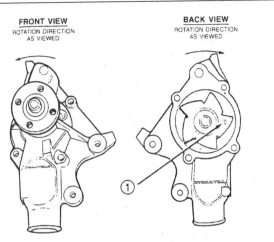

Fig. 61 Reverse rotating water pump with impeller mark (1) - 2.5L

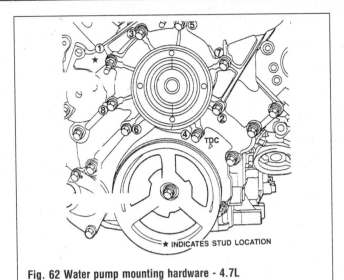

Fig. 62 Water pump mounting hardware - 4.7L

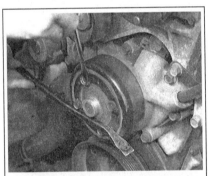

Fig. 63 Use a prybar to keep the water pump from rotating while pulley bolts are removed

Fig. 64 Bypass hose clamp (arrow) must be tended to before removing the pump

Fig. 65 Water pump bolts differ in length; note check location before installation - 3.9L, 5.2L, 5.9L gasoline engine show

9. Check accessory belt routing against the plate in the engine compartment before starting engine.
10. Allow the engine to reach operating temperature and check for leaks.

2.5L Engine
See Figure 61

> **WARNING:**
>
> The 2.5L engine has a reverse (CCW) rotating water pump. The letter "R" is stamped into the back of the pump impeller for ID purposes. Engines from previous model years may be equipped with a forward (CW) rotating water pump. Installing the wrong pump will cause engine damage. Check any replacement parts before installation.

1. Carry out the "common steps", above.
2. Remove the power steering pump.
3. Remove the four water pump bolts and remove the pump. Note that one bolt is longer than the others. Be sure it is properly installed during installation.

To install:
4. Reverse the removal procedure.
5. Install the new gasket with the silicone bead facing the water pump.
6. Tighten mounting bolts to 22 ft. lbs. (30 Nm).

3.9L, 4.7L, 5.2L, 5.9L, 8.0L Gasoline Engines
See Figures 62, 63, 64, 65, 66 and 67

1. Carry out the "common steps", above.
2. Remove the upper radiator hose.
3. Remove the fan and shroud assembly.
4. Except 4.7L: remove the four water pump pulley bolts and remove the pulley.
5. Remove the heater hose pipe mounting bolt, if fitted.
6. Except 4.7L: loosen the bypass hose clamp at the water pump.
7. Remove the seven water pump mounting bolts and, on the 4.7L, one nut.

Fig. 66 After removing the bolts, pull off the pump and place it in a safe location

> **WARNING:**
>
> Bolts may be of different lengths. Be sure to note their locations.

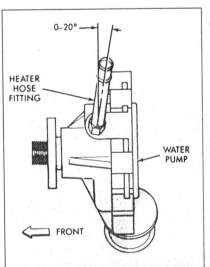

Fig. 67 Make sure the heater hose fitting is positioned as shown - V-10 engine

3-32 ENGINE AND ENGINE OVERHAUL

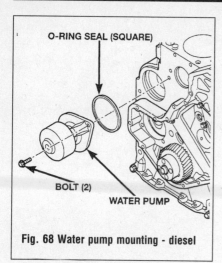

Fig. 68 Water pump mounting - diesel

Read the procedures before starting to get a thorough understanding of the tools and equipment needed. Many cylinder head components are precision machined pieces that need to be cleaned and carefully inspected before installation. Refer to ENGINE RECONDITIONING for this information.

Common Steps

The following steps are common to most models. Refer to individual procedures for particulars for each vehicle.

1. Depressurize the fuel system (see Chapter 5), then disconnect the battery negative cable(s).
2. Drain the coolant from engine and radiator.
3. Remove the upper radiator hose.
4. Remove the air cleaner assembly and duct work.
5. Disconnect control cables from the throttle body: typically, throttle, cruise control and kickdown cables.
6. Disconnect all hoses and vacuum lines from the intake and exhaust manifolds and from the valve cover(s). This will normally include crankcase ventilation and evaporative control systems.
7. If equipped with A/C, unbolt the compressor from the mounting bracket and place it aside. Unbolt and remove the compressor bracket.
8. Mark and disconnect the spark plug wires. On the 4.7L, remove the ignition coils.
9. Loosen or remove the spark plugs.
10. Remove any heat shields on the exhaust manifolds.
11. Disconnect the exhaust manifold from the head.
12. Remove the accessory drive belt.
13. Remove the alternator (except 2.5L).
14. Disconnect the heater hoses.
15. Disconnect the coolant temperature sending unit(s) on the intake manifold.
16. Disconnect sensor wires from intake manifold sensors.
17. Disconnect throttle body electrical connectors.
18. Detach the fuel line from the fuel rail.
19. Disconnect fuel system and injector electrical connectors.
20. Remove the intake manifold complete with fuel system components.
21. Remove cylinder head cover(s).
22. Remove the rocker arms and pushrods, if fitted.
23. Check for and disconnect ground straps, wiring harness or other electrical connections on the head(s).

To install:

24. Use new gaskets throughout.

25. Do not use sealants or gasket compounds on head gasket.
26. Observe cylinder head tightening sequences and proper fastener torque. Head bolts are tightened in a cross pattern starting near the center of the head and working outwards. On some engines, a multiple-step tightening sequence is recommended. Bolt tightness should always get a final check after the sequence is completed.
27. Be careful when installing rocker arms that the piston is not at TDC.
28. In some cases, installation may be made easier by fabricating locating dowels from old head bolts. Cut off the hex, slot the end to accept a screwdriver, and fit at each end of the block to locate the head during installation.

2.5L Engine
See Figure 70

1. Carry out the "common steps", above.
2. Unbolt the power steering pump bracket and set pump and bracket aside. It is not necessary to disconnect the hoses.
3. Remove the cylinder head bolts.

Note: If the bolts have a dab of paint on the heads, they have been removed before. Use new ones when assembling. If they do not have the dab of paint, mark them after installation.

4. Remove the cylinder head and gasket.

To install:

5. Remove carbon, gasket material, etc., from mating surfaces.
6. Decarbonize piston crowns, exhaust ports and clean intake ports and water passages.
7. Refer to the illustration of the cylinder head bolt tightening sequence. Be sure to tighten head bolts in gradually and evenly according to the pattern provided.
8. Use sealant (Loctite Pipe Sealant with Teflon no. 592, or equivalent) on the threads of bolt 7.
9. Tighten all cylinder head bolts in the order shown to 22 ft. lbs. (30 Nm).
10. Tighten all cylinder head bolts to 45 ft. lbs. (61 Nm).
11. Tighten bolts 1 through 6 to 110 ft. lbs. (149 Nm).
12. Tighten bolt 7 to 100 ft. lbs. (136 Nm).
13. Tighten bolts 8, 9, 10 to 110 ft. lbs. (149 Nm).
14. Carry out a final torque check on all bolts.

To install:

8. Reverse the removal procedure.
9. The 8.0L engine uses an O-ring to seal. Use a new one and lubricate it with petroleum jelly before installation of the pump. Also, use sealant on the threads of the heater hose pipe, position the pipe as shown before installing the pump and tighten the nut to 12 ft. lbs. (16 Nm).
10. On pushrod engines, connect the bypass hose to the pump while installing it.
11. Tighten the bolts to 30 ft. lbs. (40 Nm) on pushrod engines, 40 ft. lbs. (54 Nm) on the 4.7L engine.
12. Spin pump to ensure the impeller is not rubbing against the timing chain case.
13. When installing the heater hose pipe, ensure that the slot in the pipe bracket is bearing against the mounting bolt. This will hold the pipe in as far as possible to prevent leaks.
14. Tighten pulley bolts to 20 ft. lbs. (27 Nm) on V-6 and V-8, 16 ft. lbs. (22 Nm) on the V-10.

Diesel
See Figure 68

1. Carry out the "common steps", above. Remove of radiator hoses is not required.
2. Remove the wiring harness bolt from the water pump and arrange the harness out of the way.
3. Remove the water pump mounting bolts and remove the pump.

To install:

4. Use a new O-ring and lubricate it with petroleum jelly before installation.
5. Tighten mounting bolts to 18 ft. lbs. (24 Nm).

Cylinder Head

REMOVAL & INSTALLATION

See Figure 69

WARNING:
The engine must be cold before attempting to remove the cylinder head. Loosening head bolts on a hot engine may cause component damage.

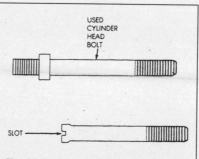

Fig. 69 Locating dowels made from old head bolts may ease head installation

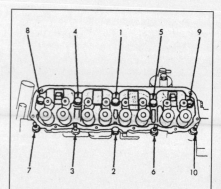

Fig. 70 Cylinder head bolt tightening sequence - 2.5L engine

ENGINE AND ENGINE OVERHAUL 3-33

Fig. 71 Removing the cylinder head - 5.2L shown

15. Mark reused head bolts with a dab of paint.
16. The remainder of the procedure is the reverse of disassembly.

V-6, V-8, V-10 OHV Engines
See Figures 71, 72, 73, 74 and 75

1. Carry out the "common steps", above.
2. Remove the distributor cap or coil packs (V-10).
3. Remove the lower intake manifold (V-10).
4. Remove the cylinder head bolts.
5. Remove the cylinder head and gasket.

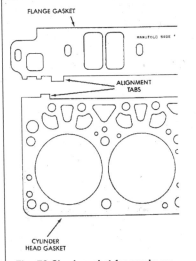

Fig. 72 Check gasket for marks or properties that aid in installation

To install:

6. Reverse the removal procedure.
7. 6 and 8 cylinder engines: tighten head bolts in the order shown in two steps: to 50 ft. lbs. (68 Nm), then 105 ft. lbs. (143 Nm) on the second.
8. V-10: tighten head bolts in the order shown in two steps: to 43 ft. lbs. (58 Nm), then 105 ft. lbs. (143 Nm) on the second.

4.7L Engine
See Figures 76, 77 and 78

A special timing chain locking tool (mfg. P/N 8515) is required for this procedure.

1. Carry out the "common steps", above.
2. Remove the engine cooling fan and shroud.
3. Remove the oil fill housing.
4. Remove the power steering pump.
5. Before disconnecting the timing chains alignment of crankshaft and cam sprocket is required for both cylinder heads
6. Rotate the crankshaft so that the damper timing mark aligns with the Top Dead Center (TDC) mark on the front cover, and the V8 marks on the camshaft sprockets are at 12 o'clock as shown.
7. Remove the crankshaft damper. Remove the timing chain cover.
8. Lock the secondary timing chains to the idler sprocket with Timing Chain Locking tool 8515.
9. Matchmark the secondary timing chains to the camshaft sprockets.
10. Remove the secondary timing chain tensioner.
11. Remove the cylinder head access plug.
12. Remove the secondary chain guide.
13. Remove the retaining bolt and the camshaft drive gear.

WARNING:
Do not allow the engine to rotate. Damage to the valve train can occur.

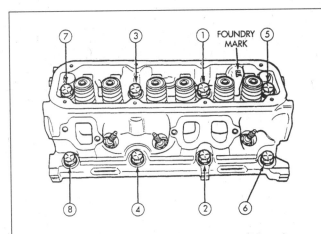

Fig. 73 Cylinder head bolt tightening sequence - V-6 engines

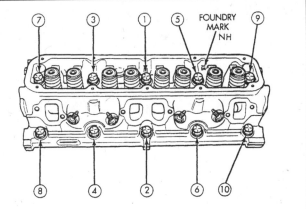

Fig. 74 Cylinder head bolt tightening sequence - 5.2L, 5.9L gasoline engines

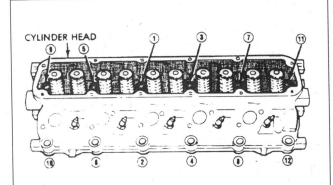

Fig. 75 Cylinder head bolt tightening order - V-10 engine

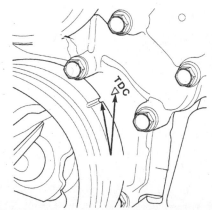

Fig. 76 Before unbolting the cam sprockets, first align the damper mark with the "TDC" mark (arrows); check alignment of cam sprockets as shown in the illustration - 4.7L

3-34 ENGINE AND ENGINE OVERHAUL

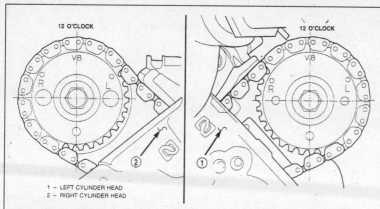

Fig. 77 After the damper mark is aligned, check that the cam sprocket "V8" marks are at 12 o'clock. If not, turn the crankshaft 360-degrees and recheck - 4.7L

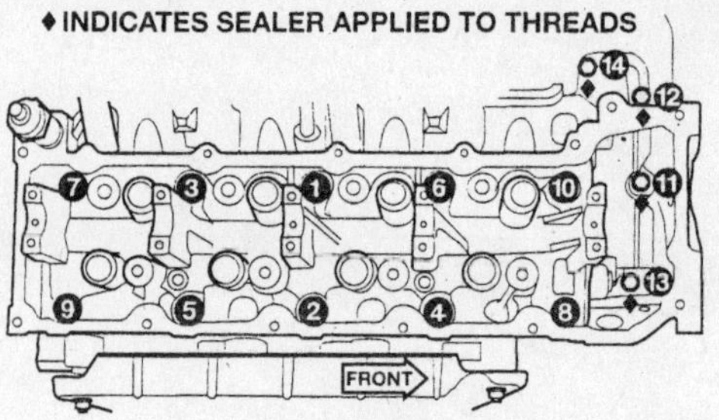

1 – LEFT CYLINDER HEAD
2 – RIGHT CYLINDER HEAD

Fig. 78 Apply thread sealer to bolts 11 - 14, oil the rest and tighten in the order shown in steps outlined in the text - 4.7L

14. Remove the cylinder head bolts, head and gasket.

Note: Each cylinder head is retained by ten 11mm bolts and four 8mm bolts. Do not neglect the four smaller bolts at the front of the cylinder head.

15. Repeat the procedure with the other cylinder head.

To install:

16. Check the cylinder head bolts for signs of stretching and replace as necessary.
17. Lubricate the threads of the 11mm bolts with clean engine oil.
18. Coat the threads of the 8mm bolts with thread locking compound.

Note: Refer to the illustration for the cylinder head torque sequence illustration.

19. Install the cylinder heads. Use new gaskets and tighten the bolts, in sequence, as follows:
    a. Step 1: Bolts 1 - 10 to 15 ft. lbs. (20 Nm)
    b. Step 2: Bolts 1 - 10 to 35 ft. lbs. (47 Nm)
    c. Step 3: Bolts 11 - 14 to 18 ft. lbs. (25 Nm)
    d. Step 4: Bolts 1 - 10 plus 1/4 (90-degrees) turn
    e. Step 5: Bolts 11 - 14 to 22 ft. lbs. (30 Nm)

20. Install the camshaft sprockets. Align the crankshaft with the TDC mark and the secondary chain matchmarks and tighten the bolts to 90 ft. lbs. (122 Nm).
21. Install the secondary timing chain guides, head access plugs and secondary chain tensioners. Refer to the timing chain procedure.
22. Remove the timing chain locking tool P/N 8515.
23. Tighten crankshaft damper to 130 ft. lbs. (175 Nm).

Diesel
See Figures 79, 80, 81 and 82

CAUTION:

A crane or other dedicated lifting device designed for the job is required to remove the cylinder head.

1. Carry out the "common steps", above.
2. Remove the turbocharger.
3. Remove the engine harness-to-cylinder head attaching bolt at the front of the head.
4. Remove the engine harness ground fastener at the front of the head below the thermostat housing.
5. Remove the throttle linkage cover. Remove the six Accelerator Pedal Position Sensor (APPS) assembly-to-cylinder head bracket bolts and secure the entire assembly out of the way.
6. Disconnect the APPS connector.
7. Remove the intake air grid heater wires from the grid heater.
8. Remove the oil level indicator tube attaching bolt from the air inlet housing.
9. Remove the charge air cooler-to-air inlet housing pipe.
10. Remove the air inlet housing and intake grid heater from the intake manifold cover.
11. Remove the engine lift bracket at the rear of the cylinder head.
12. Remove the high pressure fuel lines: see "Fuel System" for details.
13. Remove the fuel filter-to-injection pump low pressure line.

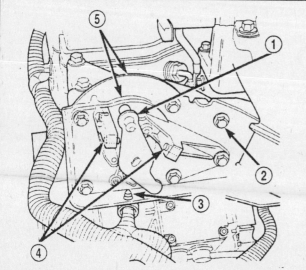

Fig. 79 APPS Assembly (5) showing lever (1), mounting bolts (2), wiring harness clip (3), and calibration screws (4)

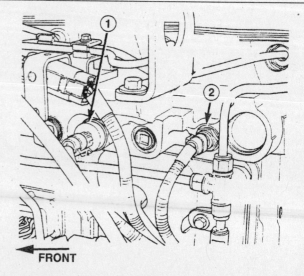

Fig. 80 MAP (1) and IAT (2) sensor locations

ENGINE AND ENGINE OVERHAUL 3-35

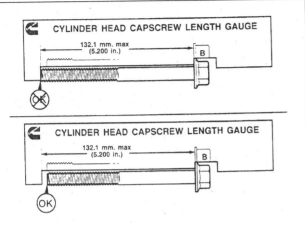

Fig. 81 Check head bolt stretch against the gauge, if included in the head gasket kit

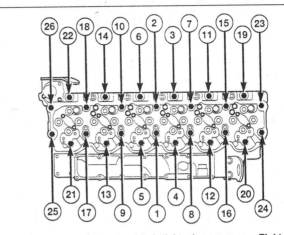

Fig. 82 Diesel cylinder head bolt tightening sequence. Tighten in steps: see Torque Specifications charts for details

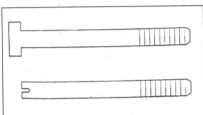

Fig. 83 Alignment dowels made from bolts will make pan installation easier. Diameter and TPI will depend on engine

14. Disconnect the water-in-fuel and fuel heater connectors.
15. Remove the fuel filter assembly-to-manifold cover bolts and remove the filter assembly from the truck.
16. Disconnect the IAT and MAP sensors.
17. Remove the cylinder head cover(s).
18. Remove the rocker arms and pushrods.
19. Remove the fuel return line banjo fitting at the rear of the cylinder head.
20. Reinstall the engine lift bracket.
21. Remove the 26 cylinder head bolts.
22. With a crane or other suitable lifting device, remove the cylinder head from the vehicle.

To install:

23. Reverse the removal procedure.
24. Cylinder head bolts should be replaced with new ones if they have been removed before or if stretched. A cylinder head capscrew length gauge may be included with replacement head gaskets.
25. Install head gasket with P/N up.
26. Tighten head bolts in the order shown. Tighten in steps with torque given in the specifications charts. Note that 12-valve and 24-valve engines differ in this regard.

Oil Pan

REMOVAL & INSTALLATION

Common Steps

See Figures 83 and 84

1. New pan gaskets must be used. Obtain these

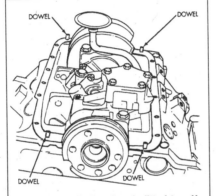

Fig. 84 Installation dowels fitted to a V-8 engine

before beginning.
2. Disconnect the negative battery cable(s).
3. Raise the vehicle for access.
4. Drain the engine oil.
5. Remove the engine oil dipstick.
6. When installing, clean all mating surfaces thoroughly. Use a quality adhesive sealant on gasket(s) and split lines.
7. Installation can be made easier by making alignment dowels from bolts. Diameter and TPI will depend on engine. Cut off the heads and cut in a screwdriver slot. Install one "dowel" in each corner of the pan mating surface and slide the pan over them when fitting.
8. Tighten pan bolts gradually and evenly.

2.5L Engine

See Figure 85

1. Perform "common steps", above.
2. Disconnect the exhaust pipe at the manifold.
3. Disconnect the exhaust hanger at the catalytic converter and lower the pipe.
4. Remove the starter motor.
5. Remove the flywheel/torque converter housing access cover.
6. Position a jack beneath the vibration damper, protecting the damper with a block of wood.
7. Remove the engine mount throughbolts.
8. Raise the engine until the pan can be removed.

9. Installation is the reverse of removal. Apply sealer/adhesive where shown. Tighten 1/4" bolts to 84 inch lbs (9.5 Nm); tighten 5/16" bolts to 132 inch lbs. (15 Nm).

RAM With V-6, V-8 Engines

1. Perform "common steps", above.
2. Remove the exhaust pipe.
3. Remove the left engine-to-transmission strut.
4. Loosen the right side engine support bracket cushion throughbolt nut and raise the engine slightly.
5. Remove the oil pan bolts and remove the pan by sliding it back and out.
6. Installation is the reverse of removal. Tighten bolts to 18 ft. lbs. (24 Nm).

2WD Dakota With 3.9L, 5.2L, 5.9L Engines

1. Perform "common steps", above.
2. Disconnect the distributor cap and position away from cowl.
3. Remove the exhaust pipe.
4. Remove the engine mount insulator throughbolts.
5. Place a block of wood and jack beneath the oil pan and raise it.
6. When the engine is high enough, place mount throughbolts in the engine mount attaching points on the frame brackets.
7. Lower the engine so the bottom of engine mounts rest on the replacement bolts placed in the engine mount frame brackets.
8. Remove the oil pan.
9. Installation is the reverse of removal. Tighten bolts to 18 ft. lbs. (24 Nm).

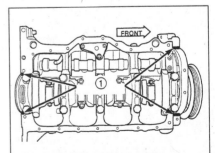

Fig. 85 On 2.5L engines, apply sealer/adhesive to locations shown

3-36 ENGINE AND ENGINE OVERHAUL

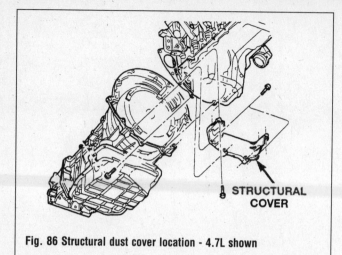

Fig. 86 Structural dust cover location - 4.7L shown

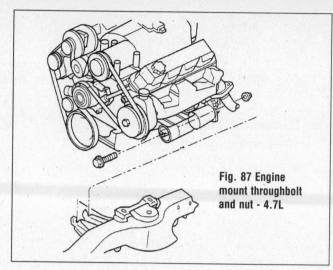

Fig. 87 Engine mount throughbolt and nut - 4.7L

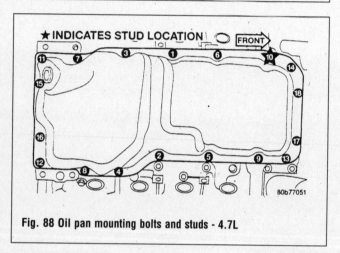

Fig. 88 Oil pan mounting bolts and studs - 4.7L

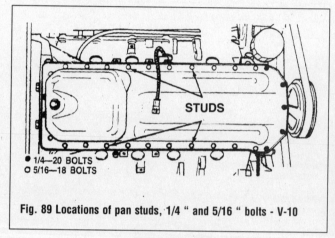

Fig. 89 Locations of pan studs, 1/4" and 5/16" bolts - V-10

Durango, 4WD Dakota With 3.9L, 5.2L, 5.9L Engines

1. Perform "common steps", above.
2. Remove the front axle.
3. Remove both engine mount support brackets.
4. Remove the transmission inspection cover.
5. Remove the oil pan.
6. Installation is the reverse of removal. Tighten pan bolts to 18 ft. lbs. (24 Nm).

4.7L With 2WD

See Figures 86, 87 and 88

1. Perform "common steps", above.
2. Drain the cooling system.
3. Remove the upper fan shroud.
4. Remove the throttle body resonator and air inlet hose.
5. Remove the intake manifold.
6. Disconnect the exhaust pipe at the manifold.
7. Remove the structural dust cover.
8. Support the engine.
9. Remove the left and right engine mount throughbolts.
10. Raise the engine for clearance to remove the oil pan. Note that the gasket will remain on the engine.
11. Installation is the reverse of removal. Tighten the pan bolts to 11 ft. lbs. (15 Nm).

4.7L With 4WD

See Figures 86 and 88

1. Perform "common steps", above.
2. Remove the front axle.
3. Remove the structural dust cover.
4. Remove the oil pan. Note that the gasket will remain on the engine.
5. Installation is the reverse of removal. Tighten pan bolts to 11 ft. lbs. (15 Nm).

8.0L Engine

See Figure 89

1. Perform "common steps", above.
2. Remove the left engine-to-transmission strut.
3. Remove the oil pan mounting bolts, pan and gasket. On 2WD vehicles, it may be necessary to raise the engine slightly.
4. Installation is the reverse of removal. Tighten 1/4" bolts to 96 inch lbs. (11 Nm), studs and 5/16" bolt to 144 inch lbs. (16 Nm).

Diesel

1. Perform "common steps", above.
2. Remove the transmission and transfer case, if fitted.
3. Remove the flywheel.
4. Disconnect the starter cables from the starter motor.
5. Remove the starter motor and transmission adapter plate assembly.
6. Remove the pan bolts. Break the seal with a plastic mallet or pry bar. Remove the pan.
7. Installation is the reverse of removal. Tighten pan bolts to 18 ft. lbs. (24 Nm).

Oil Pump

REMOVAL & INSTALLATION

2.5L Engine

See Figure 90

1. Remove the oil pan.
2. Remove the two pump attaching bolts. Remove the pump and gasket.

WARNING:

If the pump is not to be serviced, DO NOT disturb the position of the oil inlet tube and strainer assembly in the pump body. If the tube is moved within the pump body, a replacement tube and strainer assembly must be installed to ensure an airtight seal.

3. Installation is the reverse of removal. Use a new gasket and tighten the pump attaching bolts to 17 ft. lbs. (23 Nm).

ENGINE AND ENGINE OVERHAUL 3-37

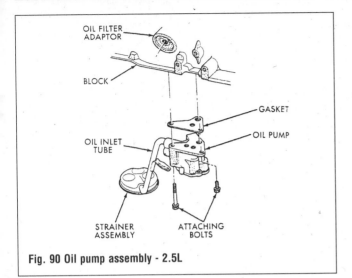

Fig. 90 Oil pump assembly - 2.5L

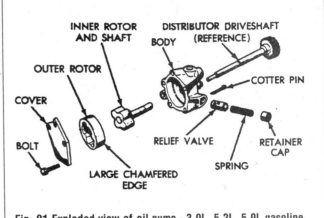

Fig. 91 Exploded view of oil pump - 3.9L, 5.2L, 5.9L gasoline engines

3.9L, 5.2L, 5.9L Gasoline Engines
See Figure 91

1. Remove the oil pan.
2. Remove the oil pump from the rear main bearing cap.

To install:

3. Install while slowly rotating the pump body to ensure drive shaft engagement.
4. Use a new gasket.
5. Hold the pump base flush against the mating surface on the main bearing cap and tighten the bolts to 30 ft. lbs. (41 Nm).

4.7L Engine
See Figure 92

1. Remove the oil pan and pickup tube.
2. Remove the timing chain cover.
3. Remove the timing chains and tensioners.
4. Remove the four bolts, primary timing chain tensioner and oil pump.

To install:

5. Position the oil pump onto the crankshaft and install two mounting bolts.
6. Position the primary timing chain tensioner and install the two retaining bolts.
7. Tighten the pump and tensioner retaining bolts to 21 ft. lbs. (28 Nm) in the sequence shown.
8. Install the secondary timing chain tensioners, chains, cover, pickup tube and oil pan.

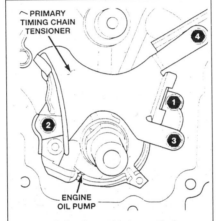

Fig. 92 Oil pump and primary timing chain tensioner tightening sequence - 4.7L

8.0L Engine
See Figure 93

1. Remove the timing chain cover.
2. Remove the mounting screws and the oil pump cover.

To install:

3. Lubricate pump internals with petroleum jelly to prime it.

4. Install the pump and tighten the pump cover to 10 ft. lbs. (14 Nm).
5. Make sure that the inner ring moves freely after the cover is installed.
6. Install the timing chain cover.
7. Squirt oil into the relief valve hole until oil runs out.
8. Install the relief valve and spring.
9. Using a new pressure relief valve gasket, install the relief valve plug. Tighten it to 15 ft. lbs. (20 Nm).
10. Remember to fill the filter with oil before installation.

Diesel Engines
See Figure 94

1. Disconnect the battery negative cables.
2. On 1997-99 models, remove the radiator.
3. Remove the oil fill tube, if necessary.
4. Remove the fan/drive assembly.
5. Remove the accessory drive belt.
6. Remove the fan support/hub assembly.
7. Remove the crankshaft damper.
8. Remove the gear cover-to-housing bolts and pry the cover away from the housing, taking care not to mar the gasket surface.
9. Remove the oil pump mounting bolts and remove the pump.

To install:

10. Fill the pump with clean engine oil.

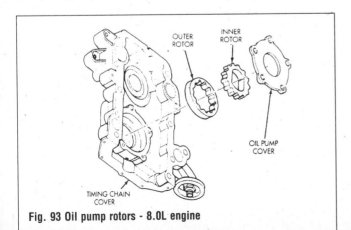

Fig. 93 Oil pump rotors - 8.0L engine

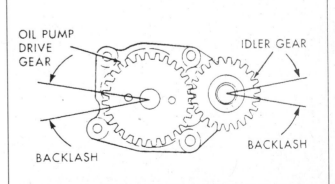

Fig. 94 Check backlash at each oil pump gear in turn - diesel

3-38 ENGINE AND ENGINE OVERHAUL

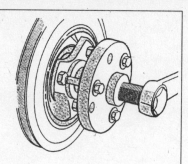

Fig. 95 Using the damper puller on the 2.5L engine

Fig. 96 The crankshaft pulley is attached to the damper by retaining bolts

Fig. 97 After unfastening the retaining bolts, remove the crankshaft pulley from the damper and set it aside

Fig. 98 Unfasten the damper retaining bolt and washer. To do so, it may be necessary to keep the crankshaft from turning

Fig. 99 Withdraw the bolt and washer that retain the damper to the crankshaft

Fig. 100 Attach a puller to the damper, then tighten the center bolt to press off the damper

11. Check that the idler gear pin is installed in the locating bore in the cylinder block.
12. Tighten the pump mounting bolts in two steps: first to 44 inch lbs. (5 Nm), then to 18 ft. lbs. (24 Nm).
13. The back plate on the pump seats against the bottom of the bore in the cylinder block. When the pump is correctly installed, the flange on the pump will not touch the cylinder block.
14. Check the idler gear to pump drive gear backlash and the idler gear to crankshaft gear backlash. Replace the pump assembly if it exceeds 0.003 - 0.013 in. (0.080 - 0.330mm).
15. Tighten cover bolts to 18 ft. lbs. (24 Nm). Tighten vibration damper bolts to 92 ft. lbs. (125 Nm).

Crankshaft Damper

REMOVAL & INSTALLATION

2.5L Engine

See Figure 95

Note: Crankshaft Vibration Damper Puller 7697 or equivalent is required.

1. Disconnect the negative battery cable.
2. Remove the accessory drivebelt.
3. Remove the fan shroud.
4. Remove the vibration damper bolt and washer.
5. Use the puller to remove the damper from the crankshaft.

Fig. 101 When sufficiently loosened, remove the damper from the crankshaft

To install:

6. Apply silicone sealer to the keyway and install the key.
7. Fit the damper, aligning key and keyway.
8. Tighten bolt to 80 ft. lbs. (108 Nm).

3.9L, 5.2L, 5.9L Gasoline Engines

See Figures 96, 97, 98, 99, 100, 101 and 102

Note: Crankshaft Vibration Damper Puller/Installer Kit, Tool C-3688 or equivalent is required.

1. Disconnect the negative battery cable.
2. Remove the fan shroud.
3. Remove the engine fan.
4. Remove the accessory drivebelt.
5. Remove the crankshaft pulley.
6. Remove the vibration damper bolt and washer from the crankshaft.

Fig. 102 Inspect the damper for wear on its front seal surface. If worn, be sure to replace it, or the engine may leak oil

7. Install the puller assembly on the vibration damper, and thread the forcing screw into the crankshaft. Install two bolts and washers through the puller and into the damper. Remove the vibration damper by turning the forcing screw.
8. When loosened, pull the damper off of the crankshaft

To install:

9. Using tool C-3688, or equivalent, press the damper onto the shaft. Remove the tool and install the damper bolt. Tighten the bolt to 135 ft. lbs. (183 Nm).
10. Install the crankshaft pulley. Tighten pulley bolts to 17 ft. lbs. (23 Nm). Use same torque for the fan bolts.
11. The remainder of the procedure is the reverse of removal.

ENGINE AND ENGINE OVERHAUL 3-39

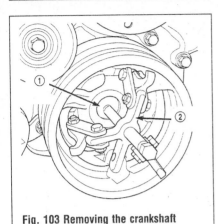

Fig. 103 Removing the crankshaft damper - typical

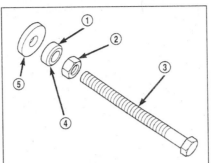

Fig. 104 Proper assembly of special tool 8512: bearing (1) with hardened surface (4) facing nut (2); threaded shaft (3) and hardened washer (5) - 4.7L engine

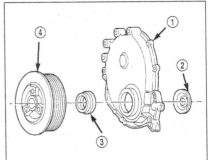

Fig. 105 Timing case cover components - 2.5L engine
1 Cover
2 Oil slinger
3 Oil seal
4 Crankshaft damper

4.7 L Engine
See Figures 103 and 104

Note: Special Tools 8512, 8513 and 1026, or equivalents, are required.

1. Disconnect the battery negative cable.
2. Remove the accessory drivebelt.
3. Drain the cooling system. Remove the upper radiator hose. Remove the upper fan shroud.
4. Remove the belt-driven fan.
5. Disconnect the electric fan connector.
6. Remove the crankshaft damper bolt.
7. Remove the damper with the special tools.

To install:

8. Clean the damper bore and crankshaft nose thoroughly before installation.
9. Align slot and key and install damper onto crankshaft.
10. Assemble special tool 8512 as follows: thread nut onto shaft, then bearing with hardened surface facing nut, then the hardened washer. Coat threads with anti-seize compound.
11. Press damper home with the special tool
12. Tighten damper bolt to 130 ft. lbs. (175 Nm).
13. The remainder of the procedure is the reverse of removal.

8.0L Engine
See Figure 103

Note: Special Tools 8513 and 1026, or equivalents, are required.

1. Disconnect the battery negative cable.

2. Remove the accessory drivebelt.
3. Remove the fan.
4. Drain the cooling system. Remove the radiator.
5. Remove the crankshaft pulley/damper bolt and washer from the crankshaft.
6. Using the special tool, remove the damper from the crankshaft.

To install:
7. Reverse the removal procedure.
8. Tighten the pulley/damper bolt to 230 ft. lbs. (312 Nm).

Diesel

1. Remove the accessory drivebelt.
2. Remove the four damper bolts. Remove the damper.
3. Tighten bolts to 92 ft. lbs. (125 Nm) when installing.

Timing Cover and Seal (Gasoline Engines)

REMOVAL & INSTALLATION

2.5L Engine
See Figure 105

1. Disconnect the negative battery cable.
2. Remove the accessory drivebelt.
3. Remove any accessory bracket attached to the timing case cover.
4. Remove the fan and hub assembly and fan shroud.

5. Remove the A/C compressor (if fitted) and alternator bracket from the head and place them to one side.
6. Remove the vibration damper.
7. Remove the oil pan-to-timing case cover bolts and timing case cover-to-cylinder block bolts.
8. Remove the timing case cover and gasket.
9. Pry out the oil seal if replacement is required.

To install:
10. The open end of the seal should face the engine.
11. A special alignment tool (6139) is available to ease assembly.
12. Tighten 1/4" cover-to-block bolts to 60 inch lbs. (7 Nm). Tighten 5/16" front cover-to-block bolts 192 inch lbs. (22 Nm). Tighten the oil pan-to-cover bolts to 84 inch lbs. (9.5 Nm).
13. The remainder of the procedure is the reverse of removal.

3.9L, 5.2L, 5.9L, 8.0L Gasoline Engines
See Figures 106, 107 and 108

1. Disconnect the negative battery cable.
2. Drain and recycle the engine coolant.
3. Remove the accessory drivebelt.
4. Remove the fan and shroud.
5. Remove the alternator. Unbolt the A/C compressor and place it to one side if required for access to the water pump.
6. Remove the air pump and bracket (8.0L engine)
7. Remove the water pump.
8. Remove the power steering pump.

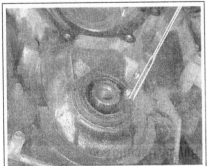

Fig. 106 You can use a seal puller, such as this one from Lisle®, to remove the front seal from the timing chain cover

Fig. 107 Make sure the seal area of the cover is clean and smooth before installing a new seal

Fig. 108 A large socket can be used if the proper tool is unavailable; however, this is not the recommended method

3-40 ENGINE AND ENGINE OVERHAUL

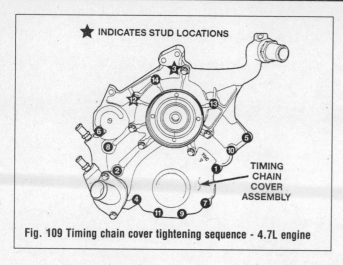

Fig. 109 Timing chain cover tightening sequence - 4.7L engine

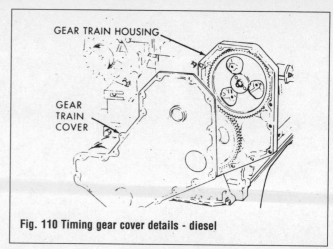

Fig. 110 Timing gear cover details - diesel

9. Remove the crankshaft pulley.
10. Remove the vibration damper.
11. If replacing the seal only: use a suitable tool to pry the seal from out of the timing chain cover. Use caution as to not damage the seal surface of the timing chain cover. Proceed to the installation steps.
12. Loosen the oil pan bolts and remove the two front oil pan bolts on each side of the crankshaft that retain the cover.
13. Remove the bolts that retain the timing chain cover to the engine block.
14. Remove the cover, using caution to avoid damaging the oil pan gasket.
15. Remove the seal from the timing chain cover using a suitable tool. Take care not to damage the seal surface of the timing cover.

To install:

16. The oil seal can be installed before or after installation of the cover. Use a suitable seal installation tool or large socket to press seal into place. Spring side goes towards the engine.
17. Using a new gasket, place the cover on the engine block.
18. Install the timing chain cover bolts to the engine block. Tighten to 30 ft. lbs. (41 Nm) on V-6 and V-8 engines, 35 ft. lbs. (47 Nm) on V-10 engines.
19. Install the front two oil pan bolts, tighten to 215 inch lbs. (24 Nm).
20. the reset of the procedure is the reverse of removal.

4.7L Engine

See Figure 109

1. Disconnect the negative battery cable.
2. Drain and recycle the engine coolant.
3. Remove the belt-driven fan and radiator shroud.
4. Disconnect both heater hoses at the timing cover.
5. Disconnect the lower radiator hose at the engine.
6. Remove the crankshaft damper.
7. Remove the accessory belt tensioner assembly.
8. Remove the alternator and A/C compressor.
9. Remove the timing chain cover and gasket.

To install:

10. Reverse the removal procedure.
11. Tighten cover hardware in order shown to 40 ft. lbs. (54 Nm).

12. Tighten accessory drive belt tensioner to 40 ft. lbs. (54 Nm).

Timing Gear Cover and Seal (Diesels)

REMOVAL & INSTALLATION

See Figure 110

This procedure applies only to diesel engines.
1. Disconnect the battery negative cable(s).
2. Drain the coolant from the block.
3. Remove the upper radiator hose.
4. Remove the coolant and windshield washer bottles from the fan shroud.
5. Remove the fan assembly (left-hand thread).
6. Remove the fan and shroud from the vehicle.
7. Remove the accessory belt.
8. Remove the fan support/hub from the engine.
9. Remove the crankshaft damper.
10. Remove the gear cover-to-housing bolts and pry the cover off.

To install:

11. Check seal condition and replace it if damaged. A pilot eases assembly.
12. Reverse the removal procedure.
13. Tighten cover bolts to 18 ft. lbs. (24 Nm).

Timing Chain and Sprockets

REMOVAL & INSTALLATION

2.5L Engine

See Figures 111, 112, 113 and 114

1. Remove the timing case cover. See procedures above.
2. Turn the engine until the camshaft sprocket mark and the crankshaft sprocket mark are as close together as possible and aligned as shown.
3. Remove the crankshaft oil slinger.
4. Remove the camshaft sprocket bolt.
5. Remove the sprockets and chain together.

To install:

6. Turn the chain tensioner lever to the unlocked (down) position.
7. Pull the tensioner block towards the tensioner lever to compress the spring. Hold the block and turn the tensioner lever to the lock position.
8. Fit the crankshaft key, holding it in place with silicone adhesive, if needed.
9. Install the sprockets and chain, ensuring that alignment is maintained.
10. Tighten the camshaft sprocket bolt to 80 ft. lbs. (108 Nm).
11. To verify alignment, turn the crankshaft to po-

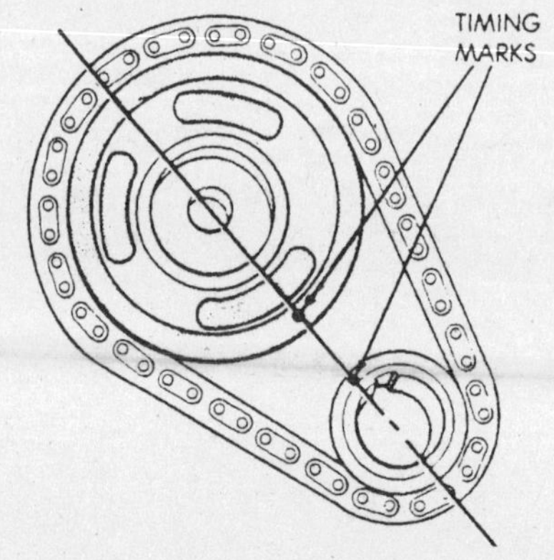

Fig. 111 Camshaft and crankshaft sprocket alignment marks must align for removal and installation - 2.5L engine

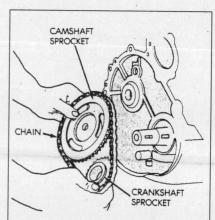

Fig. 112 Removing the sprocket and timing chain as an assembly - 2.5L engine

ENGINE AND ENGINE OVERHAUL 3-41

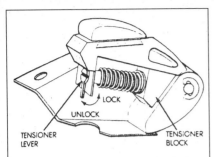

Fig. 113 Be sure to install and unlock the cam chain tensioner - 2.5L engine

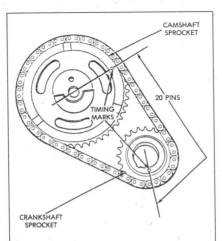

Fig. 114 Verify timing by positioning sprocket as shown and counting 20 pins between the marks - 2.5L engine

Fig. 115 View of the timing chain and sprockets, with the cover removed - 5.2L shown

sition the timing marks as shown in the illustration and count the number of chain pins between the marks. There must be 20 pins.

12. Unlock the chain tensioner lever.
13. Install the oil slinger.
14. The remainder of the installation procedure is the reverse of removal.

3.9L, 5.2L, 5.9L, 8.0L Engines
See Figures 115, 116 and 117

1. Remove the timing chain cover. See procedures above.
2. Using the vibration damper bolt, turn the engine until the camshaft sprocket mark is at 6 o'clock and the crankshaft sprocket mark is at 12 o'clock as shown in the illustration.
3. Remove the camshaft sprocket bolt; take off both sprockets with the timing chain. Sprocket puller may be required on the V-10 crankshaft sprocket.
4. If a tensioner is fitted, replace the crankshaft sprocket, Pry up between sprocket and tensioner with a large screwdriver until the hole in the tensioner shoe lines up with the hole in the bracket. Slip in a pin to keep the tensioner out of the way.

To install:

5. Align the sprocket marks as on removal.
6. Install the chain and sprockets together, ensuring that mark alignment is maintained after installation.
7. Remove the tensioner pin, if installed.
8. Tighten the camshaft sprocket bolt to 50 ft. lbs. (68 Nm) on V-6 and V-8; 45 ft. lbs. (61 Nm) on the V-10.

4.7L Engine
See Figures 118, 119, 120, 121 and 122

1. Remove the timing chain cover. See procedures above.
2. Remove cylinder head covers. See Procedures above.
3. Rotate engine to line up crankshaft damper mark and TDC mark. Check that camshaft sprocket V8 marks are at 12 o'clock. If not, rotate engine 360-degrees and recheck.
4. Compress and pin the primary chain tensioner.

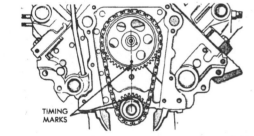

Fig. 116 Be sure to maintain proper alignment of the timing marks

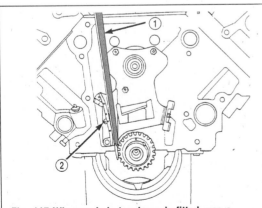

Fig. 117 When a chain tensioner is fitted, use a screwdriver (1) to pry back the tensioner and install a keeper pin (2)

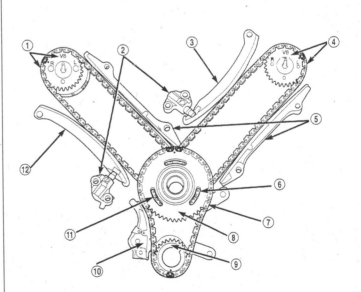

1 – RIGHT CAMSHAFT SPROCKET AND SECONDARY CHAIN
2 – SECONDARY TIMING CHAIN TENSIONER
3 – SECONDARY TENSIONER ARM
4 – LEFT CAMSHAFT SPROCKET AND SECONDARY CHAIN
5 – CHAIN GUIDE
6 – TWO PLATED LINKS ON RIGHT CAMSHAFT CHAIN
7 – PRIMARY CHAIN
8 – IDLER SPROCKET
9 – CRANKSHAFT SPROCKET
10 – PRIMARY CHAIN TENSIONER
11 – TWO PLATED LINKS ON LEFT CAMSHAFT CHAIN
12 – SECONDARY TENSIONER ARM

Fig. 118 Timing chain assembly - 4.7L engine

3-42 ENGINE AND ENGINE OVERHAUL

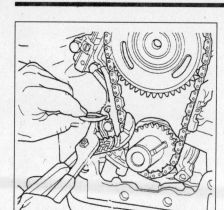

Fig. 119 Insert a pin to hold the primary chain tensioner - 4.7L engine

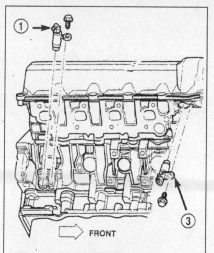

Fig. 120 Crankshaft position sensor (1) and camshaft position sensor (3) - 4.7L engine

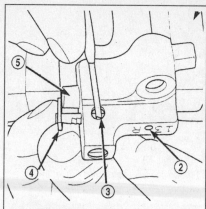

Fig. 121 Resetting the secondary chain tensioners: vise (1), insert lock pin (2), ratchet pawl (3), ratchet (4), piston (5) - 4.7L engine

5. Cover the oil pan opening.
6. remove the secondary chain tensioners.
7. Remove the camshaft position sensor from the right cylinder head.
8. Remove both cam sprocket bolts.
9. Hold the left camshaft steel tube with adjustable pliers and remove the sprocket. Slowly rotate the camshaft about 15-degrees CW to a neutral position.
10. Hold the right camshaft with adjustable pliers and remove the sprocket. Slowly rotate the camshaft about 45-degrees CCW to a neutral position.
11. Remove the idler sprocket assembly bolt.
12. Slide the idler sprocket assembly and crank sprocket forward at the same time to remove the primary and secondary chains.
13. Remove both pivoting tensioner arms and chain guides.

To install:

14. Compress the secondary chain tensioner piston until the piston step is flush with the tensioner body. Using a pin or suitable tool, release the ratchet pawl by pulling the pawl back against spring force through the access hole on the side of the tensioner. While holding the pawl back, push ratchet device to about 0.08 in. (2mm) from the tensioner body. Install special tool 8514 lock pin into hole on front of ten-

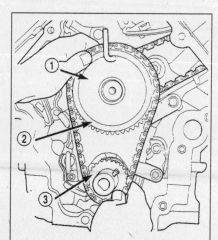

Fig. 122 Installing the timing chains: special tool 8515 (1), primary chain idler sprocket (2), crankshaft sprocket (3) - 4.7L engine

sioner. Slowly open vise to transfer piston spring force to lock pin.
15. Position primary chain tensioner over oil pump and insert bolts into lower two holes on tensioner bracket. Tighten bolts to 250 inch lbs. (28 Nm).
16. Install the right side chain tensioner arm. Apply thread locking compound to Torx, bolt and tighten to 150 inch lbs. (17 Nm).

Note: The silver bolts retain the guides to the cylinder heads and the black bolts retain the guides to the engine block.

17. Install the left side chain guide. Tighten the bolts to 250 inch lbs. (28 Nm).
18. Install the left side chain tensioner arm. Apply thread locking compound to the Torx, bolt and tighten to 150 inch lbs. (17 Nm).

WARNING:

Do not overtighten tensioner bolts. Tighten to specified torque only.

19. Install both secondary chains to idler sprocket. Align two plated links on the secondary chains to be visible through the two lower openings on the idler sprocket (4 o'clock and 8 o'clock). Once the secondary timing chains are installed, position special tool 8515 to hold the chains in place for installation.
20. Align the primary chain double plated links with the timing mark at 12 o'clock on the idler sprocket. Align the primary chain single plated link with the timing mark at 6 o'clock on the crankshaft sprocket.
21. Lube the idler shaft with clean engine oil.
22. Install all chains, crankshaft sprocket and idler sprocket as an assembly. After guiding both secondary chains through the block and cylinder head openings, tie up chains with rubber strap or equivalent to maintain tension.

Note: It is necessary to rotate the camshafts slightly for sprocket installation.

23. Align the left cam sprocket L dot with the plated link on the chain.

24. Align the right cam sprocket R dot with the plated link on the chain.
25. Remove excess oil from the cam sprocket bolts to ensure proper torque.
26. Remove special tool 8515. Attach both sprocket to camshafts. Install, but do not torque sprocket bolts.
27. Verify that all plated links are aligned with marks on all sprocket and that the V8 marks on the sprockets are at 12 o'clock.

WARNING:

Ensure the plate between the left secondary chain tensioner and block is correctly installed.

28. Install both secondary chain tensioners and tighten bolts to 250 inch lbs. (28 Nm).
29. Lubricate the idler sprocket bolt washer with oil and tighten bolt to 25 ft. lbs. (34 Nm).
30. Remove all locking pins (3) from tensioners. Do not manually extend the tensioner ratchets. This will over-tension the chains.
31. Prevent the engine from rotating and tighten the camshaft sprocket bolts to 90 ft. lbs. (122 Nm).
32. Rotate the engine two full revolutions. Verify alignment of all timing marks:
  • primary chain idler sprocket dot at 12 o'clock
  • primary chain crankshaft sprocket dot at 6 o'clock
  • secondary chain camshaft sprockets' V8 marks at 12 o'clock
33. Lube the chains with clean engine oil.
34. Check idler gear end play. It must be 0.004 - 0.010 in. (0.10 - 0.25 mm). If not within spec, replace the idler gear.
35. The remainder of the procedure is the reverse of removal.

INSPECTION

Timing Chain Slack (3.9L, 5.2L, 5.9L, 8.0L)

See Figure 123

1. If a chain tensioner is fitted, it must be removed before this check is carried out.
2. Position a scale (ruler or straightedge) next to the timing chain to detect any movement in the chain.

ENGINE AND ENGINE OVERHAUL

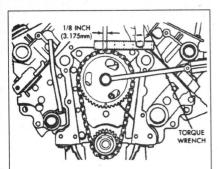

Fig. 123 Measuring timing chain slack with a torque wrench and ruler

3. Place a torque wrench and socket on the camshaft sprocket attaching bolt. Apply 30 ft. lbs. (41 Nm) of torque if the cylinder heads are installed on the engine, or 15 ft. lbs. (20 Nm) if the cylinder heads are removed, and rotate the bolt in the direction of crankshaft rotation in order to remove all slack from the chain.
4. While applying torque to the camshaft sprocket bolt, the crankshaft should not be allowed to rotate. It may be necessary to block the crankshaft to prevent rotation.
5. Position the scale over the edge of a timing chain link and apply an equal amount of torque in the opposite direction. If the movement of the chain exceeds 1/8 in. (3.175mm), replace the chain.
6. If camshaft end-play exceeds 0.010 in., install a new thrust plate. It should be 0.002 - 0.006 in. with the new plate.

Chain Guides (4.7L)

Check for grooves. Replace parts if grooves are more than 0.039 in. (1 mm) deep.

Camshaft, Bearings and Lifters

REMOVAL & INSTALLATION

2.5L, 3.9L, 5.2L, 5.9L, 8.0L

See Figures 124, 125, 126, 127, 128, 129, 130 and 131

1. Disconnect the negative battery cable.
2. Remove the fan, shrouds and radiator.
3. Unbolt the A/C compressor and place it aside.
4. Remove the intake manifold.

Fig. 127 Remove the aligning yoke for each lifter you are servicing, noting its location

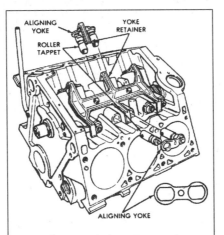

Fig. 124 Exploded view of a set of tappets (lifters), aligning yoke, and yoke retainer - 5.9L engine shown

5. Remove the valve covers.
6. Remove and label the rocker arms and pivots.
7. Remove and label the pushrods.
8. On 2.5L engines, remove the tappets.
9. On other models,
   a. Remove the yoke retainer and label the aligning yokes for each cylinder, keeping them in order to ensure installation in their original locations.
   b. Slide a hydraulic valve tappet remover/installer into the bore and remove each lifter keeping them in order to ensure the installation in their original locations.
   c. If the lifter or bore in the cylinder block is scored, ream the bore to the next oversize and replace with oversize lifters (leave this job to an automotive machine shop).
10. Remove the vibration damper, timing chain (or case) cover, timing chain and sprockets.
11. Remove the distributor and lift out the oil pump and distributor driveshaft. Mark position first to ease reassembly.
12. Remove the camshaft thrust plate, if fitted.
13. On the 2.5L engine, remove the camshaft.
14. On "V" engines, install a long bolt into the front of the camshaft and remove the cam with caution.
15. See camshaft bearing section if replacing camshaft and bearings.

To install:

16. If a new camshaft is fitted, all the tappets

Fig. 128 Inspect the lifter for wear after removing; also inspect the bore for wear

Fig. 125 Remove the bolts that hold the yoke retainer to the engine block - 5.2 engine shown

Fig. 126 Lift the yoke retainer off of the engine block

should be replaced as well.
17. Prior to installation, lubricate the camshaft lobes and bearings journals. It is recommended that 1 pt. of Crankcase Conditioner be added to the initial crankcase oil fill.
18. Insert the camshaft into the engine block within 2 in. (51mm) of its final position in the block.
19. Have an assistant support the camshaft with a suitable tool to prevent the camshaft from contacting the plug in the rear of the engine block.
20. Position the suitable tool against the rear side of the cam gear and be careful not to damage the cam lobes.
21. Replace the camshaft thrust plate, if fitted. If camshaft end-play exceeds 0.010 in. (0.254mm), install a new thrust plate. It should be 0.002 - 0.006 in. (0.051 - 0.152mm) with the new plate.
22. Install the timing chain and sprockets, timing chain cover, vibration damper, and pulley.

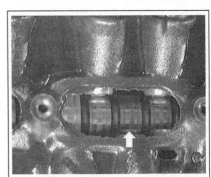

Fig. 129 The camshaft, as seen through the top of the engine block

3-44 ENGINE AND ENGINE OVERHAUL

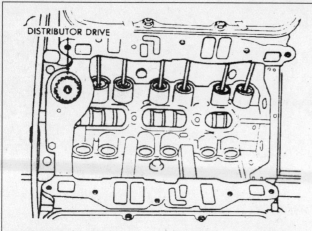

Fig. 130 The distributor drive gear is located in the distributor bore and is driven by the camshaft; the distributor must first be removed in order to access the drive gear

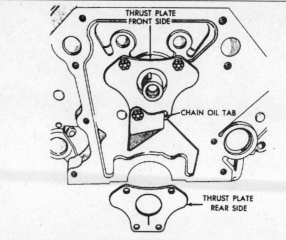

Fig. 131 The camshaft thrust plate is located on the front of the engine block, under the camshaft sprocket

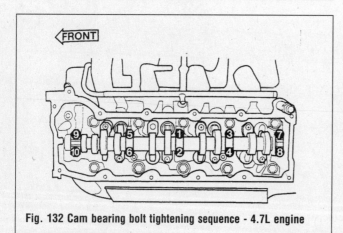

Fig. 132 Cam bearing bolt tightening sequence - 4.7L engine

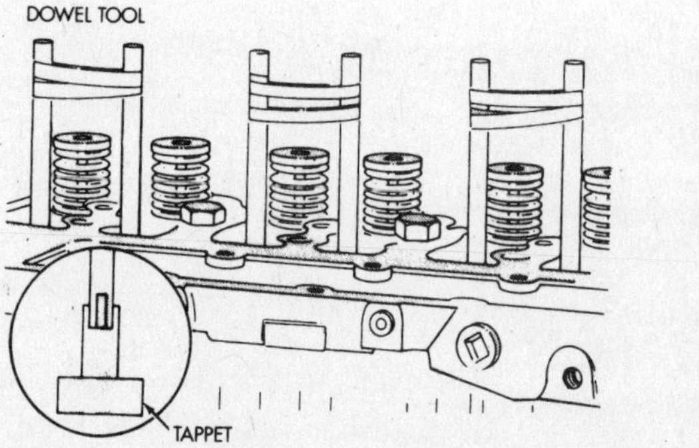

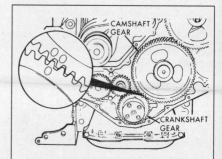

Fig. 133 Insert the tappet dowels to hold the tappets up securely - diesel engine

23. To install the lifters:
 a. Lubricate the lifters.
 b. Install the lifters in their original locations if installing the old ones.
 c. Ensure that the oil feed hole in the side of the lifter faces up (away from the crankshaft).
 d. Install the yokes and yoke retainer. Tighten the bolts to 200 inch lbs. (23 Nm).
24. The remainder of the procedure is the reverse of removal.

4.7L Engine
See Figure 132

WARNING:

Cams can be removed after the cylinder head covers are taken off if special tool 8350 (timing chain wedge) is available. If not, removal of the timing chain cover is required in order to reset chain tensioners.

1. Remove the cylinder head covers.
2. align the crankshaft damper mark with the TDC mark on the cover and both cam sprocket V8 marks to the 12 o'clock position.
3. Mark secondary timing chains relative to the sprocket.
4. Loosen but do not remove the cam sprocket bolts.

5. Use timing chain wedge 8350 between the timing chain strands to keep the tensioners from extending when the sprocket are removed.
6. Remove the sprocket. See procedures above.
7. Starting at the ends of the cam and working inward, loosen the cam bearing cap bolts 1/2 turn at a time until loose.
8. Mark the rocker arms for position.
9. Remove the cam bearing caps and camshaft.

To install:

10. Lube all parts with clean engine oil.
11. Install the left side camshaft so the sprocket dowel is near the 1 o'clock position.
12. Tighten the bearing cap bolts in 1/2 turn increments, starting in the middle and working out (see illustration). Final torque is 100 inch lbs. (11 Nm).
13. Assemble and install the timing chain sprocket assembly as outlined in the relevant section.

Diesel Engine
See Figures 133, 134 and 135

1. Disconnect the negative battery cable.
2. Remove the cylinder head covers.
3. Remove the rocker pedestal and arm assemblies.
4. Remove the pushrods.
5. Remove the drive belt.
6. Drain the cooling system. Remove the fan assembly, radiator and all related parts.

7. Remove the crankshaft pulley.
8. Remove the front gear cover.
9. Remove the fuel pump.
10. Insert the special dowels into the pushrod holes and onto the top of each lifter. When properly installed, the dowels can be used to hold the tappets up securely. Wrap rubber bands around the top of the dowels to prevent them from dropping down.
11. Rotate the crankshaft to align the crankshaft to camshaft timing marks.
12. Remove the bolts from the thrust plate.
13. Remove the camshaft and thrust plate.
14. Press the gear from the camshaft and remove the key.

Fig. 134 Align the crankshaft to the camshaft on the diesel engine as shown

ENGINE AND ENGINE OVERHAUL 3-45

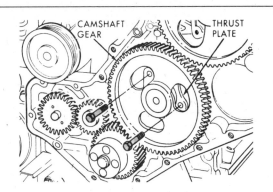

Fig. 135 Insert the thrust plate bolts on the diesel engine in the location shown

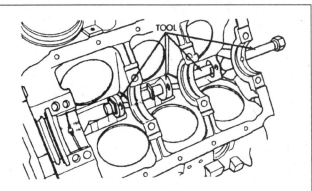

Fig. 136 Removal/installation of the camshaft bearings using a bearing installer/remover tool kit

To install:
15. Install the key on the camshaft.
16. Heat the camshaft gear to 250-degrees F (121-degrees C) for 45 minutes. Lubricate the gear mount surface with Lubriplate 105,. Install the gear to the camshaft with the timing marks facing away from the shaft.
17. Lubricate the camshaft bores, lobes, journals and thrust washer with Lubriplate 105,.

Note: Do not push the camshaft in too far or it may dislodge the plug in the rear of the camshaft bore, possibly creating a leak.

18. Install the camshaft and thrust washer so the ""E" timing mark on the injection pump gear aligns with the "C" timing mark on the camshaft gear and the timing mark on the crankshaft gear align with those on the camshaft gear.
19. Install the thrust washer bolts and tighten to 18 ft. lbs. (24 Nm).
20. Check the end-play of the camshaft. The specification is 0.006-0.010 in. (0.152-0.254mm).
21. Check the backlash of the camshaft gear. The specification is 0.003-0.013 in. (0.080-0.330mm).
22. Install the tappets and pushrods.
23. Install the rocker pedestal and arm assemblies.
24. Install the front cover and crankshaft pulley.
25. Install the drive belt and fan assembly.
26. Install the fuel pump.
27. Adjust the valves.
28. Install the cylinder head covers.
29. Connect the negative battery cable and check for leaks.

INSPECTION

Camshaft Lobe Lift (3.9L, 5.2L, 5.9L Engines)

Note: Check the lift of each lobe in consecutive order and make a note of the reading.

1. Disconnect the negative battery cable.
2. Remove the engine cover.
3. Remove the valve cover(s).
4. Remove the intake manifold.
5. Remove the rocker arm stud nut or fulcrum bolts, fulcrum seat and rocker arm.
6. Make sure the pushrod is in the valve tappet socket. Install a dial indicator so that the actuating point of the indicator is in the push rod socket (or the indicator ball socket adapter is on the end of the pushrod) and is in the same plane as the pushrod movement.
7. Disconnect the I terminal and the S terminal at the starter relay. Install an auxiliary starter switch between the battery and S terminals of the start relay.
8. Crank the engine with the ignition switch off. Turn the crankshaft over until the tappet is on the base circle of the camshaft lobe. At this position, the push rod will be in its lowest position.
9. Zero the dial indicator. Continue to rotate the crankshaft slowly until the push rod is in the fully raised position.
10. Compare the total lift recorded on the dial indicator with the specification shown on the Camshaft Specification chart.
11. To check the accuracy of the original indicator reading, continue to rotate the crankshaft until the indicator reads zero. If the left on any lobe is below specified wear limits listed, the camshaft and the valve lifter operating on the worn lobe(s) must be replaced.
12. Remove the dial indicator and auxiliary starter switch.
13. Install the rocker arm, fulcrum seat and stud nut or fulcrum bolts. Check the valve clearance. Adjust if required.
14. Install the intake manifold.
15. Install the valve cover(s).
16. Install the engine cover.
17. Connect the negative battery cable.

Camshaft Bearings
See Figure 136

1. Remove the engine.
2. Remove the camshaft, flywheel and crankshaft.
3. Push the pistons to the top of the cylinder.
4. Remove the camshaft rear bearing bore plug.
5. Remove the camshaft bearings with a bearing removal tool.
   a. Select the proper size expanding collet and back-up nut and assemble on the mandrel. With the expanding collet collapsed, install the collet assembly in the camshaft bearing and tighten the back-up nut on the expanding mandrel until the collet fits the camshaft bearing.
   b. Assemble the puller screw and extension (if necessary) and install on the expanding mandrel. Wrap a cloth around the threads of the puller screw to protect the front bearing or journal.
   c. Tighten the pulling nut against the thrust bearing and pulling plate to remove the camshaft bearing. Be sure to hold a wrench on the end of the puller screw to prevent it from turning.
6. To remove the front bearing, install the puller from the rear of the cylinder block.

To install:
7. Position the new bearings at the bearing bores, and press them in place. Be sure to center the pulling plate and puller screw to avoid damage to the bearing. Failure to use the correct expanding collet can cause severe bearing damage. Align the oil holes in the bearings with the oil holes in the cylinder block before pressing bearings into place.
8. Install the camshaft rear bearing bore plug.
9. Install the camshaft, crankshaft, flywheel and related parts.
10. Install the engine.

Rear Main Seal

REMOVAL & INSTALLATION

CAUTION:

The EPA warns that prolonged contact with used engine oil may cause a number of skin disorders, including cancer! You should make every effort to minimize your exposure to used engine oil. Protective gloves should be worn when changing the oil. Wash your hands and any other exposed skin areas as soon as possible after exposure to used engine oil. Soap and water, or waterless hand cleaner should be used.

2.5L Engine

1. Disconnect the negative battery cable.
2. Raise and support the vehicle.
3. Drain the engine oil.
4. Remove the flywheel or converter driveplate. Discard the old bolts.
5. Pry out the seal from around the crankshaft flange, taking care not to scratch the crank.

To install:
6. Clean the contact areas.
7. Coat the outer lip of the replacement seal with clean oil.
8. Position the seal and drive the seal in until flush with the engine block.

3-46 ENGINE AND ENGINE OVERHAUL

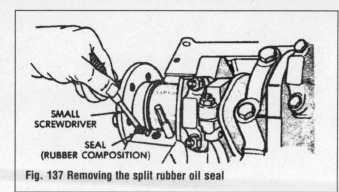

Fig. 137 Removing the split rubber oil seal

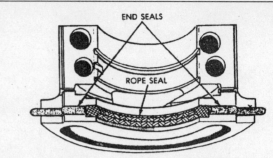

Fig. 138 Assembled view of the rear main bearing cap and seals

WARNING:

The felt lip must be located inside the flywheel mounting surface.

9. Use new bolts on the flywheel or converter plate. Tighten to 50 ft. lbs. (68 Nm).

3.9L, 5.2L, 5.9L Gasoline Engines
See Figures 137, 138 and 139

1. Disconnect the negative battery cable.
2. Raise and support the vehicle.
3. Drain the engine oil.
4. Remove the oil pan.
5. Remove the oil pump.
6. Loosen all the main bearing cap bolts, thereby lowering the crankshaft slightly but not to exceed 1/32 in. (0.8mm).
7. Remove the rear main bearing cap, and remove the oil seal from the bearing cap and cylinder block. On the block half of the seal use a seal removal tool, or install a small metal screw in one end of the seal, and pull on the screw to remove the seal. Exercise caution to prevent scratching or damaging the crankshaft seal surfaces.
8. Remove the oil seal retaining pin from the bearing cap (if equipped). The pin is not used with the split-lip seal.

To install:

9. Thoroughly clean the seal groove in the cap and block.
10. Dip the split lip-type seal halves in clean engine oil.
11. If the replacement seals are marked with white paint, the marks must face the REAR of the engine.

12. Carefully install the upper seal (cylinder block) into its groove with undercut side of the seal (rubber type) toward the FRONT of the engine, by rotating it on the seal journal of the crankshaft until approximately 3/8 in. (9.5mm) protrudes below the parting surface. On rope type, pull into position with the seal installing tool. Be sure no rubber has been shaved from the outside diameter of the seal by the bottom edge of the groove. Do not allow oil to get on the sealer area.
13. Tighten the main bearing cap bolts.
14. On the rope type seal, trim the upper seal ends flush with the block surface. Install the lower seal (rubber type) in the rear main bearing cap under undercut side of seal toward the FRONT of the engine, allow the seal to protrude approximately 3/8 in. (9.5mm) above the parting surface to mate with the upper seal when the cap is installed. With rope type seals, press the seal full and firmly into the cap groove. Trim the ends flush with the cap.
15. Install the side seals into the bearing cap. Apply an even 1/16 in. (1.6mm) bead of RTV silicone sealer at the bearing cap to block joint to provide oil pan end sealing.

Note: This sealer sets up in 15 minutes.

16. Install the rear main bearing cap.
17. Tighten the cap bolts to 85 ft. lbs. (115 Nm).
18. Install the oil pump.
19. Install the oil pan.
20. Refill the engine oil.

WARNING:

Operating the engine without the proper amount and type of engine oil will result in severe engine damage.

21. Start the engine and check for leaks.
22. Turn the engine off, recheck the engine oil.
23. Lower the vehicle.

4.7L Engine

1. Remove the transmission.
2. Remove the driveplate.
3. Pry out the old seal.
4. To install, drive a new seal in with a suitable driver.

8.0L Engine
See Figure 140

1. Remove the transmission.
2. Remove the oil pan.
3. Remove the rear seal retainer. Discard the oil seal and gasket.

To install:

4. Wash all parts in a safe solvent and inspect carefully for wear or damage.
5. Position the new seal in the retainer.
6. Position the retainer and oil seal over the crankshaft using special tool No. 6687, or equivalent. Install the bolts, tightening them to 16 ft. lbs. (22 Nm).
7. Check that the seal surface is within 0.020 in. (0.508mm) full indicator movement relative to the rear face of the crankshaft. If it is out of limits, use a soft face hammer to gently tap the high side into the retainer.
8. Seal the split line with a small amount of a suitable silicon rubber adhesive sealant.
9. Install the oil pan.
10. Install the transmission.

Diesel Engine
See Figure 141

The rear crankshaft seal is mounted in a housing that is bolted to the rear of the block. A double-lipped

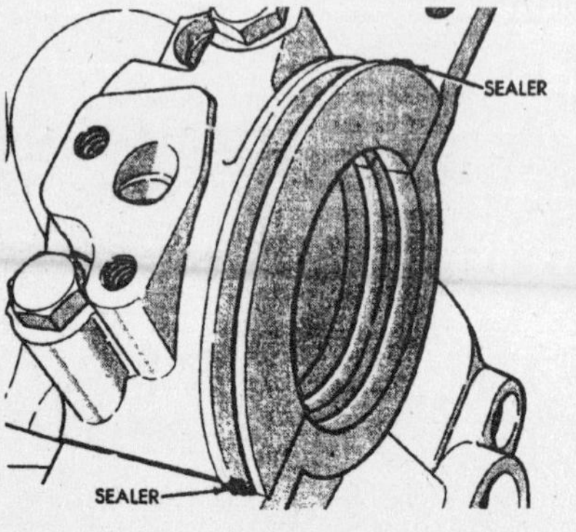

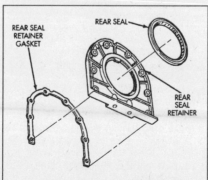

Fig. 140 Exploded view of the rear crankshaft main seal assembly - V10 engine

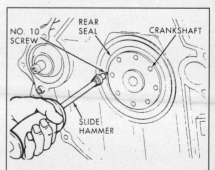

Fig. 141 Use a slide hammer to remove the rear main oil seal - diesel engine

Fig. 139 Oil pan end main bearing cap sealer points

ENGINE AND ENGINE OVERHAUL 3-47

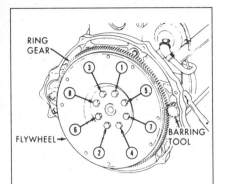

Fig. 142 Flywheel mounting bolt loosening/tightening sequence (diesel shown, others similar)

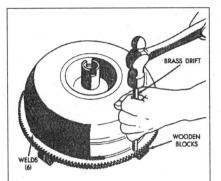

Fig. 143 Removing the starter ring gear with a cold chisel

Teflon, seal is used.
1. Disconnect the negative battery cable.
2. Remove the transmission.
3. Remove the clutch cover and plate, if equipped.
4. Remove the flywheel/driveplate.
5. Drill two 0.118 in. (3mm) holes 180-degrees apart into the seals. Be extremely careful not to drill against the crankshaft.
6. Insert two No. 10 sheet metal screws and, using a slide hammer, remove the rear seal.

To install:
7. Thoroughly clean and dry the crankshaft surface. Do not oil the crankshaft or seal prior to installation or the seal will leak.
8. Install the seal pilot included in the replacement seal kit, on the crankshaft. Push the seal on the pilot and crankshaft. Remove the pilot.
9. If the new seal has a rubber outer diameter, lubricate it with soapy water. If the seal does not have a rubber outer diameter, use Loctite, 277, or equivalent on the outer diameter.
10. Use the alignment tool to install the seal to the proper depth in the housing. Drive the seal in gradually and evenly until the alignment tool stops against the housing.

Flywheel/Driveplate

REMOVAL & INSTALLATION

See Figures 142 and 143

Note: Driveplate is the term for a flywheel mated with an automatic transmission.

The flywheel serves as the forward clutch engagement surface. It also serves as the ring gear with which the starter pinion engages to crank the engine. The most common reasons to replace the flywheel are:
- Broken teeth on the flywheel ring gear
- Excessive driveline chatter when engaging the clutch
- Excessive wear, scoring or cracking of the clutch surface

3.9L, 5.2L, 5.9L Gasoline Engines

Note: The ring gear is replaceable only on engines mated with a manual transmission, or automatic transmissions with a non lock-up torque converter. On engines with an automatic transmission and lock-up torque converter, the ring gear is not replaceable, so you must replace the torque converter assembly.

1. Disconnect the negative battery cable.
2. Raise and support the vehicle.
3. Remove the transmission (see Chapter 7).
4. If your vehicle has a manual transmission:
   a. Remove the clutch assembly.

Note: The bolts should be loosened a little at a time in a crisscross pattern to avoid warping the pressure plate.

   b. Remove the flywheel bolts.

Note: The flywheel bolts should be loosened a little at a time in a crisscross pattern to avoid warping the flywheel.

   c. Replace the pilot bearing in the end of the crankshaft after removing the flywheel.
   d. The flywheel should be checked for cracks and glazing. It can be resurfaced by a machine shop.

5. If the ring gear is to be replaced, drill a hole in the gear between two teeth, being careful not to contact the flywheel surface. Using a cold chisel at this point, crack the ring gear and remove it.
Polish the inner surface of the new ring gear and heat it in an oven to about 600-degrees F (316-degrees C). Quickly place the ring gear on the flywheel and tap it into place, making sure that it is fully seated.

WARNING:

Never heat the ring gear past 800-degrees F (426-degrees C), or the tempering will be destroyed.

6. If your vehicle has an automatic transmission:
   a. Remove the torque converter from the transmission. This is done by removing the c-clamp from the edge of the bell-housing, and carefully sliding the torque converter off the input shaft of the transmission.

CAUTION:

The torque converter is usually rather heavy, be prepared for the weight when the converter comes off of the input shaft.

   b. The converter is mated to the engine by a driveplate. To remove the driveplate, unscrew the retaining bolts from the crankshaft.

To install:
7. If your vehicle has a manual transmission:
   a. Position the flywheel on the end of the crankshaft. Tighten the bolts a little at a time, in a cross pattern, to 55 ft. lbs. (75 Nm).
   b. Install the clutch assembly.
8. If your vehicle has an automatic transmission:
   a. Install the torque converter on the input shaft. Turn the converter while pushing in to make sure it engages completely.
   b. Install the c-clamp that retains the converter to the bell housing.
9. Install the transmission.
10. Lower the vehicle.
11. Connect the negative battery cable.
12. Check the transmission fluid level.

4.7L Engine

1. Remove the transmission.
2. Remove the bolts and driveplate.

To install:
3. Tighten bolts in a cross pattern to 45 ft. lbs. (60 Nm).

EXHAUST SYSTEM

See Figures 144, 145, 146, 147, 148, 149, 150, 151, 152, 153 and 154

CAUTION:

Safety glasses should be worn at all times when working on or near the exhaust system.

Older exhaust systems will almost always be covered with loose rust particles which will become airborne when disturbed. These particles are more than a nuisance - they could injure your eyes.

n working on the exhaust system always keep the following in mind:

- Check the complete exhaust system for open seams, holes, loose connections, or other deterioration which could permit exhaust fumes to seep into the passenger compartment.
- The exhaust system may be supported by free-hanging rubber mounts which permit some movement of the exhaust system, but block the transfer of noise and vibration into the passenger com-

3-48 ENGINE AND ENGINE OVERHAUL

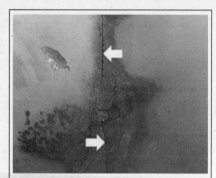

Fig. 144 Cracks in the muffler are a guaranteed leak

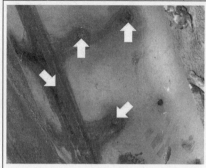

Fig. 145 Check the muffler for rotted spot welds and seams

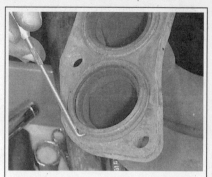

Fig. 146 Clean around the edges of the flange gasket with a sharp pick to help with the removal of the sealing ring

partment. Do not replace rubber mounts with solid ones.
- Before removing any component of the exhaust system, ALWAYS squirt a penetrating lubricant or rust dissolving agent onto the fasteners for ease of removal. A lot of knuckle skin will be saved by following this rule.

CAUTION:

Because some penetrating fluids are flammable, never use them on a hot exhaust system or near an open flame.

- Annoying rattles and noise vibrations in the exhaust system are usually caused by misalignment of the parts. When aligning the system, leave mounting hardware loose until all parts are properly aligned, then tighten, working from front to rear.
- When installing exhaust system parts, make sure there is enough clearance between the hot exhaust system components and other pipes and hoses that would be adversely affected by excessive heat. Also make sure there is adequate clearance from the floor pan to avoid possible overheating of the floor.

Safety Precautions

For a number of reasons, exhaust system work can be among the most dangerous type of work you can do on your truck. Always observe the following precautions:
- Support the truck extra securely. Not only will you often be working directly under it, but you'll frequently be using a lot of force, say, heavy hammer blows, to dislodge rusted parts. This can cause a truck that's improperly supported to shift and possibly fall.
- Wear goggles. Exhaust system parts are always rusty. Metal chips can be dislodged, even when you're only turning rusted bolts. Attempting to pry pipes apart with a chisel makes the chips fly even more frequently.
- If you're using a cutting torch, keep it a great distance from either the fuel tank or lines. Stop what you're doing and feel the temperature of the fuel bearing pipes on the tank frequently. Even slight heat can expand and/or vaporize fuel, resulting in accumulated vapor, or even a liquid leak, near your torch.
- Watch where your hammer blows fall and make sure you hit squarely. You could easily tap a brake or fuel line when you hit an exhaust system part with a glancing blow. Inspect all lines and hoses in the area where you've been working.

CAUTION:

Be very careful when working on or near the catalytic converter. External temperatures can reach 1,500-degrees F (816-degrees C) and more, capable of causing severe burns on contact. Headers also get extremely hot in use and require adequate cool-down time. Removal or installation should be performed only on a cold exhaust system.

Component Replacement

A number of special exhaust system tools can be rented from auto supply houses or local stores that rent special equipment. A common one is a tail pipe expander, designed to enable you to join pipes of identical diameter

System components may be welded or clamped together. The system consists of a head pipe, catalytic converter, intermediate pipe, muffler and tail pipe, in that order from the engine to the back of the truck. Some extensions may be fitted between major components to compensate for variations in wheelbase, transmissions, or other chassis obstructions.

The head pipe is bolted to the exhaust manifold, on one end, and the catalytic converter on the other. Various hangers suspend the system from the floor pan. When assembling exhaust system parts, the relative clearances around all system parts is extremely critical. Observe all clearances during assembly.

In the event that the system is welded, the various parts will have to be cut apart for removal. In these cases, the cut parts may not be reused. To cut the parts, a hacksaw is the best choice. An oxy-acetylene cutting torch may be faster but the sparks are DANGEROUS near the fuel tank, and, at the very least, accidents could happen, resulting in damage to other under-truck parts, not to mention yourself!

The following replacement steps relate to clamped parts:

1. Raise and support the truck on jackstands. It's much easier on you if you can get the truck up on four jackstands. Some pipes need lots of clearance for removal and installation
2. Remove the nuts from the U-bolts. Don't be surprised if the U-bolts break while removing the nuts. Age and rust account for this. Besides, you shouldn't reuse old U-bolts. When unbolting the headpipe from the exhaust manifold, make sure that the bolts are free before trying to remove them. If you snap a stud in the exhaust manifold, the stud will have to be removed with a bolt extractor, which often necessitates the removal of the manifold itself.

3. After the clamps are removed from the joints, first twist the parts at the joints to break loose rust and scale, then pull the components apart with a twisting motion. If the parts twist freely but won't pull apart, check the joint. The clamp may have been installed so tightly that it has caused a slight crushing of the joint. In this event, the best thing to do is secure a chisel designed for the purpose and, using the chisel and a hammer, peel back the female pipe end until the parts are freed.

4. Once the parts are freed, check the condition of the pipes which you had intended keeping. If their condition is at all in doubt, replace them too. You went to a lot of work to get one or more components out. You don't want to have to go through that again in the near future. If you are retaining a pipe, check the pipe end. If it was crushed by a clamp, it can be restored to its original diameter using a pipe expander.

Check the condition of the exhaust system hangers. If ANY deterioration is noted, replace them.

Use only parts designed for your truck. Don't use "universal" parts or flex pipes. "Universal" parts rarely fit like originals and flex pipes don't last very long.

5. When installing the new parts, coat the pipe ends with high temperature lubricant. It makes fitting the parts much easier. It's also a good idea to assemble all the parts in position before clamping them. This will ensure a good fit, detect any problems and allow you to check all clearances between the parts and surrounding frame and floor members.

6. When you are satisfied with all fits and clearances, install the clamps. The headpipe-to-manifold nuts should be torqued to 20 ft. lbs. (27 Nm). If the studs were rusty, wire-brush them clean, and spray them with a lubricant. This will ensure a proper torque reading. Position the clamps on the slip points. The slits in the female pipe ends should be under the U-bolt, not under the clamp. Tighten the U-bolt nuts securely, without crushing the pipe. The pipe fit should be tight, so that you can't swivel the pipe by hand. Don't forget: always use new clamps.

7. When the system is tight, recheck all clearances. Start the engine and check the joints for leaks. A leak can be felt by hand.

CAUTION:

MAKE CERTAIN THAT THE TRUCK IS SECURE BEFORE GETTING UNDER IT WITH THE ENGINE RUNNING!!

ENGINE AND ENGINE OVERHAUL 3-49

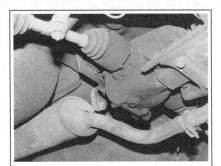
Fig. 147 Make sure the exhaust components are not contacting the body or suspension

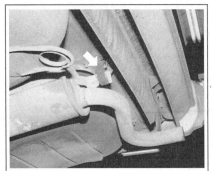
Fig. 148 Check for overstretched or torn exhaust hangers

Fig. 149 Example of a badly deteriorated exhaust pipe

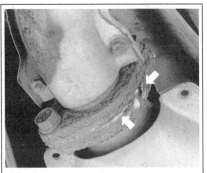
Fig. 150 Inspect flanges for gaskets that have deteriorated and need replacement

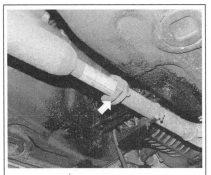
Fig. 151 Some systems, like this one, use large O-rings (doughnuts) in between the flanges

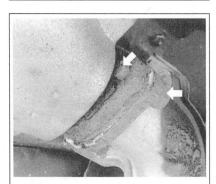
Fig. 152 Nuts and bolts will be extremely difficult to remove when deteriorated with rust

8. If any leaks are detected, tighten the clamp until the leak stops. If the pipe starts to deform before the leak stops, reposition the clamp and tighten it. If that still doesn't stop the leak, it may be that you don't have enough overlap on the pipe fit. Shut off the engine and try pushing the pipe together further. Be careful; the pipe gets hot quickly.

9. When everything is tight and secure, lower the truck and take it for a road test. Make sure there are no unusual sounds or vibration. Most new pipes are coated with a preservative, so the system will be pretty smelly for a day or two while the coating burns off.

Fig. 153 Example of a flange type exhaust system joint

Fig. 154 Example of a common slip joint type system

ENGINE RECONDITIONING

Determining Engine Condition

Anything that generates heat and/or friction will eventually burn or wear out (for example, a light bulb generates heat, therefore its life span is limited). With this in mind, a running engine generates tremendous amounts of both; friction is encountered by the moving and rotating parts inside the engine and heat is created by friction and combustion of the fuel. However, the engine has systems designed to help reduce the effects of heat and friction and provide added longevity. The oiling system reduces the amount of friction encountered by the moving parts inside the engine, while the cooling system dissipates the heat created by friction and combustion. If either system is not maintained, a break-down will be inevitable. Therefore, you can see how regular maintenance can affect the service life of your vehicle. If you do not drain, flush and refill your cooling system at the proper intervals, deposits will begin to accumulate in the radiator, thereby reducing the amount of heat it can extract from the coolant. The same applies to your oil and filter; if oil is not changed often enough it becomes laden with contaminates and is unable to properly lubricate the engine. This increases friction and wear.

There are a number of methods for evaluating the condition of your engine. A compression test can reveal the condition of your pistons, piston rings, cylinder bores, head gasket(s), valves and valve seats. An oil pressure test can warn you of possible engine bearing, or oil pump failures. Excessive oil consumption, evidence of oil in the engine air intake area and/or bluish smoke from the tailpipe may indicate worn piston rings, worn valve guides and/or valve seals. As a general rule, an engine that uses no more than one quart of oil every 1000 miles is in good condition. Engines that use one quart of oil or more in less than 1000 miles should first be checked for oil leaks. If any oil leaks are present, have them fixed before determining how much oil is consumed by the engine, especially if blue smoke is not visible at the tailpipe.

Fig. 155 A screw-in type compression gauge is more accurate and easier to use without an assistant

COMPRESSION TEST

A noticeable lack of engine power, excessive oil consumption and/or poor fuel mileage measured over an extended period are all indicators of internal engine wear. Worn piston rings, scored or worn cylinder bores, blown head gaskets, sticking or burnt valves, and worn valve seats are all possible culprits. A check of each cylinder's compression will help locate the problem.

Gasoline Engines
See Figure 155

Note: A screw-in type compression gauge is more accurate than the type you simply hold against the spark plug hole. Although it takes slightly longer to use, it's worth the effort to obtain a more accurate reading. Moreover, the limited space in some engine compartments may make hold-in gauges impossible to use in some plug holes.

1. Make sure that the proper amount and viscosity of engine oil is in the crankcase, then ensure the battery is fully charged.
2. Warm-up the engine to normal operating temperature, then shut the engine OFF.
3. Disable the ignition system.
4. Label and disconnect all of the spark plug wires from the plugs. On the 4.7L engine, remove the coils.
5. Thoroughly clean the cylinder head area around the spark plug ports, then remove the spark plugs.
6. Set the throttle plate to the fully open (wide-open throttle) position. You can block the accelerator linkage open for this, or you can have an assistant fully depress the accelerator pedal.
7. Install a screw-in type compression gauge into the No. 1 spark plug hole until the fitting is snug.

WARNING:

Be careful not to crossthread the spark plug hole.

8. According to the tool manufacturer's instructions, connect a remote starting switch to the starting circuit.
9. With the ignition switch in the OFF<B/> position, use the remote starting switch to crank the engine through at least five compression strokes (approximately 5 seconds of cranking) and record the highest reading on the gauge.
10. Repeat the test on each cylinder, cranking the engine approximately the same number of compression strokes and/or time as the first.
11. Compare the highest readings from each cylinder to that of the others. The indicated compression pressures are considered within specifications if the lowest reading cylinder is within 75 percent of the pressure recorded for the highest reading cylinder. For example, if your highest reading cylinder pressure was 150 psi (1034 kPa), then 75 percent of that would be 113 psi (779 kPa). So the lowest reading cylinder should be no less than 113 psi (779 kPa).
12. If a cylinder exhibits an unusually low compression reading, pour a tablespoon of clean engine oil into the cylinder through the spark plug hole and repeat the compression test. If the compression rises after adding oil, it means that the cylinder's piston rings and/or cylinder bore are damaged or worn. If the pressure remains low, the valves may not be seating properly (a valve job is needed), or the head gasket may be blown near that cylinder. If compression in any two adjacent cylinders is low, and if the addition of oil doesn't help raise compression, there is leakage past the head gasket. Oil and coolant in the combustion chamber, combined with blue or constant white smoke from the tailpipe, are symptoms of this problem. However, don't be alarmed by the normal white smoke emitted from the tailpipe during engine warm-up or from cold weather driving. There may be evidence of water droplets on the engine dipstick and/or oil droplets in the cooling system if a head gasket is blown.

Diesel Engines

Checking cylinder compression on diesel engines is basically the same procedure as on gasoline engines except for the following:

1. A special compression gauge suitable for diesel engines (because these engines have much greater compression pressures) **MUST** be used.
2. Remove the injector tubes and remove the injectors from each cylinder.

WARNING:

Do not forget to remove the washer underneath each injector. Otherwise, it may get lost when the engine is cranked.

3. When fitting the compression gauge adapter to the cylinder head, make sure the bleeder of the gauge (if equipped) is closed.
4. When reinstalling the injector assemblies, install new washers underneath each injector.

OIL PRESSURE TEST

Check for proper oil pressure at the sending unit passage with an externally mounted mechanical oil pressure gauge (as opposed to relying on a factory installed dash-mounted gauge). A tachometer may also be needed, as some specifications may require running the engine at a specific rpm.

1. With the engine cold, locate and remove the oil pressure sending unit.
2. Following the manufacturer's instructions, connect a mechanical oil pressure gauge and, if necessary, a tachometer to the engine.
3. Start the engine and allow it to idle.
4. Check the oil pressure reading when cold and record the number. You may need to run the engine at a specified rpm, so check the specifications.
5. Run the engine until normal operating temperature is reached (upper radiator hose will feel warm).
6. Check the oil pressure reading again with the engine hot and record the number. Turn the engine OFF.
7. Compare your hot oil pressure reading to that given in the chart. If the reading is low, check the cold pressure reading against the chart. If the cold pressure is well above the specification, and the hot reading was lower than the specification, you may have the wrong viscosity oil in the engine. Change the oil, making sure to use the proper grade and quantity, then repeat the test.
8. Low oil pressure readings could be attributed to internal component wear, pump related problems (such as a clogged screen), a low oil level, or oil viscosity that is too low. High High pressure readings could be caused by an overfilled crankcase, too high of an oil viscosity or a faulty pressure relief valve. Gasoline or water leaking into the lubrication system will also cause readings above normal.

Buy or Rebuild?

Now that you have determined that your engine is worn out, you must make some decisions. The question of whether or not an engine is worth rebuilding is largely a subjective matter and one of personal worth. Is the engine a popular one, or is it an obsolete model? Are parts available? Will it get acceptable gas mileage once it is rebuilt? Is the car it's being put into worth keeping? Would it be less expensive to buy a new engine, have your engine rebuilt by a pro, rebuild it yourself or buy a used engine from a salvage yard? Or would it be simpler and less expensive to buy another car? If you have considered all these matters and more, and have still decided to rebuild the engine, then it is time to decide how you will rebuild it.

Note: The editors at Chilton feel that most engine machining should be performed by a professional machine shop. Don't think of it as wasting money, rather, as an assurance that the job has been done right the first time. There are many expensive and specialized tools required to perform such tasks as boring and honing an engine block or having a valve job done on a cylinder head. Even inspecting the parts requires expensive micrometers and gauges to properly measure wear and clearances. Also, a machine shop can deliver to you clean, and ready to assemble parts, saving you time and aggravation. Your maximum savings will come from performing the removal, disassembly, assembly and installation of the engine and purchasing or renting only the tools required to perform the above tasks. Depending on the particular circumstances, you may save 40 to 60 percent of the cost doing these yourself.

A complete rebuild or overhaul of an engine involves replacing all of the moving parts (pistons, rods, crankshaft, camshaft, etc.) with new ones and machining the non-moving wearing surfaces of the block and heads. Unfortunately, this may not be cost effective. For instance, your crankshaft may have been damaged or worn, but it can be machined un-

ENGINE AND ENGINE OVERHAUL 3-51

Fig. 156 Use a gasket scraper to remove the old gasket material from the mating surfaces

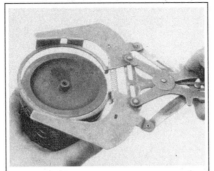

Fig. 157 Use a ring expander tool to remove the piston rings

Fig. 158 Clean the piston ring grooves using a ring groove cleaner tool, or...

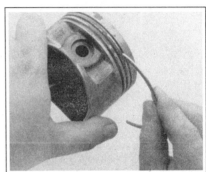

Fig. 159 ... use a piece of an old ring to clean the grooves. Be careful, the ring can be quite sharp

dersize for a minimal fee.

So, as you can see, you can replace everything inside the engine, but, it is wiser to replace only those parts which are really needed, and, if possible, repair the more expensive ones. Later in this section, we will break the engine down into its two main components: the cylinder head and the engine block. We will discuss each component, and the recommended parts to replace during a rebuild on each.

Engine Overhaul Tips

Most engine overhaul procedures are fairly standard. In addition to specific parts replacement procedures and specifications for your individual engine, this section is also a guide to acceptable rebuilding procedures. Examples of standard rebuilding practice are given and should be used along with specific details concerning your particular engine.

Competent and accurate machine shop services will ensure maximum performance, reliability and engine life. In most instances it is more profitable for the do-it-yourself mechanic to remove, clean and inspect the component, buy the necessary parts and deliver these to a shop for actual machine work.

Much of the assembly work (crankshaft, bearings, piston rods, and other components) is well within the scope of the do-it-yourself mechanic's tools and abilities. You will have to decide for yourself the depth of involvement you desire in an engine repair or rebuild.

TOOLS

The tools required for an engine overhaul or parts replacement will depend on the depth of your involvement. With a few exceptions, they will be the tools found in a mechanic's tool kit (see Section 1 of this manual). More in-depth work will require some or all of the following:
- A dial indicator (reading in thousandths) mounted on a universal base
- Micrometers and telescope gauges
- Jaw and screw-type pullers
- Scraper
- Valve spring compressor
- Ring groove cleaner
- Piston ring expander and compressor
- Ridge reamer
- Cylinder hone or glaze breaker
- Plastigage,
- Engine stand

The use of most of these tools is illustrated in this section. Many can be rented for a one-time use from a local parts jobber or tool supply house specializing in automotive work.

Occasionally, the use of special tools is called for. See the information on Special Tools and the Safety Notice in the front of this book before substituting another tool.

OVERHAUL TIPS

Aluminum has become extremely popular for use in engines, due to its low weight. Observe the following precautions when handling aluminum parts:
- Never "hot tank" aluminum parts (the caustic hot tank solution will eat the aluminum).
- Remove all aluminum parts (identification tag, etc.) from engine parts prior to the tanking.
- Always coat threads lightly with engine oil or anti-seize compounds before installation, to prevent seizure.
- Never overtighten bolts or spark plugs especially in aluminum threads.

When assembling the engine, any parts that will be exposed to frictional contact must be prelubed to provide lubrication at initial start-up. Any product specifically formulated for this purpose can be used, but engine oil is not recommended as a prelube in most cases.

When semi-permanent (locked, but removable) installation of bolts or nuts is desired, threads should be cleaned and coated with Loctite® or another similar, commercial non-hardening sealant.

CLEANING

See Figures 156, 157, 158 and 159

Before the engine and its components are inspected, they must be thoroughly cleaned. You will need to remove any engine varnish, oil sludge and/or carbon deposits from all of the components to ensure an accurate inspection. A crack in the engine block or cylinder head can easily become overlooked if hidden by a layer of sludge or carbon.

Most of the cleaning process can be carried out with common hand tools and readily available solvents or solutions. Carbon deposits can be chipped away using a hammer and a hard wooden chisel. Old gasket material and varnish or sludge can usually be removed using a scraper and/or cleaning solvent. Extremely stubborn deposits may require the use of a power drill with a wire brush. If using a wire brush, use extreme care around any critical machined surfaces (such as the gasket surfaces, bearing saddles, cylinder bores, etc.). Use of a wire brush is NOT RECOMMENDED on any aluminum components. Always follow any safety recommendations given by the manufacturer of the tool and/or solvent. You should always wear eye protection during any cleaning process involving scraping, chipping or spraying of solvents.

An alternative to the mess and hassle of cleaning the parts yourself is to drop them off at a local garage or machine shop. They will, more than likely, have the necessary equipment to properly clean all of the parts for a nominal fee.

CAUTION:

Always wear eye protection during any cleaning process involving scraping, chipping or spraying of solvents.

Remove any oil galley plugs, freeze plugs and/or pressed-in bearings and carefully wash and degrease all of the engine components including the fasteners and bolts. Small parts such as the valves, springs, etc., should be placed in a metal basket and allowed to soak. Use pipe cleaner type brushes, and clean all passageways in the components. Use a ring expander and remove the rings from the pistons. Clean the piston ring grooves with a special tool or a piece of broken ring. Scrape the carbon off of the top of the piston. You should never use a wire brush on the pistons. After preparing all of the piston assemblies in this manner, wash and degrease them again.

WARNING:

Use extreme care when cleaning around the cylinder head valve seats. A mistake or slip may cost you a new seat.

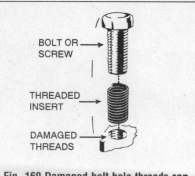

Fig. 160 Damaged bolt hole threads can be replaced with thread repair inserts

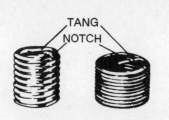

Fig. 161 Standard thread repair insert (left), and spark plug thread insert

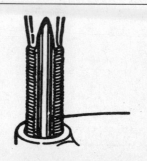

Fig. 163 Using the kit, tap the hole in order to receive the thread insert. Keep the tap well oiled and back it out frequently to avoid clogging the threads

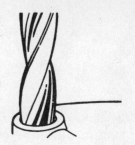

Fig. 162 Drill out the damaged threads with the specified size bit. Be sure to drill completely through the hole or to the bottom of a blind hole

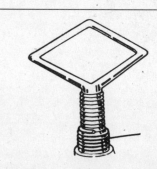

Fig. 164 Screw the insert onto the installer tool until the tang engages the slot. Thread the insert into the hole until it is 1/4 - 1/2 turn below the top surface, then remove the tool and break off the tang using a punch

When cleaning the cylinder head, remove carbon from the combustion chamber with the valves installed. This will avoid damaging the valve seats.

REPAIRING DAMAGED THREADS

See Figures 160, 161, 162, 163 and 164

Several methods of repairing damaged threads are available. Heli-Coil® (shown here), Keenserts® and Microdot® are among the most widely used. All involve basically the same principle - drilling out stripped threads, tapping the hole and installing a prewound insert - making welding, plugging and oversize fasteners unnecessary.

Two types of thread repair inserts are usually supplied: a standard type for most inch coarse, inch fine, metric course and metric fine thread sizes and a spark plug type to fit most spark plug port sizes. Consult the individual tool manufacturer's catalog to determine exact applications. Typical thread repair kits will contain a selection of prewound threaded inserts, a tap (corresponding to the outside diameter threads of the insert) and an installation tool. Spark plug inserts usually differ because they require a tap equipped with pilot threads and a combined reamer/tap section. Most manufacturers also supply blister-packed thread repair inserts separately in addition to a master kit containing a variety of taps and inserts plus installation tools.

Before attempting to repair a threaded hole, remove any snapped, broken or damaged bolts or studs. Penetrating oil can be used to free frozen threads. The offending item can usually be removed with locking pliers or using a screw/stud extractor. After the hole is clear, the thread can be repaired, as shown in the series of accompanying illustrations and in the kit manufacturer's instructions.

Engine Preparation

To properly rebuild an engine, you must first remove it from the vehicle, then disassemble and diagnose it. Ideally you should place your engine on an engine stand. This affords you the best access to the engine components. Follow the manufacturer's directions for using the stand with your particular engine. Remove the flywheel or driveplate before installing the engine to the stand.

Now that you have the engine on a stand, and assuming that you have drained the oil and coolant from the engine, it's time to strip it of all but the necessary components. Before you start disassembling the engine, you may want to take a moment to draw some pictures, or fabricate some labels or containers to mark the locations of various components and the bolts and/or studs which fasten them. Modern day engines use a lot of little brackets and clips which hold wiring harnesses and such, and these holders are often mounted on studs and/or bolts that can be easily mixed up. The manufacturer spent a lot of time and money designing your vehicle, and they wouldn't have wasted any of it by haphazardly placing brackets, clips or fasteners on the vehicle. If it's present when you disassemble it, put it back when you assemble, you will regret not remembering that little bracket which holds a wire harness out of the path of a rotating part.

You should begin by unbolting any accessories still attached to the engine, such as the water pump, power steering pump, alternator, etc. Then, unfasten any manifolds (intake or exhaust) which were not removed during the engine removal procedure. Finally, remove any covers remaining on the engine such as the rocker arm, front or timing cover and oil pan. Some front covers may require the vibration damper and/or crank pulley to be removed beforehand. The idea is to reduce the engine to the bare necessities (cylinder head(s), valve train, engine block, crankshaft, pistons and connecting rods), plus any other 'in block' components such as oil pumps, balance shafts and auxiliary shafts.

Finally, remove the cylinder head(s) from the engine block and carefully place on a bench. Disassembly instructions for each component follow later in this section.

Cylinder Head

There are two basic types of cylinder heads used on today's automobiles: the Overhead Valve (OHV) and the Overhead Camshaft (OHC). The latter can also be broken down into two subgroups: the Single Overhead Camshaft (SOHC) and the Dual Overhead Camshaft (DOHC). Generally, if there is only a single camshaft on a head, it is just referred to as an OHC head. Also, an engine with an OHV cylinder head is also known as a pushrod engine.

All engines covered in this manual are OHV with the exception of the SOHC 4.7L V-8.

Some cylinder heads these days are made of an aluminum alloy due to its light weight, durability and heat transfer qualities. However, cast iron was the material of choice in the past, and is still used on most vehicles today.

All engines covered in this manual have iron heads with the exception of the aluminum head 4.7L V-8.

Whether made from aluminum or iron, all cylinder heads have valves and seats. Some use two valves per cylinder, while the more hi-tech engines will utilize a multi-valve configuration using 3, 4 and even 5 valves per cylinder.

All the engines covered in this manual have two valves per cylinder with the exception of the 24-valve, 6-cylinder diesel.

When the valve contacts the seat, it does so on precision machined surfaces, which seals the combustion chamber. All cylinder heads have a valve guide for each valve. The guide centers the valve to the seat and allows it to move up and down within it. The clearance between the valve and guide can be critical. Too much clearance and the engine may consume oil, lose vacuum and/or damage the seat. Too little, and the valve can stick in the guide causing the

ENGINE AND ENGINE OVERHAUL 3-53

Fig. 165 When removing an OHV valve spring, use a compressor tool to relieve the tension from the retainer

Fig. 166 A small magnet will help in removal of the valve locks

Fig. 167 Be careful not to lose the small valve locks (keepers)

Fig. 168 Remove the valve seal from the valve stem - O-ring type seal shown

Fig. 169 Removing an umbrella/positive type seal

Fig. 170 Invert the cylinder head and withdraw the valve from the valve guide bore

engine to run poorly if at all, and possibly causing severe damage. The last component all cylinder heads have are valve springs. The spring holds the valve against its seat. It also returns the valve to this position when the valve has been opened by the valve train or camshaft. The spring is fastened to the valve by a retainer and valve locks (sometimes called keepers).

An ideal method of rebuilding the cylinder head would involve replacing all of the valves, guides, seats, springs, etc. with new ones. However, depending on how the engine was maintained, often this is not necessary. A major cause of valve, guide and seat wear is an improperly tuned engine. An engine that is running too rich, will often wash the lubricating oil out of the guide with gasoline, causing it to wear rapidly. Conversely, an engine which is running too lean will place higher combustion temperatures on the valves and seats allowing them to wear or even burn. Springs fall victim to the driving habits of the individual. A driver who often runs the engine rpm to the redline will wear out or break the springs faster then one that stays well below it. Unfortunately, mileage takes it toll on all of the parts. Generally, the valves, guides, springs and seats in a cylinder head can be machined and re-used, saving you money. However, if a valve is burnt, it may be wise to replace all of the valves, since they were all operating in the same environment. The same goes for any other component on the cylinder head. Think of it as an insurance policy against future problems related to that component.

Unfortunately, the only way to find out which components need replacing, is to disassemble and carefully check each piece. After the cylinder head(s) are disassembled, thoroughly clean all of the components.

DISASSEMBLY

OHV Heads

See Figures 165, 166, 167, 168, 169, 170 and 171

Before disassembling the cylinder head, you may want to fabricate some containers to hold the various parts, as some of them can be quite small (such as keepers) and easily lost. Also keeping yourself and the components organized will aid in assembly and reduce confusion. Where possible, try to maintain a components original location; this is especially important if there is not going to be any machine work performed on the components.

1. If you haven't already removed the rocker arms completely do so now.
2. Position the head so that the springs are easily accessed.
3. Use a valve spring compressor tool, and relieve spring tension from the retainer.

Note: Due to engine varnish, the retainer may stick to the valve locks. A gentle tap with a hammer may help to break it loose.

4. Remove the valve locks from the valve tip and/or retainer. A small magnet may help in removing the locks.
5. Lift the valve spring, tool and all, off of the valve stem.
6. If equipped, remove the valve seal. If the seal is difficult to remove with the valve in place, try removing the valve first, then the seal. Follow the steps below for valve removal.
7. Position the head to allow access for withdrawing the valve.

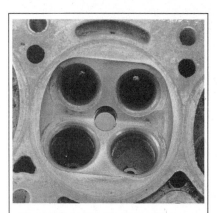

Fig. 171 Example of a multi-valve cylinder head. Note how it has 2 intake and 2 exhaust valve ports

Note: Cylinder heads that have seen a lot of miles and/or abuse may have mushroomed the valve lock grove and/or tip, causing difficulty in removal of the valve. If this has happened, use a metal file to carefully remove the high spots around the lock grooves and/or tip. Only file it enough to allow removal.

8. Remove the valve from the cylinder head.
9. If equipped, remove the valve spring shim. A small magnetic tool or screwdriver will aid in removal.
10. Repeat Steps 3 though 9 until all of the valves have been removed.

3-54 ENGINE AND ENGINE OVERHAUL

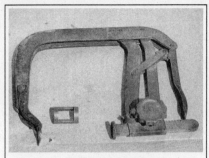

Fig. 172 A C-clamp type spring compressor is easier to use on OHC heads

Fig. 173 The 4.7L OHC head uses rocker arms under the camshafts

Fig. 174 Before the camshaft can be removed, all of the followers must first be removed…

Fig. 175 … then the camshaft can be removed by unbolting the bearing caps

Fig. 176 Remove the valve spring and retainer from the cylinder head

Fig. 177 Remove the valve seal from the guide. Some gentle prying or pliers may help to remove stubborn ones

4.7L Heads

See Figures 172, 173, 174, 175, 176, 177 and 178

Be sure to label the position of all components before disassembly.

Although this is an OHC cylinder head, it can be disassembled using a standard valve spring compressor. A C-clamp style compressor tool is recommended.

1. If not already removed, remove the rocker arms and the camshaft. Mark their positions for assembly.
2. Position the cylinder head to allow access to the valve spring.
3. Use a valve spring compressor tool to relieve the spring tension from the retainer.

Note: Due to engine varnish, the retainer may stick to the valve locks. A gentle tap with a hammer may help to break it loose.

4. Remove the valve locks from the valve tip and/or retainer. A small magnet may help in removing the small locks.
5. Lift the valve spring, tool and all, off of the valve stem.
6. If equipped, remove the valve seal. If the seal is difficult to remove with the valve in place, try removing the valve first, then the seal. Follow the steps below for valve removal.
7. Position the head to allow access for withdrawing the valve.

Note: Cylinder heads that have seen a lot of miles and/or abuse may have mushroomed the valve lock grove and/or tip, causing difficulty in removal of the valve. If this has happened, use a metal file to carefully remove the high spots around the lock grooves and/or tip. Only file it enough to allow removal.

8. Remove the valve from the cylinder head.
9. If equipped, remove the valve spring shim. A small magnetic tool or screwdriver will aid in removal.
10. Repeat Steps 3 though 9 until all of the valves have been removed.

INSPECTION

Now that all of the cylinder head components are clean, it's time to inspect them for wear and/or damage. To accurately inspect them, you will need some specialized tools:
- A 0 - 1 in. micrometer for the valves
- A dial indicator or inside diameter gauge for the valve guides
- A spring pressure test gauge

If you do not have access to the proper tools, you may want to bring the components to a shop that does.

Valves

See Figures 179 and 180

The first thing to inspect are the valve heads. Look closely at the head, margin and face for any cracks, excessive wear or burning. The margin is the best place to look for burning. It should have a squared edge with an even width all around the diameter. When a valve burns, the margin will look melted and the edges rounded. Also inspect the valve head for any signs of "tulipping." This will show as a lifting of the edges or dishing in the center of the head and will usually not occur to all of the valves. All of the heads should look the same, any that seem dished more than others are probably bad. Next, inspect the valve lock grooves and valve tips. Check for any burrs

Fig. 178 Most heads will have these valve spring shims. Remove all of them as well

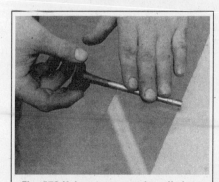

Fig. 179 Valve stems may be rolled on a flat surface to check for bends

ENGINE AND ENGINE OVERHAUL   3-55

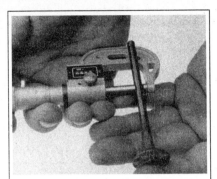

Fig. 180 Use a micrometer to check the valve stem diameter

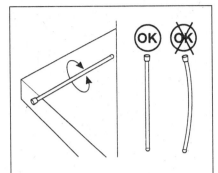

Fig. 181 Roll each pushrod on a flat surface and check for a bent condition

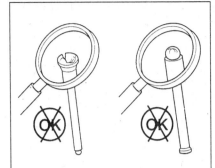

Fig. 182 Check pushrod ends for damage or wear

Fig. 183 Pushrod ball end must not have flat spots or other visible damage

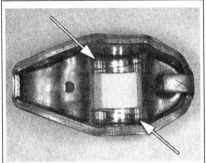

Fig. 184 Check rocker arm wear surfaces (arrows)

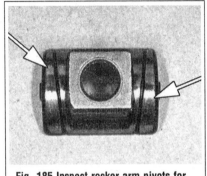

Fig. 185 Inspect rocker arm pivots for wear or damage

around the lock grooves, especially if you had to file them to remove the valve. Valve tips should appear flat, although slight rounding with high mileage engines is normal. Slightly worn valve tips will need to be machined flat. Last, measure the valve stem diameter with the micrometer. Measure the area that rides within the guide, especially towards the tip where most of the wear occurs. Take several measurements along its length and compare them to each other. Wear should be even along the length with little to no taper. If no minimum diameter is given in the specifications, then the stem should not read more than 0.001 in. (0.025mm) below the unworn area of the valve stem. Any valves that fail these inspections should be replaced.

Pushrods
See Figures 181, 182 and 183

1. Clean the pushrods thoroughly.
2. Roll each pushrod on a flat surface. A bent pushrod will make itself known in short order. Replace bent pushrods. They cannot be straightened.
3. Check the cup and ball ends of each rod for damage, flat spots, wear, cracks, etc. Replace any pushrods showing signs of damage.

Rocker Arms
See Figures 184 and 185

4. Most damage to these components would be caused by lack of lubrication, very high mileage or a combination of the two.
5. Check for bluing which is a result of overheating. Replace any parts thus damaged.
6. Check the wear surfaces on rocker arm and pivots. If normal, high-mileage wear is obvious on any rocker arm, replacing all of them might be a good idea.
7. On diesels, check rocker arm shafts and bores for excessive play and obvious physical damage.

Springs, Retainers and Valve Locks
See Figures 186, 187 and 188

The first thing to check is the most obvious, broken springs. Next check the free length and squareness of each spring. If applicable, ensure to distinguish between intake and exhaust springs. Use a ruler and/or carpenter's square to measure the length. A carpenter's square should be used to check the springs for squareness. If a spring pressure test gauge is available, check each springs rating and compare to the specifications chart. Check the readings against the specifications given. Any springs that fail these inspections should be replaced.

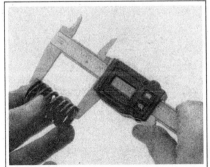

Fig. 186 Use a caliper to check the valve spring free-length

Fig. 187 Check the valve spring for squareness on a flat surface; a carpenter's square can be used

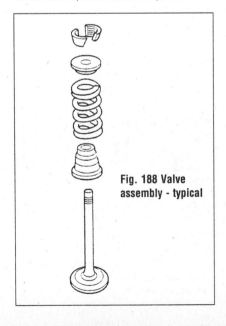

Fig. 188 Valve assembly - typical

3-56 ENGINE AND ENGINE OVERHAUL

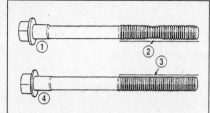

Fig. 189 Check head bolts for rounded heads (1, 4), necking (2), or thread damage (3)

Fig. 190 A dial gauge may be used to check valve stem-to-guide clearance; read the gauge while moving the valve stem

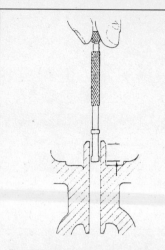

Fig. 191 Measuring valve guide bore with an inside micrometer

The spring retainers rarely need replacing, however they should still be checked as a precaution. Inspect the spring mating surface and the valve lock retention area for any signs of excessive wear. Also check for any signs of cracking. Replace any retainers that are questionable.

Valve locks should be inspected for excessive wear on the outside contact area as well as on the inner notched surface. Any locks which appear worn or broken and its respective valve should be replaced.

Cylinder Head

There are several things to check on the cylinder head: bolts, valve guides, seats, cylinder head surface flatness, cracks and physical damage.

CYLINDER HEAD BOLTS

See Figure 189

Using new head bolts after a rebuild is good practice. Engines in heavy-duty applications or operating under severe service conditions should always be given new head bolts.

No head bolts should be used (removed/installed) more than twice. Mark reused head bolts with a dab of paint so the next rebuilder will know to use new parts.

Check head bolts for thread damage, a stretched condition, or "necking".

VALVE GUIDES

See Figures 190 and 191

Now that you know the valves are good, you can use them to check the guides, although a new valve, if available, is preferred. Before you measure anything, look at the guides carefully and inspect them for any cracks, chips or breakage. Also if the guide is a removable style (as in most aluminum heads), check them for any looseness or evidence of movement. All of the guides should appear to be at the same height from the spring seat. If any seem lower (or higher) from another, the guide has moved. Mount a dial indicator onto the spring side of the cylinder head. Lightly oil the valve stem and insert it into the cylinder head. Position the dial indicator against the valve stem near the tip and zero the gauge. Grasp the valve stem and wiggle towards and away from the dial indicator and observe the readings. Mount the dial indicator 90 degrees from the initial point and zero the gauge and again take a reading. Compare the two readings for a out of round condition. Check the readings against the specifications given. An Inside Diameter (I.D.) gauge designed for valve guides will give you an accurate valve guide bore measurement. If the I.D. gauge is used, compare the readings with the specifications given. Any guides that fail these inspections should be replaced or machined.

VALVE SEATS

A visual inspection of the valve seats should show a slightly worn and pitted surface where the valve face contacts the seat. Inspect the seat carefully for severe pitting or cracks. Also, a seat that is badly worn will be recessed into the cylinder head. A severely worn or recessed seat may need to be replaced. All cracked seats must be replaced. A seat concentricity gauge, if available, should be used to check the seat run-out. If run-out exceeds specifications the seat must be machined (if no specification is given use 0.002 in. or 0.051mm).

CYLINDER HEAD SURFACE FLATNESS

See Figures 192 and 193

After you have cleaned the gasket surface of the cylinder head of any old gasket material, check the head for flatness.

Place a straightedge across the gasket surface. Using feeler gauges, determine the clearance at the center of the straightedge and across the cylinder head at several points. Check along the centerline and diagonally on the head surface. If the warpage exceeds 0.003 in. (0.076mm) within a 6.0 in. (15.2cm) span, or 0.006 in. (0.152mm) over the total length of the head, the cylinder head must be resurfaced. After resurfacing the heads of a V-type engine, the intake manifold flange surface should be checked, and if necessary, milled proportionally to allow for the change in its mounting position.

CRACKS AND PHYSICAL DAMAGE

Generally, cracks are limited to the combustion chamber, however, it is not uncommon for the head to crack in a spark plug hole, port, outside of the head or in the valve spring/rocker arm area. The first

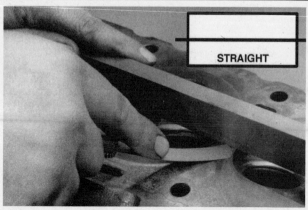

Fig. 192 Check the head for flatness across the center of the head surface using a straightedge and feeler gauge

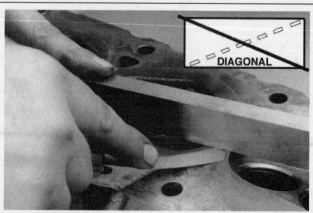

Fig. 193 Checks should also be made along both diagonals of the head surface

ENGINE AND ENGINE OVERHAUL 3-57

area to inspect is always the hottest: the exhaust seat/port area.

A visual inspection should be performed, but just because you don't see a crack does not mean it is not there. Some more reliable methods for inspecting for cracks include Magnaflux,, a magnetic process or Zyglo,, a dye penetrant. Magnaflux, is used only on ferrous metal (cast iron) heads. Zyglo, uses a spray-on fluorescent mixture along with a black light to reveal the cracks. It is strongly recommended to have your cylinder head checked professionally for cracks, especially if the engine was known to have overheated and/or leaked or consumed coolant. Contact a local shop for availability and pricing of these services.

Physical damage is usually very evident. For example, a broken mounting ear from dropping the head or a bent or broken stud and/or bolt. All of these defects should be fixed or, if unrepairable, the head should be replaced.

CAMSHAFTS AND TAPPETS

Refer to the camshaft section for OHV engines under "Engine Block", following.

REFINISHING & REPAIRING

Many of the procedures given for refinishing and repairing the cylinder head components must be performed by a machine shop. Certain steps, if the inspected part is not worn, can be performed yourself inexpensively. However, you spent a lot of time and effort so far, why risk trying to save a couple bucks if you might have to do it all over again?

Valves

Any valves that were not replaced should be refaced and the tips ground flat. Unless you have access to a valve grinding machine, this should be done by a machine shop. If the valves are in extremely good condition, as well as the valve seats and guides, they may be lapped in without performing machine work.

It is a recommended practice to lap the valves even after machine work has been performed and/or new valves have been purchased. This ensures a positive seal between the valve and seat.

LAPPING THE VALVES

Note: Before lapping the valves to the seats, read the rest of the cylinder head section to ensure that any related parts are in acceptable enough condition to continue.

Note: Before any valve seat machining and/or lapping can be performed, the guides must be within factory recommended specifications.

1. Invert the cylinder head.
2. Lightly lubricate the valve stems and insert them into the cylinder head in their numbered order.
3. Raise the valve from the seat and apply a small amount of fine lapping compound to the seat.
4. Moisten the suction head of a hand-lapping tool and attach it to the head of the valve.
5. Rotate the tool between the palms of both hands, changing the position of the valve on the valve seat and lifting the tool often to prevent grooving.
6. Lap the valve until a smooth, polished circle is evident on the valve and seat.

7. Remove the tool and the valve. Wipe away all traces of the grinding compound and store the valve to maintain its lapped location.

WARNING:

Do not get the valves out of order after they have been lapped. They must be put back with the same valve seat with which they were lapped.

Springs, Retainers and Valve Locks

There is no repair or refinishing possible with the springs, retainers and valve locks. If they are found to be worn or defective, they must be replaced with new (or known good) parts.

Cylinder Head

Most refinishing procedures dealing with the cylinder head must be performed by a machine shop. Read the sections below and review your inspection data to determine whether or not machining is necessary.

VALVE GUIDES

Note: If any machining or replacements are made to the valve guides, the seats must be machined.

Unless the valve guides need machining or replacing, the only service to perform is to thoroughly clean them of any dirt or oil residue.

There are only two types of valve guides used on automobile engines: the replaceable type (all aluminum heads) and the cast-in integral type (most cast iron heads). There are four recommended methods for repairing worn guides.
- Knurling
- Inserts
- Reaming oversize
- Replacing

Knurling is a process in which metal is displaced and raised, thereby reducing clearance, giving a true center, and providing oil control. It is the least expensive way of repairing the valve guides. However, it is not necessarily the best, and in some cases, a knurled valve guide will not stand up for more than a short time. It requires a special knurlizer and precision reaming tools to obtain proper clearances. It would not be cost effective to purchase these tools, unless you plan on rebuilding several of the same cylinder head.

Installing a guide insert involves machining the guide to accept a bronze insert. One style is the coil-type which is installed into a threaded guide. Another is the thin-walled insert where the guide is reamed oversize to accept a split-sleeve insert. After the insert is installed, a special tool is then run through the guide to expand the insert, locking it to the guide. The insert is then reamed to the standard size for proper valve clearance.

Reaming for oversize valves restores normal clearances and provides a true valve seat. Most cast-in type guides can be reamed to accept an valve with an oversize stem. The cost factor for this can become quite high as you will need to purchase the reamer and new, oversize stem valves for all guides which were reamed. Oversizes are generally 0.003 to 0.030 in. (0.076 to 0.762mm), with 0.015 in. (0.381mm) being the most common.

To replace cast-in type valve guides, they must be drilled out, then reamed to accept replacement guides. This must be done on a fixture which will allow centering and leveling off of the original valve seat or guide, otherwise a serious guide-to-seat misalignment may occur making it impossible to properly machine the seat.

Replaceable-type guides are pressed into the cylinder head. A hammer and a stepped drift or punch may be used to install and remove the guides. Before removing the guides, measure the protrusion on the spring side of the head and record it for installation. Use the stepped drift to hammer out the old guide from the combustion chamber side of the head. When installing, determine whether or not the guide also seals a water jacket in the head, and if it does, use the recommended sealing agent. If there is no water jacket, grease the valve guide and its bore. Use the stepped drift, and hammer the new guide into the cylinder head from the spring side of the cylinder head. A stack of washers the same thickness as the measured protrusion may help the installation process.

VALVE SEATS

Note: Before any valve seat machining can be performed, the guides must be within factory recommended specifications. If any machining or replacements were made to the valve guides, the seats must be machined.

If the seats are in good condition, the valves can be lapped to the seats, and the cylinder head assembled. See the valves section for instructions on lapping.

If the valve seats are worn, cracked or damaged, they must be serviced by a machine shop. The valve seat must be perfectly centered to the valve guide, which requires very accurate machining.

CYLINDER HEAD SURFACE

If the cylinder head is warped, it must be machined flat. If the warpage is extremely severe, the head may need to be replaced. In some instances, it may be possible to straighten a warped head enough to allow machining. In either case, contact a professional machine shop for service.

Note: Any OHC cylinder head that shows excessive warpage should have the camshaft bearing journals align bored after the cylinder head has been resurfaced.

WARNING:

Failure to align bore the camshaft bearing journals could result in severe engine damage including but not limited to: valve and piston damage, connecting rod damage, camshaft and/or crankshaft breakage.

CRACKS AND PHYSICAL DAMAGE

Certain cracks can be repaired in both cast iron and aluminum heads. For cast iron, a tapered threaded insert is installed along the length of the crack. Aluminum can also use the tapered inserts, however welding is the preferred method. Some physical damage can be repaired through brazing or welding. Contact a machine shop to get expert advice for your particular dilemma.

3-58 ENGINE AND ENGINE OVERHAUL

ASSEMBLY

The first step for any assembly job is to have a clean area in which to work. Next, thoroughly clean all of the parts and components that are to be assembled. Finally, place all of the components onto a suitable work space and, if necessary, arrange the parts to their respective positions.

OHV Engines

1. Lightly lubricate the valve stems and insert all of the valves into the cylinder head. If possible, maintain their original locations.
2. If equipped, install any valve spring shims which were removed.
3. If equipped, install the new valve seals, keeping the following in mind:
   - If the valve seal presses over the guide, lightly lubricate the outer guide surfaces.
   - If the seal is an O-ring type, it is installed just after compressing the spring but before the valve locks.
4. Place the valve spring and retainer over the stem.
5. Position the spring compressor tool and compress the spring.
6. Assemble the valve locks to the stem.
7. Relieve the spring pressure slowly and ensure that neither valve lock becomes dislodged by the retainer.
8. Remove the spring compressor tool.
9. Repeat Steps 2 through 8 until all of the springs have been installed.

4.7L Engine

1. Lightly lubricate the valve stems and insert all of the valves into the cylinder head. If possible, maintain their original locations.
2. If equipped, install any valve spring shims which were removed.
3. If equipped, install the new valve seals, keeping the following in mind:
   - If the valve seal presses over the guide, lightly lubricate the outer guide surfaces.
   - If the seal is an O-ring type, it is installed just after compressing the spring but before the valve locks.
4. Place the valve spring and retainer over the stem.
5. Position the spring compressor tool and compress the spring.
6. Assemble the valve locks to the stem.
7. Relieve the spring pressure slowly and ensure that neither valve lock becomes dislodged by the retainer.
8. Remove the spring compressor tool.
9. Repeat Steps 2 through 8 until all of the springs have been installed.
10. Install the camshaft(s), rockers, shafts and any other components that were removed for disassembly.

Engine Block

GENERAL INFORMATION

A thorough overhaul or rebuild of an engine block would include replacing the pistons, rings, bearings, timing belt/chain assembly and oil pump. For OHV engines also include a new camshaft and lifters. The block would then have the cylinders bored and honed and the crankshaft would be cut undersize to provide new wearing surfaces and perfect clearances. However, your particular engine may not have everything worn out. What if only the piston rings have worn out and the clearances on everything else are still within factory specifications? Well, you could just replace the rings and put it back together, but this would be a very rare example. Chances are, if one component in your engine is worn, other components are sure to follow, and soon. At the very least, you should always replace the rings, bearings and oil pump. This is what is commonly called a "freshen up".

Cylinder Ridge Removal

Because the top piston ring does not travel to the very top of the cylinder, a ridge is built up between the end of the travel and the top of the cylinder bore.

Pushing the piston and connecting rod assembly past the ridge can be difficult, and damage to the piston ring lands could occur. If the ridge is not removed before installing a new piston or not removed at all, piston ring breakage and piston damage may occur.

Note: It is always recommended that you remove any cylinder ridges before removing the piston and connecting rod assemblies. If you know that new pistons are going to be installed and the engine block will be bored oversize, you may be able to forego this step. However, some ridges may actually prevent the assemblies from being removed, necessitating its removal.

There are several different types of ridge reamers on the market, none of which are inexpensive. Unless a great deal of engine rebuilding is anticipated, borrow or rent a reamer.

1. Turn the crankshaft until the piston is at the bottom of its travel.
2. Cover the head of the piston with a rag.
3. Follow the tool manufacturer's instructions and cut away the ridge, exercising extreme care to avoid cutting too deeply.
4. Remove the ridge reamer, the rag and as many of the cuttings as possible. Continue until all of the cylinder ridges have been removed.

DISASSEMBLY

See Figures 194 and 195

The following engine disassembly instructions assume that you have the engine mounted on an engine stand. If not, it is easiest to disassemble the engine on a bench or the floor with it resting on the bell housing or transmission mounting surface. You must be able to access the connecting rod fasteners and turn the crankshaft during disassembly. Also, all engine covers (timing, front, side, oil pan, whatever) should have already been removed. Engines which are seized or locked up may not be able to be completely disassembled, and a core (salvage yard) engine should be purchased.

Pushrod Engines

If not done during the cylinder head removal, remove the pushrods and lifters, keeping them in order for assembly. Remove the timing gears and/or timing chain assembly, then remove the oil pump drive assembly and withdraw the camshaft from the engine block. Remove the oil pick-up and pump assembly. If equipped, remove any balance or auxiliary shafts. If necessary, remove the cylinder ridge from the top of the bore. See the cylinder ridge removal procedure earlier in this section.

4.7L Engine

If not done during the cylinder head removal, remove the timing chain and/or sprocket assembly. If necessary, remove the cylinder ridge from the top of the bore. See the cylinder ridge removal procedure earlier in this section.

All Engines

Rotate the engine over so that the crankshaft is exposed. Use a number punch or scribe and mark each connecting rod with its respective cylinder number. Refer to the illustrations of firing orders under "Engine Electrical" for cylinder numbers. On in-line engines, the No. 1 cylinder is the one closest to the radiator. On "V" engines it's the one closest to the radiator on the driver's side. Use a number punch or scribe and also mark the main bearing caps from front to rear with the front most cap being number 1 (if there are five caps, mark them 1 through 5, front to rear).

Fig. 194 Place rubber hose over the connecting rod studs to protect the crankshaft and cylinder bores from damage

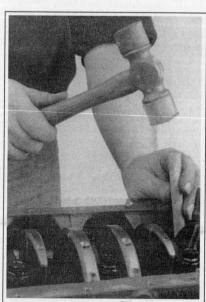

Fig. 195 Carefully tap the piston out of the bore using a wooden dowel

ENGINE AND ENGINE OVERHAUL 3-59

WARNING:

Take special care when pushing the connecting rod up from the crankshaft because the sharp threads of the rod bolts/studs will score the crankshaft journal. Ensure that special plastic caps are installed over them, or cut two pieces of rubber hose to do the same.

Again, rotate the engine, this time to position the number one cylinder bore (head surface) up. Turn the crankshaft until the number one piston is at the bottom of its travel, this should allow the maximum access to its connecting rod. Remove the number one connecting rods fasteners and cap and place two lengths of rubber hose over the rod bolts/studs to protect the crankshaft from damage. Using a sturdy wooden dowel and a hammer, push the connecting rod up about 1 in. (25mm) from the crankshaft and remove the upper bearing insert. Continue pushing or tapping the connecting rod up until the piston rings are out of the cylinder bore. Remove the piston and rod by hand, put the upper half of the bearing insert back into the rod, install the cap with its bearing insert installed, and hand-tighten the cap fasteners. If the parts are kept in order in this manner, they will not get lost and you will be able to tell which bearings came from what cylinder if any problems are discovered and diagnosis is necessary. Remove all the other piston assemblies in the same manner. On V-style engines, remove all of the pistons from one bank, then reposition the engine with the other cylinder bank head surface up, and remove the piston assemblies from that bank.

The only remaining component in the engine block should now be the crankshaft. Mark bearing caps for orientation and position. Loosen the main bearing caps evenly until the fasteners can be turned by hand, then remove them and the caps. Remove the crankshaft from the engine block. Thoroughly clean all of the components.

WARNING:

Bearing caps are not interchangeable. Be sure to mark them before removal and install them in their original locations during assembly.

INSPECTION

Now that the engine block and all of its components are clean, it's time to inspect them for wear and/or damage. To accurately inspect them, you will need some specialized tools:
- Two or three separate micrometers to measure the pistons and crankshaft journals
- A dial indicator
- Telescoping gauges for the cylinder bores
- A rod alignment fixture to check for bent connecting rods

If you do not have access to the proper tools, you may want to bring the components to a shop that does.

Generally, you shouldn't expect cracks in the engine block or its components unless it was known to leak, consume or mix engine fluids, it was severely overheated, or there was evidence of bad bearings and/or crankshaft damage. A visual inspection should be performed on all of the components, but just because you don't see a crack does not mean it is not there. Some more reliable methods for inspecting for cracks include Magnaflux,, a magnetic process or Zyglo,, a dye penetrant. Magnaflux, is used only on ferrous metal (cast iron). Zyglo, uses a spray on fluorescent mixture along with a black light to reveal the cracks. It is strongly recommended to have your engine block checked professionally for cracks, especially if the engine was known to have overheated and/or leaked or consumed coolant. Contact a local shop for availability and pricing of these services.

Engine Block

ENGINE BLOCK BEARING ALIGNMENT

Remove the main bearing caps and, if still installed, the main bearing inserts. Inspect all of the main bearing saddles and caps for damage, burrs or high spots. If damage is found, and it is caused from a spun main bearing, the block will need to be align-bored or, if severe enough, replaced. Any burrs or high spots should be carefully removed with a metal file.

Place a straightedge on the bearing saddles, in the engine block, along the centerline of the crankshaft. If any clearance exists between the straightedge and the saddles, the block must be align-bored.

Align-boring consists of machining the main bearing saddles and caps by means of a flycutter that runs through the bearing saddles.

DECK FLATNESS

The top of the engine block where the cylinder head mounts is called the deck. Ensure that the deck surface is clean of dirt, carbon deposits and old gasket material. Place a straightedge across the surface of the deck along its centerline and, using feeler gauges, check the clearance along several points. Repeat the checking procedure with the straightedge placed along both diagonals of the deck surface. If the reading exceeds 0.003 in. (0.076mm) within a 6.0 in. (15.2cm) span, or 0.006 in. (0.152mm) over the total length of the deck, it must be machined.

CYLINDER BORES

See Figure 196

The cylinder bores house the pistons and are slightly larger than the pistons themselves. A common piston-to-bore clearance is 0.0015 - 0.0025 in.

Fig. 196 Use a telescoping gauge to measure the cylinder bore diameter - take several readings within the same bore

(0.0381mm - 0.0635mm). Inspect and measure the cylinder bores. The bore should be checked for out-of-roundness, taper and size. The results of this inspection will determine whether the cylinder can be used in its existing size and condition, or a rebore to the next oversize is required (or in the case of removable sleeves, have replacements installed).

The amount of cylinder wall wear is always greater at the top of the cylinder than at the bottom. This wear is known as taper. Any cylinder that has a taper of 0.0012 in. (0.305mm) or more, must be rebored. Measurements are taken at a number of positions in each cylinder: at the top, middle and bottom and at two points at each position; that is, at a point 90 degrees from the crankshaft centerline, as well as a point parallel to the crankshaft centerline. The measurements are made with either a special dial indicator or a telescopic gauge and micrometer. If the necessary precision tools to check the bore are not available, take the block to a machine shop and have them mike it. Also if you don't have the tools to check the cylinder bores, chances are you will not have the necessary devices to check the pistons, connecting rods and crankshaft. Take these components with you and save yourself an extra trip.

For our procedures, we will use a telescopic gauge and a micrometer. You will need one of each, with a measuring range which covers your cylinder bore size.

1. Position the telescopic gauge in the cylinder bore, loosen the gauges lock and allow it to expand.

Note: Your first two readings will be at the top of the cylinder bore, then proceed to the middle and finally the bottom, making a total of six measurements.

2. Hold the gauge square in the bore, 90 degrees from the crankshaft centerline, and gently tighten the lock. Tilt the gauge back to remove it from the bore.
3. Measure the gauge with the micrometer and record the reading.
4. Again, hold the gauge square in the bore, this time parallel to the crankshaft centerline, and gently tighten the lock. Again, you will tilt the gauge back to remove it from the bore.
5. Measure the gauge with the micrometer and record this reading. The difference between these two readings is the out-of-round measurement of the cylinder.
6. Repeat steps 1 through 5, each time going to the next lower position, until you reach the bottom of the cylinder. Then go to the next cylinder, and continue until all of the cylinders have been measured.

The difference between these measurements will tell you all about the wear in your cylinders. The measurements which were taken 90 degrees from the crankshaft centerline will always reflect the most wear. That is because at this position is where the engine power presses the piston against the cylinder bore the hardest. This is known as thrust wear. Take your top, 90 degree measurement and compare it to your bottom, 90 degree measurement. The difference between them is the taper. When you measure your pistons, you will compare these readings to your piston sizes and determine piston-to-wall clearance.

Crankshaft

Inspect the crankshaft for visible signs of wear or damage. All of the journals should be perfectly round and smooth. Slight scores are normal for a used crankshaft, but you should hardly feel them with your fingernail. When measuring the crankshaft with a micrometer, you will take readings at the front and rear of each journal, then turn the micrometer 90 degrees and take two more readings, front and rear. The difference between the front-to-rear readings is the journal taper and the first-to-90 degree reading is the out-of-round measurement. Generally, there should be no taper or out-of-roundness found, however, up to 0.0005 in. (0.0127mm) for either can be overlooked. Also, the readings should fall within the factory specifications for journal diameters.

If the crankshaft journals fall within specifications, it is recommended that it be polished before being returned to service. Polishing the crankshaft ensures that any minor burrs or high spots are smoothed, thereby reducing the chance of scoring the new bearings.

Pistons and Connecting Rods

PISTONS
See Figure 197

The piston should be visually inspected for any signs of cracking or burning (caused by hot spots or detonation), and scuffing or excessive wear on the skirts. The wrist pin attaches the piston to the connecting rod. The piston should move freely on the wrist pin, both sliding and pivoting. Grasp the connecting rod securely, or mount it in a vise, and try to rock the piston back and forth along the centerline of the wrist pin. There should not be any excessive play evident between the piston and the pin. If there are C-clips retaining the pin in the piston then you have wrist pin bushings in the rods. There should not be any excessive play between the wrist pin and the rod bushing. Normal clearance for the wrist pin is approx. 0.001 - 0.002 in. (0.025mm - 0.051mm).

WARNING:

Mark pistons and rods for cylinder and orientation before removal.

Except 2.5L engine: Use a micrometer and measure the diameter of the piston, perpendicular to the wrist pin, on the skirt. Compare the reading to its original cylinder measurement obtained earlier. The difference between the two readings is the piston-to-wall clearance. If the clearance is within specifications, the piston may be used as is. If the piston is out of specification, but the bore is not, you will need a new piston. If both are out of specification, you will need the cylinder rebored and oversize pistons installed. Generally if two or more pistons/bores are out of specification, it is best to rebore the entire block and purchase a complete set of oversize pistons.

Fig. 197 Measure the piston's outer diameter, perpendicular to the wrist pin, with a micrometer

2.5L engine: Piston skirts are coated with a friction-fighting moly compound and cannot be measured accurately. Use an inside micrometer on the cylinder bore only and compare readings against specifications given.

On many engines, pistons are graded for size, usually with a letter (e.g.: "A" through "E"). This will not normally be a factor if new pistons are fitted, since the cylinders will be bored to match the measured diameters of the replacement pistons.

CONNECTING ROD

You should have the connecting rod checked for straightness at a machine shop. If the connecting rod is bent, it will unevenly wear the bearing and piston, as well as place greater stress on these components. Any bent or twisted connecting rods must be replaced. If the rods are straight and the wrist pin clearance is within specifications, then only the bearing end of the rod need be checked. Place the connecting rod into a vice, with the bearing inserts in place, install the cap to the rod and torque the fasteners to specifications. Use a telescoping gauge and carefully measure the inside diameter of the bearings. Compare this reading to the rods original crankshaft journal diameter measurement. The difference is the oil clearance. If the oil clearance is not within specifications, install new bearings in the rod and take another measurement. If the clearance is still out of specifications, and the crankshaft is not, the rod will need to be reconditioned by a machine shop.

Note: You can also use Plastigage, to check the bearing clearances. The assembling section has complete instructions on its use.

Camshaft
See Figure 198

1. Mount the camshaft on v-blocks and check run-out. Max allowable in 0.001 in. (0.00004 mm).

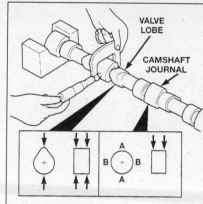

Fig. 198 Camshaft lobe and bearing measuring points

2. Check for obvious signs of wear or damage, such as pitting or bluing of the lobes or bearings. Bluing, indicating heat damage, often damages the hardened surfaces, resulting in rapid wear. Imperfect camshafts should be replaced.
3. Check diameters of cam journals and bearing surfaces on the engine against the specifications given. Replace bearing inserts if not within specification.

Tappets
See Figure 199

These engine use hydraulic tappets. Check that the tappet's roller or camshaft contact pad is unworn. The tappet can be disassembled after prying out the spring clip at the top, but is normally not necessary. Usually tappets can be satisfactorily cleaned with thin oil.

Bearings

All of the engine bearings should be visually inspected for wear and/or damage. The bearing should look evenly worn all around with no deep scores or pits. If the bearing is severely worn, scored, pitted or heat blued, then the bearing, and the components that use it, should be brought to a machine shop for inspection. Full-circle bearings (used on most camshafts) require specialized tools for removal and installation, and should be brought to a machine shop for service.

Oil Pump
See Figure 200

Note: The oil pump is responsible for providing constant lubrication to the whole engine

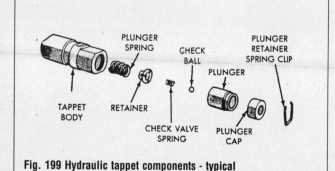

Fig. 199 Hydraulic tappet components - typical

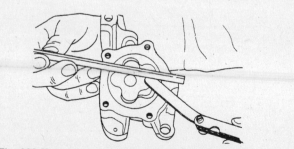

Fig. 200 Measuring clearance over installed rotors on a gear-type oil pump

ENGINE AND ENGINE OVERHAUL 3-61

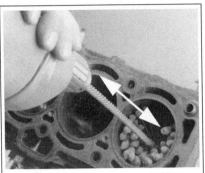

Fig. 201 Use a ball type cylinder hone to remove any glaze and provide a new surface for seating the piston rings

Fig. 202 Most pistons are marked to indicate positioning in the engine (usually a mark means the side facing the front)

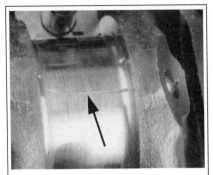

Fig. 203 Apply a strip of gauging material to the bearing journal, then install and torque the cap

Fig. 204 After the cap is removed again, use the scale supplied with the gauging material to check the clearance

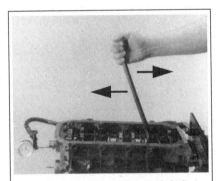

Fig. 205 Carefully pry the crankshaft back and forth while reading the dial gauge for end-play

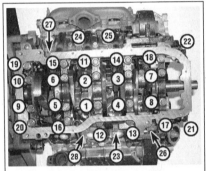

Fig. 206 On the 4.7L V8, the main bearing cap/bedplate bolts must be tightened in a very specific sequence (see the torque specifications chart at the end of this Chapter).

and so it is recommended that a new oil pump be installed when rebuilding the engine.

1. Completely disassemble the oil pump and thoroughly clean all of the components. Inspect the oil pump gears and housing for wear and/or damage. Ensure that the pressure relief valve operates properly and there is no binding or sticking due to varnish or debris.
2. Using a micrometer, check dimensions of the pump components against the specifications given. Typical checks include:
   - Clearance between rotors
   - Outer rotor-to-housing clearance
   - Inner and outer rotor thickness
   - Clearance over installed rotors
   - Cover flatness
3. If all of the parts are in proper working condition, lubricate the gears and relief valve, and assemble the pump.

REFINISHING

See Figure 201

Almost all engine block refinishing must be performed by a machine shop. If the cylinders are not to be rebored, then the cylinder glaze can be removed with a ball hone. When removing cylinder glaze with a ball hone, use a light or penetrating type oil to lubricate the hone. Do not allow the hone to run dry as this may cause excessive scoring of the cylinder bores and wear on the hone. If new pistons are required, they will need to be installed to the connecting rods. This should be performed by a machine

shop as the pistons must be installed in the correct relationship to the rod or engine damage can occur.

Pistons and Connecting Rods
See Figure 202

Only pistons with the wrist pin retained by C-clips are serviceable by the home mechanic. Press fit pistons require special presses and/or heaters to remove/install the connecting rod and should only be performed by a machine shop.

All pistons will have a mark indicating the direction to the front of the engine and the must be installed into the engine in that manner. Usually it is a notch or arrow on the top of the piston, or it may be the letter F cast or stamped into the piston. Make your own marks to indicate proper cylinder and piston and rod orientation to be sure.

C-CLIP TYPE PISTONS

1. Note the location of the forward mark on the piston and mark the connecting rod in relation.
2. Remove the C-clips from the piston and withdraw the wrist pin.

Note: Varnish build-up or C-clip groove burrs may increase the difficulty of removing the wrist pin. If necessary, use a punch or drift to carefully tap the wrist pin out.

3. Ensure that the wrist pin bushing in the connecting rod is usable, and lubricate it with assembly lube.
4. Remove the wrist pin from the new piston and lubricate the pin bores on the piston.
5. Align the forward marks on the piston and the

connecting rod and install the wrist pin.
6. The new C-clips will have a flat and a rounded side to them. Install both C-clips with the flat side facing out.
7. Repeat all of the steps for each piston being replaced.

ASSEMBLY

Before you begin assembling the engine, first give yourself a clean, dirt free work area. Next, clean every engine component again. The key to a good assembly is cleanliness.

Mount the engine block into the engine stand and wash it one last time using water and detergent (dishwashing detergent works well). While washing it, scrub the cylinder bores with a soft bristle brush and thoroughly clean all of the oil passages. Completely dry the engine and spray the entire assembly down with an anti-rust solution such as WD-40, or similar product. Take a clean lint-free rag and wipe up any excess anti-rust solution from the bores, bearing saddles, etc. Repeat the final cleaning process on the crankshaft. Replace any freeze or oil galley plugs which were removed during disassembly.

Crankshaft
See Figures 203, 204, 205 and 206

1. Remove the main bearing inserts from the block and bearing caps.
2. If the crankshaft main bearing journals have

3-62 ENGINE AND ENGINE OVERHAUL

Fig. 207 Checking the piston ring-to-ring groove side clearance using the ring and a feeler gauge

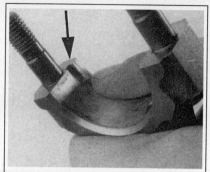

Fig. 208 The notch on the side of the bearing cap matches the tang on the bearing insert

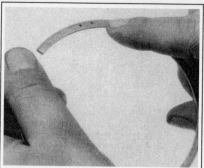

Fig. 209 Most rings are marked to show which side of the ring should face up when installed to the piston. In this application, two marks indicates a lower compression ring

been refinished to a definite undersize, install the correct undersize bearing. Be sure that the bearing inserts and bearing bores are clean. Foreign material under inserts will distort bearing and cause failure.

3. Place the upper main bearing inserts in bores with tang in slot.

Note: The oil holes in the bearing inserts must be aligned with the oil holes in the cylinder block.

4. Install the lower main bearing inserts in bearing caps.
5. Clean the mating surfaces of block and rear main bearing cap.
6. Carefully lower the crankshaft into place. Be careful not to damage bearing surfaces.
7. Check the clearance of each main bearing by using the following procedure:
  a. Place a piece of Plastigage, or its equivalent, on bearing surface across full width of bearing cap and about 1/4 in. off center.
  b. Install cap and tighten bolts to specifications. Do not turn crankshaft while Plastigage is in place.
  c. Remove the cap. Using the supplied Plastigage scale, check width of Plastigage, at widest point to get maximum clearance. Difference between readings is taper of journal.
  d. If clearance exceeds specified limits, try a 0.001 in. or 0.002 in. undersize bearing in combination with the standard bearing. Bearing clearance must be within specified limits. If standard and 0.002 in. undersize bearing does not bring clearance within desired limits, refinish crankshaft journal, then install undersize bearings.
8. After the bearings have been fitted, apply a light coat of engine oil to the journals and bearings. Install the rear main bearing cap. Install all bearing caps except the thrust bearing cap. Be sure that main bearing caps are installed in original locations. Tighten the bearing cap bolts to specifications.
9. Install the thrust bearing cap with bolts finger-tight.
10. Pry the crankshaft forward against the thrust surface of upper half of bearing.
11. Hold the crankshaft forward and pry the thrust bearing cap to the rear. This aligns the thrust surfaces of both halves of the bearing.
12. Retain the forward pressure on the crankshaft. Tighten the cap bolts to specifications.
13. Measure the crankshaft end-play as follows:
  a. Mount a dial gauge to the engine block and position the tip of the gauge to read from the crankshaft end.

  b. Carefully pry the crankshaft toward the rear of the engine and hold it there while you zero the gauge.
  c. Carefully pry the crankshaft toward the front of the engine and read the gauge.
  d. Confirm that the reading is within specifications. If not, install a new thrust bearing and repeat the procedure. If the reading is still out of specifications with a new bearing, have a machine shop inspect the thrust surfaces of the crankshaft, and if possible, repair it.
14. Rotate the crankshaft so as to position the first rod journal to the bottom of its stroke.
15. Install the rear main seal.

Pistons and Connecting Rods
See Figures 207, 208, 209, 210, 211, 212, 213 and 214

16. Before installing the piston/connecting rod assembly, oil the pistons, piston rings and the cylinder walls with light engine oil. Install connecting rod bolt protectors or rubber hose onto the connecting rod bolts/studs. Also perform the following:
  a. Select the proper ring set for the size cylinder bore.
  b. Position the ring in the bore in which it is going to be used.
  c. Push the ring down into the bore area where normal ring wear is not encountered.
  d. Use the head of the piston to position the ring in the bore so that the ring is square with the cylinder wall. Use caution to avoid damage to the ring or cylinder bore.
  e. Measure the gap between the ends of the ring with a feeler gauge. Ring gap in a worn cylinder is normally greater than specification. If the ring gap is greater than the specified limits, try an oversize ring set.
  f. If ring gap is less than specified, carefully file the ends until proper gap is achieved. This should be done using a jig to hold the ring, otherwise it will be difficult to keep ring ends parallel.
  g. Check the ring side clearance of the compression rings with a feeler gauge inserted between the ring and its lower land according to specification. The gauge should slide freely around the entire ring circumference without binding. Any wear that occurs will form a step at the inner portion of the lower land. If the lower lands have high steps, the piston should be replaced.
17. Unless new pistons are installed, be sure to install the pistons in the cylinders from which they were removed. The numbers on the connecting rod

and bearing cap must be on the same side when installed in the cylinder bore. If a connecting rod is ever transposed from one engine or cylinder to another, new bearings should be fitted and the connecting rod should be numbered to correspond with the new cylinder number. The notch on the piston head goes toward the front of the engine.

18. Install all of the rod bearing inserts into the rods and caps.
19. Rings are purchased in sets and are normally identified for type and position, although marking systems vary.
  • On the 2.5L engine, the top compression ring is unmarked; the lower compression ring has one punch mark
  • On most other engines, the top compression ring has ONE mark, the lower compression ring has TWO
  • Rings may also be marked "TOP" or "0", or have an oval depression
  • Rings are always installed with marks UP
  • Some rings may have a chamfer on the inner side. This is usually installed facing UP on top compression rings and DOWN on second compression - but ring marks, if fitted, determine orientation
20. Install the oil control ring first, then the second compression ring and finally the top compression ring. Use a piston ring expander tool to aid in installation and to help reduce the chance of breakage. With a multi-piece oil ring, install the lower rail first, winding it on until beneath the ring land. Then install the spacer, and finally the upper rail. Then install the rails on either side of the spacer.

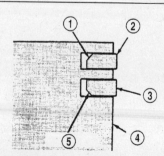

Fig. 210 Proper installation of chamfered compression rings: top compression (2), lower compression (3), piston (4), ring chamfers (1,5)

ENGINE AND ENGINE OVERHAUL 3-63

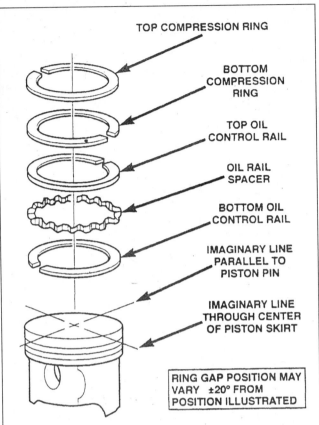

Fig. 211 Stagger compression ring end gaps 180 degrees apart on the 2.5L engine. Place the oil rail spacer on the centerline of the wrist pin bore and the two rails 90 degrees on either side.

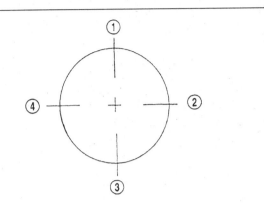

Fig. 212 Ring end gap positions on the 3.9L, 5.2L and 5.9L "V" engines: oil ring spacer (1), lower compression ring AND top oil rail (2), lower oil rail (3), top compression ring (4)

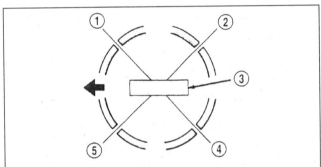

Fig. 213 Ring end gap positions - 4.7L engine: arrow indicates front of engine; upper oil rail (1), top compression (2), piston pin (3), lower oil rail (4), lower compression and oil spacer (5)

21. Make sure the ring gaps are properly spaced around the circumference of the piston. See illustrations. Manufacturer's recommendations differ for different engines. The important thing is that the end gaps of adjacent rings do not align.
- Diesels have a one-piece oil ring and end gaps should be 120 degrees apart
- On the 8.0L engine, position compression ring gaps 180 degrees apart, position the oil ring spacer gap at the rear of the piston and the two oil rail end gaps 90 degrees on either side with the top rail beneath the top compression ring end gap
- See illustrations for other engines

22. Fit a piston ring compressor around the piston and slide the piston and connecting rod assembly down into the cylinder bore, pushing it in with the wooden hammer handle. Push the piston down until it is only slightly below the top of the cylinder bore. Guide the connecting rod onto the crankshaft bearing journal carefully, to avoid damaging the crankshaft.

23. Check the bearing clearance of all the rod bearings, fitting them to the crankshaft bearing journals. Follow the procedure in the crankshaft installation above.

24. After the bearings have been fitted, apply a light coating of assembly oil to the journals and bearings.

25. Turn the crankshaft until the appropriate bearing journal is at the bottom of its stroke, then push the piston assembly all the way down until the connecting rod bearing seats on the crankshaft journal. Be careful not to allow the bearing cap screws to strike the crankshaft bearing journals and damage them.

26. After the piston and connecting rod assemblies have been installed, check the connecting rod side clearance on each crankshaft journal.

27. Prime and install the oil pump and the oil pump intake tube.

OHV Engines

CAMSHAFT, LIFTERS AND TIMING ASSEMBLY

1. Install the camshaft.
2. Install the lifters/followers into their bores.
3. Install the timing gears/chain assembly.

CYLINDER HEAD(S)

1. Install the cylinder head(s) using new gaskets.
2. Assemble the rest of the valve train (pushrods and rocker arms and/or shafts).

4.7L Engine

CYLINDER HEAD(S)

1. Install the cylinder head(s) using new gaskets.
2. Install the timing sprockets/gears and the belt/chain assemblies.

Engine Covers and Components

Install the timing cover(s) and oil pan. Refer to your notes and drawings made prior to disassembly and install all of the components that were removed. Install the engine into the vehicle.

Fig. 214 Install the piston and rod assembly into the block using a ring compressor and the handle of a hammer

Engine Start-up and Break-in

STARTING THE ENGINE

Now that the engine is installed and every wire and hose is properly connected, go back and double check that all coolant and vacuum hoses are connected. Check that your oil drain plug is installed and properly tightened. If not already done, install a new oil filter onto the engine. Fill the crankcase with the proper amount and grade of engine oil. Fill the cooling system with a 50/50 mixture of coolant/water.

1. Connect the vehicle battery.

ENGINE AND ENGINE OVERHAUL

2. Start the engine. Keep your eye on your oil pressure indicator; if it does not indicate oil pressure within 10 seconds of starting, turn the vehicle off.

WARNING:

Damage to the engine can result if it is allowed to run with no oil pressure. Check the engine oil level to make sure that it is full. Check for any leaks and if found, repair the leaks before continuing. If there is still no indication of oil pressure, you may need to prime the system.

3. Confirm that there are no fluid leaks (oil or other).

4. Allow the engine to reach normal operating temperature (the upper radiator hose will be hot to the touch).

5. At this point you can perform any necessary checks or adjustments, such as checking the ignition timing.

6. Install any remaining components or body panels which were removed.

BREAKING IT IN

Make the first miles on the new engine easy ones. Vary the speed but do not accelerate hard. Most importantly, do not lug the engine, and avoid sustained high speeds until at least 100 miles. Check the engine oil and coolant levels frequently. Expect the engine to use a little oil until the rings seat. Change the oil and filter at 500 miles, 1500 miles, then every 3000 miles past that.

KEEP IT MAINTAINED

Now that you have just gone through all of that hard work, keep yourself from doing it all over again by thoroughly maintaining it. Not that you may not have maintained it before; heck, you could have had one to two hundred thousand miles on it before doing this. However, you may have bought the vehicle used, and the previous owner did not keep up on maintenance. Which is why you just went through all of that hard work. See?

ENGINE AND ENGINE OVERHAUL 3-65

2.5L TORQUE SPECIFICATIONS

Components	English	Metric
A/C compressor bracket-to-engine		
1997-99	25 ft. lbs.	34 Nm
2000	35 ft. lbs.	47 Nm
A/C compressor mounting bolts	20 ft. lbs.	27 Nm
Alternator		
Mounting bolts	42 ft. lbs.	57 Nm
Mounting bracket-to-engine	35 ft. lbs.	47 Nm
Block heater nut	16 in. lbs.	1.8 Nm
Camshaft sprocket bolt	80 ft. lbs.	108 Nm
Clutch cover-to-flywheel bolts	23 ft. lbs.	31 Nm
Connecting rod nuts	33 ft. lbs.	45 Nm
Cylinder block drain plugs	30 ft. lbs.	41 Nm
Cylinder head bolts**		
Nos. 1-6	110 ft. lbs.	149 Nm
No. 7	100 ft. lbs.	135 Nm
Cylinder head cover	115 inch. lbs.	13 Nm
Distributor clamp bolt	204 inch. lbs.	23 Nm
Engine shock damper stud nuts	17 ft. lbs.	23 Nm
Engine mounts - front		
Insulator bracket bolts	60 ft. lbs.	81 Nm
Insulator bracket nuts	35 ft. lbs.	47 Nm
Insulator thru-bolt	60 ft. lbs.	81 Nm
Engine mounts - rear		
Support cushion/crossmember nuts	192 inch. lbs.	22 Nm
Support cushion/bracket nuts	34 ft. lbs.	46 Nm
Transmission support bracket bolts	32 ft. lbs.	43 Nm
Transmission support bracket/cushion bolt	55 ft. lbs.	75 Nm
Transmission support adaptor bracket bolts	55 ft. lbs.	75 Nm
Exhaust manifold/pipe nuts	20 ft. lbs.	27 Nm
Exhaust manifold		
Bolt #1	30 ft. lbs.	41 Nm
Bolts #2-5	23 ft. lbs.	31 Nm
Nuts #6 & 7	11 ft. lbs.	14 Nm
Flywheel/converter housing bolts	28 ft. lbs.	38 Nm
Flywheel/crankshaft bolts	105 ft. lbs.	143 Nm
Front cover-to-block		
1/4-20	60 inch. lbs.	7 Nm
5/16-18	192 inch. lbs.	22 Nm
Main bearing bolts	80 ft. lbs.	108 Nm
Oil filter	13 ft. lbs	18 Nm
Oil filter connector	40 ft. lbs.	54 Nm
Oil gallery plug	30 ft. lbs.	41 Nm
Oil pan bolts		
1/4-20 bolts	129 inch. lbs.	14 Nm
5/16-18 bolts	156 inch. lbs.	18 Nm
Oil pump	30 ft. lbs.	41 Nm
Short	17 ft. lbs.	23 Nm
Long	17 ft. lbs.	23 Nm
Cover	70 inch. lbs.	8 Nm
P/S Pump pressure hose	38 ft. lbs.	52 Nm
Rocker arm ass'y-to-head	21 ft. lbs.	28 Nm
Spark plugs	27 ft. lbs.	37 Nm
Starter motor mounting bolts	33 ft. lbs.	45 Nm
Tensioner bracket at cylinder block	14 ft. lbs.	19 Nm
Thermostat housing	13 ft. lbs.	18 Nm
Throttle body bolts	90 inch lbs.	10 Nm
Vibration damper	80 ft. lbs.	108 Nm
Water pump bolts	23 ft. lbs.	31 Nm

** See Procedure

ENGINE AND ENGINE OVERHAUL

3.9L TORQUE SPECIFICATIONS

Components	English	Metric
Alternator M\mounting bolts	30 ft. lbs.	41 Nm.
Camshaft sprocket bolt	50 ft. lbs.	68 Nm.
Camshaft thrust plate bolts	210 inch lbs.	24 Nm.
Connecting rod cap bolts	45 ft. lbs.	61 Nm
Crankshaft pulley bolts	210 inch lbs.	24 Nm.
Cylinder head bolts (see procedure)	105 ft. lbs.	143 Nm.
Step 1	50 ft. lbs.	68 Nm
Step 2	105 ft. lbs.	143 Nm.
Cylinder head cover bolts	95 inch lbs.	11 Nm
Engine Support Bracket-to-Block (4WD)	30 ft. lbs.	41 Nm.
Exhaust manifold	25 ft. lbs.	34 Nm.
Flywheel Bolts	55 ft. lbs.	75 Nm
Front Insulator Through Bolts	70 ft. lbs.	95 Nm
Front Insulator-To-Support Bracket (4WD)		
Stud Nut	30 ft. lbs.	41 Nm.
Through Bolt/Nut	75 ft. lbs.	102 Nm.
Front Insulator-To-Block (2WD)	70 ft. lbs.	95 Nm
Main bearing cap bolts	85 ft. lbs.	115 Nm.
Oil pan bolts	215 inch lbs.	24 Nm.
Oil pan drain plug	25 ft. lbs.	34 Nm.
Oil pump bolts	30 ft. lbs.	41 Nm.
Oil Pump Cover	95 in. lbs.	11 Nm
Rear Insulator-To-Bracket Through Bolt (2WD)	50 ft. lbs.	68 Nm
Rear Insulator-To-Crossmember		
Support Bracket (2WD)	30 ft. lbs.	41 Nm.
Rear Insulator-To-Crossmember Nuts (4WD)	50 ft. lbs.	68 Nm
Rear Insulator-To-Transmission (4WD)	50 ft. lbs.	68 Nm
Rear Insulator Bracket Bolts (4WD/Auto)	50 ft. lbs.	68 Nm
Rear Support Bracket, Plate	30 ft. lbs.	41 Nm.
Rocker arm retaining bolts	21 ft. lbs.	28 Nm.
Spark Plugs	30 ft. lbs.	41 Nm.
Starter Motor Bolts	50 ft. lbs.	68 Nm.
Thermostat housing bolts	225 inch lbs.	25 Nm.
Throttle body-to-intake manifold	200 inch lbs.	23 Nm
Torque converter-to-drive plate	270 inch lbs.	31 Nm
Transfer Case-To-Insulator Mount	150 ft. lbs.	204 Nm
Transmission support bolts	50 ft. lbs.	68 Nm
Vibration Damper Bolt	135 ft. lbs.	183 Nm
Water pump-to-timing chain case cover bolts	30 ft. lbs.	41 Nm.

4.7L TORQUE SPECIFICATIONS

Components	English	Metric
Alternator Mounting Bolts		
M10	40 ft. lbs.	54 Nm
M8	250 inch lbs.	28 Nm
Camshaft sprocket bolt		
Sprocket bolt (non-oiled)	90 ft. lbs.	122 Nm
Bearing cap bolts	100 inch lbs.	11 Nm
Connecting rod cap bolts	20 ft. lbs. plus 1/4 turn	27 Nm plus 1/4 turn
Crankshaft damper bolt	130 ft. lbs.	175 Nm
Cylinder head bolts	105 ft. lbs.	143 Nm
M11	60 ft. lbs.	81 Nm
M8	250 inch lbs.	28 Nm
Cylinder head cover bolts	105 inch lbs.	12 Nm
Engine mount bracket-to-block	45 ft. lbs.	61 Nm
Rear mount-to-transmission	34 ft. lbs.	46 Nm
Exhaust manifold	18 ft. lbs.	25 Nm
Exhaust manifold heat shield nuts	72 inch lbs. minus 3/4 turn	8 Nm minus 3/4 turn
Flexplate bolts	45 ft. lbs.	60 Nm
Intake manifold bolts	105 inch lbs.	12 Nm
Main bearing cap (bedplate) bolts		
Step 1		
Tighten bolts 1-12	40 ft. lbs.	54 Nm
Step 2		
Tighten bolts 13-22	25 inch lbs.	2.8 Nm
Step 3		
Tighten bolts 13-22	Turn an additional 90-degrees	
Step 4		
Tighten bolts 23-28	20 ft. lbs	27 Nm
Oil pan bolts		
Oil pan bolts	130 inch lbs.	15 Nm
Oil pan drain plug	25 ft. lbs.	34 Nm
Oil pump bolts	250 inch lbs.	28 Nm
Oil Pump Cover	105 inch lbs.	12 Nm
Oil pickup tube	250 inch lbs.	28 nm
Oil fill tube	105 inch lbs.	12 Nm
Timing chain tensioner arm	150 inch lbs.	17 Nm
Hydraulic tensioner bolts	250 inch lbs.	28 Nm
Timing chain primary tensioner	250 inch lbs.	28 Nm
Timing drive idler sprocket	25 ft. lbs.	34 Nm
Thermostat housing bolts	105 inch lbs.	12 Nm
Water pump bolts	40 ft. lbs.	54 Nm

5.2L, 5.9L Gasoline Engine TORQUE SPECIFICATIONS

Components	English	Metric
Alternator Mounting Bolts	30 ft. lbs.	41 Nm.
Camshaft sprocket bolt	50 ft. lbs.	68 Nm.
Camshaft Thrust Plate Bolts	210 in. lbs.	24 Nm.
Chain Case Cover Bolts	30 ft. lbs.	41 Nm.
Connecting Rod Cap Bolts	45 ft. lbs.	61 Nm
Crankshaft Pulley Bolts	210 in. lbs.	24 Nm.
Cylinder head bolts	105 ft. lbs.	143 Nm.
Step 1	50 ft. lbs.	68 Nm
Step 2	105 ft. lbs.	143 Nm.
Cylinder Head Cover Bolts	95 in. lbs.	11 Nm
Engine Support Bracket-to-Block (4WD)	30 ft. lbs.	41 Nm.
Exhaust manifold	25 ft. lbs.	34 Nm.
Flywheel Bolts	55 ft. lbs.	75 Nm
Front Insulator Through Bolts	70 ft. lbs.	95 Nm
Front Insulator-To-Support Bracket (4WD)		
Stud Nut	30 ft. lbs.	41 Nm.
Through Bolt/Nut	75 ft. lbs.	102 Nm
Front Insulator-To-Block (2WD)	70 ft. lbs.	95 Nm
Main bearing cap bolts	85 ft. lbs.	115 Nm.
Oil pan bolts	215 inch lbs.	24 Nm.
Oil pan drain plug	25 ft. lbs.	34 Nm.
Oil pump bolts	30 ft. lbs.	41 Nm.
Oil Pump Cover	95 in. lbs.	11 Nm
Rear Insulator-To-Bracket Through Bolt (2WD)	50 ft. lbs.	68 Nm
Rear Insulator-To-Crossmember		
Support Bracket (2WD)	30 ft. lbs.	41 Nm.
Rear Insulator-To-Crossmember Nuts (4WD)	50 ft. lbs.	68 Nm
Rear Insulator-To-Transmission (4WD)	50 ft. lbs.	68 Nm
Rear Insulator Bracket Bolts (4WD/Auto)	50 ft. lbs.	68 Nm
Rear Support Bracket-to-		
Crossmember Flange Nuts	30 ft. lbs.	41 Nm.
Rear Support Plate-to-Transfer Case	30 ft. lbs.	41 Nm.
Rocker arm bolts	21 ft. lbs.	28 Nm.
Spark Plugs	30 ft. lbs.	41 Nm.
Starter Motor Bolts	50 ft. lbs.	68 Nm.
Thermostat housing bolts	225 inch lbs.	25 Nm.
Throttle body-to-intake manifold	200 inch lbs.	23 Nm
Torque converter-to-drive plate	23 ft. lbs.	31 Nm
Transfer Case-To-Insulator Mount	150 ft. lbs.	204 Nm
Transmission support bracket (2WD)	50 ft. lbs.	68 Nm
Vibration Damper Bolt	135 ft. lbs.	183 Nm
Water pump bolts	30 ft. lbs.	41 Nm.

ENGINE AND ENGINE OVERHAUL

5.2L, 5.9L Gasoline Engine TORQUE SPECIFICATIONS

Components	English	Metric
Alternator Mounting Bolts	30 ft. lbs.	41 Nm.
Camshaft sprocket bolt	50 ft. lbs.	68 Nm.
Camshaft Thrust Plate Bolts	210 in. lbs.	24 Nm.
Chain Case Cover Bolts	30 ft. lbs.	41 Nm.
Connecting Rod Cap Bolts	45 ft. lbs.	61 Nm
Crankshaft Pulley Bolts	210 in. lbs.	24 Nm.
Cylinder head bolts	105 ft. lbs.	143 Nm.
Step 1	50 ft. lbs.	68 Nm
Step 2	105 ft. lbs.	143 Nm.
Cylinder Head Cover Bolts	95 in. lbs.	11 Nm
Engine Support Bracket-to-Block (4WD)	30 ft. lbs.	41 Nm.
Exhaust manifold	25 ft. lbs.	34 Nm.
Flywheel Bolts	55 ft. lbs.	75 Nm
Front Insulator Through Bolts	70 ft. lbs.	95 Nm
Front Insulator-To-Support Bracket (4WD)		
Stud Nut	30 ft. lbs.	41 Nm.
Through Bolt/Nut	75 ft. lbs.	102 Nm
Front Insulator-To-Block (2WD)	70 ft. lbs.	95 Nm
Main bearing cap bolts	85 ft. lbs.	115 Nm.
Oil pan bolts	215 inch lbs.	24 Nm.
Oil pan drain plug	25 ft. lbs.	34 Nm.
Oil pump bolts	30 ft. lbs.	41 Nm.
Oil Pump Cover	95 in. lbs.	11 Nm
Rear Insulator-To-Bracket Through Bolt (2WD)	50 ft. lbs.	68 Nm
Rear Insulator-To-Crossmember Support Bracket (2WD)	30 ft. lbs.	41 Nm.
Rear Insulator-To-Crossmember Nuts (4WD)	50 ft. lbs.	68 Nm
Rear Insulator-To-Transmission (4WD)	50 ft. lbs.	68 Nm
Rear Insulator Bracket Bolts (4WD/Auto)	50 ft. lbs.	68 Nm
Rear Support Bracket-to-Crossmember Flange Nuts	30 ft. lbs.	41 Nm.
Rear Support Plate-to-Transfer Case	30 ft. lbs.	41 Nm.
Rocker arm bolts	21 ft. lbs.	28 Nm.
Spark Plugs	30 ft. lbs.	41 Nm.
Starter Motor Bolts	50 ft. lbs.	68 Nm.
Thermostat housing bolts	225 inch lbs.	25 Nm.
Throttle body-to-intake manifold	200 inch lbs.	23 Nm
Torque converter-to-drive plate	23 ft. lbs.	31 Nm
Transfer Case-To-Insulator Mount	150 ft. lbs.	204 Nm
Transmission support bracket (2WD)	50 ft. lbs.	68 Nm
Vibration Damper Bolt	135 ft. lbs.	183 Nm
Water pump bolts	30 ft. lbs.	41 Nm.

ENGINE AND ENGINE OVERHAUL

12-Valve Diesel TORQUE SPECIFICATIONS

Components	English	Metric
Air fuel control fitting	72 inch lbs.	8 Nm
Battery cable to block bolt	57 ft. lbs.	77 Nm
Belt tensioner bolt	32 ft. lbs.	43 Nm
Block Heater Element	108 inch lbs.	12 Nm
Cab heater hose clamp screw	35 inch lbs.	4 Nm
Cab heater tubing bracket bolt	84 inch lbs.	9 Nm
Camshaft thrust plate bolts	18 ft. lbs.	24 Nm
Clutch cover-to-flywheel bolts	17 ft. lbs.	23 Nm
Connecting rod nuts		
Step 1	26 ft. lbs.	35 Nm
Step 2	51 ft. lbs.	70 Nm
Step 3	73 ft. lbs.	100 Nm
Cooling fan to fan clutch	15 ft. lbs.	20 Nm
Main bearing bolts		
Step 1	44 ft. lbs.	60 Nm
Step 2	88 ft. lbs.	119 Nm
Step 3	129 ft. lbs.	176 Nm
Cylinder head bolts		
Step 1	66 ft. lbs.	90 Nm
Step 2	66 ft. lbs.	90 Nm
Step 3 (Long bolts)	90 ft. lbs.	120 Nm
Step 4 (Long bolts)	90 ft. lbs.	120 Nm
Step 5	Another 1/4 turn	Another 1/4 turn
Cylinder head cover	18 ft. lbs.	24 Nm
Exhaust manifold bolts	32 ft. lbs.	43 Nm
Fan clutch mounting-to-fan hub (LH)	42 ft. lbs.	57 Nm
Fan hub bracket bolts	18 ft. lbs.	24 Nm
Fan hub bearing bolt	57 ft. lbs.	77 Nm
Fan pulley-to-fan hub bolts	84 inch lbs.	9 Nm
Fan housing bolts	44 ft. lbs.	60 Nm
Fan shroud mounting bolts	95 inch lbs.	11 Nm
Flywheel bolts	101 ft. lbs.	137 Nm
Generator mounting bolts	30 ft. lbs.	41 Nm
Generator pulley nut	59 ft. lbs.	80 Nm
Generator support bolts	18 ft. lbs.	24 Nm
Gear cover bolts	18 ft. lbs.	24 Nm
Gear cover housing bolts	18 ft. lbs.	24 Nm
Intake manifold cover bolts	18 ft. lbs.	24 Nm
Intercooler attaching bolts	17 inch lbs.	2 Nm
Intercooler duct clamp nuts	72 inch lbs.	8 Nm
Lift bracket (rear) bolts	57 ft. lbs.	77 Nm
Oil cooler bolts	18 ft. lbs.	24 Nm
Oil fill tube bracket bolt	32 ft. lbs.	43 Nm
Oil pan drain plug	44 ft. lbs.	60 Nm
Oil pan bolts	18 ft. lbs.	24 Nm
Oil pump bolts	18 ft. lbs.	24 Nm
Oil pressure regulator plug	60 ft. lbs.	80 Nm
Oil suction tube flange bolts	18 ft. lbs.	24 Nm
Oil suction tube brace bolt	18 ft. lbs.	24 Nm
Oil pressure sender switch	12 ft. lbs.	16 Nm
Oil supply to vacuum pump nut	89 inch lbs.	10 Nm
Rear mount		
Support cushion-to-crossmember nut	35 ft. lbs.	47 Nm
Support cushion-to-support bracket nuts	35 ft. lbs.	47 Nm
Support bracket-to-transmission bolts	75 ft. lbs.	102 Nm
Rear support plate to transfer case bolts	30 ft. lbs.	41 Nm
Rocker arm retaining bolts	18 ft. lbs.	24 Nm
Starter bolts	50 ft. lbs.	68 Nm
Torque converter-to-drive plate	35 ft. lbs.	47 Nm
Transfer case-to-insulator mounting plate	150 ft. lbs.	204 Nm
Transmission support bracket bolts (2WD)	50 ft. lbs.	68 Nm
Transmission support spacer bolts (4WD)	50 ft. lbs.	68 Nm
Transmission support		
spacer-to-mounting plate (4WD)	150 ft. lbs.	204 Nm
Vacuum pump-to-adapter nuts	18 ft. lbs.	24 Nm
Vacuum pump adapter-to-P/S pump nuts	18 ft. lbs.	24 Nm
Vacuum pump-to-gear housing bolts	57 ft. lbs.	77 Nm
Vacuum pump oil supply line fitting	89 inch lbs.	10 Nm
Vibration damper bolts	92 ft. lbs.	125 Nm
Water pump bolts	18 ft. lbs.	24 Nm

24-Valve Diesel TORQUE SPECIFICATIONS

Components	English	Metric
Battery cable to block bolt	57 ft. lbs.	77 Nm
Belt tensioner bolt	32 ft. lbs.	43 Nm
Block Heater Element	32 ft. lbs.	43 Nm
Camshaft thrust plate bolts	18 ft. lbs.	24 Nm
Charge air cooler bolts	17 inch lbs.	2 Nm
Clutch cover-to-flywheel bolts	17 ft. lbs.	23 Nm
Connecting rod bolts		
Step 1	26 ft. lbs.	35 Nm
Step 2	51 ft. lbs.	70 Nm
Step 3	73 ft. lbs.	100 Nm
Cooling fan to fan clutch	15 ft. lbs.	20 Nm
Main bearing bolts		
Step 1	44 ft. lbs.	60 Nm
Step 2	88 ft. lbs.	119 Nm
Step 3	129 ft. lbs.	176 Nm
Crankshaft damper bolts	92 ft. lbs.	125 Nm
Crankshaft rear seal retainer bolts	80 inch lbs.	9 Nm
Crankshaft tone wheel bolts	71 inch lbs.	8 Nm
Cylinder head bolts		
Step 1	59 ft. lbs.	80 Nm
Step 2	77 ft. lbs.	105 Nm
Step 3	77 ft. lbs.	105 Nm
Step 4	1/4 turn more	1/4 turn more
Cylinder head cover	18 ft. lbs.	24 Nm
Exhaust manifold bolts	32 ft. lbs.	43 Nm
Fan clutch mounting-to-fan hub	42 ft. lbs.	57 Nm
Fan hub bracket bolts	18 ft. lbs.	24 Nm
Fan hub bearing bolt	57 ft. lbs.	77 Nm
Fan pulley-to-fan hub bolts	84 inch lbs.	9 Nm
Fan housing bolts	44 ft. lbs.	60 Nm
Fan shroud mounting bolts	95 inch lbs.	11 Nm
Flywheel housing access plate	18 ft. lbs.	24 Nm
Flywheel bolts	101 ft. lbs.	137 Nm
Fuel delivery lines (high pressure)		
At pump	18 ft. lbs.	24 Nm
At cylinder head	28 ft. lbs.	38 Nm
Fuel drain line, banjo (rear of head)	18 ft. lbs.	24 Nm
Fuel filter cannister nut	10 ft. lbs.	14 Nm
Fuel injection pump gear retaining nut	125 ft. lbs.	170 Nm
Fuel injection pump support bracket bolts	18 ft. lbs.	24 Nm
Low pressure fuel line banjos	18 ft. lbs.	24 Nm
Generator mounting bolts	30 ft. lbs.	41 Nm
Generator pulley nut	59 ft. lbs.	80 Nm
Generator support bolts	18 ft. lbs.	24 Nm
Gear housing-to-block bolts	18 ft. lbs.	24 Nm
Gear housing cover bolts	18 ft. lbs.	24 Nm
Injector clamp bolts	89 inch lbs.	10 Nm
Injector pump-to-gear housing nuts	32 ft. lbs.	43 Nm
Intake manifold cover bolts	18 ft. lbs.	24 Nm
Charge air cooler pipe clamp nuts	72 inch lbs.	8 Nm
Lift bracket (rear) bolts	57 ft. lbs.	77 Nm
Lift pump mounting nuts	9 ft. lbs.	12 Nm
Lift pump mounting bracket bolts	18 ft. lbs.	24 Nm

ENGINE AND ENGINE OVERHAUL

Components	English	Metric
Oil cooler bolts	18 ft. lbs.	24 Nm
Oil pan drain plug	44 ft. lbs.	60 Nm
Oil pan bolts	18 ft. lbs.	24 Nm
Oil pump bolts	18 ft. lbs.	24 Nm
Oil pressure regulator plug	60 ft. lbs.	80 Nm
Oil suction tube flange bolts	18 ft. lbs.	24 Nm
Oil suction tube brace bolt	18 ft. lbs.	24 Nm
Oil pressure sender switch	12 ft. lbs.	16 Nm
Oil supply to vacuum pump nut	89 inch lbs.	10 Nm
Rear mount		
Support cushion-to-crossmember nut	35 ft. lbs.	47 Nm
Support cushion-to-support bracket nuts	35 ft. lbs.	47 Nm
Support bracket-to-transmission bolts	75 ft. lbs.	102 Nm
Rear support plate to transfer case bolts	30 ft. lbs.	41 Nm
Rocker arm/pedestal bolts	27 ft. lbs.	36 Nm
Rocker arm retaining bolts	21 ft. lbs.	28 Nm
Starter bolts	32 ft. lbs.	43 Nm
Thermostat housing bolts	18 ft. lbs.	24 Nm
Throttle control bracket-to-cylinder	40 ft. lbs.	56 Nm
Torque converter-to-drive plate	35 ft. lbs.	47 Nm
Transfer case-to-insulator mounting plate nuts	150 ft. lbs.	204 Nm
Transmission support bracket bolts (2WD)	50 ft. lbs.	68 Nm
Transmission support spacer bolts (4WD)	50 ft. lbs.	68 Nm
Transmission support spacer-to-mounting plate (4WD)	150 ft. lbs.	204 Nm
Turbocharger/CAC clamp(s) nut	71 inch lbs.	8 Nm
Turbocharger oil supply line nut	15 ft. lbs.	20 Nm
Turbocharger oil drain pipe bolts	20 ft. lbs.	27 Nm
Turbocharger-to-exhaust manifold nuts	33 ft. lbs.	45 Nm
Vacuum pump-to-adapter nuts	18 ft. lbs.	24 Nm
Vacuum pump adapter-to-P/S pump nuts	18 ft. lbs.	24 Nm
Vacuum pump-to-gear housing bolts	57 ft. lbs.	77 Nm
Vacuum pump oil supply line fitting	89 inch lbs.	10 Nm
Valve cover bolts	95 inch lbs.	11 Nm
Water pump bolts	18 ft. lbs.	24 Nm
Water-in-fuel sensor	18 ft. lbs.	24 Nm

AIR POLLUTION
AUTOMOTIVE POLLUTANTS 4-2
INDUSTRIAL POLLUTANTS 4-2
NATURAL POLLUTANTS 4-2

AUTOMOTIVE EMISSIONS
CRANKCASE EMISSIONS 4-4
EVAPORATIVE EMISSIONS 4-4
EXHAUST GASES 4-3

CLEARING CODES
CONTINUOUS MEMORY CODES 4-26

ELECTRONIC ENGINE CONTROLS
CAMSHAFT POSITION (CMP) SENSOR 4-19
CRANKSHAFT POSITION (CKP) SENSOR 4-21
ENGINE COOLANT TEMPERATURE (ECT)
    SENSOR 4-15
IDLE AIR CONTROL MOTOR (IAC) 4-13
INTAKE AIR TEMPERATURE (IAT) SENSOR 4-16
MANIFOLD ABSOLUTE PRESSURE (MAP)
    SENSOR 4-16
OXYGEN (O2) SENSOR 4-12
POWERTRAIN CONTROL MODULE (PCM) 4-12
THROTTLE POSITION SENSOR (TPS) 4-18

EMISSION CONTROLS
AIR INJECTION SYSTEM 4-11
CATALYTIC CONVERTER 4-10
CRANKCASE VENTILATION SYSTEM 4-6
EVAPORATIVE EMISSION CONTROLS 4-6
EXHAUST GAS RECIRCULATION SYSTEM 4-9
MALFUNCTION INDICATOR LAMP 4-11
POSITIVE CRANKCASE 4-4

TROUBLE CODES
CLEARING CODES 4-26
DIAGNOSTIC CONNECTOR 4-22
GENERAL INFORMATION 4-21
READING CODES 4-22

VACUUM DIAGRAMS 4-27

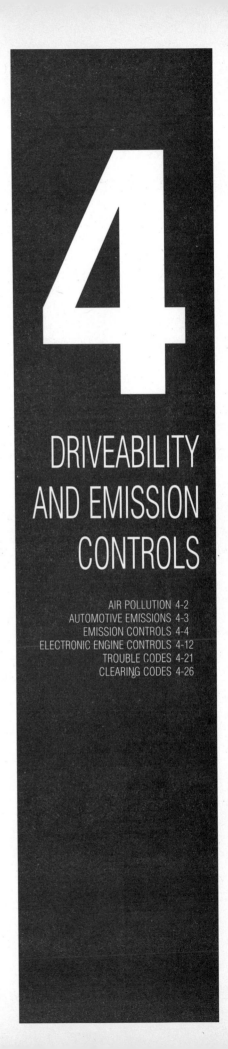

4

DRIVEABILITY AND EMISSION CONTROLS

AIR POLLUTION 4-2
AUTOMOTIVE EMISSIONS 4-3
EMISSION CONTROLS 4-4
ELECTRONIC ENGINE CONTROLS 4-12
TROUBLE CODES 4-21
CLEARING CODES 4-26

4-2 DRIVEABILITY AND EMISSION CONTROLS

AIR POLLUTION

The earth's atmosphere, at or near sea level, consists approximately of 78 percent nitrogen, 21 percent oxygen and 1 percent other gases. If it were possible to remain in this state, 100 percent clean air would result. However, many varied sources allow other gases and particulates to mix with the clean air, causing our atmosphere to become unclean or polluted.

Some of these pollutants are visible while others are invisible, with each having the capability of causing distress to the eyes, ears, throat, skin and respiratory system. Should these pollutants become concentrated in a specific area and under certain conditions, death could result due to the displacement or chemical change of the oxygen content in the air. These pollutants can also cause great damage to the environment and to the many man-made objects that are exposed to the elements.

To better understand the causes of air pollution, the pollutants can be categorized into 3 separate types, natural, industrial and automotive.

Natural Pollutants

Natural pollution has been present on earth since before man appeared and continues to be ignored as a factor when discussing air pollution, although it causes a significant percentage of the so-called pollution problem. It is the direct result of decaying organic matter, wind-borne smoke and particulates from such natural events as plain and forest fires (ignited by heat or lightning), volcanic ash, sand and dust which can spread over a large area of the countryside.

Such a phenomenon of natural pollution has been seen in the form of volcanic eruptions, with the resulting plume of smoke, steam and volcanic ash blotting out the sun's rays as it spreads and rises higher into the atmosphere. As it travels into the atmosphere the upper air currents catch and carry the smoke and ash, while condensing the steam back into water vapor. As the water vapor, smoke and ash travel on their journey, the smoke dissipates into the atmosphere while the ash and moisture settle back to earth in a trail hundreds of miles long. In some cases, lives are lost and millions of dollars of property damage result.

Industrial Pollutants

Industrial pollution is caused primarily by industrial processes, the burning of coal, oil and natural gas, which in turn produce smoke and fumes. Because the burning fuels contain large amounts of sulfur, the principal ingredients of smoke and fumes are sulfur dioxide and particulate matter. This type of pollutant occurs most severely during still, damp and cool weather, such as at night. Even in its less severe form, this pollutant is not confined to just cities. Because of air movements, the pollutants move for miles over the surrounding countryside.

Working with Federal, State and Local mandated regulations and by carefully monitoring emissions, business has greatly reduced the amount of pollutant introduced from its industrial sources, striving to obtain an acceptable level. Because of the industrial emission clean up, many land areas and streams in and around the cities have now begun to move back in the direction of nature's intended balance.

Automotive Pollutants

The third major source of air pollution is automotive emissions. The emissions from the internal combustion engines were not an appreciable problem years ago because of the small number of registered vehicles and the nation's small highway system. However, during the early 1950's, the trend of the American people was to move from the cities to the surrounding suburbs. This caused an immediate problem in transportation because the majority of suburbs were not afforded mass transit conveniences. This lack of transportation created an attractive market for the automobile manufacturers, which resulted in a dramatic increase in the number of vehicles produced and sold, along with a marked increase in highway construction between cities and the suburbs. Multi-vehicle families emerged with a growing emphasis placed on an individual vehicle per family member. As the increase in vehicle ownership and usage occurred, so did pollutant levels in and around the cities, as suburbanites drove daily to their businesses and employment, returning at the end of the day to their homes in the suburbs. It was noted that a smoke and fog type haze was being formed and at times, remained in suspension over the cities, taking time to dissipate. At first this "smog," derived from the words "smoke" and "fog," was thought to result from industrial pollution but it was determined that automobile emissions shared the blame. It was discovered that when normal automobile emissions were exposed to sunlight for a period of time, complex chemical reactions would take place.

It is now known that smog is a photo-chemical layer which develops when certain oxides of nitrogen (NOx) and unburned hydrocarbons (HC) from automobile emissions are exposed to sunlight. Pollution was more severe when smog would become stagnant over an area in which a warm layer of air settled over the top of the cooler air mass, trapping and holding the cooler mass at ground level. The trapped cooler air would keep the emissions from being dispersed and diluted through normal air flows. This type of air stagnation was given the name "Temperature Inversion."

TEMPERATURE INVERSION

In normal weather situations, surface air is warmed by heat radiating from the earth's surface and the sun's rays. This causes it to rise upward, into the atmosphere. Upon rising it will cool through a convection type heat exchange with the cooler upper air. As warm air rises, the surface pollutants are carried upward and dissipated into the atmosphere.

When a temperature inversion occurs, we find the higher air is no longer cooler, but is warmer than the surface air, causing the cooler surface air to become trapped. This warm air blanket can extend from above ground level to a few hundred or even a few thousand feet into the air. As the surface air is trapped, so are the pollutants, causing a severe smog condition. Should this stagnant air mass extend to a few thousand feet high, enough air movement with the inversion takes place to allow the smog layer to rise above ground level but the pollutants still cannot dissipate. This inversion can remain for days over an area, with the smog level only rising or lowering from ground level to a few hundred feet high. Meanwhile, the pollutant levels increase, causing eye irritation, respiratory problems, reduced visibility, plant damage and in some cases, even disease.

This inversion phenomenon was first noted in the Los Angeles, California area. The city lies in terrain resembling a basin and with certain weather conditions, a cold air mass is held in the basin while a warmer air mass covers it like a lid. Because this type of condition was first documented as prevalent in the Los Angeles area, this type of trapped pollution was named Los Angeles Smog, although it occurs in other areas where a large concentration of automobiles are used and the air remains stagnant for any length of time.

HEAT TRANSFER

Consider the internal combustion engine as a machine in which raw materials must be placed so a finished product comes out. As in any machine operation, a certain amount of wasted material is formed. When we relate this to the internal combustion engine, we find that through the input of air and fuel, we obtain power during the combustion process to drive the vehicle. The by-product or waste of this power is, in part, heat and exhaust gases with which we must dispose.

The heat from the combustion process can rise to over 4000-degrees F (2204-degrees C). The dissipation of this heat is controlled by a ram air effect, the use of cooling fans to cause air flow and a liquid coolant solution surrounding the combustion area to transfer the heat of combustion through the cylinder walls and into the coolant. The coolant is then directed to a thin-finned, multi-tubed radiator, from which the excess heat is transferred to the atmosphere by 1 of the 3 heat transfer methods, conduction, convection or radiation.

The cooling of the combustion area is an important part in the control of exhaust emissions. To understand the behavior of the combustion and transfer of its heat, consider the air/fuel charge. It is ignited and the flame front burns progressively across the combustion chamber until the burning charge reaches the cylinder walls. Some of the fuel in contact with the walls is not hot enough to burn, thereby snuffing out or quenching the combustion process. This leaves unburned fuel in the combustion chamber. This unburned fuel is then forced out of the cylinder and into the exhaust system, along with the exhaust gases.

Many attempts have been made to minimize the amount of unburned fuel in the combustion chambers due to quenching, by increasing the coolant temperature and lessening the contact area of the coolant around the combustion area. However, design limitations within the combustion chambers prevent the complete burning of the air/fuel charge, so a certain amount of the unburned fuel is still expelled into the exhaust system, regardless of modifications to the engine.

DRIVEABILITY AND EMISSION CONTROLS 4-3

AUTOMOTIVE EMISSIONS

Before emission controls were mandated on internal combustion engines, other sources of engine pollutants were discovered along with the exhaust emissions. It was determined that engine combustion exhaust produced approximately 60 percent of the total emission pollutants, fuel evaporation from the fuel tank and carburetor vents produced 20 percent, with the final 20 percent being produced through the crankcase as a by-product of the combustion process.

Exhaust Gases

The exhaust gases emitted into the atmosphere are a combination of burned and unburned fuel. To understand the exhaust emission and its composition, we must review some basic chemistry.

When the air/fuel mixture is introduced into the engine, we are mixing air, composed of nitrogen (78 percent), oxygen (21 percent) and other gases (1 percent) with the fuel, which is 100 percent hydrocarbons (HC), in a semi-controlled ratio. As the combustion process is accomplished, power is produced to move the vehicle while the heat of combustion is transferred to the cooling system. The exhaust gases are then composed of nitrogen, a diatomic gas (N_2), the same as was introduced in the engine, carbon dioxide (CO_2), the same gas that is used in beverage carbonation, and water vapor (H_2O). The nitrogen (N_2), for the most part, passes through the engine unchanged, while the oxygen (O_2) reacts (burns) with the hydrocarbons (HC) and produces the carbon dioxide (CO_2) and the water vapors (H_2O). If this chemical process would be the only process to take place, the exhaust emissions would be harmless. However, during the combustion process, other compounds are formed which are considered dangerous. These pollutants are hydrocarbons (HC), carbon monoxide (CO), oxides of nitrogen (NOx) oxides of sulfur (SOx) and engine particulates.

HYDROCARBONS

Hydrocarbons (HC) are essentially fuel which was not burned during the combustion process or which has escaped into the atmosphere through fuel evaporation. The main sources of incomplete combustion are rich air/fuel mixtures, low engine temperatures and improper spark timing. The main sources of hydrocarbon emission through fuel evaporation on most vehicles used to be the vehicle's fuel tank and carburetor float bowl.

To reduce combustion hydrocarbon emission, engine modifications were made to minimize dead space and surface area in the combustion chamber. In addition, the air/fuel mixture was made more lean through the improved control which feedback carburetion and fuel injection offers and by the addition of external controls to aid in further combustion of the hydrocarbons outside the engine. Two such methods were the addition of air injection systems, to inject fresh air into the exhaust manifolds and the installation of catalytic converters, units that are able to burn traces of hydrocarbons without affecting the internal combustion process or fuel economy.

To control hydrocarbon emissions through fuel evaporation, modifications were made to the fuel tank to allow storage of the fuel vapors during periods of engine shut-down. Modifications were also made to the air intake system so that at specific times during engine operation, these vapors may be purged and burned by blending them with the air/fuel mixture.

CARBON MONOXIDE

Carbon monoxide is formed when not enough oxygen is present during the combustion process to convert carbon (C) to carbon dioxide (CO_2). An increase in the carbon monoxide (CO) emission is normally accompanied by an increase in the hydrocarbon (HC) emission because of the lack of oxygen to completely burn all of the fuel mixture.

Carbon monoxide (CO) also increases the rate at which the photo-chemical smog is formed by speeding up the conversion of nitric oxide (NO) to nitrogen dioxide (NO_2). To accomplish this, carbon monoxide (CO) combines with oxygen (O_2) and nitric oxide (NO) to produce carbon dioxide (CO_2) and nitrogen dioxide (NO_2). ($CO + O_2 + NO = CO_2 + NO_2$). The dangers of carbon monoxide, which is an odorless and colorless toxic gas, are many. When carbon monoxide is inhaled into the lungs and passed into the blood stream, oxygen is replaced by the carbon monoxide in the red blood cells, causing a reduction in the amount of oxygen supplied to the many parts of the body. This lack of oxygen causes headaches, lack of coordination, reduced mental alertness and, should the carbon monoxide concentration be high enough, death could result.

NITROGEN

Normally, nitrogen is an inert gas. When heated to approximately 2500-degrees F (1371-degrees C) through the combustion process, this gas becomes active and causes an increase in the nitric oxide (NO) emission.

Oxides of nitrogen (NOx) are composed of approximately 97 - 98 percent nitric oxide (NO). Nitric oxide is a colorless gas but when it is passed into the atmosphere, it combines with oxygen and forms nitrogen dioxide (NO_2). The nitrogen dioxide then combines with chemically active hydrocarbons (HC) and when in the presence of sunlight, causes the formation of photo-chemical smog.

Ozone

To further complicate matters, some of the nitrogen dioxide (NO_2) is broken apart by the sunlight to form nitric oxide and oxygen. (NO_2 + sunlight = NO + O). This single atom of oxygen then combines with diatomic (meaning 2 atoms) oxygen (O_2) to form ozone (O_3). Ozone is one of the smells associated with smog. It has a pungent odor (like around copy machines), irritates the eyes and lung tissues, affects the growth of plant life and causes rapid deterioration of rubber products. Ozone can be formed by sunlight as well as electrical discharge into the air.

The most common discharge area on the automobile engine is the secondary ignition electrical system, especially when inferior quality spark plug cables are used. As the surge of high voltage is routed through the secondary cable, the circuit builds up an electrical field around the wire, which acts upon the oxygen in the surrounding air to form the ozone. The faint glow along the cable with the engine running that may be visible on a dark night, is called the "corona discharge." It is the result of the electrical field passing from a high along the cable, to a low in the surrounding air, which forms the ozone gas. The combination of corona and ozone has been a major cause of cable deterioration. Recently, different and better quality insulating materials have lengthened the life of the electrical cables.

Although ozone at ground level can be harmful, there is a concentrated ozone layer called the "ozonosphere," between 10 and 20 miles (16 - 32 km) up in the atmosphere. Much of the ultraviolet radiation from the sun's rays are absorbed and screened. The effect of this ozone layer on life is a matter of much debate.

OXIDES OF SULFUR

Oxides of sulfur (SOx) were initially ignored in the exhaust system emissions, since the sulfur content of gasoline as a fuel is less than 1/10 of 1 percent. Because of this small amount, it was felt that it contributed very little to the overall pollution problem. However, because of the difficulty in solving the sulfur emissions in industrial pollution and the introduction of catalytic converters to automobile exhaust systems, a change was mandated. The automobile exhaust system, when equipped with a catalytic converter, changes the sulfur dioxide (SO_2) into sulfur trioxide (SO_3).

When this combines with water vapors (H_2O), a sulfuric acid mist (H_2SO_4) is formed and is a very difficult pollutant to handle since it is extremely corrosive. This sulfuric acid mist that is formed, is the same mist that rises from the vents of an automobile battery when an active chemical reaction takes place within the battery cells.

When a large concentration of vehicles equipped with catalytic converters are operating in an area, this acid mist may rise and be distributed over a large ground area causing land, plant, crop, paint and building damage.

PARTICULATE MATTER

A certain amount of particulate matter is present in the burning of any fuel, with carbon constituting the largest percentage of the particulates. In gasoline, the remaining particulates are the burned remains of the various other compounds used in its manufacture. When a gasoline engine is in good internal condition, the particulate emissions are low but as the engine wears internally, the particulate emissions increase. By visually inspecting the tail pipe emissions, a determination can be made as to where an engine defect may exist. An engine with light gray or blue smoke emitting from the tail pipe normally indicates an increase in the oil consumption through burning due to internal engine wear. Black smoke would indicate a defective fuel delivery system, causing the engine to operate in a rich mode. Regardless of the color of the smoke, the internal part of the engine or the fuel delivery system should be repaired to prevent excess particulate emissions.

Diesel and turbine engines emit a darkened plume of smoke from the exhaust system because of the type of fuel used. Emission control regulations are mandated for this type of emission and more strin-

4-4 DRIVEABILITY AND EMISSION CONTROLS

gent measures are being used to prevent excess emission of the particulate matter. Electronic components are being introduced to control the injection of the fuel at precisely the proper time of piston travel, to achieve the optimum in fuel ignition and fuel usage. Other particulate after-burning components are being tested to achieve a cleaner emission.

Good grades of engine lubricating oils should be used, which meet the manufacturer's specification. Cut-rate oils can contribute to the particulate emission problem because of their low flash or ignition temperature point. Such oils burn prematurely during the combustion process causing emission of particulate matter.

The cooling system is an important factor in the reduction of particulate matter. The optimum combustion will occur with the cooling system operating at a temperature specified by the manufacturer. The cooling system must be maintained in the same manner as the engine oiling system, as each system is required to perform properly in order for the engine to operate efficiently for a long time.

Crankcase Emissions

Crankcase emissions are made up of water, acids, unburned fuel, oil fumes and particulates. These emissions are classified as hydrocarbons (HC) and are formed by the small amount of unburned, compressed air/fuel mixture entering the crankcase from the combustion area (between the cylinder walls and piston rings) during the compression and power strokes. The heat caused by compression and combustion help to form the remaining crankcase emissions.

Since the first engines, crankcase emissions were allowed into the atmosphere through a road draft tube, mounted on the lower side of the engine block. Fresh air came in through an open oil filler cap or breather. The air passed through the crankcase mixing with blow-by gases. The motion of the vehicle and the air blowing past the open end of the road draft tube caused a low pressure area (vacuum) at the end of the tube. Crankcase emissions were simply drawn out of the road draft tube into the air.

To control the crankcase emission, the road draft tube was deleted. A hose and/or tubing was routed from the crankcase to the intake manifold so the blow-by emission could be burned with the air/fuel mixture. However, it was found that intake manifold vacuum, used to draw the crankcase emissions into the manifold, would vary in strength at the wrong time and not allow the proper emission flow. A regulating valve was needed to control the flow of air through the crankcase.

Testing showed that the removal of the blow-by gases from the crankcase as quickly as possible was most important to the longevity of the engine. Should large accumulations of blow-by gases remain and condense, dilution of the engine oil would occur to form water, soot, resins, acids and lead salts, resulting in the formation of sludge and varnishes. This condensation of the blow-by gases occurs more frequently on vehicles used in numerous starting and stopping conditions, excessive idling and when the engine is not allowed to attain normal operating temperature through short runs.

Evaporative Emissions

Gasoline fuel is a major source of pollution, before and after it is burned in the automobile engine. From the time the fuel is refined, stored, pumped and transported, again stored until it is pumped into the fuel tank of the vehicle, the gasoline gives off unburned hydrocarbons (HC) into the atmosphere. Through the redesign of storage areas and venting systems, the pollution factor was diminished, but not eliminated, from the refinery standpoint. However, the automobile still remained the primary source of vaporized, unburned hydrocarbon (HC) emissions.

Fuel pumped from an underground storage tank is cool but when exposed to a warmer ambient temperature, will expand. Before controls were mandated, an owner might fill the fuel tank with fuel from an underground storage tank and park the vehicle for some time in warm area, such as a parking lot. As the fuel would warm, it would expand and should no provisions or area be provided for the expansion, the fuel would spill out of the filler neck and onto the ground, causing hydrocarbon (HC) pollution and creating a severe fire hazard. To correct this condition, the vehicle manufacturers added overflow plumbing and/or gasoline tanks with built in expansion areas or domes.

However, this did not control the fuel vapor emission from the fuel tank. It was determined that most of the fuel evaporation occurred when the vehicle was stationary and the engine not operating. Most vehicles carry 5 - 25 gallons (19 - 95 liters) of gasoline. Should a large concentration of vehicles be parked in one area, such as a large parking lot, excessive fuel vapor emissions would take place, increasing as the temperature increases.

To prevent the vapor emission from escaping into the atmosphere, the fuel systems were designed to trap the vapors while the vehicle is stationary, by sealing the system from the atmosphere. A storage system is used to collect and hold the fuel vapors from the carburetor (if equipped) and the fuel tank when the engine is not operating. When the engine is started, the storage system is then purged of the fuel vapors, which are drawn into the engine and burned with the air/fuel mixture.

EMISSION CONTROLS

Positive Crankcase Ventilation System (PCV)

The 3.9L, 4.7L, 5.2L and 5.9L gasoline engines are fitted with a Positive Crankcase Ventilation (PCV) system.

OPERATION

See Figures 1, 2, 3 and 4

When the engine is running, a small portion of the gases which are formed in the combustion chamber leak by the piston rings and enter the crankcase. Since these gases are under pressure they tend to escape from the crankcase and enter into the atmosphere. If these gases are allowed to remain in the crankcase for any length of time, they would contaminate the engine oil and cause sludge to build up. If the gases are allowed to escape into the atmosphere, they would pollute the air, as they contain unburned

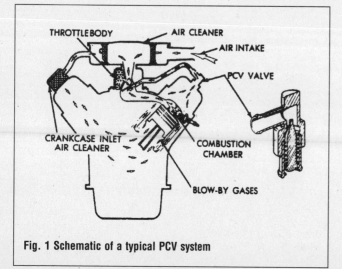

Fig. 1 Schematic of a typical PCV system

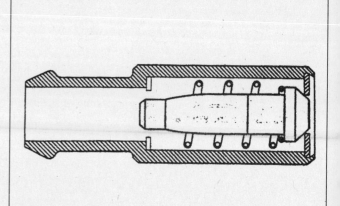

Fig. 2 PCV valve position with no vapor flow - engine off or backfire

DRIVEABILITY AND EMISSION CONTROLS 4-5

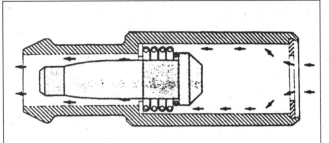

Fig. 3 PCV valve position with minimal vapor flow - high intake manifold vacuum

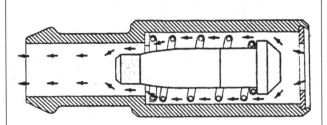

Fig. 4 PCV valve position with maximum vapor flow - moderate intake manifold vacuum

hydrocarbons. The crankcase emission control equipment recycles these gases back into the engine combustion chamber, where they are burned.

Crankcase gases are recycled in the following manner. While the engine is running, clean filtered air is drawn into the crankcase through the intake air filter and then through a hose leading to the oil filler cap or the valve cover. Early models covered in this manual may have a filter here which requires maintenance. As the air passes through the crankcase it picks up the combustion gases and carries them out of the crankcase, up through the PCV valve and into the intake manifold. After they enter the intake manifold they are drawn into the combustion chamber and are burned.

The most critical component of the system is the PCV valve. This vacuum-controlled valve regulates the amount of gases which are recycled into the combustion chamber. At low engine speeds the valve is partially closed, limiting the flow of gases into the intake manifold. As engine speed increases, the valve opens to admit greater quantities of the gases into the intake manifold. If the valve should become blocked or plugged, the gases will be prevented from escaping the crankcase by the normal route. Since these gases are under pressure, they will find their own way out of the crankcase. This alternate route is usually a weak oil seal or gasket in the engine. As the gas escapes by the gasket, it also creates an oil leak. Besides causing oil leaks, a clogged PCV valve also allows these gases to remain in the crankcase for an extended period of time, promoting the formation of sludge in the engine.

COMPONENT TESTING

See Figures 5, 6, 7, 8 and 9

PCV Valve

1. The PCV valve is located in the valve cover on all engines except the 4.7L, where it is located on the oil filler tube housing.
2. With the engine running, pull the PCV valve

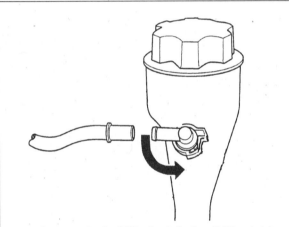

Fig. 5 On the 4.7L, the PCV valve is in the oil filler: twist and pull to remove

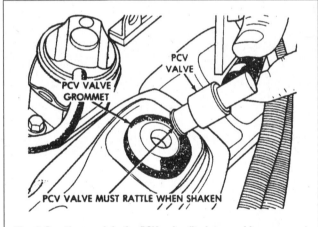

Fig. 6 On other models the PCV valve fits into a rubber grommet on the valve cover; replace the grommet if it leaks

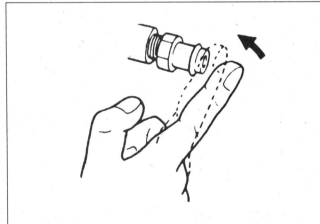

Fig. 7 Testing a PCV valve: vacuum should be strong and steady

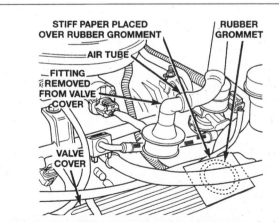

Fig. 8 Checking PCV suction with a piece of paper over the fresh air inlet

4-6 DRIVEABILITY AND EMISSION CONTROLS

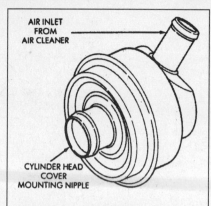

Fig. 9 The crankcase inlet air cleaner (early models) should be cleaned in solvent

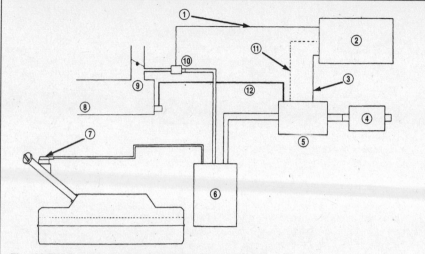

Fig. 11 EVAP system incorporating leak detection pump

1. Duty Cycle Purge Solenoid (DCPS) driver
2. Powertrain Control Module (PCM)
3. 3-port solenoid driver
4. Remote filter
5. Combined canister vent valve and leak detection pump
6. Canister
7. Tank rollover valve and vapor flow control orifice
8. Intake manifold
9. Throttle body
10. DCPS
11. Switch signal input to the PCM
12. Engine vacuum line

and hose from the valve cover rubber grommet (or oil filler housing on the 4.7L). On this engine, twist it 90-degrees and then pull it out.

3. If the valve is working properly, a hissing noise should be heard as air passes through the valve and a strong vacuum should be felt when you place a finger over the valve inlet.

4. While you have your finger over the PCV valve inlet, check for vacuum leaks in the hose and at the connections.

5. When the PCV valve is removed, a metallic clicking noise should be heard when it is shaken. This indicates that the metal check ball inside the valve is still free and is not gummed up. If not operating properly, replace the valve.

6. If no vacuum is felt at the PCV valve when it is removed from the engine, remove the valve from the hose and check the vacuum supply in the hose.

7. To check the PCV valve in operation, remove the fresh air fitting from the valve cover. Place a stiff piece of paper over the fitting grommet. After about a minute, the paper should be drawn against the grommet with considerable force. If not, replace the PCV valve.

Crankcase Inlet Air Cleaner

1. Some early models may be fitted with a crankcase inlet air cleaner on the valve cover (the valve cover opposite the one with the PCV valve, if there are two). This supplies air to the engine to replace that vented through the PCV valve.

2. Wash the valve in solvent and dry it thoroughly before refitting.

3. Maintenance should be performed more often

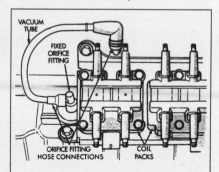

Fig. 10 CCV fixed orifice and hose - 8.0L engine

if the vehicle is used for short trips, extended idling periods or very dusty conditions.

Hoses

1. Check the breather hose for restrictions.
2. Check the intake manifold fittings for sludge buildup; this can reduce the flow of the system.

REMOVAL & INSTALLATION

Refer to "General Information And Maintenance" for removal and installation of the PCV valve.

Crankcase Ventilation System (CCV)

The 2.5L and 8.0L engines are fitted with a Crankcase Ventilation (CCV) system. It performs the same function as the PCV system described above, but does not use a vacuum-controlled valve.

OPERATION

See Figure 10

A molded vacuum tube connects a fitting on the intake manifold to a fixed orifice fitting of a calibrated size. This fitting meters the amount of crankcase vapors drawn out of the engine. The fixed orifice fitting is located on the side of the valve cover on 2.5L engines, the right valve cover on 8.0L engines. A fresh air supply hose from the air cleaner housing is connected to a fitting at the top/rear of the cylinder head cover on 2.5L engines, on the left valve cover on 8.0L engines. When the engine is running, fresh air enters the engine through the fresh air supply hose and mixes with crankcase vapors. Engine manifold vacuum draws the vapor/air mixture through the fixed orifice and into the intake manifold. The vapors are then consumed during engine combustion.

COMPONENT TESTING

No maintenance is required. Check hoses for condition and replace if cracked or abraded. On the 8.0L engine, only the right valve cover hose has a fixed orifice in the system (gray in color). The left side (fresh air) hose has a black connector. Do not interchange left and right side hose connections, as this would restrict fresh air supply.

Evaporative Emission Controls

OPERATION

See Figures 11, 12, 13, 14, 15, 16, 17 and 18

Changes in atmospheric temperature cause fuel tanks to breathe, that is, the air within the tank expands and contracts with outside temperature changes. If an unsealed system was used, when the temperature rises, air would escape through the tank vent tube or the vent in the tank cap. The air which escapes contains gasoline vapors.

The Evaporative Emission Control System provides a sealed fuel system with the capability to store and condense fuel vapors. When the fuel evaporates in the fuel tank, the vapor passes through vent hoses or tubes to a charcoal-filled evaporative canister. When the engine is operating the vapors are drawn into the intake manifold.

The PCM determines when these vapors will be passed into the intake manifold.

EVAP Canister

A sealed, maintenance-free evaporative canister is used (some RAM trucks have two). The canister is mounted under the vehicle on the frame rail behind the cab. It may be on either side (or both sides) of the vehicle depending on model and year.

DRIVEABILITY AND EMISSION CONTROLS 4-7

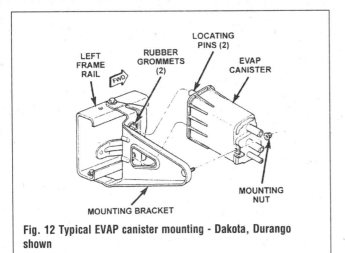

Fig. 12 Typical EVAP canister mounting - Dakota, Durango shown

Fig. 13 EVAP canister mounted on a 97 RAM truck

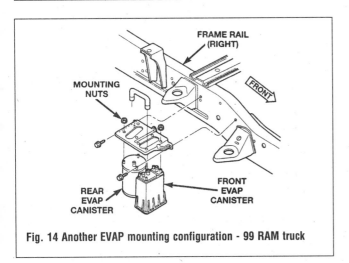

Fig. 14 Another EVAP mounting configuration - 99 RAM truck

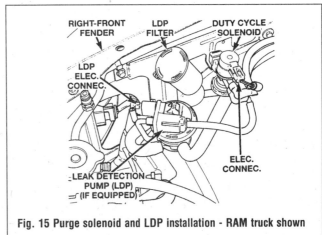

Fig. 15 Purge solenoid and LDP installation - RAM truck shown

The canister is filled with granules of an activated carbon mixture which absorb fuel vapors.

Purge Solenoid

All gasoline engines use a duty cycle purge system. The PCM controls vapor flow by operating the duty cycle EVAP purge solenoid. This regulates the rate of vapor flow from the EVAP canister to the intake manifold.

The PCM regulates the solenoid by switching the ground circuit on and off based on engine operating conditions. When energized, the solenoid prevents vacuum from reaching the canister. When not energized the solenoid allows vacuum to flow through to the canister.

During warm up and for a specified time after hot starts, the PCM energizes (grounds) the solenoid preventing vacuum from reaching the canister. When the engine temperature reaches the operating level of about 120-degrees F (49-degrees C), the PCM removes the ground from the solenoid allowing vacuum to flow through the canister and purges vapors through the throttle body. During certain idle conditions, the purge solenoid may be grounded to control fuel mix calibrations.

After the engine reaches operating temperature, the PCM energizes and de-energizes the solenoid 5 or 10 times per second, depending on operating conditions. The PCM varies the vapor flow rate by changing solenoid pulse width, which is the amount of time the solenoid energizes.

Fuel Tank Cap

The fuel tank is sealed with a pressure-vacuum relief filler cap. Under normal operating conditions, the filler cap operates as a check valve, allowing air to

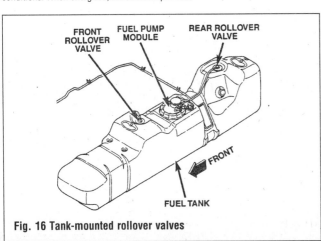

Fig. 16 Tank-mounted rollover valves

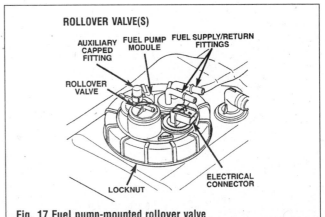

Fig. 17 Fuel pump-mounted rollover valve

4-8 DRIVEABILITY AND EMISSION CONTROLS

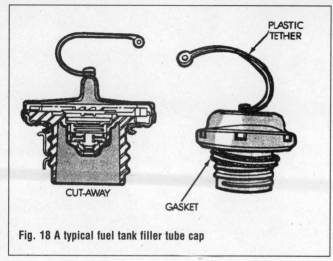

Fig. 18 A typical fuel tank filler tube cap

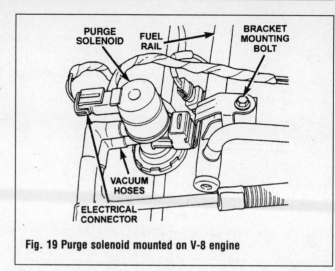

Fig. 19 Purge solenoid mounted on V-8 engine

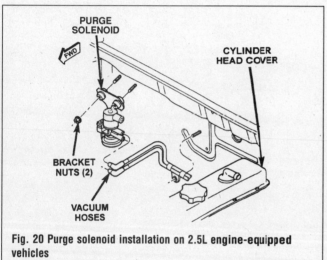

Fig. 20 Purge solenoid installation on 2.5L engine-equipped vehicles

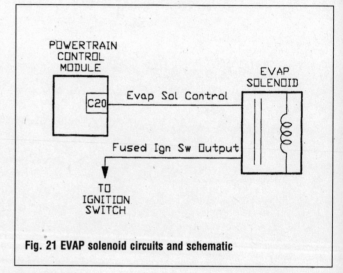

Fig. 21 EVAP solenoid circuits and schematic

enter the tank to enable fuel flow. At the same time, it prevents fuel vapors from escaping through the cap to the atmosphere. These vapors are therefore forced into the EVAP canister.

The relief valves in the cap are a safety feature, preventing excessive pressure or vacuum in the fuel tank. If the cap is malfunctioning, and needs to be replaced, ensure that the replacement is the identical cap to ensure correct system operation.

Leak Detection Pump

Certain emissions packages have a Leak Detection Pump (LDP) is used to monitor the EVAP system for leaks. On most models so equipped, this unit is located beneath the battery tray. A test port for pressurizing the EVAP system is included. The test port is used to pressurize the system with a special gas and serious precautions must be taken to avoid damage to the EVAP system and the fuel tank. This is a procedure best suited to a professional shop, due to the precautions and the equipment needed to test this system. The PCM can store trouble codes for EVAP system performance.

Rollover Valves

Two-door models have one rollover valve located on top of the fuel tank. Four-door models have two valves also located on top of the fuel tank, although configuration may differ by model. Some vehicles have valves mounted at front and rear of the tank, while others may have one of the valves located atop the fuel pump.

Rollover valves prevent fuel flow through the EVAP hoses in the event of an accidental vehicle rollover. The EVAP canister(s) draw vapor from the gas tank through the valve(s).

Tank-mounted rollover valves are not serviceable and can only be replaced with the tank. Pump-mounted valves can be serviced separately.

COMPONENT TESTING

Note: To relieve fuel tank pressure, the filler cap must be removed before disconnecting any fuel system component.

Canister Purge Solenoid
See Figures 19, 20, 21 and 22

1. Locations vary depending on model and year. Typical locations include the firewall (left or right side) and the rear of the engine near the firewall.
2. With the ignition off, unplug the connector on the EVAP solenoid.
3. Turn ignition on, measure the voltage at the ignition switch output line, voltage should be 10.0v or more. If voltage is not 10.0v or more, repair circuit from ignition switch to EVAP solenoid.
4. Disconnect the negative battery cable.
5. Disconnect the PCM harness from the PCM.
6. Check the resistance of the EVAP solenoid control circuit between the PCM harness connector and the EVAP solenoid connector. Resistance should be less than 5.0 ohms; if not, repair the opening in the circuit.
7. Connect the negative battery cable.

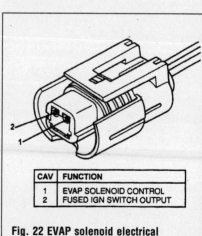

CAV	FUNCTION
1	EVAP SOLENOID CONTROL
2	FUSED IGN SWITCH OUTPUT

Fig. 22 EVAP solenoid electrical connector

DRIVEABILITY AND EMISSION CONTROLS

Fig. 23 Mark the lines connected to the EVAP canister before removal to ease installation

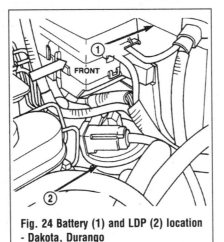

Fig. 24 Battery (1) and LDP (2) location - Dakota, Durango

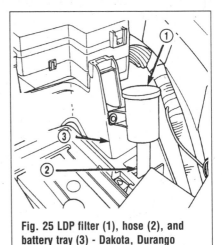

Fig. 25 LDP filter (1), hose (2), and battery tray (3) - Dakota, Durango

REMOVAL & INSTALLATION

Note: To relieve fuel tank pressure, the filler cap must be removed before disconnecting any fuel system component.

Evaporative Canister

See Figures 12, 13, 14 and 23

1. Disconnect the negative battery cable.
2. Raise and support the vehicle.
3. Label and disconnect the hoses on the top of the canister.
4. Remove the mounting bolts or nut(s).

To install:

5. Install and tighten the mounting hardware to 7 ft. lbs. (9 Nm) if two nuts or bolts are used; tighten to 18 ft. lbs. (24 Nm) for single fasteners.
6. Install the hoses in their proper locations.
7. Lower the vehicle.
8. Connect the negative battery cable.

Canister Purge Solenoid

See Figures 19 and 20

1. Disconnect the negative battery cable.
2. Disconnect the electrical wiring from the solenoid.
3. Disconnect the vacuum lines from the solenoid.
4. Remove the solenoid (or solenoid with support bracket as an assembly, if this is easier).

To install:

5. Install the solenoid or support bracket.
6. Connect the vacuum lines.
7. Connect the wiring harness.
8. Connect the negative battery cable.

Leak Detection Pump (LDP)

RAM TRUCKS

1. The Leak Detection Pump (LDP) and filter are attached to a bracket mounted to the right side inner fender. LDP and filter are serviced as a single assembly.
2. Remove the LDP filter hose.
3. Remove the filter mounting bolt and remove the filter from the vehicle.
4. Disconnect the vacuum lines at the LDP.

5. Disconnect the LDP wiring.
6. Remove the mounting screws and remove the LDP from the vehicle.

To install:

7. Reverse the removal procedure. Be sure all connections are TIGHT.

DAKOTA, DURANGO

See Figures 24 and 25

1. The Leak Detection Pump (LDP) is located in the engine compartment under the battery tray and PDC. The LDP filter is attached to the outside of the battery tray. LDP and filter are serviced as a single assembly.
2. Disconnect and remove the battery.
3. Disconnect the hose from the bottom of the LDP filter.
4. Disconnect the battery temperature sensor pigtail wiring harness at the bottom of the battery tray.
5. Remove the PDC-to-fender mounting screw at the rear of the PDC. Unsnap the PDC from the battery tray. To prevent damage to the PDC wiring, carefully position the PDC to gain access to the LDP.
6. Remove the battery tray.
7. Disconnect the vacuum lines from the LDP.
8. Disconnect the LDP wiring.
9. Remove the three mounting screws and remove the LDP.

To install:

10. Reverse the removal procedure. Be sure all connections are TIGHT.

Rollover Valve

See Figure 17

1. Tank-mounted rollover valves are molded into the gas tank and are not replaceable. The following procedure refers to pump-mounted valves only.
2. Disconnect the battery negative cable(s).
3. Drain the fuel tank.
4. Remove the fuel tank from the truck.
5. Disconnect the line at the rollover valve.
6. The valve is kept by a rubber grommet. Pry one side upward and roll it out.

To install:

7. Use a new grommet.
8. Press the valve in using finger pressure only.

Exhaust Gas Recirculation System

An Exhaust Gas Recirculation (EGR) system is fitted to 1997 - 98 California diesels.

OPERATION

See Figures 26, 27, 28 and 29

The EGR system reduces oxides of nitrogen in the engine exhaust. This is done by allowing a predetermined amount of hot exhaust gas to recirculate and dilute the incoming fuel/air mixture.

The following are the major components of the system:

- An EGR valve assembly located at the front of the intake manifold. The EGR valve is a poppet style valve (on/off only) and is controlled by an internal diaphragm
- An EGR valve vacuum regulator solenoid located at the front/top of the cylinder head which controls the on-time and off-time of the EGR valve
- The Powertrain Control Module (PCM) operates the EGR valve vacuum regulator solenoid
- The Engine Coolant Temperature (ECT) sensor supplies temperature input to the PCM
- The Intake Manifold Air Temperature (IAT) sensor supplies air temperature input to the PCM

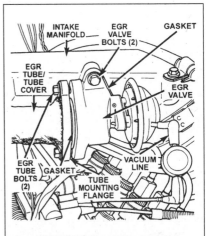

Fig. 26 EGR location and associated components - California diesel

4-10 DRIVEABILITY AND EMISSION CONTROLS

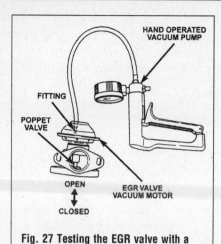

Fig. 27 Testing the EGR valve with a vacuum pump

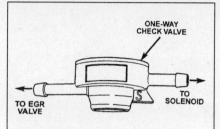

Fig. 28 One-way check valve and connection details

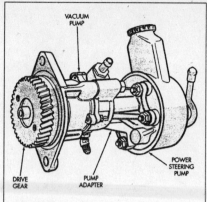

Fig. 29 Power steering and vacuum pump assembly

- The Throttle Position Sensor (TPS) supplies a voltage reference input to the PCM
- A vacuum pump supplies vacuum for the EGR valve vacuum regulator solenoid and EGR valve. The pump is located at the front of the engine and is attached to the power steering pump
- A quick-release one-way check valve provides fast release of engine vacuum from the EGR valve diaphragm when the system is shut down

When the PCM supplies a ground signal to the EGR valve vacuum regulator solenoid, the EGR system operation starts to occur. The PCM will monitor and determine when to supply and remove this ground signal. This will depend on inputs from the engine coolant temperature, throttle position and intake manifold air temperature sensors.

When the ground signal is supplied to the EGR solenoid, vacuum from the vacuum pump is allowed to pass through the EGR solenoid and on to the EGR valve with a connecting hose.

The EGR valve is normally open when the engine is running at idle or above with the coolant at normal operating temperature. Certain changes to engine status will close the valve, including rapid acceleration, wide-open throttle or after about two minutes without a change in throttle position.

COMPONENT TESTING

EGR
See Figure 27

1. Remove the EGR valve from the manifold (see procedure below).
2. Examine the head of the poppet valve at base opening on the bottom of the valve. Look for heavy carbon build-up. Some carbon build-up is normal.
3. Shine a bright light through the valve opening and examine the valve seat. No light should be evident at the valve edge. Replace the EGR valve if either condition exists.
4. Connect a vacuum pump to the EGR's vacuum pump fitting. Slowly apply about 10 inches Hg of vacuum. The valve should start to open. Vacuum should hold steady. If it doesn't, replace the valve.
5. The valve should be fully open at 20 inches Hg of vacuum. If it isn't, replace it.
6. Check passages for carbon clogging. Cleaning is not practical. Replace any clogged valve.

Check Valve Test

1. A quick-release, one-way check valve is located in the vacuum line between the EGR valve and the vacuum regulator solenoid. Failure may result in the EGR not opening or remaining open when it should not.
2. Attach a vacuum gauge with a "T" fitting into the vacuum line at the EGR valve between the EGR valve and EGR vacuum regulator solenoid.
3. Run the engine until operating temperature is reached.
4. Drive at a steady speed. Vacuum should be observed at the gauge.
5. Quickly open the throttle. Vacuum should drop to "0".
6. The check-valve can also be tested with a hand-held vacuum gauge connected to the inlet side of the valve (marked with an "S"). It should hold 20 inches Hg of vacuum or more without leakage.
7. The other side of the valve should not hold any vacuum.

Vacuum Supply Test

1. Disconnect the vacuum supply line at the EGR valve vacuum regulator solenoid and attach a vacuum gauge.
2. Start the engine and check vacuum. It should range from 8 - 25 inches Hg as rpm varies.

If vacuum does not fall within this range, check for leaks in lines and connections. If the system is sound, replace the vacuum pump.

REMOVAL & INSTALLATION

EGR
See Figure 26

1. Disconnect the vacuum line at the EGR.
2. Remove the two bolts securing the EGR tube.
3. Remove the EGR mounting bolts and remove the valve.

To install:

4. Clean off old gasket material.
5. Tighten bolts gradually and evenly. Torque to 18 inch lbs. (24 Nm).

Check Valve
See Figure 28

1. Pull off vacuum lines on both sides of the valve. A bit of penetrating fluid safe for rubber and plastic will aid removal.
2. Check lines for cracking or damage and replace as required.
3. When installing, be sure connections are clean and tight.

4. Be sure the vacuum lines are properly connected. The inlet side of the valve is marked "S".

Vacuum Pump
See Figure 29

1. Disconnect the power steering hoses and the vacuum line.
2. Remove the sender unit from the engine block and plug the hole.
3. Remove and cap the oil feed line from the bottom of the vacuum pump.
4. Remove the lower bolt that attaches the pump assembly to the engine block.
5. Remove the nut from the steering pump attaching bracket.
6. Remove the upper bolt from the assembly and remove the assembly.
7. Remove the steering pump to vacuum pump bracket attaching nuts.

To install:

8. Reverse the removal procedure.
9. Rotate the drive gear until the steering pump and vacuum pump drive dogs align. Install the steering pump onto the vacuum pump bracket. Use care to avoid damaging the oil seal.
10. The steering pump housing and spacers must mate completely with the vacuum pump bracket.
11. Tighten the bracket-to-steering pump nuts to 18 ft. lbs. (24 Nm). Tighten the pump-to-engine block attaching bolts to 57 ft. lbs. (77 Nm). Tighten the bracket nut to 18 ft. lbs. (24 Nm).

Catalytic Converter

OPERATION

The catalytic converter, mounted in the exhaust system, is a muffler-shaped device containing a ceramic honeycomb shaped material coated with alumina and impregnated with catalytically active precious metals such as platinum, palladium and rhodium.

The catalyst's job is to reduce air pollutants by oxidizing hydrocarbons (HC) and carbon monoxide (CO). Catalysts containing palladium and rhodium also oxidize nitrous oxides (NOx).

On some trucks, the catalyst is also fed by the secondary air system, via a small supply tube in the side of the catalyst.

DRIVEABILITY AND EMISSION CONTROLS 4-11

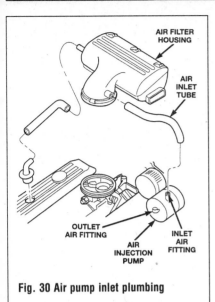

Fig. 30 Air pump inlet plumbing

No maintenance is possible on the converter, other than keeping the heat shield clear of flammable debris, such as leaves and twigs.

Other than external damage, the only significant damage possible to a converter is through the use of leaded gasoline, or by way of a too rich fuel/air mixture. Both of these problems will ruin the converter through contamination of the catalyst and will eventually plug the converter causing loss of power and engine performance.

When this occurs, the catalyst must be replaced. For catalyst replacement, see the Exhaust System procedures under "Engine And Engine Overhaul".

Air Injection System

5.9L V-8 and 8.0L V-10 engines with HDC packages are equipped with an air injection system.

The system consists of:
- A belt-driven air injection (AIR) pump mounted on the front of the engine and driven by the accessory belt
- Two air pressure relief valves
- Two one-way check valves located on each of the air injection downstream tubes
- A replaceable pump air filter on the V-10
- Tubes and hoses

OPERATION

See Figure 30

The air injection system adds a controlled amount of air to the exhaust gases aiding oxidation of hydrocarbons and carbon monoxide in the exhaust stream.

On the 5.9L HDC engine air is drawn into the pump through a rubber tube that is connected to a fitting on the air cleaner housing.

On the 8.0L engine, air is drawn into the pump through a rubber tube that is connected to a fitting on the air injection pump filter housing. Air is drawn into the filter housing from the front of the vehicle with rubber tube. This tube is used as a silencer to help prevent air intake noise at the opening to the pump filter housing.

Air is then compressed by the air injector pump. It is expelled from the pump and routed into a rubber tube where it reaches the air pressure relief valve. Pressure relief holes in the relief valve will prevent excess downstream pressure. If excess downstream pressure occurs at the relief valve, it will be vented to the atmosphere.

Air is then routed from the relief valve, through a tube, down to a "Y" connector, through the two, one-way check valves and injected at both of the catalytic converters.

The two one-way check valves protect the hoses, pump and tubes from hot exhaust gases backing up into the system. Air is allowed to flow through these valves in one direction only (towards the catalytic converters).

Downstream air flow assists the oxidation process in the catalyst, but does not interfere with EGR operation (if fitted).

COMPONENT TESTING

Air Pump

1. The air injection system and pump is not completely noiseless. Under normal conditions, noise rises in pitch as engine speed increases.
2. To determine if the pump is the source of excessive noise, disconnect the accessory drive belt and temporarily run the engine.
3. Run the engine to normal operating temperature and let it idle.
4. Disconnect the output hose.
5. If the pump is operating properly, airflow should be felt at the pump outlet. The flow should increase as you increase the engine speed. The pump is not serviceable and should be replaced if it is not functioning properly.

WARNING:

Do not attempt to lubricate the air pump. Oil in the pump will quickly destroy it.

One-Way Check Valves

1. Remove the rubber air tube from the inlet side of each check valve.
2. Start the engine.
3. If exhaust gas is escaping through the inlet side of the check valve, it must be replaced. Valves are not repairable.

REMOVAL & INSTALLATION

Air Injection Pump

1. The AIR pump does not have any internal serviceable parts.
2. Disconnect both hoses at the pump.
3. Loosen, but do not remove yet, the three pump pulley mounting bolts.
4. Relax the accessory belt tension and remove the accessory belt.
5. Remove the pulley bolts and remove the pulley.
6. Remove the two air pump mounting bolts and remove the pump from the mounting bracket.

To install:

7. Reverse the removal procedure.
8. Tighten the pump mounting bolts to 30 ft. lbs. (40 Nm).
9. Tighten the pulley bolts to 8 ft. lbs. (11 Nm).

Air Injection Pump Filter

On 8.0L engines, the air filter for the pump is located inside a housing in the right front side of the engine compartment. Connection is via a rubber hose.

The filter should be replaced every 24,000 miles/24 months under normal operating conditions.

1. Remove the rubber tubes at the filter housing.
2. Remove the housing mounting nut and remove the housing.
3. Remove the housing lid which snaps off.
4. Remove the filter.

To install:

5. Clean the inside of the housing.
6. Tighten the mounting nut to 8 ft. lbs. (11 Nm).
7. The reset of the procedure is the reverse of removal.

One-Way Check Valves

1. Remove the hose clamp at the inlet side of the valve.
2. Remove the hose from the valve.
3. Unscrew the valve from the catalyst tube. To prevent damage, use a backup wrench on the tube.

To install:

4. Reverse the removal procedure.
5. Tighten the valve in the catalyst tube to 25 ft. lbs. (33 Nm).

Malfunction Indicator Lamp (MIL)

See Figure 31

The Malfunction Indicator Lamp (MIL) is located on the instrument panel. It is displayed as an engine icon (graphic).

The MIL lamp lights up when the ignition key is turned on as a bulb test.

Under other conditions the MIL lamp will be illuminated when the PCM sets a Diagnostic Trouble Code (DTC).

DTCs can be retrieved with a scan tool via the 16-way data link connector. This connector is located on the lower edge of the instrument panel near the steering column. On 1997 models, the DTCs can be retrieved with the ignition key. See "Trouble Codes", following.

RESETTING

Some scan tools have a function to clear trouble codes. However, the problem should be attended to and resolved or the trouble code(s) will simply reappear.

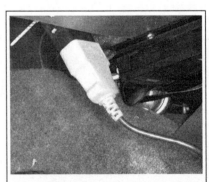

Fig. 31 Scan tool connected to 16-way data link connector beneath the instrument panel.

4-12 DRIVEABILITY AND EMISSION CONTROLS

ELECTRONIC ENGINE CONTROLS

Powertrain Control Module (PCM)

OPERATION

The Powertrain Control Module (PCM) performs many functions on your vehicle. The module accepts information from various engine sensors and computes the required fuel flow rate necessary to maintain the correct amount of air/fuel ratio and ignition timing throughout the entire engine operational range.

Based on the information that is received and programmed into the PCM's memory, the PCM generates output signals to control relays, actuators and solenoids. The PCM also sends out a command to the fuel injectors that meters the appropriate quantity of fuel. The module automatically senses and compensates for any changes in altitude when driving your vehicle.

Included among its many outputs, the PCM manages EVAP and other emissions components, ignition timing, alternator output, fuel injectors, electric fans (where fitted), cruise control, tachometer, fuel pump, and other systems.

REMOVAL & INSTALLATION

See Figures 32 and 33

The PCM is located on the right side of the engine compartment, either on the fender inner shield or on the firewall.

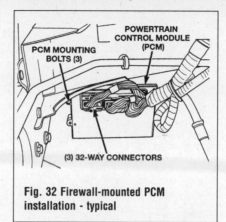

Fig. 32 Firewall-mounted PCM installation - typical

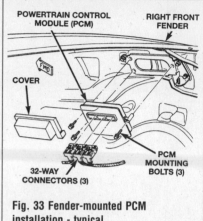

Fig. 33 Fender-mounted PCM installation - typical

1. Disconnect the battery negative cable(s).
2. Remove the connector cover, if fitted.
3. Disconnect the wiring from the PCM.
4. Remove the mounting screws and take the unit out of the vehicle.
5. Installation is the reverse of removal.

Oxygen (O2) Sensor

OPERATION

See Figures 34 and 35

The oxygen sensor (O2) is a galvanic battery which produces an electrical voltage when exposed to the oxygen present in the exhaust gases. Some oxygen sensors are electrically heated internally for faster switching when the engine is running. The oxygen sensor produces a voltage within 0 and 1 volt, providing this signal to the PCM which interprets it to learn how much oxygen is in the exhaust. When there is a large amount of oxygen present (lean mixture), the sensor produces a low voltage (less than 0.4v). When there is a lesser amount present (rich mixture) it produces a higher voltage (0.6 - 1.0v). The stoichiometric or correct air-to-fuel ratio will read between 0.4 and 0.6v. By monitoring the oxygen content and converting it to electrical voltage, the sensor acts as a rich-lean switch. The PCM signals the power module to trigger the fuel injector and maintains the 14.7:1 air:fuel ratio necessary for proper engine operation and emissions control.

Two or four sensors may be fitted, depending on engine type and emissions package. Sensors are positioned before and after the catalytic converter and may be installed on both sides of the exhaust system. The one before the catalyst measures the exhaust emissions right out of the engine, and sends the signal to the PCM about the state of the mixture as previously discussed. The second sensor reports the difference in the emissions after the exhaust gases have gone through the catalyst. This sensor reports to the PCM the amount of emissions reduction the catalyst is performing.

Engines equipped with either a downstream sensor or a post-catalytic sensor will monitor catalytic converter efficiency. If efficiency is below emission standards, the MIL will be illuminated and a DTC will be set.

The oxygen sensor will not work until a predetermined temperature is reached. Until this time the PCM is running in what is known as OPEN LOOP operation. OPEN LOOP means that the PCM has not yet begun to correct the air-to-fuel ratio by reading the oxygen sensor. After the engine reaches operating temperature, the PCM will monitor the oxygen sensor and correct the air/fuel ratio from the sensor's readings. This is what is known as CLOSED LOOP operation.

A heated oxygen sensor has a heating element that keeps the sensor at proper operating temperature during all operating modes. Maintaining correct sensor temperature at all times allows the system to enter into CLOSED LOOP operation sooner.

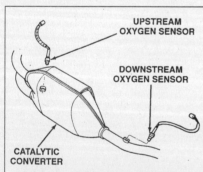

Fig. 34 Most models have O2 sensors to monitor emissions before and after the catalytic converter

Fig. 35 O2 sensor fitted upstream of the catalytic converter on a RAM truck

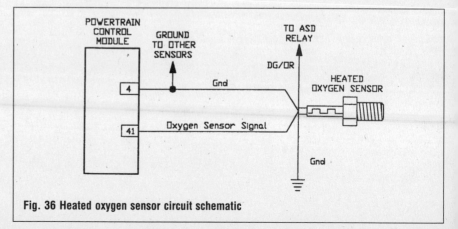

Fig. 36 Heated oxygen sensor circuit schematic

DRIVEABILITY AND EMISSION CONTROLS 4-13

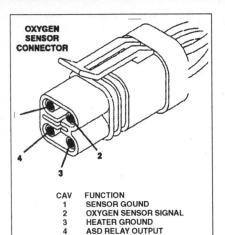

Fig. 37 Heated oxygen sensor's electrical connector terminal identification

CAV	FUNCTION
1	SENSOR GOUND
2	OXYGEN SENSOR SIGNAL
3	HEATER GROUND
4	ASD RELAY OUTPUT

The O2 sensor uses a Positive Thermal Co-efficient (PTC) heater element. As temperature increases, resistance increases. At temperatures around 70-degrees F, the resistance of the heating element is about 4.5 ohms. As the sensor's temperature increases, the resistance in the heater element increases. This allows the heater to maintain the optimum operating temperature of about 930-degrees - 1100-degrees F (500-degrees - 600-degrees C). Although the sensors operate the same, there are physical differences, due to the environment they operate in, that keep them from being interchangeable.

In CLOSED LOOP operation the PCM monitors the sensor input (along with other inputs) and adjusts the injector pulse width accordingly. During OPEN LOOP operation the PCM ignores the sensor input and adjusts the injector pulse to a preprogrammed value based on other inputs.

WARNING:
The O2 sensor must have a source of oxygen from outside the exhaust stream for comparison. Current sensors receive their fresh oxygen supply through the wiring harness. This is why it is important to never solder an O2 sensor connector or pack the connector with grease.

Four wires are used on each sensor: a 12-volt feed circuit for the sensor heating element, a ground circuit for the heater element, a low-noise sensor return circuit to the PCM, and an input circuit from the sensor back to the PCM to detect sensor operation.

TESTING

See Figures 36 and 37

1. To test the O2 sensor in a static mode, disconnect the wiring connector and connect an ohmmeter across the sensor's two (white) wires. Resistance should be 4 - 7 ohms. If an open circuit is detected, replace the O2 sensor.
2. Start the engine and bring it up to operating temperature.
3. Raise and support the vehicle.

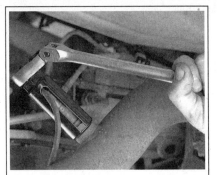

Fig. 38 Sensor sockets, like this one from Lisle®, are slotted on one side to accommodate wiring

CAUTION:
The exhaust pipe gets extremely hot during engine operation, and if touched, severe burns can occur. If servicing the oxygen sensor, avoid contacting the exhaust system.

4. Backprobe the O2 sensor between the O2 sensor output wire and ground with a suitable high impedance voltmeter.
5. The O2 sensor should be rapidly switching between 0 and 1v. If working properly, it should be switching from a lean mixture (less than 0.4v) to a rich mixture (0.6 - 1.0v), and back. The average voltage should fall between 0.4 - 0.6v.
6. If the sensor switches slowly, or is stuck in the middle of the range, the O2 sensor may be faulty.
7. If the sensor is stuck rich or lean, it most likely indicates a problem with the engine; for example, a vacuum leak would cause the sensor to read a lean mixture, and a malfunctioning fuel pressure regulator would cause a rich mixture.
8. If the O2 sensor is above or below the specified range (0 - 1v), a wiring or computer problem is most likely the cause.
9. Lower the vehicle.
10. Turn the engine off.

REMOVAL & INSTALLATION

See Figures 38 and 39

WARNING:
Never apply any type of grease to the O2 sensor electrical connector or attempt any soldering of the wiring harness. This will influence the operation of the sensor.

1. Disconnect the negative battery cable.
2. Raise and support the vehicle.

CAUTION:
The exhaust pipe gets extremely hot during engine operation, and if touched, severe burns can occur. If servicing the oxygen sensor, avoid contacting the exhaust system.

3. Disconnect the wiring harness from the oxygen sensor.
4. Remove the sensor using the appropriate tool.

Fig. 39 Removing the O2 sensor from the exhaust pipe

Note: The oxygen sensor threads are coated with an anti-seize compound. The compound must be removed from the mounting boss threads, either in the exhaust manifold or Y-pipe. An 18mm x 1.5 x 6E tap is required.

To install:
5. Clean the threads of the mount to remove any old anti-seize compound.
6. If the old sensor is to be reused, apply anti-seize compound to its threads. New sensors come with the compound already applied.
7. Install and tighten the sensor to 22 ft. lbs. (30 Nm). Connect the wiring harness.
8. Lower the vehicle.
9. Connect the negative battery cable.

Idle Air Control Motor (IAC)

OPERATION

See Figures 40 and 41

The Idle Air Control (IAC) system consists of a stepper motor and pintle. It is mounted to the throttle body and regulates the amount of air bypassing the throttle plate. It is operated by the PCM. The throttle body has an air control passage that provides air for the engine at idle (when the throttle plate is closed). The pintle protrudes into the air control passage and regulates air flow through it. Based on various sensor inputs, the PCM adjusts engine speed by moving the pintle in and out of the air control passage. The IAC motor is positioned when the ignition is turned to the ON position.

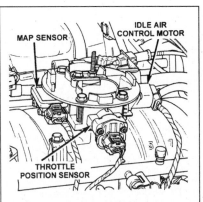

Fig. 40 MAP sensor, TPS and IAC motor locations - typical

4-14 DRIVEABILITY AND EMISSION CONTROLS

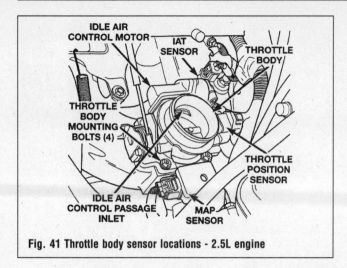

Fig. 41 Throttle body sensor locations - 2.5L engine

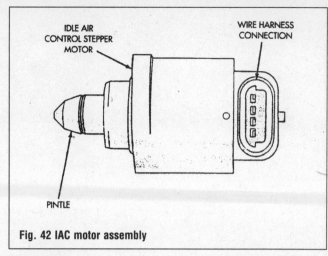

Fig. 42 IAC motor assembly

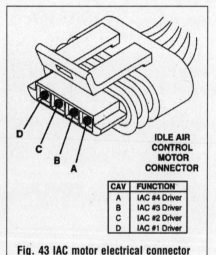

Fig. 43 IAC motor electrical connector pinouts

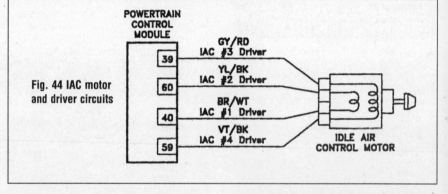

Fig. 44 IAC motor and driver circuits

At idle, engine speed can be increased by retracting the IAC motor pintle and allowing more air to pass through the port, or it can be decreased by restricting the passage with the pintle.

The PCM uses the IAC motor to control idle speed (along with timing) and to keep the engine from stalling during deceleration.

The IAC motor has four wires with four circuits. Two of the wires are for 12 volts and ground to supply electrical current to the motor windings and to operate the stepper motor in one direction. The other two wires are also for 12 volts and ground to supply electrical current to operate the stepper motor in the opposite direction. The PCM reverses polarity to move the pintle in one direction or the other.

When engine rpm is above idle speed, the IAC is used for the following:
- Off-idle dashpot (throttle blade will close quickly but idle speed will not stop quickly)
- Deceleration air flow
- A/C compressor load control (so the engine does not lose rpm when the A/C engages)
- Power steering load control

TESTING

See Figures 42, 43 and 44

Conclusive testing of this component requires the use of special, factory-supplied equipment and is not included here. The tests below will provide valuable information regarding condition of the IAC motor and require only a VOM.
1. Ignition must be OFF.
2. Disconnect the IAC motor connector.
3. Disconnect the PCM harness connector(s).
4. Measure the resistance of the four IAC motor wires between the motor connector and the PCM connector. Refer to the illustrations for wire colors.
5. Resistance should be under 5.0 ohms. Replace any wire with a higher resistance.
6. Check resistance between each motor lead and ground. Replace any wire with resistance under 5.0 ohms.
7. Remove the ASD relay from the PDC (location is shown on the underside of the PDC cover. Connect a jumper between cavity 30 and 87. Turn the ignition ON. Measure the voltage at each of the driver wires. Voltage should not exceed 1.0 v.
8. With the ignition OFF and the IAC and PCM connectors detached, measure the resistance between each driver wire and ground. Resistance must be above 5.0 ohms in each case.

REMOVAL & INSTALLATION

See Figure 45

The IAC motor is located on the rear of the throttle body secured by two Torx® screws.
1. Disconnect the negative battery cable.
2. Remove the air cleaner assembly.
3. Unplug the IAC motor connector.
4. Remove the two mounting Torx® screws.
5. Remove the IAC motor from the throttle body.

WARNING:
When the IAC motor is removed from the throttle body, do not extend the pintle more than 0.250 inch (6.35mm). If the pintle is extended more than this amount, it may separate from the IAC motor and the motor will have to be replaced.

To install:
6. Install the IAC motor in the throttle body.
7. Tighten the mounting bolts to 60 inch lbs. (7 Nm).
8. Plug the IAC electrical connector in.
9. Install the air cleaner assembly.
10. Connect the negative battery cable.

Fig. 45 IAC motor mounting screws (arrows)

DRIVEABILITY AND EMISSION CONTROLS 4-15

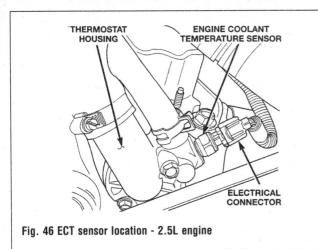

Fig. 46 ECT sensor location - 2.5L engine

Fig. 47 ECT sensor location - 3.9L, 5.2L, 5.9L engines. Thermostat housing removed for clarity

Engine Coolant Temperature (ECT) Sensor

OPERATION

See Figures 46, 47 and 48

The Engine Coolant Temperature (ECT) is located on or near the thermostat housing.

The ECT sensor's resistance changes in response to engine coolant temperature. The sensor resistance decreases as the coolant temperature increases, and vice-versa. This provides a reference signal to the PCM, which indicates engine coolant temperature. The signal sent to the PCM by the ECT sensor helps the PCM to determine spark advance, EGR flow rate, and air/fuel ratio.

Some vehicles may have two ECTs fitted: one for input to the PCM and the other for input to the coolant temperature gauge or warning light.

TESTING

See Figures 49, 50 and 51

1. Check resistance across the sensor terminals. Using an input impedance (digital) VOM is recommended.
2. Resistance will vary with coolant temperature. Generally, resistance will be about 18K ohms at 50-degrees F (10-degrees C), decreasing to about 600 ohms at operating temperature. If the measured resistance is at or near zero (shorted) or at or near infinity (open), replace the sensor.
3. Bring the engine up to or near operating temperature.
4. Disconnect the negative battery cable.
5. Unplug the connector on the sensor.

Note: Be sure you are unplugging and checking the ECT, and not the temperature gauge sending unit; they are two different parts.

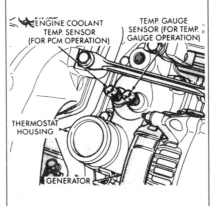

Fig. 48 Location of the ECT and temperature gauge sensors - early V-10 engine

Fig. 49 The coolant temperature sensor can be tested using the Auto Xray, or equivalent tool's data display feature

6. Connect an ohmmeter across the sensor connectors and observe the reading over time as the engine cools.
7. As the engine cools the resistance in the sensor should increase slowly.
8. If it doesn't, replace the sensor.
9. If a scan tool is available, it can be used to check the sensor. Call up sensor voltage in the "live" mode. Start and run the engine observing the sensor voltage. It should decrease and increase smoothly as engine temperature increase and then decreases.

REMOVAL & INSTALLATION

1. Disconnect the negative battery cable.

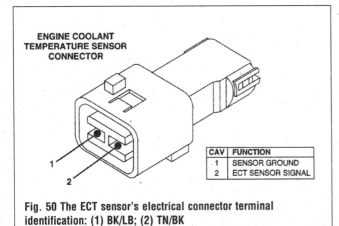

Fig. 50 The ECT sensor's electrical connector terminal identification: (1) BK/LB; (2) TN/BK

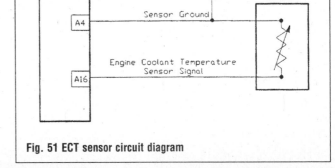

Fig. 51 ECT sensor circuit diagram

4-16 DRIVEABILITY AND EMISSION CONTROLS

CAUTION:

Never open, service or drain the radiator or cooling system when hot; serious burns can occur from the steam and hot coolant.

2. Drain the engine coolant below the level of the thermostat housing.
3. Remove the air cleaner assembly if necessary for access to the sensor.
4. Remove the connector from the sensor. Vehicles with air conditioning may benefit if you fabricate an L-shaped hook tool from a wire coat hanger about 8 in. long. Use the hook for removal rather than pulling on the wiring harness.
5. Remove the sensor from the intake manifold.

To install:

6. Install the sensor in the manifold and tighten to 8 ft. lbs. (11 Nm).
7. Install the connector on the sensor. There is no polarity: the connector cannot be accidentally reversed.
8. Add the proper mixture of coolant and water to the cooling system.
9. Connect the negative battery cable.
10. Inspect for leaks.
11. Install the engine cover.

Intake Air Temperature (IAT) Sensor

OPERATION

See Figure 52

The two-wire Intake Manifold Air Temperature (IAT) sensor is installed in the intake manifold with the sensor element extending into the air stream.

1. This is a Negative Thermal Coefficient (NTC), meaning that as the manifold air temperature increases, the resistance (voltage) in the sensor decreases, and vice-versa.
2. The IAT sensor provides an input voltage to the PCM indicating the density of the air entering the intake manifold based on intake manifold air temperature. When the ignition is turned ON, a 5-volt power circuit is supplied to the sensor from the PCM. The sensor is grounded at the PCM through a low-noise, sensor-return circuit.
3. The PCM uses this input to determine:
- Injector pulse-width (air:fuel ratio)
- Ignition timing (to prevent spark knock)

Fig. 52 IAT sensor installed in the intake manifold of a 5.2L engine. Others similar

TESTING

See Figures 53, 54 and 55

1. Be sure the ignition is OFF.
2. Disconnect the connector at the IAT sensor.
3. Check resistance across the sensor terminals. Using an input impedance (digital) VOM is recommended.
4. Resistance will vary with temperature, but should be about 18K ohms at 50-degrees F (10-degrees C), decreasing rapidly as temperature increases to about 1K ohms at 176-degrees F (80-degrees C) which is near engine operating temperature. If the measured resistance is at or near zero (shorted) or at or near infinity (open), replace the sensor.
5. To test the harness, proceed as follows:
   a. Shut the ignition OFF and disconnect the battery negative cable.
   b. Unplug the PCM, test the harness between the proper PCM connector terminal and the sensor connector. Also check the harness between the proper PCM connector terminal and the sensor connector. If resistance is more than 1 ohm, repair the wiring harness.

WARNING:

Testing the wiring harness with the PCM still connected can cause serious damage to the processor. ALWAYS disconnect the PCM before testing the wiring harness, unless instructed otherwise.

   c. Connect the negative battery cable.

Fig. 53 Testing the IAT sensor with an ohmmeter connected across the two sensor terminals

REMOVAL & INSTALLATION

See Figures 52 and 56

1. Disconnect the negative battery cable.
2. Unplug the electrical connector from the sensor.
3. Unscrew the sensor from the intake manifold.

To install:

4. Install the sensor into the intake manifold and tighten it to 20 ft. lbs. (28 Nm).
5. Plug the connector into the sensor.
6. Connect the negative battery cable.

Manifold Absolute Pressure (MAP) Sensor

OPERATION

See Figures 40 and 57

The Manifold Absolute Pressure (MAP) sensor measures the pressure inside the intake manifold, by measuring the vacuum level. It sends a voltage signal to the PCM in relation to the pressure inside the manifold, which varies according to engine load and altitude. The PCM uses this signal to adjust the air/fuel ratio (injector pulse width) and spark advance.

A 5v reference is supplied from the PCM and returns a voltage signal to the PCM that reflects manifold pressure. The zero-pressure reading is 0.5v and full scale is 4.5v. For a pressure swing of 0 - 15 psi,

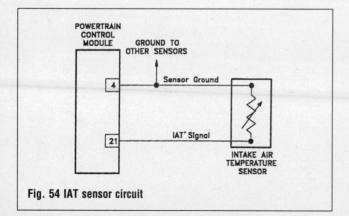

Fig. 54 IAT sensor circuit

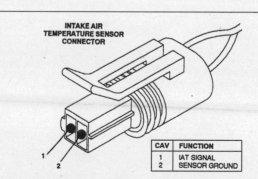

Fig. 55 IAT sensor's electrical connector terminal identification: (1)BK/LB; (2) BK/RD

DRIVEABILITY AND EMISSION CONTROLS 4-17

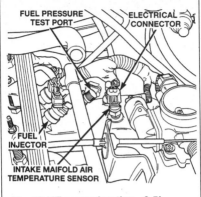

Fig. 56 IAT sensor location - 2.5L engine shown

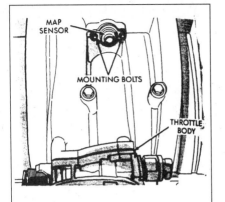

Fig. 57 MAP sensor location - 8.0L engine

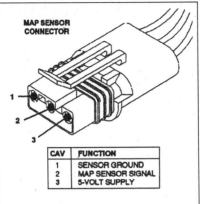

Fig. 58 MAP sensor electrical connector - except 4.7L and 8.0L engines

the voltage changes 4.0v. To operate the sensor, it is supplied a regulated 4.8 - 5.1v. Ground is provided through the low-noise, sensor return circuit at the PCM.

MAP sensor input is the primary contributor to fuel injector pulse width. The most important function of this sensor is to determine barometric pressure. The PCM needs to know if the vehicle is at sea level or high altitude because air density changes with altitude. As altitude goes up, barometric pressure goes down.

Readings are taken every 12 msec.

The PCM uses the MAP sensor input to aid in calculating the following parameters:
- Manifold pressure
- Barometric pressure
- Engine load
- Injector pulse width
- Spark advance
- Shift-point strategies
- Idle speed
- Decel fuel shutoff

TESTING

See Figures 58, 59, 60 and 61

The MAP sensor is fitted to the front of the throttle body except on 4.7L and 8.0L engines where it is installed on the intake manifold.

1. Remove the air cleaner assembly.
2. On throttle-body mounted MAP sensors, inspect the L-shaped tube from the throttle body to the sensor for cracks, blockage, and damage. Repair as necessary.
3. Unplug the MAP sensor connector.
4. Test the MAP sensor output voltage at the

Fig. 59 Testing the MAP sensor

MAP sensor connector between terminals "sensor ground" and "sensor signal". With ignition ON and the engine OFF, the voltage should be between 4 - 5 volts.

5. Check voltage across the same terminals with the engine running at idle at operating temperature. Voltage should be 1.5 - 2.1v.
6. Test MAP sensor supply voltage at sensor connector between terminals "sensor ground" and "5v supply" with the ignition ON and engine off, voltage should be 4.5 - 5.0 volts.
7. Test the MAP sensor ground circuit at the "sensor ground" of the MAP connector, voltage should be less than 0.2 volts. If not inspect for open harness from pin 4 of PCM harness and terminal A. If no voltage is present, proceed to the next step.
8. Turn the ignition OFF.
9. Disconnect the PCM harness from the PCM.

WARNING:

Testing the wiring harness with the PCM still connected can cause serious damage to the processor. ALWAYS disconnect the PCM before testing the wiring harness, unless instructed otherwise.

10. Test the MAP sensor output voltage at the PCM connector. At the PCM cavity A-27, the voltage should be about 5.0v.
11. Check the voltage at PCM cavity A-17. It should be 5.0v.

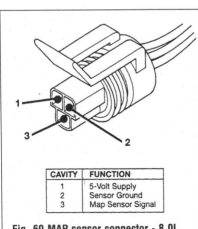

Fig. 60 MAP sensor connector - 8.0L engine

12. If all of the above tests pass, plug in PCM wiring harness and MAP sensor connector.
13. With the ignition in the ON position, and the engine OFF, remove the tube from the throttle body to the MAP sensor.
14. Connect a vacuum pump to the nipple on the MAP sensor, and pump the sensor to 20 - 27 in. Hg. of vacuum. Check the sensor output voltage, it should be below 1.8 volts. If not, replace the MAP sensor. If the voltage is OK, proceed to next step.
15. Relieve vacuum pressure on the sensor, then check the output voltage. The voltage should be 4 - 5 volts; if not, replace the MAP sensor.
16. Install the air cleaner assembly.

REMOVAL & INSTALLATION

See Figure 62

2.5L, 3.9L, 5.2L, 5.9L Engines

1. Disconnect the negative battery cable.
2. Remove the air cleaner assembly.

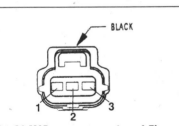

Fig. 61 MAP sensor connector - 4.7L engine. (1) 5V supply, (2) sensor ground, (3) sensor signal

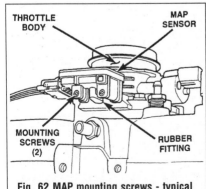

Fig. 62 MAP mounting screws - typical

4-18 DRIVEABILITY AND EMISSION CONTROLS

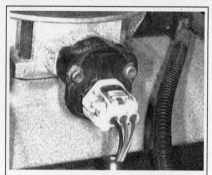

Fig. 63 TPS location on the throttle body - typical

3. Remove the two MAP sensor retaining bolts.
4. Slide the L-shaped tube to the MAP sensor from the throttle body off while removing the MAP sensor from the throttle body.

To install:
5. Install the L-shaped tube onto the MAP sensor.
6. Position the sensor to the throttle body while guiding the rubber fitting over the throttle body vacuum nipple.
7. Install the MAP sensor onto the throttle body assembly. Tighten the retaining bolts to 25 inch lbs. (3 Nm). If applicable, install the throttle body.
8. Install the air cleaner assembly.
9. Connect the negative battery cable.

4.7L, 8.0L Engines

1. Clean the area around the sensor before removal to keep debris out of the intake tract.

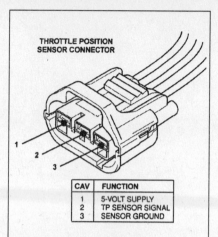

Fig. 64 The Throttle Position Sensor's electrical connector terminal identification

2. Remove the electrical connector.
3. Remove the sensor mounting bolts, then remove the sensor.

To install:
4. Make sure the sensor seal is in good condition.
5. Clean the sensor and lubricate the rubber gasket with clean engine oil.
6. Clean the sensor opening in the intake manifold.
7. Install the sensor, tighten the bolts to 20 inch lbs. (2 Nm).
8. Connect the wires to the sensor.

Throttle Position Sensor (TPS)

OPERATION

See Figure 63

The Throttle Position sensor (TPS) is a potentiometer that provides a voltage signal to the PCM that is directly proportional to the throttle plate position. The sensor is mounted on the side of the throttle body and is connected to the throttle plate shaft. The TPS monitors throttle plate movement and position, and transmits an appropriate electrical signal to the PCM. These signals are used by the PCM to adjust the air/fuel mixture, spark timing and EGR operation according to engine load at idle, part throttle, or full throttle. The TPS is not adjustable.

The TPS receives a 5-volt reference signal and a ground circuit from the PCM. A return signal circuit is connected to a wiper that runs on a resistor internally on the sensor. The further the throttle is opened, the wiper moves along the resistor, at wide open throttle, the wiper essentially creates a loop between the reference signal and the signal return returning the full or nearly full 5 volt signal back to the PCM. At idle the signal return should be approximately 0.9 volts.

TESTING

See Figures 64, 65, 66, 67 and 68

1. With the engine OFF and the ignition ON, check the voltage at the signal return circuit of the TP sensor by carefully backprobing the connector using a DVOM.
2. Voltage should be between 0.2 and 1.4 volts at idle.
3. Slowly move the throttle pulley to the wide open throttle (WOT) position and watch the voltage on the DVOM. The voltage should slowly rise to slightly less than 4.5v at Wide Open Throttle (WOT).
4. If no voltage is present, check the wiring harness for supply voltage (5.0v) and ground (0.3v or less), by referring to your corresponding wiring guide. If supply voltage and ground are present, but no output voltage from TPS, replace the TPS. If supply voltage and ground do not meet specifications, make necessary repairs to the harness or PCM.

REMOVAL & INSTALLATION

See Figures 69 and 70

1. Disconnect the negative battery cable.

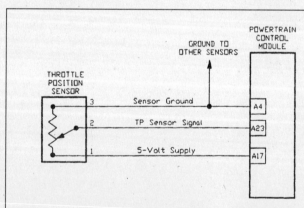

Fig. 65 The Throttle Position Sensor circuit schematic

Fig. 66 Using a voltmeter to check the TPS for signal voltage

Fig. 67 Using a voltmeter to check the TPS for reference (supply) voltage

Fig. 68 You can use the data display function of the Auto Xray, or other scan tool to get the TPS readings

DRIVEABILITY AND EMISSION CONTROLS 4-19

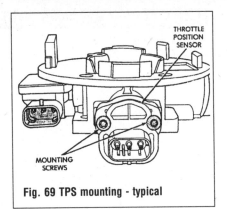

Fig. 69 TPS mounting - typical

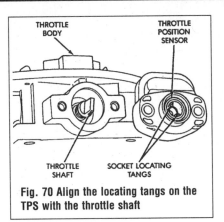

Fig. 70 Align the locating tangs on the TPS with the throttle shaft

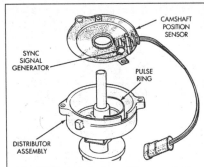

Fig. 71 Exploded view of the camshaft position sensor mounting - 2.5L, 3.9L, 5.2L, 5.9L engines

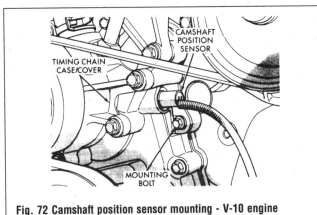

Fig. 72 Camshaft position sensor mounting - V-10 engine

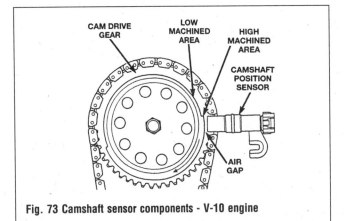

Fig. 73 Camshaft sensor components - V-10 engine

2. Remove the air cleaner assembly.
3. Unplug the TPS connector.
4. Remove the two retaining bolts.
5. Remove the sensor from the throttle body.

To install:

6. Install the sensor on the throttle body.
7. Tighten the sensor bolts to 60 inch lbs. (7 Nm)
8. Open the throttle to WOT and back again several times to check for binding.
9. Plug in the connector.
10. Install the air cleaner assembly.
11. Connect the negative battery cable.

Camshaft Position (CMP) Sensor

OPERATION

See Figures 71, 72 and 73

The camshaft position sensor is required for all vehicles that utilize a sequential multi-port fuel injection system. It serves this system by providing a fuel sync signal to the engine controller. The sync signal is used in conjunction with the signal sent by the crankshaft signal to maintain the correct injector firing order.

- On 2.5L, 3.9L, 5.2L and 5.9L gasoline engines, the camshaft position sensor is located in the distributor
- On the 4.7L engine, the camshaft sensor is bolted to the front top of the right-hand cylinder head
- On V-10 engines, the camshaft position sensor is found in the timing chain case/cover on the left-front side of the engine

- On diesel engines, the sensor is located below the fuel injection pump and is attached to the back of the timing gear cover housing

TESTING

See Figures 74 and 75

To completely test this sensor and circuitry, you need the DRBII scan tool, or equivalent. This is a test of the camshaft position sensor only.

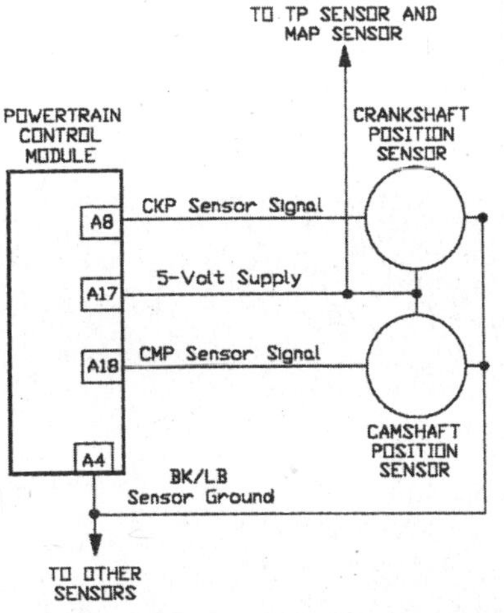

Fig. 74 Camshaft Position Sensor circuit schematic

For this test you will need an analog (non-digital) voltmeter. Do not remove the distributor connector. Using small paper clips, insert them into the backside of the distributor wire harness connector to make contact with the terminals. Do not damage the connector when inserting the paper clips. Attach the voltmeter leads to these clips.

1. Connect the positive voltmeter lead to the sensor output wire.
2. Connect the negative voltmeter lead to the ground wire.
3. Turn the ignition ON. Rotate the engine. The

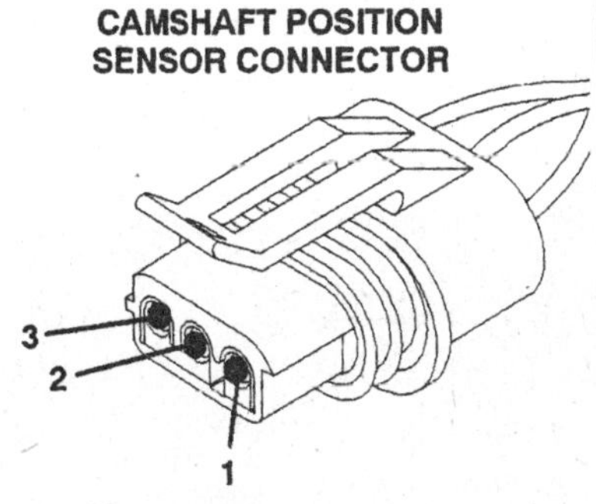

Fig. 75 Camshaft Position Sensor wires: (1) OR (5v supply), (2) BK/LB (sensor ground), (3) TN/YL (camshaft position sensor signal)

4-20 DRIVEABILITY AND EMISSION CONTROLS

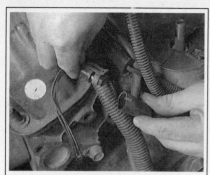

Fig. 76 Detach the CMP sensor connector - 5.2L engine shown

Fig. 77 The CMP sensor pickup removed from the distributor. Note the pulse ring on the distributor shaft

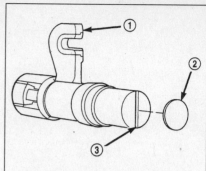

Fig. 78 Sensor depth positioning rib - 8.0L engine: sensor (1), paper spacer (2), rib for sensor depth positioning (3)

meter should show a reading of less than 1 volt and a high voltage reading of 5.0 volts as the high and low points on the wheel pass the sensor.

4. If there is no voltage reading at all, check the meter connections.

5. If voltage is still not present, check for voltage at the supply wire.

6. If 5.0 volts is not found at the supply wire, check for voltage at cavity A-17 of the 32-way connector. Leave the engine controller harness connected for this test.

Note: If voltage is not found at the PCM connector, you will need to diagnose the system using a DRBII scan tool, or equivalent, or take the vehicle to a qualified shop.

7. If voltage is present at the engine controller connector, but not the supply wire:
   a. Check continuity between the sensor and supply wire. This is checked between the distributor connector and cavity A-17. If continuity is not present, repair the wire harness.
   b. Check for continuity between the camshaft position sensor output wire and cavity A-18 at the engine controller. If continuity is not present, repair the wire harness.
   c. Check for continuity between the ground circuit wire at the distributor connector and ground. If continuity is not present, repair the wire harness.

8. Crank the engine while observing the voltmeter. The voltmeter needle should fluctuate 0 - 5 volts, approximately. This will verify the camshaft position sensor is operating properly and a sync pulse signal is being generated.

9. If a sync pulse signal is not detected, and all other variables have been eliminated, replace the camshaft position sensor.

REMOVAL & INSTALLATION

2.5L, 3.9L, 5.2L, 5.9L Gasoline Engines
See Figures 71, 76 and 77

1. Remove the air cleaner assembly.
2. Disconnect the negative battery cable.
3. Remove the distributor cap and unplug the camshaft position sensor wire.
4. Remove the rotor from the shaft, then lift the camshaft position sensor assembly from the distributor.

To install:

5. Install the camshaft position sensor, aligning the sensor notch to the housing.
6. Connect the wire, install the rotor and cap.
7. Install the air cleaner assembly, if removed.

Other Models

8. Disconnect the negative battery cable.
9. Locate the sensor. Disconnect the wiring.
10. Remove the mounting screws and pull off the sensor.
11. Check condition of any seals or O-rings.
12. Installation is the reverse of removal.

Sensor Replacement (8.0L) - Old Sensor
See Figure 78

A thin plastic rib is molded into the face of the camshaft position sensor to position the depth of the sensor to the upper cam gear (sprocket). This rib is found on original and replacement sensors. The first time the engine is operated with the new sensor, part of the rib may be sheared (ground) off. Depending on parts tolerances, some of the plastic rib may still be observed after removal.

Note: This procedure will require a special peel-and-stick paper spacer of a calibrated thickness for setting the tolerance of the sensor upon installation. This special paper should be available from the dealer parts department.

If the original camshaft position sensor is to be removed and installed, such as when servicing the timing chain, timing gears or timing chain cover, use this procedure.

1. Unplug the sensor harness connector from the engine harness.
2. Remove the sensor mounting bolt.
3. Carefully pry the sensor from the timing chain case/cover in a rocking action with two small pry-tools.
4. Remove the sensor. Check the condition of the sensor O-ring.

To install:

Note: When installing a used camshaft position sensor, the depth must be adjusted.

5. Inspect the face of the sensor. If any of the original rib material remains, cut it flush to the face of the sensor with a razor knife. Remove only enough of the rib material until the face is flat.

6. Apply the special peel-and-stick paper spacer to the (clean and flat) face of the sensor.
7. Apply a small amount of clean engine oil to the O-ring.
8. Using a slight rocking action, install the sensor into the timing case/cover until the paper spacer just contacts the gear. Do not twist the sensor or you may tear the paper or O-ring. Do not install the mounting bolt.
9. Use a scribe to scratch a line into the timing case/cover to indicate the depth of the sensor, then remove the sensor.
10. Remove the paper spacer so it will not remain in the engine and get lost in the engine oil passages.
11. Again, apply a small amount of oil to the sensor O-ring and install the sensor until it aligns with the scribe mark.
12. Install the mounting bolt, tightening it to 50 inch lbs. (6 Nm).
13. Plug-in the engine wire harness connector to the sensor connector.

Sensor Replacement (8.0L) - New Sensor

Use this procedure if a NEW camshaft position sensor is to be installed.

1. Unplug the wire connectors and unbolt the sensor.
2. Carefully pry the sensor from the timing chain case/cover in a rocking action with two small pry-tools.
3. Remove the sensor. Check the condition of the sensor O-ring.

To install:

4. Lightly lubricate the O-ring with fresh engine oil.
5. Using a slight rocking action, install the sensor into the timing case/cover until the paper spacer just contacts the gear. Do not twist the sensor or you may tear the O-ring. Push the sensor all the way into the cover until the rib material on the sensor contacts the camshaft gear.
6. Install the mounting bolt, tightening it to 50 inch lbs. (6 Nm).
7. Connect the sensor wiring.

WARNING:

Do not remove more rib material than necessary or damage to the sensor may result. Never use an electric grinder to remove rib material, as the magnetic field from the electric motor may cause electrical damage to the sensor.

DRIVEABILITY AND EMISSION CONTROLS 4-21

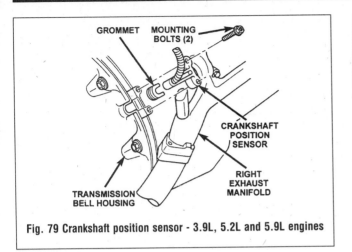

Fig. 79 Crankshaft position sensor - 3.9L, 5.2L and 5.9L engines

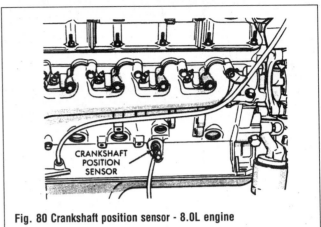

Fig. 80 Crankshaft position sensor - 8.0L engine

Crankshaft Position (CKP) Sensor

OPERATION

See Figures 79, 80 and 81

The crankshaft position sensor generates pulses and sends them to the PCM. The PCM interprets this input signal to determine the crankshaft position (and engine speed). With this and other information, the engine controller determines injector sequence and ignition timing. The sensor is a hall-effect device combined with an internal magnet. It is sensitive to steel within a certain distance from it.

On the 4-cylinder engine the crankshaft position sensor is mounted to the transmission bell housing. On the V6 and V8 engines, the crankshaft position sensor is bolted to the cylinder block near the rear of the right cylinder head. On the V10 engine, the crankshaft position sensor is located on the right-lower side of the cylinder block, just above the oil pan rail.

TESTING

See Figure 82

On the 2.5L engine, the Crankshaft Position (CMK) sensor is located near the outer edge of the flywheel (or starter ring gear). On other models it is fitted to the right side of the engine block.

The only test you can perform without a DRBII scan tool, or equivalent, is a basic check of the sensor.

1. Disconnect the sensor harness connector from the main wire harness connector.
2. Place an ohmmeter across terminals 1 and 2 (see the illustration). The meter reading should be open (infinite resistance). If a low resistance is read, replace the camshaft position sensor.

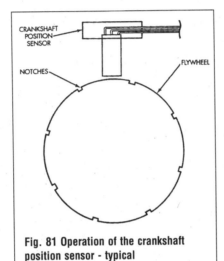

Fig. 81 Operation of the crankshaft position sensor - typical

REMOVAL & INSTALLATION

All Except V-10 Engine

1. Remove the air cleaner assembly if necessary for access to the sensor.
2. Disconnect the pigtail harness from the sensor.
3. Remove the nut holding the sensor wire clip to the fuel rail mounting stud.
4. Remove the sensor mounting hardware, then the sensor.
5. Remove the clip from the sensor wire harness.

To install:

6. Install the sensor flush against the opening.
7. Install and tighten the bolts or nuts, as applicable.
8. Install the electrical connector.
9. Install the clip on the sensor wire harness.

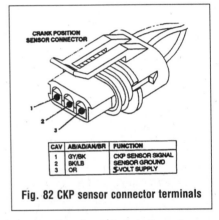

Fig. 82 CKP sensor connector terminals

10. Install the clip over the fuel rail mounting stud. Install the clip mounting nut.

V-10 Engine

1. Raise and safely support the vehicle.
2. Unplug the sensor harness connector from the engine harness.
3. Remove the sensor mounting bolt.
4. Carefully pry the sensor from the timing chain case/cover in a rocking action with two small pry-tools.
5. Remove the sensor. Check the condition of the sensor O-ring.

To install:

6. Apply a small amount of fresh engine oil to the sensor O-ring.
7. Using a slight rocking action, install the sensor into the timing case/cover until the paper spacer just contacts the gear. Do not twist the sensor or you may tear the O-ring.
8. Install the mounting bolt, tightening it to 70 inch lbs. (8 Nm).
9. Connect the sensor wiring.

TROUBLE CODES

General Information

The Powertrain Control Module (PCM) is given responsibility for the operation of the emission control devices, cooling fans, ignition and advance and in some cases, automatic transmission functions. Because the PCM oversees both the ignition timing and the fuel injector operation, a precise air/fuel ratio will be maintained under all operating conditions. The PCM is a microprocessor or small computer which receives electrical inputs from several sensors, switches and relays on and around the engine.

Based on combinations of these inputs, the PCM controls outputs to various devices concerned with engine operation and emissions. The PCM relies on the signals to form a correct picture of current vehicle operation. If any of the input signals are incorrect, the PCM reacts to what ever picture is painted for it. For example, if the coolant temperature sensor is inaccurate and reads too low, the PCM may see a picture of the engine never warming up. Consequently, the engine settings will be maintained as if the engine were cold. Because so many inputs can affect one output,

4-22 DRIVEABILITY AND EMISSION CONTROLS

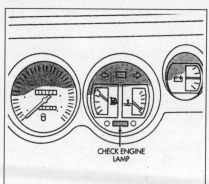

Fig. 83 Check engine lamp location

Fig. 84 Plug the scan tool right into the connector to retrieve DTC's or to perform other tests

Fig. 85 Although sometimes no DTCs are found, that does not rule out a potential problem

Fig. 86 When you find a DTC, proceed to diagnostics

correct diagnostic procedures are essential on these systems.

One part of the PCM is devoted to monitoring both input and output functions within the system. This ability forms the core of the self-diagnostic system. If a problem is detected within a circuit, the control module will recognize the fault, assign it a Diagnostic Trouble Code (DTC), and store the code in memory. The stored code(s) may be retrieved during diagnosis.

While the PCM is capable of recognizing many internal faults, certain faults will not be recognized. Because the PCM sees only electrical signals, it cannot sense or react to mechanical or vacuum faults affecting engine operation. Some of these faults may affect another component which will set a code. For example, the PCM monitors the output signal to the fuel injectors, but cannot detect a partially clogged injector. As long as the output driver responds correctly, the computer will read the system as functioning correctly. However, the improper flow of fuel may result in a lean mixture. This would, in turn, be detected by the oxygen sensor and noticed as a constantly lean signal by the PCM. Once the signal falls outside the pre-programmed limits, the PCM would notice the fault and set a trouble code.

Additionally, the PCM employs adaptive fuel logic. This process is used to compensate for normal wear and variability within the fuel system. Once the engine enters steady-state operation, the PCM watches the oxygen sensor signal for a bias or tendency to run slightly rich or lean. If such a bias is detected, the adaptive logic corrects the fuel delivery to bring the air/fuel mixture towards a centered or 14.7:1 ratio. This compensating shift is stored in a non-volatile memory which is retained by battery power even with the ignition switched OFF. The correction factor is then available the next time the vehicle is operated.

Note: If the battery cable(s) are disconnected for longer than 5 minutes, the adaptive fuel factor will be lost. After repair it will be necessary to drive the truck at least 10 miles to allow the processor to relearn the correct factors. The driving period should include steady-throttle open road driving if possible. During the drive, the vehicle may exhibit driveability symptoms not noticed before. These symptoms should clear as the control module computes the correction factor.

The electronic engine control format is called OBD-II. OBD-II vehicles are able to monitor certain systems of the vehicle through various input sensors and output hardware. These monitors include the Catalyst Efficiency, Engine Misfire Detection, Comprehensive Component, EGR System Flow (where fitted), EVAP System Integrity, Secondary Air (if equipped), Fuel System, and Heated O2 Sensor Monitors.

Some of the hardware included with these monitors are an O2 sensor behind the catalyst, a Leak Detection Pump (LDP) on the EVAP system, and a High Data Rate CKP sensor to help monitor engine misfires. All monitors in the OBD-II system have separate trouble codes and are diagnosed by standard diagnostic methods.

MALFUNCTION INDICATOR LAMP (MIL)

See Figure 83

The Malfunction Indicator Lamp (MIL) is located on the instrument panel. The lamp is connected to the control unit and will alert the driver to certain malfunctions detected by the PCM. When the lamp is illuminated, the PCM has detected a fault and stored a DTC in memory.

The light will stay illuminated as long as the fault is present. Should the fault self-correct, the MIL will extinguish but the stored code will remain in memory.

Under normal operating conditions, the MIL should illuminate briefly when the ignition key is turned ON. This is commonly known as a bulb check. As soon as the PCM receives a signal that the engine is cranking, the lamp should extinguish. The lamp should remain extinguished during the normal operating cycle.

Diagnostic Connector

See Figure 84

The diagnostic connector is located under the dashboard, on the driver's side, next to the steering column.

Reading Codes

See Figures 85 and 86

There are two kinds of codes stored in the PCM: hard faults and continuous faults. A hard fault is a fault detected that is malfunctioning at the time of testing. A continuous fault is a fault that was detected by the PCM, but is not malfunctioning at the moment. The PCM will erase this code if the fault does not recur within a set amount of time; different year models take different amounts of time to clear these code, however it usually is around 40 starts.

When the MIL lamp (Check Engine light) illuminates while the engine is running, it indicates a detected fault by the PCM.

On 1998 and later models, you must use the DRB-II scan tool, or equivalent, to retrieve this information from the PCM.

On 1997 models, you can use the scan tool, or use the MIL as follows: Turn the ignition key ON - OFF - ON - OFF - ON within five seconds. The MIL lamp will then flash the DTC(s). Each code, if there is more than one, will be preceded by a four-second pause to distinguish it from the previous code. An example of a DTC flashed by the MIL would be:

1. Lamp on for two seconds, then turns off.
2. Lamp flashes two times, then pauses, then flashes seven times.
3. Lamp pauses for four seconds.
4. Lamp flashes three times, then pauses, then flashes one time.
5. Lamp pauses for four seconds.
6. Lamp flashes five times, then pauses, then flashes five times.

In this instance, the DTCs stored were 27, and 31. The PCM will then flash code 55 after it has reached the last code stored in its memory.

DRIVEABILITY AND EMISSION CONTROLS 4-23

The following is a list of OBD-II codes. The two-digit number in parentheses after the problem description is the MIL lamp code (where applicable and for 97 models only):

P0000 No Failures
      Test Complete (55)
P0100 Mass or Volume Air Flow Circuit Malfunction
P0101 Mass or Volume Air Flow Circuit Range/Performance Problem
P0102 Mass or Volume Air Flow Circuit Low Input
P0103 Mass or Volume Air Flow Circuit High Input
P0104 Mass or Volume Air Flow Circuit Intermittent
P0105 Manifold Absolute Pressure/Barometric Pressure Circuit Malfunction
P0106 Manifold Absolute Pressure/Barometric Pressure Circuit Range/Performance Problem
P0107 Manifold Absolute Pressure/Barometric Pressure Circuit Low Input (14)
P0108 Manifold Absolute Pressure/Barometric Pressure Circuit High Input (14)
P0109 Manifold Absolute Pressure/Barometric Pressure Circuit Intermittent
P0110 Intake Air Temperature Circuit Malfunction
P0111 Intake Air Temperature Circuit Range/Performance Problem
P0112 Intake Air Temperature Circuit Low Input (23)
P0113 Intake Air Temperature Circuit High Input (23)
P0114 Intake Air Temperature Circuit Intermittent
P0115 Engine Coolant Temperature Circuit Malfunction
P0116 Engine Coolant Temperature Circuit Range/Performance Problem
P0117 Engine Coolant Temperature Circuit Low Input (22)
P0118 Engine Coolant Temperature Circuit High Input (22)
P0119 Engine Coolant Temperature Circuit Intermittent
P0120 Throttle/Pedal Position Sensor/Switch "A" Circuit Malfunction
P0121 TPS Voltage Does Not Agree With MAP (24)
P0122 Throttle/Pedal Position Sensor/Switch "A" Circuit Low Input (24)
P0123 Throttle/Pedal Position Sensor/Switch "A" Circuit High Input (24)
P0124 Throttle/Pedal Position Sensor/Switch "A" Circuit Intermittent (24)
P0125 Insufficient Coolant Temperature For Closed Loop Fuel Control
P0126 Insufficient Coolant Temperature For Stable Operation
P0130 O2 Circuit Malfunction (Bank no. 1 Sensor no. 1)
P0131 O2 Sensor Circuit Low Voltage (Bank no. 1 Sensor no. 1) (21)
P0132 O2 Sensor Circuit High Voltage (Bank no. 1 Sensor no. 1) (21)
P0133 O2 Sensor Circuit Slow Response (Bank no. 1 Sensor no. 1) (21)
P0134 O2 Sensor Circuit No Activity Detected (Bank no. 1 Sensor no. 1)
P0135 O2 Sensor Heater Circuit Malfunction (Bank no. 1 Sensor no. 1) (21)
P0136 O2 Sensor Circuit Malfunction (Bank no. 1 Sensor no. 2) (21)
P0137 O2 Sensor Circuit Low Voltage (Bank no. 1 Sensor no. 2) (21)
P0138 O2 Sensor Circuit High Voltage (Bank no. 1 Sensor no. 2) (21)
P0139 O2 Sensor Circuit Slow Response (Bank no. 1 Sensor no. 2) (21)
P0140 O2 Sensor Circuit No Activity Detected (Bank no. 1 Sensor no. 2) (21)
P0141 O2 Sensor Heater Circuit Malfunction (Bank no. 1 Sensor no. 2) (21)
P0142 O2 Sensor Circuit Malfunction (Bank no. 1 Sensor no. 3) (21)
P0143 O2 Sensor Circuit Low Voltage (Bank no. 1 Sensor no. 3) (21)
P0144 O2 Sensor Circuit High Voltage (Bank no. 1 Sensor no. 3)(21)
P0145 O2 Sensor Circuit Slow Response (Bank no. 1 Sensor no. 3)(21)
P0146 O2 Sensor Circuit No Activity Detected (Bank no. 1 Sensor no. 3) (21)
P0147 O2 Sensor Heater Circuit Malfunction (Bank no. 1 Sensor no. 3) (21)
P0150 O2 Sensor Circuit Malfunction (Bank no. 2 Sensor no. 1)(21)
P0151 O2 Sensor Circuit Low Voltage (Bank no. 2 Sensor no. 1)(21)
P0152 O2 Sensor Circuit High Voltage (Bank no. 2 Sensor no. 1)(21)
P0153 O2 Sensor Circuit Slow Response (Bank no. 2 Sensor no. 1) (21)
P0154 O2 Sensor Circuit No Activity Detected (Bank no. 2 Sensor no. 1) (21)
P0155 O2 Sensor Heater Circuit Malfunction (Bank no. 2 Sensor no. 1) (21)
P0156 O2 Sensor Circuit Malfunction (Bank no. 2 Sensor no. 2) (21)
P0157 O2 Sensor Circuit Low Voltage (Bank no. 2 Sensor no. 2) (21)
P0158 O2 Sensor Circuit High Voltage (Bank no. 2 Sensor no. 2) (21)
P0159 O2 Sensor Circuit Slow Response (Bank no. 2 Sensor no. 2) (21)
P0160 O2 Sensor Circuit No Activity Detected (Bank no. 2 Sensor no. 2) (21)
P0161 O2 Sensor Heater Circuit Malfunction (Bank no. 2 Sensor no. 2) (21)
P0162 O2 Sensor Circuit Malfunction (Bank no. 2 Sensor no. 3) (21)
P0163 O2 Sensor Circuit Low Voltage (Bank no. 2 Sensor no. 3) (21)
P0164 O2 Sensor Circuit High Voltage (Bank no. 2 Sensor no. 3) (21)
P0165 O2 Sensor Circuit Slow Response (Bank no. 2 Sensor no. 3) (21)
P0166 O2 Sensor Circuit No Activity Detected (Bank no. 2 Sensor no. 3) (21)
P0167 O2 Sensor Heater Circuit Malfunction (Bank no. 2 Sensor no. 3) (21)
P0170 Fuel Trim Malfunction (Bank no. 1/left)
P0171 System Too Lean (Bank no. 1/left)(51)
P0172 System Too Rich (Bank no. 1/left) (52)
P0173 Fuel Trim Malfunction (Bank no. 2/right)
P0174 System Too Lean (Bank no. 2/right)(51)
P0175 System Too Rich (Bank no. 2/right)(52)
P0176 Fuel Composition Sensor Circuit Malfunction
P0177 Fuel Composition Sensor Circuit Range/Performance
P0178 Fuel Composition Sensor Circuit Low Input
P0179 Fuel Composition Sensor Circuit High Input
P0180 Fuel Temperature Sensor "A" Circuit Malfunction
P0181 Fuel Temperature Sensor "A" Circuit Range/Performance
P0182 Fuel Temperature Sensor "A" Circuit Low Input
P0183 Fuel Temperature Sensor "A" Circuit High Input
P0184 Fuel Temperature Sensor "A" Circuit Intermittent
P0185 Fuel Temperature Sensor "B" Circuit Malfunction
P0186 Fuel Temperature Sensor "B" Circuit Range/Performance
P0187 Fuel Temperature Sensor "B" Circuit Low Input
P0188 Fuel Temperature Sensor "B" Circuit High Input
P0189 Fuel Temperature Sensor "B" Circuit Intermittent
P0190 Fuel Rail Pressure Sensor Circuit Malfunction
P0191 Fuel Rail Pressure Sensor Circuit Range/Performance
P0192 Fuel Rail Pressure Sensor Circuit Low Input
P0193 Fuel Rail Pressure Sensor Circuit High Input

4-24 DRIVEABILITY AND EMISSION CONTROLS

P0194 Fuel Rail Pressure Sensor Circuit Intermittent
P0195 Engine Oil Temperature Sensor Malfunction
P0196 Engine Oil Temperature Sensor Range/Performance
P0197 Engine Oil Temperature Sensor Low
P0198 Engine Oil Temperature Sensor High
P0199 Engine Oil Temperature Sensor Intermittent
P0200 Injector Circuit Malfunction
P0201 Injector Circuit Malfunction - Cylinder no. 1 (27)
P0202 Injector Circuit Malfunction - Cylinder no. 2 (27)
P0203 Injector Circuit Malfunction - Cylinder no. 3 (27)
P0204 Injector Circuit Malfunction - Cylinder no. 4 (27)
P0205 Injector Circuit Malfunction - Cylinder no. 5 (27)
P0206 Injector Circuit Malfunction - Cylinder no. 6 (27)
P0207 Injector Circuit Malfunction - Cylinder no. 7 (27)
P0208 Injector Circuit Malfunction - Cylinder no. 8 (27)
P0209 Injector Circuit Malfunction - Cylinder no. 9 (27)
P0210 Injector Circuit Malfunction - Cylinder no. 10 (27)
P0213 Cold Start Injector no. 1 Malfunction
P0214 Cold Start Injector no. 2 Malfunction
P0215 Engine Shutoff Solenoid Malfunction
P0216 Injection Timing Control Circuit Malfunction
P0217 Engine Over Temperature Condition
P0218 Transmission Over Temperature Condition
P0219 Engine Over Speed Condition
P0220 Throttle/Pedal Position Sensor/Switch "B" Circuit Malfunction
P0221 Throttle/Pedal Position Sensor/Switch "B" Circuit Range/Performance Problem
P0222 Throttle/Pedal Position Sensor/Switch "B" Circuit Low Input
P0223 Throttle/Pedal Position Sensor/Switch "B" Circuit High Input
P0224 Throttle/Pedal Position Sensor/Switch "B" Circuit Intermittent
P0225 Throttle/Pedal Position Sensor/Switch "C" Circuit Malfunction
P0226 Throttle/Pedal Position Sensor/Switch "C" Circuit Range/Performance Problem
P0227 Throttle/Pedal Position Sensor/Switch "C" Circuit Low Input
P0228 Throttle/Pedal Position Sensor/Switch "C" Circuit High Input
P0229 Throttle/Pedal Position Sensor/Switch "C" Circuit Intermittent
P0230 Fuel Pump Primary Circuit Malfunction
P0231 Fuel Pump Secondary Circuit Low
P0232 Fuel Pump Secondary Circuit High
P0233 Fuel Pump Secondary Circuit Intermittent
P0234 Engine Over Boost Condition
P0236 MAP Sensor Voltage Too High Too Long (Turbocharger Wastegate)
P0238 MAP Sensor Voltage Too Low
P0251 Fuel Injection Pump Problem
P0261 Cylinder no. 1 Injector Circuit Low
P0262 Cylinder no. 1 Injector Circuit High
P0263 Cylinder no. 1 Contribution/Balance Fault
P0264 Cylinder no. 2 Injector Circuit Low
P0265 Cylinder no. 2 Injector Circuit High
P0266 Cylinder no. 2 Contribution/Balance Fault
P0267 Cylinder no. 3 Injector Circuit Low
P0268 Cylinder no. 3 Injector Circuit High

P0269 Cylinder no. 3 Contribution/Balance Fault
P0270 Cylinder no. 4 Injector Circuit Low
P0271 Cylinder no. 4 Injector Circuit High
P0272 Cylinder no. 4 Contribution/Balance Fault
P0273 Cylinder no. 5 Injector Circuit Low
P0274 Cylinder no. 5 Injector Circuit High
P0275 Cylinder no. 5 Contribution/Balance Fault
P0276 Cylinder no. 6 Injector Circuit Low
P0277 Cylinder no. 6 Injector Circuit High
P0278 Cylinder no. 6 Contribution/Balance Fault
P0279 Cylinder no. 7 Injector Circuit Low
P0280 Cylinder no. 7 Injector Circuit High
P0281 Cylinder no. 7 Contribution/Balance Fault
P0282 Cylinder no. 8 Injector Circuit Low
P0283 Cylinder no. 8 Injector Circuit High
P0284 Cylinder no. 8 Contribution/Balance Fault
P0285 Cylinder no. 9 Injector Circuit Low
P0286 Cylinder no. 9 Injector Circuit High
P0287 Cylinder no. 9 Contribution/Balance Fault
P0288 Cylinder no. 10 Injector Circuit Low
P0289 Cylinder no. 10 Injector Circuit High
P0290 Cylinder no. 10 Contribution/Balance Fault
P0291 Cylinder no. 11 Injector Circuit Low
P0292 Cylinder no. 11 Injector Circuit High
P0293 Cylinder no. 11 Contribution/Balance Fault
P0294 Cylinder no. 12 Injector Circuit Low
P0295 Cylinder no. 12 Injector Circuit High
P0296 Cylinder no. 12 Contribution/Balance Fault
P0300 Random/Multiple Cylinder Misfire Detected (43)
P0301 Cylinder no. 1 - Misfire Detected (43)
P0302 Cylinder no. 2 - Misfire Detected (43)
P0303 Cylinder no. 3 - Misfire Detected (43)
P0304 Cylinder no. 4 - Misfire Detected (43)
P0305 Cylinder no. 5 - Misfire Detected (43)
P0306 Cylinder no. 6 - Misfire Detected (43)
P0307 Cylinder no. 7 - Misfire Detected (43)
P0308 Cylinder no. 8 - Misfire Detected (43)
P0309 Cylinder no. 9 - Misfire Detected (43)
P0310 Cylinder no. 10 - Misfire Detected (43)
P0320 Crankshaft Position Sensor Problem
P0321 Ignition/Distributor Engine Speed Input Circuit Range/Performance
P0322 Ignition/Distributor Engine Speed Input Circuit No Signal
P0323 Ignition/Distributor Engine Speed Input Circuit Intermittent
P0325 Knock Sensor no. 1 - Circuit Malfunction (Bank no. 1 or Single Sensor)
P0336 Crankshaft Position Sensor Problem
P0340 Camshaft Position Sensor Circuit Malfunction (54)
P0341 Camshaft Position Sensor Circuit Range/Performance
P0342 Camshaft Position Sensor Circuit Low Input
P0343 Camshaft Position Sensor Circuit High Input
P0344 Camshaft Position Sensor Circuit Intermittent
P0350 Ignition Coil Primary/Secondary Circuit Malfunction

DRIVEABILITY AND EMISSION CONTROLS

Code	Description
P0351 P0358	Ignition Coil #1 - #8 Primary/Secondary Circuit Malfunction (43)
P0370	Timing Reference High Resolution Signal "A" Malfunction
P0371	Timing Reference High Resolution Signal "A" Too Many Pulses
P0372	Timing Reference High Resolution Signal "A" Too Few Pulses
P0373	Timing Reference High Resolution Signal "A" Intermittent/Erratic Pulses
P0374	Timing Reference High Resolution Signal "A" No Pulses
P0375	Timing Reference High Resolution Signal "B" Malfunction
P0376	Timing Reference High Resolution Signal "B" Too Many Pulses
P0377	Timing Reference High Resolution Signal "B" Too Few Pulses
P0378	Timing Reference High Resolution Signal "B" Intermittent/Erratic Pulses
P0379	Timing Reference High Resolution Signal "B" No Pulses
P0380	Glow Plug/Heater Circuit "A" Malfunction
P0381	Glow Plug/Heater Indicator Circuit Malfunction
P0382	Glow Plug/Heater Circuit "B" Malfunction
P0385	Crankshaft Position Sensor "B" Circuit Malfunction
P0386	Crankshaft Position Sensor "B" Circuit Range/Performance
P0387	Crankshaft Position Sensor "B" Circuit Low Input
P0388	Crankshaft Position Sensor "B" Circuit High Input
P0389	Crankshaft Position Sensor "B" Circuit Intermittent
P0400	Exhaust Gas Recirculation Flow Malfunction (68)
P0401	Exhaust Gas Recirculation Flow Insufficient Detected
P0402	Exhaust Gas Recirculation Flow Excessive Detected
P0403	Exhaust Gas Recirculation Circuit Malfunction (32)
P0404	Exhaust Gas Recirculation Circuit Range/Performance
P0405	Exhaust Gas Recirculation Sensor "A" Circuit Low
P0406	Exhaust Gas Recirculation Sensor "A" Circuit High
P0412	Secondary Air Injection System Solenoid Circuit Malfunction
P0420	Catalyst System Efficiency Below Threshold (left side)(72)
P0432	Main Catalyst Efficiency Below Threshold (right side)(72)
P0440	Evaporative Emission Control System Malfunction
P0441	Evaporative Emission Control System Incorrect Purge Flow (31)
P0442	Evaporative Emission Control System Leak Detected (Small Leak)
P0443	Evaporative Emission Control System Purge Control Valve Circuit Malfunction (31)
P0455	Evaporative Emission Control System Leak Detected (Gross Leak)
P0460	Fuel Level Sensor Circuit Malfunction
P0461	Fuel Level Sensor Circuit Range/Performance
P0462	Fuel Level Sensor Circuit Low Input
P0463	Fuel Level Sensor Circuit High Input
P0500	Vehicle Speed Sensor Malfunction (15)
P0501	Vehicle Speed Sensor Range/Performance
P0502	Vehicle Speed Sensor Circuit Low Input
P0503	Vehicle Speed Sensor Intermittent/Erratic/High
P0505	Idle Control System Malfunction (25)
P0506	Idle Control System RPM Lower Than Expected
P0522	Engine Oil Pressure Sensor/Switch Low Voltage
P0523	Engine Oil Pressure Sensor/Switch High Voltage
P0524	Oil Pressure Too Low
P0530	A/C Refrigerant Pressure Sensor Circuit Malfunction
P0531	A/C Refrigerant Pressure Sensor Circuit Range/Performance
P0532	A/C Refrigerant Pressure Sensor Circuit Low Input
P0533	A/C Refrigerant Pressure Sensor Circuit High Input
P0534	A/C Refrigerant Charge Loss
P0545	A/C Clutch Problem
P0550	Power Steering Pressure Sensor Circuit Malfunction
P0551	Power Steering Switch Failure
P0552	Power Steering Pressure Sensor Circuit Low Input
P0553	Power Steering Pressure Sensor Circuit High Input
P0554	Power Steering Pressure Sensor Circuit Intermittent
P0560	Charging System Voltage Malfunction
P0561	Charging System Voltage Unstable
P0562	System Voltage Low
P0563	Charging System Voltage High
P0600	PCM Serial Communication Link Malfunction
P0601	Internal Control Module Memory Check Sum Error (53)
P0602	Engine Control Module Programming Error
P0604	Internal Control Module Random Access Memory (RAM) Error
P0605	Internal Control Module Read Only Memory (ROM) Error
P0606	PCM Processor Fault
P0615	Starter Relay Control Circuit
P0620	Alternator Control Circuit Malfunction
P0621	Alternator Lamp "L" Control Circuit Malfunction
P0622	Alternator Field "F" Control Circuit Malfunction
P0645	A/C Clutch Relay Open Or Shorted
P0700	Transmission Control System Malfunction
P0701	Transmission Control System Range/Performance
P0702	Transmission Control System Electrical
P0703	Brake Switch Stuck Or Released
P0704	Clutch Switch Input Circuit Malfunction
P0705	Transmission Range Sensor Circuit Malfunction (PRNDL Input)
P0706	Transmission Range Sensor Circuit Range/Performance
P0707	Transmission Range Sensor Circuit Low Input
P0708	Transmission Range Sensor Circuit High Input
P0709	Transmission Range Sensor Circuit Intermittent
P0710	Transmission Fluid Temperature Sensor Circuit Malfunction (45)
P0711	Transmission Fluid Temperature Sensor Circuit Range/Performance (45)
P0712	Transmission Fluid Temperature Sensor Circuit Low Input (45)
P0713	Transmission Fluid Temperature Sensor Circuit High Input (45)
P0714	Transmission Fluid Temperature Sensor Circuit Intermittent
P0715	Input/Turbine Speed Sensor Circuit Malfunction
P0716	Input/Turbine Speed Sensor Circuit Range/Performance
P0717	Input/Turbine Speed Sensor Circuit No Signal
P0718	Input/Turbine Speed Sensor Circuit Intermittent
P0719	Torque Converter/Brake Switch "B" Circuit Low
P0720	Output Speed Sensor
P0740	Torque Converter Clutch Circuit Malfunction
P0741	Torque Converter Clutch Circuit Performance or Stuck Off
P0742	Torque Converter Clutch Circuit Stuck On
P0743	Torque Converter Clutch Circuit Electrical (37)
P0744	Torque Converter Clutch Circuit Intermittent
P0745	Pressure Control Solenoid Malfunction

DRIVEABILITY AND EMISSION CONTROLS

Code	Description
P0746	Pressure Control Solenoid Performance or Stuck Off
P0747	Pressure Control Solenoid Stuck On
P0748	Pressure Control Solenoid Electrical
P0749	Pressure Control Solenoid Intermittent
P0750	Shift Solenoid "A" Malfunction
P0751	Shift Solenoid "A" Performance or Stuck Off
P0752	Shift Solenoid "A" Stuck On
P0753	Shift Solenoid "A" Electrical (37)
P0754	Shift Solenoid "A" Intermittent
P0755	Shift Solenoid "B" Malfunction
P0756	Shift Solenoid "B" Performance or Stuck Off
P0757	Shift Solenoid "B" Stuck On
P0758	Shift Solenoid "B" Electrical
P0759	Shift Solenoid "B" Intermittent
P0760	Shift Solenoid "C" Malfunction
P0761	Shift Solenoid "C" Performance Or Stuck Off
P0762	Shift Solenoid "C" Stuck On
P0763	Shift Solenoid "C" Electrical
P0764	Shift Solenoid "C" Intermittent
P0765	Shift Solenoid "D" Malfunction
P0766	Shift Solenoid "D" Performance Or Stuck Off
P0767	Shift Solenoid "D" Stuck On
P0768	Shift Solenoid "D" Electrical
P0769	Shift Solenoid "D" Intermittent
P0770	Shift Solenoid "E" Malfunction
P0771	Shift Solenoid "E" Performance Or Stuck Off
P0772	Shift Solenoid "E" Stuck On
P0773	Shift Solenoid "E" Electrical
P0774	Shift Solenoid "E" Intermittent
P0780	Shift Malfunction
P0781	1 - 2 Shift Malfunction
P0782	2 - 3 Shift Malfunction
P0783	3 - 4 Shift Malfunction
P0784	4 - 5 Shift Malfunction
P0785	Shift/Timing Solenoid Malfunction
P0786	Shift/Timing Solenoid Range/Performance
P0787	Shift/Timing Solenoid Low
P0788	Shift/Timing Solenoid High
P0789	Shift/Timing Solenoid Intermittent
P0790	Normal/Performance Switch Circuit Malfunction
P0801	Reverse Inhibit Control Circuit Malfunction
P0803	1 - 4 Upshift (Skip Shift) Solenoid Control Circuit Malfunction
P0804	1 - 4 Upshift (Skip Shift) Lamp Control Circuit Malfunction
P0830	Clutch Switch Circuit Problem
P0833	Clutch Switch Circuit Problem
P1110	Decreased Engine Performance Due To High Intake Air Temp
P1180	Decreased Engine Performance Due To High Fuel Temp
P1195	Slow O2 Sensor (2/1)
P1195	Slow O2 Sensor (1/2)
P1198	Radiator Temperature Sensor Volts Too High
P1199	Radiator Temperature Sensor Volts Too Low
P1281	Engine Cold For Too Long (Thermostat Malfunction) (17)
P1282	Defective Fuel Pump Relay Control Circuit
P1283	Fuel Injection System Fault
P1291	No Temp Rise From Intake Air Heaters
P1294	Target Idle Not Reached. Possible Vacuum Leak (25)
P1295	Low Voltage To TPS (14)
P1296	No 5v To MAP Sensor
P1297	No Change In MAP From Start To Run (13)
P1298	Lean Operation At Wide Open Throttle
P1299	Vacuum Leak Found (IAC Fully Seated)
P1388	ASD Circuit Problem
P1390	Timing Belt Skipped One Tooth Or More
P1391	Intermittent Loss Of CMP Or CKP Signal (11)
P1398	Defective Crankshaft Position Sensor (11)
P1399	Wait-To-Start Lamp Circuit Open Or Shorted
P1403	No 5v To EGR Sensor
P1476	Too Little Secondary Air
P1477	Too Much secondary Air
P1478	Battery Temp Sensor Out Of Range
P1479	Open Or Shorted Transmission Fan Relay
P1480	Open Or Shorted Condition In The PCV Solenoid Circuit
P1486	LDP Has Found A Pinched EVAP Hose
P1491	Radiator Fan Relay Circuit Open Or Shorted
P1492	Ambient/Batt Temp Sensor Voltage Too High (44)
P1493	Ambient/Batt Temp Sensor Voltage Too Low (44)
P1607	Powertrain Control Module Internal Circuit Failure A
P1618	Serial Peripheral Interface Communication Error
P1640	Output Driver Module 'A' Fault
P1682	Charging system Voltage Too Low
P1684	The Battery Has Been Disconnected In The Last 50 Starts
P1691	Fuel Injection Pump Calibration Error
P1765	Transmission 12v Supply Relay (37)
P1899	P/N Switch Stuck (37)

Clearing Codes

CONTINUOUS MEMORY CODES

These codes are retained in memory for usually around 40 warm-up cycles. To clear the codes for the purposes of testing or confirming repair, perform the code reading procedure. Use the DRB-II scan tool or equivalent to erase the memory. Disconnecting the negative battery cable will not erase the code completely from the memory; it will remain as a continuous fault. The MIL lamp will remain out if you disconnect the battery, and the malfunction that lit the MIL lamp is repaired. After 40 warm-up cycles or so are reached, the code will be erased from the PCM's memory.

DRIVEABILITY AND EMISSION CONTROLS 4-27

VACUUM DIAGRAMS

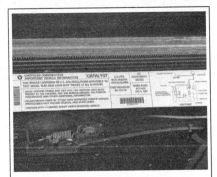

Fig. 87 The Vehicle Emission Control Information (VECI) label contains information critical to the truck's performance

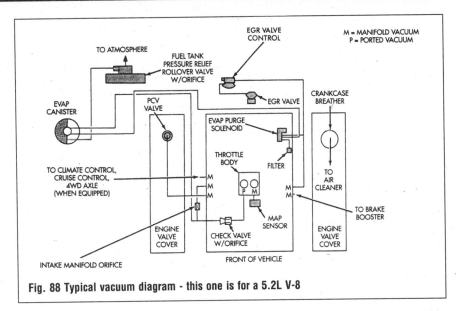

Fig. 88 Typical vacuum diagram - this one is for a 5.2L V-8

See Figures 87, 88 and 89

By law, each vehicle is equipped with a Vehicle Emission Control Information (VECI) label which is permanently affixed. On the models covered in this manual, the label will be found in the engine compartment with typical locations being on the underside of the hood, the top of the radiator, the alternator bracket and the air cleaner box.

The VECI label provides a vacuum schematic for the vehicle in question. Because vacuum circuits will vary based on year, various engine and vehicle options, always refer first to the vehicle emission control information label.

VECI labels are designed to be permanently attached to the vehicle and cannot be removed without being destroyed. Absence of the label in your vehicle may indicate undocumented repair work has been performed.

If you wish to obtain a replacement emissions label, most manufacturers make the labels available for purchase.

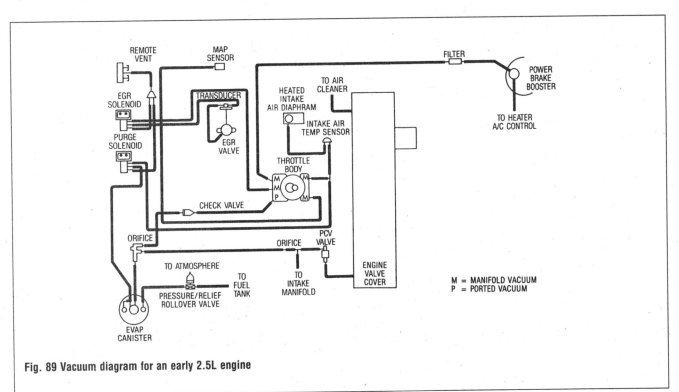

Fig. 89 Vacuum diagram for an early 2.5L engine

Notes

BASIC FUEL SYSTEM DIAGNOSIS
PRECAUTIONS 5-2
DIESEL FUEL SYSTEM
DIESEL INJECTION PUMP 5-14
DIESEL INJECTION PUMP INJECTION PUMP
   TIMING 5-16
DIESEL INJECTION PUMP REMOVAL &
   INSTALLATION 5-14
FUEL TRANSFER PUMP 5-12
FUEL/WATER SEPARATOR FILTER 5-16
FUEL/WATER SEPARATOR FILTER DRAINING
   WATER FROM THE SYSTEM 5-17
IDLE SPEED ADJUSTMENT (1997 - 98
   MODELS) 5-17
INJECTION LINES 5-10
INJECTORS 5-10
FUEL LINES AND FITTINGS
CONVENTIONAL TYPE FUEL FITTINGS
FUEL LINES AND HOSES 5-2
QUICK-CONNECT FUEL FITTINGS 5-2
FUEL TANK
TANK ASSEMBLY 5-17
GASOLINE FUEL INJECTION SYSTEM
FUEL FILTER/PRESSURE REGULATOR
   ASSEMBLY 5-6
FUEL GAUGE SENDING UNIT 5-6
FUEL INJECTORS 5-10
FUEL PUMP 5-5
FUEL RAIL ASSEMBLY 5-7
GENERAL INFORMATION 5-4
RELIEVING FUEL SYSTEM PRESSURE 5-4
ROLLOVER VALVE 5-6
THROTTLE BODY 5-7

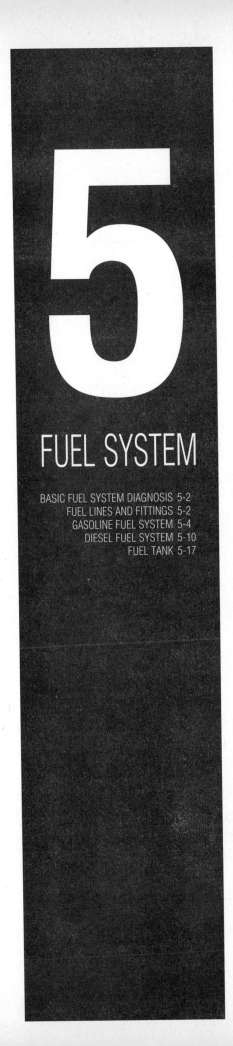

5

FUEL SYSTEM

BASIC FUEL SYSTEM DIAGNOSIS 5-2
FUEL LINES AND FITTINGS 5-2
GASOLINE FUEL SYSTEM 5-4
DIESEL FUEL SYSTEM 5-10
FUEL TANK 5-17

5-2 FUEL SYSTEM

BASIC FUEL SYSTEM DIAGNOSIS

When there is a problem starting or driving a vehicle, two of the most important checks involve the ignition and the fuel systems. The questions most mechanics attempt to answer first, "is there spark?" and "is there fuel?" will often lead to solving most basic problems. For ignition system diagnosis and testing, refer to the information on engine electrical components and ignition systems found in this manual. If the ignition system checks out (there is spark), then you must determine if the fuel system is operating properly (is there fuel?).

Precautions

- Disconnect the negative battery terminal(s), except for testing when battery voltage is required
- Always use a flashlight instead of a drop light to inspect fuel system components or connections
- Keep all open flames and smoking material out of the area and make sure there is adequate ventilation to remove fuel vapors
- Use a clean shop cloth to catch fuel when opening a fuel system. Dispose of gasoline-soaked rags properly
- Relieve the fuel system pressure before any service procedures are attempted that require disconnecting a fuel line
- Use eye protection
- Always keep a dry chemical (class B) fire extinguisher near the area

FUEL LINES AND FITTINGS

CAUTION:

Observe all applicable safety precautions when working around fuel. Whenever servicing the fuel system, always work in a well-ventilated area. Do not allow fuel spray or vapors to come in contact with a spark or open flame. Keep a dry chemical fire extinguisher near the work area. Always keep fuel in a container specifically designed for fuel storage; also, always properly seal fuel containers to avoid the possibility of fire or explosion.

CAUTION:

The fuel injection system is under constant pressure (up to approximately 50 psi). Before servicing any part of the fuel injection system, the pressure must be released. Observe all applicable and common sense safety precautions during these procedures.

Fuel Lines and Hoses

The hoses used on fuel injected vehicles are of special construction to prevent contamination of the fuel system. The hose clamps used are of a special construction. The clamps have a rolled edge design to prevent the edges from cutting the hose when tightened down. Only these type of clamps may be used on the fuel system hoses.

Conventional Type Fuel Fittings

REMOVAL & INSTALLATION

1. Remove the air cleaner assembly, if needed for access.
2. Perform the fuel system pressure release.
3. Disconnect the negative battery cable(s).
4. Loosen the fuel intake and return hose clamps. Wrap a shop towel around each hose, twist and pull off each hose.
5. Remove each fitting and note the inlet diameter. Remove the copper washers.

To install:

6. Clean both ends of the fuel line with a clean, lint-free cloth.

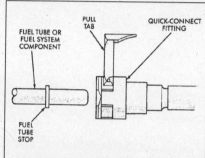

Fig. 1 Disengaged view of a single-tab style quick-connect fitting

7. Replace the copper washers with new washers.
8. Install the fuel fittings in the proper ports and tighten to 175 inch lbs. (20 Nm).
9. Using new original equipment type hose clamps, install the fuel return and supply hoses.
10. Reconnect the negative battery cable(s).
11. Start the engine and check for leaks.
12. Install the air cleaner assembly.

Quick-Connect Fuel Fittings

REMOVAL & INSTALLATION

Single-Tab Type

See Figure 1

1. Remove the air cleaner assembly.
2. Properly relieve the fuel system pressure.
3. Disconnect the negative battery cable(s).
4. Clean the fitting of any foreign material before disassembly.
5. Press the release tab on the side of the fitting to release the pull tab.

WARNING:

The release tab must be pressed prior to releasing the pull tab or the tab will be damaged.

6. While pressing the release tab on the side of the fitting, use a small pry tool to pry up the pull tab.
7. Raise the pull tab until it separates from the quick-disconnect fitting. Discard the old pull tab.
8. Disconnect the quick-connect fitting.

9. Inspect the fitting body and the fuel system component for damage. Replace as necessary.

To install:

10. Clean both ends of the fuel line with a clean, lint-free cloth.
11. Lubricate fittings with a drop of clean engine oil.
12. Insert the quick-connect fitting into the fuel tube or component until the built-on stop on the fuel tube or component rests against the back of the fitting.
13. Obtain a new pull tab. Push the new tab down until it locks into place in the quick-connect fitting.
14. Pull firmly on the fitting to verify the fitting is secure.
15. Connect the negative battery cable(s).
16. Start the engine and check for leaks.
17. Install the air cleaner assembly.

Two-Tab Type

See Figures 2 and 3

1. Remove the air cleaner assembly.
2. Properly relieve the fuel system pressure.
3. Disconnect the negative battery cable(s).
4. Clean the fitting of any foreign material before disassembly.
5. Squeeze the plastic retaining tabs against the sides of the fitting with your fingers.
6. Pull the fitting from the component or tube being disconnected. The plastic retainer will stay on the component or tube. The O-rings and spacer will remain in the fitting connector body.
7. Inspect the fitting body and the fuel system component for damage. Replace as necessary.

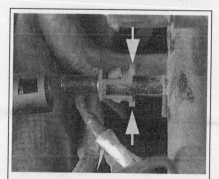

Fig. 2 Remove the fuel line connection by pressing down on its tabs (arrows) and pulling the line outward

FUEL SYSTEM 5-3

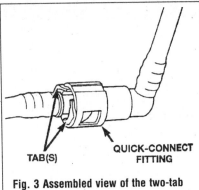

Fig. 3 Assembled view of the two-tab style quick-connect fitting

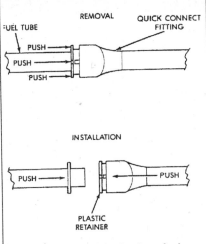

Fig. 4 Plastic retainer ring type fuel fitting

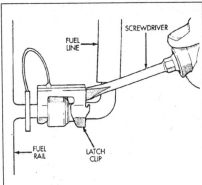

Fig. 5 Use a screwdriver to carefully remove the latch

To install:

8. Clean both ends of the fuel line with a clean, lint-free cloth.
9. Lubricate fittings with a drop of clean engine oil.
10. Insert the quick-connect fitting to the component or fuel tube and into the plastic retainer. When a connection is made, a click will be heard.
11. Pull firmly on the fitting to verify the fitting is secure.
12. Connect the negative battery cable(s).
13. Start the engine and check for leaks.
14. Install the air cleaner assembly.

Plastic Retainer Ring Type
See Figure 4

1. Remove the air cleaner assembly.
2. Properly relieve the fuel system pressure.
3. Disconnect the negative battery cable(s).
4. Clean the fitting of any foreign material before disassembly.
5. To release the fitting, firmly push the fitting towards the component it's attached to while firmly pushing the plastic retainer ring into the fitting.
6. Pull the fitting from the component or tube being disconnected. The plastic retainer will stay on the component or tube.
7. Inspect the fitting body and the fuel system component for damage. Replace as necessary.

To install:

8. Clean both ends of the fuel line with a clean, lint-free cloth.
9. Lubricate fittings with a drop of clean engine oil.
10. Insert the quick-connect fitting to the component or fuel tube and into the plastic retainer. When a connection is made, a click will be heard.
11. Pull firmly on the fitting to verify the fitting is secure.
12. Connect the negative battery cable(s).
13. Start the engine and check for leaks.
14. Install the air cleaner assembly.

Latch Clip Type 1
See Figures 5 and 6

1. Remove the air cleaner assembly.
2. Properly relieve the fuel system pressure.
3. Disconnect the negative battery cable(s).
4. Clean the fitting of any foreign material before disassembly.
5. Pry up on the latch clip with a screwdriver and

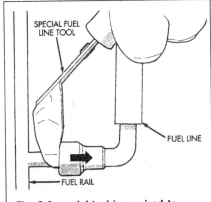

Fig. 6 A special tool is required to disengage this type of connection

remove the clip from the quick-connect fitting.
6. Insert a proper fuel line removal tool, available from such companies as Lisle,, into the fuel line.
7. With the tool inserted inside the line releasing the locking fingers, pull the line releasing the connection.
8. Inspect the fitting, locking fingers and fuel line for damage, repair as necessary.

To install:

9. Clean both ends of the fuel line with a clean, lint-free cloth.
10. Lubricate fittings with a drop of clean engine oil.
11. Insert the fitting end onto the line or component until a click is heard.
12. Pull firmly on the fitting to verify the fitting is secure.
13. Install the latch clip onto the fitting. It snaps into position. If the clip will not fit, it indicates that the fuel line is not properly installed.
14. Connect the negative battery cable(s).
15. Start the engine and check for leaks.
16. Install the air cleaner assembly.

Latch Clip Type 2
See Figure 7

1. Remove the air cleaner assembly.
2. Properly relieve the fuel system pressure.
3. Disconnect the negative battery cable(s).
4. Clean the fitting of any foreign material before disassembly.
5. Separate and unlatch the two small arms on the end of the clip and swing away from fuel line.
6. Slide the latch clip toward the fuel rail while lifting with a screwdriver.
7. Insert a proper fuel line removal tool, available from such companies as Lisle®, into the fuel line.
8. With the tool inserted inside the line releasing the locking fingers, pull the line releasing the connection.
9. Inspect the fitting, locking fingers and fuel line for damage, repair as necessary.

To install:

10. Clean both ends of the fuel line with a clean, lint-free cloth.
11. Lubricate fittings with a drop of clean engine oil.
12. Insert the fitting end onto the line or component until a click is heard.
13. Pull firmly on the fitting to verify the fitting is secure.
14. Install the latch clip onto the fitting. It snaps into position. If the clip will not fit, it indicates that the fuel line is not properly installed.
15. Connect the negative battery cable(s).
16. Start the engine and check for leaks.
17. Install the air cleaner assembly.

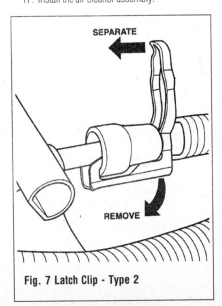

Fig. 7 Latch Clip - Type 2

5-4 FUEL SYSTEM

GASOLINE FUEL INJECTION SYSTEM

General Information

See Figure 8

A Multi-port Fuel Injection (MFI) system is used on all engines. The MFI system is controlled by a pre-programmed digital computer known as the Powertrain Control Module (PCM). The PCM controls ignition timing, air/fuel ratio, emission control devices, charging system and idle speed. The PCM constantly varies timing, fuel delivery and idle speed to meet changing engine operating conditions.

Various sensors provide the input necessary for the PCM to correctly regulate the fuel flow at the fuel injectors. These include the manifold absolute pressure, throttle position, oxygen sensor, coolant temperature, intake air temperature, and camshaft and crankshaft position sensors. In addition to the sensors, various switches also provide important information. These include the neutral safety, air conditioning, air conditioning clutch, and brake light switches.

All inputs to the PCM are converted into signals which are used to calculate and adjust the fuel flow at the injectors or ignition timing or both. The PCM accomplishes this by varying the pulse width of the injectors to adjust the fuel/air ratio, or advancing or retarding timing. The PCM tests many of its own input and output circuits. If a fault is found in a major system, this information is stored in the PCM as a Diagnostic Trouble Code (DTC). Information on this fault can be displayed to a technician by means of the grounding a terminal and reading the check engine lamp flashes or by connecting a scan tool and reading the DTCs (see "Driveability And Emission Controls" for a more complete procedure).

The primary variables that the engine controller uses to determine pulse width are manifold absolute pressure (air density) and engine rpm (speed). In addition to manifold absolute pressure (MAP) and engine speed (rpm), the engine controller also considers input from the following sensors to determine the pulse width:

- Exhaust gas content
- Coolant temperature
- Throttle position
- Battery voltage
- Air conditioning selection
- Transmission gear selection
- Speed control

The fuel injection delivery system consists of the following major components:

- Fuel tank
- Fuel pump
- Pressure regulator
- Fuel injectors
- Fuel rails and lines

The fuel tank incorporates a fuel pump assembly and may have one or more rollover valves.

The fuel pump module is located in the top of the fuel tank and incorporates a pressure regulator. Two fuel filters are also used. One is incorporated into the pump/regulator assembly and the other is at the bottom of the module. The filters do not need routine maintenance.

The pressure regulator is a mechanical device. It maintains a constant pressure of about 49 psi (339 kPa).

Fuel is delivered through the supply line to the metal fuel rails. Injector nozzles (one for each cylinder) are installed on the rails. The injectors are electrical solenoids and are fired by the PCM. When activated, voltage is applied across the injector terminals. The injector opens, releasing the fuel built up behind the injector. The pulse width, or amount of time the injector remains open, is determined by the inputs to the PCM and is modified depending on the needs of the engine, air pollution standards and so on.

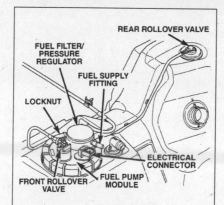

Fig. 8 Fuel tank pump module components - typical

Relieving Fuel System Pressure

CAUTION:

The fuel injection system is under constant pressure (up to approximately 50 psi). Before servicing any part of the fuel injection system, the pressure must be released. Observe all applicable and common sense safety precautions during these procedures.

The system may or may not be fitted with a pressure test port. Check first, then follow the appropriate procedure.

CAUTION:

Whenever servicing the fuel system, always work in a well ventilated area. Do not allow fuel spray or vapors to come in contact with a spark or open flame. Keep a dry chemical fire extinguisher near the work area. Always keep fuel in a container specifically designed for fuel storage; also, always properly seal fuel containers to avoid the possibility of fire or explosion. Fuel system work is made more hazardous by a hot engine. Work on a cold engine whenever possible.

USING THE PRESSURE TEST PORT

See Figure 9

This procedure requires the use of factory fuel pressure gauge tool set No. 5069, or equivalent. It relieves pressure through the pressure test port.

Fig. 9 Fuel pressure test port is protected by a plastic cap. This 5.2L engine has the port on the driver's side fuel rail

1. Disconnect the negative battery cable.
2. Remove the fuel filler cap.
3. Remove the air cleaner assembly, if needed for access.
4. Remove the protective cap from the fuel pressure test port on the fuel rail.
5. Remove the gauge from the fuel pressure gauge tool set and place the gauge end of the hose into a suitable container. Screw the other end of the hose onto the fuel pressure test port and relieve the fuel pressure by venting the pressure out the hose.
6. Remove the gauge tool set hose.
7. Install protective cap on fuel pressure test port.
8. Install removed components after servicing the fuel system.

BYPASSING THE PRESSURE TEST PORT

This procedure requires the use of a pair of jumper wires with alligator clips. It can be used on systems with or without a pressure test port. It relieves pressure by having the engine consume fuel in the lines and manually discharging an injector nozzle to remove the rest.

1. Remove the fuel filler cap.
2. Remove the fuel pump relay from the Power Distribution Center (PDC). The location of the relay is provided on the underside of the PDC cover.

WARNING:

This may cause one or more DTCs to be set in memory.

3. Start and run the engine until it stalls.
4. Attempt restarting until the engine will not run at all.
5. Turn the ignition key OFF.
6. This should have eliminated high pressure in the fuel lines. To relieve the rest of the pressure, the procedure involves manually activating an injector nozzle to empty the fuel in the rail(s).

WARNING:

Do not carry out the following steps unless the system has first been relieved as described above.

FUEL SYSTEM 5-5

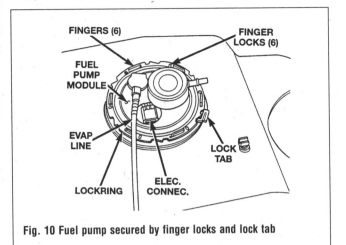

Fig. 10 Fuel pump secured by finger locks and lock tab

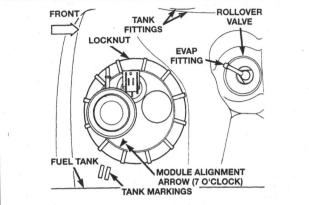

Fig. 11 Fuel pump and related components. Two-door models align at 7 o'clock; four-door models align at 4 o'clock

7. Unplug the electrical connector from any injector. Place a suitable container beneath the injector to catch the fuel when it discharges.
8. Attach one end of a jumper wire (18 ga. or smaller) to either of the injector terminals.
9. Connect the other end of the jumper wire to the battery positive (+) terminal.
10. Connect a second jumper wire to the injector and ground it for two seconds.
11. The voltage will discharge any residual fuel pressure through the injector.

WARNING:

Do not leave the jumper connected for more than two seconds. To do so will damage the injector.

12. Disconnect the negative battery cable before performing work on the fuel system.
13. Replace the fuel pump relay and gas cap. Clear any stored DTCs from memory.

Fuel Pump

The fuel pump is electric and is fitted to the top of the fuel tank. The pump module may also incorporate the filter/pressure regulator and rollover valve. The fuel gauge sending unit and float are also a part of this module.

The pump has a permanent magnet electric motor. Fuel is pulled through a filter at the bottom of the module and pushed through the electric motor gearset to the pump outlet.

The pump outlet contains a one-way check valve to prevent fuel flow back into the tank and to maintain fuel supply line pressure when the pump is not running. It is also used to keep the fuel supply line full of gasoline.

When the engine has not been run for some time, fuel pressure may drop to zero. This is normal.

REMOVAL & INSTALLATION

See Figures 10, 11, 12 and 13

1. Depressurize the fuel system, then disconnect the negative battery cable.
2. Drain the fuel tank.
3. Remove the fuel tank from the vehicle.
4. Disconnect any lines which may remain attached to the pump.
5. Thoroughly clean the area around the pump so that foreign matter won't fall into the tank when the unit is removed.
6. Check the top of the pump for arrows or alignment marks. Check for corresponding marks on the tank itself. Alignment of the pump is important for routing of wiring and lines. Matchmark pump and tank for assembly.
7. If the pump is equipped with "fingers" and "finger locks", apply some lubricant to the assembly.
8. Remove the pump locknut. Some units are fitted with a lock tab. Pry it back while removing the locknut.

To install:

9. Clean the filter at the bottom of the module.
10. Reverse the removal procedure.
11. Use a new gasket.
12. Align pump and tank marks to ensure correct routing of wires and lines.
13. If the locknut is the type without fingers and finger locks, tighten the nut to 40 ft. lbs. (54 Nm).

TESTING

See Figure 14

This test requires the use of special factory fuel pressure gauge tool set No. 5069, or equivalent (0 - 60 psi/0 - 414 kPa).

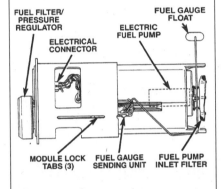

Fig. 12 Fuel pump module components - typical

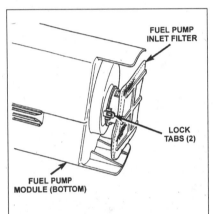

Fig. 13 Clean the filter at the bottom of the fuel pump module

1. Remove the cap from the fuel rail test port. Connect the gauge.
2. Start and run the engine.
3. Fuel pressure at idle should be 44.2 - 54.2 psi (305 - 373 kPa).
4. If pressure is too low, check for kinked hoses. If none are found, suspect the pump.
5. If pressure is too high, suspect the regulator.
6. Be sure to replace the test port cap after completing the test.

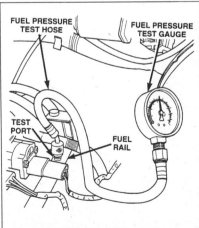

Fig. 14 Fuel pressure test gauge installed on fuel rail

5-6 FUEL SYSTEM

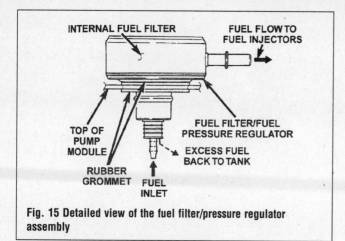

Fig. 15 Detailed view of the fuel filter/pressure regulator assembly

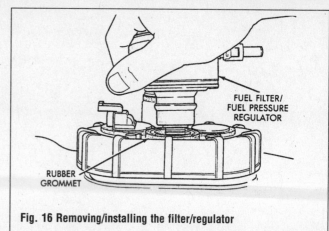

Fig. 16 Removing/installing the filter/regulator

Fuel Filter/Pressure Regulator Assembly

A combination filter and pressure regulator is used on all models. The regulator is a mechanical device not controlled by the PCM or engine vacuum.

It is calibrated to maintain a pressure a pressure of about 49 psi (339 kPa) at the fuel injectors. The unit contains a diaphragm, calibrated springs and a return valve.

The regulator's pressure maintenance feature permits easier starting. If pressure exceeds the specified amount, the unit routs the excess fuel back to the tank.

REMOVAL & INSTALLATION

See Figures 15, 16, 17 and 18

1. Remove the fuel tank filler cap.
2. Properly relieve the fuel system pressure.
3. Disconnect the negative battery cable.
4. Raise and safely support the vehicle.
5. Remove the fuel tank assembly.
6. The fuel filter/pressure regulator assembly is located on the top of the fuel pump assembly. It is not necessary to remove the fuel pump to remove the filter/regulator assembly. Matchmark the direction the hose fitting points. Unit must be reinstalled pointing in the same direction.
7. Twist the filter/regulator assembly out of its grommet on the fuel pump.
8. Remove the snap ring that retains the convoluted tube to the filter/regulator. Slide the tube down the plastic fuel tube to access the fuel tube clamp.
9. Gently cut the old fuel tube clamp off, taking care not to damage the fuel tube or drop the clamp inside the tank.
10. Remove the fuel tube from the filter/regulator assembly by gently pulling downward.
11. Remove the filter/regulator assembly from the fuel pump module.

To install:

12. Install a new clamp over the plastic fuel tube.
13. Install the filter/regulator assembly to the fuel tube. Rotate the filter/regulator until it is pointed to line up with the matchmark previously made.
14. Tighten the clamp to fuel line using Hose Clamp Pliers C-4124 or equivalent.
15. Slide the convoluted plastic tube up to the bottom of the filter/regulator and install the snap ring.

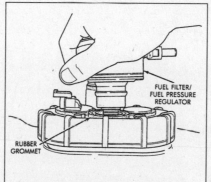

Fig. 17 Filter/regulator fuel tube and clamp

16. Press the filter/regulator assembly into the rubber grommet making sure it is pointed in the 10 o'clock position.
17. Install the fuel tank.
18. Lower the vehicle.
19. Connect the negative battery cable.
20. Install the fuel filler cap.

Fuel Gauge Sending Unit

See Figure 12

The fuel gauge sending unit (fuel level sensor) is attached to the side of the fuel pump module. The sending unit consists of a float, an arm, and a variable resistor track. The resistor track is used to send electrical signals to the PCM for fuel gauge and for OBD II emission requirements.

As fuel level increases the float and arm move up. This decreases the sending unit resistance, causing the fuel gauge to move towards "F". As fuel level decreases, the float and arm move down, and the gauge moves to "E".

When this signal is sent to the PCM, the PCM will transmit the data across the CCD bus circuits to the instrument panel. Here it is translated to an appropriate fuel gauge level reading.

For OBD II emission monitor requirements, a voltage signal is sent from the resistor track on the sending unit to the PCM to indicate fuel level. The purpose of this feature is to prevent the OBD II from recording/setting false misfire and fuel system monitor trouble codes. The feature is activated if the level in the tank is less than approximately 15 percent of

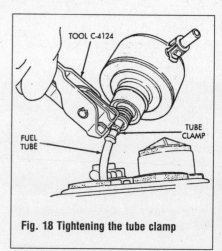

Fig. 18 Tightening the tube clamp

its rated capacity. If equipped with a leak detection pump, this feature will also be activated if the fuel level in the tank is more than about 85 percent of its rated capacity.

REMOVAL & INSTALLATION

The unit is removed with the fuel pump module. Refer to that procedure.

TESTING

1. Measure the resistance across the sending unit terminals.
2. With the float in the "up" position, resistance should be 14 - 26 ohms.
3. With the float in the "down" position, resistance should be 214 - 226 ohms.
4. Resistance should increase and decrease gradually as the float arm is moved. If odd resistance readings are evident at any point in the movement, replace the sending unit.

Rollover Valve

REMOVAL & INSTALLATION

See Figure 19

1. Tank-mounted rollover valves are molded into the gas tank and are not replaceable. The following

FUEL SYSTEM

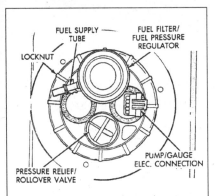

Fig. 19 Top view of the fuel pump module, including the rollover valve

Fig. 20 Disconnecting the throttle cable

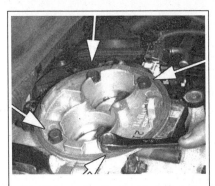

Fig. 21 Removing the throttle body bolts

procedure refers to pump module-mounted valves only.

2. Disconnect the battery negative cable(s).
3. Drain the fuel tank.
4. Remove the fuel tank from the truck.
5. Disconnect the line at the rollover valve.
6. The valve is kept by a rubber grommet. Pry one side upward and roll it out.

To install:

7. Use a new grommet.
8. Press the valve in using finger pressure only.

Throttle Body

WARNING:

A factory-adjusted setscrew is used to limit the position of the throttle plate. NEVER attempt to adjust idle speed with this screw. Idle speed is controlled by the PCM based on sensor inputs.

REMOVAL & INSTALLATION

See Figures 20, 21, 22, 23, 24 and 25

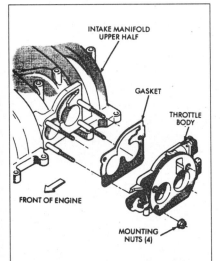

Fig. 23 Throttle body mounting - 8.0L engine

CAUTION:

Whenever servicing the fuel system, always work in a well ventilated area. Do not allow fuel spray or vapors to come in contact with a spark or open flame. Keep a dry chemical fire extinguisher near the work area. Always keep fuel in a container specifically designed for fuel storage; also, always properly seal fuel containers to avoid the possibility of fire or explosion. Fuel system work is made more hazardous by a hot engine. Work on a cold engine whenever possible.

1. Remove the air cleaner assembly.
2. Perform the fuel system depressurization procedure.
3. Disconnect the battery negative cable.
4. Disconnect all vacuum lines and electrical connectors from the throttle body. Tag for location, if necessary. Electrical connectors include TPS, MAP and IAC motor.
5. Disconnect cables: throttle, cruise control and others fitted.
6. Remove the throttle body bolts: there are three on the 4.7L engine, four on the others.
7. Lift the throttle body from the intake manifold.
8. Put a clean rag into the intake manifold to prevent the entry of foreign matter.

To install:

9. Always use a new gasket.
10. Reverse the removal procedure.
11. Tighten the throttle body bolts to the following torque:
    - 2.5L, 8.0L engines: 8 ft. lbs. (11 Nm)
    - 4.7L: 9 ft. lbs. (12 Nm)
    - 3.9L, 5.2L, 5.8L: 17 ft. lbs. (23 Nm)

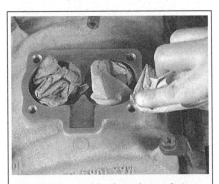

Fig. 24 Thoroughly clean the gasket surface before reinstalling the throttle body

Fig. 22 Removing the throttle body from the manifold

Fuel Rail Assembly

REMOVAL & INSTALLATION

CAUTION:

Whenever servicing the fuel system, always work in a well ventilated area. Do not allow fuel spray or vapors to come in contact with a spark or open flame. Keep a dry chemical fire extinguisher near the work area. Always keep fuel in a container specifically designed for fuel storage; also, always properly seal fuel containers to avoid the possibility of fire or explosion. Fuel system work is made more hazardous by a hot engine. Work on a cold engine whenever possible.

Fig. 25 Install a new gasket on the intake manifold. Don't forget to remove the rag before installing the throttle body

5-8 FUEL SYSTEM

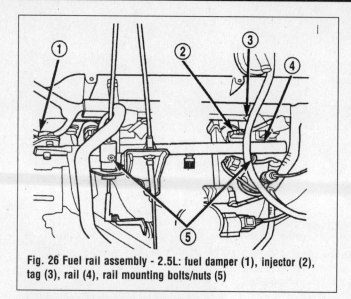

Fig. 26 Fuel rail assembly - 2.5L: fuel damper (1), injector (2), tag (3), rail (4), rail mounting bolts/nuts (5)

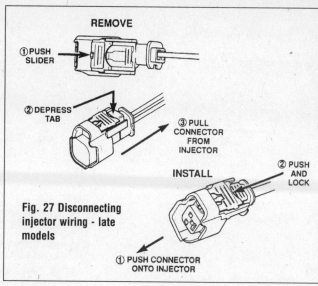

Fig. 27 Disconnecting injector wiring - late models

WARNING:

On "V" engines, the left and right fuel rails are joined by a connecting hose. Do not attempt to separate the two rails. The hose does not require clamps. Never install any kind of clamp on this hose. Be careful not to bend or kink the hose. The hose is not serviceable. If there is damage to the hose or either rail, the entire assembly must be replaced.

2.5L Engine
See Figure 26

1. Remove the gas cap.
2. Depressurize the fuel system.
3. Disconnect the battery negative cable.
4. Remove the air tube at the top of the throttle body. Some vehicles may require removal of the air cleaner ducts at the throttle body.
5. Disconnect the wiring at each injector. They should be tagged by the factory, but check first and tag the leads for cylinder before disconnecting if necessary.
6. Disconnect the fuel supply line latch clip and fuel line at the rail.
7. Disconnect the throttle cable at the throttle body. Disconnect any other cables fitted, such as cruise control, transmission, etc.
8. Remove the cable routing bracket.
9. Remove the nut securing the crankshaft position sensor harness to the fuel rail mounting stud. Remove the clamp and harness from the fuel rail mounting stud.
10. Clean any dirt or debris from around the injectors on the intake manifold.
11. Remove the rail mounting nuts and bolts.
12. Gently remove the fuel rail by rocking until all the injectors are clear of the manifold.

To install:

13. Clean the injector bores.
14. Oil the injector O-rings.
15. Position the injectors over their bores and install the rail assembly.
16. Tighten rail mounting bolts to 100 inch lbs. (11 Nm).
17. The remainder of the procedure is the reverse of removal. Check for leaks before driving the vehicle.

4.7L Engine
See Figure 27

1. Remove the gas cap.
2. Depressurize the fuel system.
3. Disconnect the battery negative cable.
4. Remove the air cleaner duct at the throttle body air box, then the air box itself.
5. Disconnect the wiring at the alternator.
6. Disconnect the wiring at each injector. They should be tagged by the factory, but check first and tag the leads for cylinder before disconnecting if necessary. To disconnect, push the red colored slider away from the injector while depressing the tab.
7. Disconnect the fuel supply line latch clip and fuel line at the rail.
8. Disconnect the throttle body vacuum lines.
9. Disconnect electrical connectors at the throttle body. Disconnect the TPS, MAP and IAT sensors.
10. Remove the first three ignition coils on each bank of cylinders.
11. Clean any dirt or debris from around the injectors on the intake manifold.
12. Remove the rail mounting bolts.
13. Gently remove the fuel rail by rocking until all the injectors are clear of the manifold. Free one side at a time, then remove the assembly.

To install:

14. Clean the injector bores.
15. Oil the injector O-rings.
16. Position the injectors over their bores and install the rail assembly. Install one side at a time.
17. Tighten rail mounting bolts to 20 ft. lbs. (27 Nm).
18. The remainder of the procedure is the reverse of removal. Check for leaks before driving the vehicle.

3.9L, 5.2L, 5.9L Engines
See Figures 28, 29, 30, 31, 32, 33, 34 and 35

1. Remove the air cleaner assembly.

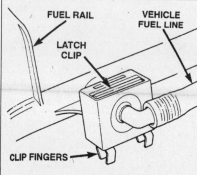

Fig. 28 Fuel feed line connector: compress the clip fingers and push up to disconnect

Fig. 29 Disconnect the fuel line by removing the latch

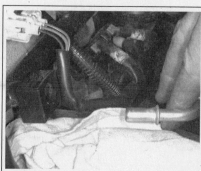

Fig. 30 Use a rag to catch any residual fuel in the line

FUEL SYSTEM 5-9

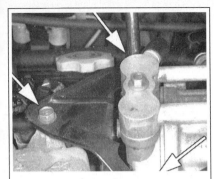

Fig. 31 The A/C compressor support bracket must be removed to access the fuel rail on some vehicles

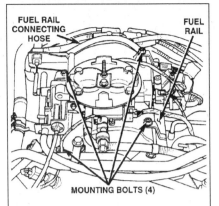

Fig. 32 Location of the fuel rail mounting bolts on 3.9L, 5.2L and 5.9L engines

Fig. 33 Remove the fuel rail retaining bolts

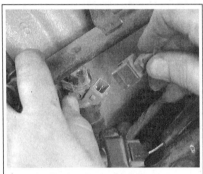

Fig. 34 Unplug the fuel injector connectors

2. Properly relieve the fuel system pressure.
3. Disconnect the negative battery cable.
4. Remove the throttle body assembly.
5. If equipped with A/C, remove the A-shaped A/C compressor-to-intake manifold bracket.
6. Disconnect the fuel line at the fuel rail.
7. Disconnect and label all the injector wiring harness connectors. They should be tagged by the factory, but check first and tag the leads for cylinder before disconnecting if necessary. To disconnect late models, push the red colored slider away from the injector while depressing the tab.
8. On the 3.9L, disconnect the IAT sensor.
9. Remove the fuel rail attaching bolts.
10. Gently rock and pull the driver's side fuel rail until the injectors unseat themselves from the intake manifold. Repeat this procedure for the passenger's side and then remove the fuel rail from the engine with the injectors attached.

To install:

11. Apply a small amount of engine oil to each injector O-ring.
12. Install the fuel rail onto intake aligning injectors into the openings. Guide the injectors into intake, taking care not to tear O-rings.
13. Push down passenger's side of fuel rail until the injectors have bottomed on injector shoulder. Repeat for the driver's side.
14. Install the fuel rail mounting bolts.
15. Plug in all the injector wiring harness connectors.

16. Connect the fuel line onto fuel rail.
17. Install the EVAP canister purge solenoid and bracket onto intake manifold.
18. On the 3.9L engine, plug in the IAT sensor connector.
19. On vehicles with A/C, install the A/C compressor support bracket.
20. Install the throttle body assembly.
21. Install the air cleaner assembly.
22. Connect the negative battery cable.
23. Start the engine and check for leaks.

8.0L Engine

See Figure 36

1. Disconnect the negative battery cable.
2. Remove the air cleaner tube and housing.
3. Release the fuel system pressure.
4. Remove the throttle body from the intake manifold.
5. Remove the ignition coil pack and bracket assembly at the intake manifold and right valve cover.
6. Remove the upper half of the intake manifold.
7. Unplug the electrical connectors at all of the injectors. The factory harness should already be numerically tagged (INJ 1, INJ 2, etc.) to facilitate reconnection. If not, tag as needed.
8. Disconnect the main fuel line at the rear of the engine.

9. Remove the six fuel rail mounting bolts from the lower half of the intake manifold.
10. Gently rock and pull the LEFT fuel rail until the injectors just start to clear the intake manifold. Then do the same for the right. Alternate each side progressively until all left and right injectors have cleared the intake manifold.
11. With the injectors attached, remove the fuel rail.

To install:

12. Apply a small amount of engine oil to each fuel injector O-ring.
13. With the injectors attached, position the fuel rail/fuel injector assembly to the injector openings on the intake manifold.
14. Guide each injector into the manifold. Be

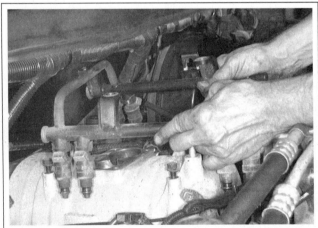

Fig. 35 Pull straight up on the injector rail to unseat it from the intake manifold

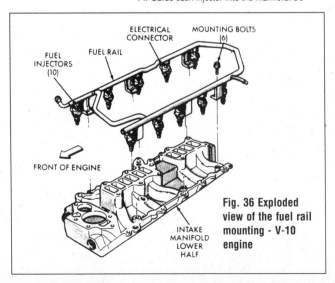

Fig. 36 Exploded view of the fuel rail mounting - V-10 engine

5-10 FUEL SYSTEM

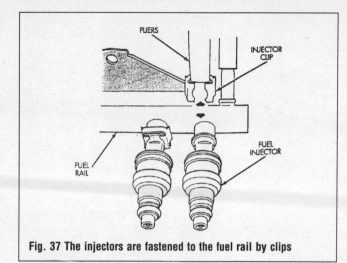

Fig. 37 The injectors are fastened to the fuel rail by clips

Fig. 38 Check condition of the injector O-rings. Apply a drop of clean engine oil before installation

careful not to tear any O-rings.

15. Push the RIGHT fuel rail down until the injectors bottom on the injector shoulder. Then do the same for the left side injectors.

Note: The fuel injector electrical connectors on all 10 injectors should be facing to the right (passenger) side of the vehicle.

16. Install the six fuel rail mounting bolts. Tighten them to 136 inch lbs. (15 Nm).
17. Connect all removed wires.
18. Install the upper half of the intake manifold.
19. Install the ignition coil pack.
20. Install the throttle body and the throttle linkage.
21. Install the main fuel line at the fuel rail.
22. Install the air cleaner tube and housing.
23. Connect the negative battery cable.
24. Start the engine and check for leaks. If any are detected, shut the engine off immediately and correct.

Fuel Injectors

REMOVAL & INSTALLATION

See Figures 37, 38 and 39

1. Disconnect the negative battery cable.
2. Remove the air cleaner assembly.
3. Remove the fuel rail assembly.
4. Remove the clip(s) retaining the injector(s) to the rail.
5. Remove the injector(s) from the fuel rail.

To install:

6. Apply a small amount of engine oil to each injector O-ring.
7. Install the injector(s) onto the fuel rail and install the clip(s).
8. Install the fuel rail assembly.
9. Install the air cleaner assembly.
10. Connect the negative battery cable.
11. Start the engine and check for leaks.

TESTING

A scan tool is the recommended method of troubleshooting injector-related problems, but the following tests can provide useful information in many applications.

1. With the engine running, disconnect a wire of each injector, in turn. If no change in running condition is observed when one of the injectors is disconnected, the problem has been isolated to that cylinder.
2. Check the resistance across the injector terminals. It should be about 12 ohms. If open or shorted,

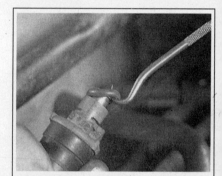

Fig. 39 Use a suitable tool to remove O-rings from the injector

replace the injector.

3. After removing an injector, apply 12 VDC across the terminals. A "click" should be heard. If not, replace the injector.

WARNING:

Do not leave the injector connected to voltage for more than a couple a seconds. To do so will destroy the unit.

DIESEL FUEL SYSTEM

See Figure 40

Injection Lines

The high pressure fuel lines deliver the diesel fuel under pressure of approximately 18,000 psi (120,000 kPa) from the injection pump to the fuel injectors. The lines expand and contract with the high pressure fuel pulses generated during the injection process. All the high pressure fuel lines are of the same length and inside diameter. Correct line usage and installation is critical.

The high pressure fuel lines must be retained securely in their holders. The lines must not be allowed to contact each other or other components along side of them. If a line fails, it must be replaced with the correct replacement. Repair is not possible.

CAUTION:

If examining for leaks, use a sheet of cardboard to show any spraying fuel. Do not use your hand and wear safety goggles. The extremely high pressure can cause serious personal injury, especially to your skin and eyes.

Injectors

GENERAL INFORMATION

See Figure 41

The fuel injectors are mounted on the left side of the cylinder head on 97 - 98 models, on the top, center of the head on 99 - 00 models, and are connected to the fuel pump by the high pressure fuel lines.

The injectors consist of the nozzle holder, O-ring, water seal, shims, spring, needle valve and nozzle. Fuel enters the injector through the top of the injector (fuel inlet) and is routed to the needle valve bore. The injector fires when fuel pressure rises to an amount sufficient to overcome the needle valve spring tension. The pressure needed to overcome the needle valve spring tension is about 4,000 psi (28,000 kPa). This pressure is commonly known as the "pop" pressure.

As the needle valve opens, fuel flows rapidly through the spray holes in the nozzle tip into the combustion chamber. After this injection, the fuel pressure drops and the needle valve is closed preventing further fuel flow, and conversely, exhaust flow into the injector.

FUEL SYSTEM 5-11

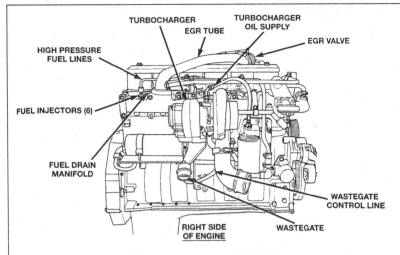

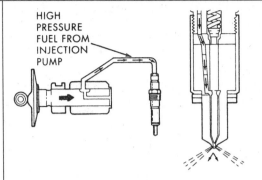

Fig. 41 The injector is a precise one-way valve that sprays a metered amount of fuel at high pressure into the combustion chamber

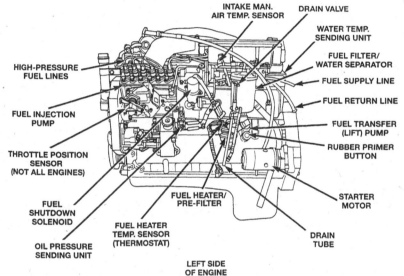

Fig. 40 Common diesel fuel system components - 98 shown, others similar

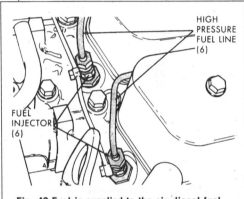

Fig. 42 Fuel is supplied to the six diesel fuel injectors by individual high pressure lines

REMOVAL & INSTALLATION

See Figures 42, 43, 44, 45 and 46

1997 - 98 Models

1. Disconnect the negative battery cable(s). Remove the throttle linkage and bracket if necessary.
2. Disconnect the high pressure fuel supply line to the injector.
3. Disconnect the fuel drain manifold.
4. Clean the area around the injector.

Note: Certain types of injectors MAY have an O-ring located above the hold-down nut.

5. Using the correct deepwell socket, remove the injector from the cylinder head.

To install:

6. Clean the injector bore with a bore brush.
7. Assemble the injector and 1 new copper sealing washer. Never use more than 1 copper washer.
8. Apply a thin coat of anti-seize compound to the threads of the injector hold-down nut and between the top of the nut and the injector body.
9. Align the protrusion in the injector with the notch in the bore and install the injector. Tighten the injector retainer nut to 44 ft. lbs. (60 Nm).
10. Push the O-ring into the groove at the top of the injector, if applicable.
11. Using new sealing washers, assemble the fuel drain manifold and high pressure lines. Tighten the

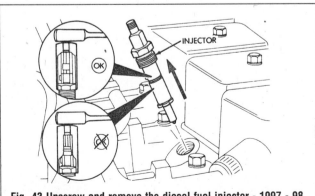

Fig. 43 Unscrew and remove the diesel fuel injector - 1997 - 98 models shown

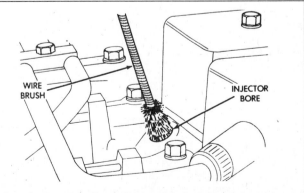

Fig. 44 Use a wire brush to clean the injector bore as shown

5-12 FUEL SYSTEM

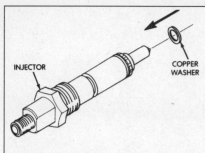

Fig. 45 Use only one copper washer on the diesel injector

banjo fitting bolt to 6 ft. lbs. (8 Nm). Leave the high pressure line loose temporarily.

CAUTION:
Do not place any part of the hand near the base of the high pressure line. A fuel leak from a high pressure fuel line has sufficient pressure to penetrate the skin and cause serious bodily harm. Do not bleed the lines if the engine is hot. Fuel spilling onto a hot exhaust manifold creates the danger of fire.

12. To bleed air from the system, run or crank the engine and tighten the fitting after the air has expelled. If more than 1 injector was replaced, tighten each fitting after the air has expelled before going on to the next injector fitting. Tighten the fittings to 22 ft. lbs. (30 Nm). The operation is complete when the engine runs smoothly. If the air cannot be removed, check the pump and supply line for suction leaks.

13. Install the throttle linkage and bracket if they were removed.

1999 - 2000 Models
See Figure 47

1. Disconnect the negative battery cable(s). Remove the throttle linkage and bracket if necessary.
2. To access the injectors at cylinders #1 and #2, remove the intake manifold heater assembly. To access the injector at cylinder #5, remove the engine lifting bracket.
3. Clean the area around the injector.
4. Disconnect the high pressure fuel supply line to the injector.
5. Remove the valve cover.
6. Thread special tool 8324 onto the end of the injector tube and pull injector tube from cylinder head. Remove and discard the old O-ring.
7. Remove the fuel injector hold down clamp bolt at front end of clamp. Do not loosen or remove special bolt at rear end of clamp. Remove injector by sliding it out.
8. Thread rod from special tool 8318 into the top of the fuel injector. Remove injector from cylinder head.

To install:
9. Reverse the removal procedure.
10. Use new copper sealing washer and O-ring. Copper washer thickness must be 0.060 in. (1.5mm).
11. Oil all parts before installation.
12. Position inlet hole towards connector tube.
13. Tighten opposite clamp bolt to 89 inch lbs. (10 Nm).

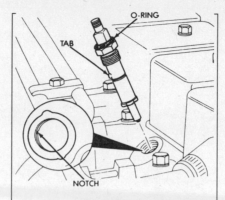

Fig. 46 Install the diesel injector so that its tab is aligned with the notch in the bore - 1997 - 98 models

TESTING (ALL MODELS)
See Figures 48 and 49

CAUTION:
If examining for leaks, use a sheet of cardboard to show any spraying fuel. Do not use your hand and do wear safety goggles. The extremely high pressure can cause serious personal injury, especially to your skin and eyes.

A leaking fuel injector can cause fuel knock, poor performance, black smoke, poor fuel economy and rough engine idle. If the needle valve does not operate properly, the engine may misfire and produce low power.

A leak in the injection pump-to-injector high-pressure fuel line can cause many of the same symptoms as a malfunctioning injector. First check the lines before the injectors.

1. To check the injectors, start the engine and loosen the high-pressure line nut one-at-a-time at each injector. Listen for a decrease in engine speed.
2. If the engine speed drops, the injector is operating normally. If the engine speed remains the same, the injector is malfunctioning. After testing each injector, tighten the line nuts to 22 ft. lbs. (30 Nm) on

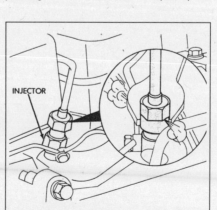

Fig. 48 Loosen the fittings one at a time, to check the effect on engine rpm - no change in rpm indicates a faulty injector

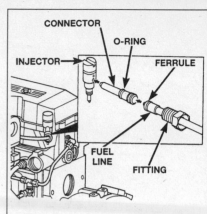

Fig. 47 Fuel injector connections - 1999 - 2000

1997 - 98 models, 30 ft. lbs. (40 Nm) on later units.

3. If an injector is found to be malfunctioning, remove it and test it on a standard bench-mount injector tester. Follow the manufacturer's instructions for bench testing the injector.
4. If the opening (pop) pressure is below specifications, replace the injector. The pop pressure is 3,822 psi (26,252 kPa) for 1997 - 98 models, 4,500 psi (31,026 kPa) for 1999 - 2000 models.

Fuel Transfer Pump

REMOVAL & INSTALLATION

See Figures 50 and 51

The fuel transfer pump is located on the left side of the engine and above the starter motor.

1. Disconnect both negative battery cables.
2. Thoroughly clean the area around the pump to prevent fuel contamination.
3. Remove the starter motor.
4. Place a drain pan beneath the pump.
5. Remove the fuel line fittings.
6. Disconnect the fuel line banjos, or clamps and hose, as fitted.
7. Disconnect the electrical connector.
8. Unfasten the mounting bolts or nuts, then re-

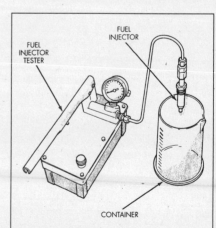

Fig. 49 Use a tester like the one shown to measure pop pressure for the diesel fuel injector

FUEL SYSTEM 5-13

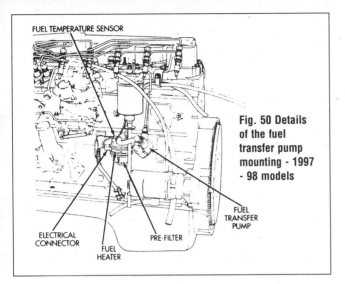

Fig. 50 Details of the fuel transfer pump mounting - 1997 - 98 models

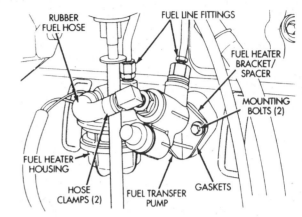

Fig. 51 Disconnect the fuel lines and the rubber hose, then unbolt and remove the fuel pump and heater housing as a unit

move the fuel pump (and fuel heater assembly, as fitted) from the engine.

To install:

9. Clean the mating surfaces of the fuel heater mounting bracket, the fuel pump and the engine block of any gasket material.
10. Position the new gaskets, the fuel heater housing mounting bracket and the fuel pump to the engine.
11. Assemble, but do not tighten banjos, yet.
12. Install mounting nuts, if fitted, but do not tighten yet.
13. Install bracket, if fitted.
14. Install the bolts if fitted, tightening them ALTERNATELY to 18 ft. lbs. (24 Nm). If mounting nuts are used, tighten to 9 ft. lbs. (12 Nm). Bracket torque is the same.

> **WARNING:**
>
> As these bolts are tightened, the fuel pump plunger will be compressed. Tighten the fuel pump mounting bolts alternately to prevent damage to the fuel pump housing.

15. Install the fuel line fittings to the pump and fuel heater, tightening them to 18 ft. lbs. (24 Nm).
16. Install the starter motor.
17. Connect the negative battery cables.
18. Bleed the fuel system of air.

LOW PRESSURE BLEEDING

A certain amount of air becomes trapped in the fuel system when components on the supply and/or high pressure side are serviced or replaced.

> **CAUTION:**
>
> Do not bleed air from the fuel system of a hot engine.

1997 - 98 Models
See Figures 52 and 53

1. Loosen the low pressure bleed bolt.
2. Operate the rubber push-button primer on the fuel transfer pump. Do this until the fuel exiting the bleed screw is free of air. If the primer button feels as if it is not pumping, rotate (crank) the engine approximately 90-degrees, then continue pumping as described.
3. Tighten the low pressure bleed screw to 6 ft. lbs. (8 Nm).

1999 - 2000 Models
See Figure 54

Primary air bleeding is accomplished using the electric fuel transfer (lift) pump. If the vehicle has been allowed to run completely out of fuel, the fuel injectors must also be bled as the pump is not self-bleeding.

1. Loosen, but do not remove, the banjo bolt holding the low pressure fuel supply line to the side of the fuel injection pump. Place a rag around the banjo fitting to catch excess fuel.
2. Turn the key to the CRANK position and quickly release it to the ON position before the engine starts. This will operate the fuel transfer pump for about 25 seconds.
3. Repeat the procedure, if needed, until fuel appears at the supply line.
4. Tighten the banjo bolt to 18 ft. lbs. (24 Nm).
5. If the engine will not start, crank it for 30 seconds at a time until it starts.
6. If necessary, loosen the high pressure fuel

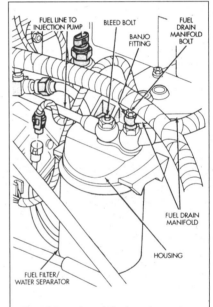

Fig. 52 Location of the low pressure bleed bolt - 1997 - 98 models

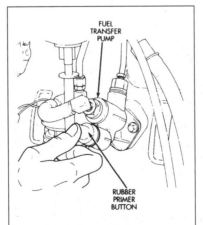

Fig. 53 Repeatedly press the rubber primer button to bleed the system of air - 1997 - 98 models

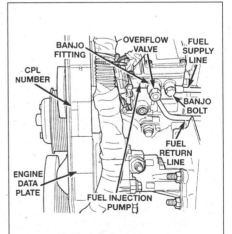

Fig. 54 Fuel supply and return lines at pump - 1999 - 2000 models

5-14 FUEL SYSTEM

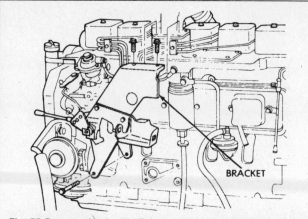

Fig. 55 Remove the throttle linkage and bracket from the engine - 1997 - 98 models

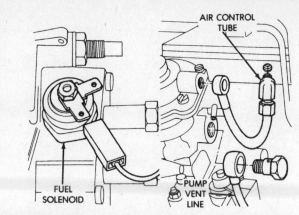

Fig. 56 Detach the connector and lines shown when removing the injection pump

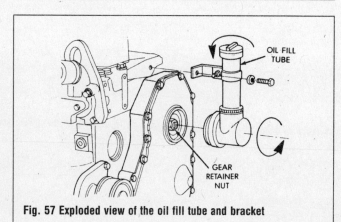

Fig. 57 Exploded view of the oil fill tube and bracket

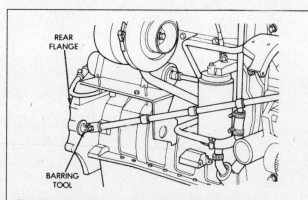

Fig. 58 Use the barring tool to rotate the diesel engine

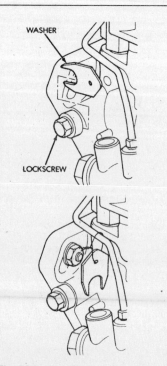

Fig. 59 Loosen the injection pump shaft lockscrew and remove the special washer shown

line fittings at the three rear-most cylinders. Continue until engine runs smoothly. Tighten fittings to 28 ft. lbs. (38 Nm).

Diesel Injection Pump

REMOVAL & INSTALLATION

1997 - 98 Models

See Figures 55, 56, 57, 58, 59, 60 and 61

Note: The Bosch VE lever is indexed to the shaft during pump calibration. Do not remove it from the pump during removal.

1. Disconnect the negative battery cables.
2. Remove the throttle linkage and bracket.
3. Disconnect the fuel drain manifold.
4. Remove the injection pump supply line.
5. Remove the high pressure lines.
6. Disconnect the electrical wire to the fuel shut off valve.
7. Remove the fuel air control tube.
8. Remove the pump support bracket.
9. Remove the oil fill tube bracket and adapter from the front gear cover.
10. Place a shop towel in the gear cover opening in a position that will prevent the nut and washer from falling into the gear housing. Remove the gear retaining nut and washer.
11. Install the turning tool into the flywheel hous-

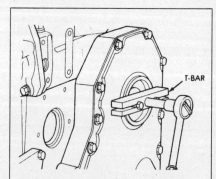

Fig. 60 Use a puller tool as shown to remove the injection pump timing gear

ing opening on the exhaust side of the engine. Place a 1/2 in. drive universal joint in the turning tool and attach enough extensions to the joint to make it convenient to turn the tool.

12. Using a ratchet to turn the barring tool, turn the engine until the keyway on the fuel pump shaft is pointing approximately in the six o'clock position.
13. Locate TDC for cylinder No. 1 by turning the engine slowly while pushing in on the TDC pin. Stop turning the engine as soon as the pin engages with the gear timing hole. Disengage the pin after locating TDC and remove the turning equipment.
14. Loosen the lockscrew, remove the special washer from the injection pump and wire it to the line above it so it will not get misplaced. Retighten the

FUEL SYSTEM 5-15

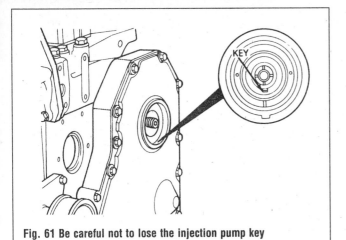

Fig. 61 Be careful not to lose the injection pump key

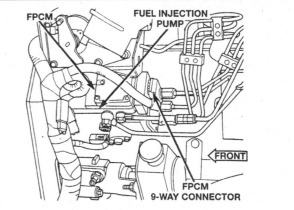

Fig. 62 Fuel pump control module (FPCM) 9 - way connector - 1999 - 2000 models

lockscrew to 22 ft. lbs. (30 Nm) to lock the driveshaft.

15. Using a suitable puller, pull the pump drive gear from the driveshaft.

Note: Be careful not to drop the drive gear key into the front cover when removing or installing the pump. If it does drop in, it must be removed before proceeding.

16. Remove the 3 mounting nuts and remove the injection pump from the vehicle.
17. Remove the gasket and clean the mounting surface.

To install:

18. Install a new gasket.

Note: The shaft of a new or reconditioned pump is locked so the key aligns with the drive gear keyway with cylinder No. 1 at TDC.

19. Install the pump and finger-tighten the mounting nuts; the pump must be free to move in the slots.
20. Install the pump drive gear, washer and nut to the driveshaft. The pump will rotate slightly because of gear helix and clearance. This is acceptable providing the pump is free to move on the flange slots and the crankshaft does not move. Tighten the nut to 11 - 15 ft. lbs. (15 - 20 Nm). This is not the final torque; do not overtighten.
21. If installing the original pump, rotate the pump to align the original timing marks and tighten the mounting nuts to 18 ft. lbs. (24 Nm).
22. If installing a replacement pump, take up gear lash by rotating the pump counterclockwise toward the cylinder head, and tighten the mounting nuts to 18 ft. lbs. (24 Nm). Permanently mark the new injection pump flange to match the mark on the gear housing.
23. Loosen the lockscrew and install the special washer under the lockscrew; tighten to 13 ft. lbs. (18 Nm). Disengage the TDC pin.
24. Install the injection pump support bracket. Finger-tighten the bolts initially, then tighten them to 18 ft. lbs. (24 Nm) in the following sequence:
    a. Bracket-to-block bolts.
    b. Bracket-to-injection pump bolts.
    c. Throttle support bracket bolts.
25. Now perform the final tightening of the pump drive gear retaining nut to 48 ft. lbs. (65 Nm).
26. Install the oil filler tube assembly and clamp. Tighten the bolts to 32 ft. lbs. (43 Nm).
27. Install all fuel lines and the electrical connector to the fuel shut off valve. Tighten the high pressure lines to 18 ft. lbs. (24 Nm).
28. Install the fuel air control tube. Tighten the banjo fitting bolt to 9 ft. lbs. (12 Nm).
29. Install the throttle bracket and linkage. When connecting the cable to the control lever, adjust the length so the lever has stop-to-stop movement.
30. Connect the negative battery cables.

CAUTION:

Do not place your hand near the base of the high pressure line. A fuel leak from a high pressure fuel line has sufficient pressure to penetrate the skin and cause serious bodily harm. Do not bleed the lines if the engine is hot. Fuel spilling onto a hot exhaust manifold creates the danger of fire.

31. To bleed air from the system, run or crank the engine and carefully loosen the high pressure fitting from each injector one at a time. Retighten the fitting after the air has expelled before going on to the next injector fitting. The operation is complete when the engine runs smoothly. If the air cannot be removed, check the pump and supply line for suction leaks.
32. Adjust the idle speed if necessary.

1999 - 2000 Models
See Figures 54, 62, 63 and 64

1. Disconnect the battery negative cables.
2. Clean all fuel line and fitting ends to prevent possible contamination.
3. Disconnect the 9 - way electrical connector at the fuel pump control module.
4. Remove the fuel line return line at the side of the injection pump by removing the overflow valve.
5. Remove the supply line banjo bolt.
6. Remove high pressure lines, intake air tube, throttle sensor, air intake housing, dipstick tube, cables and clips, and engine lifting bracket.
7. Unscrew the plastic access cap at the front gear cover.

WARNING:

To prevent pump/gear keyway from falling into gear housing, the engine must be rotated until the keyway is at 12 o'clock.

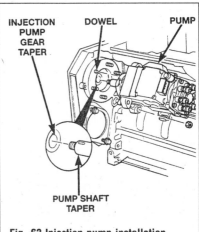

Fig. 63 Injection pump installation - 1999 - 2000

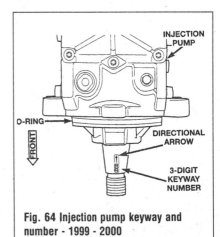

Fig. 64 Injection pump keyway and number - 1999 - 2000

8. Remove the nut and washer retaining injection pump gear to shaft.
9. Rotate the engine until the keyway is at 12 o'clock.
10. Use a puller to separate the gear from the shaft. Attach two M8 x 1.24mm screws through the puller and into the threaded holes in the pump gear. Pull gear forward until loose. Move only enough to free it.

5-16 FUEL SYSTEM

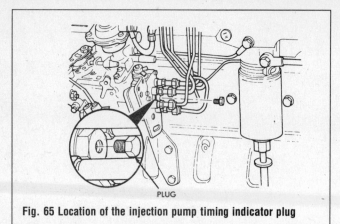

Fig. 65 Location of the injection pump timing indicator plug

Fig. 66 Check the injection pump timing with the timing gauge installed

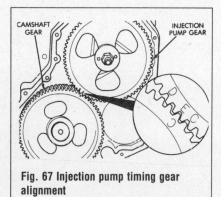

Fig. 67 Injection pump timing gear alignment

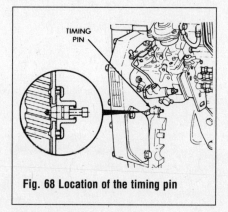

Fig. 68 Location of the timing pin

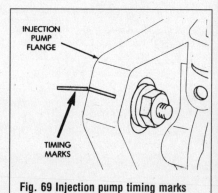

Fig. 69 Injection pump timing marks

11. Remove the three lower pump bracket bolts and the bracket. Loosen, but do not remove, the two engine bracket bolts.
12. Remove the four injection pump-to-gear housing mounting nuts. Remove the injection pump.

To install:

13. Use new O-rings and gaskets. Machined tapers must be clean and dry before installation.
14. The pump/gear keyway has an arrow and a three-digit number stamped at the top edge. Position the keyway into the pump shaft with the arrow pointing to the rear of the pump. Be sure keyway and pump data plate have the same number.
15. Position pump assembly to mounting flange while aligning pump shaft through back of injection pump gear. Dowel on mounting flange must align with hole in front of pump.
16. Install and tighten four pump mounting nuts finger tight only.
17. Tighten injection pump shaft washer and nut finger tight.
18. Position lower pump bracket and install three bolts finger tight only.
19. Tighten shaft nut to 15 - 22 ft. lbs. (30 Nm). This is not the final torque.
20. Tighten the four pump mounting nuts to 32 ft. lbs. (43 Nm).
21. Tighten the three lower pump bracket-to-pump bolts to 18 ft. lbs. (24 Nm).
22. Tighten the two engine bracket-to-engine bolts to 18 ft. lbs. (24Nm).
23. Final tighten the injection pump shaft nut to 125 ft. lbs. (170 Nm).
24. The remainder of the procedure is the reverse or removal.

INJECTION PUMP TIMING

1997 - 98 Models

See Figures 65, 66, 67, 68 and 69

1. Install the barring tool into the flywheel housing opening on the exhaust side of the engine. Place a 1/2 in. drive universal joint in the tool and attach enough extensions to the joint to make it convenient to turn the tool.
2. Using a ratchet to turn the barring tool, turn the engine until the keyway on the fuel pump shaft is pointing approximately in the six o'clock position.
3. Locate TDC for cylinder No. 1 by turning the engine slowly while pushing in on the TDC pin. Stop turning the engine as soon as the pin engages with the gear timing hole. Disengage the pin after locating TDC.
4. Remove the plug from the end of the pump.
5. Install the special timing indicator, allowing for adequate indicator pin travel. It may be necessary to disconnect one or more fuel lines to properly install the indicator.

Note: The indicator is marked in increments of 0.01mm. One revolution of the indicator needle is equal to 0.050mm.

6. Turn the engine counterclockwise until the indicator needle stops moving. Adjust the indicator face to read zero.
7. Rotate the engine back to TDC and count the number of revolutions of the indicator needle. The reading shown when the engine timing pin engages is the amount of the pump's plunger lift at that point.
8. Readjust the indicator face to read zero.

Loosen the pump mounting nuts and rotate the pump clockwise toward the cylinder head until the indicator reads the correct value for plunger lift (the reading in Step 7). Tighten the mounting nuts to 18 ft. lbs. (24 Nm).
9. Remove the engine turning equipment and timing indicator. Install the timing plug and tighten to 7.5 ft. lbs. (10 Nm). Connect any fuel lines that were disconnected.
10. Road test the vehicle.

1999 - 2000 Models
See Figure 70

1. All timing and fuel adjustments are made by the Engine Control Module. No manual adjustments are required unless a DTC has been displayed indicating "engine sync error" or "static timing error". If this appears after installation of a new or rebuilt injection pump, the keyway has probably been installed backwards.
2. Refer to the illustration of timing marks. Be sure the keyway has been installed with the arrow pointing to the rear of the pump.

Fuel/Water Separator Filter

See Figure 71

REMOVAL & INSTALLATION

1. Disconnect the negative battery cable(s).
2. Disconnect the Water In Filter (WIF) sensor connector.
3. Remove the separator filter assembly from the

FUEL SYSTEM 5-17

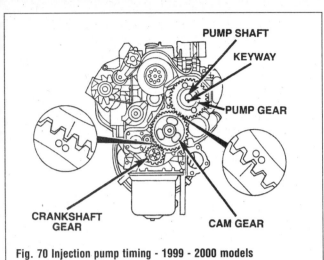

Fig. 70 Injection pump timing - 1999 - 2000 models

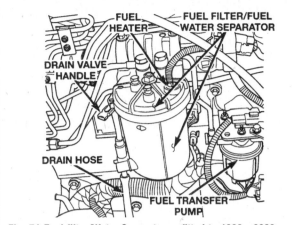

Fig. 71 Fuel filter/Water Separator as fitted to 1999 - 2000 models

filter head with a standard oil filter wrench or loosen the filter canister nut at top, depending on type fitted.
4. Remove the O-ring from the filter mounting bushing, if equipped.
5. Drain the fuel/water separator filter and remove the assembly from the fuel filter.

To install:

6. Install a new O-ring to the WIF assembly and install to the new separator filter.
7. Install a new O-ring to the mounting bushing.
8. Fill the fuel/water separator filter with clean diesel fuel.
9. Apply a light coat of oil to the sealing surface of the separator filter.
10. Install the assembly and tighten it 1/2 turn after the seal contacts the filter head, or if a nut is used, tighten it to 30 - 40 inch lbs. (3 - 5 Nm).
11. Reconnect the WIF sensor connector.
12. Connect the negative battery cable(s), start the engine and check for leaks.

DRAINING WATER FROM THE SYSTEM

Filtration and separation of water from the fuel is

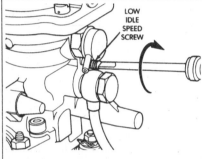

Fig. 72 Turn the screw shown to adjust the low idle speed

important for trouble free operation and long life of the fuel system. Regular maintenance, including draining moisture from the fuel/water separator filter is essential to keep water out of the fuel pump. To remove the collected water, simply unscrew the drain at the bottom of the WIF assembly located at the bottom of the filter separator, or turn the drain handle to OPEN, depending on type fitted.

Idle speed adjustment (1997 - 98 models)

See Figure 72

WARNING:

This procedure applies to 1997 - 98 models only. It applies only to low-speed idle. High-speed idle cannot be adjusted.

1. Start the engine and run until at normal operating temperature.
2. An optical tachometer must be used to read engine speed; a conventional tachometer connected to the coil is useless in this instance.
3. Turn the air conditioning ON if equipped.
4. Turn the idle speed screw until the desired idle speed is obtained. The specification for a vehicle equipped with automatic transmission is 750 - 800 rpm in Drive with A/C ON. The specification for a vehicle equipped with manual transmission is 780 rpm in Neutral with A/C ON.

FUEL TANK

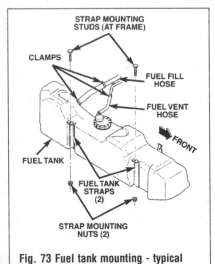

Fig. 73 Fuel tank mounting - typical

Tank Assembly

REMOVAL & INSTALLATION

See Figures 73, 74, 75 and 76

CAUTION:

Observe all applicable safety precautions when working around fuel. Whenever servicing the fuel system, always work in a well ventilated area. Do not allow fuel spray or vapors to come in contact with a spark or open flame. Keep a dry chemical fire extinguisher near the work area. Always keep fuel in a container specifically designed for fuel storage; also, always properly seal fuel containers to avoid the possibility of fire or explosion.

1. Disconnect the negative battery cable(s).
2. Remove the fuel tank filler cap.
3. Properly relieve the fuel system pressure.
4. Pump all fuel from the tank into an approved holding tank.
5. Raise and safely support the vehicle.
6. Some models are fitted with a ground strap connecting fuel filler tube assembly to the body. If equipped, disconnect the ground strap.
7. Disconnect the fuel line and wire lead to the gauge unit.
8. Open the fuel filler door and remove the screws mounting the fuel filler tube assembly to the body. Do not disconnect the rubber fuel filler or vent hoses at this time.
9. Place a transmission jack or other proper support device under the center of the tank and apply sufficient pressure to support the tank.
10. Remove the nuts which secure the tank mounting straps. Remove the fuel tank shield bolts, if fitted.

5-18 FUEL SYSTEM

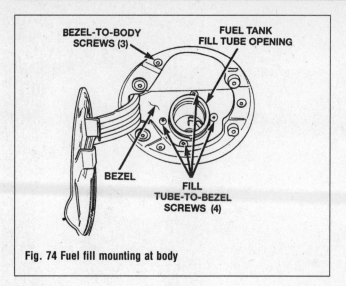

Fig. 74 Fuel fill mounting at body

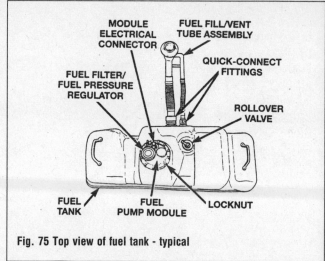

Fig. 75 Top view of fuel tank - typical

11. Lower the tank just enough for access to the remaining tubes and lines.
12. Disconnect the vent hose and the fill hose.
13. Disconnect the electrical lines.
14. Disconnect the EVAP line(s).
15. Disconnect the fuel supply line, which may have a quick-connect fitting.
16. Remove the tank.

To install:

17. Inspect the fuel pump filter and if it is clogged or damaged, replace it.
18. Install fuel pump assembly.
19. Position the tank on a transmission jack and hoist it into place, feeding the hoses through the grommets on the way up, if necessary.
20. Connect any vacuum and electrical lines or hoses possible before bolting in the tank.

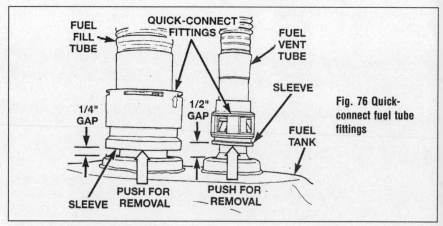

Fig. 76 Quick-connect fuel tube fittings

21. Connect the retaining straps, and tighten to 30 ft. lbs. (41 Nm). Remove the jack.
22. Connect the remaining hoses.
23. Connect ground strap.
24. Refill the tank and inspect it for leaks.
25. Reconnect the battery cable ground cable(s).

AIR BAG (SUPPLEMENTAL RESTRAINT SYSTEM)
GENERAL INFORMATION, 6-7
BATTERY CABLES
DISCONNECTING THE CABLES, 6-7
CIRCUIT PROTECTION
FUSES, 6-26
CRUISE CONTROL
GENERAL INFORMATION, 6-11
ENTERTAINMENT SYSTEMS
RADIO RECEIVER/AMPLIFIER/TAPE PLAYER/CD PLAYER, 6-12
SPEAKERS, 6-13
FUSES
CARTRIDGE FUSE, 6-27
CIRCUIT BREAKERS, 6-27
FLASHERS, 6-28
HEATING AND AIR CONDITIONING, 6-9
AIR CONDITIONING COMPONENTS, 6-10
BLOWER MOTOR, 6-10
CONTROL PANEL, 6-11
HEATER - A/C HOUSING, 6-9
HEATER CORE, 6-10
TEMPERATURE CONTROL CABLE, 6-10
INSTRUMENTS AND SWITCHES
FOG LAMP SWITCH, 6-18
GAUGES & BULBS, 6-17
HEADLIGHT SWITCH, 6-18
INSTRUMENT CLUSTER, 6-17
WINDSHIELD WIPER SWITCH, 6-17
LIGHTING
FOG/DRIVING LIGHTS, 6-23
HEADLIGHTS, 6-18
SIGNAL AND MARKER LIGHTS, 6-20
TRAILER WIRING, 6-25
UNDERSTANDING AND TROUBLESHOOTING ELECTRICAL SYSTEMS
BASIC ELECTRICAL THEORY, 6-2
ELECTRICAL COMPONENTS, 6-2
TEST EQUIPMENT, 6-4
TESTING, 6-6
TROUBLESHOOTING ELECTRICAL SYSTEMS, 6-5
WIRE AND CONNECTOR REPAIR, 6-7
WINDSHIELD WIPER SYSTEM
WASHER FLUID LEVEL SENSOR, 6-15
WINDSHIELD WASHER PUMP, 6-15
WINDSHIELD WASHER RESERVOIR, 6-16
WINDSHIELD WIPER MOTOR, 6-14
WIPER ARMS, 6-14
WIPER BLADES, 6-14

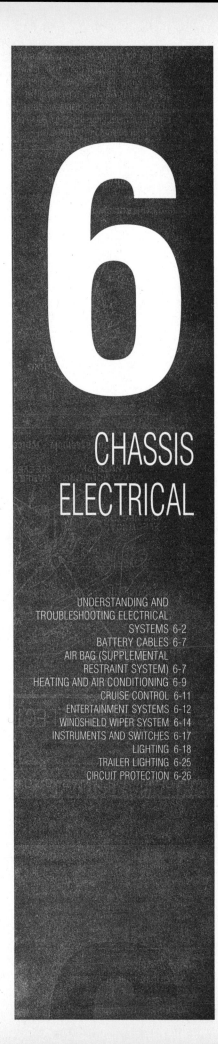

6

CHASSIS ELECTRICAL

UNDERSTANDING AND TROUBLESHOOTING ELECTRICAL SYSTEMS 6-2
BATTERY CABLES 6-7
AIR BAG (SUPPLEMENTAL RESTRAINT SYSTEM) 6-7
HEATING AND AIR CONDITIONING 6-9
CRUISE CONTROL 6-11
ENTERTAINMENT SYSTEMS 6-12
WINDSHIELD WIPER SYSTEM 6-14
INSTRUMENTS AND SWITCHES 6-17
LIGHTING 6-18
TRAILER LIGHTING 6-25
CIRCUIT PROTECTION 6-26

6-2 CHASSIS ELECTRICAL

UNDERSTANDING AND TROUBLESHOOTING ELECTRICAL SYSTEMS

Basic Electrical Theory

See Figure 1

For any 12 volt, negative ground, electrical system to operate, the electricity must travel in a complete circuit. This simply means that current (power) from the positive (+) terminal of the battery must eventually return to the negative (−) terminal of the battery. Along the way, this current will travel through wires, fuses, switches and components. If, for any reason, the flow of current through the circuit is interrupted, the component fed by that circuit will cease to function properly.

Perhaps the easiest way to visualize a circuit is to think of connecting a light bulb (with two wires attached to it) to the battery - one wire attached to the negative (−) terminal of the battery and the other wire to the positive (+) terminal. With the two wires touching the battery terminals, the circuit would be complete and the light bulb would illuminate. Electricity would follow a path from the battery to the bulb and back to the battery. It's easy to see that with longer wires on our light bulb, it could be mounted anywhere. Further, one wire could be fitted with a switch so that the light could be turned on and off.

The normal automotive circuit differs from this simple example in two ways. First, instead of having a return wire from the bulb to the battery, the current travels through the frame of the vehicle. Since the negative (−) battery cable is attached to the frame (made of electrically conductive metal), the frame of the vehicle can serve as a ground wire to complete the circuit. Secondly, most automotive circuits contain multiple components which receive power from a single circuit. This lessens the amount of wire needed to power components on the vehicle.

HOW DOES ELECTRICITY WORK: THE WATER ANALOGY

Electricity is the flow of electrons - the subatomic particles that constitute the outer shell of an atom. Electrons spin in an orbit around the center core of an atom. The center core is comprised of protons (positive charge) and neutrons (neutral charge). Electrons have a negative charge and balance out the positive charge of the protons. When an outside force causes the number of electrons to unbalance the charge of the protons, the electrons will split off the atom and look for another atom to balance out. If this imbalance is kept up, electrons will continue to move and an electrical flow will exist.

Many people have been taught electrical theory using an analogy with water. In a comparison with water flowing through a pipe, the electrons would be the water and the wire is the pipe.

The flow of electricity can be measured much like the flow of water through a pipe. The unit of measurement used is amperes, frequently abbreviated as amps (a). You can compare amperage to the volume of water flowing through a pipe. When connected to a circuit, an ammeter will measure the actual amount of current flowing through the circuit. When relatively few electrons flow through a circuit, the amperage is low. When many electrons flow, the amperage is high.

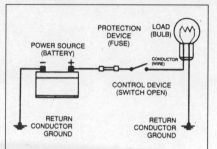

Fig. 1 This example illustrates a simple circuit. When the switch is closed, power from the positive (+) battery terminal flows through the fuse and the switch, and then to the light bulb. The light illuminates and the circuit is completed through the ground wire back to the negative (−) battery terminal. In reality, the two ground points shown in the illustration are attached to the metal frame of the vehicle, which completes the circuit back to the battery

Water pressure is measured in units such as pounds per square inch (psi); The electrical pressure is measured in units called volts (v). When a voltmeter is connected to a circuit, it is measuring the electrical pressure.

The actual flow of electricity depends not only on voltage and amperage, but also on the resistance of the circuit. The higher the resistance, the higher the force necessary to push the current through the circuit. The standard unit for measuring resistance is an ohm. Resistance in a circuit varies depending on the amount and type of components used in the circuit. The main factors which determine resistance are:

• Material - some materials have more resistance than others. Those with high resistance are said to be insulators. Rubber materials (or rubber-like plastics) are some of the most common insulators used in vehicles as they have a very high resistance to electricity. Very low resistance materials are said to be conductors. Copper wire is among the best conductors. Silver is actually a superior conductor to copper and is used in some relay contacts, but its high cost prohibits its use as common wiring. Most automotive wiring is made of copper.

• Size - the larger the wire size being used, the less resistance the wire will have. This is why components which use large amounts of electricity usually have large wires supplying current to them.

• Length - for a given thickness of wire, the longer the wire, the greater the resistance. The shorter the wire, the less the resistance. When determining the proper wire for a circuit, both size and length must be considered to design a circuit that can handle the current needs of the component.

• Temperature - with many materials, the higher the temperature, the greater the resistance (positive temperature coefficient). Some materials exhibit the opposite trait of lower resistance with higher temperatures (negative temperature coefficient). These principles are used in many of the sensors on the engine.

OHM'S LAW

There is a direct relationship between current, voltage and resistance. The relationship between current, voltage and resistance can be summed up by a statement known as Ohm's law.

Voltage (E) is equal to amperage (I) times resistance (R): $E = I \times R$

Other forms of the formula are $R = E/I$ and $I = E/R$

In each of these formulas, E is the voltage in volts, I is the current in amps and R is the resistance in ohms. The basic point to remember is that as the resistance of a circuit goes up, the amount of current that flows in the circuit will go down, if voltage remains the same.

The amount of work that the electricity can perform is expressed as power. The unit of power is the watt (w). The relationship between power, voltage and current is expressed as:

Power (w) is equal to amperage (I) times voltage (E): $W = I \times E$

This is only true for direct current (DC) circuits; The alternating current formula is a tad different, but since the electrical circuits in most vehicles are DC type, we need not get into AC circuit theory.

Electrical Components

POWER SOURCE

Power is supplied to the vehicle by two devices: The battery and the alternator. The battery supplies electrical power during starting or during periods when the current demand of the vehicle's electrical system exceeds the output capacity of the alternator. The alternator supplies electrical current when the engine is running. The alternator doesn't just supply the current needs of the vehicle, but it recharges the battery.

The Battery

In most modern vehicles, the battery is a lead/acid electrochemical device consisting of six 2 volt subsections (cells) connected in series, so that the unit is capable of producing approximately 12 volts of electrical pressure. Each subsection consists of a series of positive and negative plates held a short distance apart in a solution of sulfuric acid and water.

The two types of plates are of dissimilar metals. This sets up a chemical reaction, and it is this reaction which produces current flow from the battery when its positive and negative terminals are connected to an electrical load. The power removed from the battery is replaced by the alternator, restoring the battery to its original chemical state.

The Alternator

Alternators are devices that consist of coils of wires wound together making big electromagnets. One group of coils spins within another set and the interaction of the magnetic fields causes a current to flow. This current is then drawn off the coils and fed into the vehicle's electrical system.

CHASSIS ELECTRICAL 6-3

Fig. 2 Most vehicles use one or more fuse panels. This one is located on the left side of the instrument panel

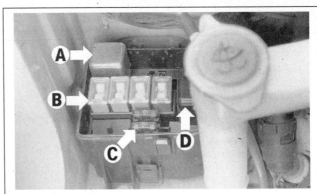

Fig. 3 The underhood fuse and relay panel usually contains fuses, relays, flashers and fusible links

GROUND

Two types of grounds are used in automotive electric circuits. Direct ground components are grounded to the frame through their mounting points. All other components use some sort of ground wire which is attached to the frame or chassis of the vehicle. The electrical current runs through the chassis of the vehicle and returns to the battery through the ground (–) cable; if you look, you'll see that the battery ground cable connects between the battery and the frame or chassis of the vehicle.

Note: It should be noted that a good percentage of electrical problems can be traced to bad grounds.

PROTECTIVE DEVICES

See Figure 2

It is possible for large surges of current to pass through the electrical system of your vehicle. If this surge of current were to reach the load in the circuit, the surge could burn it out or severely damage it. It can also overload the wiring, causing the harness to get hot and melt the insulation. To prevent this, fuses, circuit breakers and/or fusible links are connected into the supply wires of the electrical system. These items are nothing more than a built-in weak spot in the system. When an abnormal amount of current flows through the system, these protective devices work as follows to protect the circuit:

• Fuse - when an excessive electrical current passes through a fuse, the fuse "blows" (the conductor melts) and opens the circuit, preventing the passage of current.

• Circuit Breaker - a circuit breaker is basically a self-repairing fuse. It will open the circuit in the same fashion as a fuse, but when the surge subsides, the circuit breaker can be reset and does not need replacement.

• Fusible Link - a fusible link (fuse link or main link) is a short length of special, high temperature insulated wire that acts as a fuse. When an excessive electrical current passes through a fusible link, the thin gauge wire inside the link melts, creating an intentional open to protect the circuit. To repair the circuit, the link must be replaced. Some newer type fusible links are called "maxi fuses" and are housed in plug-in modules, which are simply replaced like a fuse, while older type fusible links must be cut and spliced if they melt. Since this link is very early in the electrical path, it's the first place to look if nothing on the vehicle works, yet the battery seems to be charged and is properly connected.

CAUTION:

Always replace fuses and circuit breakers with identically rated components. Under no circumstances should a component of higher or lower amperage rating be substituted.

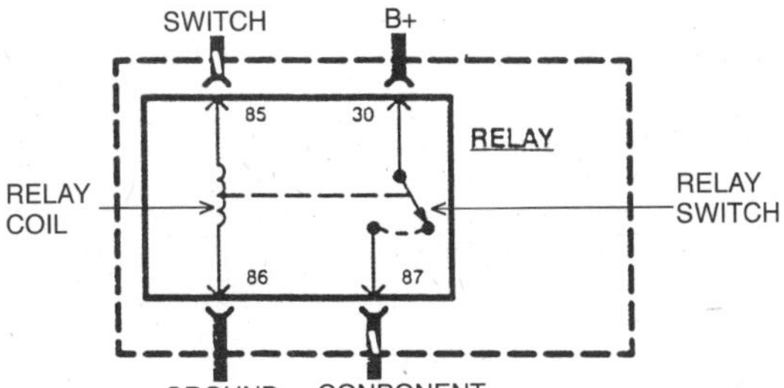

SWITCHES & RELAYS

See Figures 3 and 4

Switches are used in electrical circuits to control the passage of current. The most common use is to open and close circuits between the battery and the various electric devices in the system. Switches are rated according to the amount of amperage they can handle. If a sufficient amperage rated switch is not used in a circuit, the switch could overload and cause damage.

Some electrical components which require a large amount of current to operate use a special switch called a relay. Since these circuits carry a large amount of current, the thickness of the wire in the circuit is also greater. If this large wire were connected from the load to the control switch, the switch would have to carry the high amperage load and the fairing or dash would be twice as large to accommodate the increased size of the wiring harness. To prevent these problems, a relay is used.

Relays are composed of a coil and a set of contacts. When the coil has a current passed though it, a magnetic field is formed and this field causes the contacts to move together, completing the circuit. Most relays are normally open, preventing current from passing through the circuit, but they can take any electrical form depending on the job they are intended to do. Relays can be considered "remote control switches." They allow a smaller current to operate devices that require higher amperages. When a small current operates the coil, a larger current is allowed to pass by the contacts. Some common circuits which may use relays are the horn, headlights, starter, electric fuel pump and other high draw circuits.

LOAD

Every electrical circuit must include a "load" (something to use the electricity coming from the source). Without this load, the battery would attempt to deliver its entire power supply from one pole to another. This is called a "short circuit." All this electricity would take a shortcut to ground and cause a great amount of damage to other components in the circuit by developing a tremendous amount of heat. This condition could develop sufficient heat to melt the insulation on all the surrounding wires and reduce a multiple wire cable to a lump of plastic and copper.

Fig. 4 Relays are composed of a coil and a switch. These two components are linked together so that when one operates, the other operates at the same time. The large wires in the circuit are connected from the battery to one side of the relay switch (B+) and from the opposite side of the relay switch to the load (component). Smaller wires are connected from the relay coil to the control switch for the circuit and from the opposite side of the relay coil to ground

6-4 CHASSIS ELECTRICAL

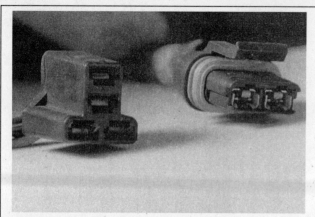

Fig. 5 Hard shell (left) and weatherproof (right) connectors have replaceable terminals

Fig. 6 Weatherproof connectors are most commonly used in the engine compartment or where the connector is exposed to the elements

WIRING & HARNESSES

The average vehicle contains meters and meters of wiring, with hundreds of individual connections. To protect the many wires from damage and to keep them from becoming a confusing tangle, they are organized into bundles, enclosed in plastic or taped together and called wiring harnesses. Different harnesses serve different parts of the vehicle. Individual wires are color coded to help trace them through a harness where sections are hidden from view.

Automotive wiring or circuit conductors can be either single strand wire, multi-strand wire or printed circuitry. Single strand wire has a solid metal core and is usually used inside such components as alternators, motors, relays and other devices. Multi-strand wire has a core made of many small strands of wire twisted together into a single conductor. Most of the wiring in an automotive electrical system is made up of multi-strand wire, either as a single conductor or grouped together in a harness. All wiring is color coded on the insulator, either as a solid color or as a colored wire with an identification stripe. A printed circuit is a thin film of copper or other conductor that is printed on an insulator backing. Occasionally, a printed circuit is sandwiched between two sheets of plastic for more protection and flexibility. A complete printed circuit, consisting of conductors, insulating material and connectors for lamps or other components is called a printed circuit board. Printed circuitry is used in place of individual wires or harnesses in places where space is limited, such as behind instrument panels.

Since automotive electrical systems are very sensitive to changes in resistance, the selection of properly sized wires is critical when systems are repaired. A loose or corroded connection or a replacement wire that is too small for the circuit will add extra resistance and an additional voltage drop to the circuit.

The wire gauge number is an expression of the cross-section area of the conductor. Vehicles from countries that use the metric system will typically describe the wire size as its cross-sectional area in square millimeters. In this method, the larger the wire, the greater the number. Another common system for expressing wire size is the American Wire Gauge (AWG) system. As gauge number increases, area decreases and the wire becomes smaller. An 18 gauge wire is smaller than a 4 gauge wire. A wire with a higher gauge number will carry less current than a wire with a lower gauge number. Gauge wire size refers to the size of the strands of the conductor, not the size of the complete wire with insulator. It is possible, therefore, to have two wires of the same gauge with different diameters because one may have thicker insulation than the other.

It is essential to understand how a circuit works before trying to figure out why it doesn't. An electrical schematic shows the electrical current paths when a circuit is operating properly. Schematics break the entire electrical system down into individual circuits. In a schematic, usually no attempt is made to represent wiring and components as they physically appear on the vehicle; switches and other components are shown as simply as possible. Face views of harness connectors show the cavity or terminal locations in all multi-pin connectors to help locate test points.

CONNECTORS

See Figures 5 and 6

Three types of connectors are commonly used in automotive applications - weatherproof, molded and hard shell.

- **Weatherproof** - these connectors are most commonly used where the connector is exposed to the elements. Terminals are protected against moisture and dirt by sealing rings which provide a weathertight seal. All repairs require the use of a special terminal and the tool required to service it. Unlike standard blade type terminals, these weatherproof terminals cannot be straightened once they are bent. Make certain that the connectors are properly seated and all of the sealing rings are in place when connecting leads.

- **Molded** - these connectors require complete replacement of the connector if found to be defective. This means splicing a new connector assembly into the harness. All splices should be soldered to insure proper contact. Use care when probing the connections or replacing terminals in them, as it is possible to create a short circuit between opposite terminals. If this happens to the wrong terminal pair, it is possible to damage certain components. Always use jumper wires between connectors for circuit checking and NEVER probe through weatherproof seals.

- **Hard Shell** - unlike molded connectors, the terminal contacts in hard-shell connectors can be replaced. Replacement usually involves the use of a special terminal removal tool that depresses the locking tangs (barbs) on the connector terminal and allows the connector to be removed from the rear of the shell. The connector shell should be replaced if it shows any evidence of burning, melting, cracks, or breaks. Replace individual terminals that are burnt, corroded, distorted or loose.

Test Equipment

Pinpointing the exact cause of trouble in an electrical circuit is most times accomplished by the use of special test equipment. The following describes different types of commonly used test equipment and briefly explains how to use them in diagnosis. In addition to the information covered below, the tool manufacturer's instructions booklet (provided with the tester) should be read and clearly understood before attempting any test procedures.

JUMPER WIRES

CAUTION:

Never use jumper wires made from a thinner gauge wire than the circuit being tested. If the jumper wire is of too small a gauge, it may overheat and possibly melt. Never use jumpers to bypass high resistance loads in a circuit. Bypassing resistances, in effect, creates a short circuit. This may, in turn, cause damage and fire. Jumper wires should only be used to bypass lengths of wire or to simulate switches.

Jumper wires are simple, yet extremely valuable, pieces of test equipment. They are basically test wires which are used to bypass sections of a circuit. Although jumper wires can be purchased, they are usually fabricated from lengths of standard automotive wire and whatever type of connector (alligator clip, spade connector or pin connector) that is required for the particular application being tested. In cramped, hard-to-reach areas, it is advisable to have insulated boots over the jumper wire terminals in order to prevent accidental grounding. It is also advisable to include a standard automotive fuse in any jumper wire. This is commonly referred to as a "fused jumper". By inserting an in-line fuse holder between a set of test leads, a fused jumper wire can be used for bypassing open circuits. Use a 5 amp fuse to provide protection against voltage spikes.

CHASSIS ELECTRICAL 6-5

Jumper wires are used primarily to locate open electrical circuits, on either the ground (–) side of the circuit or on the power (+) side. If an electrical component fails to operate, connect the jumper wire between the component and a good ground. If the component operates only with the jumper installed, the ground circuit is open. If the ground circuit is good, but the component does not operate, the circuit between the power feed and component may be open. By moving the jumper wire successively back from the component toward the power source, you can isolate the area of the circuit where the open is located. When the component stops functioning, or the power is cut off, the open is in the segment of wire between the jumper and the point previously tested.

You can sometimes connect the jumper wire directly from the battery to the "hot" terminal of the component, but first make sure the component uses 12 volts in operation. Some electrical components, such as fuel injectors or sensors, are designed to operate on about 4 to 5 volts, and running 12 volts directly to these components will cause damage.

TEST LIGHTS

See Figure 7

The test light is used to check circuits and components while electrical current is flowing through them. It is used for voltage and ground tests. To use a 12 volt test light, connect the ground clip to a good ground and probe wherever necessary with the pick. The test light will illuminate when voltage is detected. This does not necessarily mean that 12 volts (or any particular amount of voltage) is present; it only means that some voltage is present. It is advisable before using the test light to touch its ground clip and probe across the battery posts or terminals to make sure the light is operating properly.

WARNING:

Do not use a test light to probe electronic ignition, spark plug or coil wires. Never use a pick-type test light to probe wiring on computer controlled systems unless specifically instructed to do so. Any wire insulation that is pierced by the test light probe should be taped and sealed with silicone after testing.

Like the jumper wire, the 12 volt test light is used to isolate opens in circuits. But, whereas the jumper wire is used to bypass the open to operate the load, the 12 volt test light is used to locate the presence of voltage in a circuit. If the test light illuminates, there is power up to that point in the circuit; if the test light does not illuminate, there is an open circuit (no power). Move the test light in successive steps back toward the power source until the light in the handle illuminates. The open is between the probe and a point which was previously probed.

The self-powered test light is similar in design to the 12 volt test light, but contains a 1.5 volt penlight battery in the handle. It is most often used in place of a multimeter to check for open or short circuits when power is isolated from the circuit (continuity test).

The battery in a self-powered test light does not provide much current. A weak battery may not provide enough power to illuminate the test light even when a complete circuit is made (especially if there is high resistance in the circuit). Always make sure that the test battery is strong. To check the battery, briefly

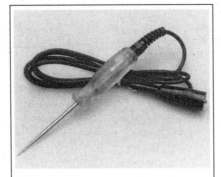

Fig. 7 A 12 volt test light is used to detect the presence of voltage in a circuit

touch the ground clip to the probe; if the light glows brightly, the battery is strong enough for testing.

Note: A self-powered test light should not be used on any computer controlled system or component. The small amount of electricity transmitted by the test light is enough to damage many electronic automotive components.

MULTIMETERS

Multimeters are an extremely useful tool for troubleshooting electrical problems. They can be purchased in either analog or digital form and have a price range to suit any budget. A multimeter is a voltmeter, ammeter and ohmmeter (along with other features) combined into one instrument. It is often used when testing solid state circuits because of its high input impedance (usually 10 mega-ohms or more). A brief description of the multimeter main test functions follows:

- Voltmeter - the voltmeter is used to measure voltage at any point in a circuit, or to measure the voltage drop across any part of a circuit. Voltmeters usually have various scales and a selector switch to allow the reading of different voltage ranges. The voltmeter has a positive and a negative lead. To avoid damage to the meter, always connect the negative lead to the negative (–) side of the circuit (to ground or nearest the ground side of the circuit) and connect the positive lead to the positive (+) side of the circuit (to the power source or the nearest power source). Note that the negative voltmeter lead will always be black and that the positive voltmeter will always be some color other than black (usually red).
- Ohmmeter - the ohmmeter is designed to read resistance (measured in ohms) in a circuit or component. Most ohmmeters will have a selector switch which permits the measurement of different ranges of resistance (usually the selector switch allows the multiplication of the meter reading by 10, 100, 1,000 and 10,000). Some ohmmeters are "auto-ranging" which means the meter itself will determine which scale to use. Since the meters are powered by an internal battery, the ohmmeter can be used like a self-powered test light. When the ohmmeter is connected, current from the ohmmeter flows through the circuit or component being tested. Since the ohmmeter's internal resistance and voltage are known values, the amount of current flow through the meter depends on the resistance of the circuit or component being tested. The ohmmeter can also be used to perform a continuity test for suspected open circuits. In using the meter for making continuity checks, do not be concerned with the actual resistance readings. Zero resistance, or any ohm reading, indicates continuity in the circuit. Infinite resistance indicates an opening in the circuit. A high resistance reading where there should be none indicates a problem in the circuit. Checks for short circuits are made in the same manner as checks for open circuits, except that the circuit must be isolated from both power and normal ground. Infinite resistance indicates no continuity, while zero resistance indicates a dead short.

WARNING:

Never use an ohmmeter to check the resistance of a component or wire while there is voltage applied to the circuit.

- Ammeter - an ammeter measures the amount of current flowing through a circuit in units called amperes or amps. At normal operating voltage, most circuits have a characteristic amount of amperes, called "current draw" which can be measured using an ammeter. By referring to a specified current draw rating, then measuring the amperes and comparing the two values, one can determine what is happening within the circuit to aid in diagnosis. An open circuit, for example, will not allow any current to flow, so the ammeter reading will be zero. A damaged component or circuit will have an increased current draw, so the reading will be high. The ammeter is always connected in series with the circuit being tested. All of the current that normally flows through the circuit must also flow through the ammeter; if there is any other path for the current to follow, the ammeter reading will not be accurate. The ammeter itself has very little resistance to current flow and, therefore, will not affect the circuit, but it will measure current draw only when the circuit is closed and electricity is flowing. Excessive current draw can blow fuses and drain the battery, while a reduced current draw can cause motors to run slowly, lights to dim and other components to not operate properly.

Troubleshooting Electrical Systems

When diagnosing a specific problem, organized troubleshooting is a must. The complexity of a modern automotive vehicle demands that you approach any problem in a logical, organized manner. There are certain troubleshooting techniques, however, which are standard:

- Establish when the problem occurs. Does the problem appear only under certain conditions? Were there any noises, odors or other unusual symptoms? Isolate the problem area. To do this, make some simple tests and observations, then eliminate the systems that are working properly. Check for obvious problems, such as broken wires and loose or dirty connections. Always check the obvious before assuming something complicated is the cause.
- Test for problems systematically to determine the cause once the problem area is isolated. Are all the components functioning properly? Is there power going to electrical switches and motors. Performing careful, systematic checks will often turn up most causes on the first inspection, without wasting time checking components that have little or no relationship to the problem.

6-6 CHASSIS ELECTRICAL

Fig. 8 The infinite reading on this multimeter indicates that the circuit is open

Fig. 9 This voltage drop test revealed high resistance (low voltage) in the circuit

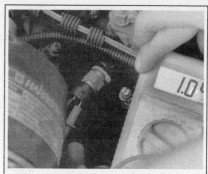

Fig. 10 Checking the resistance of a coolant temperature sensor with an ohmmeter. Reading is 1.04 k-ohms

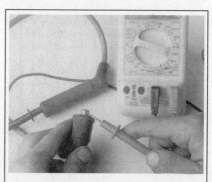

Fig. 11 Spark plug wires can be checked for excessive resistance using an ohmmeter

• Test all repairs after the work is done to make sure that the problem is fixed. Some causes can be traced to more than one component, so a careful verification of repair work is important in order to pick up additional malfunctions that may cause a problem to reappear or a different problem to arise. A blown fuse, for example, is a simple problem that may require more than another fuse to repair. If you don't look for a problem that caused a fuse to blow, a shorted wire (for example) may go undetected.

Experience has shown that most problems tend to be the result of a fairly simple and obvious cause, such as loose or corroded connectors, bad grounds or damaged wire insulation which causes a short. This makes careful visual inspection of components during testing essential to quick and accurate troubleshooting.

Testing

OPEN CIRCUITS

See Figure 8

This test already assumes the existence of an open in the circuit and it is used to help locate the open portion.

1. Isolate the circuit from power and ground.
2. Connect the self-powered test light or ohmmeter ground clip to the ground side of the circuit and probe sections of the circuit sequentially.
3. If the light is out or there is infinite resistance, the open is between the probe and the circuit ground.
4. If the light is on or the meter shows continuity, the open is between the probe and the end of the circuit toward the power source.

SHORT CIRCUITS

Note: Never use a self-powered test light to perform checks for opens or shorts when power is applied to the circuit under test. The test light can be damaged by outside power.

1. Isolate the circuit from power and ground.
2. Connect the self-powered test light or ohmmeter ground clip to a good ground and probe any easy-to-reach point in the circuit.
3. If the light comes on or there is continuity, there is a short somewhere in the circuit.
4. To isolate the short, probe a test point at either end of the isolated circuit (the light should be on or the meter should indicate continuity).

5. Leave the test light probe engaged and sequentially open connectors or switches, remove parts, etc. until the light goes out or continuity is broken.
6. When the light goes out, the short is between the last two circuit components which were opened.

VOLTAGE

This test determines voltage available from the battery and should be the first step in any electrical troubleshooting procedure after visual inspection. Many electrical problems, especially on computer controlled systems, can be caused by a low state of charge in the battery. Excessive corrosion at the battery cable terminals can cause poor contact that will prevent proper charging and full battery current flow.

1. Set the voltmeter selector switch to the 20V position.
2. Connect the multimeter negative lead to the battery's negative (–) post or terminal and the positive lead to the battery's positive (+) post or terminal.
3. Turn the ignition switch ON to provide a load.
4. A well charged battery should register over 12 volts. If the meter reads below 11.5 volts, the battery power may be insufficient to operate the electrical system properly.

VOLTAGE DROP

See Figure 9

When current flows through a load, the voltage beyond the load drops. This voltage drop is due to the resistance created by the load and also by small resistances created by corrosion at the connectors and damaged insulation on the wires. The maximum allowable voltage drop under load is critical, especially if there is more than one load in the circuit, since all voltage drops are cumulative.

1. Set the voltmeter selector switch to the 20 volt position.
2. Connect the multimeter negative lead to a good ground.
3. Operate the circuit and check the voltage prior to the first component (load).
4. There should be little or no voltage drop in the circuit prior to the first component. If a voltage drop exists, the wire or connectors in the circuit are suspect.
5. While operating the first component in the circuit, probe the ground side of the component with the positive meter lead and observe the voltage readings.

A small voltage drop should be noticed. This voltage drop is caused by the resistance of the component.

6. Repeat the test for each component (load) down the circuit.
7. If a large voltage drop is noticed, the preceding component, wire or connector is suspect.

RESISTANCE

See Figures 10 and 11

WARNING:

Never use an ohmmeter with power applied to the circuit. The ohmmeter is designed to operate on its own power supply. The normal 12 volt electrical system voltage could damage the meter!

1. Isolate the circuit from the vehicle's power source.
2. Ensure that the ignition key is OFF when disconnecting any components or the battery.
3. Where necessary, also isolate at least one side of the circuit to be checked, in order to avoid reading parallel resistances. Parallel circuit resistances will always give a lower reading than the actual resistance of either of the branches.
4. Connect the meter leads to both sides of the circuit (wire or component) and read the actual measured ohms on the meter scale. Make sure the selector switch is set to the proper ohm scale for the circuit being tested, to avoid misreading the ohmmeter test value.

CHASSIS ELECTRICAL 6-7

Wire and Connector Repair

Almost anyone can replace damaged wires, as long as the proper tools and parts are available. Wire and terminals are available to fit almost any need. Even the specialized weatherproof, molded and hard shell connectors are now available from aftermarket suppliers.

Be sure the ends of all the wires are fitted with the proper terminal hardware and connectors. Wrapping a wire around a stud is never a permanent solution and will only cause trouble later. Replace wires one at a time to avoid confusion. Always route wires exactly the same as the factory.

Note: If connector repair is necessary, only attempt it if you have the proper tools. Weatherproof and hard shell connectors require special tools to release the pins inside the connector. Attempting to repair these connectors with conventional hand tools will damage them.

BATTERY CABLES

Disconnecting the Cables

When working on any electrical component on the vehicle, it is always a good idea to disconnect the negative (–) battery cable (some diesel-engined vehicles have two batteries, and therefore two cables which must be disconnected). This will prevent potential damage to many sensitive electrical components such as the Powertrain Control Module (PCM), radio, alternator, etc.

Note: Any time you disengage the battery cables, it is recommended that you disconnect the negative (–) battery cable first. This will prevent your accidentally grounding the positive (+) terminal to the body of the vehicle when disconnecting it, thereby preventing damage to the above mentioned components.

Before you disconnect the cable(s), first turn the ignition to the OFF position. This will prevent a draw on the battery which could cause arcing (electricity trying to ground itself to the body of a vehicle, just like a spark plug jumping the gap) and, of course, damaging some components such as the alternator diodes. When the battery cable(s) are reconnected (negative cable last), be sure to check that your lights, windshield wipers and other electrically operated safety components are all working correctly. If your vehicle contains an Electronically Tuned Radio (ETR), don't forget to also reset your radio stations. Ditto for the clock.

WARNING:

Check polarity before reconnecting battery cables. On batteries with top post connectors, note that the posts are not the same size.

Anytime the battery cables have been disconnected and then reconnected, some abnormal drive symptoms could occur. The is due to the PCM losing the memory voltage and its learned adaptive strategy. The vehicle will need to be driven for 10 miles (16 km) or more until the PCM relearns its adaptive strategy, and acclimates the engine and transmission functions to your driving style.

AIR BAG (SUPPLEMENTAL RESTRAINT SYSTEM)

General Information

CAUTION:

This system is a sensitive, complex electro-mechanical unit. Before attempting to diagnose or service, you must first disconnect the negative battery cable and wait at least 2 minutes for the system capacitor to discharge. Failure to do so could result in accidental deployment which could cause personal injury.

SYSTEM OPERATION

The air bag or Supplemental Restraint System (SRS) is a safety device designed to be used in conjunction with the seat belt. Its purpose is to help protect the driver in a frontal impact exceeding a certain set limit. The system consists of the air bag module, three impact sensors, a clockspring and a dedicated air bag control module.

The air bag is a fabric bag or balloon with an explosive inflator unit attached. The system employs impact sensors and a safing sensor, as well as an inflator circuit and control module.

When the control unit receives the sensor signals, power is supplied to the inflator circuit, either from the battery or backup system. A small heater causes a chemical reaction in the igniter; the non-toxic gas from the chemical mixture expands very rapidly (in milliseconds), filling the bag and forcing it through the cover pad. Since all this is happening very rapidly, the expanding bag should reach the occupant before he/she reaches the steering wheel/dashboard during a frontal collision. The chemical reaction is complete by the time the air bag is fully inflated; as the occupant hits the bag, the gas is allowed to escape slowly through vents in the back of the bag.

SYSTEM COMPONENTS

Air Bag Module

The driver's air bag module is mounted directly to the steering wheel beneath a protective cover. Under the air bag module protective cover, the air bag cushion and its supporting components are contained. The air bag module contains a housing to which the cushion and inflator are attached and sealed. The air bag module is non-repairable. If it is dropped or damaged, it must be replaced.

The inflator assembly is mounted to the back of the module. The inflator seals the hole in the air bag cushion so it can discharge the gas it produces directly into the cushion when supplied with the proper electrical signal. Upon deployment, the protective cover will split horizontally.

The passenger's side air bag door on the instrument panel above the glove box is the most visible part of the passenger side air bag system. This system includes an extruded aluminum housing within which the cushion and inflator are mounted.

The module cannot be repaired. Following a deployment, the air bag module and instrument panel assembly must be replaced.

A passenger side air bag on/off switch is fitted to all models except the quad cab.

Impact Sensors

The three impact sensors used in the SRS system verify the direction and severity of an impact. One of the sensors is called the safing sensor. It is located in the Air Bag Control Module (ACM), which is mounted to a bracket under the instrument panel, on top of the floor pan transmission tunnel. The other two are impact sensors and are mounted on the left and right inner fender extension panels behind the grille. The sensors are calibrated for the particular vehicle that they serve.

The impact sensors are threshold-sensitive switches that complete an electrical circuit when an impact provides a sufficient deceleration force to close the switch. The safing sensor is an accelerometer that senses the rate of deceleration. The microprocessor in the ACM monitors the sensor signals. A pre-programmed decision algorithm in the microprocessor determines when the deceleration rate indicates an impact that is severe enough to require air bag system protection.

The two impact sensors are available for service replacement. The safing sensor is only serviced as part of the ACM.

Clockspring

See Figure 12

The clockspring is mounted on the steering column behind the steering wheel. Its purpose is to maintain a continuous electrical circuit between the

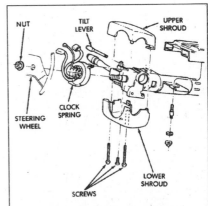

Fig. 12 The clockspring assembly is mounted to the end of the column, behind the steering wheel

6-8 CHASSIS ELECTRICAL

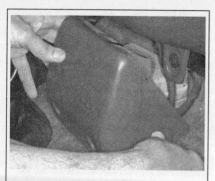

Fig. 13 Removing the ACM cover

Fig. 14 The air bag control module is located on the floorboard beneath the dash

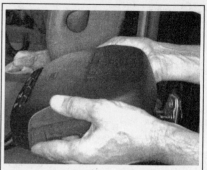

Fig. 15 Always carry undeployed air bags with the padded side facing away from your body

wiring harness and the driver's side air bag module. This assembly consists of a flat, ribbon-like electrically conductive tape that winds and unwinds with the steering wheel rotation.

Air Bag Control Module (ACM)
See Figures 13 and 14

The air bag control module (ACM) is mounted on the transmission tunnel beneath the dashboard.

The ACM contains the safing sensor, and a microprocessor that monitors the air bag system to determine readiness. It also monitors the impact sensors to determine when the proper conditions exist to provide the electrical signal that deploys the air bag. The ACM contains On-Board Diagnostics (OBD), and will light the air bag warning lamp on the instrument panel if a (monitored) air bag system fault occurs. If the light does not come on, does not go out or comes on when driving, the system must be diagnosed and repaired by your dealer or reputable shop. The system is NOT repairable at home.

The ACM also contains an energy-storage capacitor. The capacitor stores enough electrical energy to deploy the air bag for up to two minutes following a battery disconnect or failure. The purpose of the capacitor is to provide air bag system protection in a severe secondary impact if the initial impact somehow damaged or disconnected the battery, but did not deploy the air bag.

This unit is not serviceable and must be replaced if damaged or defective.

SERVICE PRECAUTIONS

Note: This manual does not cover SRS repairs or replacement as such work should be left to a trained professional. The following precautions then, are only to inform the do-it-yourselfer and give him or her a greater appreciation for the system when required to work in proximity to SRS components.

- Replace air bag system components only with parts specified in the Chrysler Mopar parts catalog. Substitute parts may appear interchangeable, but internal differences may result in inferior occupant protection.
- The fasteners, screws and bolts have special coatings and are specially designed for use in the air bag system. Never replace them with substitutes. Always use the correct replacement fasteners as supplied by the Chrysler Mopar parts catalog.
- No SRS component should be used if it shows any sign of being dropped, dented or otherwise damaged.
- SRS components formerly installed in another vehicle should never be used. Only new components should be installed.
- Whenever working on SRS components (except for electrical inspections), always disconnect the negative battery cable and then wait at least two minutes before beginning (and taking other precautions as necessary). Once the air bag has been deployed, replace the SRS unit.
- Whenever the ignition switch is ON or has been turned OFF for less than two minutes, be careful not to bump the SRS unit; the air bag could accidentally deploy and do damage or cause injuries.
- Do not try to take apart the air bag assembly. Once deployed, it cannot be re-used or repaired.
- For temporary storage of the air bag assembly while servicing the vehicle, place it with the pad surface UP. Store the air bag assembly on a secure, clean, flat surface away from heat, oil, grease, water or detergent.

CAUTION:

If the air bag is stored face down, it could cause serious injury and damage if it accidentally deploys.

- Take extra care when doing paint or body work near the air bag assembly and keep heat guns, welding and spray equipment away from the air bag assembly.
- Make sure SRS wiring harnesses are not pinched, and that all ground contacts are clean. Poor grounding can cause intermittent problems that are difficult to diagnose.

DISARMING THE SYSTEM

1. First read the system precautions.
2. Disconnect and isolate the negative battery cable(s).
3. If the air bag module is undeployed, wait at least two minutes for the system capacitor to discharge.

ARMING THE SYSTEM

Assuming that the system components (air bag control module, sensors, air bag, etc.) are installed correctly and are in good working order, the system is armed whenever the battery's positive and negative battery cables are connected.

WARNING:

If you have disarmed the air bag system for any reason, and are re-arming the system, make sure no one is in the vehicle (as an added safety measure), then connect the negative battery cable(s)1.

HANDLING A LIVE MODULE

See Figure 15

At no time should any source of electricity be permitted near the inflator on the back of the module. When carrying a live module (such as when removing the steering wheel), the trim cover should be pointed away from the body to minimize injury in the event of accidental deployment. In addition, if the module is placed on a bench or other surface, the plastic trim cover should be face up to minimize movement in case of accidental deployment.

When handling a steering column with an air bag module attached, never place the column on the floor or other surface with the steering wheel or module face down.

DEPLOYED MODULE

The vehicle interior may contain a very small amount of sodium hydroxide powder, a by-product of air bag deployment. Since this powder can irritate the skin, eyes, nose or throat, be sure to wear safety glasses, rubber gloves and long sleeves during cleanup.

If you find that the cleanup is irritating your skin,

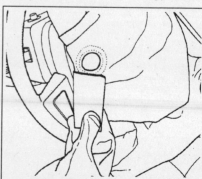

Fig. 16 Applying tape to the vent holes in a deployed module

CHASSIS ELECTRICAL 6-9

run cool water over the affected area. Also, if you experience nasal or throat irritation, exit the vehicle for fresh air until the irritation ceases. If irritation continues, see a physician.

Clean-Up Procedure
See Figure 16

Begin the clean-up by putting tape over the two air bag exhaust vents so that no additional powder will find its way into the vehicle interior. Then, remove the air bag and air bag module from the vehicle.

Use a vacuum cleaner to remove any residual powder from the vehicle interior. Work from the outside in so that you avoid kneeling or sitting in a unclean area.

Be sure to vacuum the heater and A/C outlets as well; in fact, it's a good idea to run the blower on LOW and to vacuum up any powder expelled from the plenum. You may need to vacuum the interior of the vehicle a second time to recover all of the powder.

Servicing a Deployed Air Bag

After an air bag has been deployed, the air bag module and clockspring must be replaced because they cannot be reused. Other air bag system components must also be replaced if damaged.

HEATING AND AIR CONDITIONING

Heater - A/C Housing

REMOVAL & INSTALLATION

See Figure 17

On Ram, the heater - A/C housing contains the heater core. On Dakota and Durango, this housing contains the heater core and the blower motor. For access to these components, the housing must be removed from the vehicle. To remove the housing, the dash panel must be removed and the A/C lines disconnected.

Note: According to law, refrigerant must be captured and reused which requires the use of special equipment and training. This procedure must be performed by a certified MVAC technician.

Dakota, Durango
See Figure 18

1. Remove the instrument panel using the following steps
2. Disconnect the battery negative cable.
3. Remove the trim from the sills of both doors.
4. Remove the trim from the cowl side inner panels on both sides.
5. Remove the steering column opening cover.
6. Remove the inside hood release latch.
7. Disconnect the overdrive lockout switch, if fitted.
8. Remove the steering column. Do not remove the air bag, steering wheel or switches. Be sure the steering wheel is locked and secured from rotation to prevent loss of clockspring centering.
9. Detach the parking brake cable and switch.
10. Disconnect all accessible wiring harness connectors, ground wires, headlight wiring, sound system, etc., from the instrument panel.
11. Disconnect the vacuum harness connector on the left side of the heater - A/C housing.
12. Remove the center support bracket from the instrument panel.
13. Remove the ACM ground line. Disconnect the ACM harness.
14. Remove the glove box.
15. Disconnect the radio antenna.
16. Disconnect the blower motor wiring harness.
17. Disconnect the speakers.
18. Disconnect the radio ground wire.
19. Loosen the right and left instrument panel cowl side roll-down bracket screw about a quarter of an inch.
20. Remove the screws that secure the top of the instrument panel to the top of the dash panel, removing the center screw last.
21. Pull the lower instrument panel rearward until the right and left cowl side roll-down bracket screws are in the roll-down slot position of both brackets.
22. Roll down the instrument panel and install a temporary hook in the center hole. Support the panel.
23. Disconnect the remaining wiring.
24. Disconnect remaining hoses.
25. Remove the panel.
26. If the vehicle is equipped with A/C, recover the refrigerant. See "Note" above.
27. Disconnect the liquid line refrigerant line fitting from the evaporator inlet tube.
28. Disconnect the accumulator inlet tube.
29. Drain the cooling system.
30. Disconnect the heater hoses from the heater core tubes.
31. Remove the nuts from the heater - A/C mounting studs on the engine compartment side of the dash panel.
32. Remove the nut that secures the heater - A/C housing mounting brace to the stud on the passenger compartment side of the dash board.
33. Pull the housing rearward far enough for the mounting studs and evaporator condensate drain tube to clear the dash panel holes.
34. Remove the housing from the vehicle.
35. Installation is the reverse of removal.

Ram

1. Remove the instrument panel using the following steps
2. Disconnect the battery negative cable(s).
3. Remove the ACM from the floor.
4. Remove the trim from the cowl side inner panels on both sides.
5. Remove the steering column opening cover.
6. Remove the inside hood release latch.
7. Disconnect the driver's air bag module wire harness connector.
8. Disconnect the overdrive lockout switch, if fitted.
9. Detach the parking brake cable and switch.
10. Remove the steering column. Do not remove the air bag, steering wheel or switches. Be sure the steering wheel is locked and secured from rotation to prevent loss of clockspring centering.
11. Disconnect all accessible wiring harness connectors, ground wires, headlight wiring, sound system, etc., from the instrument panel.
12. Disconnect the vacuum harness connector on the left side of the heater - A/C housing.
13. Disconnect the radio antenna.
14. Loosen the right and left instrument panel cowl side roll-down bracket screw about half an inch.
15. Remove the screws that secure the top of the instrument panel to the top of the dash panel, removing the center screw last.
16. Roll down the instrument panel and install a

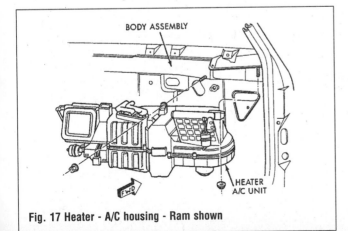

Fig. 17 Heater - A/C housing - Ram shown

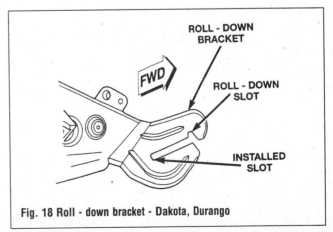

Fig. 18 Roll - down bracket - Dakota, Durango

6-10 CHASSIS ELECTRICAL

temporary hook in the center hole. Support the panel.
17. Disconnect the remaining wiring.
18. Disconnect remaining hoses.
19. Remove the panel.
20. If the vehicle is equipped with A/C, recover the refrigerant. See "Note" above.
21. Disconnect the liquid line refrigerant line fitting from the evaporator inlet tube.
22. Disconnect the accumulator.
23. Drain the cooling system.
24. Disconnect the heater hoses from the heater core tubes.
25. Remove the powertrain control module from the dash panel. It is not necessary to disconnect the wiring.
26. Remove the nuts from the heater - A/C mounting studs on the engine compartment side of the dash panel.
27. Remove the nuts that secure the heater - A/C housing mounting brace to the studs on the passenger compartment side of the dash board.
28. Pull the housing rearward far enough for the mounting studs and evaporator condensate drain tube to clear the dash panel holes.
29. Remove the housing from the vehicle.
30. Installation is the reverse of removal.

Blower Motor

REMOVAL & INSTALLATION

Dakota, Durango

Note: Removal of the blower motor on these models requires that the A/C lines be disconnected. According to law, refrigerant must be captured and reused which requires the use of special equipment and training. This procedure must be performed by a certified MVAC technician.

1. Disconnect the battery negative cable.
2. Remove the instrument panel.
3. Recover the refrigerant. See "Note" above.
4. Disconnect the liquid line refrigerant line fitting from the evaporator inlet tube.
5. Disconnect the accumulator inlet tube.
6. Drain the cooling system.
7. Disconnect the heater hoses from the heater core tubes.
8. Remove the four nuts from the heater - A/C mounting studs on the engine compartment side of the dash panel.
9. Remove the nut that secures the heater - A/C housing mounting brace to the stud on the passenger compartment side of the dash board.
10. Pull the housing rearward far enough for the mounting studs and evaporator condensate drain tube to clear the dash panel holes.
11. Remove the housing from the vehicle.
12. Remove the three screws that secure the motor to the housing.
13. Remove the wheel, if desired, by removing the clip from the motor shaft.
14. Installation is the reverse of removal. Note the following points:
    a. When installing the wheel onto the shaft, be sure to line up the flats.
    b. The ears of the retaining clip must be on the shaft flat.
    c. Be sure the blower motor seal is in place.

Fig. 19 Removing the blower motor - Ram

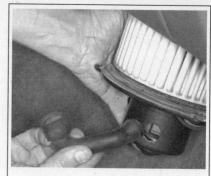

Fig. 20 Blower motor cooling tube - Ram

Ram

See Figures 19 and 20

The blower motor is located on the passenger's side beneath the dashboard.
1. Disconnect the battery negative cable(s).
2. Disconnect the blower wiring after disengaging the wiring harness from the retainer clip.
3. Remove the blower motor cooling tube.
4. Remove the hardware securing the blower motor and wheel assembly to the heater - A/C housing.
5. Remove the blower motor assembly.
6. Remove the wheel, if desired, by removing the clip from the motor shaft.
7. Installation is the reverse of removal. Note the following points:
    a. When installing the wheel onto the shaft, be sure to line up the flats.
    b. The ears of the retaining clip must be on the shaft flat.
    c. Be sure the blower motor seal is in place.

Heater Core

REMOVAL & INSTALLATION

See Figure 21

1. Disconnect the battery negative cable.
2. Remove the instrument panel from the vehicle.
3. If equipped with air conditioning, recover the refrigerant.
4. Disconnect the liquid line refrigerant line fitting from the evaporator inlet tube.
5. Ram: remove the accumulator.
6. Dakota, Durango: disconnect the accumulator inlet tube.
7. Drain the cooling system.
8. Disconnect the heater hoses from the heater core tubes.
9. Ram: remove the powertrain control module from the dash panel. Disconnecting wiring is not necessary.
10. Remove the four nuts from the heater - A/C mounting studs on the engine compartment side of the dash panel.
11. Remove the nut(s) that secure(s) the heater - A/C housing mounting brace to the stud(s) on the passenger compartment side of the dash board.
12. Pull the housing rearward far enough for the mounting studs and evaporator condensate drain tube to clear the dash panel holes.

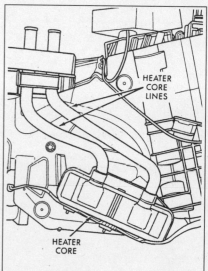

Fig. 21 Heater core removal/installation

13. Remove the housing from the vehicle.
14. Lift the heater core out of the housing.
15. Installation is the reverse of removal.

Air Conditioning Components

Repair or service of air conditioning components is not covered by this manual, because of the risk of personal injury or death, and because of the legal ramifications of servicing these components without the proper EPA certification and experience. Cost, personal injury or death, environmental damage, and legal considerations (such as the fact that it is a federal crime to vent refrigerant into the atmosphere), dictate that the A/C components on your vehicle should be serviced only by a Motor Vehicle Air Conditioning (MVAC) trained, and EPA certified automotive technician.

Temperature Control Cable

REMOVAL & INSTALLATION

See Figure 22

On some models, the heating system uses a cable-controlled door for air management.
1. Disconnect the battery negative cable(s).
2. Roll down the instrument panel assembly, but do not remove it.

CHASSIS ELECTRICAL 6-11

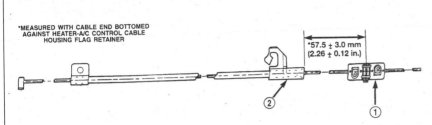

Fig. 22 Temperature control cable self-adjuster clip (1) adjustment relative to the flag retainer-housing end (2)

Control Panel

REMOVAL & INSTALLATION

See Figures 23 and 24

1. Disconnect the battery negative cable(s).
2. Reach under the instrument panel on the driver's side near the transmission tunnel and unplug the heater - A/C control-to-housing vacuum harness connector.
3. Disengage the retainer on the heater - A/C control half of the vacuum harness from the hole in the center of the distribution duct.
4. Remove the cluster bezel from the instrument panel.
5. Remove the screws that secure the control panel to the instrument panel.
6. Pull out the assembly far enough to access the connections at the back of the control.
7. Unplug the wiring.
8. Disconnect the cable, if fitted.
9. If a vacuum unit is fitted, remove the two stamped nuts that secure the vacuum harness connector and unplug the connector.
10. Installation is the reverse of removal.

3. Disconnect the cable from the control panel.
4. Disconnect the cable housing flag retainer from the receptacle on the top of the heater - A/C housing.
5. Pull the control cable core self-adjuster clip off the pin on the end of the blend-air door lever.
6. Remove the cable.

To install:

7. Be sure that the self-adjuster clip is properly positioned. See illustration. The measurement is taken from the end of the flag retainer on the housing end of the cable to the self - adjuster clip. Reposition the clip as needed to achieve the proper distance.
8. Connect the cable.
9. Route the cable making sure it is not kinked or pinched.
10. Push the cable clip onto the pin on the end of the door.
11. Snap the flag retainer into the receiver on top of the housing.
12. The remainder of the procedure is the reverse of disassembly.

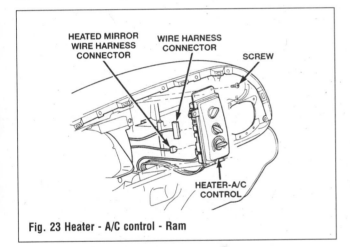

Fig. 23 Heater - A/C control - Ram

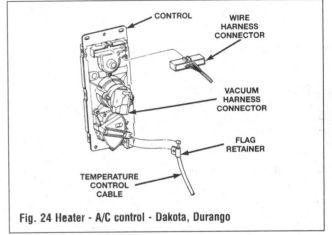

Fig. 24 Heater - A/C control - Dakota, Durango

CRUISE CONTROL

General Information

See Figures 25, 26, 27, 28, 29 and 30

The cruise (speed) control system is electrically actuated and vacuum operated. Two separate switch pods on the steering column cover all necessary functions: ON , OFF , RESUME , ACCELERATE , SET DECEL , and CANCEL . To operate the speed control system, consult your owner's manual.

The speed control system includes the multi-function switches on the steering column, the servo and solenoids, the speed control cable, and a vacuum reservoir. When the speed control servo receives a SET input from the switch, the vent solenoid is actu-

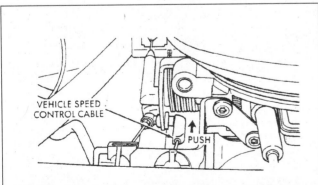

Fig. 25 The speed control cable connection at the throttle body

Fig. 26 The vacuum reservoir is attached on the passenger side beneath the grille or on the firewall

6-12 CHASSIS ELECTRICAL

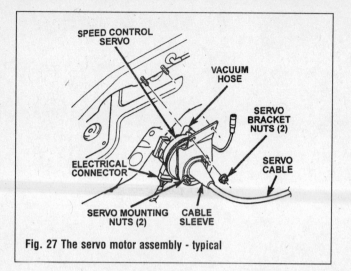

Fig. 27 The servo motor assembly - typical

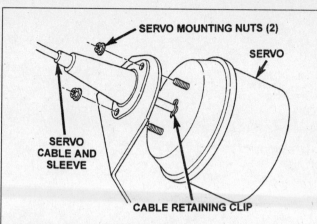

Fig. 28 A retaining clip secures the speed control cable to the servo

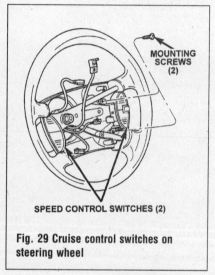

Fig. 29 Cruise control switches on steering wheel

CRUISE CONTROL TROUBLESHOOTING

Problem	Possible Cause
Will not hold proper speed	Incorrect cable adjustment Binding throttle linkage Leaking vacuum servo diaphragm Leaking vacuum tank Faulty vacuum or vent valve Faulty stepper motor Faulty transducer Faulty speed sensor Faulty cruise control module
Cruise intermittently cuts out	Clutch or brake switch adjustment too tight Short or open in the cruise control circuit Faulty transducer Faulty cruise control module
Vehicle surges	Kinked speedometer cable or casing Binding throttle linkage Faulty speed sensor Faulty cruise control module
Cruise control inoperative	Blown fuse Short or open in the cruise control circuit Faulty brake or clutch switch Leaking vacuum circuit Faulty cruise control switch Faulty stepper motor Faulty transducer Faulty speed sensor Faulty cruise control module

Note: Use this chart as a guide. Not all systems will use the components listed.

Fig. 30 CRUISE CONTROL TROUBLESHOOTING

ated while the vacuum solenoid is "duty-cycled". This action causes the diaphragm inside the servo to move, pulling on the speed control cable and opening the throttle. When the vehicle reaches its target speed, the vacuum solenoid is deactivated so that the diaphragm holds, and the cable is held. When the vehicle is above target speed, the vent solenoid is duty-cycled, while the vacuum solenoid is still deactivated to maintain the target speed.

ENTERTAINMENT SYSTEMS

Fig. 31 Removing the cup holder - 97 Ram

Fig. 32 Removing the ashtray bracket screws - 97 Ram

Radio Receiver/Amplifier/Tape Player/CD Player

REMOVAL & INSTALLATION

See Figures 31, 32, 33, 34, 35 and 36

1. Disconnect the negative battery cable(s).
2. On automatic transmission models, be sure the parking brake is set, then lower the shift lever to L
3. On models with tilt wheels, lower the steering wheel.
4. On 97 Ram models: remove the ashtray and the two screws holding the ashtray bracket. Pull out the cup holder and remove the two screws holding

CHASSIS ELECTRICAL 6-13

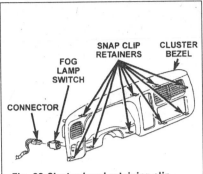

Fig. 33 Cluster bezel retaining clip - typical

Fig. 34 Removing the cluster bezel

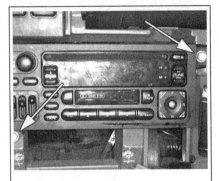

Fig. 35 Radio mounting screws (arrows)

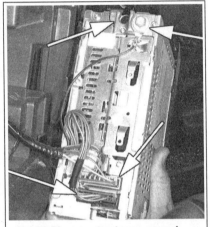

Fig. 36 Disconnect antenna, ground wire, connectors (arrows) - typical

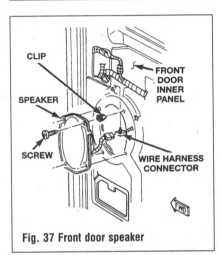

Fig. 37 Front door speaker

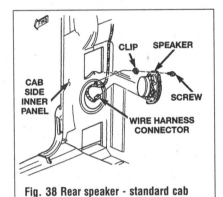

Fig. 38 Rear speaker - standard cab

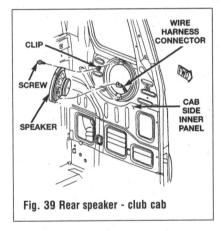

Fig. 39 Rear speaker - club cab

the assembly. Unscrew and remove the ashtray bracket and cup holder.

5. Using a trim stick or other wide, flat-bladed tool, carefully pry around the perimeter of the cluster bezel to disengage the snap clips. Remove the bezel.
6. Remove the two screws which secure the radio.
7. Pull the radio out as far as possible until you can disconnect the wiring, antenna and ground wire (if fitted).
8. To install, reverse the removal process.

Speakers

REMOVAL & INSTALLATION

See Figures 37, 38, 39, 40 and 41

Speakers are installed in various locations in the cab depending on model, year and option package. Refer to the illustrations provided for removal/installation details. Remove the speaker cover or trim for access to the mounting screws. On most door-mounted speakers, the door panel must be removed first.

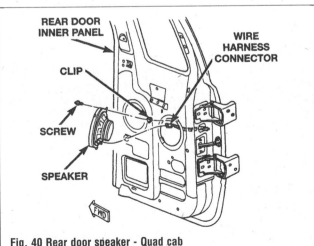

Fig. 40 Rear door speaker - Quad cab

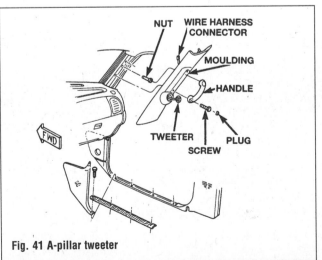

Fig. 41 A-pillar tweeter

6-14 CHASSIS ELECTRICAL

WINDSHIELD WIPER SYSTEM

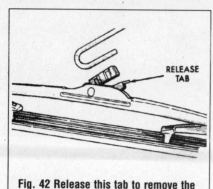

Fig. 42 Release this tab to remove the wiper blade

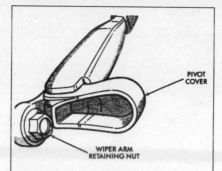

Fig. 43 Dakotas and Durangos secure the wiper arms with a nut beneath a plastic cap or cover

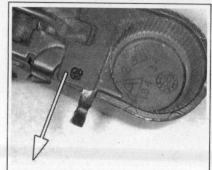

Fig. 44 RAM wiper arm. Lift off windshield and pull latch to release

Wiper Blades

REMOVAL & INSTALLATION

See Figure 42

1. Turn the wiper switch ON, then turn the ignition ON and OFF. This brings the wipers to a more convenient working position.
2. Lift the wiper arm to raise the wiper blade and element off of the windshield.
3. Push the release tab under the arm tip and slide the blade away from the tip (towards the pivot), and remove the blade from the arm.

To install:

4. Slide the blade retainer into the U-shaped formation on the tip of the wiper arm until the release tab snaps into position.
5. Place the arm and blade assembly back onto the windshield.

Wiper Arms

REMOVAL & INSTALLATION

Dakota, Durango
See Figure 43

1. Wiper arms are secured on their pivots by nuts. A small puller (like that used for battery terminals) is used to remove the arms.
2. "Park" the wiper arms in the normal OFF position.
3. Front wipers: mark the locations of the wiper blade on the windshield to ease installation. This is important! Installation even one spline off may result in serious malfunction of the wiper system.
4. Rear wipers have a wiper alignment line concealed in the upper margin of the lower rear window blackout area.
5. Open the hood.
6. Remove the plastic pivot or nut cover.
7. Remove the pivot nut.
8. Use a small puller to remove the wiper arm from its shaft.

To install:

9. Fit the arm onto the pivot ensuring that the wiper blade lines up with the mark made in the removal sequence. Push the blade fully home, taking care to first align the splines.
10. Tighten the nut to 18 ft. lbs. (24 Nm).
11. Refit the cover.
12. Operate the wipers and put them in PARK. Check for proper wiper arm positioning and adjust if necessary.

Ram
See Figure 44

1. Wiper arms are secured by a latch.
2. "Park" the wiper arms in the normal OFF position.
3. Mark the locations of the wiper blade on the windshield to ease installation. This is important! Installation even one spline off may result in serious malfunction of the wiper system.
4. Open the hood.
5. Lift the arm off the windshield as far as possible to permit the latch to be pulled out to the holding position. Pull out the latch. The arm will remain off the windshield.
6. Pull the arm off its shaft using a rocking motion.

WARNING:

The use of a screwdriver or other pry tool on the arm may distort it. This may allow it to come off the pivot shaft during operation.

To install:

7. Mount the arm(s) on the pivot shafts.
8. Lift the wiper arm away from the windshield slightly to relieve the spring tension on the latch.
9. Push the latch into the locked position and slowly release the arm until the wiper blade rests on the windshield.
10. Check that the wiper blade is aligned with the mark made in the removal sequence.
11. Operate the wipers and put them in PARK. Check for proper wiper arm positioning and adjust if necessary.

Windshield Wiper Motor

See Figures 45, 46, 47, 48, 49 and 50

REMOVAL & INSTALLATION

Front

1. Disconnect the negative battery cable(s).

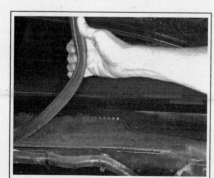

Fig. 45 Removing the weather stripping from the grille

Fig. 46 Remove the grill fasteners - RAM shown

Fig. 47 Lift the grille for access to the windshield washer hoses

CHASSIS ELECTRICAL 6-15

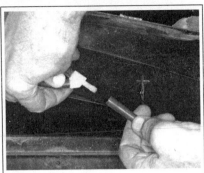

Fig. 48 Disconnecting the windshield washer hose

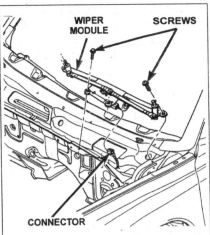

Fig. 49 The wiper module (motor/crank) is held by four screws

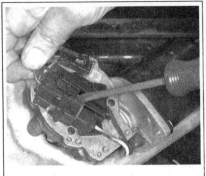

Fig. 50 Disconnecting the motor wiring

2. Remove the wiper arms.
3. Peel off the weather stripping from the front of the grille
4. RAM: Remove the four screws that hold the grille.
5. Dakota, Durango: remove the four plastic nuts along the length of the grille, and the plastic rivets and push-in retainers at the corners.
6. Raise the grille enough to gain access to the windshield washer hoses.
7. Disconnect the washer hose at the connector on the right side of the grille.
8. Dakota, Durango: Disconnect the vacuum supply hose from the vacuum reservoir on the underside of the grille.
9. Remover the grille.
10. Remove the four screws holding the motor/crank assembly: two near the motor and one on each end.
11. Turn the assembly over for access to the motor wiring. Disconnect the motor wiring at the motor.
12. Installation is the reverse of removal. Be sure that hoses are correctly routed and installed in retainers. Tighten the motor/crank screws to 72 inch lbs. (8 Nm).

Rar

See Figure 51

1. Disconnect the negative battery cable.
2. Remove the wiper arm.
3. With a door trim panel removal tool, gently pry at the base of the nut cover where it meets the wiper motor shaft bezel and grommet on the outer liftgate panel until it unsnaps. Be sure to use caution to protect the paint finish.

4. Remove the nut that secures the wiper motor shaft to the outer liftgate panel.
5. Pull the bezel and grommet off of the motor shaft.
6. Remove the liftgate trim panel.
7. Unplug the wiring harness at the connectors.
8. While supporting the motor with a free hand, remove the two screws that secure the motor mounting bracket.
9. Remove the motor and bracket from the liftgate.

To install:

10. Position the motor and bracket to the liftgate and install the two screws, but do not tighten them yet.
11. From the outside, center the wiper motor shaft in the outer panel mounting hole and secure it with the shaft retaining nut. Tighten the nut to 43 inch lbs. (4.8 Nm).
12. The remainder of the procedure is the removal. Tighten the motor bracket mounting screws to 72 inch lbs. (8.1 Nm).

Windshield Washer Pump

See Figure 52

REMOVAL & INSTALLATION

The windshield washer pumps are fitted to the bottom of the fluid reservoir. Procedures vary depending on the type of reservoir fitted, and whether front or rear types. The pumps can be pried out of the reservoir if accessible with the reservoir in place. Otherwise, the reservoir will have to be removed first. Note the following:

1. Disconnect the hose from the pump and drain the washer fluid into a suitable container.
2. Disconnect any electrical wiring connected to the pump.
3. Disconnect the washer fluid level sensor, if so equipped.
4. Disconnect any other wiring attached to the pump.
5. Using a trim tool or other wide, flat-bladed instrument, pry the barbed inlet nipple of the washer pump out of the rubber grommet seal.

To install:

6. Reverse the removal procedure. Use a new rubber grommet seal.

Washer Fluid Level Sensor

See Figure 52

REMOVAL & INSTALLATION

1. Disconnect the hose from the pump and drain the washer fluid into a suitable container.
2. Disconnect the electrical wiring at the pump.
3. Disconnect the washer fluid level sensor.
4. The level sensor float must be positioned horizontally relative to the reservoir in order to be removed. This will be obtained when the reservoir is empty.

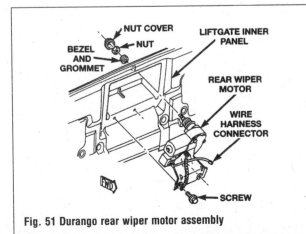

Fig. 51 Durango rear wiper motor assembly

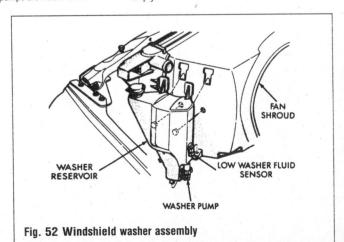

Fig. 52 Windshield washer assembly

6-16 CHASSIS ELECTRICAL

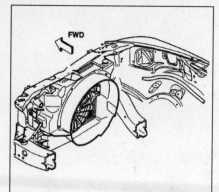

Fig. 53 Fan shroud reservoir - Dakota, Durango

5. Using a trim tool or other wide, flat-bladed instrument, pry the barbed inlet nipple of the sensor out of the rubber grommet seal.

To install:

6. Reverse the removal procedure. Use a new rubber grommet seal.

Windshield Washer Reservoir

REMOVAL & INSTALLATION

RAM

See Figure 52

1. Disconnect the battery negative cable(s).
2. Drain the coolant below the level of the upper radiator hose.
3. Disconnect the upper radiator hose, taking care to catch any coolant left in the hose.
4. Disconnect the electrical connectors at pump and fluid level sensor, if equipped.
5. Disconnect the supply line and drain the fluid into a suitable container.
6. Pull the reservoir away from the fan shroud and lift the unit enough to disengage the mounting tabs from the shroud.
7. Installation is the reverse of removal.

Dakota, Durango (Fan Shroud Reservoir)
See Figure 53

For models with the reservoir incorporated into the upper fan shroud, proceed as follows:

1. Disconnect the electrical connectors at pump and fluid level sensor, if equipped.
2. Disconnect the supply line and drain the fluid into a suitable container.
3. Remove the upper fan shroud.
4. Installation is the reverse of removal.

Dakota, Durango (Fender-Mounted Reservoir)
See Figure 54

1. Disconnect the battery negative cable.

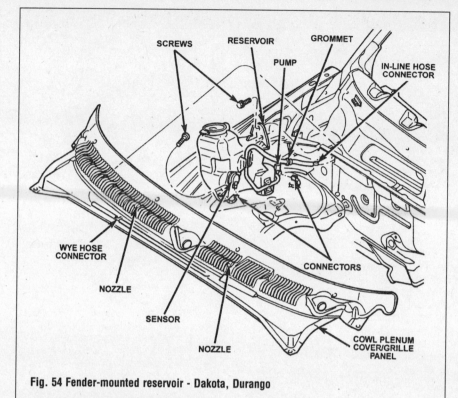

Fig. 54 Fender-mounted reservoir - Dakota, Durango

2. Remove the air filter housing.
3. Remove the screw that holds the reservoir to the inner fender.
4. Remove the two screws that hold the reservoir to the cowl plenum panel.
5. Move the reservoir forward to access the pump wiring. Unplug wires from the pump and fluid level sensor.
6. Disconnect the supply hose and drain the reservoir.
7. Remove the reservoir from the vehicle.
8. Installation is the reverse of removal. Tighten mounting screws to 24 inch lbs. (2.7 Nm).

Durango (Rear)
See Figure 55

1. Disconnect the battery negative cable.
2. Remove the right quarter trim panel.
3. Disconnect the washer supply hose from the check valve located on the inside surface of the right liftgate opening pillar near the liftgate opening header.
4. Disengage the washer supply hose from the retainers on the right liftgate opening pillar. Use the supply hose to siphon the washer fluid from the reservoir through the washer pump and into a clean container for reuse.
5. Remove the washer supply hose from the washer pump.
6. Unplug the wire harness connector.
7. Disconnect the vent hose.
8. Remove the two screws that secure the reser-

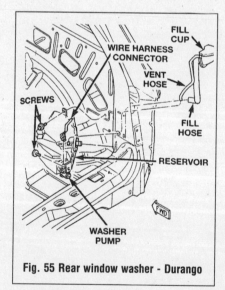

Fig. 55 Rear window washer - Durango

voir to the quarter inner panel and the liftgate opening sill panel.

9. Pull the washer reservoir away from the right liftgate opening pillar far enough to access and disconnect the fill hose.
10. Remove the reservoir.
11. Installation is the reverse of removal. Tighten the mounting screws to 20 inch lbs. (2.2 Nm).

CHASSIS ELECTRICAL 6-17

INSTRUMENTS AND SWITCHES

Fig. 56 Lower instrument cluster mounting screws

Fig. 57 Removing the instrument cluster

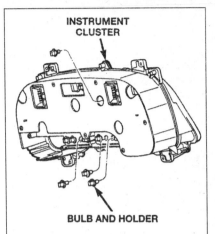

Fig. 58 Bulbs and holders in instrument cluster - typical

Instrument Cluster

REMOVAL & INSTALLATION

See Figures 56 and 57

1. Disconnect the battery negative cable(s).

CAUTION:

On air bag-equipped vehicles, wait at least two minutes for the system to discharge before proceeding. Failure to take proper precautions could result in accidental deployment and possible personal injury.

2. On vehicles with automatic transmission, set the parking brake and place the lever in L for clearance.
3. On models with a tilt wheel, set the wheel to the lowest position for clearance.
4. On 97 Ram models: remove the ashtray and the two screws holding the ashtray bracket. Pull out the cup holder and remove the two screws holding the assembly. Unscrew and remove the ashtray bracket and cup holder.
5. Using a trim stick or other wide, flat-bladed tool, carefully pry around the perimeter of the cluster bezel to disengage the snap clips.
6. Remove the cluster bezel.
7. Remove the four screws that secure the instru-

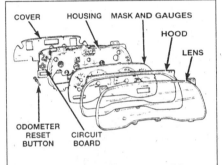

Fig. 59 Instrument cluster assembly

ment cluster to the instrument panel.

8. On models with an automatic transmission, place the transmission in P.
9. Pull the instrument cluster rearward far enough to disengage the two self-docking instrument panel wire harness connectors.
10. On models with an automatic transmission, remove the gear selector indicator or disconnect the gear indicator cable.
11. Remove the instrument cluster.
12. Installation is the reverse of removal.

Note: Certain indicator lamps in the cluster are programmable. If a new cluster is being installed, these lamps must be activated using the scan tool.

Gauges & Bulbs

See Figures 58, 59, 60 and 61

REMOVAL & INSTALLATION

1. Remove the instrument cluster.
2. To replace bulbs, turn the bulb holder CCW and pull out.
3. Instruments can be accessed by separating the lens, hood and housing from the mask/gauge assembly.
4. Installation is the reverse of removal.

Windshield Wiper Switch

Removal/installation of this switch is covered under "Suspension And Steering".

Ignition Switch

Removal/installation of this switch is covered under "Suspension And Steering".

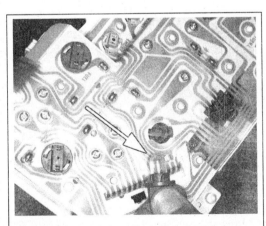

Fig. 60 Turn instrument bulb CCW to remove

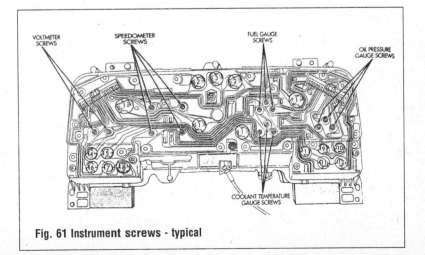

Fig. 61 Instrument screws - typical

6-18 CHASSIS ELECTRICAL

Headlight Switch

REMOVAL & INSTALLATION

See Figure 62

1. Disconnect the battery negative cable(s).
2. Remove the cluster bezel.
3. Remove the three screws that secure the switch.
4. Pull out the switch and disconnect the wires.
5. Installation is the reverse of removal.

Fog Lamp Switch

REMOVAL & INSTALLATION

See Figure 63

1. Disconnect the battery negative cable(s).
2. Remove the cluster bezel.

Fig. 62 Headlight switch mounting screws

3. On some models the switch is attached to the cluster bezel. If so, squeeze the lock tabs on the switch and disconnect it from the panel. If not, proceed to the next step.
4. Remove the three screws that secure the switch mounting plate to the instrument panel.

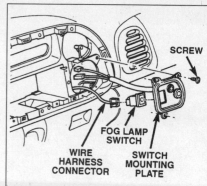

Fig. 63 Fog lamp switch mounting assembly - typical

5. Pull out the switch and disconnect the wires.
6. Squeeze the tabs on the back of the switch and remove it from the mounting plate.
7. Installation is the reverse of removal.

LIGHTING

See Figure 64

Headlights

All models use halogen headlights. Bulbs can be replaced without disturbing the lens assembly (headlamp module). Some models have a single combination high/low bulb, while others use separate bulbs for each application.

BULB REMOVAL & INSTALLATION

See Figures 65, 66 and 67

1. Open the hood.
2. Unfasten the locking ring (turn CCW) which secures the bulb and socket assembly, then withdraw the assembly rearward.
3. If necessary, gently pry the socket's retaining clip over the projection on the bulb (use care not to break the clip.) Pull the bulb from the socket.

To install:

4. Before installing a light bulb into the socket, ensure that all electrical contact surfaces are free of corrosion or dirt.
5. Line up the replacement headlight bulb with the socket. Firmly push the bulb onto the socket until the spring clip latches over the bulb's projection.

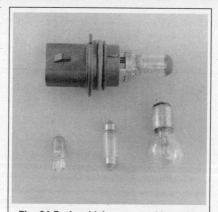

Fig. 64 Each vehicle uses a wide variety of different light bulbs

WARNING:

Do not touch the glass bulb with your fingers. Oil from your fingers can severely shorten the life of the bulb. If necessary, wipe off any dirt or oil from the bulb with rubbing alcohol before completing installation.

Fig. 65 Remove the bulb assembly after turning the lockring CCW

6. To ensure that the replacement bulb functions properly, activate the applicable switch to illuminate the bulb which was just replaced. (If this is a combination low and high beam bulb, be sure to check both intensities.) If the replacement light bulb does not illuminate, either it too is faulty or there is a problem in the bulb circuit or switch. Correct if necessary.
7. Position the headlight bulb and secure it with the locking ring (turn CW).
8. Close the hood.

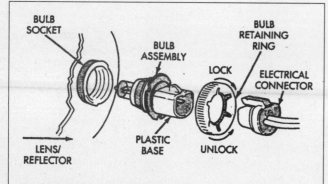

Fig. 66 Headlamp bulb assembly

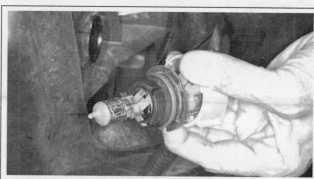

Fig. 67 Halogen bulbs run very HOT. Clean any fingerprints or foreign matter from the glass before installing

CHASSIS ELECTRICAL 6-19

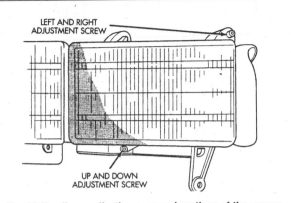

Fig. 68 Headlamp adjusting screws. Locations of the screws may vary depending on model

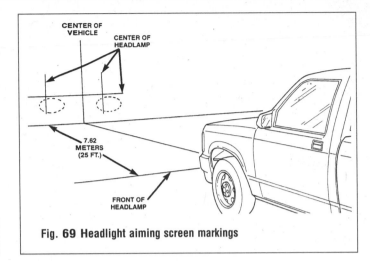

Fig. 69 Headlight aiming screen markings

AIMING THE HEADLIGHTS

See Figures 68 and 69

The headlights must be properly aimed to provide the best, safest road illumination. The lights should be checked for proper aim and adjusted as necessary. Certain state and local authorities have requirements for headlight aiming; these should be checked before adjustment is made.

CAUTION:

About once a year, when the headlights are replaced or any time front end work is performed on your vehicle, the headlight should be accurately aimed by a reputable repair shop using the proper equipment. Headlights not properly aimed can make it virtually impossible to see and may blind other drivers on the road, possibly causing an accident. Note that the following procedure is a temporary fix, until you can take your vehicle to a repair shop for a proper adjustment.

Headlight adjustment may be temporarily made using a wall, as described below, or on the rear of another vehicle. When adjusted, the lights should not glare in oncoming car or truck windshields, nor should they illuminate the passenger compartment of vehicles driving in front of you. These adjustments are rough and should always be fine-tuned by a repair shop which is equipped with headlight aiming tools. Improper adjustments may be both dangerous and illegal.

Horizontal and vertical aiming of each unit is provided by two adjusting screws which move the retaining ring and adjusting plate against the tension of a coil spring.

The vertical (up and down) adjustment screw will be near the center of the headlight at the top or bottom of the assembly. The horizontal (left and right) adjustment screw will be on the side of the headlight at the top or bottom of the assembly.

1. Park the vehicle on a level surface, with the fuel tank full and with the vehicle with an average load if one is normally carried.
2. Be sure all tires are properly inflated.
3. Bounce the truck up and down to center the suspension.
4. The vehicle should be facing a wall which is no less than 6 feet (1.8m) high and 12 feet (3.7m) wide. The front of the vehicle should be about 25 feet from the wall.
5. Make headlight alignment marks on the wall according to the following parameters:
   a. On the wall, mark the exact the center of the vehicle with a vertical line.
   b. Measure the distance from the center of the vehicle to the center of each headlight. Transfer these measurements to the wall with vertical lines.
   c. Measure the distance from the floor to the center of the headlights and mark this on the wall with a horizontal line.
   d. Where the headlamp vertical lines and horizontal line meet are the "headlamp centerlines" and will be the basis point for the adjustment.
6. If aiming is to be performed outdoors, it is advisable to wait until dusk in order to properly see the headlight beams on the wall. If done in a garage, darken the area around the wall as much as possible by closing shades or hanging cloth over the windows.
7. Turn the headlights ON and put on the low beam. Note the position of the pattern.
   - A properly aimed low beam will put the top edge of the high intensity pattern ("hot spot") about 2 in. (50mm) above or below the horizontal headlamp centerline.
   - A properly aimed low beam will put the outboard edge of the high intensity pattern about 2 in. (50mm) to the left or right of the vertical headlamp centerlines.
   - The preferred headlamp up-and-down alignment is 1 in. (25mm) below the horizontal line and the preferred aside-to-side alignment is dead on the vertical lines.
8. Use the vertical and horizontal adjustment screws to move the pattern as required. Adjusting the low beam correctly ensures that high beam adjustment will be correct.
9. Be sure that both headlamps have the exact same adjustment.

HEADLAMP MODULE REMOVAL & INSTALLATION

See Figures 70 and 71

Bulb replacement does not require removal of the headlamp module. Removal will only be required if the module has been damaged. Several types of assembly are fitted depending on model and year. Refer

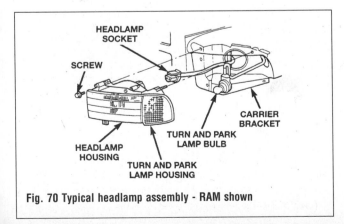

Fig. 70 Typical headlamp assembly - RAM shown

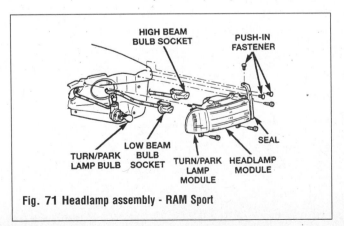

Fig. 71 Headlamp assembly - RAM Sport

6-20 CHASSIS ELECTRICAL

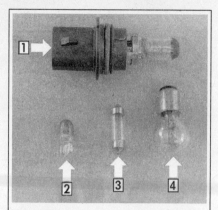

Fig. 72 Popular bulb designs and some of their more common applications: headlight and fog light (1), sidemarker and interior (2), dome and license plate (3), brake, turn signal, park (4). But each may be found in other places!

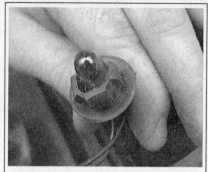

Fig. 73 Here is an example of a burned-out bulb: note the dark, silvery color

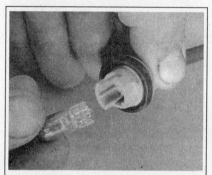

Fig. 75 Simply pull this side marker light bulb straight from its socket

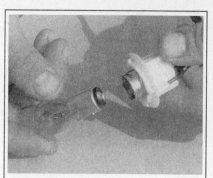

Fig. 74 Push this type of bulb in, then twist CCW to remove

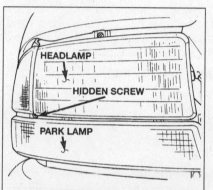

Fig. 76 Parking light/turn signal module - Dakota, Durango

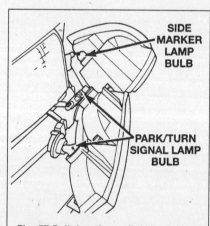

Fig. 77 Bulb locations - Dakota, Durango

to the illustrations for assembly details.
1. Raise the hood.
2. Remove the push-in fastener attaching the seal to the radiator closure panel, if fitted.
3. RAM: remove the screw securing the turn/park lamp module to the headlamp module. Remove the turn/park lamp module.
4. All: remove the screws attaching the headlamp module and remove the module.
5. Disconnect the bulb wiring.
6. Installation is the reverse of removal.

Signal and Marker Lights

See Figure 72

Note: A wide variety of lighting components have been and are being used on this line of trucks and of course a wide range of bulbs. Compare those in your truck against the illustrations.

REMOVAL & INSTALLATION

See Figures 73, 74 and 75

Front Turn Signal and Parking Lights
See Figures 76, 77, 78, 79, 80 and 81

1. On some models, the bulb may be accessible from inside the engine compartment. On others, the lamp module must be removed after taking off the securing screw. On Dakota and Durango models, the module is secured by a screw and incorporates the side marker bulb as well.
2. After the bulb is accessed, turn it CCW 1/4 turn and remove it and the connector from the module.
3. Pull the bulb out of the connector.
4. Installation is the reverse of removal.
5. Before installing the light bulb into the metal contacts, ensure that all electrical conducting surfaces are free of corrosion or dirt.
6. Clean the bulb glass surfaces.
7. To ensure that the replacement bulb functions properly, activate the applicable switch to illuminate the bulb which was just replaced. If the replacement light bulb does not illuminate, either it is faulty or there is a problem in the bulb circuit or switch. Correct as necessary.
8. Install the module, noting that on some mod-

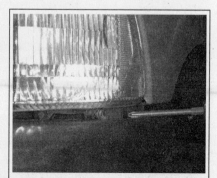

Fig. 78 Removing the turn signal/parking light module screw - RAM shown

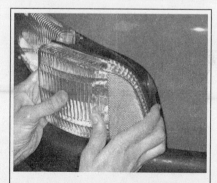

Fig. 79 Removing the lamp module

Fig. 80 Turn the bulb CCW to remove from the module. Then pull the bulb out of the connector

CHASSIS ELECTRICAL 6-21

Fig. 81 Be aware of alignment and mounting devices for the module (arrows)

Fig. 82 Removing the brake light module mounting screw

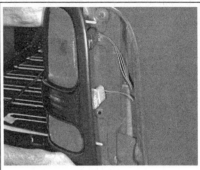
Fig. 83 Pulling off the brake light module

els it is located with slots and tabs which must be properly aligned.

9. Install the securing screw.

Taillight Module
See Figures 82, 83, 84 and 85

Rear modules incorporate taillight, brake, turn signal and back-up lights. Configuration varies depending on model and year.

1. Open the tailgate.
2. Remove the screw which secures the lamp module.
3. Pull the module away from the truck body.
4. Remove the bulb holder from the module by twisting 1/4 turn CCW and pulling out.
5. Remove the bulb from the connector.
6. Installation is the reverse of removal.
7. Before installing the light bulb into the metal contacts, ensure that all electrical conducting surfaces are free of corrosion or dirt.
8. Clean the bulb glass surfaces.
9. Check condition of the retaining studs before refitting the module.
10. To ensure that the replacement bulb functions properly, activate the applicable switch to illuminate the bulb which was just replaced. If the replacement light bulb does not illuminate, either it is faulty or there is a problem in the bulb circuit or switch. Correct as necessary.

Side Marker Light
See Figures 76, 77, 86 and 87

1. On Dakota and Durango models, the front

Fig. 84 Twist the bulb holders 1/4 turn CCW and pull them out of the module

side marker is incorporated into the turn signal/parking light module. Refer to that section, if necessary. The module is secured by a single screw.

2. Some side marker lights are fitted to the fenders. On this version, pry off the bulb holder assembly using a flat-bladed screwdriver at the clip provided.
3. Disengage the bulb and socket assembly from the lens housing.
4. Gently grasp the light bulb and pull it straight out of the socket.

To install:

5. Before installing the light bulb into the socket, ensure that all electrical contact surfaces are free of corrosion or dirt.
6. Line up the base of the light bulb with the socket, then insert the light bulb into the socket until it is fully seated.
7. Clean the glass surface to remove fingerprints or foreign matter.
8. To ensure that the replacement bulb functions properly, activate the applicable switch to illuminate

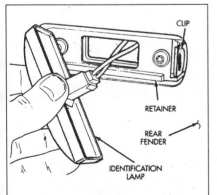
Fig. 86 Side marker bulb holder (early) - pry off with a screwdriver at the clip

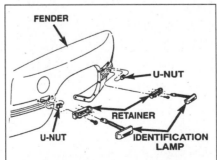
Fig. 87 Side market light assembly - typical

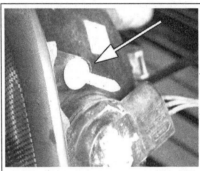
Fig. 85 Retaining studs (arrow) must be in serviceable condition

the bulb which was just replaced. If the replacement light bulb does not illuminate, either it too is faulty or there is a problem in the bulb circuit or switch. Correct as necessary.

9. Install the socket and bulb assembly into the lens housing.

Dome Light
See Figure 88

1. Using a small pry tool, carefully remove the cover lens from the lamp assembly. Pry slots are provided.
2. Remove the bulb by pulling it straight out.

To install:

3. Before installing the light bulb into the metal contacts, ensure that all electrical conducting surfaces are free of corrosion or dirt.
4. Push the bulb into place. Clean the glass surface.
5. To ensure that the replacement bulb functions

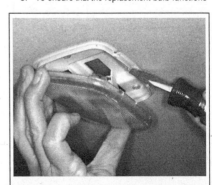
Fig. 88 Removing the dome light lens

6-22 CHASSIS ELECTRICAL

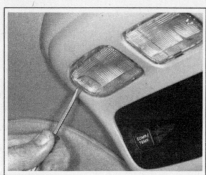
Fig. 89 Removing the reading lamp bulb lens

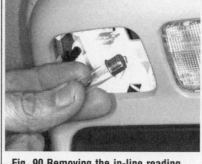
Fig. 90 Removing the in-line reading lamp bulb

Fig. 91 Removing the screws of the high-mount brake light lens assembly

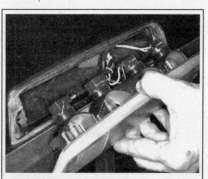
Fig. 92 Removing the high-mount brake light lens

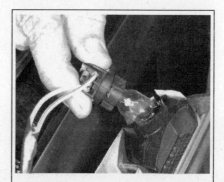
Fig. 93 Remove the brake light bulb

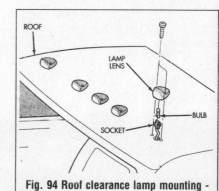
Fig. 94 Roof clearance lamp mounting - typical

properly, activate the applicable switch to illuminate the bulb which was just replaced. If the replacement light bulb does not illuminate, either it is faulty or there is a problem in the bulb circuit or switch. Correct as necessary.

6. Install the cover lens until its retaining tabs are properly engaged.

Reading Lamp Bulb
See Figures 89 and 90

1. Using a small pry tool, remove the reading lamp bulb lens. A pry slot is provided.
2. Pry out the in-line bulb.
3. Installation is the reverse of removal.
4. Clean the bulb glass surfaces.
5. Before installing the light bulb into the metal contacts, ensure that all electrical conducting surfaces are free of corrosion or dirt.
6. To ensure that the replacement bulb functions properly, activate the applicable switch to illuminate the bulb which was just replaced. If the replacement light bulb does not illuminate, either it is faulty or there is a problem in the bulb circuit or switch. Correct as necessary.
7. Install the cover lens until its retaining tabs are properly engaged.

Central High - Mounted Stop Lamp
See Figures 91, 92 and 93

This assembly incorporates two brake light bulbs in the center with cargo area bulbs on each end.
1. Remove the two screws securing the brake light lens assembly.
2. Remove the bulb holder from the assembly.
3. Remove the bulb by pushing in and turning CCW.
4. Installation is the reverse of removal.

5. Clean the bulb glass surfaces.
6. Before installing the light bulb into the metal contacts, ensure that all electrical conducting surfaces are free of corrosion or dirt.
7. To ensure that the replacement bulb functions properly, activate the applicable switch to illuminate the bulb which was just replaced. If the replacement light bulb does not illuminate, either it is faulty or there is a problem in the bulb circuit or switch. Correct as necessary.

Cab Mounted Roof Marker Lights
See Figure 94

1. For each lens: remove the screws holding the clearance lamp to the roof panel.
2. Rotate the socket about 1/4 turn CCW and remove the lamp from the socket.

To install:
3. Installation is the reverse of removal.
4. Clean the bulb glass surfaces.
5. Before installing the light bulb into the metal contacts, ensure that all electrical conducting surfaces are free of corrosion or dirt.
6. Install the lamp in the socket and rotate CW 1/4 turn.
7. Position the lamp on the roof.
8. Install the screws, tightening each to 13 inch lbs. (1 Nm).
9. To ensure that the replacement bulb functions properly, activate the applicable switch to illuminate the bulb which was just replaced. If the replacement light bulb does not illuminate, either it is faulty or there is a problem in the bulb circuit or switch. Correct as necessary.

License Plate Lights
See Figures 95, 96 and 97

1. On models with dual lights (one on each side of the plate), remove the cap over the bulb.
2. On models with a single light over the plate, remove the two screws securing the lens.
3. Pull the wiring harness and connector out for

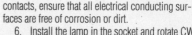

Fig. 95 Removing the license plate bulb cap

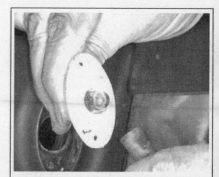
Fig. 96 Slip a piece of rubber tubing over the bulb to pull it out of its socket

CHASSIS ELECTRICAL 6-23

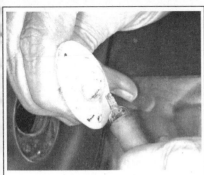

Fig. 97 Using tubing to remove or install a small bulb

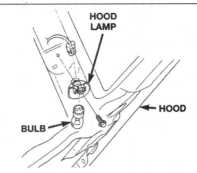

Fig. 98 Early type underhood lamp: push in and twist bulb CCW to remove

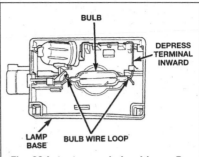

Fig. 99 Later type underhood lamp. Pry off lens (slot provided), then press terminal in to release bulb

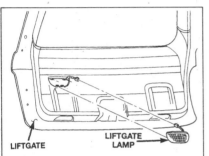

Fig. 100 Liftgate lamp replacement - pry off the lens with a small screwdriver to get to the bulb. If removal of the entire assembly is desired, pry it out of the trim panel

Fig. 101 Removing the fog light bulb assembly from the lens. Arrow indicates vertical lamp adjuster

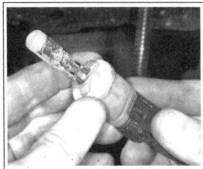

Fig. 102 Removing the bulb from the connector

access to the bulb.
4. Slip a piece of rubber tubing over the bulb and pull it straight out.

To install:
5. Clean the bulb glass surfaces.
6. Before installing the light bulb into the metal contacts, ensure that all electrical conducting surfaces are free of corrosion or dirt.
7. Use the tubing to install a new bulb.
8. To ensure that the replacement bulb functions properly, activate the applicable switch to illuminate the bulb which was just replaced. If the replacement light bulb does not illuminate, either it is faulty or there is a problem in the bulb circuit or switch. Correct as necessary.
9. Install the cap, noting that it cannot be fitted wrong way up.

Underhood Lamp
See Figures 98 and 99

Different types are fitted depending on model and year. They range from a simple bulb in a socket to a more elaborate set-up. Refer to illustrations for construction details.
1. Early models have a simple, uncovered bulb: twist it CCW to remove.
2. Later models have a lens. Remove the lens, if fitted, by prying it off at the slot provided.
3. On this type, press on the terminal to remove the bulb.

To install:
4. Installation is the reverse of removal.
5. Clean the bulb glass surfaces.
6. Before installing the light bulb into the metal contacts, ensure that all electrical conducting surfaces are free of corrosion or dirt.
7. To ensure that the replacement bulb functions properly, activate the applicable switch to illuminate the bulb which was just replaced. If the replacement light bulb does not illuminate, either it is faulty or there is a problem in the bulb circuit or switch. Correct as necessary.

Liftgate Lamp
See Figure 100

1. Open the liftgate.
2. Using a small screwdriver, pry the lens off.
3. Pull out the bulb.

To install:
4. Installation is the reverse of removal.
5. Clean the bulb glass surfaces.
6. Before installing the light bulb into the metal contacts, ensure that all electrical conducting surfaces are free of corrosion or dirt.
7. Push the bulb home.
8. To ensure that the replacement bulb functions properly, activate the applicable switch to illuminate the bulb which was just replaced. If the replacement light bulb does not illuminate, either it is faulty or there is a problem in the bulb circuit or switch. Correct as necessary.

Fog/Driving Lights

BULB REMOVAL & INSTALLATION

See Figures 101 and 102

1. Fog lights have replaceable bulbs.
2. Reach behind the fog lamp housing. Turn the bulb assembly 1/4 turn CCW and then pull it out.
3. Pull the bulb out of the connector

To install:
4. Clean the bulb glass surfaces.
5. Before installing the light bulb into the metal contacts, ensure that all electrical conducting surfaces are free of corrosion or dirt.
6. Push the new bulb into the connector.

WARNING:

Do not touch the glass bulb with your fingers. Oil from your fingers can severely shorten the life of the bulb. If necessary, wipe off any dirt or oil from the bulb with rubbing alcohol before completing installation.

7. Insert the bulb assembly into the lens and turn CW to lock.
8. To ensure that the replacement bulb functions properly, activate the applicable switch to illuminate the bulb which was just replaced. If the replacement light bulb does not illuminate, either it is faulty or there is a problem in the bulb circuit or switch. Correct as necessary.

AIMING FACTORY FOG LIGHTS

See Figure 101

Fog lights have a vertical adjustment screw (see photo). Fog light adjustment may be made using a wall, as described below
1. Park the vehicle on a level surface, with the fuel tank full and with the vehicle empty of all extra

6-24 CHASSIS ELECTRICAL

cargo (unless normally carried). Be sure all tires are properly inflated. The vehicle should be facing a wall which is no less than 6 feet (1.8m) high and 12 feet (3.7m) wide. The front of the vehicle should be about 25 feet from the wall.

2. Measure the distance from the floor to the center of the fog lights and mark this on the wall with a horizontal line.

3. If aiming is to be performed outdoors, it is advisable to wait until dusk in order to properly see the beams on the wall. If done in a garage, darken the area around the wall as much as possible by closing shades or hanging cloth over the windows.

4. Turn the fog lights ON. Note the position of the pattern.

5. A properly aimed fog light beam will put the high intensity area about 4 inches below the horizontal line made on the wall.

6. Use the adjustment screw to move the pattern up or down as required.

7. Be sure that both fog lamps have the exact same adjustment.

FOG LAMP MODULE REMOVAL & INSTALLATION

See Figure 103

1. The module is secured to the bumper or fascia by two bolts or screws.
2. Remove the bulb.
3. Remove the mounting hardware and take off the lamp module assembly.
4. Installation is the reverse of removal.

INSTALLING AFTERMARKET AUXILIARY LIGHTS

Note: Before installing any aftermarket light, make sure it is legal for road use. Most acceptable lights will have a DOT approval number. Also check your local and regional inspection regulations. In certain areas, aftermarket lights must be installed in a particular manner or they may not be legal for inspection.

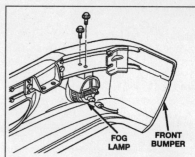

Fig. 103 Fog lamp module mounting - typical. On some models, mounting hardware is accessible from the front of the bumper. Some optional units use screws mounting the unit to the fascia from the back

1. Disconnect the negative battery cable.
2. Unpack the contents of the light kit purchased. Place the contents in an open space where you can

Bulb Application Chart - Durango

Lamp	Bulb
Back-up	3057
Center High Mounted Stop Lamp	
98	921
99-On	922
Fog lamp	893
Front Side Marker	194
Front Park/Turn Signal	3157
Headlights	9007
License Plate	168
Tail, Brake Turn Signal, Side Marker	3057
Dome	579
Dome/Reading	579
Glove Compartment	194
Overhead Console	212-2
Underhood	105
Underpanel Courtesy	904
A/C Control	74
Airbag	PC194
Anti-lock Brake	PC194
Ashtray	161
Brake Warning	PC194
Check Engine	PC194
Cigarette Lighter	161
Engine Oil Pressure	PC194
Fasten Seat Belts	PC194
Four-Wheel Drive	161
Heater Control	158
High Beam	PC194
Ignition Key	53
Illumination	PC194
Low Fuel	PC194
Low Washer Fluid	PC194
Maintenance Required	PC194
Radio	ASC
Temperature Indicator	PC194

93256c01

Bulb Application Chart - Dakota

Lamp	Bulb
Back-up	3157
Cargo	921
Center High Mounted Stop Lamp	921
Fog Lamp	
97-99	893
00	899
Front Side Marker	194
Park/Turn Signal	3157
Headlights	9007
License Plate	168
Tail, Brake	3157
Dome	579
Glove Compartment	194
Overhead Console	212-2
Underhood	
97-98	105
99-00	561
A/C Heater Control	6233137
Ashtray	161
Headlamp Switch	158
Heater Control	6233137
Instrument Cluster	PC194
Overhead Console	
97-98	579
99-00	578
Radio	ASC
Stepwell	904
Airbag	LED
Anti-lock Brake	PC74
Brake Warning	LED
Check Engine	PC74
Check Gauges	LED
Cruise	PC74
Engine Oil Pressure	PC74
Four-Wheel Drive	PC194
High Beam	PC74
Ignition Key	53
Low Fuel	PC74
Low Washer Fluid	PC74
Overdrive Off	PC74
Seat Belt	LED
Turn Signal	PC194
Security	PC74
RWAL	PC74

93256c02

CHASSIS ELECTRICAL 6-25

easily retrieve a piece if needed.

3. Choose a location for the lights. If you are installing fog lights, below the bumper and apart from each other is desirable. Most fog lights are mounted below or very close to the headlights. If you are installing driving lights, above the bumper and close together is desirable. Most driving lights are mounted between the headlights.

4. Drill the needed hole(s) to mount the light. Install the light, and secure using the supplied retainer nut and washer. Tighten the light mounting hardware, but not the light adjustment nut or bolt.

5. Install the relay that came with the light kit in the engine compartment, in a rigid area, such as a fender. Always install the relay with the terminals facing down. This will prevent water from entering the relay assembly.

6. Using the wire supplied, locate the ground terminal on the relay, and connect a length of wire from this terminal to a good ground source. You can drill a hole and screw this wire to an inside piece of metal; just scrape the paint away from the hole to ensure a good connection.

7. Locate the light terminal on the relay; and attach a length of wire between this terminal and the fog/driving lamps.

8. Locate the ignition terminal on the relay, and connect a length of wire between this terminal and the light switch.

9. Find a suitable mounting location for the light switch and install. Some examples of mounting areas are a location close to the main light switch, auxiliary light position in the dash panel, if equipped, or in the center of the dash panel.

10. Depending on local and regional regulations, the other end of the switch can be connected to a constant power source such as the battery, an ignition opening in the fuse panel, or a parking or headlight wire.

11. Locate the power terminal on the relay, and connect a wire with an in-line fuse of at least 10 amperes between the terminal and the battery.

12. With all the wires connected and tied up neatly, connect the negative battery cable.

13. Turn the lights ON and adjust the light pattern, if necessary.

AIMING

1. Park the vehicle on level ground, so it is perpendicular to and, facing a flat wall about 25 ft. (7.6m) away.

2. Remove any stone shields, if equipped, and switch ON the lights.

3. Loosen the mounting hardware of the lights so you can aim them as follows:

   a. The horizontal distance between the light beams on the wall should be the same as between the lights themselves.

   b. The vertical height of the light beams above the ground should be 4 in. (10cm) less than the distance between the ground and the center of the lamp lenses for fog lights. For driving lights, the vertical height should be even with the distance between the ground and the center of the lamp.

4. Tighten the mounting hardware.

5. Test to make sure the lights work correctly, and the light pattern is even.

Bulb Application Chart - RAM

Lamp	Bulb
Back-up	3157
Cargo	921
Center High Mounted Stop Lamp	921
Clearance	
97	194
98-00	168
Fog lamp	893
Front Side Marker	194
Front Park/Turn Signal	3157
Headlights	
97	9004
98-00	9004LL
Sport, HI	9004LL
Sport, LO	9007
License Plate	1155
License Plate, Step Bumper	168
Park/Turn Signal	3157NA
Snow Plow Control	161
Tail, Brake Turn Signal	3157
Tail/Stop (Chassis Cab)	1157
Dome	1004
Glove Compartment	1891
Underhood	105
A/C Heater Control	158
Ashtray	161
Cigarette Lighter	161
Headlamp Switch	158
Heater Control	158
Instrument Cluster	PC194
Radio	ASC
Airbag High Line	PC194
Airbag Low Line	PC74
Anti-lock Brake	PC74
Battery Voltage	PC194
Brake Warning	PC194
Check Engine	PC74
Engine Oil Pressure	PC74
Fasten Seat Belts	PC74
Four-Wheel Drive	PC194
High Beam	PC194
Low Fuel	PC194
Low Washer Fluid	PC74
Upshift	PC74
Maintenance Required	PC74
Message Center	PC194
Turn Signal	PC194

93256c03

TRAILER WIRING

Wiring the vehicle for towing is fairly easy. There are a number of good wiring kits available and these should be used, rather than trying to design your own.

All trailers will need brake lights and turn signals as well as tail lights and side marker lights. Most areas require extra marker lights for overwide trailers. Also, most areas have recently required back-up lights for trailers, and most trailer manufacturers have been building trailers with back-up lights for several years.

Additionally, some Class I, most Class II and just about all Class III and IV trailers will have electric brakes. Add to this number an accessories wire, to operate trailer internal equipment or to charge the trailer's battery, and you can have as many as seven wires in the harness.

Determine the equipment on your trailer and buy the wiring kit necessary. The kit will contain all the wires needed, plus a plug adapter set which includes the female plug, mounted on the bumper or hitch, and the male plug, wired into, or plugged into the trailer harness.

When installing the kit, follow the manufacturer's instructions. The color coding of the wires is usually standard throughout the industry. One point to note: some domestic vehicles, and most imported vehicles, have separate turn signals. On most domestic vehicles, the brake lights and rear turn signals operate with the same bulb. For those vehicles without separate turn signals, you can purchase an isolation unit so that the brake lights won't blink whenever the turn signals are operated.

One final point, the best kits are those with a spring loaded cover on the vehicle mounted socket. This cover prevents dirt and moisture from corroding the terminals. Never let the vehicle socket hang loosely; always mount it securely to the bumper or hitch.

6-26 CHASSIS ELECTRICAL

CIRCUIT PROTECTION

CAUTION:
On vehicles equipped with air bags, disconnect the battery negative cable(s) and wait for two minutes before doing any work on the PDC and junction block. This will eliminate possibility of an accidental deployment.

Fuses

The vehicle's electrical components are protected with a large number of fuses connected at critical junctions of the wiring system.

"Maxi fuses", located in the Power Distribution Center (PDC), have high amperage ratings and replace the "fusible links" of previous years.

"Mini fuses", with lower amperage ratings, are located in the PDC and on the junction block (or fuse panel) located on the left side of the dash panel.

Both the PDC and the junction block provide location for relays as well.

JUNCTION BLOCK

See Figures 104 and 105

The electrical junction block is concealed behind the left end of the instrument panel cover. This block serves to simplify and centralize numerous electrical components and to distribute electrical current to many of the accessory systems in the vehicle. It also eliminates the need for numerous splice connections and serves in place of a bulkhead connector between many of the engine compartment, instrument panel, and chassis wire harnesses.

The junction block houses up to nineteen blade-type fuses (two standard and seventeen mini), up to two blade-type automatic resetting circuit breakers, and two ISO relays (one standard, one micro).

A fuse puller and spare fuse holders are located on the back of the fuse access cover.

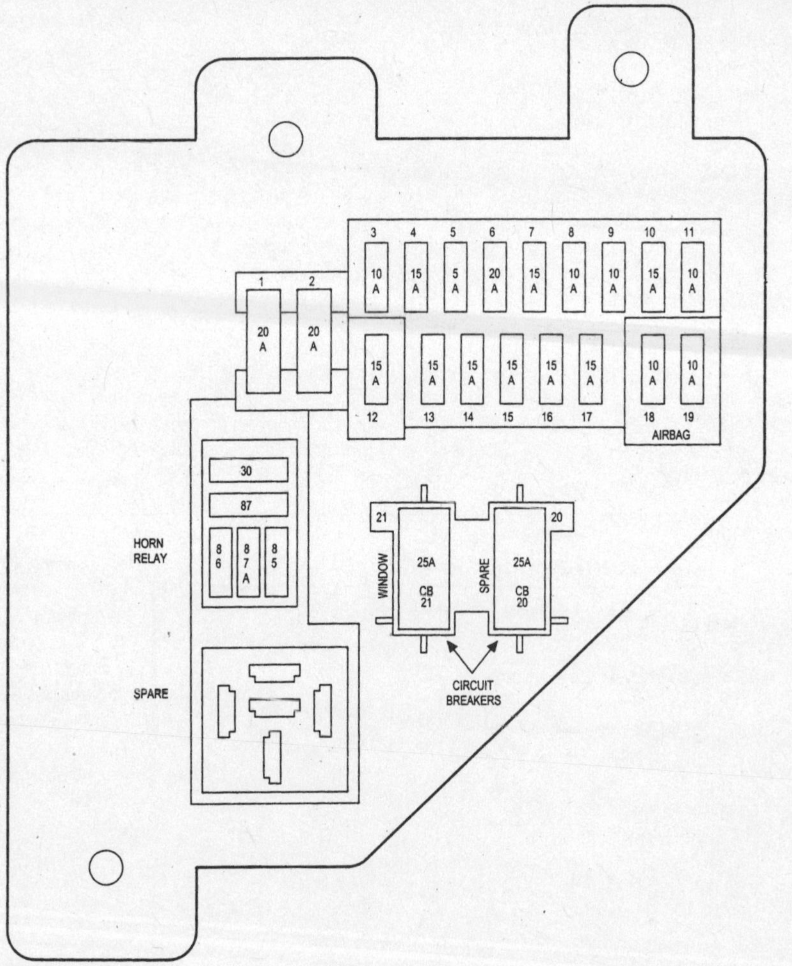

POWER DISTRIBUTION CENTER

See Figures 106, 107 and 108

Located in the engine compartment, the PDC houses the alternator cartridge fuse, up to ten maxi fuses, up to seven blade-type mini fuses, and as many as 13 ISO relays (standard and micro).

The inside of the PDC cover has a map identifying each element.

Fig. 104 Typical junction block layout. Fuse ratings and applications on your vehicle may differ. Refer to the information on the junction block and cover for your vehicle

REPLACEMENT

Excessive current draw is what causes a fuse to blow. Observing the condition of the fuse will provide insight as to what caused this to occur.

Fig. 105 Fuse panel (junction block) location on left side of instrument panel cover: lid identifies function

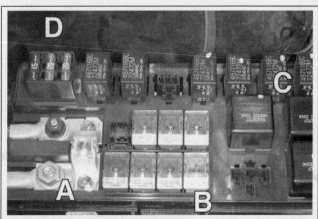

Fig. 106 PDC components on a 97 RAM: cartridge fuse (A), maxi fuses (B), relays (C), mini fuses (D)

CHASSIS ELECTRICAL 6-27

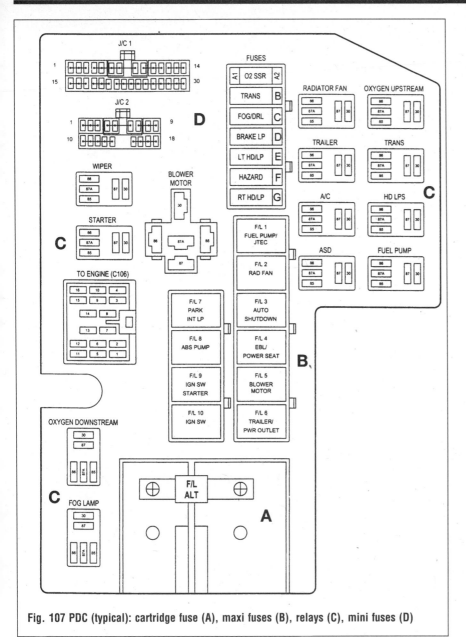

Fig. 107 PDC (typical): cartridge fuse (A), maxi fuses (B), relays (C), mini fuses (D)

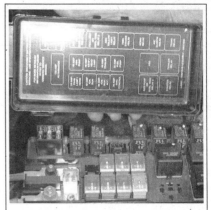

Fig. 108 PDC in the engine compartment has relays and maxi fuses and standard mini fuses: lid identifies function

A fuse with signs of burns, melting of the plastic shell, or little to no trace of the wire that once served as the conductor indicates that a direct short to ground exists.

1. On air bag-equipped vehicles, disconnect the battery negative cable(s) and wait two minutes for the air bag capacitor to discharge.
2. Remove the fuse or PDC cover.
3. Inspect the fuses to determine which is faulty.
4. Unplug and discard the blown fuse.
5. Inspect the box terminals and clean if corroded. If any terminals are damaged, replace the terminals.
6. Plug in a new fuse of the same amperage rating.

WARNING:

Never exceed the amperage rating of a blown fuse. If the replacement fuse also blows, check for a problem in the circuit.

7. Check for proper operation of the affected component or circuit.

Cartridge Fuse

See Figures 106 and 107

The truck is equipped with a cartridge fuse in the Power Distribution Center (PDC) to protect the electrical system from excessive output from the alternator resulting from a defective alternator or regulator circuits. This fuse is rated at 120 amps on 97 RAM trucks and 140 amps on the other models.

If this fuse fails, a thorough inspection of the charging system must be carried out before replacement.

REMOVAL & INSTALLATION

1. Disconnect the battery negative cable(s).
2. Remove the PDC cover.

3. Remove the two screws which secure the cartridge fuse to the B+ terminal stud bus bars.
4. Remove the fuse.
5. Installation is the reverse of removal. Tighten the two fuse screws to 30 inch lbs. (3.4 Nm). Proper torque is important.

Circuit Breakers

See Figure 104

The junction block at the left end of the instrument panel can accommodate up to two circuit breakers. These are used to protect power windows and power seats, if fitted.

Circuit breakers will interrupt the circuit in the event of a short circuit or an overload condition caused by an obstruction to free movement of the window or seat which would tend to overheat the motor.

The circuit breakers are automatically reset when the failure corrects itself, however, that does not mean that the problem is repaired. If a circuit has a motor that is going bad, the resistance will be high and eventually the circuit breaker trips and power is shut off to the circuit. When the circuit cools down, and the circuit breaker works again, so will the windows or seats, however the breaker will trip again when the resistance builds.

Circuits that might be needed in case of an emergency or to operate the vehicle (like the power door locks) have circuit breakers instead of fuses.

RESETTING AND/OR REPLACEMENT

Circuit breakers are automatically resetting. No attention is necessary in the event of activation.

A circuit breaker, if defective, is replaced exactly like a fuse.

1. On air bag-equipped vehicles, disconnect the battery negative cable(s) and wait two minutes for the air bag capacitor to discharge.
2. Locate the correct circuit breaker in the junction block.
3. Pull the circuit breaker out slightly, but leave the circuit breaker terminals connected to the terminals in the junction block cavity.
4. Connect the negative lead of a 12-volt DC voltmeter to a good ground.
5. With the voltmeter's positive lead, check both

6-28 CHASSIS ELECTRICAL

exposed circuit breaker terminals for voltage.
6. If only one terminal has voltage, the breaker must be replaced.
7. If neither terminal has voltage, the problem may lie in the circuit or the PDC.

Flashers

See Figures 109, 110 and 111

REPLACEMENT

Early models (97) have separate flashers for the turn signals and hazard flashers. They are located on the junction (fuse) block under the cover on the left side of the instrument panel. The hazard warning flasher is the upper one; the turn signal flasher is the lower one.

The flasher contains sophisticated internal circuitry which cannot be tested with conventional equipment.

Models from 98 on use a combination flasher.

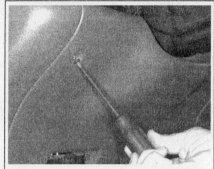

Fig. 109 Removing the steering column cover for access to the flasher

This flasher is mounted either on the front or on the back of the junction block.

The combination flasher is a smart relay that functions as both the turn signal system and the hazard warning system flasher. It is designed to handle factory-fitted equipment. If aftermarket equipment is added to this circuit (such as trailers with lights), the flasher will automatically try to compensate to keep the flashing rate the same.

1. The flasher contains sophisticated internal circuitry which cannot be tested with conventional equipment. Substituting a new unit for a suspected defective version is the easiest test.
2. On air bag-equipped vehicles, disconnect the battery negative cable(s) and wait two minutes for the air bag capacitor to discharge.
3. Remove the junction block cover. If the flasher is accessible, grasp and pull the flasher from the fuse panel.
4. If the flasher is not mounted on the outside of the junction block, remove the steering column cover from the instrument panel. Reach through the opening to access the fuse block on the back of junction block. Pull off the flasher.
5. Inspect the socket for corrosion or any other signs of a bad contact.

To install:

6. Install a new flasher in the connector.
7. Install the fuse panel cover.

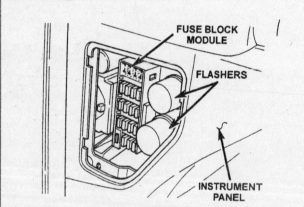

Fig. 110 Early style flasher system: dual flashers mounted on junction (fuse) block. The turn signal flasher is the lower of the two; hazard flasher is the higher one

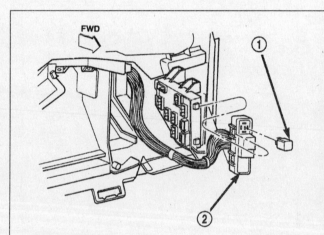

Fig. 111 Combination flasher location (1) on fuse block (2) (later models)

CHASSIS ELECTRICAL 6-29

WIRING DIAGRAMS

INDEX OF WIRING DIAGRAMS

DIAGRAM #	DESCRIPTION
DIAGRAM 1	Sample Diagram: How To Read & Interpret Wiring Diagrams
DIAGRAM 2	Sample Diagram: Wiring Diagram Symbols
DIAGRAM 3	1997-98 Dakota 2.5L/3.9L/5.2L Engine Schematic
DIAGRAM 4	1998 Dakota 5.9L Engine Schematic
DIAGRAM 5	1999-00 Dakota 2.5L/3.9L/5.2L, 2000 5.9L Engine Schematic
DIAGRAM 6	1999 Dakota 5.9L Engine Schematic
DIAGRAM 7	2000 Dakota 4.7L (Except California) Engine Schematic
DIAGRAM 8	2000 Dakota 4.7L (California) Engine Schematic
DIAGRAM 9	1998 Durango 3.9L/5.2L/5.9L Engine Schematic
DIAGRAM 10	1999-00 Durango 3.9L/5.2L/5.9L (Except California) Engine Schematic
DIAGRAM 11	1999-00 Durango 3.9L/5.2L/5.9L (California) Engine Schematic
DIAGRAM 12	1999-00 Durango 4.7L (Except California) Engine Schematic
DIAGRAM 13	1999-00 Durango 4.7L (California) Engine Schematic
DIAGRAM 14	1997 Ram 1500-3500 3.9L/5.2L/5.9L Engine Schematic
DIAGRAM 15	1997 Ram 1500-3500 5.9L Diesel Engine Schematic
DIAGRAM 16	1997 Ram 1500-3500 8.0L Engine Schematic
DIAGRAM 17	1998 Ram 1500-3500 3.9L/5.2L/5.9L (Except Heavy Duty) Engine Schematic
DIAGRAM 18	1998 Ram 1500-3500 5.9L (Heavy Duty) Engine Schematic
DIAGRAM 19	1998 Ram 1500-3500 5.9L Diesel Engine Schematic
DIAGRAM 20	1998 Ram 1500-3500 8.0L Engine Schematic
DIAGRAM 21	1999-00 Ram 1500-3500 3.9L/5.2L/5.9L (Except HD California) Engine Schematic
DIAGRAM 22	1999-00 Ram 1500-3500 5.9L (Heavy Duty California) Engine Schematic
DIAGRAM 23	1999-00 Ram 1500-3500 5.9L Diesel Engine Schematic
DIAGRAM 24	1999-00 Ram 1500-3500 8.0L Engine Schematic
DIAGRAM 25	1997-00 Dakota,1998 Durango Starting, 1997-00 Dakota, 1998-00 Durango Charging Chassis Schematics
DIAGRAM 26	1997-98 Dakota Headlights, Fog Lights Chassis Schematics
DIAGRAM 27	1999-00 Dakota Headlights Chassis Schematics
DIAGRAM 28	1997-00 Dakota Parking/Marker Lights, Back-up Lights, Stop Lights, 1999-00 Fog Lights Chassis Schematics
DIAGRAM 29	1997-00 Dakota, 1997-98 Ram 1500-3500 Turn/Hazard Lights Chassis Schematics

6-30 CHASSIS ELECTRICAL

DIAGRAM #	DESCRIPTION
DIAGRAM 30	1997-00 Dakota 2 Door, Ram 1500-3500 Power Windows, 1997-00 Dakota 2 Door Power Door Locks Chassis Schematics
DIAGRAM 31	2000 Dakota 4 Door Power Windows, Power Door Locks Chassis Schematics
DIAGRAM 32	1997-00 Dakota, 1998-00 Durango Washer/Wipers, 1997-00 Dakota Horns Chassis Schematics
DIAGRAM 33	1999-00 Durango Starting, Charging, 1998-00 Horns Chassis Schematics
DIAGRAM 34	1998 Durango Headlights, Fog Lights Chassis Schematics
DIAGRAM 35	1999-00 Durango Headlights, Fog Lights Chassis Schematics
DIAGRAM 36	1998 Durango Parking/Marker Lights, Stop Lights, 1998-00 Back-up Lights Chassis Schematics
DIAGRAM 37	1999-00 Durango Parking/Marker Lights, Stop Lights Chassis Schematics
DIAGRAM 38	1998-00 Durango Turn/Hazard Lights Chassis Schematic
DIAGRAM 39	1998-00 Durango Power Door Locks Chassis Schematic
DIAGRAM 40	1998-00 Durango Power Windows Chassis Schematic
DIAGRAM 41	1997-00 Ram 1500-3500 Starting, Charging, Horns Chassis Schematics
DIAGRAM 42	1997-98 Ram 1500-3500 Headlights, Fog Lights Chassis Schematics
DIAGRAM 43	1999-00 Ram 1500-3500 Headlights Chassis Schematics
DIAGRAM 44	1997-00 Ram 1500-3500 Parking/Marker Lights Chassis Schematics
DIAGRAM 45	1997-00 Ram 1500-3500 Turn/Hazard Lights Chassis Schematics
DIAGRAM 46	1997 Ram 1500-3500 Power Door Locks Chassis Schematics
DIAGRAM 47	1998-00 Ram 1500-3500 Power Door Locks Chassis Schematics
DIAGRAM 48	1997-00 Ram 1500-3500 Washer/Wipers, Back-up Lights, Stop Lights Chassis Schematics
DIAGRAM 49	1997-00 Dakota, Durango Cooling Fan, 1997-00 Dakota, Durango, Ram 1500-3500 Fuel Pump Chassis Schematics

CHASSIS ELECTRICAL 6-31

SAMPLE DIAGRAM: HOW TO READ & INTERPRET WIRING DIAGRAMS

BLACK	B	PINK	PK
BROWN	BR	PURPLE	P
RED	R	GREEN	G
ORANGE	O	WHITE	W
YELLOW	Y	LIGHT BLUE	LBL
GRAY	GY	LIGHT GREEN	LG
BLUE	BL	DARK GREEN	DG
VIOLET	V	DARK BLUE	DBL
TAN	T	NO COLOR AVAILABLE-	NCA

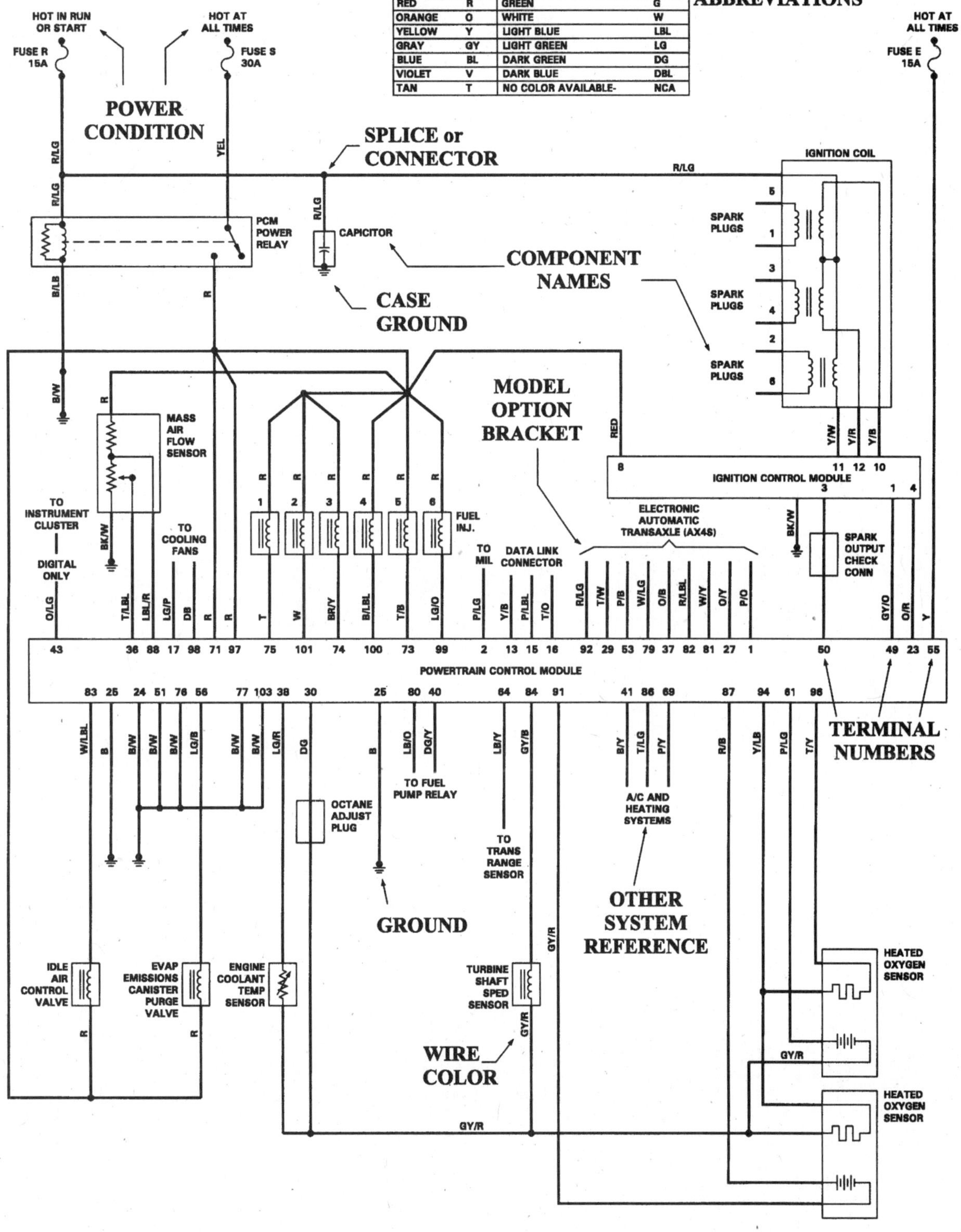

WIRE COLOR ABBREVIATIONS

DIAGRAM 1

WIRING DIAGRAM SYMBOLS

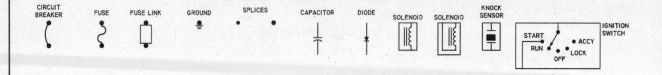

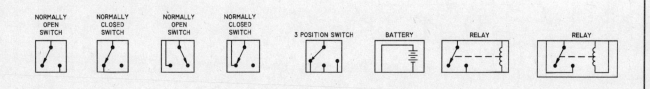

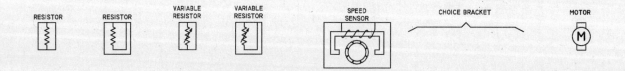

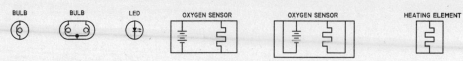
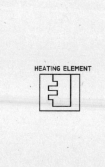

DIAGRAM 2

CHASSIS ELECTRICAL 6-33

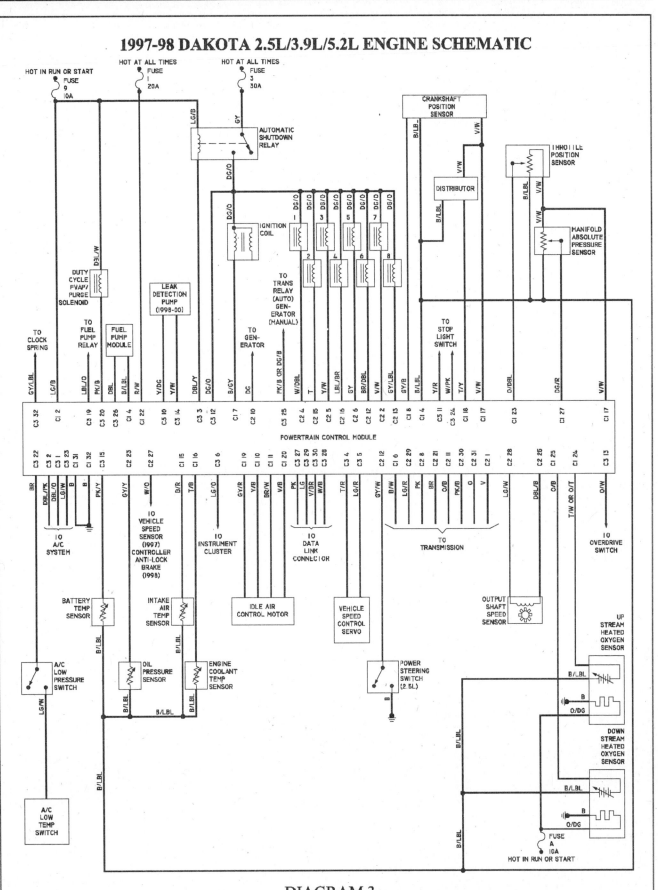

DIAGRAM 3

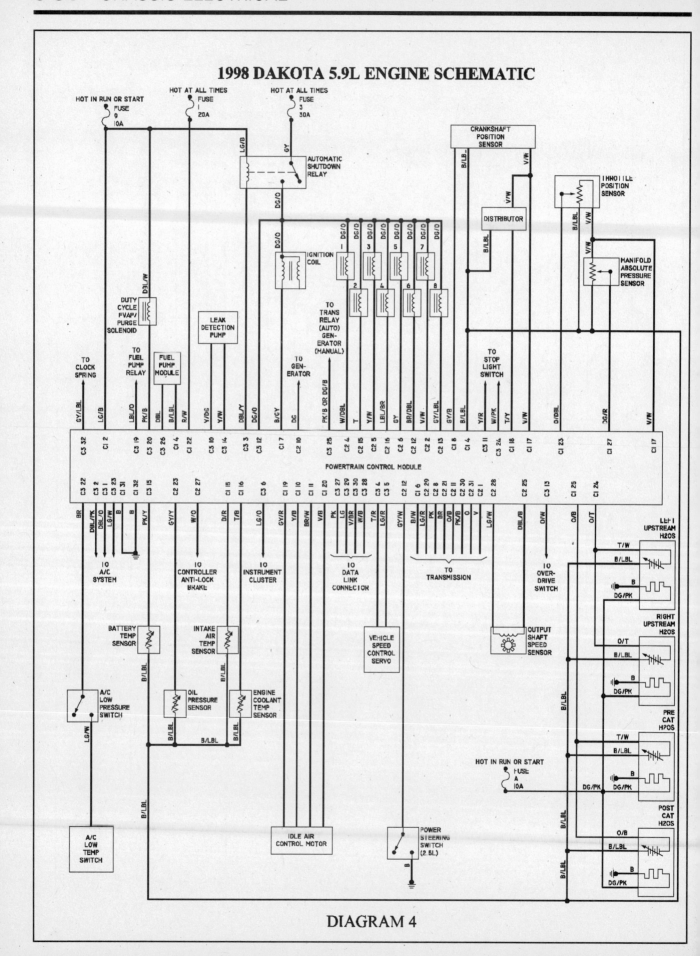

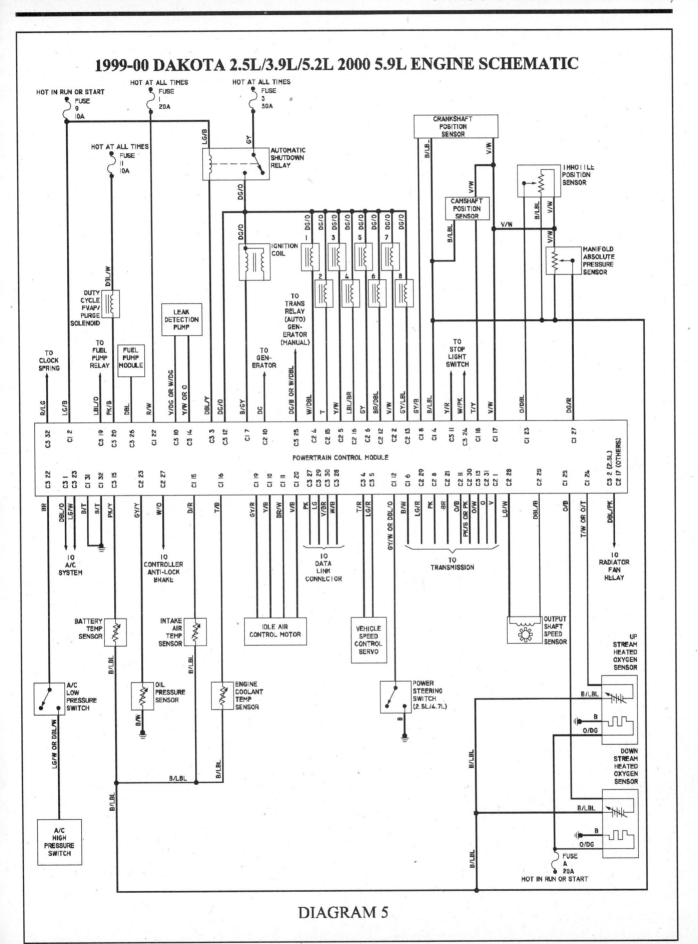

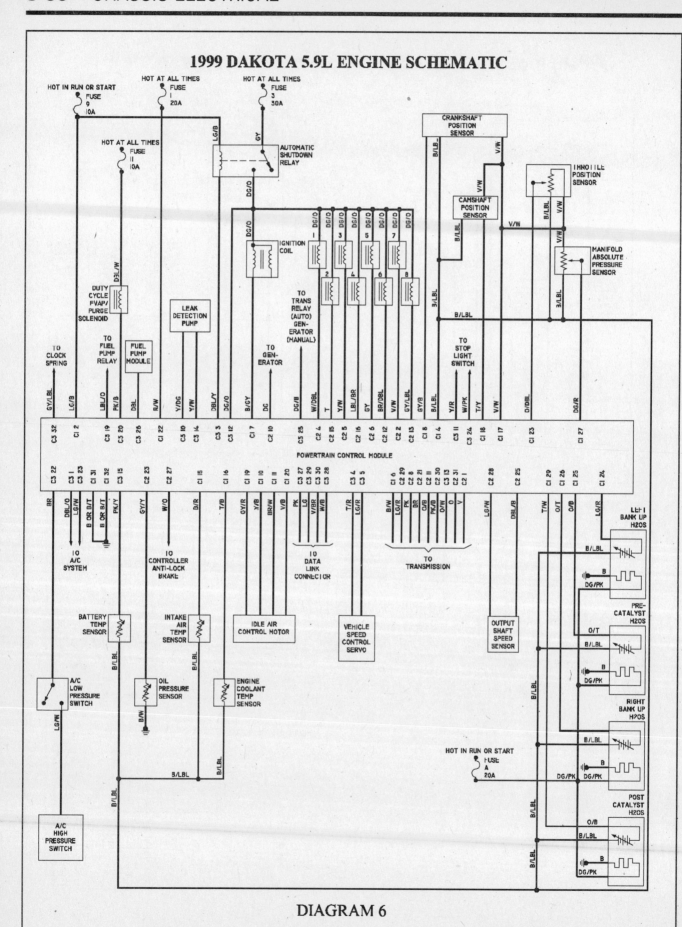

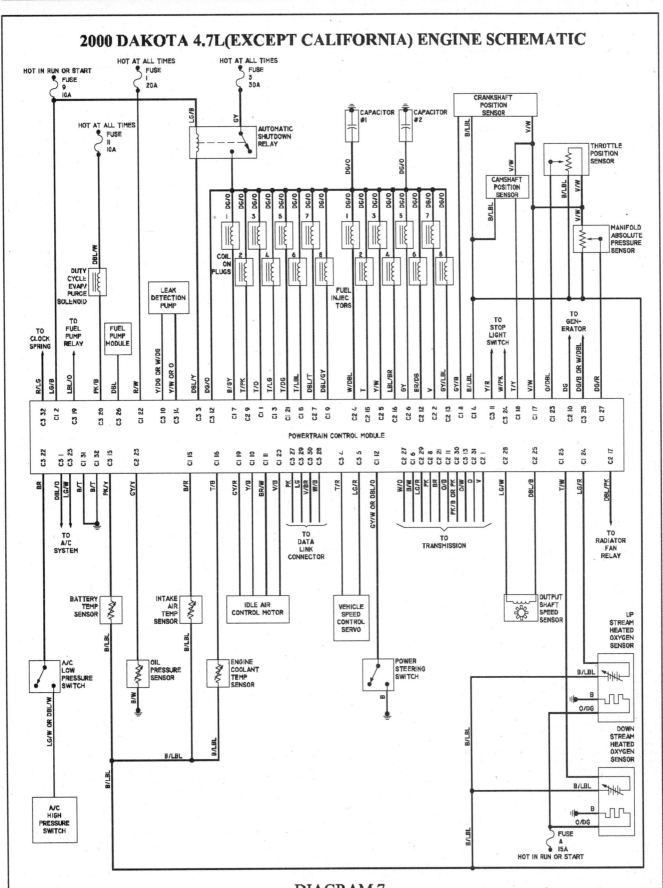

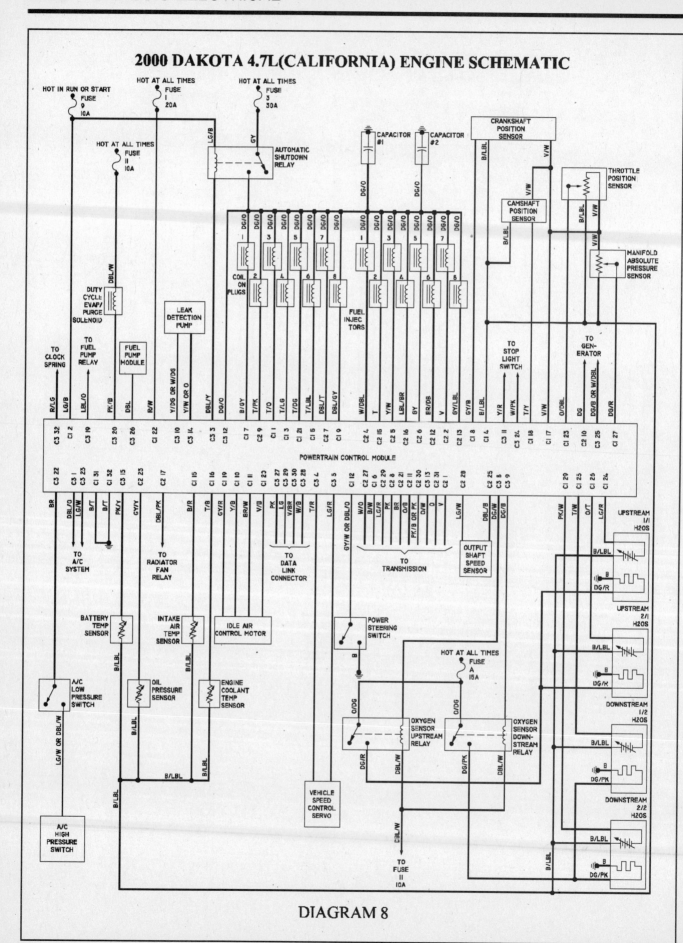

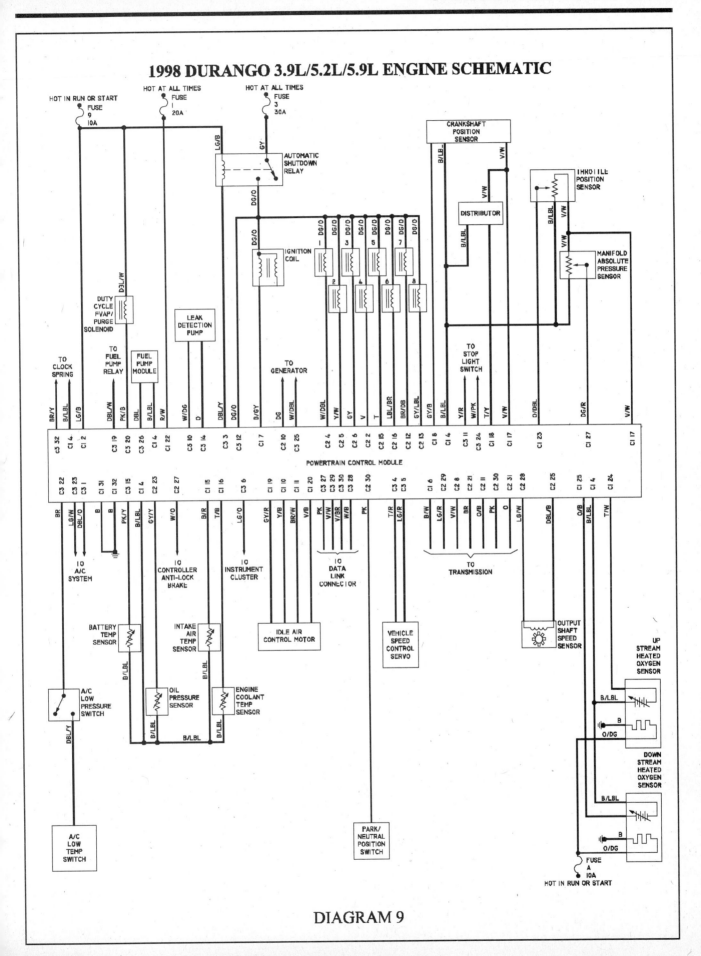

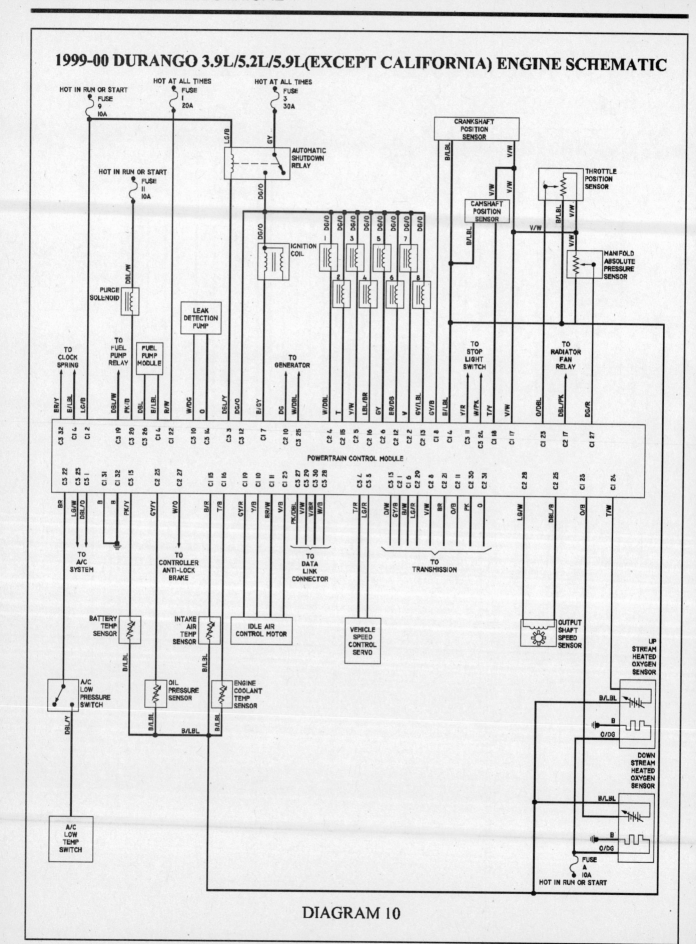

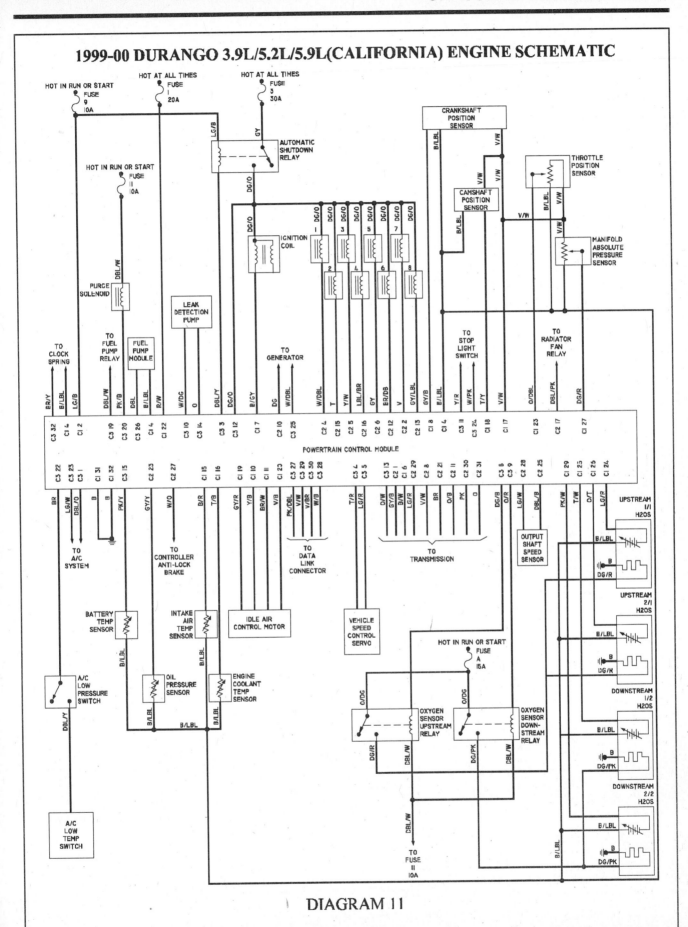

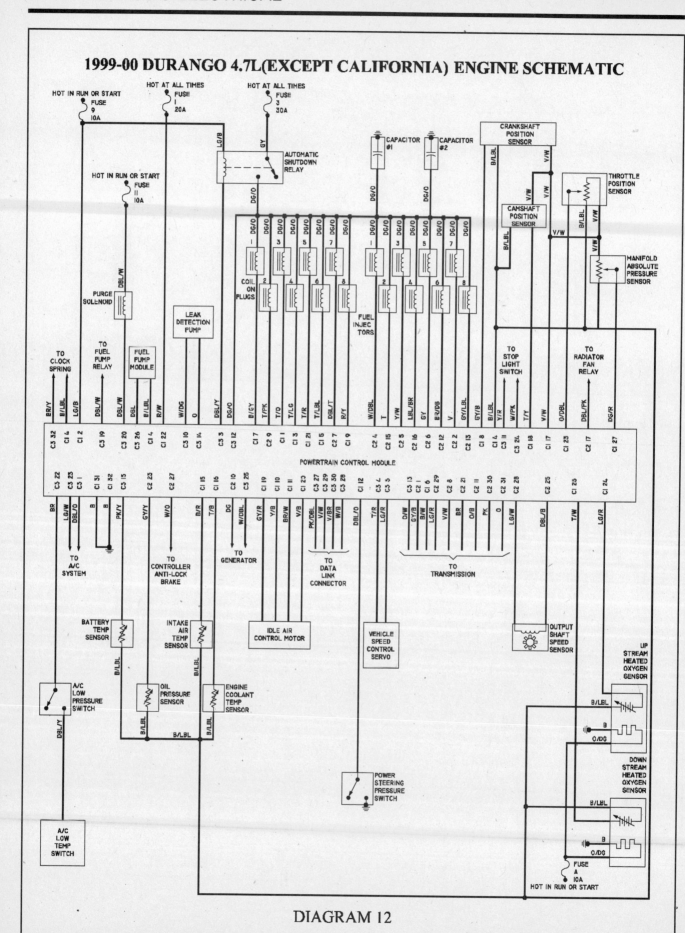

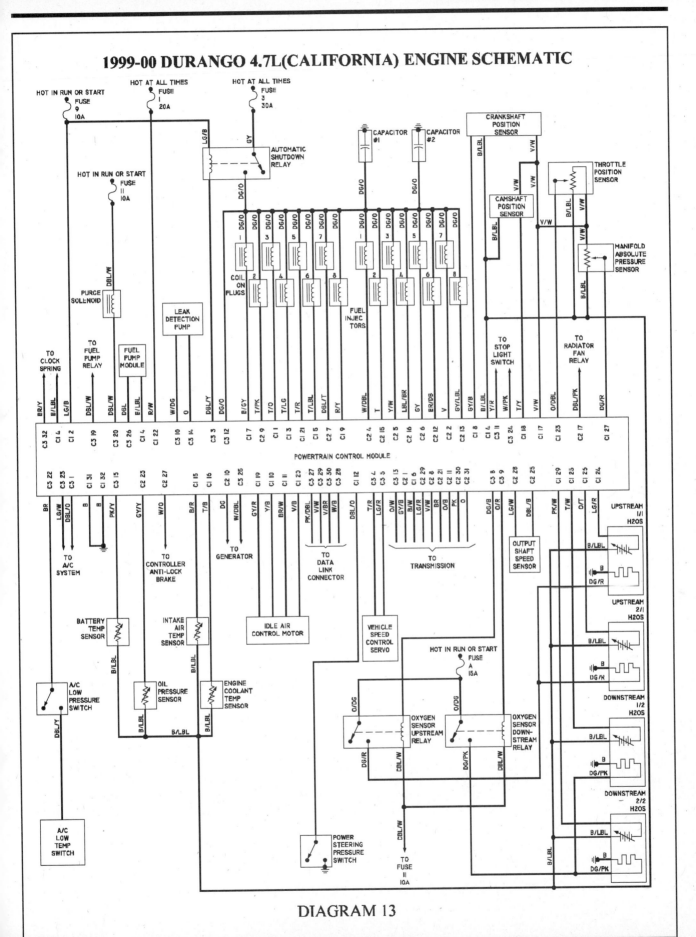

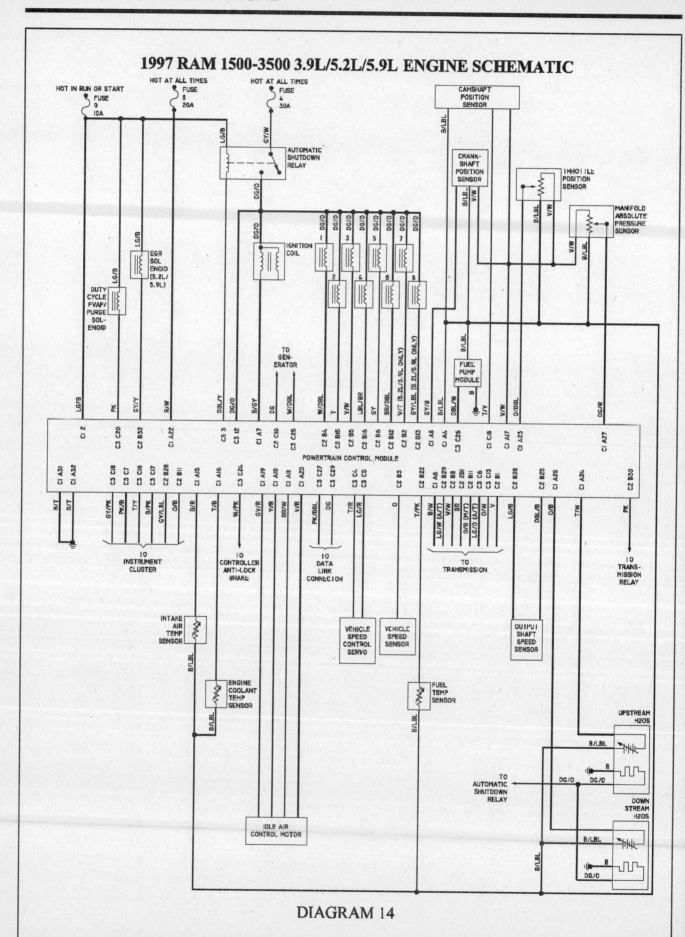

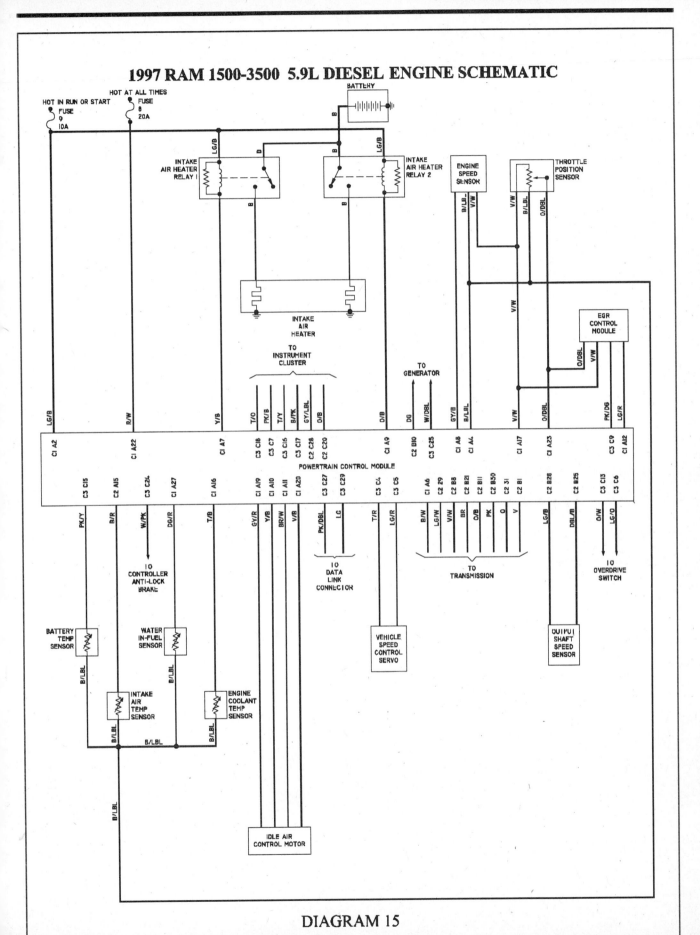

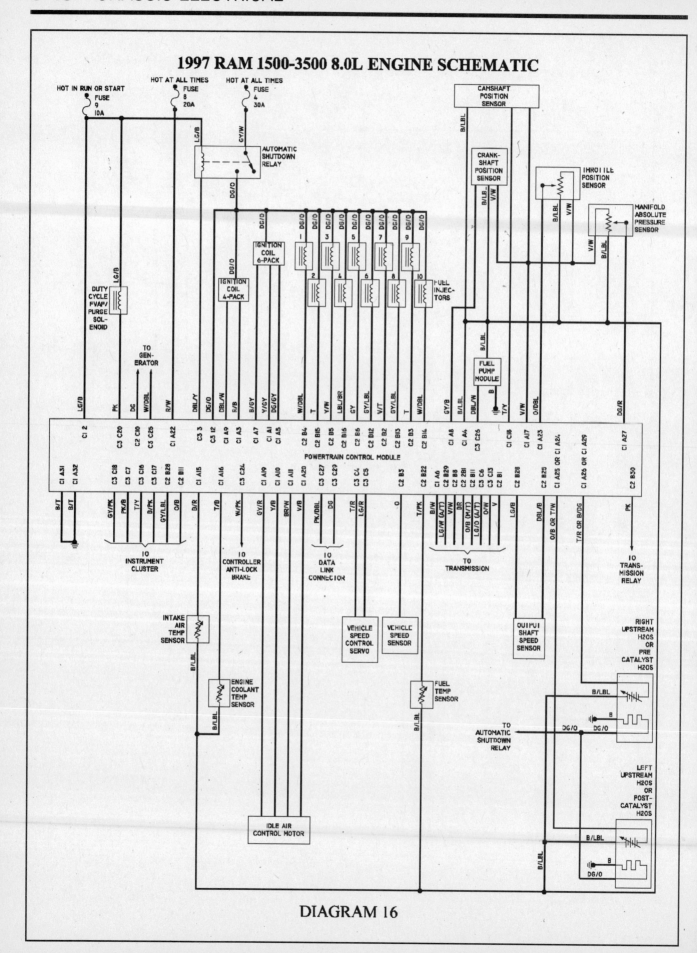

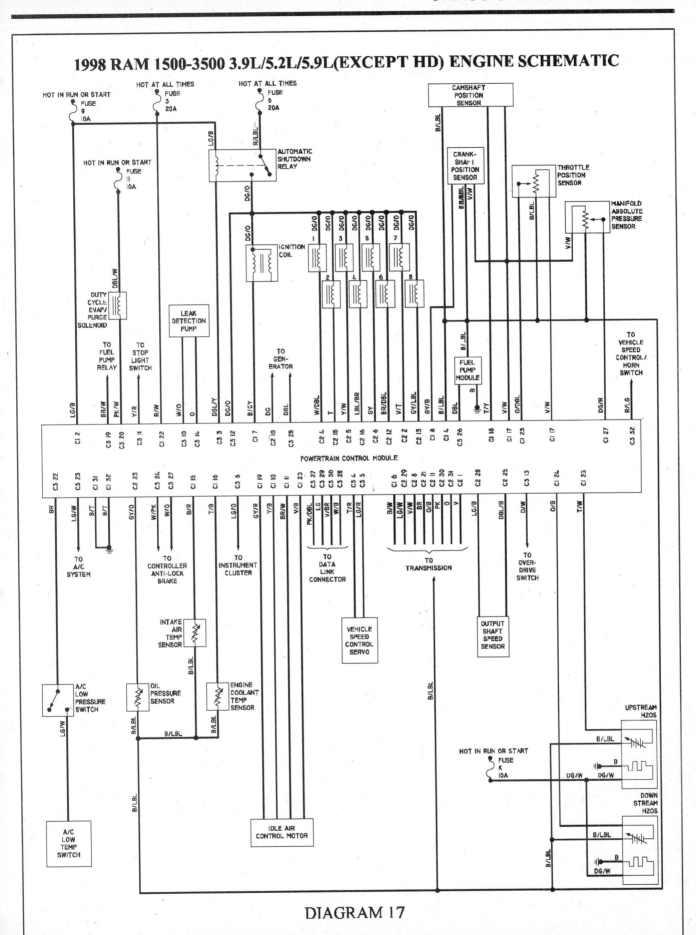

6-48 CHASSIS ELECTRICAL

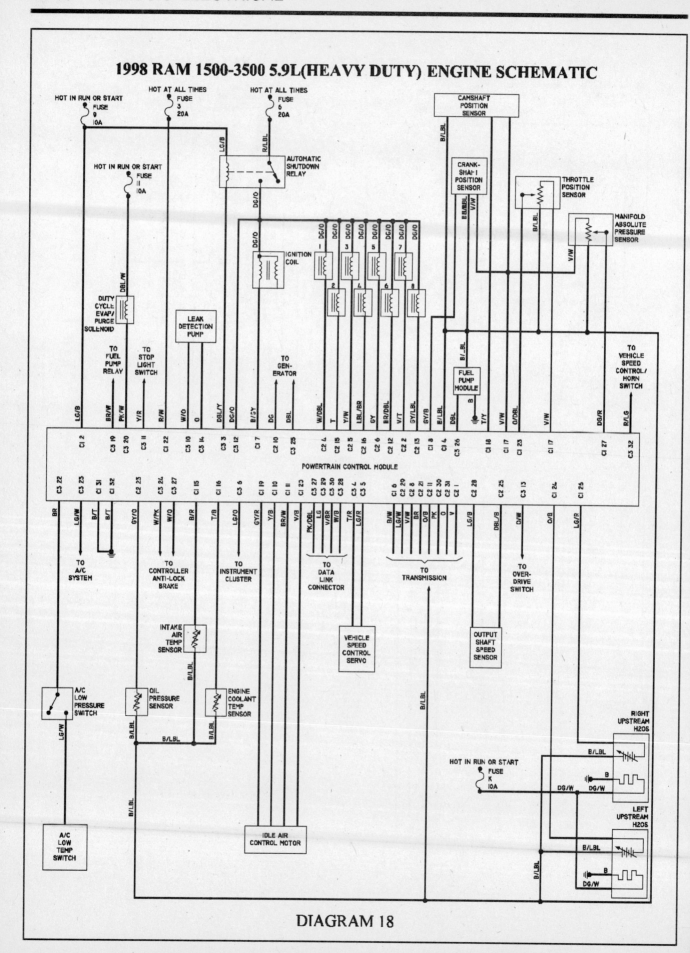

DIAGRAM 18

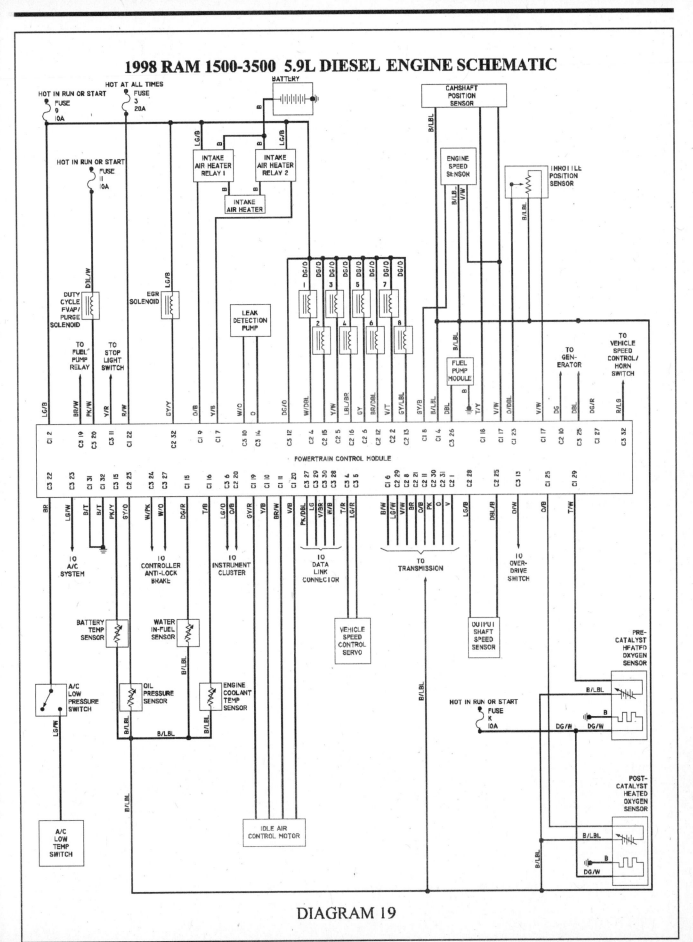

6-50 CHASSIS ELECTRICAL

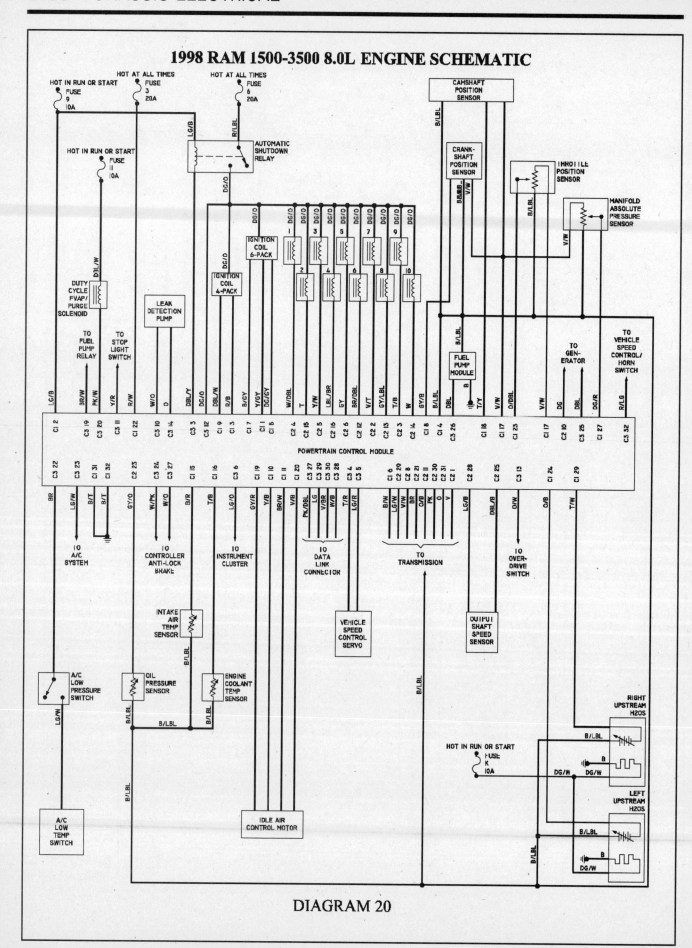

DIAGRAM 20

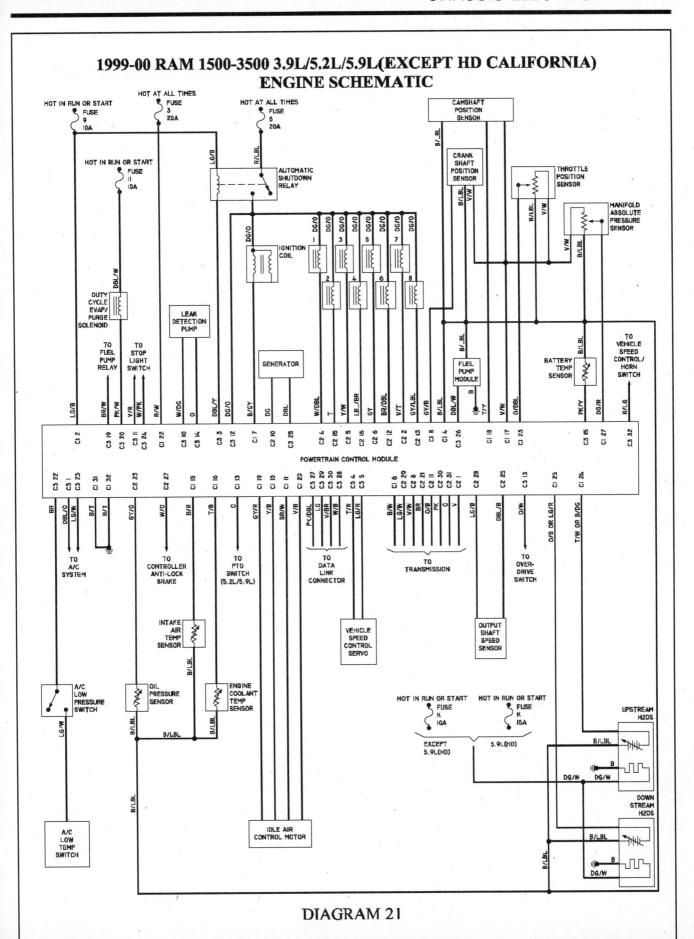

6-52 CHASSIS ELECTRICAL

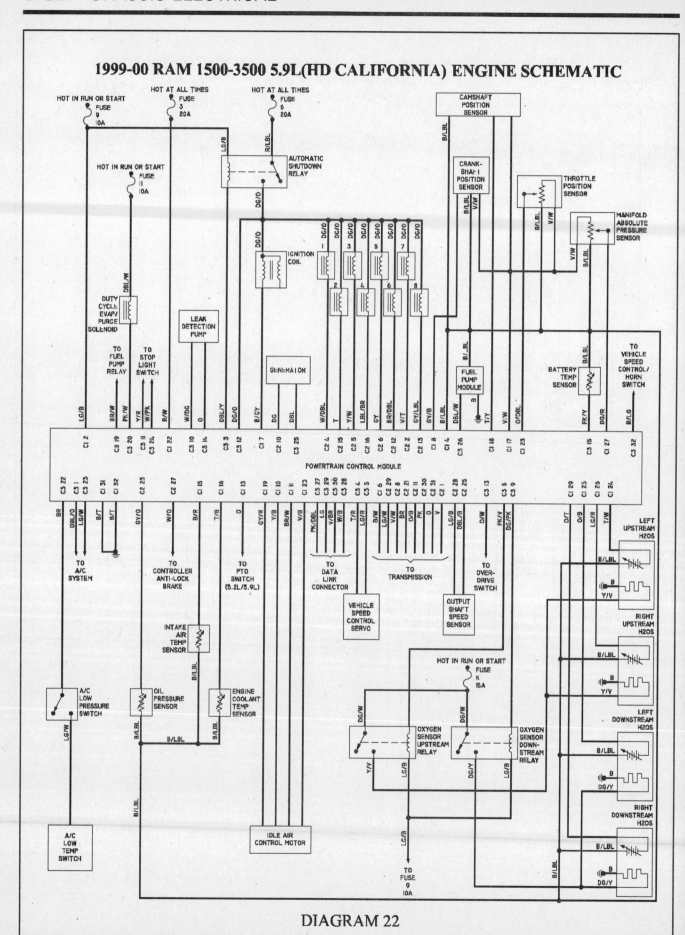

DIAGRAM 22

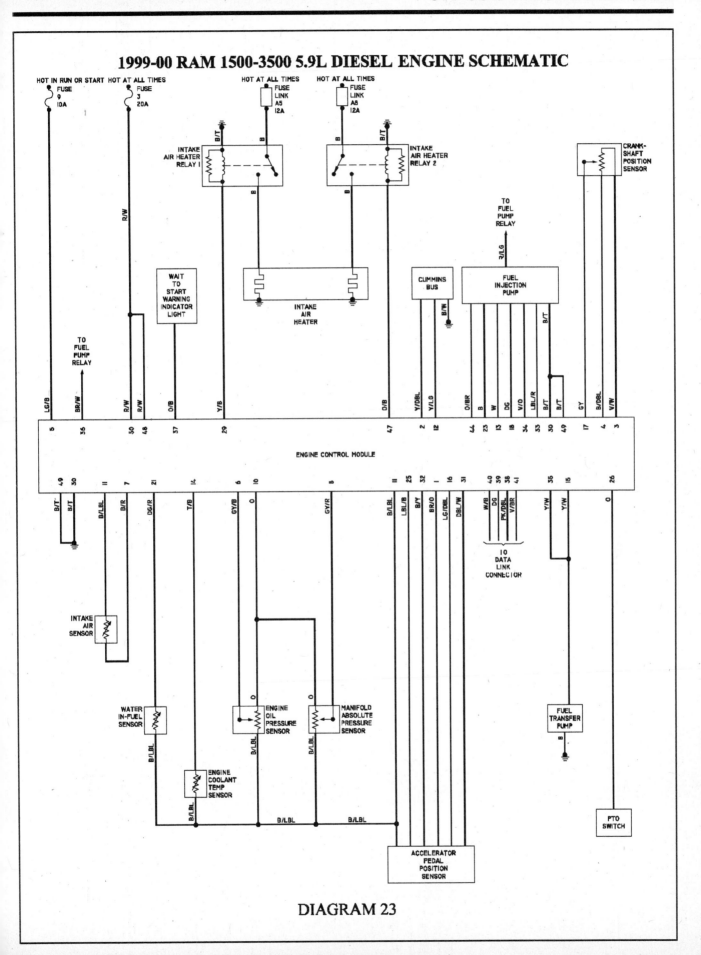

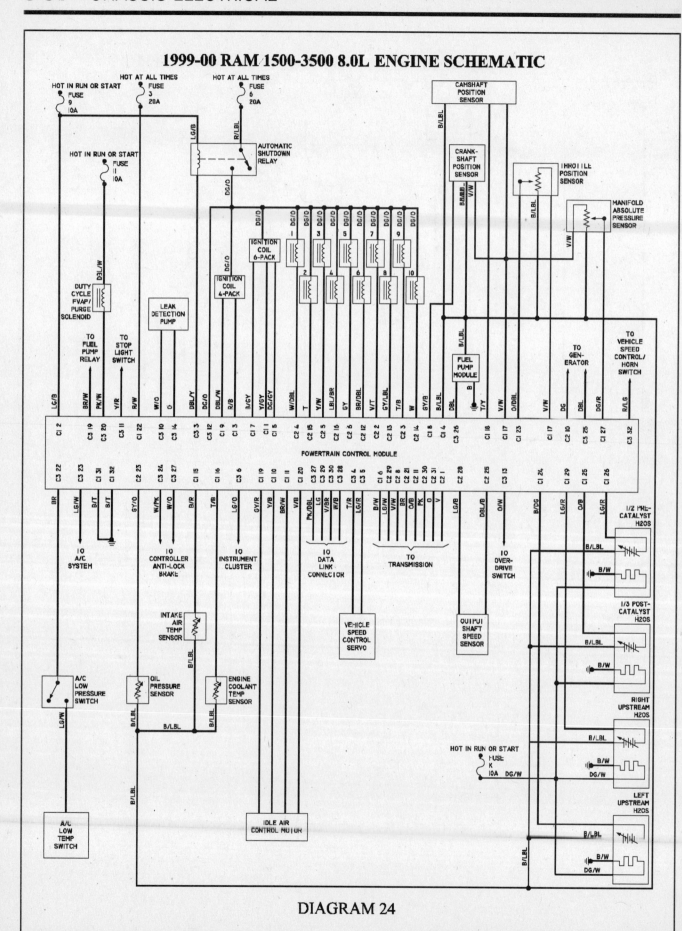

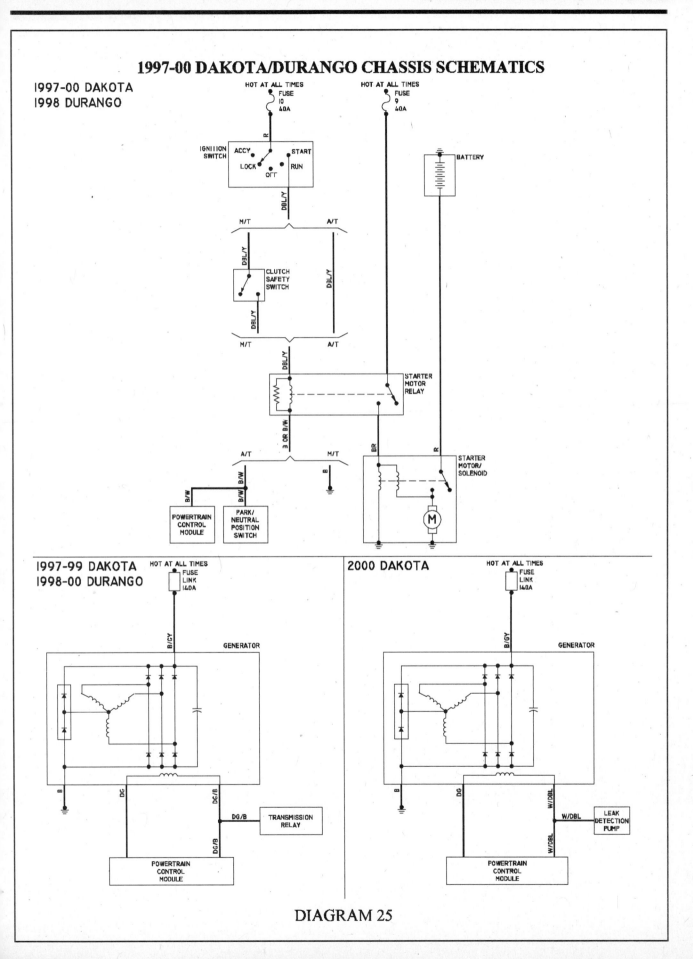

DIAGRAM 25

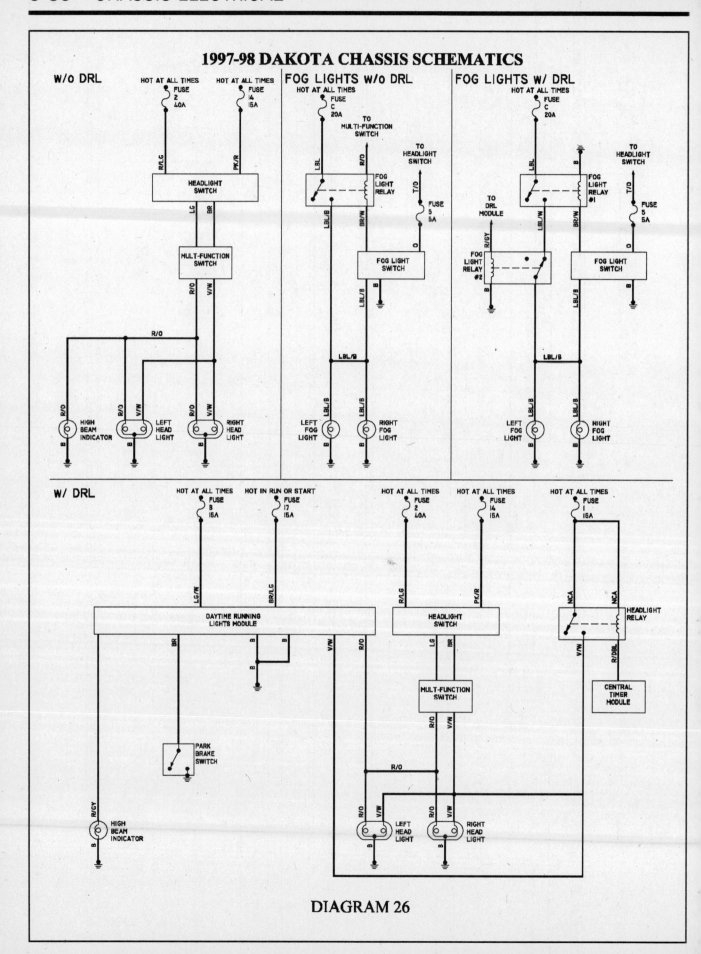

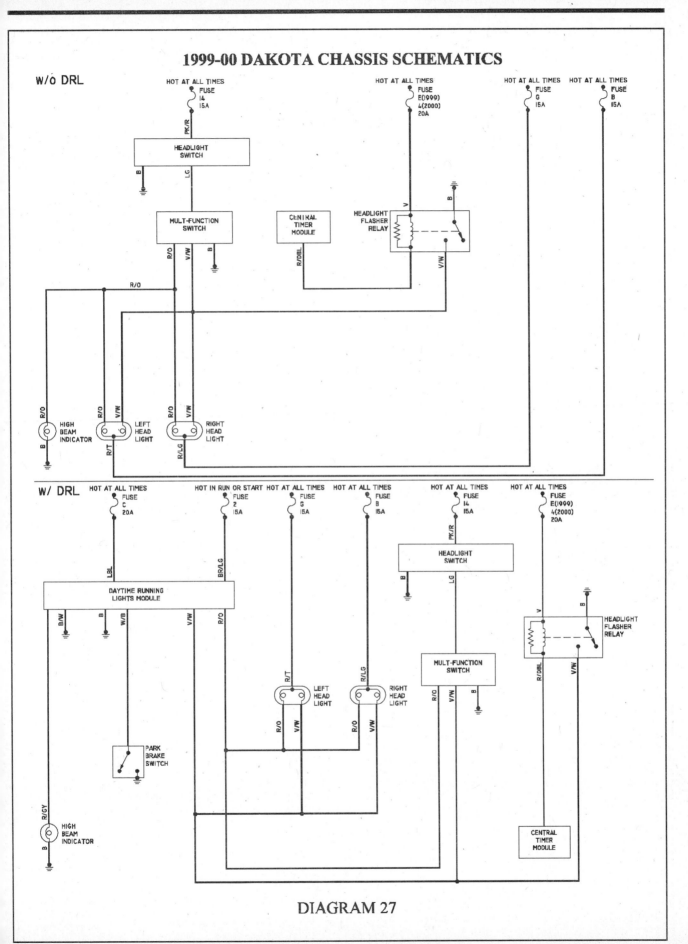

6-58 CHASSIS ELECTRICAL

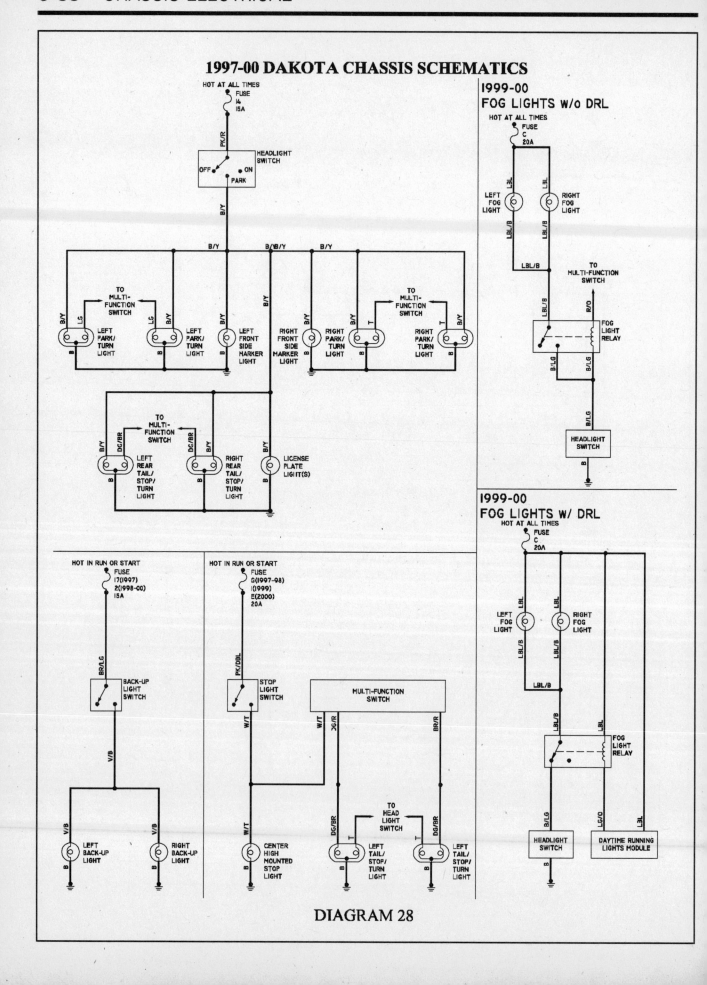

DIAGRAM 28

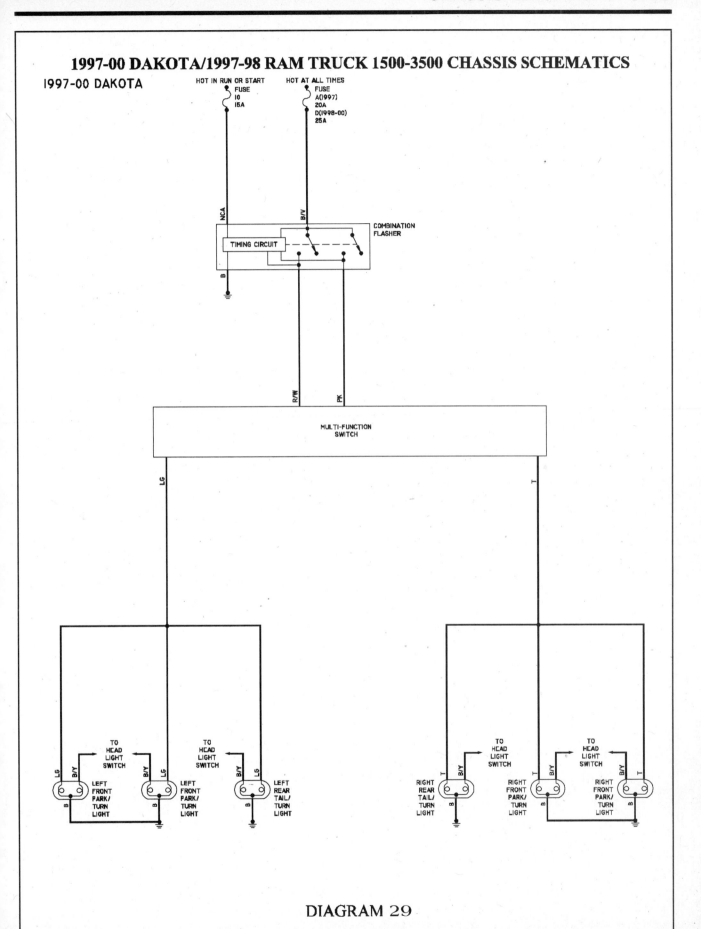

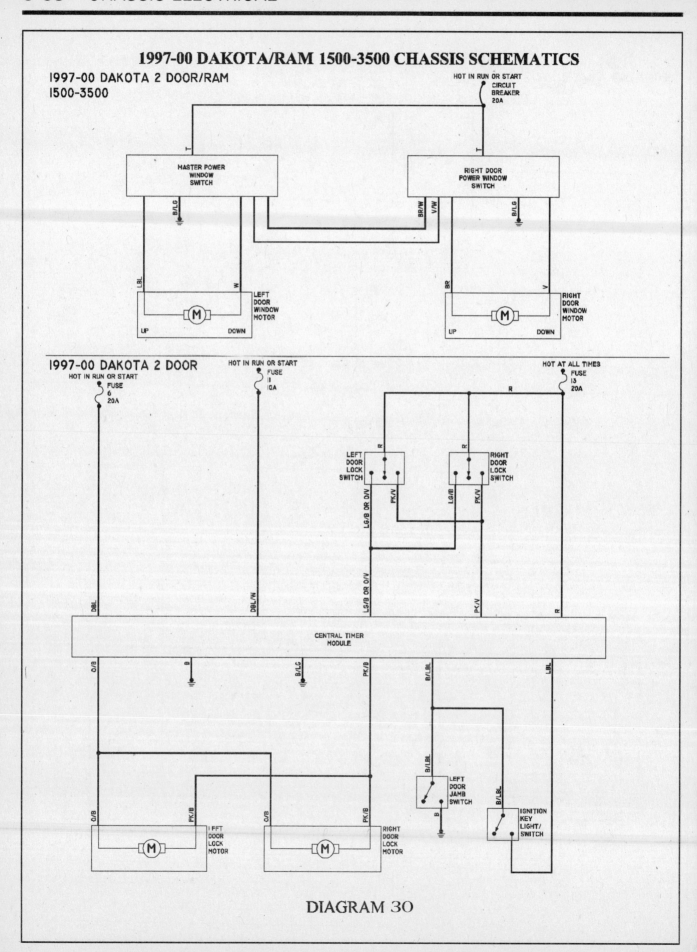

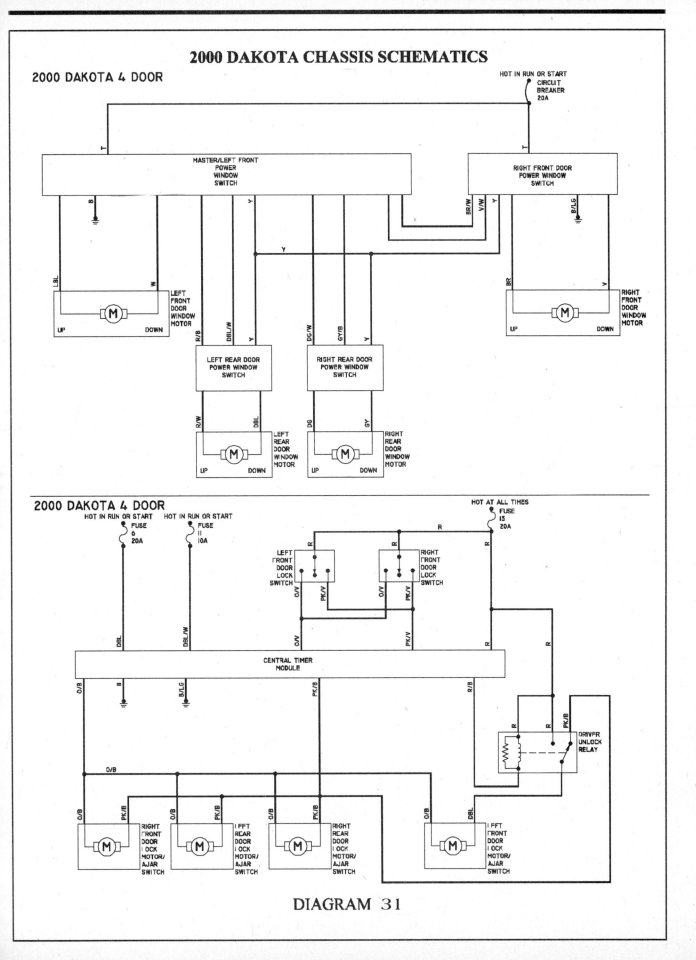

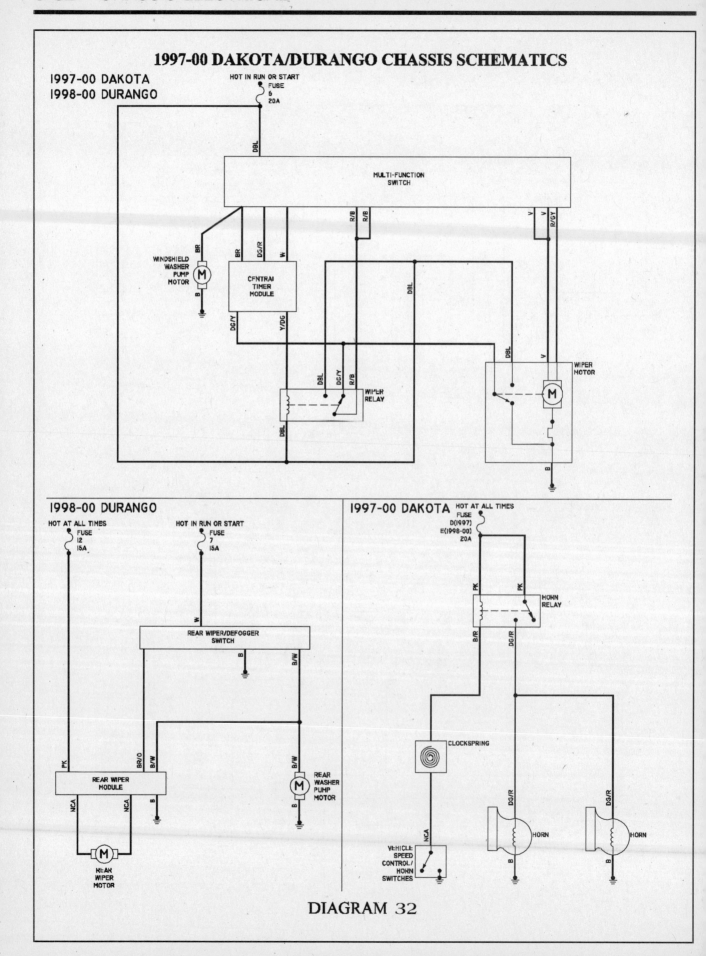

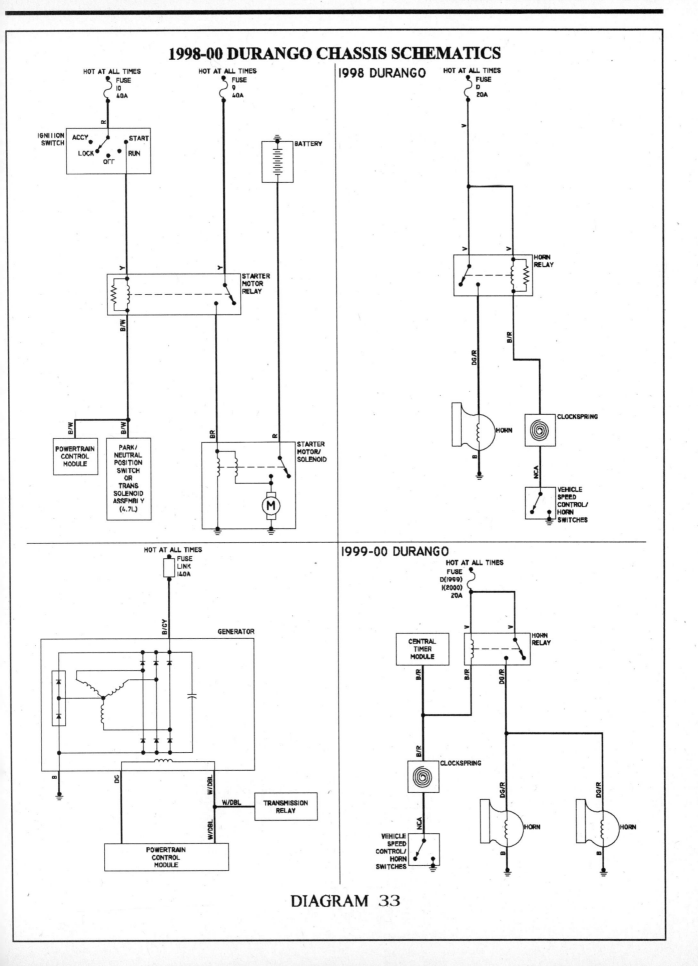

6-64 CHASSIS ELECTRICAL

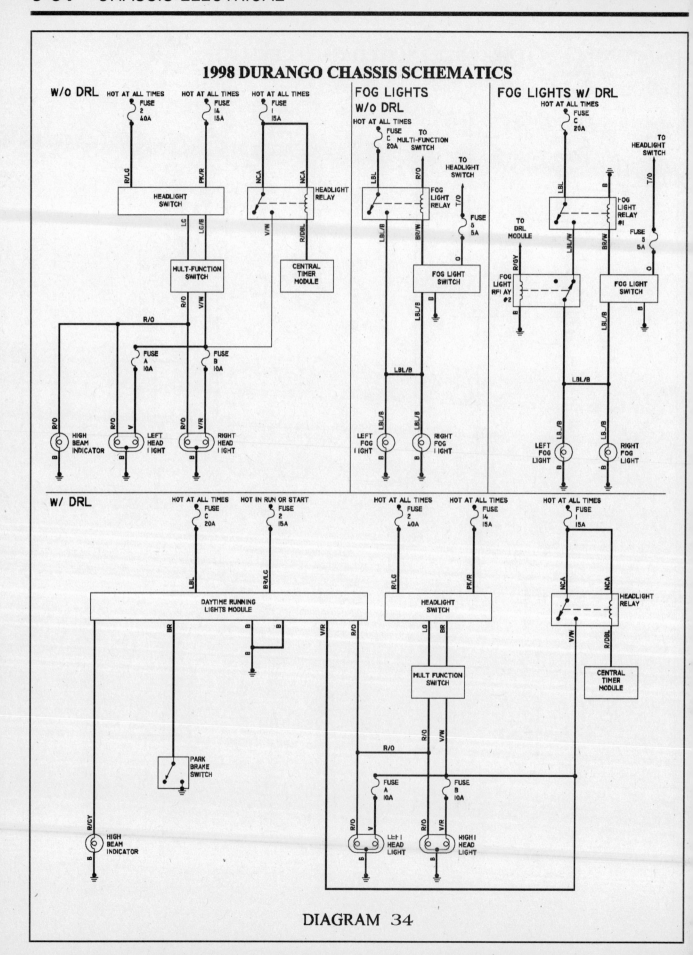

DIAGRAM 34

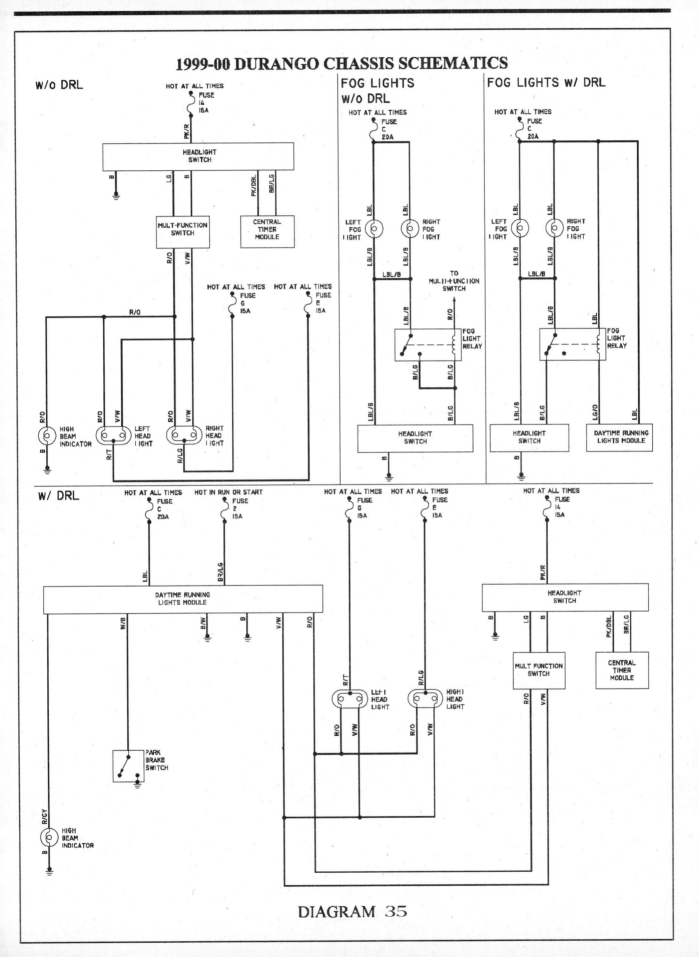

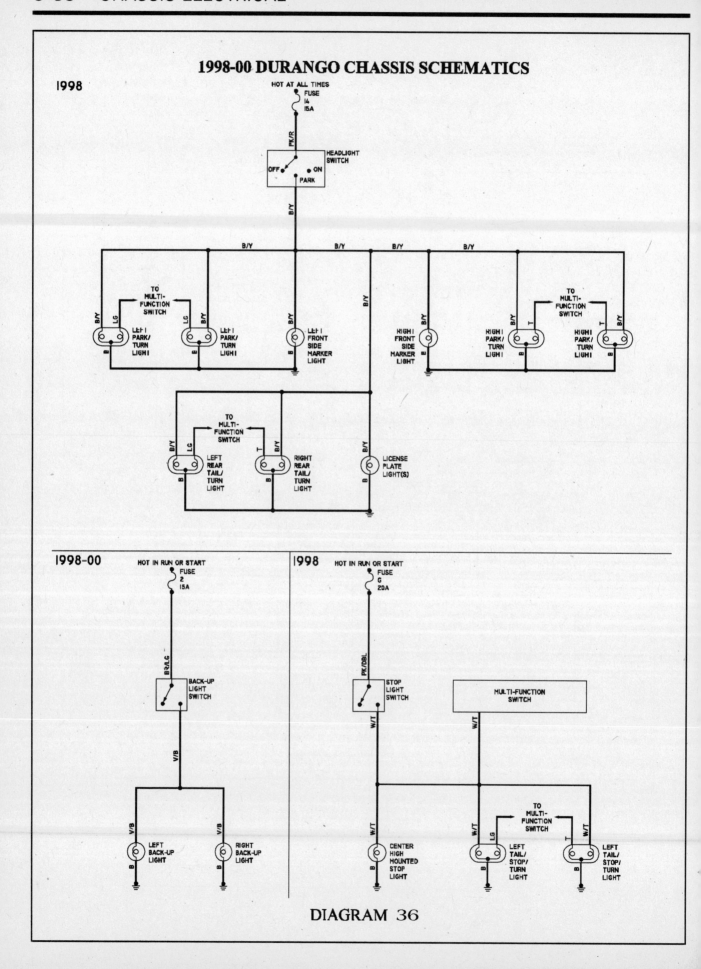

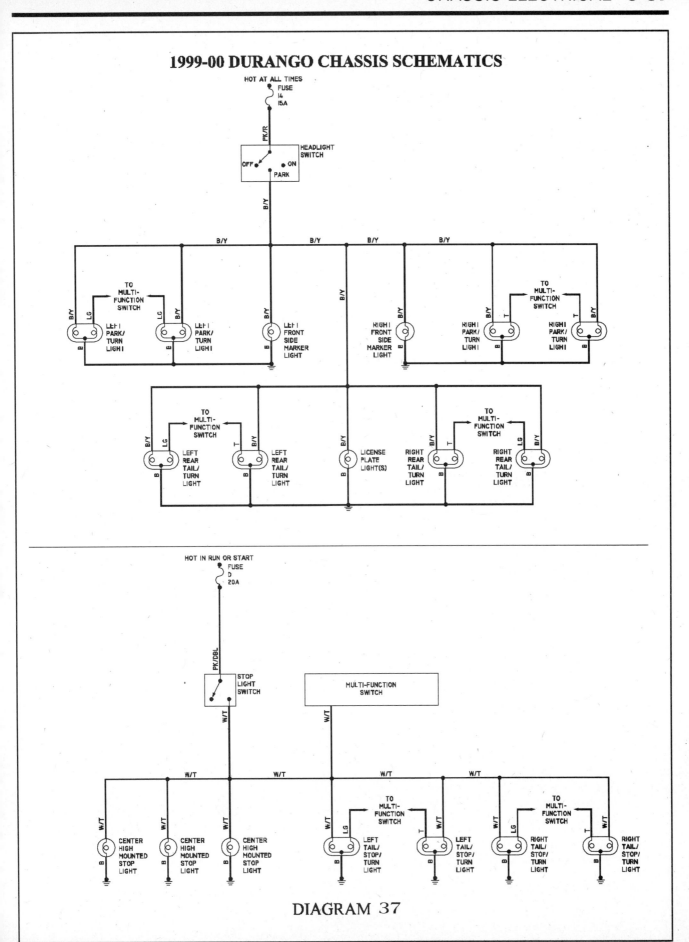

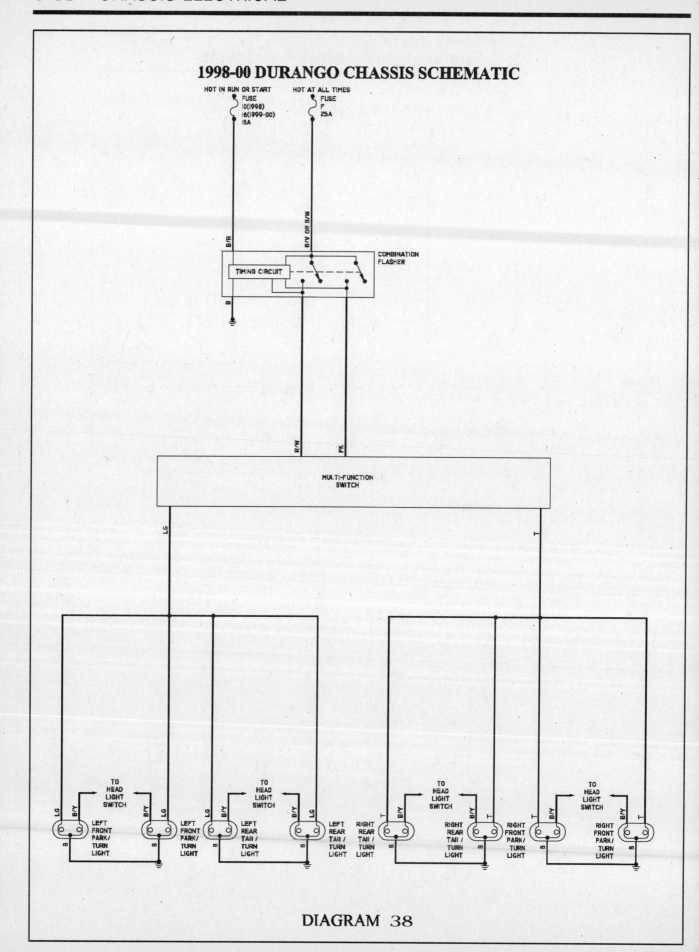

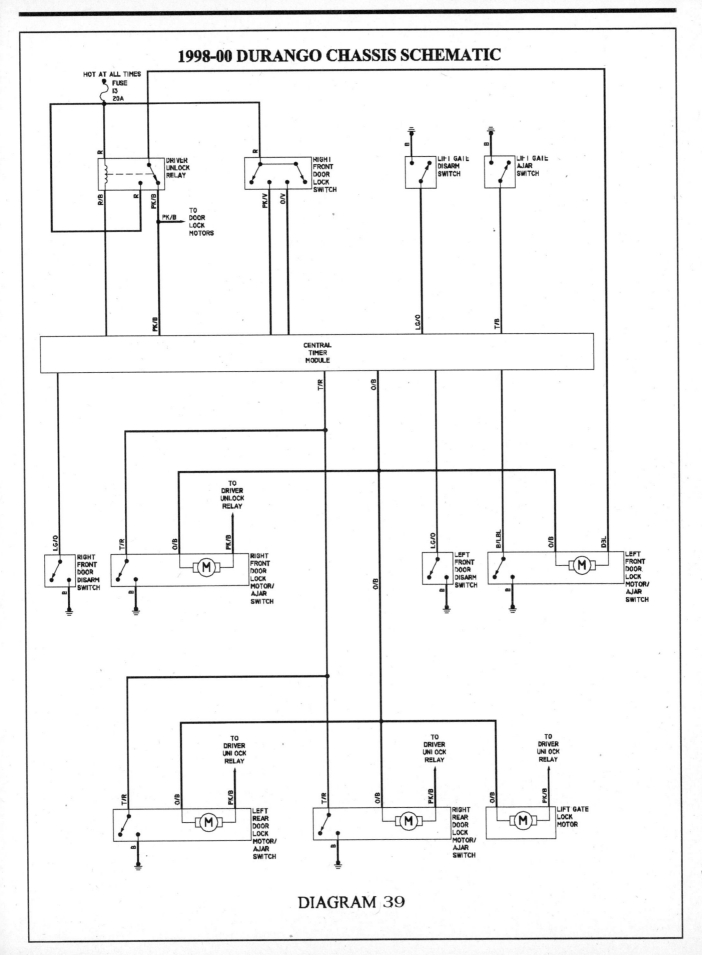

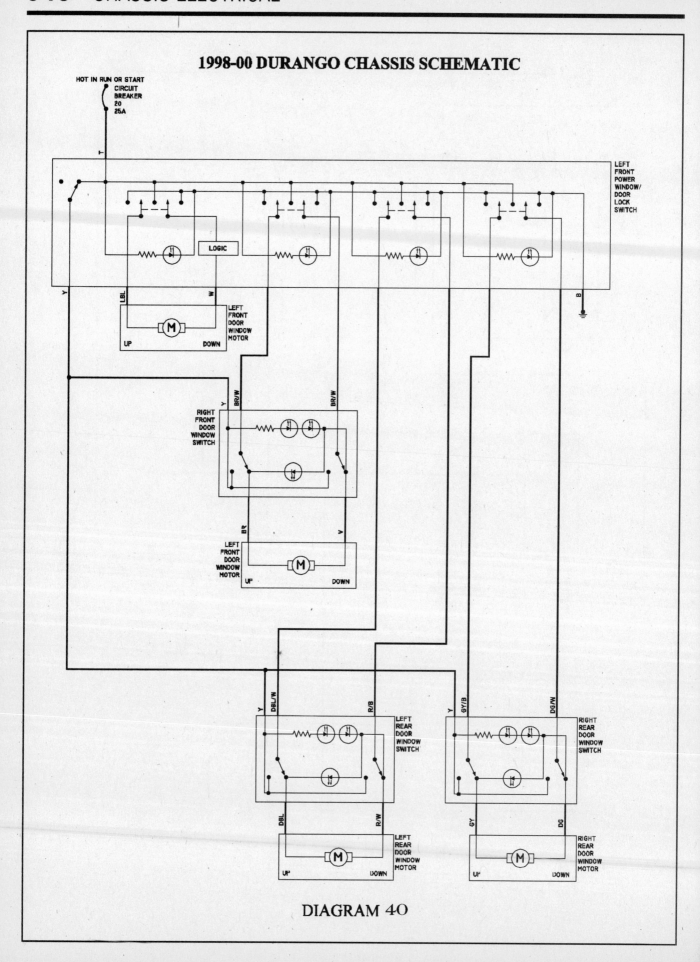

CHASSIS ELECTRICAL 6-71

1997-00 RAM 1500-3500 CHASSIS SCHEMATICS

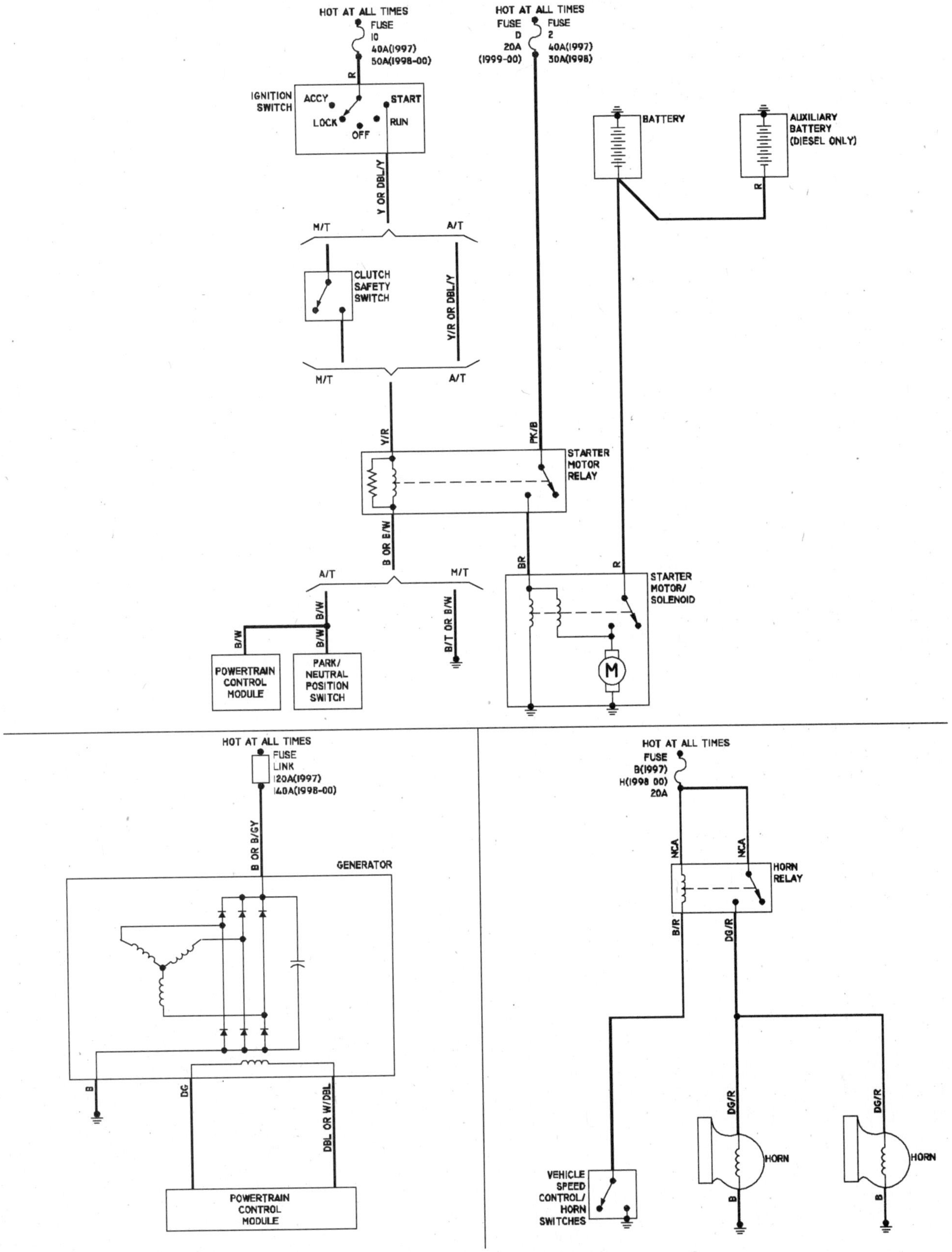

DIAGRAM 41

6-72 CHASSIS ELECTRICAL

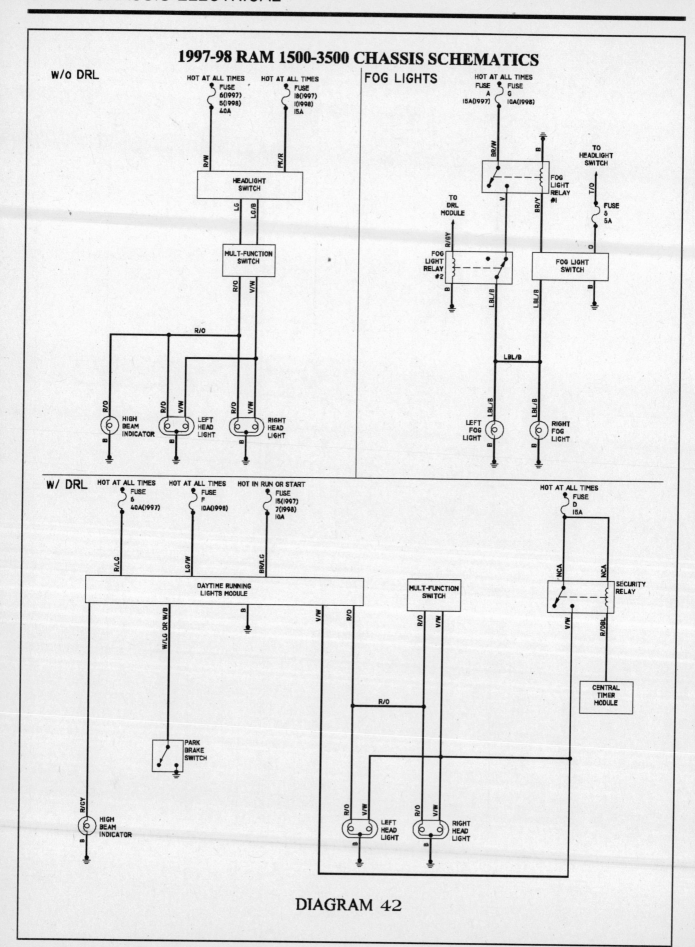

DIAGRAM 42

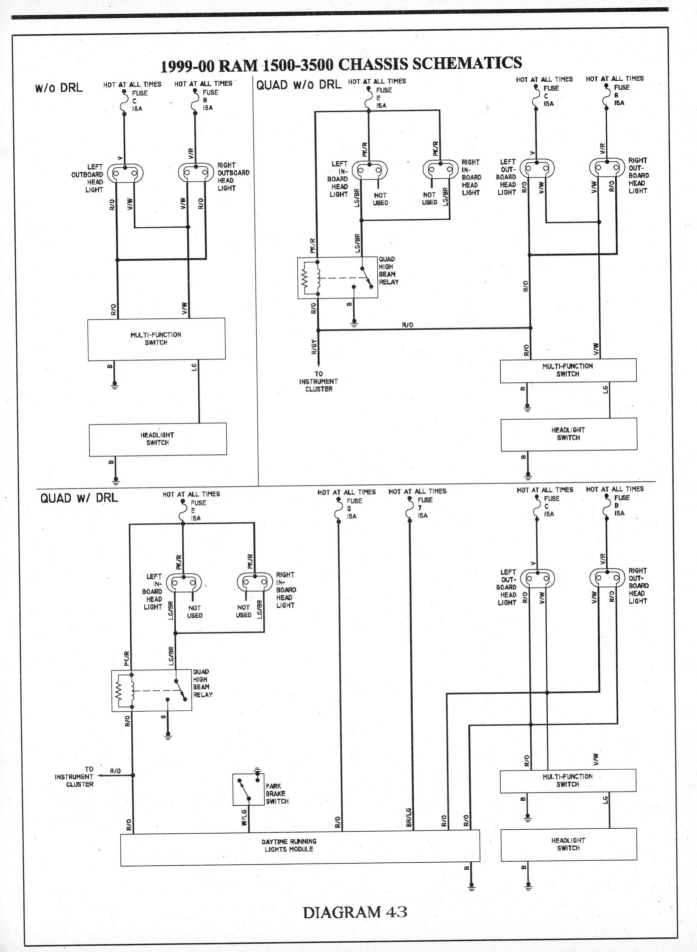

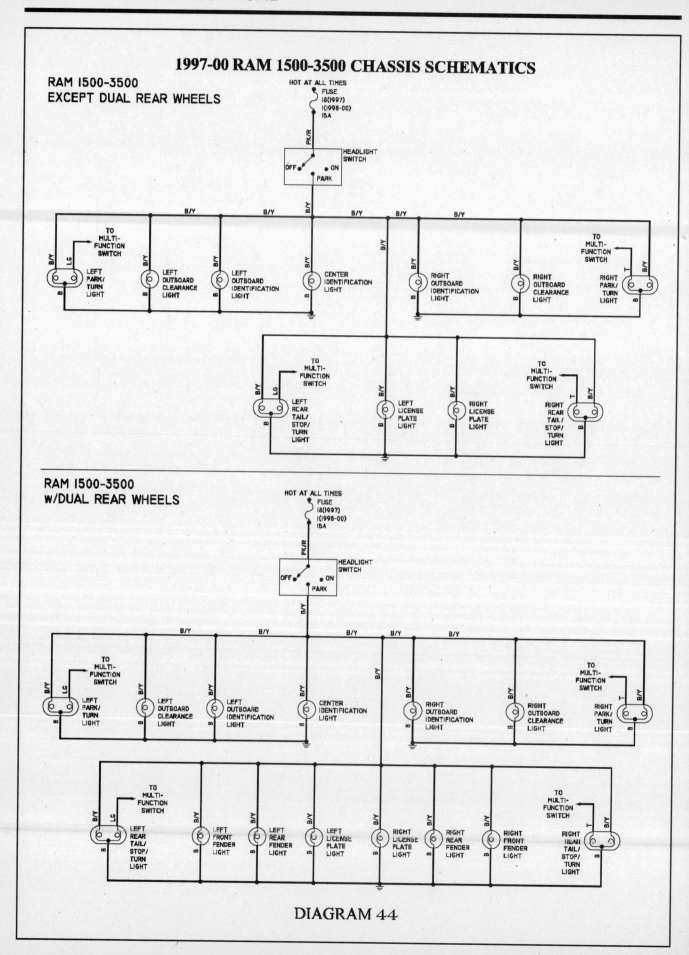

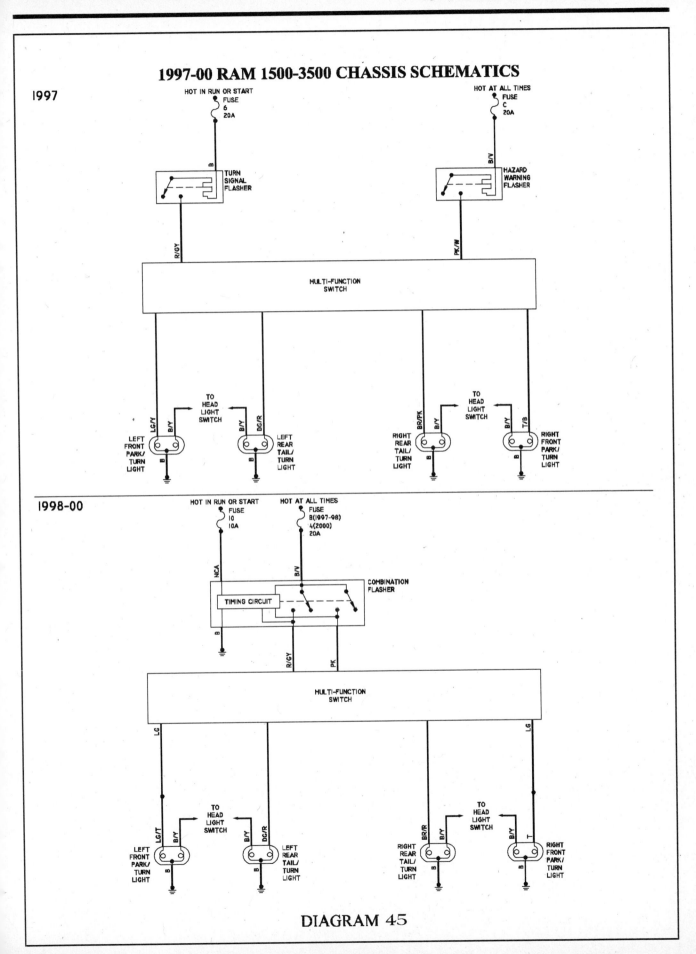

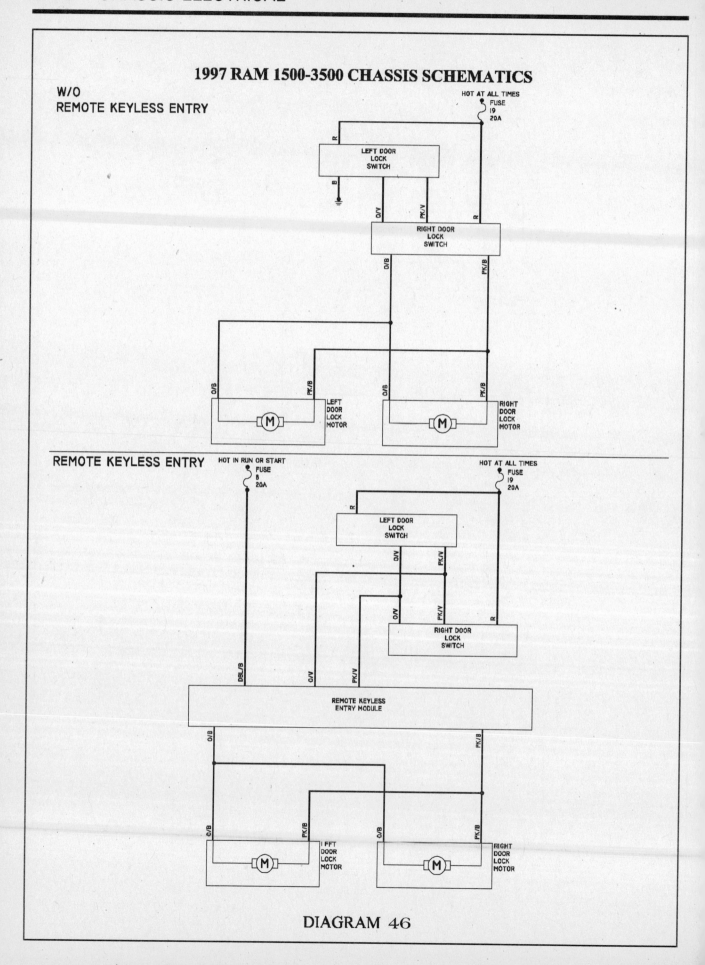

CHASSIS ELECTRICAL 6-77

1998-00 RAM 1500-3500 CHASSIS SCHEMATICS

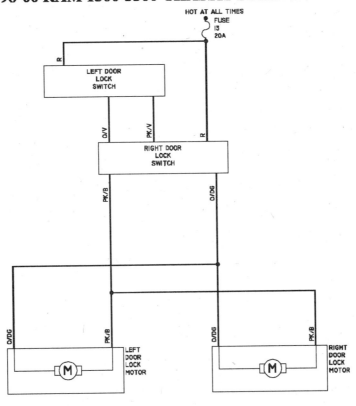

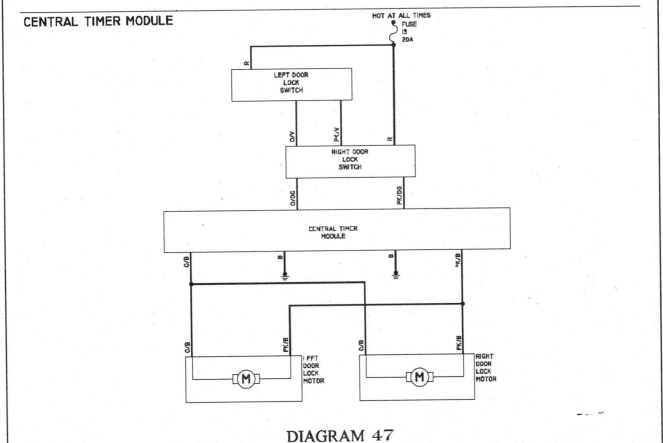

DIAGRAM 47

6-78 CHASSIS ELECTRICAL

1997-00 RAM 1500-3500 CHASSIS SCHEMATICS

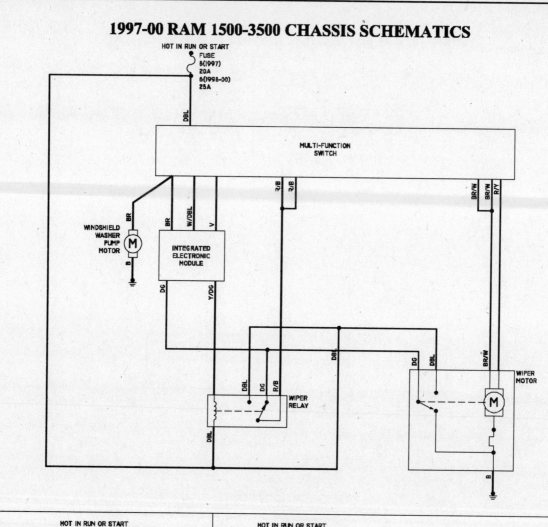

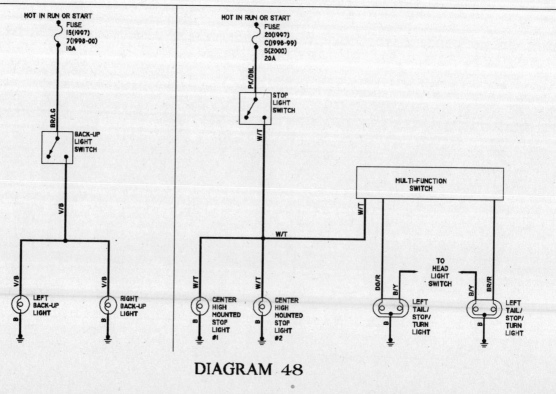

DIAGRAM 48

CHASSIS ELECTRICAL 6-79

1997-00 DAKOTA/DURANGO/RAM 1500-3500 CHASSIS SCHEMATICS

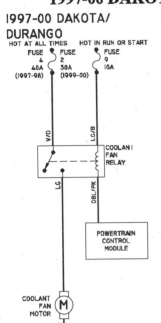

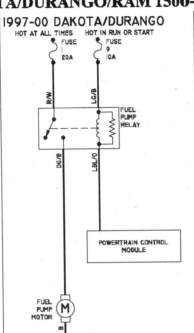

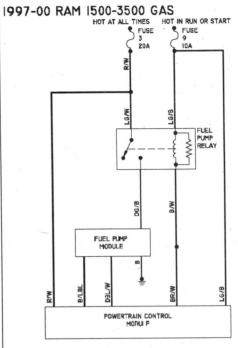

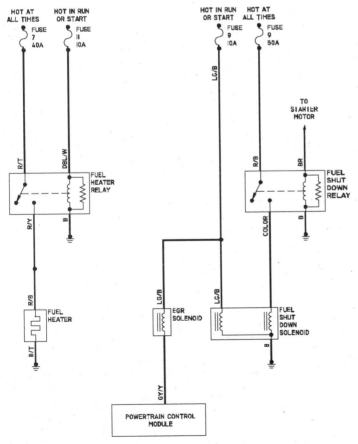

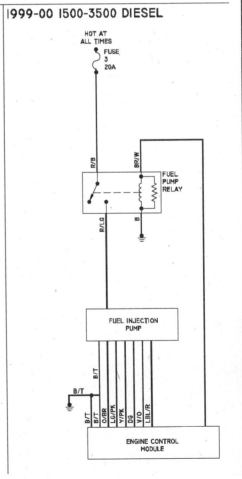

DIAGRAM 49

Notes

AUTOMATIC TRANSMISSION
AUTOMATIC TRANSMISSION ASSEMBLY 7-10
EXTENSION HOUSING SEAL 7-10
NEUTRAL SAFETY SWITCH 7-9
UNDERSTANDING THE AUTOMATIC
 TRANSMISSION 7-8

CLUTCH
CLUTCH HYDRAULICS 7-7
DRIVEN DISC AND PRESSURE PLATE 7-4
DRIVEN DISC AND PRESSURE PLATE
 ADJUSTMENTS 7-7
PILOT BEARING 7-7
RELEASE BEARING 7-7
UNDERSTANDING THE CLUTCH 7-4

DRIVELINE
CENTER BEARING 7-18
FRONT DRIVESHAFT AND U-JOINTS 7-14
REAR DRIVESHAFT AND U-JOINTS 7-16
TROUBLESHOOTING 7-14

FRONT DRIVE AXLE 7-19
AXLE SHAFT BEARING AND SEAL 7-22
CV JOINTS 7-20
FRONT AXLE HOUSING ASSEMBLY 7-24
FRONT DRIVESHAFT 7-19
PINION SEAL 7-23
TROUBLESHOOTING 7-19

MANUAL TRANSMISSION
BACK-UP LIGHT SWITCH 7-2
EXTENSION HOUSING SEAL 7-3
MANUAL TRANSMISSION ASSEMBLY 7-3
SHIFT HANDLE 7-2
UNDERSTANDING THE MANUAL
 TRANSMISSION 7-2

REAR AXLE 7-25
AXLE SHAFT BEARING AND SEAL 7-26
PINION SEAL 7-28
REAR AXLE HOUSING ASSEMBLY 7-29
TROUBLESHOOTING 7-25

TRANSFER CASE
FRONT OUTPUT SHAFT SEAL 7-13
REAR OUTPUT SHAFT SEAL 7-12
SHIFT LINKAGE 7-12
TRANSFER CASE ASSEMBLY 7-13

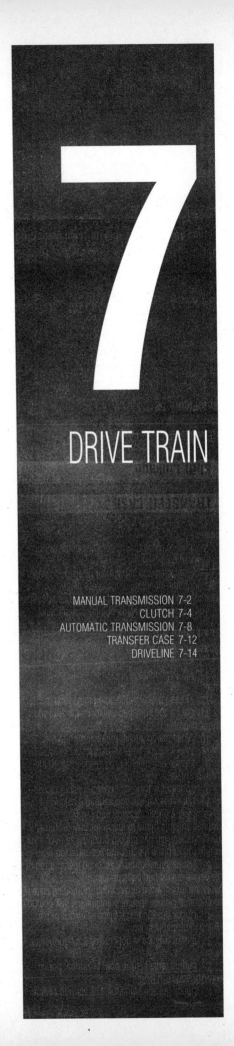

7

DRIVE TRAIN

MANUAL TRANSMISSION 7-2
CLUTCH 7-4
AUTOMATIC TRANSMISSION 7-8
TRANSFER CASE 7-12
DRIVELINE 7-14

7-2 DRIVE TRAIN

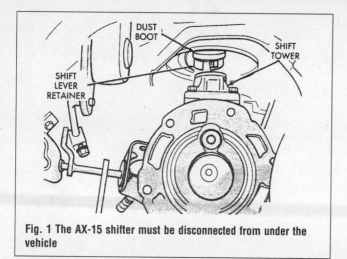

Fig. 1 The AX-15 shifter must be disconnected from under the vehicle

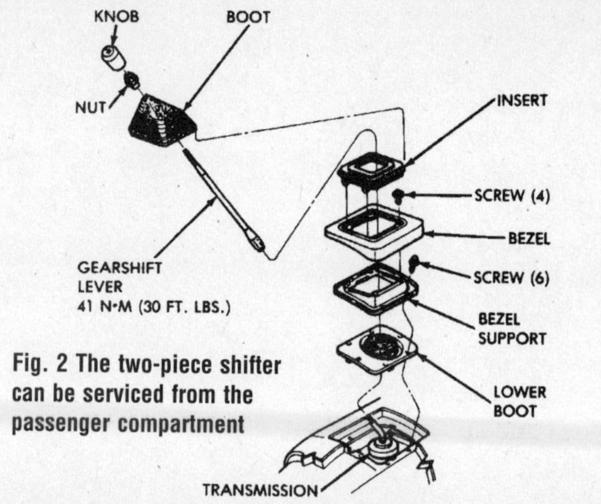

Fig. 2 The two-piece shifter can be serviced from the passenger compartment

MANUAL TRANSMISSION

Understanding the Manual Transmission

Because of the way an internal combustion engine breathes, it can produce torque (or twisting force) only within a narrow speed range. Most overhead valve pushrod engines must turn at about 2500 rpm to produce their peak torque. Often by 4500 rpm, they are producing so little torque that continued increases in engine speed produce no power increases.

The torque peak on overhead camshaft engines is, generally, much higher, but much narrower.

The manual transmission and clutch are employed to vary the relationship between engine RPM and the speed of the wheels so that adequate power can be produced under all circumstances. The clutch allows engine torque to be applied to the transmission input shaft gradually, due to mechanical slippage. The vehicle can, consequently, be started smoothly from a full stop.

The transmission changes the ratio between the rotating speeds of the engine and the wheels by the use of gears. 4-speed or 5-speed transmissions are most common. The lower gears allow full engine power to be applied to the rear wheels during acceleration at low speeds.

The clutch driveplate is a thin disc, the center of which is splined to the transmission input shaft. Both sides of the disc are covered with a layer of material which is similar to brake lining and which is capable of allowing slippage without roughness or excessive noise.

The clutch cover is bolted to the engine flywheel and incorporates a diaphragm spring which provides the pressure to engage the clutch. The cover also houses the pressure plate. When the clutch pedal is released, the driven disc is sandwiched between the pressure plate and the smooth surface of the flywheel, thus forcing the disc to turn at the same speed as the engine crankshaft.

The transmission contains a mainshaft which passes all the way through the transmission, from the clutch to the driveshaft. This shaft is separated at one point, so that front and rear portions can turn at different speeds.

Power is transmitted by a countershaft in the lower gears and reverse. The gears of the countershaft mesh with gears on the mainshaft, allowing power to be carried from one to the other. Countershaft gears are often integral with that shaft, while several of the mainshaft gears can either rotate independently of the shaft or be locked to it. Shifting from one gear to the next causes one of the gears to be freed from rotating with the shaft and locks another to it. Gears are locked and unlocked by internal dog clutches which slide between the center of the gear and the shaft. The forward gears usually employ synchronizers; friction members which smoothly bring gear and shaft to the same speed before the toothed dog clutches are engaged.

Shift Handle

REMOVAL & INSTALLATION

AX-15

See Figure 1

The gearshift lever is a one-piece design which must be disconnected from under the vehicle.
1. Lower the transmission slightly (about 3 in.).
2. Reach up and around the transmission case and press the shift lever retainer downward with your fingers.
3. Turn the retainer CCW to release it.
4. Lift the lever and retainer out of the shift tower.

Note: It is not necessary to remove the shift lever from the floorpan boot. Simply leave the lever in place for later installation.

To install:

5. Position the transmission in the vehicle.
6. Reach up and around the transmission.
7. Position the lever and retainer in the shift tower.
8. Turn the retainer CW to install it.

Other Manual Transmissions
See Figure 2

The gearshift lever is a two-piece design. The upper part of the lever is threaded to the transmission stub lever. The upper lever can be removed for service without having to remove the entire shift lever assembly or the transmission.
1. Remove the boot mounting screws, the support and trim bezel for access to the lower end of the shifter.
2. Unthread and remove the upper end of the shift lever.

To install:
3. Service the lever as required.
4. Thread the upper lever to the stud end and secure the shift boot and components.

Back-up Light Switch

Manual transmissions incorporate a back-up light switch. This is normally located on the driver's side, just below and behind the shifter. The switch is activated by the internal shift mechanism.

TESTING

1. Disconnect the switch wiring.
2. Place the transmission in REVERSE.
3. Check for continuity across the switch terminals. If there is none, replace the switch.

REMOVAL & INSTALLATION

See Figure 3

1. Disconnect the switch wiring.
2. Unscrew and remove the switch.
3. Installation is the reverse of removal. Use a new gasket and tighten the switch to 25 ft. lbs. (30 Nm).

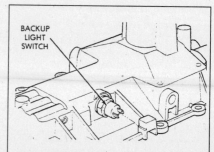

Fig. 3 Back-up switch location - typical

DRIVE TRAIN 7-3

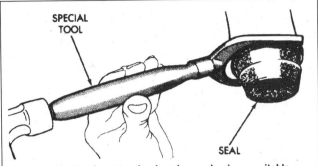

Fig. 4 Removing the extension housing seal using a suitable seal removal fork

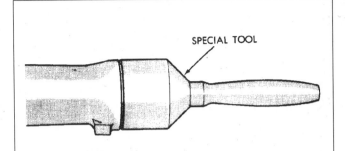

Fig. 5 Installing the extension housing seal using a suitable seal driver

Extension Housing Seal

REMOVAL & INSTALLATION

See Figures 4 and 5

Note: A special seal removal/installation tool is required.

1. Raise and safely support the vehicle securely on jackstands.
2. Place a drain pan under the end of the extension housing.
3. Mark the position of the driveshaft for installation reference.
4. Disconnect the driveshaft at the rear universal joint and carefully slide the shaft out of the transmission extension housing.
5. Remove the extension housing seal with the special tool.
6. Clean the end and inside of the extension housing.
7. Start the new seal into position and, with the installing tool, tap the seal into place.
8. Check the condition of the driveshaft slip yoke. Repair or replace as necessary.
9. Lubricate the lips of the seal and the slip yoke.
10. Align and install the driveshaft, taking care when guiding the slip yoke into the extension housing.
11. Remove the drain pan.
12. Check and correct the transmission fluid level.
13. Lower the vehicle.

Manual Transmission Assembly

REMOVAL & INSTALLATION

See Figures 6, 7, 8 and 9

CAUTION:

Significant equipment is required to remove and install the transmission safely. This includes supports and jacks for the truck, the engine, transfer case (if equipped) and transmission. Be sure all equipment is rated to safely support the weight of the component. Heavy-duty transmissions are a significant load. Scissors-style jacks are recommended for these items.

1. Disconnect the battery negative cable(s).
2. Place the transmission in NEUTRAL.

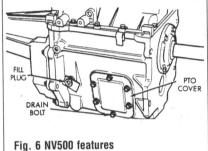

Fig. 6 NV500 features

3. Remove as much of the shift lever assembly as is accessible from inside the truck: knob, boot, shaft, tower, isolator plate, etc. On some models removal of the console and/or floor panel is required.
4. Raise and support the vehicle safely.
5. Remove the skidplate(s) if fitted.
6. Remove any frame crossmembers which may interfere with removal.
7. If the transmission is to be rebuilt, drain the oil.
8. Matchmark the driveshaft(s) and yokes or flanges for assembly alignment, then disconnect and remove them. On 4x4 vehicles, remove the front driveshaft as well.
9. Remove the exhaust system Y-pipe, then any other segments of the system that would interfere with transmission removal.
10. Support the engine with a suitable jack device and wooden block.
11. Disconnect the backup switch wires.

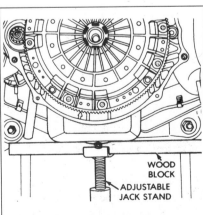

Fig. 8 Use a reliable method of supporting components during the removal process

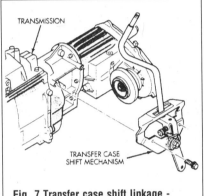

Fig. 7 Transfer case shift linkage - typical

12. Disconnect the speed sensor wires.
13. Remove the rear transmission mount hardware, if fitted.
14. If 4x4, disconnect the transfer case shift linkage at the case. Remove the transfer case shift mechanism from the transmission.
15. Support, unbolt and remove the transfer case.
16. Support the transmission with a suitable jack and secure it with safety chains.
17. Remove the transmission harness from the retaining clips on the transmission shift cover.
18. Remove the hardware attaching the transmission mount to the rear crossmember, if one is fitted.
19. Remove the clutch slave cylinder splash shield, if fitted.
20. Remove the clutch slave cylinder attaching nuts and place it aside.

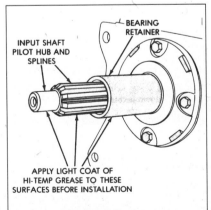

Fig. 9 Lubricate the input shaft and other contact surfaces during installation

7-4 DRIVE TRAIN

21. Dakota:
    a. Remove the starter motor.
    b. Remove the transmission splash shield, if fitted.
    c. Lower the transmission slightly.
    d. Remove the shift tower and lever assembly, if it wasn't removed before.
22. Remove the bolts attaching the transmission to the clutch housing.
23. Move the box rearward until the input shaft clears the clutch disc and release bearing.
24. Lower the transmission and remove it from the vehicle.

To install:

25. Reverse the removal procedure. Note the following points.
26. Apply a light coat of high temperature grease to the moving parts of the clutch and shifter assemblies:
    - Drive gear splines and pilot bearing hub
    - Release bearing slide surface of the front retainer
    - Pilot bearing
    - Release bearing bore
    - Release fork
    - Release fork ball stud
    - Driveshaft slip yoke
27. Tighten transmission-to-engine bolts to 80 ft. lbs. (108 Nm).
28. Tighten the transfer case-to-transmission nuts: 5/16 in. nuts to 22 - 30 ft. lbs.; 3/8 in. nuts to 30 - 35 ft. lbs. (41 - 47 Nm).
29. Frame crossmember and skidplate hardware should be torqued to 30 ft. lbs. (41 Nm).

CLUTCH

Understanding the Clutch

The purpose of the clutch is to disconnect and connect engine power at the transmission. A vehicle at rest requires a lot of engine torque to get all that weight moving. An internal combustion engine does not develop a high starting torque (unlike steam engines) so it must be allowed to operate without any load until it builds up enough torque to move the vehicle. To a point, torque increases with engine rpm. The clutch allows the engine to build up torque by physically disconnecting the engine from the transmission, relieving the engine of any load or resistance.

The transfer of engine power to the transmission (the load) must be smooth and gradual; if it weren't, drive line components would wear out or break quickly. This gradual power transfer is made possible by gradually releasing the clutch pedal. The clutch disc and pressure plate are the connecting link between the engine and transmission. When the clutch pedal is released, the disc and plate contact each other (the clutch is engaged) physically joining the engine and transmission. When the pedal is pushed in, the disc and plate separate (the clutch is disengaged) disconnecting the engine from the transmission.

Most clutch assemblies consists of the flywheel, the clutch disc, the clutch pressure plate, the throw out bearing and fork, the actuating linkage and the pedal. The flywheel and clutch pressure plate (driving members) are connected to the engine crankshaft and rotate with it. The clutch disc is located between the flywheel and pressure plate, and is splined to the transmission shaft. A driving member is one that is attached to the engine and transfers engine power to a driven member (clutch disc) on the transmission shaft. A driving member (pressure plate) rotates (drives) a driven member (clutch disc) on contact and, in so doing, turns the transmission shaft.

There is a circular diaphragm spring within the pressure plate cover (transmission side). In a relaxed state (when the clutch pedal is fully released) this spring is convex; that is, it is dished outward toward the transmission. Pushing in the clutch pedal actuates the attached linkage. Connected to the other end of this is the throw out fork, which holds the throw out bearing. When the clutch pedal is depressed, the clutch linkage pushes the fork and bearing forward to contact the diaphragm spring of the pressure plate. The outer edges of the spring are secured to the pressure plate and are pivoted on rings so that when the center of the spring is compressed by the throw out bearing, the outer edges bow outward and, by so doing, pull the pressure plate in the same direction - away from the clutch disc. This action separates the disc from the plate, disengaging the clutch and allowing the transmission to be shifted into another gear. A coil type clutch return spring attached to the clutch pedal arm permits full release of the pedal. Releasing the pedal pulls the throw out bearing away from the diaphragm spring resulting in a reversal of spring position. As bearing pressure is gradually released from the spring center, the outer edges of the spring bow outward, pushing the pressure plate into closer contact with the clutch disc. As the disc and plate move closer together, friction between the two increases and slippage is reduced until, when full spring pressure is applied (by fully releasing the pedal) the speed of the disc and plate are the same. This stops all slipping, creating a direct connection between the plate and disc which results in the transfer of power from the engine to the transmission. The clutch disc is now rotating with the pressure plate at engine speed and; because it is splined to the transmission shaft, the shaft now turns at the same engine speed.

The clutch is operating properly if:
1. It will stall the engine when released with the vehicle held stationary.
2. The shift lever can be moved freely between 1st and reverse gears when the vehicle is stationary and the clutch disengaged.

CAUTION:

The clutch driven disc may contain asbestos which has been determined to be a cancer-causing agent. Never clean clutch surfaces with compressed air! Avoid inhaling any dust from any clutch surface! When cleaning clutch surfaces, use a commercially available brake cleaning fluid.

These vehicles use a hydraulic system to engage the clutch. The system consists of a master cylinder, reservoir, slave cylinder and fluid lines.

The master cylinder push rod is connected to the clutch pedal. The slave cylinder push rod is connected to the clutch release fork. The master cylinder is mounted on the firewall adjacent to the brake master cylinder. The reservoir is mounted above it and the fluid is gravity fed to the cylinder. The slave cylinder is mounted on the clutch housing.

The system is serviced as an assembly. Individual components cannot be overhauled or replaced separately.

Driven Disc and Pressure Plate

REMOVAL & INSTALLATION

See Figures 10, 11, 12, 13, 14, 15, 16, 17, 18, 19, 20, 21, 22, 23, 24, 25, 26, 27, 28, 29, 30 and 31

1. Remove the transmission and transfer case, if fitted.
2. Remove the clutch housing.
3. Remove the clutch fork and release bearing assembly.
4. Matchmark the clutch cover and flywheel for installation reference.
5. Insert a clutch alignment tool in the clutch disc and into the pilot bushing. The tool will hold the disc in place when the cover bolts are removed.
6. Remove the pressure plate retaining bolts, loosening them evenly so the clutch cover will not be distorted.
7. Pull the pressure plate assembly clear of the flywheel and, while supporting the pressure plate, slide the clutch disc from between the flywheel and pressure plate.

To install:

8. Note the following points:
    a. Check the flywheel surface for scoring. Scuff sand with 180 grit emery cloth to remove minor scratches and glazing.
    b. Machining the flywheel is not recommended, although a surface grinder can be used. If score marks are deeper than 0.003 in. (0.076mm), replace the component.

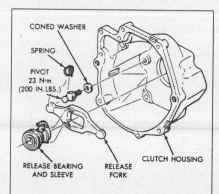

Fig. 10 Release bearing and release fork assembly

DRIVE TRAIN 7-5

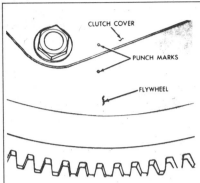

Fig. 11 Matchmark the clutch cover and flywheel

Fig. 12 Typical clutch alignment tool; note how the splines match the transmission's input shaft

Fig. 13 Loosen and remove the clutch pressure plate bolts evenly, a little at a time . . .

Fig. 14 . . . then carefully remove the clutch and pressure plate assembly from the flywheel

Fig. 15 Check across the flywheel surface; it should be flat

Fig. 16 If necessary, lock the flywheel in place and remove the retaining bolts . .

Fig. 17 . . . then remove the flywheel from the crankshaft in order to replace it or have it machined

Fig. 18 Upon installation, it is a good idea to apply a threadlocking compound to the flywheel bolts

Fig. 19 Check the pressure plate for warpage

   c. Maximum allowable flywheel runout is 0.003 in. (0.08 mm).
9. Thoroughly clean all working surfaces of the flywheel and pressure plate.
10. Check runout and free operation of new clutch disc.
11. Insert alignment tool in clutch disc hub.
12. Check that the disc hub is positioned correctly. The raised side of the hub faces away from the flywheel. "FLYWHEEL SIDE" may be stamped on the surface.
13. Lubricate pilot bearing with high temperature bearing grease.
14. Position clutch disc and plate against flywheel and insert clutch alignment tool through clutch disc hub and into main drive pilot bearing.
15. Rotate clutch cover until the punch marks on cover and flywheel line up.

Fig. 20 Be sure that the flywheel surface is clean, before installing the clutch

Fig. 21 Install a clutch alignment tool, to align the clutch assembly during installation

7-6 DRIVE TRAIN

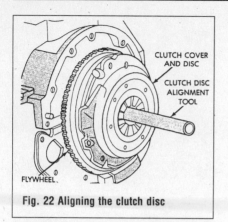

Fig. 22 Aligning the clutch disc

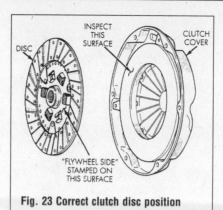

Fig. 23 Correct clutch disc position

Fig. 24 Clutch plate installed with the alignment tool in place

Fig. 25 Clutch plate and pressure plate installed with the alignment tool in place

Fig. 26 The pressure plate-to-flywheel bolt holes should align

Fig. 27 You may want to use a threadlocking compound on the clutch assembly bolts

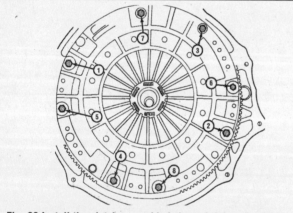

Fig. 28 Install the clutch assembly bolts and tighten in steps, using this sequence

Fig. 29 Be sure to use a torque wrench to accurately tighten all bolts

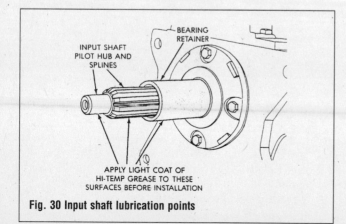

Fig. 30 Input shaft lubrication points

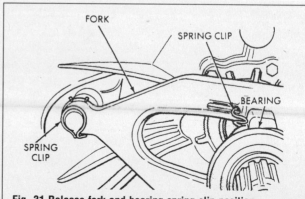

Fig. 31 Release fork and bearing spring clip position

DRIVE TRAIN 7-7

16. Bolt the cover loosely to flywheel. Tighten the bolts a few turns at a time, in progression, until tight. This is necessary to avoid distortion. Then tighten the 5/16 in. bolts to 17 ft. lbs. (23 Nm) and the 3/8 in. bolts to 30 ft. lbs. (41 Nm).
17. Remove the release lever and bearing from the clutch housing. Apply high temperature bearing grease to the bore of the release bearing and lever contact surfaces.
18. Apply a light coat of high temperature bearing grease to the transmission shaft splines and release bearing slide surface.
19. Install the release lever and bearing. Be sure that the spring clips that retain the fork on the pivot ball and release bearing on fork are properly installed. On some versions, the lever part number will be towards the bottom of the transmission and right side up. The stamped "I" goes to the pivot ball side of the transmission.
20. Install the clutch housing.
21. Install the transmission.

ADJUSTMENTS

All vehicles are equipped with a hydraulic operating linkage system. No adjustment is required. The hydraulic linkage system is serviced as an assembly only. The individual components that form the system cannot be overhauled or serviced separately.

Release Bearing

REMOVAL & INSTALLATION

See Figures 10 and 31

1. Remove the transmission and transfer case, if fitted.
2. On NV4500 equipped vehicles, remove the clutch housing.
3. Disconnect the release bearing from the release fork and remove the bearing.

To install:

4. Inspect the slide surface on transmission front for scoring or wear.
5. Lubricate the input shaft splines, bearing retainer slide surface, lever pivot ball stud and release lever pivot surface with high temperature bearing grease.
6. Install the release fork and bearing. Be sure that the fork and bearing are properly secured by spring clips. Ensure that the release fork is installed properly. The rear side of the release lever has one end with a raised area. This area goes toward the slave cylinder side of the transmission.
7. Install the clutch housing, if removed.

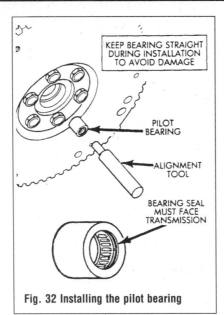

Fig. 32 Installing the pilot bearing

Pilot Bearing

REMOVAL & INSTALLATION

See Figure 32

1. Remove the transmission and transfer case, if fitted.
2. Remove the clutch housing.
3. Remove the clutch cover and disc.
4. Using a suitable blind hole puller, remove the pilot bearing.

To install:

5. Clean the bearing bore with solvent and wipe dry.
6. Install the new bearing with the clutch alignment tool. Keep the bearing straight during installation. Do not allow it to become cocked. Tap bearing into place until it is flush with the edge of the bearing bore. Do not drive it in past this point.
7. Install the clutch housing, transmission, transfer case.

Clutch Hydraulics

REMOVAL & INSTALLATION

See Figures 33 and 34

1. Raise and safely support the vehicle.
2. Remove the nuts attaching the slave cylinder to the clutch housing.
3. Remove the slave cylinder from the housing.
4. Remove any line clips.
5. Lower the vehicle.
6. Disconnect the interlock switch wires.
7. Remove the locating clip from the clutch master cylinder mounting bracket.
8. Remove the clip that secures the clutch pushrod to the clutch pedal. Remove the retaining ring, flat washer, and wave washer, if equipped. Slide the pushrod off the pedal pin.
9. Inspect the bushing on the pedal pin and replace if it is excessively worn.

Note: Verify that the cap on the clutch master cylinder reservoir is tight so that fluid will not spill during removal.

10. Some master cylinders are bolted in place. Others are locked by turning. If fastened by hardware, remove it at this time.
11. Remove the hardware attaching the reservoir and bracket to the dash panel and remove the reservoir.
12. Pull the clutch master cylinder rubber seal from the dash panel.
13. On master cylinders locked in place, rotate it 45-degrees CCW to remove it. Remove the remote reservoir, slave cylinder and connecting lines from the vehicle.

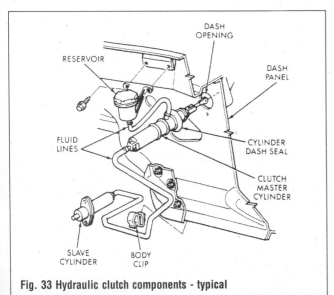

Fig. 33 Hydraulic clutch components - typical

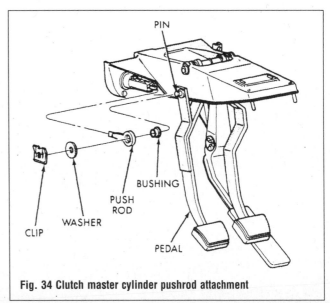

Fig. 34 Clutch master cylinder pushrod attachment

7-8 DRIVE TRAIN

To install:

14. Verify that the cap on the fluid reservoir is tight so that fluid will not spill during installation.
15. Insert the master cylinder in the dash. Rotate it 45-degrees CW to lock it in place or replace the fastening hardware.
16. Lubricate the rubber seal to ease installation.
17. Seat the seal around the cylinder in the dash.
18. Install the fluid reservoir and bracket, if equipped, to the dash panel. Tighten screws to 40 inch lbs. (5 Nm).
19. Install the master cylinder pushrod to the clutch pedal pin. Secure the rod with the wave washer, flat washer and retaining ring. Install the locating clip.

Note: Do not remove the plastic shipping stop from the pushrod until the slave cylinder has been installed.

20. Raise and safely support the vehicle securely on jackstands.
21. Insert the slave cylinder pushrod through the opening and make sure the cap on the end of the pushrod is securely engaged in the release lever before tightening the attaching nuts. Tighten the nuts to 17 ft. lbs. (23 Nm).
22. Lower the vehicle.
23. Remove the plastic shipping stop from the master cylinder pushrod.
24. Bleed the hydraulic system.
25. Operate the clutch pedal a few times to verify proper operation of the system.

HYDRAULIC SYSTEM SERVICE

The hydraulic linkage system is serviced as an assembly only. No maintenance is required. The individual components that form the system cannot be overhauled or serviced separately. The cylinders and connecting lines are sealed units. Bleeding or flushing is not required.

Fluid level will increase as the clutch wears. A level ring is provided on the reservoir. Level should not be above the ring. If it is, suspect an unacceptable amount of clutch wear.

AUTOMATIC TRANSMISSION

Understanding the Automatic Transmission

The automatic transmission allows engine torque and power to be transmitted to the rear wheels within a narrow range of engine operating speeds. It will allow the engine to turn fast enough to produce plenty of power and torque at very low speeds, while keeping it at a sensible rpm at high vehicle speeds (and it does this job without driver assistance). The transmission uses a light fluid as the medium for the transmission of power. This fluid also works in the operation of various hydraulic control circuits and as a lubricant. Because the transmission fluid performs all of these functions, trouble within the unit can easily travel from one part to another. For this reason, and because of the complexity and unusual operating principles of the transmission, a very sound understanding of the basic principles of operation will simplify troubleshooting.

TORQUE CONVERTER

See Figure 35

The torque converter replaces the conventional clutch. It has three functions:
1. It allows the engine to idle with the vehicle at a standstill, even with the transmission in gear.
2. It allows the transmission to shift from range-to-range smoothly, without requiring that the driver close the throttle during the shift.
3. It multiplies engine torque to an increasing extent as vehicle speed drops and throttle opening is increased. This has the effect of making the transmission more responsive and reduces the amount of shifting required.

The torque converter is a metal case which is shaped like a sphere that has been flattened on opposite sides. It is bolted to the rear end of the engine's crankshaft. The entire metal case rotates at engine speed and serves as the engine's flywheel.

The case contains three sets of blades. One set is attached directly to the case. This set forms the torus or pump. Another set is directly connected to the output shaft, and forms the turbine. The third set is mounted on a hub which, in turn, is mounted on a stationary shaft through a one-way clutch. This third set is known as the stator.

A pump, which is driven by the converter hub at engine speed, keeps the torque converter full of

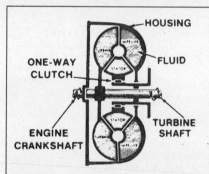

Fig. 35 The torque converter housing is rotated by the engine's crankshaft, and turns the impeller. The impeller then spins the turbine, which gives motion to the turbine shaft, driving the gears

transmission fluid at all times. Fluid flows continuously through the unit to provide cooling.

Under low speed acceleration, the torque converter functions as follows:

The torus is turning faster than the turbine. It picks up fluid at the center of the converter and, through centrifugal force, slings it outward. Since the outer edge of the converter moves faster than the portions at the center, the fluid picks up speed.

The fluid then enters the outer edge of the turbine blades. It then travels back toward the center of the converter case along the turbine blades. In impinging upon the turbine blades, the fluid loses the energy picked up in the torus.

If the fluid was now returned directly into the torus, both halves of the converter would have to turn at approximately the same speed at all times, and torque input and output would both be the same.

In flowing through the torus and turbine, the fluid picks up two types of flow, or flow in two separate directions. It flows through the turbine blades, and it spins with the engine. The stator, whose blades are stationary when the vehicle is being accelerated at low speeds, converts one type of flow into another. Instead of allowing the fluid to flow straight back into the torus, the stator's curved blades turn the fluid almost 90-degrees toward the direction of rotation of the engine. Thus the fluid does not flow as fast toward the torus, but is already spinning when the torus picks it up. This has the effect of allowing the torus to turn much faster than the turbine. This differ-

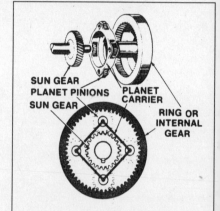

Fig. 36 Planetary gears work in a similar fashion to manual transmission gears, but are composed of three parts

ence in speed may be compared to the difference in speed between the smaller and larger gears in any gear train. The result is that engine power output is higher, and engine torque is multiplied.

As the speed of the turbine increases, the fluid spins faster and faster in the direction of engine rotation. As a result, the ability of the stator to redirect the fluid flow is reduced. Under cruising conditions, the stator is eventually forced to rotate on its one-way clutch in the direction of engine rotation. Under these conditions, the torque converter begins to behave almost like a solid shaft, with the torus and turbine speeds being almost equal.

PLANETARY GEARBOX

See Figures 36, 37 and 38

The ability of the torque converter to multiply engine torque is limited. Also, the unit tends to be more efficient when the turbine is rotating at relatively high speeds. Therefore, a planetary gearbox is used to carry the power output of the turbine to the driveshaft.

Planetary gears function very similarly to conventional transmission gears. However, their construction is different in that three elements make up one gear system, and, in that all three elements are different from one another. The three elements are: an outer gear that is shaped like a hoop, with teeth cut into the inner surface; a sun gear, mounted on a shaft

DRIVE TRAIN 7-9

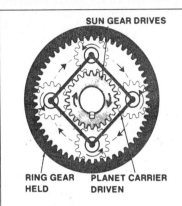

Fig. 37 Planetary gears in the maximum reduction (low) range. The ring gear is held and a lower gear ratio is obtained

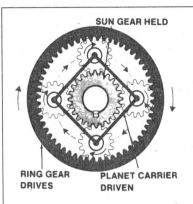

Fig. 38 Planetary gears in the minimum reduction (drive) range. The ring gear is allowed to revolve, providing a higher gear ratio

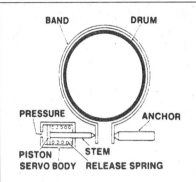

Fig. 39 Servos, operated by pressure, are used to apply or release the bands, to either hold the ring gear or allow it to rotate

and located at the very center of the outer gear; and a set of planet gears (usually four), held by pins in a ring-like planet carrier, meshing with both the sun gear and the outer gear. Either the outer gear or the sun gear may be held stationary, providing more than one possible torque multiplication factor for each set of gears. Also, if all three gears are forced to rotate at the same speed, the gearset forms, in effect, a solid shaft.

Most automatics use the planetary gears to provide various reduction ratios. Bands and clutches are used to hold various portions of the gearset to the transmission case or to the shaft on which they are mounted. Shifting is accomplished, then, by changing the portion of each planetary gearset which is held to the transmission case or to the shaft.

SERVOS AND ACCUMULATORS

See Figure 39

The servos are hydraulic pistons and cylinders. They resemble the hydraulic actuators used on many other machines, such as bulldozers. Hydraulic fluid enters the cylinder, under pressure, and forces the piston to move to engage the band or clutches.

The accumulators are used to cushion the engagement of the servos. The transmission fluid must pass through the accumulator on the way to the servo. The accumulator housing contains a thin piston which is sprung away from the discharge passage of the accumulator. When fluid passes through the accumulator on the way to the servo, it must move the piston against spring pressure, and this action smoothes out the action of the servo.

HYDRAULIC CONTROL SYSTEM

The hydraulic pressure used to operate the servos comes from the main transmission oil pump. This fluid is channeled to the various servos through the shift valves. There is generally a manual shift valve which is operated by the transmission selector lever and an automatic shift valve for each automatic upshift the transmission provides.

Note: Many new transmissions are electronically controlled. On these models, electrical solenoids are used to better control the hydraulic fluid. Usually, the solenoids are regulated by an electronic control module.

There are two pressures which affect the operation of these valves. One is the governor pressure which is affected by vehicle speed. The other is the modulator pressure which is affected by intake manifold vacuum or throttle position. Governor pressure rises with an increase in vehicle speed, and modulator pressure rises as the throttle is opened wider. By responding to these two pressures, the shift valves cause the upshift points to be delayed with increased throttle opening to make the best use of the engine's power output.

Most transmissions also make use of an auxiliary circuit for downshifting. This circuit may be actuated by the throttle linkage, the vacuum line which actuates the modulator, by a cable or by a solenoid. It applies pressure to a special downshift surface on the shift valve or valves.

The transmission modulator also governs the line pressure, used to actuate the servos. In this way, the clutches and bands will be actuated with a force matching the torque output of the engine.

Neutral Safety Switch

The neutral safety switch, otherwise known as the park/neutral position switch, prevents the vehicle from being started in any position other than PARK and NEUTRAL. It also functions as the back-up lamp switch.

The neutral safety switch is threaded into the transmission case. When the gearshift lever is placed in either the PARK or NEUTRAL position, a cam, which is attached to the transmission throttle lever

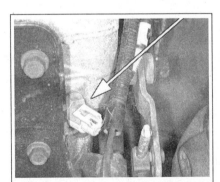

Fig. 40 Neutral safety switch location - typical

inside the transmission, contacts the neutral safety switch and provides a ground to complete the starter solenoid circuit. The back-up light switch is incorporated into the neutral safety switch.

The center terminal is for the neutral safety switch and the two outer terminals are for the back-up lights.

There is no adjustment for the switch. If a malfunction occurs, the switch must be removed and replaced.

TESTING

See Figure 40

1. Disconnect the wiring connector at the switch.
2. With an ohmmeter, check for continuity between the center terminal and ground on the transmission case. Continuity should exist only in PARK and NEUTRAL .
3. Shift the transmission into REVERSE .
4. Check for continuity across the two outer terminals. It should exist only when the transmission is in REVERSE.
5. If the switch does not conform to these specifications, check gearshift linkage adjustment to ensure that the switch operating fingers are correctly positioned.

REMOVAL & INSTALLATION

See Figures 41 and 42

1. Raise and safely support the vehicle securely on jackstands.
2. Place a drain pan under the switch. A consid-

Fig. 41 Remove the switch using an appropriately sized socket

7-10 DRIVE TRAIN

Fig. 42 The park/neutral position switch contains three terminals, and controls both the park/neutral function and the back-up lamps

erable amount of ATF will be lost unless the flow is stopped.
3. Disconnect the electrical harness.
4. Using the proper size wrench, unscrew the switch

To install:
5. Move the selector lever to the PARK and NEUTRAL positions. Verify that the switch operating lever fingers are centered in the switch opening in the transmission.
6. Using a new seal, install the new switch and tighten it to 25 ft. lbs. (34 Nm).
7. Test continuity of the switch as outlined above.
8. Connect the electrical harness.
9. Lower the vehicle.
10. Add the proper amount of ATF. If the flow was not stopped after switch removal, about four quarts may have been lost, typically.

Extension Housing Seal

REMOVAL & INSTALLATION

See Figures 3 and 4

Note: A special seal removal/installation tool is required.

1. Raise and safely support the vehicle securely on jackstands.
2. Place a drain pan under the end of the extension housing.
3. Mark the position of the driveshaft for installation reference.
4. Disconnect the driveshaft at the rear universal joint and carefully slide the shaft out of the transmission extension housing.
5. Remove the extension housing seal with the special tool.
6. Clean the end and inside of the extension housing.
7. Start the new seal into position and, with the installing tool, tap the seal into place.
8. Check the condition of the driveshaft slip yoke. Repair or replace as necessary.
9. Lubricate the lips of the seal and the slip yoke.
10. Align and install the driveshaft, taking care when guiding the slip yoke into the extension housing.
11. Remove the drain pan.
12. Check and correct the transmission fluid level.
13. Lower the vehicle.

Automatic Transmission Assembly

REMOVAL & INSTALLATION

See Figures 43, 44, 45 and 46

Note: The transmission and torque converter must be removed as an assembly to avoid component damage. The converter drive plate, pump bushing, or oil seal can be damaged if the converter is left attached to the drive plate during removal.

CAUTION:

Significant equipment is required to remove and install the transmission safely. This includes supports and jacks for the truck, the engine, transfer case (if equipped) and transmission.

1. Disconnect the battery negative cable(s).
2. Raise and support the vehicle safely.
3. Remove the skidplate(s) if fitted.
4. Remove the skidplate support crossmember, if fitted.
5. Remove any exhaust pipes and crossover pipes that would interfere with transmission removal.
6. Disconnect the fluid cooler lines at the transmission.
7. Remove the starter motor.
8. Ram: remove the engine-to-transmission struts, if fitted.
9. Dakota, Durango:
   a. Support the engine with a suitable jack device and wooden block.
   b. Remove the bolts attaching the engine-to-transmission brackets to the transmission.
   c. Remove the bolt and nut securing each engine-to-transmission bracket to the motor mounts.
   d. Remove the engine-to-transmission brackets from the front axle, if fitted.
   e. Loosen the brackets on each side of the engine block.
10. Disconnect and remove the crankshaft position sensor. Retain the sensor attaching bolts.
11. Remove the torque converter access cover.
12. If the transmission is being removed for rebuilding, remove the oil pan and drain the fluid. Reinstall the pan before removing the unit to protect internal components.
13. Remove the fill tube bracket bolts and pull the tube out of the transmission. Retain the fill tube seal.
14. On 4x4 vehicles, remove the bolt attaching the transfer case vent tube to the converter housing.
15. Mark the torque converter and drive plate for assembly alignment. Note that the bolt holes in the crankshaft flange, drive plate and torque converter all have one offset hole.
16. Rotate the crankshaft CW until the converter bolts are accessible. Then remove the bolts one at a time. Rotate the crankshaft with a socket wrench on the dampener bolt.
17. Matchmark the driveshaft(s) and yokes or flanges for assembly alignment, then disconnect and remove them. On 4x4 vehicles, remove the front driveshaft as well.
18. Disconnect the wires from the park/neutral position switch and the transmission solenoid.
19. Disconnect the gearshift rod and torque shaft assembly from the transmission.
20. Disconnect the throttle valve cable from the transmission bracket and throttle valve lever.
21. On 4x4 models, disconnect the shift rod from the transfer case shift lever.
22. Support the rear of the engine with safety stands or a heavy-duty jack and a block of wood.
23. Raise the transmission slightly with a service

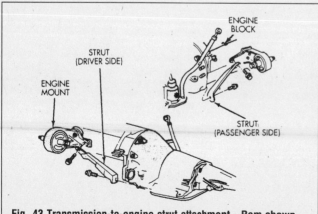

Fig. 43 Transmission-to-engine strut attachment - Ram shown

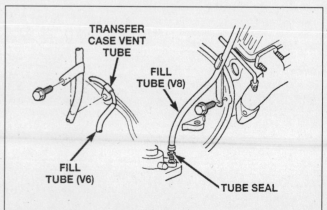

Fig. 44 Fill tube attachment - typical

DRIVE TRAIN 7-11

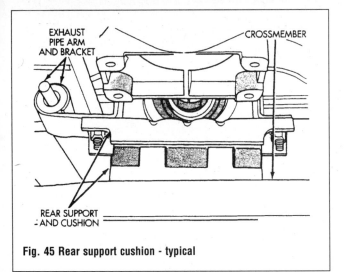

Fig. 45 Rear support cushion - typical

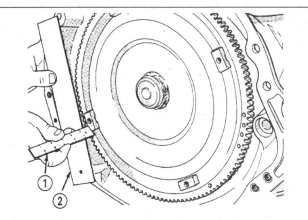

Fig. 46 Checking converter seating with scale (1) and straightedge (2). Surface of lugs should be 1/2 in. below bell housing mounting flange

jack to relieve the load on the crossmember and supports.
24. Remove the bolts securing the rear support and cushion to transmission and crossmember.
25. Raise the transmission slightly, slide the exhaust hanger arm from the bracket and remove the rear support.
26. Remove the bolts attaching the crossmember to the frame. Remove the crossmember.
27. On 4x4 vehicles, remove the transfer case.
28. Attach a small C-clamp to the edge of the bell housing to hold the converter in place during transmission removal. Otherwise the front pump bushing might be damaged.
29. Remove the converter housing bolts.
30. Carefully work the transmission and torque converter assembly rearward off the engine block dowels.
31. Lower the transmission assembly and remove it from the vehicle.
32. To remove the torque converter, remove the C-clamp from the edge of the bell housing and carefully slide the torque converter out of the transmission.

To install:
33. Reverse the removal procedure. Note the following points.
34. Check the torque converter hub and hub drive notches for sharp edges, burrs, and scratches. Polish off any imperfections with 400 grit paper. The hub must be smooth to avoid damaging the seal during installation.
35. Lubricate seal lips and engaging parts with clean ATF.
36. Check converter seating after installation with a steel scale and straightedge. The surface of the converter lugs should be 1/2 in. below bell housing mounting (see illustration).
37. Be sure the transmission dowel pins are seated in the engine block and protrude far enough to hold the transmission in alignment.
38. Don't forget to align converter and drive plate.
39. Tighten converter-to-drive plate bolts as follows:
- 10.75 in. converter: 23 ft. lbs. (31 Nm)
- 12.2 in. converter: 35 ft. lbs. (47 Nm)
40. Don't forget to fill the transmission with ATF after installation.

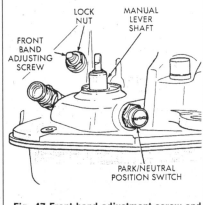

Fig. 47 Front band adjustment screw and locknut - 42/44RE shown

ADJUSTMENTS

Front Band

See Figure 47

The front (kickdown) band adjusting screw is located on the left side of the transmission case above the manual and throttle valve levers.
1. Raise the vehicle.
2. Loosen the band adjusting screw locknut. Then back off locknut 3 - 5 turns. Be sure the adjusting screw turns freely in the case. Lubricate the screw threads if necessary.
3. Tighten the band adjusting screw to 72 in. lbs. (8 Nm).

CAUTION:

Torque is critical. If an extension is used on the torque wrench to reach the adjusting screw, adjust the torque to 50 in. lbs. (5 Nm).

4. Back off the front band adjusting screw by the following amounts:
- 42RE: 3 - 5/8 turns
- 44RE: 2 - 1/4 turns
- 46RE: 2 - 7/8 turns
- 47RE: 1 - 7/8

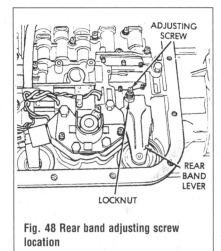

Fig. 48 Rear band adjusting screw location

5. Hold the adjuster screw in position and tighten the locknut to 30 ft. lbs. (41 Nm).

Rear Band

See Figure 48

The oil pan must be removed for access to the adjusting screw.
1. Raise the vehicle.
2. Remove the transmission oil pan.
3. Loosen the band adjusting screw locknut 5 - 6 turns. Be sure the adjusting screw turns freely in the lever.
4. Tighten the adjusting screw to 72 in. lbs. (8 Nm).
5. Back off the front band adjusting screw by the following amounts:
- 42/44RE: 4 turns
- 46RE: 2 turns
- 47RE: 3 turns
6. Hold the adjuster screw in position and tighten the locknut to 25 ft. lbs. (34 Nm).

Gearshift Linkage

See Figure 49

The following procedure applies to automatic transmission-equipped models fitted with a shift linkage. For cable-operated models, see below.

7-12 DRIVE TRAIN

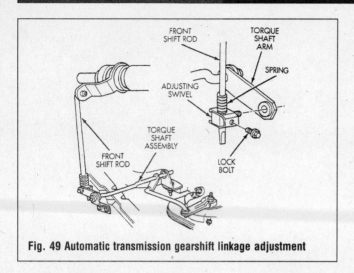

Fig. 49 Automatic transmission gearshift linkage adjustment

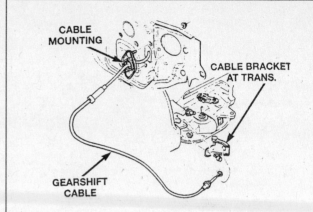

Fig. 50 Automatic transmission gearshift cable adjustment

1. Check linkage adjustment by starting engine in PARK and NEUTRAL.
2. Engine must start in both positions, but no other.
3. If performance is not correct, the park/neutral switch is probably at fault.
4. Check linkage and replace any worn or bent component. Do not attempt adjustment if any components are loose, worn or damaged. Tighten all fasteners before attempting adjustment.
5. Replace the grommet securing the shift rod or torque rod in place if either rod was removed from the grommet. Remove the old grommet as necessary and use suitable pliers to install the new grommet.
6. Shift the transmission into PARK.
7. Raise the vehicle and support it safely.
8. Loosen the lockbolt in the front shift rod adjusting swivel.
9. Ensure that the shift rod slides freely in the swivel. Lubricate the parts as necessary.
10. Move the shift lever fully rearward to the PARK detent.
11. Center the adjusting swivel on the shift rod.
12. Tighten the swivel lockbolt to 90 inch lbs. (10 Nm).
13. Verify operation.

Gearshift Cable
See Figure 50

The following procedure applies to automatic transmission-equipped models fitted with a shift linkage. For cable-operated models, see below.
1. Check linkage adjustment by starting engine in PARK and NEUTRAL.
2. Engine must start in both positions, but no other.

3. If performance is not correct, the park/neutral switch is probably at fault.
4. Shift the transmission into PARK.
5. Release the cable adjuster locknut (underneath the power brake booster) to unlock the cable.
6. Raise the vehicle and support it safely. Slide the cable eyelet off the transmission shift lever.
7. Verify the transmission shift lever is in the PARK detent by moving the lever fully rearward to the last detent position.
8. Verify the positive engagement of the transmission park lock by attempting to rotate the driveshaft. The shaft will not rotate when the park lock is engaged.
9. Slide the cable eyelet onto the transmission shift lever.
10. Lower the engine and check engine starting, which should be in PARK and NEUTRAL only.
11. Lock the shift cable by pressing the cable adjuster clamp down until it snaps into place.

Throttle Linkage
See Figure 51

1. Turn the ignition switch OFF and shift into PARK.
2. Remove the air cleaner.
3. Disconnect the cable end from the attachment stud on the throttle body. Carefully slide the cable off the stud. Do not pull or pry the cable off.
4. Verify that the transmission throttle lever is in the idle (full forward) position. Then be sure the lever on the throttle body is at curb idle position.
5. Insert a small screwdriver under the edge of the retaining clip and remove the retraining clip.
6. Center the cable end on attachment stud to within 0.04 in. (1mm).

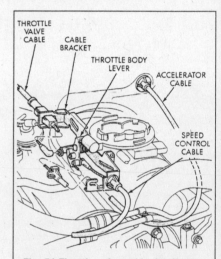

Fig. 51 Throttle valve cable details at engine

Note: Be sure that as the cable is pulled forward and centered on the throttle lever stud, the cable housing moves smoothly with the cable. Due to the angle at which the cable housing enters the spring housing, the cable housing may bind slightly and create an incorrect adjustment.

7. Install the retaining clip onto the cable housing.
8. Check the cable adjustment. Be sure the transmission throttle lever and lever on the throttle body move simultaneously and as described in the cable adjustment checking procedure.

TRANSFER CASE

Shift Linkage

REMOVAL & INSTALLATION

See Figure 52

1. Remove the shifter knob.
2. Raise and support the vehicle in a safe manner.

3. Loosen the adjusting trunnion lock bolt and slide the shift rod out of the trunnion. If there is not enough clearance, push the trunnion out of the shifter arm.
4. Remove the shifter bezel.
5. Remove the shift lever.
6. Remove the bolts holding the shifter to the vehicle floor.
7. Installation is the reverse of removal.

Rear Output Shaft Seal

REMOVAL & INSTALLATION

See Figure 53

1. Raise and safely support the vehicle securely on jackstands.
2. Remove the rear driveshaft.

DRIVE TRAIN 7-13

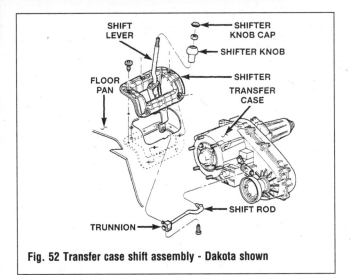

Fig. 52 Transfer case shift assembly - Dakota shown

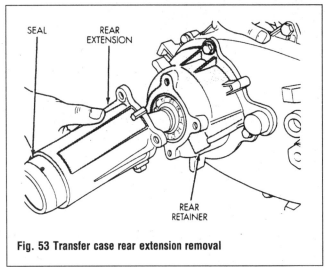

Fig. 53 Transfer case rear extension removal

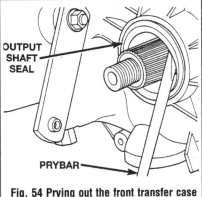

Fig. 54 Prying out the front transfer case seal - NV231, NV242

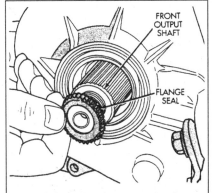

Fig. 55 Removing the companion flange

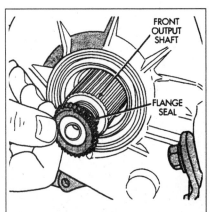

Fig. 56 Removing/installing the flange rubber seal

3. Place a drip pan beneath the transfer case.
4. Pry out the rear seal.
5. If the seal cannot be removed, the rear extension can be taken off and the seal can be punched out.
6. Grease the lips of the new seal and drive it straight in.

Front Output Shaft Seal

REMOVAL & INSTALLATION

NV231, NV242

See Figure 54

1. Raise and safely support the vehicle securely on jackstands.
2. Remove the front driveshaft.
3. Remove the driveshaft yoke from the output shaft.
4. Place a drip pan beneath the transfer case.
5. Pry out the front seal with a hooked tool or prybar.

To install:

6. Grease the lips of the new seal and drive it straight in until it is seated. The garter spring on the seal faces towards the case.
7. The remainder of the procedure is the reverse of removal.

NV231HD, NV241LD

See Figures 55, 56 and 57

1. Raise and safely support the vehicle securely on jackstands.
2. Remove the front driveshaft.
3. Remove the companion flange from the transfer case output shaft. Use a puller to free it from the shaft, if necessary.
4. Place a drip pan beneath the transfer case.
5. Remove the flange rubber seal from the shaft.
6. Pry out the front seal with a hooked tool or prybar.

To install:

7. Grease the lips of the new seal and drive it straight in. The garter spring on the seal faces towards the case.
    a. On the NV231HD, drive the seal in until it is seated.
    b. On the NV241LD, drive the seal in until it is recessed 0.080 - 0.100 in. (2.0 - 2.5mm) below the top edge of the seal bore.

WARNING:

The seal could loosen or become cocked if not fitted to the correct depth.

8. Install the flange rubber seal.
9. Install the companion flange. Tighten the nut to 130 - 200 ft. lbs. (176 - 271 Nm).

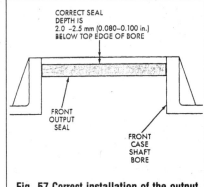

Fig. 57 Correct installation of the output shaft seal

10. The remainder of the procedure is the reverse of removal.

Transfer Case Assembly

REMOVAL & INSTALLATION

See Figures 58, 59 and 60

1. Disconnect the battery negative cable.

7-14 DRIVE TRAIN

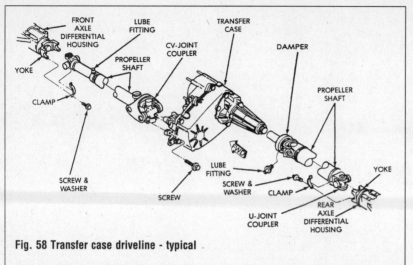

Fig. 58 Transfer case driveline - typical

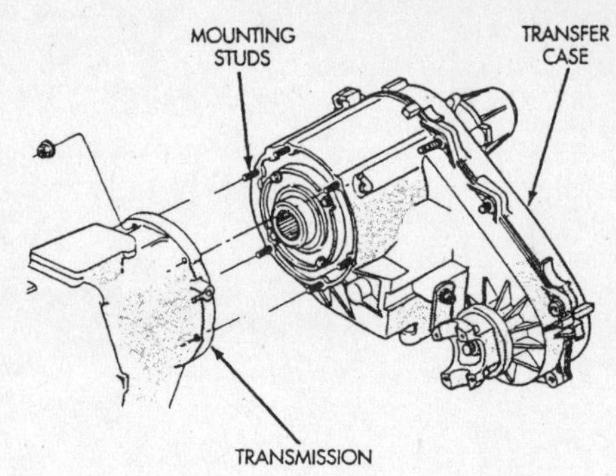

Fig. 59 Transfer case mounting arrangement - typical

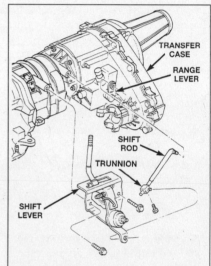

Fig. 60 Transfer case shift assembly - typical

2. Raise the vehicle and support it safely.
3. Drain the transfer case lubricant.
4. Matchmark the front and rear drive shafts on yokes or flanges so that they can be reinstalled in the original positions.
5. Remove the skidplate, if fitted, and rear frame crossmember if needed for clearance.
6. Disconnect both driveshafts from the transfer case.
7. Disconnect the PTO, if fitted.
8. Disconnect the shift linkage.
9. Disconnect the vent hose and any switch connectors fitted to the case.
10. Disconnect the wires of the vehicle speed sensor, if fitted.
11. Support the case with a transmission jack. Secure the transfer case to the jack with chains.
12. Remove the nuts securing the transfer case to the transmission.
13. Pull the assembly back towards the rear of the vehicle to disengage it.
14. Remove the transfer case from the vehicle.

To install:

15. Reverse the removal procedure.
16. Tighten the transfer case-to-transmission nuts. Tighten 5/16 in. nuts to 22 - 30 ft. lbs (30 - 40 Nm). Tighten 3/8 in. nuts to 30 - 35 ft. lbs. (41 - 47 Nm).
17. Be sure to add the proper amount and grade of fluid.

DRIVELINE

Troubleshooting

1. Low speed knock is generally caused by a worn U-joint or by worn side-gear thrust washers. A worn pinion shaft bore will also cause this.
2. Vibration may be caused by:
   - Damaged driveshaft
   - Missing driveshaft balance weights
   - Unbalanced wheels
   - Loose lug nuts
   - Worn U-joints
   - Damaged axle shaft bearings
   - Loose pinion gear nut
   - Excessive pinion yoke runout
   - Bent axle shaft
3. Check for loose or damaged front end components. Trouble here can contribute to what appears to be a rear end problem.
4. A snap or clunk when the vehicle is shifted into gear (or the clutch engaged) can be caused by:
   - high engine idle speed
   - Worn U-joints
   - Loose spring mounts
   - Loose pinion gear nut and yoke
   - Excessive ring gear backlash
   - Excessive side gear-to-case clearance

Front Driveshaft and U-Joints

REMOVAL & INSTALLATION

See Figures 61, 62 and 63

Note that two types of driveshaft bolt-ups are used. One uses small clamps to join driveshaft and yoke; the other uses a flange-and-yoke bolted directly together. Depending on model and axle fitted, these may be located at either end of the driveshaft.

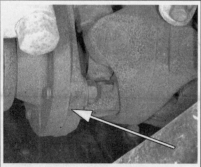

Fig. 61 Flange-and-yoke bolt-up at transfer case - typical

Note: Do not allow the driveshaft to drop or hang from any U-joint during removal. Attach the shaft to the vehicle underside with wire to prevent damage to the joints.

1. Place the transmission and transfer case in neutral.
2. Raise and safely support the vehicle securely on jackstands.
3. Remove the skidplate, if fitted.
4. Matchmark the driveshaft at both ends. Mark shaft relative to flange and/or shaft relative to cross and yoke depending on setup.
5. Remove U-joint clamp attaching bolts and both clamps from the yoke, if fitted. Some models may have a yoke bolted directly to the flange. Remove the bolts if this type is fitted.
6. Use a suitable prybar to gently pry the U-joint out of the yoke, if necessary.
7. Remove the driveshaft.
8. Wrap the U-joint caps with tape to prevent them from falling off the U-joint.
9. On models with a U-joint clamp, align the U-joint with the yoke and position the U-joint in the yoke. Install the yoke clamps.

To install:

10. Be sure the match marks will all line up when the driveshaft is finally fitted.

DRIVE TRAIN 7-15

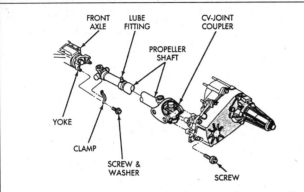

Fig. 62 One type of front driveshaft assembly: bolt torque is 14 ft. lbs. (19 Nm) at the axle, 20 ft. lbs. (27 Nm) at the transfer case

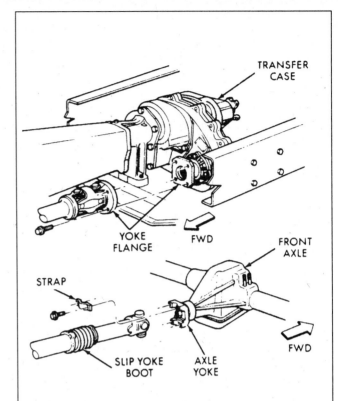

Fig. 63 Another type of front driveshaft assembly: bolt torque is 14 ft. lbs. (19 Nm) at the axle and 65 ft. lbs. (88 Nm) at the transfer case

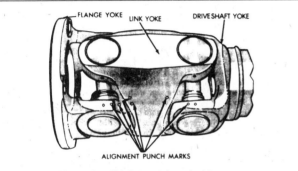

Fig. 64 Double cardan U-joint matchmarked for proper reassembly

11. Flange-and-yoke bolts are tightened to 65 ft. lbs. (88 Nm) if at the transfer case end and 80 ft. lbs. (108 Nm) if at the axle end.
12. Yoke clamp bolts are tightened to 20 ft. lbs. (27 Nm) if at the transfer case end and 14 ft. lbs. (19 Nm) if at the axle end.
13. Lower the vehicle.

U-JOINT REPLACEMENT

See Figures 64, 65, 66, 67, 68 and 69

This procedure is applicable for both single and double cardan type joints. The basic technique is to press the spider from one side, forcing a bearing cap out on the other side of the yoke. After this cap is removed, the spider is pressed in the opposite direction to remove the opposing cap. This will provide enough clearance to remove the spider from the yoke.

Sockets can be used to press the bearing caps.
1. Remove the driveshaft from the vehicle.
2. Mark the components so they can be reinstalled in their original locations.
3. Spray the assembly with penetrating fluid or rust buster and wire brush the rust off.
4. Remove any grease fittings whose position may interfere with disassembly.
5. With a soft drift, tap the outside of a bearing cap assembly in a bit to relieve tension on the snap ring, then remove its snap ring.
6. Repeat the procedure with the other snap rings.
7. Place the yoke in an arbor press or vise with a socket whose inside diameter is large enough to receive the bearing cap positioned beneath the yoke.

> **WARNING:**
> Do not clamp the driveshaft tube. Clamp only the forged portion of the welded yoke or the slip yoke. Do not overtighten the vise jaws.

8. Position the yoke with the grease fitting hole, if fitted, pointing up.
9. Place a socket with an outside diameter smaller than the bearing cap on the upper bearing cap and press the cap through the yoke to release the lower bearing cap.
10. Remove the bearing cap, if possible. If the bearing cap will not pull out of the yoke by hand after pressing, tap the yoke ear near the bearing cap to dislodge it.
11. Remove the opposite bearing cap by turning the assembly over and pressing the cross out in the opposite direction.

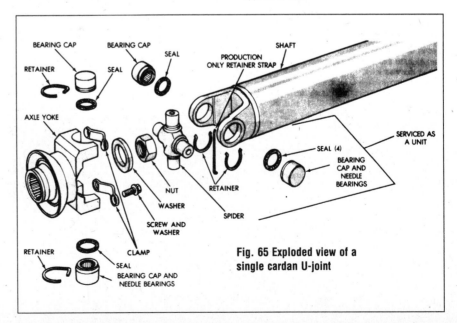

Fig. 65 Exploded view of a single cardan U-joint

7-16 DRIVE TRAIN

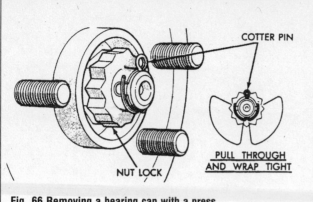

Fig. 66 Removing a bearing cap with a press

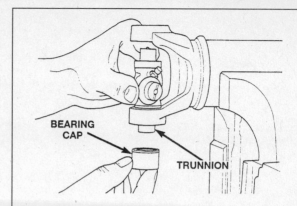

Fig. 67 Removing/installing the bearing cap on the trunnion

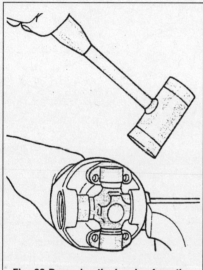

Fig. 68 Removing the bearing from the yoke - double cardan shown

WARNING:

Straighten the spider in the open hole before removal. If this is not considered, the bearing cap may score the walls of the yoke on removal.

12. Repeat the procedure with the remaining bearing caps.
13. On double cardan U-joints, remove the spider centering kit assembly and spring.

To install:

14. Be sure that matchmarks are aligned during assembly.
15. Remove the rust from the yoke bores and apply some NLGI Grade 1 or 2 EP grease.
16. Position the spider in the yoke with the grease fitting, if equipped, pointing up.
17. Place a bearing cap over the trunnion and align the cap with the yoke bore. Keep the needle bearings upright in the bearing assembly. A needle bearing lying at the bottom of the cap will prevent proper assembly.
18. Press the bearing cap into the yoke bore far enough to install the snap ring.
19. Install the snap ring and tap it lightly to en-

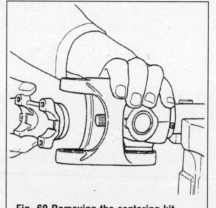

Fig. 69 Removing the centering kit (double cardan U-joint)

sure it is seated.
20. Repeat the steps for the remaining bearing assemblies.
21. If the U-joint is stiff or binding, strike the yoke with a soft-faced hammer to seat the needle bearings.
22. Double cardan joints should snap over - center in both directions when flexed beyond center.
23. Install and grease the fitting if equipped.

Rear Driveshaft and U-Joints

REMOVAL & INSTALLATION

See Figures 70, 71, 72, 73, 74, 75, 76, 77 and 78

This procedure applies to vehicles with one and

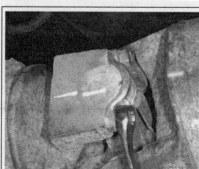

Fig. 71 Loosen the U-joint clamp bolts . . .

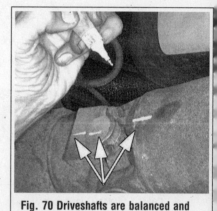

Fig. 70 Driveshafts are balanced and must be matchmarked with the differential yoke for installation reference. Mark yoke, cross and shaft

two-piece driveshafts. Also note that two types of U-joint bolt-ups are used. One uses small clamps to join driveshaft and yoke; the other uses a flange-and-yoke bolted directly together.

Note: Do not allow the shaft(s) to drop or hang from any U-joint during removal. Attach the driveshaft to the vehicle underside with wire to prevent damage to the joints.

1. Place the transmission in neutral.
2. Raise and safely support the vehicle securely on jackstands.
3. With two-piece axles, mark the position of the center support bracket on the frame crossmember. Unbolt and remove the support bracket bolts.

Fig. 72 . . . and remove the clamps from the yoke

DRIVE TRAIN 7-17

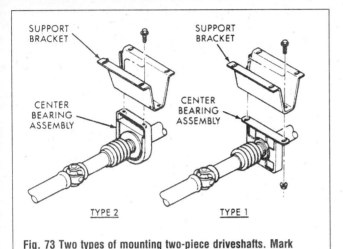

Fig. 73 Two types of mounting two-piece driveshafts. Mark bracket for position on crossmember before removal

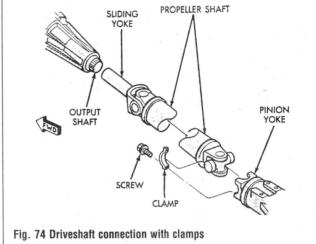

Fig. 74 Driveshaft connection with clamps

4. Matchmark the driveshaft and the rear axle drive yoke.
5. Remove the rear U-joint clamp attaching bolts and both clamps from the rear axle yoke, if fitted. Some models may have a yoke bolted directly to the flange. Remove the bolts if this type is fitted.
6. Use a suitable prybar to gently pry the U-joint out of the yoke.
7. Wrap the U-joint caps with tape to prevent them from falling off the U-joint.
8. Pull the driveshaft to the rear until the slip yoke clears the extension housing.

Note: Fluid may run from the rear of the extension housing when the shaft is removed, so position a suitable drain pan under the area. Extension housing plugs are available to prevent a constant stream of fluid from escaping the transmission (see illustration).

WARNING:

It is very important to protect the external machined surface of the slip yoke from damage during and after driveshaft removal. If the yoke is damaged, the transmission extension seal may be damaged and therefore cause a leak.

9. On two-piece driveshafts:
    a. Mark the relationship of the rear shaft to the front shaft.
    b. Slide the rear half of the shaft off the front shaft splines at the center bearing. Remove the rear half.
    c. At the transmission end of the front half, remove the bushing retaining bolts and clamps, after matchmarking. If there is a driveshaft brake, there will be flange nuts.
    d. Use a hammer and punch to tap the slinger away from the shaft to provide room for a bearing splitter.
    e. Press off the bearing.

To install:
10. Lubricate the front yoke of the driveshaft with transmission fluid and insert it into the transmission.
11. Install the center bearing support bracket bolts. Leave final tightening to the end of the procedure.
12. On models with a U-joint clamp, align the U-joint with the rear axle drive yoke and position the U-joint in the yoke. Install the yoke clamps and tighten 1/4 in. bolts to 14 ft. lbs. (19 Nm) and 5/16 in. bolts to 25 ft. lbs. (34 Nm).
13. On models with a flange-and-yoke setup, tighten the bolts to 80 ft. lbs. (108 Nm).
14. Tighten driveshaft brake flange nuts, if fitted, to 35 ft. lbs. (47 Nm).
15. Tighten the center bearing bracket bolts to 50 ft. lbs. (68 Nm), ensuring that the bracket is aligned with the marks made on removal.
16. Lower the vehicle.

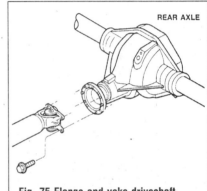

Fig. 75 Flange-and-yoke driveshaft connection

U-JOINT REPLACEMENT

Refer to the "U-Joint Replacement" information for the 'Front Driveshaft'.

DRIVESHAFT BALANCING

See Figures 79, 80 and 81

Unbalance

Driveshaft vibration increases as the vehicle speed is increased. A vibration that occurs within a specific

Fig. 76 Remove the U-joint from the rear yoke. It may be necessary to use a prybar to free the U-joint

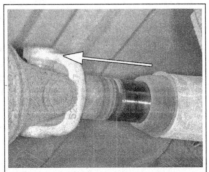

Fig. 77 Slide the driveshaft back until the sliding yoke clears the extension housing seal.

Fig. 78 This plug can be used to prevent oil flow

7-18 DRIVE TRAIN

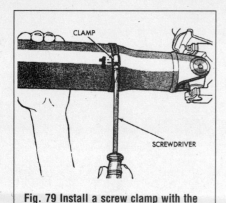

Fig. 79 Install a screw clamp with the screw at any position

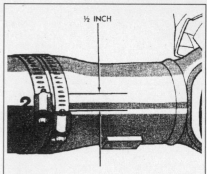

Fig. 80 If the vibration decreases, install a second clamp

Fig. 81 If the second clamp causes additional vibration, rotate the clamps in opposite directions

speed range is not usually caused by a driveshaft being unbalanced. Defective universal joints, or an incorrect driveshaft angle, are usually the cause of such a vibration.

If driveshaft is suspected of being unbalanced, it can be verified with the following procedure.

Note: Removing and re-indexing the driveshaft 180-degrees relative to the yoke may eliminate some vibrations.

1. Raise and safely support the vehicle securely on jackstands.
2. Clean all the foreign material from the driveshaft and the universal joints.
3. Inspect the driveshaft for missing balance weights, broken welds, and bent areas. If the driveshaft is dented or bent, it must be replaced.
4. Inspect the universal joints to ensure that they are not worn, are properly installed, and are correctly aligned with the shaft.
5. Check the universal joint clamp bolt torque.
6. Remove the wheels and tires. Install the wheel lug nuts to retain the brake drums or rotors.
7. Mark and number the shaft six inches (15cm) from the yoke end at four positions 90-degrees apart.
8. Run and accelerate the vehicle until vibration occurs. Note the intensity and speed the vibration occurred. Stop the engine.
9. Install a screw clamp at any position.
10. Start the engine and re-check for vibration. If there is little or no change in vibration, move the clamp to one of the other three positions. Repeat the vibration test.
11. If there is no difference in vibration at the other positions, the source of the vibration may not be driveshaft.
12. If the vibration decreased, install a second clamp and repeat the test.
13. If the additional clamp causes additional vibration, rotate the clamps (1/4 inch above and below the mark). Repeat the vibration test.
14. Increase distance between the clamp screws and repeat the test until the amount of vibration is at the lowest level. Bend the slack end of the clamps so the screws will not loosen.
15. If the vibration remains unacceptable, apply the same steps to the front end of the driveshaft.
16. Install the wheels.
17. Lower the vehicle.

Run-out

1. Remove dirt, rust, paint, and undercoating from the driveshaft surface where the dial indicator will contact the shaft.
2. The dial indicator must be installed perpendicular to the shaft surface.

Note: Measure front/rear run-out approximately 3 inches (76mm) from the weld seam at each end of the shaft tube for tube lengths over 30 inches (76cm). Under 30 inches, the maximum run-out is 0.20 inch (0.50mm) for the full length of the tube.

3. Measure run-out at the center and ends of the shaft sufficiently far away from weld areas to ensure that the effects of the weld process will not enter into the measurements.
4. Replace the driveshaft if the run-out exceeds 0.020 in. (0.50mm) at the front of shaft, 0.025 in. (0.63mm) at the center of the shaft or 0.020 in. (0.50mm) at the rear of the shaft.

Center Bearing

ADJUSTMENT

See Figure 82

1. "Drive away shudder" is a vibration that occurs at first acceleration from a stop. Shudder vibration usually peaks at the engine's highest torque output. Shudder is a symptom associated with vehicles using a two-piece driveshaft. To decrease shudder, lower the center bearing in 1/8 in. increments. Use shim stock or fabricated plates.
2. Plate stock must be able to maintain compression of the rubber insulator around the bearing. Do not use washers.
3. Replace the original bolts with longer ones if required. Remember that the nut must be 100 percent engaged with the bolt and properly torqued for safety. Proper torque is 50 ft. lbs. (68 Nm).

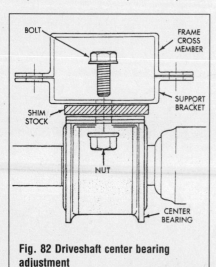

Fig. 82 Driveshaft center bearing adjustment

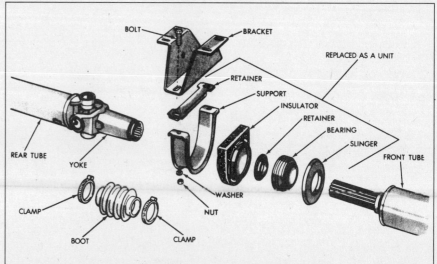

Fig. 83 Exploded view of the halfshafts, center bearing and mounting bracket

DRIVE TRAIN 7-19

4. Mark the location of the bearing bracket against the frame crossmember before removing the bolts. Mark both side-to-side and front-to-back location.

REMOVAL & INSTALLATION

See Figure 83

Two types of bearings are in use. They are not interchangeable. If replacement is required, be sure the correct item is obtained. Removal is possible without removing the bearing bracket.

1. Remove the rear driveshaft.
2. Remove the slip joint boot clamp and separate the halfshafts.
3. Use a hammer and punch to tap the slinger away from the shaft to provide room for the bearing splitter.
4. Install the bearing splitter and remove the bearing.

To install:

5. Drive the new bearing onto the shaft until seated.
6. Apply a good grade of bearing grease to the shaft splines.
7. Align the master splines and slide the front and rear halfshafts together.
8. Reposition the slip yoke boot and install a new clamp.
9. Install the driveshaft.

FRONT DRIVE AXLE

Troubleshooting

The following information should provide some guidance in pinpointing driveline faults before disassembly:

1. Axle **bearing** problems are usually caused by:
   - Insufficient or incorrect lubricant
   - Lubricant contaminated with water or foreign matter
   - Incorrect bearing preload adjustment
   - Incorrect backlash
2. Axle **gear** problems are usually caused by:
   - Insufficient lubrication
   - Lubricant contaminated with water or foreign matter
   - Overloading - excessive engine torque or vehicle/load weight
   - Incorrect clearance or backlash adjustment
3. Axle **component breakage** problems are usually caused by:
   - Severe overloading
   - Insufficient or incorrect lubricant
   - Improperly tightened components
   - Differential housing bore misalignment

Front Driveshaft

REMOVAL & INSTALLATION

See Figures 84 and 85

194 FIA

1. Remove the cotter pin, nut lock, and spring washer from the stub shaft.
2. Loosen the lug nuts and hub nut while the vehicle is on the ground.
3. Raise the vehicle and support it on jackstands.
4. Remove the skidplate, if fitted.
5. Remove the hub nut and washer from the stub shaft.
6. Remove the wheel.
7. Loosen the bolts that attach the inner CV joint to the axle shaft.
8. Remove the brake caliper and rotor.
9. Remove the ABS wheel speed sensor, if fitted.
10. Remove the bolts holding the hub bearing to the knuckle.
11. Remove the hub bearing from the axle driveshaft and steering knuckle.
12. Support the driveshaft at the CV joint housings.
13. Remove the bolts that attach the inner CV joint to the axle shaft.
14. Remove the driveshaft.

To install:

15. Insert the CV driveshaft stub into the hub bearing bore of the steering knuckle.
16. Attach the inner joint flange to the axle shaft flange. Tighten the bolts to 65 ft. lbs. (90 Nm).
17. Clean the hub bearing bore, axle driveshaft splines, and hub bearing mating surface. Apply a light coat of grease.
18. Install the hub bearing to the axle driveshaft and the steering knuckle.
19. Install the bolts that hold the hub bearing to the steering knuckle. Tighten to 123 ft. lbs. (166 Nm).
20. Install the stub shaft hub nut and washer.
21. Install the ABS sensor, caliper and rotor.
22. Tighten the hub nut to 190 ft. lbs. (258 Nm).
23. The remainder of the procedure is in the reverse of removal.

C205F

1. Remove the cotter pin, nut lock, and spring washer from the stub shaft.
2. Loosen the lug nuts and hub nut while the vehicle is on the ground.
3. Raise the vehicle.
4. Remove the skidplate, if fitted.
5. Remove the hub nut and washer from the stub shaft.
6. Remove the wheel.
7. Loosen the bolts that attach the inner CV joint to the axle shaft.
8. Remove the brake caliper and rotor.
9. Remove the ABS wheel speed sensor, if fitted.
10. Remove the bolts holding the hub bearing to the knuckle.
11. Remove the hub bearing from the axle driveshaft and steering knuckle.
12. Support the driveshaft at the CV joint housings.
13. Disengage the inner CV joint from the axle shaft. Position two pry bars between the inner CV housing and the axle housing. Apply pressure away from the differential housing. This will disengage the axle shaft snap-ring from the groove on the inside of the CV housing.
14. Remove the driveshaft.

To install:

15. Insert the CV joint stub into the hub bearing bore of the steering knuckle.
16. Apply a light coat of wheel bearing grease on the axle shaft splines.
17. Install the inner CV joint onto the axle shaft flange. Push firmly on the shaft until the shaft snap-ring engages with the groove on the inside of the joint housing.
18. Clean the hub bearing bore, axle driveshaft splines, and hub bearing mating surface. Apply a light coat of grease.
19. Install the hub bearing to the axle driveshaft and the steering knuckle.
20. Install the bolts that hold the hub bearing to the steering knuckle. Tighten to 123 ft. lbs. (166 Nm).

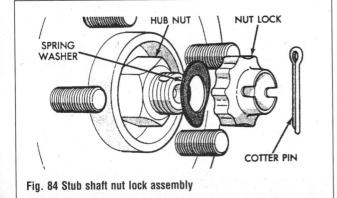

Fig. 84 Stub shaft nut lock assembly

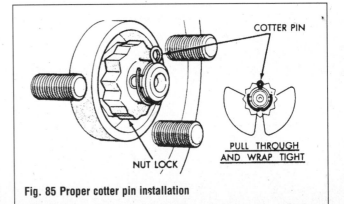

Fig. 85 Proper cotter pin installation

7-20 DRIVE TRAIN

21. Install the stub shaft hub nut and washer.
22. Install the ABS sensor, caliper and rotor.
23. Tighten the hub nut to 180 ft. lbs. (244 Nm).
24. The remainder of the procedure is in the reverse of removal.

CV Joints

CV JOINT BOOTS

Removal & Installation

See Figures 86 and 87

1. Remove the axle driveshaft.
2. Remove the outer CV joint.
3. Remove the outer CV joint small clamp. Remove boot.
4. Remove the inner CV joint boot clamps and remove boot.
5. The lubricant amounts included with replacement boots are different for inner and outer units. Apply only the included amount on each joint.
6. Clean the CV joints and shaft of old grease and foreign matter.
7. Slide the inner CV joint boot up the shaft and insert the lip located within the small diameter end of the boot into the shaft groove.
8. Retain the small diameter of the boot on the shaft with a ladder-type clamp in the boot groove. Verify that the boot and lip are properly positioned on the intermediate shaft. Position the clamp locating tabs in the slots and tighten the clamp.
9. Squeeze the clamp bridge. Take care that you don't cut through the clamp bridge or damage the boot.
10. Position the large diameter end of the boot on the CV joint housing.
11. After the inner joint boot small clamp is installed, the inboard hub must be set to a service build length.
12. Compress the inner hub down the connector shaft.
13. Use a small blunt drift between the large end and the boot seal to relive the pressure.

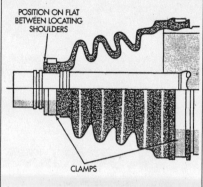

Fig. 86 Boot retaining clamp locations

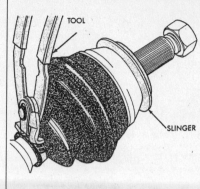

Fig. 87 Compressing the clamp bridge

14. The distance from the edge of the lip to the edge of the flange should be 7.5 in. (190mm) on 1997 models, and 7.1 in. (181mm) on other models. This will eliminate excess air that can cause a ballooning effect and possibly damage the boot.
16. Verify correct (untwisted) boot positioning.
17. Install the large ladder clamp and secure it.
18. Slide the outer CV joint boot small clamp onto the shaft.
19. Slide the outer CV joint boot onto the shaft and into position.
20. Install the small clamp to boot.
21. Install the large boot clamp over the outer CV joint.
22. Install the outer CV joint to the shaft.
23. Install the large boot clamp to boot and CV joint.
24. Install the axle driveshaft.

DISASSEMBLY & ASSEMBLY

See Figures 88, 89 and 90

Inner CV Joint

1. Remove the axle driveshaft.
2. Place the inner CV joint housing in a vise.

3. Remove the inner boot retaining clamps. Pull the inner boot back onto the interconnecting shaft. Discard the retaining clamps.
4. Pull the tripod and shaft straight out from the inner CV joint housing.
5. Remove the snap retaining ring from the groove behind the tripod. Slide the tripod toward the center of the shaft. Remove the C-clip on the outer end of the shaft.
6. Remove the tripod from the shaft. Replace the boot, if necessary.
7. Remove old lubricant.
8. Inspect the needle bearing raceways in the housing and tripod components for wear and damage. Replace the tripod as a unit only if necessary.
9. Inspect the balls for pitting, cracks, scoring and wear. A dull surface is normal.
10. Polished contact surface areas on the race-

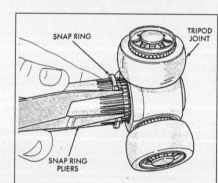

Fig. 89 Removing the snap ring – inner CV joint

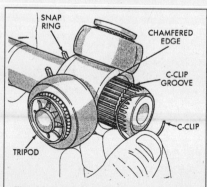

Fig. 90 Removing the C-clip – inner CV joint

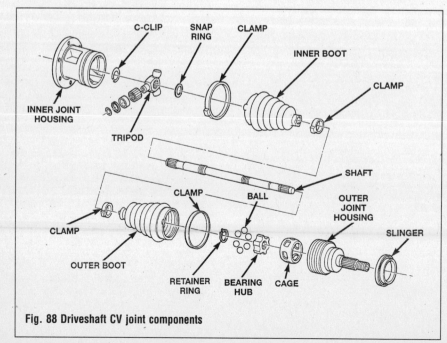

Fig. 88 Driveshaft CV joint components

DRIVE TRAIN 7-21

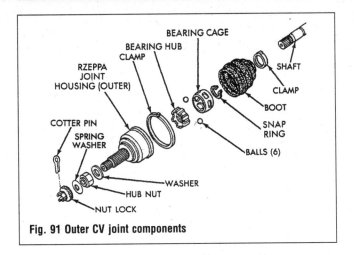

Fig. 91 Outer CV joint components

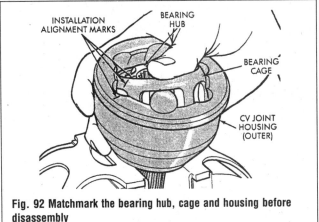

Fig. 92 Matchmark the bearing hub, cage and housing before disassembly

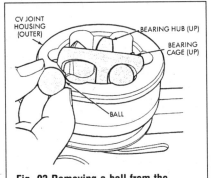

Fig. 93 Removing a ball from the assembly. Use a small pry bar, if necessary

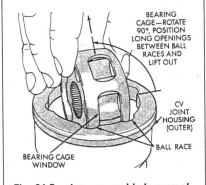

Fig. 94 Bearing cage and hub removal method

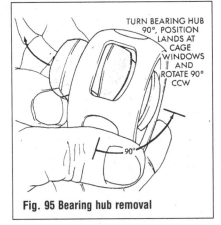

Fig. 95 Bearing hub removal

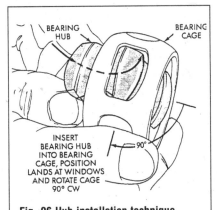

Fig. 96 Hub installation technique

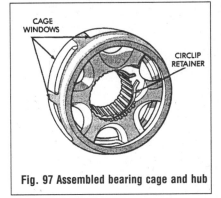

Fig. 97 Assembled bearing cage and hub

ways and on the bearing surfaces are normal. If the joints are noisy or cause vibration, they should be replaced.

11. To assemble, slide the boot down far enough for access.

12. Install the snap ring past the ring groove (towards the center of the shaft). Slide the tripod onto the end of the interconnecting shaft. Be sure the chamfered end of the tripod is adjacent to the C-clip retaining ring groove.

13. Install the C-clip in the groove. Slide the tripod out against the clip. Install the snap ring in the inner groove. Be sure they are seated.

14. Apply the required quantity of lubricant to the housing and boot. Coat the interior of the joint housing and the tripod.

15. Insert and seat the tripod and shaft in the housing.

16. Position the large diameter end of the inner CV joint boot over the edge of the housing. Insert the lip of the boot into the locating groove at the edge of the housing.

17. Install the CV joint boots as outlined in that section.

Outer CV Joint

See Figures 91, 92, 93, 94, 95, 96, 97, 98 and 99

If the outer CV joint is excessively worn, replace the entire CV joint and boot.

1. Remove the retaining clamps from the outer CV joint and discard. Slide the boot off the outer joint and down the shaft.

2. Remove the lubricant to expose the outer CV joint components.

3. Clamp the shaft in a vise with soft jaws and support the outer CV joint.

4. Remove the snap ring from the groove.

5. Slide the outer CV joint from the shaft.

6. Remove the slinger, if damaged, from the outer CV joint with a brass drift.

7. Clean the old lubricant.

8. Matchmark the bearing hub, bearing cage and housing.

9. Clamp the outer CV joint in a vertical position. Place the stub shaft in a soft-faced vise.

10. Press down on one side of the bearing cage/hub to tilt the cage. This will provide access to a ball at the opposite side. If the joint is tight, use a hammer and brass drift to loosen the hub. Do not contact the bearing cage with the drift.

11. Remove the ball from the bearing cage. If necessary, a small pry bar cab be used.

12. Repeat these steps until all balls are removed.

13. Tilt the bearing cage to a vertical position. Remove the cage from the housing. Pull the cage upwards and away from the housing.

14. Turn the bearing hub 90-degrees from the bearing cage. Align one pair of the hub lands with the cage windows. Raise and insert one of the lands with the cage windows. Raise and insert one of the lands into the adjacent cage window. Remove the bearing hub by rolling it out of the cage.

7-22 DRIVE TRAIN

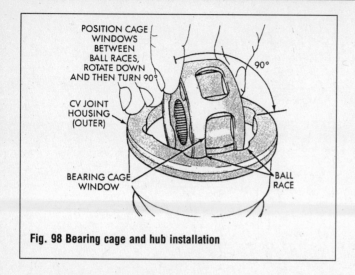

Fig. 98 Bearing cage and hub installation

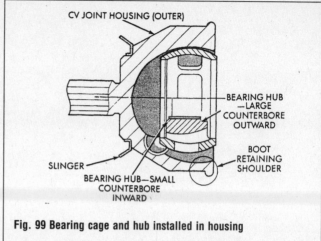

Fig. 99 Bearing cage and hub installed in housing

15. Inspect the balls for pitting, cracks, scoring and wear. A dull surface is normal.
16. Polished contact surface areas on the raceways and on the bearing surfaces are normal. If the joints are noisy or cause vibration, they should be replaced.
17. To assemble, first apply a light coat of oil to all the outer components.
18. Align the bearing hub, cage and housing with the matchmarks made during disassembly.
19. Insert one of the bearing hub lands into a bearing cage window. Roll the hub into the cage. Rotate the bearing hub 90-degrees to compete the installation.
20. Insert the bearing cage/hub into the housing. Rotate the cage/hub 90-degrees to complete the installation.
21. Apply lubricant to the ball raceways. Spread equally between all raceways.
22. Tilt the bearing hub and cage and install the balls in the raceways.
23. Apply a small amount of lubricant to the inner diameter of the slinger. Drive it squarely on.
24. Position the small diameter end of the replacement boot on the interconnecting shaft. Retain the boot with a replacement clamp.
25. Lubricate the outer CV joint and boot.
26. Align the shaft splines to the outer CV joint splines. Push the outer CV joint until the snap ring seats in the groove.
27. Ensure that the snap ring is properly seated in the housing. Pull the outer CV joint from the interconnecting shaft to test.
28. Install the boot as outlined previously.

Axle Shaft, Bearing and Seal

REMOVAL & INSTALLATION

194 FIA Axle

1. Place the transmission in NEUTRAL.
2. Raise and safely support the vehicle.
3. Remove the C/V driveshaft(s).
4. Remove the shock absorber if removing the right axle shaft.
5. Remove the skidplate if removing the left axle shaft.
6. Clean all foreign matter from the housing cover area.

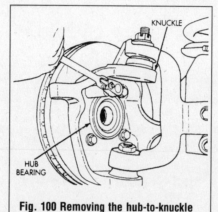

Fig. 100 Removing the hub-to-knuckle bolts

7. Remove the housing cover, draining the lubricant into a suitable container.
8. Remove the E-clip retaining the axle shaft on the side you wish to remove.
9. Remove the axle shaft. Take care to avoid damaging the bearing and seal which will remain in the axle shaft tube.
10. Pry out the axle shaft seal with a suitable tool.
11. Remove the bearing with a suitable puller.
12. Inspect the roller bearing contact surface on the axle shaft for signs of brinelling, galling or pitting. Replace the shaft and bearing if any of these conditions exist.
13. Installation is the reverse of removal. Always use a new seal and lubricate the seal lips before installing the shaft.

C205F Axle

1. Place the transmission in NEUTRAL.
2. Raise and safely support the vehicle.
3. Remove the C/V driveshaft.
4. Remove the skidplate if fitted.
5. Clean all foreign matter from the axle seal area.
6. Remove the axle shaft with a slide hammer. Take care to avoid damaging the bearing and seal which will remain in the axle shaft tube.
7. Pry out the axle shaft seal with a suitable tool.
8. Remove the bearing with a suitable puller.
9. Installation is the reverse of removal. Always use a new seal and lubricate the seal lips before installing the shaft. Push firmly on the axle shaft until the axle shaft snap-ring passes completely through the side gear and engages the snap-ring groove.

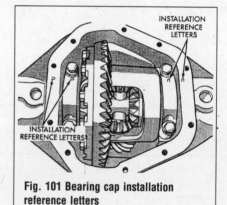

Fig. 101 Bearing cap installation reference letters

216 FBI, 248 FBI Axles
See Figures 100 and 101

1. Place the transmission in NEUTRAL.
2. Raise and safely support the vehicle.
3. Remove the wheels.
4. Remove the brake caliper and rotor.
5. Remove the ABS wheel speed sensor, if fitted.
6. Remove the axle hub nut cotter pin and the nut.
7. Remove the hub-to-knuckle bolts. Remove the hub bearing from the steering knuckle and axle shaft.
8. Remove the brake dust shield.
9. Remove the axle shaft from the housing. Be sure to avoid damaging the oil seal.
10. To service the bearings and seals, the differential must be removed. This requires the use of a spreader to enlarge the differential housing.
    a. Note the orientation of the installation reference letters stamped on the bearing caps and housing machined sealing surface.
    b. Remove the differential bearing caps.
    c. Install the spreader and enlarge the housing to remove the differential.

WARNING:

Do not spread the unit over 0.020 in. (0.50mm).

DRIVE TRAIN 7-23

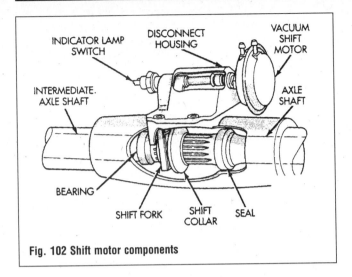

Fig. 102 Shift motor components

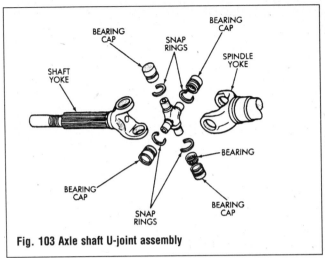

Fig. 103 Axle shaft U-joint assembly

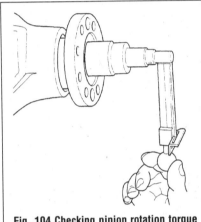

Fig. 104 Checking pinion rotation torque - C205F axle

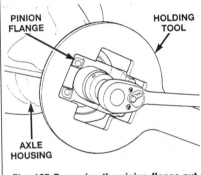

Fig. 105 Removing the pinion flange nut - typical

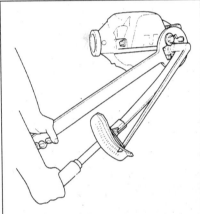

Fig. 106 Tightening the pinion nut - C205F axle

    d. Remove the bearings from the differential.
    e. Remove the oil seals from the differential housing.
  11. Installation is the reverse of removal. Tighten differential bearing cap bolts to 80 ft. lbs. (109 Nm). Other torque ratings are:
- Hub bearing-to-knuckle 125 ft. lbs. (170 Nm)
- Axle nut: 175 ft. lbs. (237 Nm)

AXLE SHIFT MOTOR

See Figure 102

  1. Disconnect the vacuum and wiring connector from the shift housing.
  2. Remove the indicator switch.
  3. Remove the shift motor housing cover, gasket and shield.

To install:

  4. Install the shift motor housing gasket and cover. Ensure the shift fork is correctly guided into the shift collar groove.
  5. Tighten the shield bolts to 8 ft. lbs. (11 Nm).

AXLE SHAFT U-JOINT

See Figure 103

This component is disassembled in the same manner as described under "Driveshaft and U-Joints".

Pinion Seal

REMOVAL & INSTALLATION

See Figures 104, 105 and 106

C205F, 194 FIA Axles

Note: C250F axles are equipped with a flange on the input shaft; 194 FIA axles have a yoke. The procedures apply to both, but with this difference.

  1. Raise and safely support the vehicle securely on jackstands.
  2. Remove the skidplate, if fitted.
  3. Scribe marks on the universal joints, pinion yokes, flanges and pinion shafts for installation reference.
  4. Disconnect the front driveshaft from the axle.
  5. Secure the driveshaft in an upright position to prevent damage to the rear universal joint.
  6. Remove the halfshafts.
  7. Rotate the pinion flange three or four times.
  8. Measure the amount of torque necessary to rotate the pinion gear with an inch pound beam or dial-type torque wrench. Record the torque reading for installation reference.
  9. Devise a method of holding the flange or yoke from turning. If the special tool is not available, the bolt holes and bolts can be utilized.
  10. Hold the flange or yoke in position with a pipe wrench or other suitable tool. Remove the pinion shaft nut. A new nut should be used on assembly.
  11. Remove the flange or yoke. A puller will probably be needed.
  12. 194 FIA axle: remove the pinion seal excluder from the yoke.
  13. Remove the pinion shaft seal with suitable pry tool or slide-hammer mounted screw.

To install:

  14. Clean the seal contact surface in the housing bore.
  15. Examine the splines on the pinion shaft for burrs or wear. Remove any burrs and clean the shaft.
  16. Inspect pinion flange or yoke for cracks, worn splines and worn seal contact surface. Replace flange or yoke if necessary.

Note: The outer perimeter of the seal is precoated with a special sealant. An additional application of sealant is not required.

  17. Apply a light coating of gear lubricant on the lip of pinion seal.
  18. Install the new pinion shaft seal. A large socket can be used.
  19. 194 FIA axle: install a new pinion seal excluder to the pinion yoke.
  20. Install the flange or yoke and nut and washer. A new nut should be used. Tighten the nut until there is zero bearing end-play.

7-24 DRIVE TRAIN

> **WARNING:**
> Do not exceed the minimum tightening torque when installing the flange nut at this point. Damage to the collapsible spacer or bearings may result.

21. Hold the pinion flange with a suitable tool and tighten shaft nut to 200 ft. lbs. (271 Nm).
22. Rotate the pinion shaft several revolutions to ensure the bearing rollers are seated.
23. Rotate the pinion shaft using an inch pound beam or dial-type torque wrench. Rotating torque should be equal to the reading recorded during removal, plus an additional 5 inch lbs. (0.56 Nm). If too low, tighten the pinion shaft nut in 5 ft. lb. (6.8 Nm) increments until the proper torque is achieved.

> **WARNING:**
> Never loosen the pinion gear nut to decrease pinion gear bearing rotating torque, and never exceed the specified preload torque. If preload torque is exceeded, a new collapsible spacer must be installed. The torque sequence will then have to be repeated.

24. The remainder of the procedure is the reverse of removal.
25. Lower the vehicle.

216 FBI, 248 FBI Axles

1. Raise and safely support the vehicle securely on jackstands.
2. Scribe marks on the universal joints, pinion yokes, and pinion shafts for installation reference.
3. Disconnect the driveshaft from the pinion yoke.
4. Secure the driveshaft in an upright position to prevent damage to the rear universal joint.
5. Remove the wheel.
6. Remove the brake calipers and rotor to prevent any drag. The drag may cause a false bearing rotating torque measurement.
7. Rotate the pinion yoke three or four times.
8. Measure the amount of torque necessary to rotate the pinion gear with an inch pound beam or dial-type torque wrench. Record the torque reading for installation reference.
9. Hold the yoke in position with a pipe wrench or other suitable tool. Remove the pinion shaft nut and washer. Note that the convex side of the washer faces outward. It must be installed the same way.
10. Remove the yoke. A puller will probably be needed.
11. Remove the pinion shaft seal with suitable pry tool or slide-hammer mounted screw.

To install:

12. Clean the seal contact surface in the housing bore.
13. Examine the splines on the pinion shaft for burrs or wear. Remove any burrs and clean the shaft.
14. Inspect pinion yoke for cracks, worn splines and worn seal contact surface. Replace yoke if necessary.

Note: The outer perimeter of the seal is pre-coated with a special sealant. An additional application of sealant is not required.

Fig. 107 Mark the suspension alignment cams for installation reference

15. Apply a light coating of gear lubricant on the lip of pinion seal.
16. Install the new pinion shaft seal. A large socket can be used.
17. Position the pinion yoke on the end of the shaft with the reference marks aligned.
18. Seat the yoke on the pinion shaft.
19. Install the pinion yoke washer. The convex side of the washer must face outward.

> **WARNING:**
> Do not exceed the minimum tightening torque when installing the pinion yoke retaining nut at this point. Damage to the collapsible spacer or bearings may result.

20. Rotate the pinion shaft several revolutions to ensure the bearing rollers are seated.
21. Rotate the pinion shaft using an inch pound beam or dial-type torque wrench. Rotating torque should be equal to the reading recorded during removal, plus an additional 5 inch lbs. (0.56 Nm). If too low, tighten the pinion shaft nut in 5 ft. lb. (6.8 Nm) increments until the proper torque is achieved.
22. The remainder of the procedure is the reverse of removal.
23. Lower the vehicle.

Front Axle Housing Assembly

REMOVAL & INSTALLATION

Ram

See Figures 107 and 108

Ram models are fitted with either 216 FBI or 248 FBI front drive axles. The removal and installation procedures are the same for both.

1. Loosen the wheel lug nuts. Raise and support the vehicle safely.
2. Remove the front wheels.
3. Remove the brake calipers and rotors.
4. Remove the ABS wheel speed sensors, if so equipped.
5. Disconnect the axle vent hose.
6. Disconnect the vacuum hose and electrical connector.
7. Matchmark the front driveshaft yokes and/or flanges for reassembly. Remove the driveshaft.
8. Disconnect the stabilizer bar links at the axle brackets.

Fig. 108 Front suspension arms and fasteners (arrows) - Ram

9. Disconnect the shock absorbers from the axle brackets.
10. Disconnect the track bar from the axle bracket.
11. Disconnect the tie rod and drag link from the steering knuckles.
12. Place a jack or other lifting device beneath the axle and attach the axle to the lifting device.
13. Mark the suspension alignment cams for installation reference.
14. Disconnect the upper and lower suspension arms from the axle bracket.
15. Lower the axle. The coil springs will drop with the axle.
16. Remove the coil springs from the axle bracket.

To install:

> **WARNING:**
> Suspension components with rubber bushings should be tightened with the weight of the vehicle on the suspension, at normal height. If the springs are not at their normal ride position, vehicle ride comfort could be affected and premature bushing wear may occur. Rubber bushings must never be lubricated.

17. After installing the axle, align the suspension cams, but do not fully tighten the nuts yet.
18. Install the track bar but do not tighten the fasteners at this time.
19. Tighten shock absorber bolts to 89 ft. lbs. (121 Nm).
20. Tighten the stabilizer bar link to 27 ft. lbs. (37 Nm).
21. Tighten the drag link and tie rod to 65 ft. lbs. (88 Nm).
22. Tighten upper suspension arm nuts at the axle to 89 ft. lbs. (121 Nm) and at the frame to 62 ft. lbs. (84 Nm).
23. Tighten the lower suspension arm nuts at the axle to 62 ft. lbs. (84 Nm) and at the frame to 88 ft. lbs. (119 Nm).
24. Tighten the track bar bolt at the axle bracket to 130 ft. lbs. (176 Nm).
25. Bleed the brake system if any hydraulic fittings were disconnected (Chapter 9).

Dakota, Durango With C205F Axle

See Figures 109 and 110

This axle is fitted to year 2000 models.

1. Raise and safely support the vehicle.
2. Remove the skidplate, if fitted.

DRIVE TRAIN 7-25

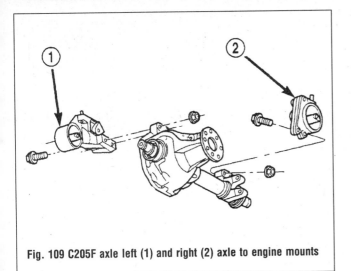

Fig. 109 C205F axle left (1) and right (2) axle to engine mounts

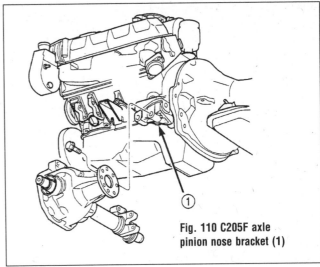

Fig. 110 C205F axle pinion nose bracket (1)

3. Remove the skidplate support crossmember, if fitted.
4. Matchmark the yokes and flanges of the front driveshaft and halfshafts so that they can be reassembled in their original locations.
5. Remove halfshafts.
6. Remove the front driveshaft.
7. Remove the axle vent tube.
8. Support the axle housing with a suitable jack.
9. Remove the bolts securing the axle to the engine mounts.
10. Remove the bolts securing the axle to the pinion nose bracket.
11. Remove the axle from the vehicle.

To install:

12. Reverse the removal procedure. Tighten all axle mounting bolts to 70 ft. lbs. (95 Nm).
13. Skidplate bolts are torqued to 17 ft. lbs. (23 Nm).

Dakota, Durango With 194 FIA Axle

See Figures 111 and 112

This axle is fitted to 1997 - 1999 models.
1. Raise and safely support the vehicle.
2. Remove the skidplate, if fitted.
3. Remove the skidplate support crossmember, if fitted.
4. Matchmark the yokes and flanges of the front driveshaft and halfshafts so that they can be reassembled in their original locations.
5. Remove halfshafts.
6. Remove the front driveshaft.

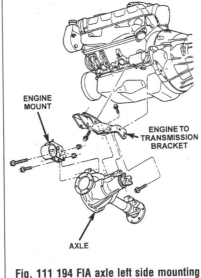

Fig. 111 194 FIA axle left side mounting devices

7. Remove the axle vent tube.
8. Support the axle housing with a suitable jack.
9. Remove the bolts securing the axle to the engine-to-transmission brackets.
10. Remove the bolts securing the axle to the engine mounts.
11. Remove the axle from the vehicle.

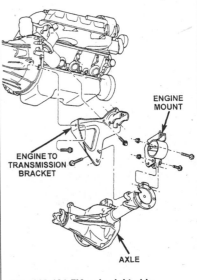

Fig. 112 194 FIA axle right side mounting devices

To install:

12. Reverse the removal procedure. Tighten the engine mount bolts to 75 ft. lbs. (102 Nm) and the transmission bracket bolts to 40 ft. lbs. (54 Nm).
13. Skidplate bolts are torqued to 17 ft. lbs. (23 Nm).

REAR AXLE

Troubleshooting

The following information should provide some guidance in pinpointing driveline faults before disassembly:
1. Axle **bearing** problems are usually caused by:
   - Insufficient or incorrect lubricant
   - Lubricant contaminated with water or foreign matter
   - Incorrect bearing preload adjustment
   - Incorrect backlash
2. Axle **gear** problems are usually caused by:
   - Insufficient lubrication
   - Lubricant contaminated with water or foreign matter
   - Overloading - excessive engine torque or vehicle/load weight
   - Incorrect clearance or backlash adjustment
3. Axle **component breakage** problems are usually caused by:
   - Severe overloading
   - Insufficient or incorrect lubricant
   - Improperly tightened components
   - Differential housing bore misalignment

7-26 DRIVE TRAIN

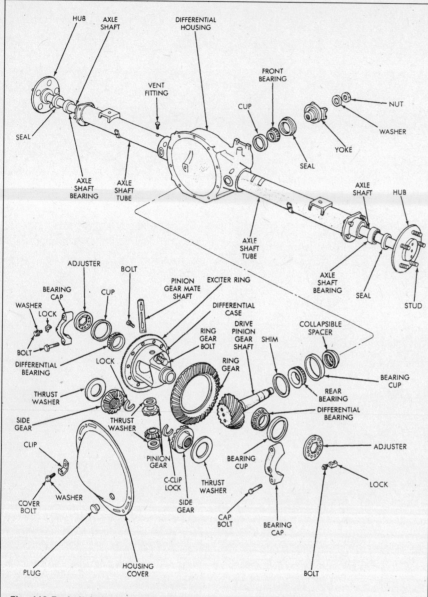

Fig. 113 Exploded view of a typical rear axle - 8 1/4, 9 1/4 shown

Fig. 114 Loosen the cover bolts and remove all of them, except two at the top and two at the bottom

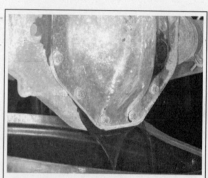

Fig. 115 Use a prybar to carefully pry the cover from the axle housing

Fig. 116 Rotate the differential case so that the pinion gear shaft lock screw is accessible, then remove it

Axle Shaft, Bearing and Seal

REMOVAL & INSTALLATION

8 1/4, 9 1/4 Axles

See Figures 113, 114, 115, 116, 117, 118, 119, 120, 121, 122, 123, 124, 125, 126, 127 and 128

Note: The following procedure is written for removal of one axle, but it applies to both.

1. Raise and safely support the vehicle securely on jackstands.
2. Ensure that the transmission is in Neutral.
3. Remove wheel.
4. Remove brake drum.
5. Clean all foreign material from housing cover area.
6. Loosen all housing cover bolts, and remove all but two bolts on the top and two on the bottom.
7. Use a prybar to pry the cover loose from the axle.
8. Allow the gear oil to drain from the housing.
9. Remove the axle housing cover.
10. Rotate the differential case so that the pinion mate gear shaft lock screw is accessible. Remove the lock screw.
11. Remove the pinion mate gear shaft from the differential case.
12. Push the axle in towards the center of the vehicle and remove the C-lock from the groove on the axle shaft.
13. Pull out and remove the axle shaft.
14. Remove the axle shaft seal from the end of the axle tube with a small prybar.

Note: The seal and bearing can be removed at the same time with the bearing removal tool.

15. Remove the axle shaft bearing from the axle tube using an appropriate puller; see Chapter 8 for the procedure.

To install:

Note: Always install a new axle bearing seal.

16. Clean the inside of the axle housing, removing any swarf or foreign matter. Some units have a magnet in the base to catch metal filings.
17. Wipe the axle tube bore clean. Remove any old sealer or burrs from the tube.
18. Install the axle shaft bearing. Ensure that the bearing part number is facing outward. Verify that the bearing is installed straight and flush with the axle tube when fully seated.
19. Inspect axle shaft seal for leakage or damage.
20. Inspect roller bearing contact surface on axle shaft for signs of brinelling, galling and pitting. If any

DRIVE TRAIN 7-27

Fig. 117 Push the pinion gear shaft up from the bottom and remove it

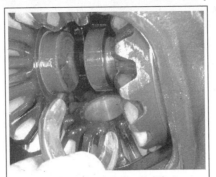

Fig. 118 Push the axle in slightly to remove the C-clip . . .

Fig. 119 . . . and the axle can then be removed from the housing

Fig. 120 Any time the differential cover is removed, inspect the gears for damage (arrow) and replace as necessary

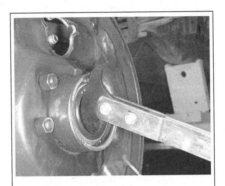

Fig. 121 Removing the axleshaft seal

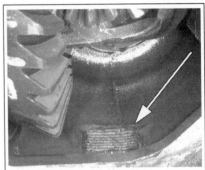

Fig. 122 Clean metal particles off the magnet, if fitted

Fig. 123 Be sure that the splines do not damage seal or bearings when installing

Fig. 124 When installing the axle shaft lock screw, always use a threadlocking compound to hold the screw in place

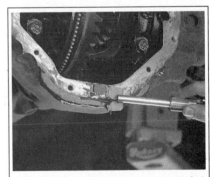

Fig. 125 Use a scraper to remove gasket material from the axle housing

of these conditions exist, the axle shaft and/or bearing and seal must be replaced.

21. Lubricate bearing bore and seal lip with gear lubricant. Insert axle shaft through seal, bearing, and engage it into side gear splines.

Note: Use care to prevent shaft splines from damaging axle shaft seal lip.

22. Insert C-clip lock in end of axle shaft.
23. Push axle shaft outward to seat C-clip lock in side gear.
24. Insert pinion mate shaft into the differential case and through the thrust washers and pinion gears.
25. Align the hole in the shaft with the hole in the differential case, and install the lock screw with Loctite, on the threads. Tighten the lock screw to 8 ft. lbs. (11 Nm).

Fig. 126 Apply sealer in a continuous bead around the cover

Fig. 127 When installing the axle cover, always replace the metal identification tag. This tag provides the only means of externally identifying the axle ratio

7-28 DRIVE TRAIN

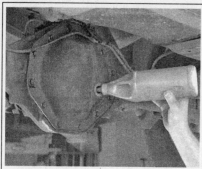

Fig. 128 Fill the axle housing through the filler hole with the proper type and amount of gear oil. Proper level is 3/8 in. (10mm) below the bottom of the hole

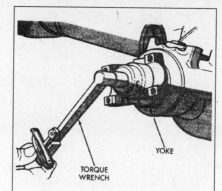

Fig. 129 Always measure bearing preload prior to loosening the pinion nut

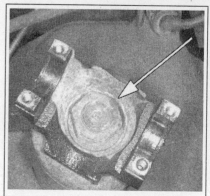

Fig. 130 Pinion shaft nut (arrow)

Fig. 131 Remove the yoke with a sturdy puller

Fig. 132 Removing the pinion seal

Fig. 133 Installing a new pinion seal with a large socket

26. Install cover and fill the differential with gear oil.
27. Install the brake drum.
28. Install the wheel.
29. Lower the vehicle.

248 RBI, 267 RBI, 286 RBI Axles

Note: The following procedure is written for removal of one axle shaft, but it applies to both.

1. Raise and safely support the vehicle securely on jackstands.
2. Ensure that the transmission is in Neutral.
3. Remove wheel.
4. Remove brake drum.
5. Clean all foreign material from housing cover area.
6. Loosen all housing cover bolts, and remove all but two bolts on the top and two on the bottom.
7. Use a prybar to pry the cover loose from the axle.
8. Allow the gear oil to drain from the housing.
9. Remove the axle housing cover.
10. Remove the axle shaft flange bolts.
11. Remove the axle.
12. Remove the lock wedge and adjustment nut.
13. Remove the hub assembly. The outer axle bearing will slide out as the hub is removed.
14. Remove the inner grease seal.
15. Remove the bearing cups.

To install:

16. Clean all parts thoroughly.
17. Install the bearing cups. Apply lubricant to the surface area.
18. Install the inner axle bearing.
19. Install a new bearing grease seal.
20. Apply a coating of multi-purpose NLGI Grade 2, EP lubricant to the axle.

WARNING:

Use care to prevent the grease seal from contacting the axle tube spindle threads during installation.

21. Carefully slide the hub onto the axle.
22. Install the outer axle bearing.
23. Install the hub bearing adjustment nut.
24. Tighten the adjustment nut to 120 - 140 ft. lbs. (163 - 190 Nm) while rotating the wheel.
25. Loosen the adjustment nut 1/8 turn at a time to provide 0.001 - 0.010 in. wheel bearing end play.
26. Tap the locking wedge into the spindle keyway and adjustment nut. Ensure that the locking wedge is installed in a new position.

Pinion Seal

REMOVAL & INSTALLATION

See Figures 129, 130, 131, 132, 133, 134 and 135

1. Loosen the wheel lug nuts. Raise and safely support the vehicle securely on jackstands.
2. Scribe a mark on the universal joint, pinion yoke, and pinion shaft for installation reference.
3. Disconnect the driveshaft from the pinion

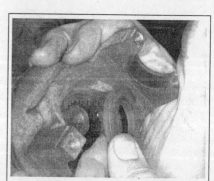

Fig. 134 The convex side of the yoke washer must face outward

yoke.
4. Secure the driveshaft in an upright position to prevent damage to the rear universal joint.
5. Remove the wheels.
6. Remove the brake drums to prevent any drag. The drag may cause a false bearing rotating torque measurement.
7. Rotate the pinion yoke three or four times.
8. Measure the amount of torque necessary to rotate the pinion gear with an inch pound beam or dial-type torque wrench. Record the torque reading for installation reference.
9. Hold the yoke in position with a pipe wrench or other suitable tool. Remove the pinion shaft nut and washer. Note that the convex side of the washer faces outward. It must be installed the same way.
10. Remove the yoke. A puller will probably be needed.

DRIVE TRAIN 7-29

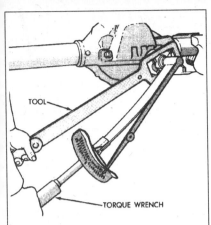

Fig. 135 Hold the pinion yoke with a suitable tool and tighten the shaft nut, then recheck the pinion preload

11. Remove the pinion shaft seal with suitable prytool or slide-hammer mounted screw.

To install:

12. Clean the seal contact surface in the housing bore.
13. Examine the splines on the pinion shaft for burrs or wear. Remove any burrs and clean the shaft.
14. Inspect pinion yoke for cracks, worn splines and worn seal contact surface. Replace yoke if necessary.

Note: The outer perimeter of the seal is pre-coated with a special sealant. An additional application of sealant is not required.

15. Apply a light coating of gear lubricant on the lip of pinion seal.
16. Install the new pinion shaft seal. A large socket can be used.

Note: The seal is correctly installed when the seal flange contacts the face of the differential housing flange.

17. Position the pinion yoke on the end of the shaft with the reference marks aligned.
18. Seat the yoke on the pinion shaft.
19. Install the pinion yoke washer. The convex side of the washer must face outward.

WARNING:

Do not exceed the minimum tightening torque when installing the pinion yoke retaining nut at this point. Damage to the collapsible spacer or bearings may result.

20. Hold the pinion yoke with pipe wrench or other suitable tool and tighten shaft nut to 210 ft. lbs. (285 Nm).
21. Rotate the pinion shaft several revolutions to ensure the bearing rollers are seated.
22. Rotate the pinion shaft using an inch pound beam or dial-type torque wrench. Rotating torque should be equal to the reading recorded during removal, plus an additional 5 inch lbs. (0.56 Nm). If too low, tighten the pinion shaft nut in 5 ft. lb. (6.8 Nm) increments until the proper torque is achieved.

WARNING:

Never loosen the pinion gear nut to decrease pinion gear bearing rotating torque, and never exceed the specified preload torque. If preload torque is exceeded, a new collapsible spacer must be installed.

23. Install the brake drums.
24. Install the wheels.
25. Align the driveshaft with the pinion yoke and install.
26. Lower the vehicle.

Rear Axle Housing Assembly

See Figure 136

REMOVAL & INSTALLATION

1. Raise and safely support the vehicle securely on jackstands.
2. Position a floor jack under the axle and secure axle to jack.
3. Remove the wheels.
4. Secure brake drums to the axle shaft using the lug nuts.
5. Remove the Rear Wheel Anti-lock Lock (RWAL) sensor from the differential housing, if necessary.

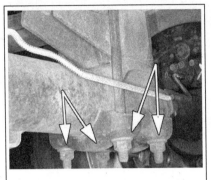

Fig. 136 Spring clamp nuts (arrows)

6. Disconnect the brake hose at the axle junction block. Do not disconnect the brake hydraulic lines at the wheel cylinders.
7. Disconnect the parking brake cables and cable brackets.
8. Disconnect the vent hose from the axle shaft tube.
9. Mark the driveshaft and yoke for installation alignment reference.
10. Remove driveshaft.
11. Disconnect the shock absorbers from the axle.
12. Remove the spring clamps and spring brackets.
13. Separate the axle from the vehicle.

To install:

14. Raise the axle with lifting device and align to the leaf spring centering bolts.
15. Install the spring clamps and spring brackets.
16. Install the shock absorbers and tighten the mounting nuts. Proper torque is 60 ft. lbs. (82 Nm)
17. Install the RWAL sensor to the differential housing, if necessary.
18. Install the brake drums.
19. Connect the brake hose to the axle junction block. Bleed the brake system (Chapter 9).
20. Install the axle vent hose.
21. Align driveshaft and pinion with reference marks.
22. Install universal joint straps and bolts. Tighten to 14 ft. lbs. (19 Nm).
23. Install the wheels.
24. Add gear lubricant, if necessary.
25. Remove the jack and lower the vehicle.

DRIVE TRAIN

TORQUE SPECIFICATIONS

Components	English	Metric
Manual Transmission		
Back-up light switch	30 ft. lbs.	41 Nm
Bearing shim cap	30 ft. lbs.	41 Nm
Bearing retainer (front)		
NV1500, 4500	30 ft. lbs.	41 Nm
NV3500	7 ft. lbs.	10 Nm
Bearing retainer (rear)	25 ft. lbs.	34 Nm
Drain/fill plugs		
NV1500	25 ft. lbs.	34 Nm
NV-3500	15 ft. lbs.	20 Nm
NV-5600	22 ft. lbs.	30 Nm
AX-15	27 ft. lbs.	37 Nm
Front-to-rear housing (NV3500)	26 ft. lbs.	35 Nm
Rear transmission mount (AX-15)	44 ft. lbs.	60 Nm
Transmission-to-clutch bell housing	50 ft. lbs.	68 Nm
Extension/adapter housing bolts		
NV4500	50 ft. lbs.	68 Nm
NV5600	35 ft. lbs.	48 Nm
Crossmember-to-frame	55 ft. lbs.	75 Nm
Crossmember-to-insulator	45 ft. lbs.	61 Nm
Transfer case attaching nuts		
5/16	22-30 ft. lbs.	30-41 Nm
3/8	30-35 ft. lbs.	41-47 Nm
Transfer case rear retainer bolt	34 ft. lbs.	46 Nm
Universal joint straps	170 inch lbs.	19 Nm
PTO Cover	40 ft. lbs.	54 Nm
Clutch		
Housing		
NV3500	45 ft. lbs.	61 Nm
NV5600	35 ft. lbs.	48 Nm
Pressure Plate		
5/16 inch bolts	20 ft. lbs.	27 Nm
3/8 inch bolts	30 ft. lbs.	41 Nm
Master Cylinder		
Fluid reservoir	40 inch lbs.	5 Nm
Slave cylinder	17 ft. lbs.	23 Nm
42/44/46/47RE Automatic Transmission		
Park/Neutral switch	25 ft. lbs.	34 Nm
Torque converter bolts		
42/44/46RE (10.75 in.)	23 ft. lbs.	31 Nm
47RE (12.2 in.)	35 ft. lbs.	47 Nm
Crossmember bolt/nut	50 ft. lbs.	68 Nm
Driveplate-to-crankshaft bolt	55 ft. lbs.	75 Nm
Pan bolt	13 ft. lbs.	17 Nm
Oil pump bolts	15 ft. lbs.	20 Nm
Overrunning clutch cam bolt	13 ft. lbs.	17 Nm
Bolt, O/D-to-transmission	25 ft. lbs.	34 Nm
Bolt, O/D piston retainer	13 ft. lbs.	17 Nm
Filter screws	35 inch lbs.	4 Nm
Bolt, reaction shaft support	15 ft. lbs.	20 Nm
Rear band locknut	30 ft. lbs.	41 Nm
Front band adjustment locknut (46/47RE	25 ft. lbs.	34 Nm
45RFE Automatic Transmission		
Park/Neutral switch	25 ft. lbs.	34 Nm
Torque converter bolt	23 ft. lbs.	31 Nm
Crossmember bolt/nut	50 ft. lbs.	68 Nm

DRIVE TRAIN 7-31

Driveplate-to-crankshaft bolt	55 ft. lbs.	75 Nm
Pan bolt	9 ft. lbs.	12 Nm
Oil pump bolts	21 ft. lbs.	28 Nm
Oil pump body-to-cover	40 inch lbs.	4.5 Nm
Reaction shaft support bolt	9 ft. lbs.	12 Nm
Valve body bolt	9 ft. lbs.	12 Nm
Extension housing bolt	40 ft. lbs.	54 Nm
Sensor bolts	9 ft. lbs.	12 Nm
Driveline		
1/4 inch bolts	14 ft. lbs.	19 Nm
Cover bolt	30 ft. lbs.	41 Nm
Bearing cap bolt		
8 1/4	70 ft lbs.	95 Nm
9 1/4	100 ft. lbs.	136 Nm
Pinion nut	210 ft. lbs.	285 Nm
Ring gear bolt		
8 1/4	75 ft. lbs.	102 Nm
9 1/4	115 ft. lbs.	157 Nm
Backing plate bolt	48 ft. lbs.	64 Nm
Threaded adjuster lock screw	90 inch lbs.	10 Nm
194 FIA, C205F Axles		
Cover bolt		
194 FIA	30 ft. lbs.	41 Nm
C205F	15 ft. lbs.	23 Nm
Bearing cap bolt	45 ft lbs.	61 Nm
Pinion nut	271-474 ft. lbs.	200-350 Nm
Ring gear bolt	70-90 ft. lbs.	95-122 Nm
216/248 FBI, 248/267/286 RBI Axles		
Fill plug	25 ft. lbs	34 Nm
Cover bolt	30 ft. lbs.	41 Nm
Bearing cap bolt	80 ft. lbs	108 Nm
Pinion nut		
216 FBI	160-200 ft. lbs.	217-271 Nm
248 FBI	215-280 ft. lbs.	291-380 Nm
248/267 RBI	215-330 ft. lbs.	292-447 Nm
286 RBI	440-500 ft. lbs.	597-678 Nm
Ring gear bolt		
216 FBI	70-90 ft. lbs.	95-122 Nm
248 FBI, 248/267 RBI	120-140 ft. lbs.	163-190 Nm
286 RBI	200-240 ft. lbs.	272-325 Nm
Shift motor bolt	8 ft. lbs.	11 Nm
Axle nut (216/248 FBI)	175 ft. lbs.	237 Nm
Axle to hub bolt (248/267/286 RBI)	90 ft. lbs.	123 Nm
Hub nut (248/267/286 RBI)	120-140 ft. lbs.	163-190 Nm
286 FBI case bolt		
Standard	65-70 ft. lbs.	89-94 Nm
Heavy duty	120-140 ft. lbs.	163-190 Nm

Notes

FRONT SUSPENSION 8-4
COIL SPRINGS 8-5
FRONT HUB AND BEARING 8-11
LOWER BALL JOINT 8-8
LOWER SUSPENSION ARM 8-10
SHOCK ABSORBERS 8-6
STABILIZER BAR 8-9
STEERING KNUCKLE 8-11
TORSION BARS 8-6
UPPER BALL JOINT 8-8
UPPER SUSPENSION ARM 8-9
WHEEL ALIGNMENT 8-14

REAR SUSPENSION
LEAF SPRINGS 8-16
REAR WHEEL BEARINGS 8-18
SHOCK ABSORBERS 8-17
STABILIZER BAR 8-17

STEERING
DAKOTA DURANGO STEERING LINKAGE 8-23
IGNITION SWITCH & LOCK CYLINDER 8-19
POWER STEERING GEAR 8-24
POWER STEERING PUMP 8-25
RACK & PINION STEERING 8-23
RAM STEERING LINKAGE 8-21
STEERING WHEEL 8-18
TURN SIGNAL (COMBINATION) SWITCH 8-19

WHEELS
INSPECTION 8-3
REMOVAL & INSTALLATION 8-2
SPECIAL NOTES 8-2
WHEEL LUG STUDS 8-3
WHEELS 8-2

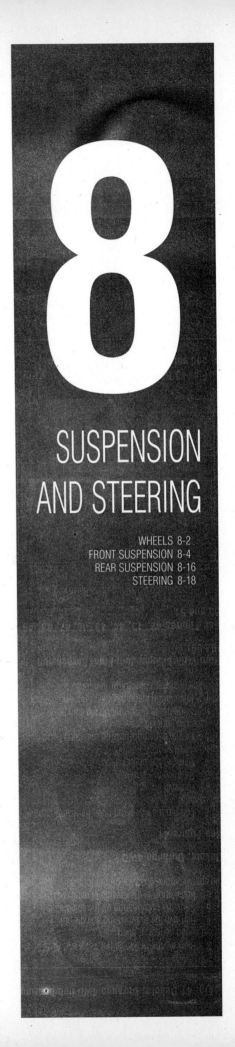

8
SUSPENSION AND STEERING

WHEELS 8-2
FRONT SUSPENSION 8-4
REAR SUSPENSION 8-16
STEERING 8-18

8-2 SUSPENSION AND STEERING

WHEELS

Wheels

REMOVAL & INSTALLATION

> **CAUTION:**
> Always refer to the owner's manual for model-specific information pertaining to your wheels and tires and vehicle jacking instructions. The safety issues involved call for attention to manufacturer-supplied information.

See Figures 1, 2, 3, 4 and 5

1. Park the vehicle on a level surface.
2. Remove the jack, tire iron and, if necessary, the spare tire from their storage compartments.
3. Check the owner's manual for model-specific information, then place the jack in the proper position.
4. If equipped with lug nut trim caps, remove them by either unscrewing or pulling them off the lug nuts, as appropriate. Consult the owner's manual, if necessary.
5. If equipped with a wheel cover or hub cap, insert the tapered end of the tire iron in the groove and pry off the cover.
6. Apply the parking brake, place the transmission in PARK for automatics or Reverse for manuals and block the diagonally opposite wheel with a wheel chock or two.

> **Note:** Wheel chocks may be purchased at your local auto parts store, or a block of wood cut into wedges may be used. If possible, keep one or two of the chocks in your tire storage compartment, in case any of the tires has to be removed on the side of the road.

7. With the tires on the ground, use the tire iron/wrench to break the lug nuts loose.

> **Note:** If a nut is stuck, never use heat to loosen it or damage to the wheel and bearings may occur. If the nuts are seized, one or two heavy hammer blows directly on the end of the nut usually loosens the rust. Be careful, as continued pounding will likely damage the brake drum or rotor. A longer breaker bar or lever usually solves all problems like this.

8. Using the jack, raise the vehicle until the tire is clear of the ground. Support the vehicle safely using jackstands, if possible.
9. Matchmark the wheel relative to the hub so that it can reinstalled in the original position to maintain balance factor if the same assembly is to be reinstalled.
10. Remove the lug nuts, then remove the tire and wheel assembly.

To install:

11. Make sure the wheel and hub mating surfaces, as well as the wheel lug studs, are clean and free of all foreign material. Always remove rust from the wheel mounting surface and the brake rotor or drum. Failure to do so may cause the lug nuts to loosen in service.
12. Never use oil or grease on the studs.
13. Install the tire and wheel assembly in its original location and hand-tighten the lug nuts with the wheel(s) OFF THE GROUND.
14. With the wheel(s) OFF THE GROUND, tighten all the lug nuts, in a crisscross pattern, until they are as close to the proper torque as possible.
15. Lower the wheel until contact with the ground will keep it from turning. Complete the tightening sequence. Refer to the specifications chart for proper torque.

SPECIAL NOTES

See Figures 6, 7 and 8

1. Lug nuts must be tightened properly to ensure efficient braking. Overtightening or torquing in the wrong pattern can cause distortion of the rotor or drums. Air wrenches are not recommended for tightening lug nuts. Always use a torque wrench.
2. Models equipped with chrome-plated wheels must NOT use chrome-plated lug nuts. The factory-supplied lug nuts used with chrome-plated wheels are not chrome-plated on purpose. Use only these lug nuts. Do not replace these items with chrome-plated lug nuts.
3. All 8800 GVW 4x4 vehicles have a factory-installed spacer behind the right front wheel.
4. Some models use a two-piece lug nut with a flat face. Do not substitute different types. Place two drops of oil (NOT MORE!) on the nut/washer interface before tightening
5. All aluminum and some steel wheels have lug nuts with an enlarged nose. This is necessary to prevent loosening.
6. Dual rear wheels use a special, heavy-duty lug nut wrench. Always use this wrench when removing or installing the wheels.
7. With dual rear wheels, be sure both wheels are off the ground when tightening the lug nuts. This will ensure correct wheel centering and maximum wheel clamping.
8. With dual rear wheels, repeat the torquing sequence twice to ensure accuracy and prevent lug nut loosening. Check tightness at 100 miles and again at 500 miles.

Fig. 1 Before jacking the vehicle, block the diagonally opposite wheel with one or, preferably, two chocks

Fig. 2 Removing the spare tire - '97 Ram

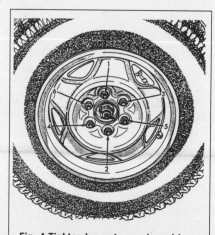

Fig. 3 Break the lug nuts loose when the tire is on the ground. But get it OFF the ground when tightening

Fig. 4 Tighten lug nuts evenly and in a cross pattern - 6-stud model shown

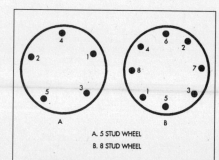

A. 5 STUD WHEEL
B. 8 STUD WHEEL

Fig. 5 Tightening patterns for 5-stud and 8-stud wheels

SUSPENSION AND STEERING 8-3

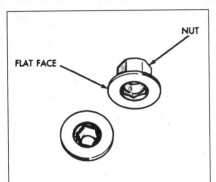

Fig. 6 Some models use two-piece lug nuts like this

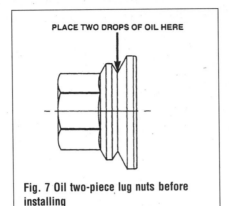

Fig. 7 Oil two-piece lug nuts before installing

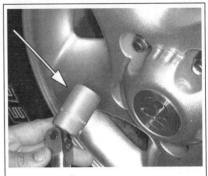

Fig. 8 Aluminum lug nut wrenches can help minimize scratching on custom lug nuts.

WARNING:

Do not overtighten the lug nuts, as this may cause the wheel studs to stretch or the brake disc (rotor) to warp.

9. If so equipped, install the wheel cover or hub cap. Make sure the valve stem protrudes through the proper opening before tapping the wheel cover into position.
10. If equipped, install the lug nut trim caps by pushing them or screwing them on, as applicable.
11. Remove the jack from under the vehicle, and place the jack and tire iron/wrench in their storage compartments. Remove the wheel chock(s).

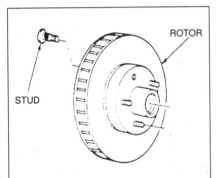

Fig. 9 View of the rotor and stud assembly - typical

12. If you have removed a flat or damaged tire, place it in the storage compartment of the vehicle and take it to your local repair station to have it fixed or replaced as soon as possible.

INSPECTION

Inspect the tires for lacerations, puncture marks, nails and other sharp objects. Repair or replace as necessary. Also check the tires for treadwear and air pressure.

Check the wheel assemblies for dents, cracks, rust and metal fatigue. Repair or replace as necessary.

Wheel Lug Studs

REMOVAL & INSTALLATION

With Disc Brakes
See Figures 9, 10, 11 and 12

1. Raise and support the appropriate end of the vehicle safely using jackstands, then remove the wheel. 2. Remove the caliper. Support the caliper aside using wire or a coat hanger.
3. Five-stud and six-stud wheels: remove the retainers on the wheel studs and remove the rotor.
4. Eight-stud wheels: remove the hub extension, if fitted. Remove the cotter pin and hub nut. Remove the ABS speed sensor, if fitted. Back off the hub/bearing mounting bolts 1/4 in. each. Then tap the bolts with a hammer to loosen the hub/bearing from the steering knuckle. Remove the hub/bearing mounting bolts and remove the hub/bearing. Remove the rotor.
5. Properly support the rotor using press bars, then drive the stud out using an arbor press.

Note: If a press is not available, CAREFULLY drive the old stud out using a blunt brass drift or a copper mallet. MAKE SURE the rotor is properly and evenly supported or it may be damaged.

To install:

6. Clean the stud hole with a wire brush and start the new stud with a hammer and drift pin. Do not use any lubricant or thread sealer.
7. Finish installing the stud with the press.

Note: If a press is not available, start the lug stud through the bore in the hub, then position about 4 flat washers over the stud and thread the lug nut. Hold the hub/rotor while tightening the lug nut, and the stud should be drawn into position. MAKE SURE THE STUD IS FULLY SEATED, then remove the lug nut and washers.

8. Install the rotor and adjust the wheel bearings.
9. Install the brake caliper and pads.
10. Install the wheel, then remove the jackstands and carefully lower the vehicle.
11. Tighten the lug nuts to the proper torque.

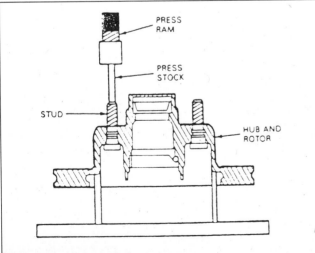

Fig. 10 Pressing the stud from the rotor

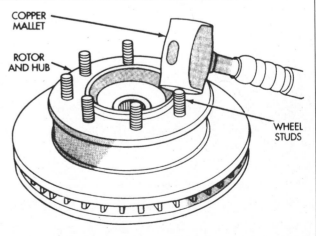

Fig. 11 Studs can be driven out carefully with a copper mallet or similar soft metal instrument

8-4 SUSPENSION AND STEERING

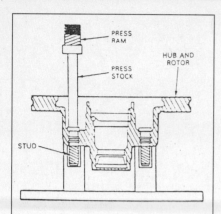

Fig. 12 Use a press to install the stud into the rotor

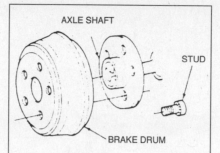

Fig. 13 Exploded view of a brake drum, axle flange and stud - typical

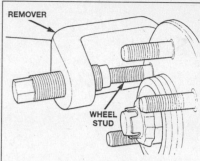

Fig. 14 One method of pressing out a rear wheel stud

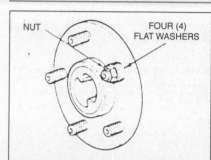

Fig. 15 Force the stud into the axle flange using washers and a lug nut

With Drum Brakes

See Figures 13, 14 and 15

1. Raise the vehicle and safely support it with jackstands, then remove the wheel.
2. Remove the brake drum.
3. If necessary to provide clearance, remove the brake shoes.
4. Using a large C-clamp and socket, press the stud from the axle flange.
5. Coat the serrated part of the stud with liquid soap and place it into the hole.

To install:

6. Position about 4 flat washers over the stud and thread the lug nut. Hold the flange while tightening the lug nut, and the stud should be drawn into position. MAKE SURE THE STUD IS FULLY SEATED, then remove the lug nut and washers.
7. If applicable, install the brake shoes.
8. Install the brake drum.
9. Install the wheel, then remove the jackstands and carefully lower the vehicle.
10. Tighten the lug nuts to the proper torque.

FRONT SUSPENSION

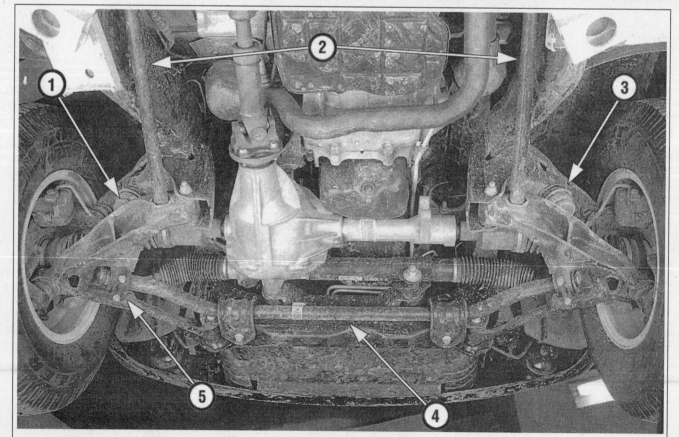

Fig. 16 Dakota, Durango 4WD Front suspension

1 Front shock absorber
2 Torsion bars
3 Upper suspension arm
4 Stabilizer bar
5 Lower suspension arm

SUSPENSION AND STEERING 8-5

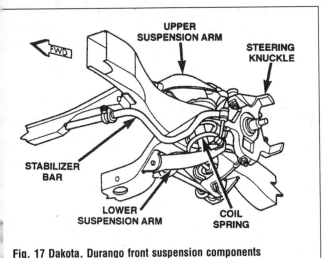

Fig. 17 Dakota, Durango front suspension components

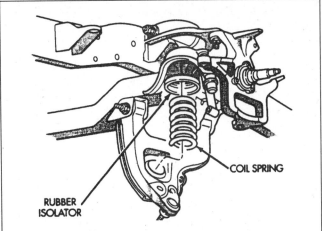

Fig. 18 Removing the coil spring - Ram with independent front suspension

Coil Springs

REMOVAL & INSTALLATION

Dakota, Durango
See Figure 17

1. Raise and support the vehicle.
2. Remove the wheel.
3. Remove the stabilizer bar link from the lower suspension arm.
4. Remove the shock absorber.
5. Compress the coil spring using either the factory tool or a generic substitute suitable for the job.
6. Remove the cotter pin and lower ball stud nut and separate the ball stud from the knuckle.
7. Pull down on the lower suspension arm to remove the spring.
8. Slowly relieve the spring tension. Remove the compressor.

To install:

9. Note that the ramped or open end of the coil spring is the BOTTOM.
10. Compress the spring.
11. Tape the isolator pad to the top of the coil spring. Position the spring in the lower suspension arm well, being sure that it is properly seated.
12. Install the lower ball stud into the knuckle and tighten the nut to 94 ft. lbs. (127 Nm). Install a new cotter pin.
13. Remove the spring compressor.
14. Install the stabilizer bar link to the lower suspension arm and tighten the nut to 35 ft. lbs. (47 Nm).
15. Install the shock absorber, wheel, and lower the vehicle.

Ram With Independent Front Suspension
See Figure 18

1. Raise and support the vehicle.
2. Remove the wheel.
3. Remove the brake caliper and rotor.
4. Disconnect the tie rod from the steering knuckle.
5. Disconnect the stabilizer bar link from the lower suspension arm.
6. Support the lower suspension arm outboard end with a suitable jack. Place the jack under the arm in front of the shock mount.
7. Remove the cotter pin and nut from the lower ball stud. Separate the ball stud assembly.
8. Remove the lower shock bolt.
9. Slowly lower the jack and suspension arm until the spring tension is relieved. Remove the spring and rubber isolator.

To install:

10. Install the rubber isolator on top of the spring.
11. Position the spring into the upper spring seat and lower suspension arm.
12. Raise the suspension arm with a jack and position shock into the suspension arm mount. Install the shock bolt and tighten it to 100 ft. lbs. (135 Nm).
13. Install the steering knuckle on the lower ball stud. Install the lower ball stud nut and tighten to:
   - Light duty: 95 ft. lbs. (129 Nm)
   - Heavy duty: 110 ft. lbs. (136 Nm)
14. Install a new cotter pin and remove jack.
15. Install the stabilizer bar link on the lower suspension arm. Install the grommet, retainer and nut and tighten to 27 ft. lbs. (37 Nm).
16. Install tie rod on steering knuckle and tighten nut to 65 ft. lbs. (88 Nm).
17. The remainder of the procedure is the reverse of removal.

Ram With Link/Coil Suspension
See Figure 19

1. Raise and support the vehicle.
2. Position a hydraulic jack under the axle to support it.
3. Paint or scribe alignment marks on the lower suspension arm cam adjusters and axle bracket for installation reference.
4. Remove the upper suspension arm and loosen the lower suspension arm bolts.
5. Matchmark and disconnect the driveshaft from the axle on 4WD models.
6. Disconnect the track bar from the frame rail bracket.
7. Disconnect the drag link from the pitman arm.
8. Disconnect the stabilizer bar link and shock absorber from the axle.
9. Lower the axle until the spring is free from the upper mount. Remove the coil spring.

To install:

10. Position the coil spring on the axle pad.
11. Raise the axle into position until the spring seats in the upper mount.
12. Connect the stabilizer bar links and shock absorbers to the axle bracket. Connect the track bar to the frame rail bracket.
13. Install the upper suspension arm.
14. If 4WD, reconnect the driveshaft.
15. Install the drag link to the pitman arm and tighten the nut to 185 ft. lbs. (251 Nm). Use a new cotter pin.
16. Lower the vehicle.
17. Tighten the remaining components to the specifications given in the chart.

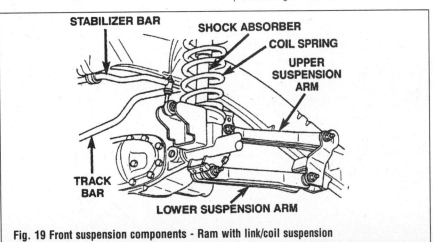

Fig. 19 Front suspension components - Ram with link/coil suspension

8-6 SUSPENSION AND STEERING

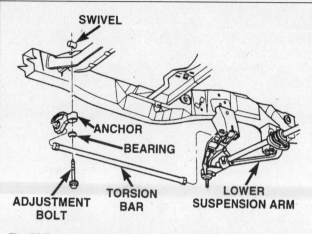

Fig. 20 Front suspension torsion bar assembly - Dakota, Durango 4WD

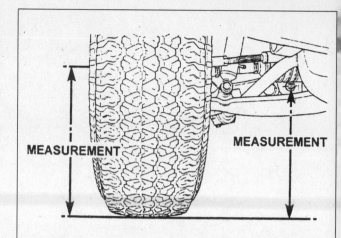

Fig. 21 Suspension height measurements - Dakota, Durango 4WD

Torsion Bars

REMOVAL & INSTALLATION

See Figure 20

Dakota and Durango 4WD trucks use torsion bars as part as the front suspension.

WARNING:

The left and right torsion bars are NOT interchangeable. The bars are identified and stamped L and R. The bars do not have a front or rear end, however, and can be installed either way as long as they are on the correct side of the vehicle.

1. Raise and support the vehicle.
2. Turn the adjustment bolt CCW to release the spring load.

Note: Count and record the number of turns for installation reference.

3. Remove the adjustment bolt.
4. Remove the torsion bar and anchor.

To install:

5. Clean torsion bar mounting in anchor and suspension arm.
6. Insert bar ends into anchor and suspension arm.
7. Position anchor and bearing in frame crossmember. Insert adjustment bolt through bearing, anchor and into swivel.
8. Turn adjustment bolt CW the recorded number of turns.
9. Lower the vehicle and adjust suspension height.

SUSPENSION HEIGHT ADJUSTMENT

See Figure 21

The vehicle suspension height should be measured and adjusted (if necessary) before wheel alignment or when front suspension components have been replaced.

The measurement should be made with the vehicle supporting its own weight and taken on both sides of the vehicle.

1. Bounce the front end of the vehicle up and down several times.
2. Measure and record the distance from the ground to the center of the lower suspension arm rear mounting bolt head.
3. Measure and record the distance from the ground to the center of the front wheel.
4. Subtract the first measurement from the second. The difference between the two measurements should be 1.85 in. (47mm) plus or minus 0.25 in. (6.4mm).
5. Use the torsion bar adjustment bolt to adjust suspension height: turn CW to raise the vehicle; turn CCW to lower the vehicle. Roll the vehicle back-and-forth 6 feet between adjustments.
6. Carry out the same procedure on the other side of the vehicle.
7. Recheck the height on both sides of the vehicle.

Shock Absorbers

REMOVAL & INSTALLATION

Ram, Independent Front Suspension

See Figure 22

1. Raise and support the vehicle.
2. Remove the shock upper nut and remove the retainer and grommet.
3. Remove the lower mounting bolt from the suspension arm.
4. Remove the shock absorber.

To install:

5. Check grommets and bushings for wear and replace if necessary.

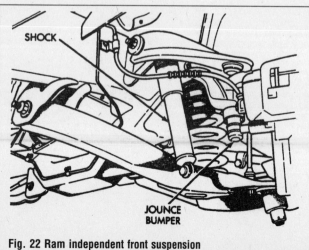

Fig. 22 Ram independent front suspension

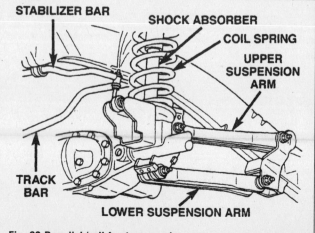

Fig. 23 Ram link/coil front suspension

SUSPENSION AND STEERING 8-7

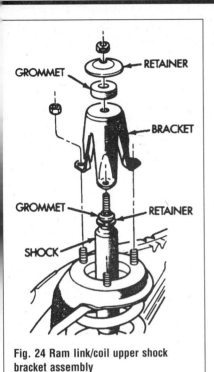

Fig. 24 Ram link/coil upper shock bracket assembly

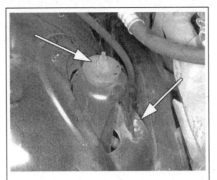

Fig. 25 Upper shock nut and bracket nut inside the engine compartment - Ram with link/coil suspension

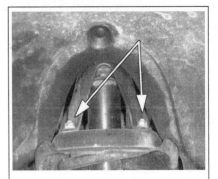

Fig. 26 Upper shock bracket nuts in the wheel well - Ram with link/coil suspension

6. Extend the shock fully, install the retainer and grommet on top of the shock.
7. Insert shock through upper suspension arm. Install top grommet, retainer and nut. Tighten the nut to 40 ft. lbs. (54 Nm).
8. Tighten the bottom bolt to 105 ft. lbs. (142 Nm).

Ram, Link/Coil Front Suspension

See Figures 23, 24, 25 and 26

1. Remove the nut, retainer and grommet from the shock absorber upper stud in the engine compartment.
2. Remove the three nuts from the upper shock bracket. Two are accessible from the wheel well, the other from the engine compartment.
3. Remove the lower bolt from the axle bracket.
4. Remove the shock absorber.

To install:

5. Check grommets and bushings for wear and replace if necessary.
6. Position the lower retainer and grommet on the upper stud. Insert the shock through the spring.
7. Install and tighten the lower bolt to 100 ft. lbs. (135 Nm).
8. Install the upper shock bracket and tighten the nuts to 55 ft. lbs. (75 Nm).
9. Install the upper grommet and retainer. Install the upper shock nut and tighten to 35 ft. lbs. (47 Nm).

Dakota, Durango, 2WD

See Figure 27

1. Remove the upper shock nut, retainer and grommet from the shock stud.
2. Raise and support the vehicle.
3. Remove the lower mounting bolts and drop the shock through the lower suspension arm.

To install:

Note: The upper shock nut must be replaced with a new one or secured with a medium-strength thread-locking compound.

4. Install the lower retainer (it's stamped with an "L") and grommet on the shock stud and install the shock.
5. Tighten the lower bolts to 21 ft. lbs. (28 Nm).

6. Lower the vehicle.
7. Install the upper grommet and retainer (it's stamped with a "U").
8. Install a new nut or secure the existing one with a medium-strength thread locking compound and tighten to 19 ft. lbs. (26 Nm).

Dakota, Durango, 4WD

See Figure 28

1. Raise and support the vehicle.
2. Remove the upper shock nut, retainer and grommet from the shock stud.
3. Remove the lower mounting bolt and remove the shock.

To install:

Note: The upper shock nut must be replaced with a new one or secured with a medium-strength thread-locking compound.

4. Install the lower retainer (it's stamped with an "L") and grommet on the shock stud and install the shock.
5. Tighten the lower bolt to 80 ft. lbs. (108 Nm).
6. Install the upper grommet and retainer (it's stamped with a "U").
7. Install a new nut or secure the existing one with a medium-strength thread locking compound and tighten to 19 ft. lbs. (26 Nm).
8. Lower the vehicle.

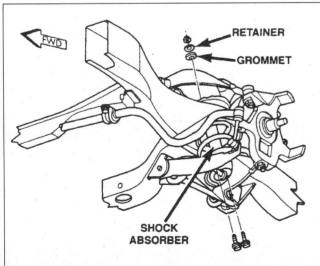

Fig. 27 Dakota, Durango 2WD front shock assembly

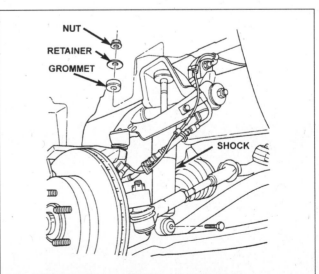

Fig. 28 Dakota, Durango 4WD front shock assembly

8-8 SUSPENSION AND STEERING

TESTING

See Figure 29

The purpose of the shock absorber is simply to dampen the motion of the spring during compression and rebound cycles. If the vehicle is not equipped with these motion dampers, the up and down motion would multiply until the vehicle was alternately trying to leap off the ground and to pound itself into the pavement.

Contrary to popular rumor, the shocks do not affect the ride height of the vehicle. This is controlled by other suspension components such as springs and tires. Worn shock absorbers can affect handling; if the front of the vehicle is rising or falling excessively, the "footprint" of the tires changes on the pavement and steering is affected.

The simplest test of the shock absorber is simply push down on one corner of the unloaded vehicle and release it. Observe the motion of the body as it is released. In most cases, it will come up beyond its original rest position, dip back below it and settle quickly to rest. This shows that the damper is controlling the spring action. Any tendency to excessive pitch (up-and-down) motion or failure to return to rest within 2-3 cycles is a sign of poor function within the shock absorber. Oil-filled shocks may have a light film of oil around the seal, resulting from normal breathing and air exchange. This should NOT be taken as a sign of failure, but any sign of thick or running oil definitely indicates failure. Gas filled shocks may also show some film at the shaft; if the gas has leaked out, the shock will have almost no resistance to motion.

While each shock absorber can be replaced individually, it is recommended that they be changed as a pair (both front or both rear) to maintain equal response on both sides of the vehicle. Chances are quite good that if one has failed, its mate is weak also.

Upper Ball Joint

INSPECTION

Front end noise, front end shimmy and difficult steering could be caused by a worn upper ball joint. Jack up the vehicle beneath the lower suspension arm. Raise the vehicle until the tire lightly contacts the floor. Position a suitable prybar under the tire and pry upward. Then grasp the top of the tire and apply force in and out. Observe any free movement in the components, especially between the upper suspension arm and the steering knuckle. Determine what piece of the suspension has excessive play. Lateral movement in the ball joint must be under 0.030 in. (0.8mm) on Rams, 0.060 in. (1.52mm) on Dakotas and Durangos.

REMOVAL & INSTALLATION

Dakota, Durango 2WD

See Figures 30 and 31

1. Raise and support the vehicle.
2. Remove the tire.
3. Remove the caliper, rotor, shield and ABS speed sensor, if fitted.
4. Remove the tie rod from the steering knuckle arm.
5. Remove the hub/bearing.
6. Remove the shock absorber.
7. Support the outer end of the lower control arm with a floor jack.
8. Remove the cotter pin and upper ball joint nut, then separate the ball joint from the knuckle.
9. Installation is the reverse of removal. Tighten the ball joint nut to 60 ft. lbs. (81 Nm). Use a new cotter pin.

Dakota, Durango 4WD

See Figures 30 and 31

1. Loosen the lug nuts. Raise and support the vehicle.
2. Remove the wheel.
3. Remove the caliper, rotor, shield and ABS speed sensor, if fitted.
4. Remove the front driveaxle.
5. Remove the tie rod from the steering knuckle arm.
6. Raise the lower suspension arm to unload the rebound bumper.
7. Remove the upper ball joint cotter pin and nut. Separate the ball joint from the knuckle.
8. Installation is the reverse of removal. Tighten the upper ball joint nut to 60 ft. lbs. (81 Nm). Install a new cotter pin.

Ram With Independent Front Suspension

1. Loosen the lug nuts. Raise and support the vehicle.
2. Remove the wheel.
3. Support the lower suspension arm outboard end with a jack.
4. Remove the cotter pin and nut from the upper ball joint. Separate the ball joint.
5. Remove the pivot bar bolts from the upper suspension arm bracket and remove the arm from the vehicle.
6. Press out the ball joint.
7. Press in a new ball joint and ensure it is fully seated.
8. The remainder of the procedure is the reverse of removal.

Lower Ball Joint

INSPECTION

Front end noise, front end shimmy and difficult steering could be caused by a worn lower ball joint. Raise the front of the vehicle until the tires lightly contact the floor. The upper suspension arms must not contact the rebound bumpers. Position a suitable prybar under the tire and pry upward. Then grasp the top of the tire and apply force in and out. Observe any

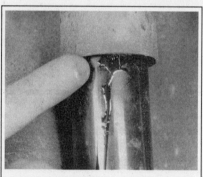

Fig. 29 When fluid is seeping out of the shock absorber, it's time to replace it

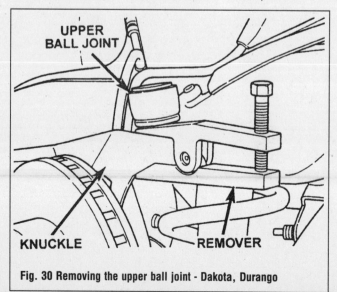

Fig. 30 Removing the upper ball joint - Dakota, Durango

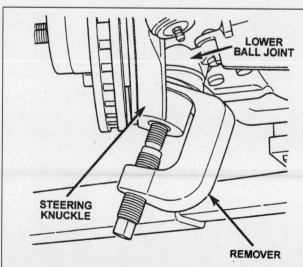

Fig. 31 Removing the lower ball joint - Dakota, Durango

SUSPENSION AND STEERING 8-9

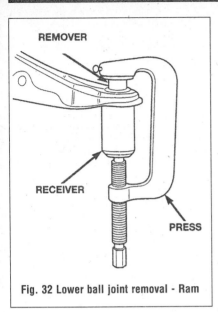

Fig. 32 Lower ball joint removal - Ram

free movement in the components and determine what piece of the suspension has excessive play. Movement in the ball joint must be under 0.030 in. (0.8mm) on Rams, 0.060 in. (1.52mm) on Dakotas and Durangos.

REMOVAL & INSTALLATION

Dakota, Durango 2WD

1. Loosen the wheel lug nuts. Raise and support the vehicle.
2. Remove the wheel.
3. Remove the caliper, rotor, shield and ABS speed sensor, if fitted.
4. Remove the tie rod from the steering knuckle arm.
5. Remove the hub/bearing.
6. Remove the shock absorber.
7. Compress the coil spring.
8. Remove the lower ball joint nut and separate the ball joint from the knuckle.
9. Installation is the reverse of removal. Tighten the ball joint nut to 94 ft. lbs. (127 Nm). Install a new cotter pin.

Dakota, Durango 4WD

1. Loosen the wheel lug nuts. Raise and support the vehicle.
2. Remove the wheel.
3. Remove the caliper, rotor, shield and ABS speed sensor, if fitted.
4. Remove the front driveshaft.
5. Remove the tie rod from the steering knuckle arm.
6. Raise the lower suspension arm to unload the rebound bumper.
7. Remove the lower ball joint cotter pin and nut. Separate the ball joint from the knuckle.
8. Installation is the reverse of removal. Tighten the lower ball joint nut to 135 ft. lbs. (183 Nm). Install a new cotter pin.

Ram With Independent Front Suspension

See Figure 32

1. Loosen the wheel lug nuts. Raise and support the vehicle.
2. Remove the wheel.
3. Remove the brake caliper assembly and rotor.
4. Disconnect the tie rod from the steering knuckle.
5. Remove the stabilizer bar link from the lower suspension arm.
6. Support the lower suspension arm outboard end with a jack. Place the jack under the arm in front of the shock mount.
7. Remove the cotter pin and nut from the lower ball joint. Separate the ball joint.
8. Remove the lower shock bolt from the suspension arm.
9. Lower the jack and suspension arm until spring tension is relieved. Remove spring and rubber isolator.
10. Remove the suspension arm to crossmember hardware and remove arm.
11. Press out the ball joint.

Note: If the ball joint is tack-welded to the arm, the suspension arm must be replaced if the ball joint is worn.

12. Press in a new ball joint and ensure it is fully seated.
13. The remainder of the procedure is the reverse of removal. Install a new cotter pin.

Stabilizer Bar

REMOVAL & INSTALLATION

See Figure 33

1. Raise and support the vehicle.
2. Remove the stabilizer bar retainer bolts from the lower suspension arms.
3. Unbolt the stabilizer bar retainers from the frame.
4. Remove the bar.
5. Installation is the reverse of removal. Be sure all hardware is properly tightened. Refer to the torque specifications charts.

Upper Suspension Arm

REMOVAL & INSTALLATION

Ram With Independent Front Suspension

See Figure 34

1. Loosen the wheel lug nuts. Raise and support the vehicle.
2. Remove the wheel.
3. Support the lower suspension arm at the outboard end with a jack stand.
4. Remove the upper ball joint cotter pin and nut and separate the joint.
5. Remove the pivot bar bolts from the upper suspension arm bracket and remove the arm from the vehicle.
6. Installation is the reverse of removal. Check the specifications charts for proper torque of mounting hardware. Install a new cotter pin.

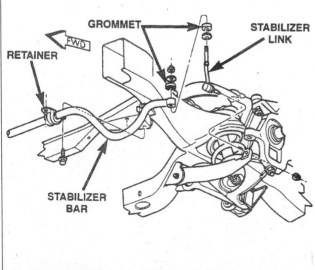

Fig. 33 Front stabilizer bar assembly - typical

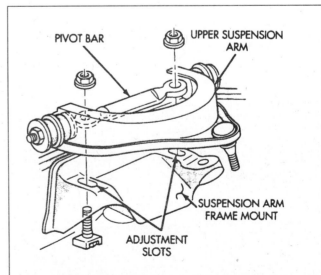

Fig. 34 Upper suspension arm assembly - Ram with independent front suspension

8-10 SUSPENSION AND STEERING

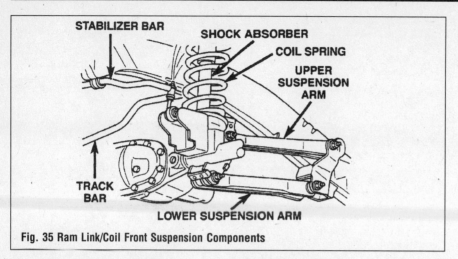

Fig. 35 Ram Link/Coil Front Suspension Components

Ram With Link/Coil Front Suspension

See Figure 35

1. Loosen the wheel lug nuts. Raise and support the vehicle.
2. Remove the upper suspension arm nut and bolt at the axle bracket.
3. Remove the nut and bolt at the frame rail and remove the upper suspension arm.
4. Installation is the reverse of removal. Check the specifications charts for proper torque of mounting hardware. Install a new cotter pin.

Dakota, Durango 2WD

See Figure 36

1. Loosen the wheel lug nuts. Raise and support the vehicle.
2. Remove the wheel.
3. Remove the brake hose bracket from the suspension arm.
4. Position a jack beneath the arm and raise it to unload the rebound bumper.
5. Remove the cotter pin and nut from the upper ball joint and separate the ball joint.
6. Remove the suspension arm pivot bar mounting nuts and remove the suspension arm.
7. Installation is the reverse of removal. Check the specifications charts for proper torque of mounting hardware. Install a new cotter pin.

Dakota, Durango 4WD

See Figure 37

1. Loosen the wheel lug nuts. Raise and support the vehicle.
2. Remove the wheel.
3. Remove the brake hose bracket from the suspension arm.
4. Position a jack beneath the arm and raise it to unload the rebound bumper.
5. Remove the shock absorber.
6. Remove the cotter pin and nut from the upper ball joint and separate the ball joint.
7. Remove the suspension arm pivot bar mounting bolts and remove the suspension arm.
8. Installation is the reverse of removal. Check the specifications charts for proper torque of mounting hardware. Install a new cotter pin.

Lower Suspension Arm

REMOVAL & INSTALLATION

Ram With Independent Front Suspension

1. Loosen the wheel lug nuts. Raise and support the vehicle.
2. Remove the wheel.
3. Remove the brake caliper assembly and rotor.
4. Remove the cotter pin and nut from the tie rod. Remove the tie rod end from the steering knuckle.
5. Remove the stabilizer bar link from the lower suspension arm.
6. Support the lower suspension arm outboard end. Place jack under the arm in front of shock mount.
7. Remove the cotter pin and nut from the lower ball joint. Separate the ball joint.
8. Remove the lower shock bolt.
9. Lower the jack and suspension arm until the spring tension is relieved. Remove spring and rubber isolator.
10. Remove the bolts mounting the suspension arm to the crossmember and remove the arm.
11. Installation is the reverse of removal. Check the specifications charts for proper torque of mounting hardware. Install a new cotter pin.

Ram With Link/Coil Suspension

1. Loosen the wheel lug nuts. Raise and support the vehicle.
2. Scribe alignment marks on the cam adjusters and suspension arm for installation reference.
3. Remove the lower suspension arm nut, cam and cam bolt from the axle.
4. Remove the nut and bolt from the frame rail bracket and remove the lower suspension arm.
5. Installation is the reverse of removal. Check the specifications charts for proper torque of mounting hardware. Install a new cotter pin.

Dakota, Durango 2WD

See Figure 38

1. Loosen the wheel lug nuts. Raise and support the vehicle.
2. Remove the wheel.
3. Remove the caliper and rotor.
4. Remove the shock absorber.
5. Remove the stabilizer bar link from the lower suspension arm.
6. Remove the coil spring.
7. Remove the lower suspension arm mounting bolts from the frame mounts and remove the arm.
8. Installation is the reverse of removal. Check the specifications charts for proper torque of mounting hardware. Install a new cotter pin.

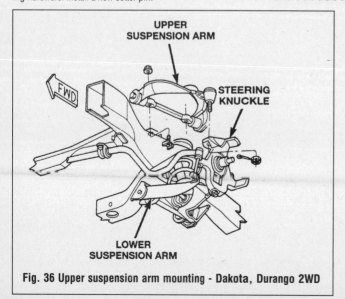

Fig. 36 Upper suspension arm mounting - Dakota, Durango 2WD

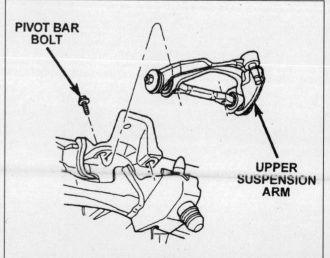

Fig. 37 Upper suspension arm mounting - Dakota, Durango 4WD

SUSPENSION AND STEERING 8-11

Dakota, Durango 4WD

See Figure 39

1. Loosen the wheel lug nuts. Raise and support the vehicle.
2. Remove the wheel.
3. Remove the front driveshaft.
4. Remove the torsion bar.
5. Remove the shock absorber lower bolt.
6. Remove the stabilizer bar.
7. Remove the cotter pin and nut from the lower ball joint. Separate the ball joint.
8. Remove the suspension arm pivot bolts and suspension arm from frame rail brackets.
9. Installation is the reverse of removal. Check the specifications charts for proper torque of mounting hardware. Install a new cotter pin.

Steering Knuckle

REMOVAL & INSTALLATION

Dakota, Durango 2WD

1. Loosen the wheel lug nuts. Raise and support the vehicle.
2. Remove the wheel.
3. Remove the caliper, rotor, shield and ABS speed sensor.
4. Remove the tie rod from the steering knuckle arm.
5. Remove the hub/bearing.
6. Compress the coil spring or support the outer end of the lower control arm with a floor jack..
7. Remove the lower ball joint nut and separate the joint.
8. Remove the upper ball joint nut and separate the ball joint.
9. Remove the steering knuckle.
10. Installation is the reverse of removal. Check the specifications charts for proper torque of mounting hardware. Install new cotter pins.

Dakota, Durango 4WD

1. Loosen the wheel lug nuts. Raise and support the vehicle.
2. Remove the wheel.
3. Remove the caliper, rotor, shield and ABS speed sensor.
4. Remove the front driveshaft.
5. Remove the tie rod end cotter pin and nut. Separate the tie rod from the knuckle.
6. Support the lower suspension arm with a floor jack and raise the jack to unload the rebound bumper.
7. Remove the upper ball joint cotter pin and nut. Separate the ball joint.
8. Remove the lower ball joint cotter pin and nut. Separate the ball joint.
9. Installation is the reverse of removal. Check the specifications charts for proper torque of mounting hardware. Be sure to use new cotter pins.

Ram With Independent Front Suspension

1. Loosen the wheel lug nuts. Raise and support the vehicle.
2. Remove the wheel.
3. Remove the brake caliper and rotor.
4. Remove the cotter pin and nut from the tie rod. Disconnect the tie rod from the steering knuckle.
5. Compress the coil spring or support the outer end of the lower control arm with a floor jack. The jack must remain in this position throughout the entire procedure.
6. Remove the cotter pins and nuts from the upper and lower ball joints. Separate the upper and lower ball joints.
7. Installation is the reverse of removal. Check the specifications charts for proper torque of mounting hardware. Be sure to use new cotter pins.

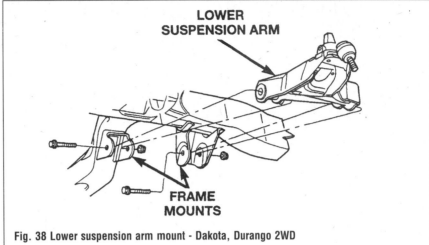

Fig. 38 Lower suspension arm mount - Dakota, Durango 2WD

Ram With Link/Coil Front Suspension

1. Loosen the wheel lug nuts. Raise and support the vehicle.
2. Remove the wheel.
3. Remove the hub bearing and axle shaft.
4. Remove the tie rod or drag link end from the steering knuckle arm.
5. Remove the ABS sensor wire and bracket from the knuckle.
6. Remove the cotter pin from the upper ball stud nut. Remove the upper and lower ball stud nuts.
7. Strike the steering knuckle with a brass hammer to loosen.
8. Remove the knuckle from the axle tube yokes.
9. Installation is the reverse of removal. Check the specifications charts for proper torque of mounting hardware.

Front Hub and Bearing

REMOVAL & INSTALLATION

Dakota, Durango 2WD

See Figure 40

1. Loosen the wheel lug nuts. Raise and support the vehicle.
2. Remove the wheel.

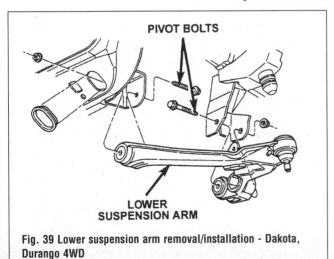

Fig. 39 Lower suspension arm removal/installation - Dakota, Durango 4WD

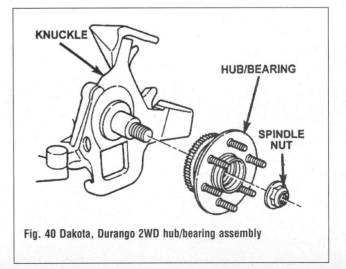

Fig. 40 Dakota, Durango 2WD hub/bearing assembly

8-12 SUSPENSION AND STEERING

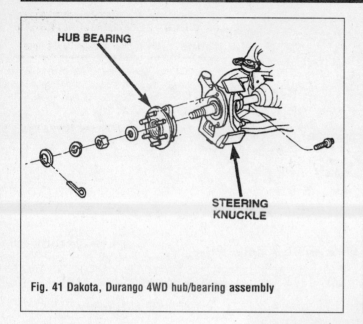

Fig. 41 Dakota, Durango 4WD hub/bearing assembly

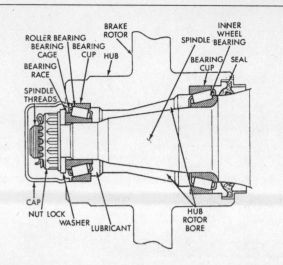

Fig. 42 Wheel bearing assembly - 97-99 Ram with independent front suspension

3. Remove the brake caliper, rotor and ABS wheel speed sensor, if fitted.
4. Remove the hub/bearing spindle nut.
5. Slide the hub/bearing off the spindle.
6. Installation is the reverse of removal. Check the specifications chart for proper torques.

Dakota, Durango 4WD

See Figure 41

1. Loosen the wheel lug nuts. Raise and support the vehicle.
2. Remove the wheel.
3. Remove the axle nut.
4. Remove the brake caliper, rotor and ABS wheel speed sensor, if fitted.
5. Remove the hub/bearing mounting bolts from the steering knuckle.
6. Slide the hub/bearing off the spindle.
7. Installation is the reverse of removal. Check the specifications chart for proper torques.

Ram With Independent Front Suspension (97-99)

See Figures 42, 43, 44, 45, 46, 47, 48, 49, 50 and 51

1. Loosen the wheel lug nuts. Raise and support the vehicle.

Fig. 43 Remove the dust cap to access the rotor hub nut

2. Remove the wheel.
3. Remove the caliper.
4. Remove the dust cap, cotter pin, nut lock, nut and washer from the spindle.
5. Remove the outer wheel bearing.
6. Slide the hub/rotor off.

WARNING:
Use care to prevent the inner wheel bearing and seal from contacting the spindle threads.

Fig. 44 After the nut is removed, place it and the washer in a safe place

7. Remove the seal and inner wheel bearing from the hub/rotor.
8. Remove the inner and outer bearing cups from the hub/rotor with a pin punch, if desired. Cups cannot be reused once removed. Check for damage or wear in place.

To install:

9. Bearing and races must be replaced as a set, if worn or damaged.

Fig. 45 Remove the outer wheel bearing and set it aside for later inspection

Fig. 46 Use a suitable tool, such as this seal remover from Lisle®, to remove the grease seal from the rotor

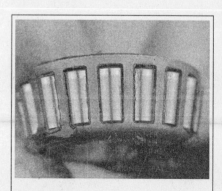

Fig. 47 Inspect the bearing rollers for wear

SUSPENSION AND STEERING 8-13

Fig. 48 This bearing is an example of one that wasn't properly maintained. Notice the sludge build-up from the old grease

Fig. 49 Always spin the rotor while tightening the retaining nut to ensure proper wheel bearing adjustment

10. Remove any burrs or roughness from the spindle with emery cloth. Clean thoroughly afterwards.
11. Clean bearings and cups with a safe solvent. Repack with a quality wheel bearing grease formulated for disc brakes.
12. Drive in the new bearing cups.
13. Grease all contact surfaces.
14. Install the inner wheel bearing with new bearing seal.
15. Slide the hub/rotor onto the shaft.
16. Install the outer wheel bearing, washer and retaining nut.
17. Tighten the nut to 30 - 40 ft. lbs. (41 - 54 Nm) to preload the bearing while rotating the hub/rotor. Stop the rotation, then loosen the nut completely.

Then tighten the nut finger tight and install the nut lock. Use a new cotter pin.

Note: Endplay should be 0.001 - 0.003 in. (0.025 - 0.076mm).

18. The remainder of the procedure is the reverse of removal.

Ram With Independent Front Suspension (00)

See Figure 52

1. Loosen the wheel lug nuts. Raise and support the vehicle.
2. Remove the wheel.
3. Remove the caliper adapter bolts from the steering knuckle and remove the assembly.

WARNING: Do not allow the caliper to hang by the hose.

4. Remove the rotor from the hub/bearing wheel studs.
5. Remove the hub/bearing nut and slide the hub/bearing off the spindle.

Note: The nut cannot be reused. Have a new one on hand before removal.

To install:

6. Reverse the removal procedure. Use a NEW spindle nut. Tighten the nut to the following torque specifications:
   - LD 1500: 185 ft. lbs. (251 Nm)
   - HD 2500/3500: 280 ft. lbs. (380 Nm)
7. Tighten the caliper adapter assembly to the following torque specifications:
   - LD 1500: 130 ft. lbs. (176 Nm)
   - HD 2500/3500: 210 ft. lbs. (285 Nm)

Ram With Coil/Link Suspension

5-STUD HUB/BEARING

See Figures 53, 54 and 57

1. Loosen the wheel lug nuts. Raise and support the vehicle.
2. Remove the wheel.
3. Remove the cotter pin and axle hub nut.
4. Remove the caliper with adapter and rotor.
5. Remove the ABS speed sensor, if fitted.

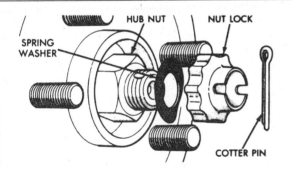

Fig. 50 This exploded view shows the correct order of installation for the spring washer, nut lock and cotter pin

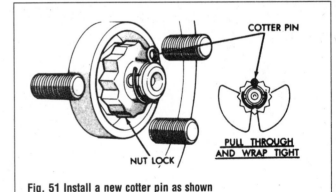

Fig. 51 Install a new cotter pin as shown

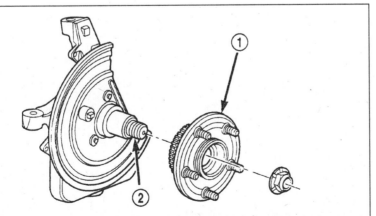

Fig. 52 Hub/bearing assembly (1) and spindle (2) - 00 Ram with independent front suspension

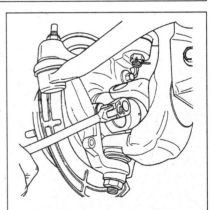

Fig. 53 Loosening the hub/bearing mounting bolts - Ram with coil/link suspension

8-14 SUSPENSION AND STEERING

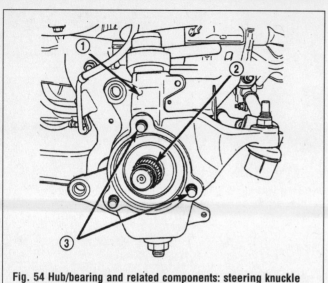

Fig. 54 Hub/bearing and related components: steering knuckle (1), axle (2) and hub/bearing bolts (3)

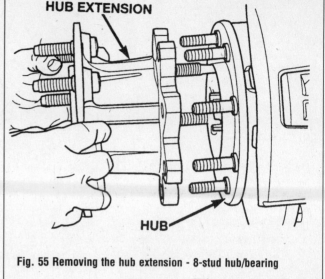

Fig. 55 Removing the hub extension - 8-stud hub/bearing

6. Back the hub/bearing mounting bolts off 1/4 in. each.
7. Tap the bolts with a hammer to loosen the hub/bearing from the steering knuckle.
8. Remove the hub/bearing mounting bolts and then the hub/bearing.

To install:
9. Apply anti-seize compound to the splines of the drive shaft.
10. The remainder of the procedure is the reverse of removal. Tighten all hardware to the proper torque specifications. Refer to the charts. Use a new cotter pin.
11. After installation, apply the brake several times to seat the brake pads and piston. Do not drive the vehicle until a firm pedal is achieved.

8-STUD HUB/BEARING

See Figures 55, 56 and 57

1. Loosen the wheel lug nuts. Raise and support the vehicle.
2. Remove the wheel.
3. Remove the hub extension mounting nuts and remove the extension from the rotor, if equipped.
4. Remove the caliper.
5. Remove the cotter pin and hub nut.

6. Remove the ABS speed sensor.
7. Back the hub/bearing mounting bolts off 1/4 in. each.
8. Tap the bolts with a hammer to loosen the hub/bearing from the steering knuckle.
9. Remove the hub/bearing mounting bolts and then the hub/bearing.

To install:
10. Apply anti-seize compound to the splines of the drive shaft.
11. The remainder of the procedure is the reverse of removal. Tighten all hardware to the proper torque specifications. Refer to the charts. Use a new cotter pin.
12. After installation, apply the brake several times to seat the brake shoes and piston. Do not drive the vehicle until a firm pedal is achieved.

To install:
13. Apply anti-seize compound to the splines of the drive shaft.
14. The remainder of the procedure is the reverse of removal. Tighten all hardware to the proper torque specifications. Refer to the charts.
15. After installation, apply the brake several times to seat the brake pads and piston. Do not drive the vehicle until a firm pedal is achieved.

Wheel Alignment

INDEPENDENT FRONT SUSPENSION

If the tires are worn unevenly, if the vehicle is not stable on the highway or if the handling seems uneven in spirited driving, the wheel alignment should be checked. If an alignment problem is suspected, first check for improper tire inflation and other possible causes. These can be worn suspension or steering components, accident damage or even unmatched tires. If any worn or damaged components are found, they must be replaced before the wheels can be properly aligned. Wheel alignment requires very expensive equipment and involves minute adjustments which must be accurate; it should only be performed by a trained technician. Take your vehicle to a properly equipped shop.

Following is a description of the alignment angles which are adjustable on most vehicles and how they affect vehicle handling. Although these angles can apply to both the front and rear wheels, only the front suspension is adjustable.

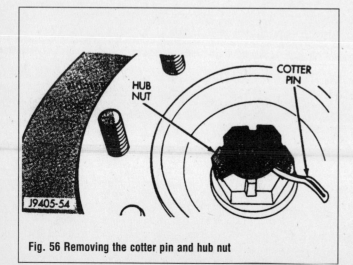

Fig. 56 Removing the cotter pin and hub nut

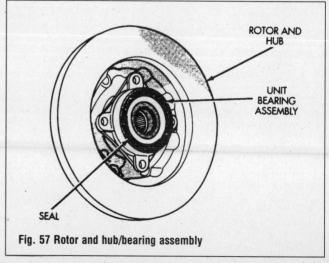

Fig. 57 Rotor and hub/bearing assembly

SUSPENSION AND STEERING 8-15

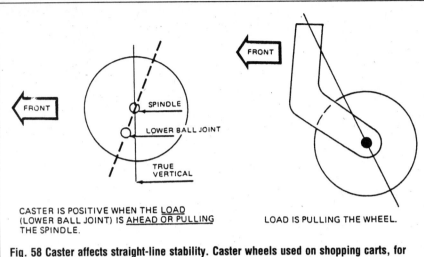

Fig. 58 Caster affects straight-line stability. Caster wheels used on shopping carts, for example, employ positive caster

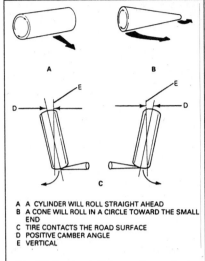

Fig. 59 Camber influences tire contact with the road

Caster

See Figure 58

Looking at a vehicle from the side, caster angle describes the steering axis rather than a wheel angle. The steering knuckle is attached to a control arm or strut at the top and a control arm at the bottom. The wheel pivots around the line between these points to steer the vehicle. When the upper point is tilted back, this is described as positive caster. Having a positive caster tends to make the wheels self-centering, increasing directional stability. Excessive positive caster makes the wheels hard to steer, while an uneven caster will cause a pull to one side. Overloading the vehicle or sagging rear springs will affect caster, as will raising the rear of the vehicle. If the rear of the vehicle is lower than normal, the caster becomes more positive.

Camber

See Figure 59

Looking from the front of the vehicle, camber is the inward or outward tilt of the top of wheels. When the tops of the wheels are tilted in, this is negative camber; if they are tilted out, it is positive. In a turn, a slight amount of negative camber helps maximize contact of the tire with the road. However, too much negative camber compromises straight-line stability, increases bump steer and torque steer.

Toe

See Figure 60

Looking down at the wheels from above the vehicle, toe angle is the distance between the front of the wheels, relative to the distance between the back of the wheels. If the wheels are closer at the front, they are said to be toed-in or to have negative toe. A small amount of negative toe enhances directional stability and provides a smoother ride on the highway.

SOLID FRONT AXLE FRONT SUSPENSION

If the tires are worn unevenly, if the vehicle is not stable on the highway or if the handling seems poor, the wheel alignment should be checked. If an alignment problem is suspected, first check for improper tire inflation and other possible causes. These can be worn suspension or steering components, accident damage or even unmatched tires. If any worn or damaged components are found, they must be replaced before the wheels can be properly aligned. Wheel alignment requires very expensive equipment and involves minute adjustments which must be accurate; it should only be performed by a trained technician. Take your vehicle to a properly equipped shop.

Following is a description of the alignment angles which are adjustable on most vehicles and how they affect vehicle handling. Although these angles can apply to both the front and rear wheels, only the front suspension is adjustable.

Caster

See Figure 61

Looking at a vehicle from the side, caster angle describes the steering axis rather than a wheel angle. The steering knuckle is attached to the axle yoke through ball joints or king pins. The wheel pivots around the line between these points to steer the vehicle. When the upper point is tilted back, this is described as positive caster. Having a positive caster tends to make the wheels self-centering, increasing directional stability. Excessive positive caster makes the wheels hard to steer, while an uneven caster will cause a pull to one side. Overloading the vehicle or sagging rear springs will affect caster, as will raising the rear of the vehicle. If the rear of the vehicle is lower than normal, the caster becomes more positive.

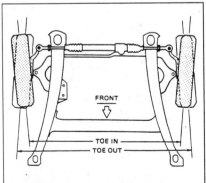

Fig. 60 With toe-in, the distance between the wheels is closer at the front than at the rear

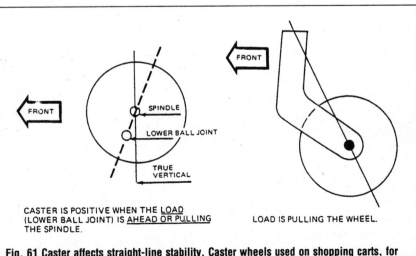

Fig. 61 Caster affects straight-line stability. Caster wheels used on shopping carts, for example, employ positive caster

8-16 SUSPENSION AND STEERING

Camber

See Figure 62

Looking from the front of the vehicle, camber is the inward or outward tilt of the top of wheels. When the tops of the wheels are tilted in, this is negative camber; if they are tilted out, it is positive. In a turn, a slight amount of negative camber helps maximize contact of the tire with the road. However, too much negative camber compromises straight-line stability, increases bump steer and torque steer.

Toe

See Figure 63

Looking down at the wheels from above the vehicle, toe angle is the distance between the front of the wheels relative to the distance between the back of the wheels. If the wheels are closer at the front, they are said to be toed-in or to have negative toe. A small amount of negative toe enhances directional stability and provides a smoother ride on the highway.

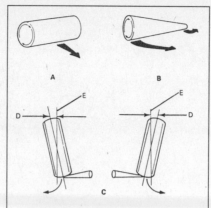

A A CYLINDER WILL ROLL STRAIGHT AHEAD
B A CONE WILL ROLL IN A CIRCLE TOWARD THE SMALL END
C TIRE CONTACTS THE ROAD SURFACE
D POSITIVE CAMBER ANGLE
E VERTICAL

Fig. 62 Camber influences tire contact with the road

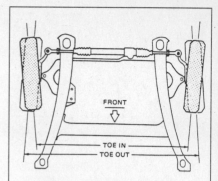

Fig. 63 With toe-in, the distance between the wheels is closer at the front than at the rear

REAR SUSPENSION

See Figure 64

Leaf Springs

REMOVAL & INSTALLATION

See Figures 65, 66 and 67

WARNING:

On Dakota and Durango, the rear of the vehicle must be lifted only with a jack or hoist. The lift must be placed under the frame rail crossmember located aft of the rear axle. Use care to avoid bending the side rail flange.

Note: On Ram models, removal of the left side spring requires removal of the fuel tank.

1. Loosen the wheel lug nuts. Raise the vehicle and support it on jackstands placed under the frame.
2. Use a hydraulic jack to support the axle to relieve the weight from the springs.
3. Remove the wheels.
4. Remove the shock absorbers if they are bolted to the spring plates.
5. Remove the spring U-bolt nuts, spring plate and U-bolts.
6. Remove the spring mounting hardware at front and rear of the spring. On Ram models, removal of the left front spring eye bolt requires fuel tank removal.
7. Remove the spring from the truck.

To install:

8. Reverse the removal procedure.
9. Install all hardware, but perform final torque after the vehicle has been lowered and the springs are supporting its weight. Note the torque specifications for the spring mounting hardware:
   - Dakota, Ram: 120 ft. lbs. (163 Nm)
   - Durango: 90 ft. lbs. (122 Nm)

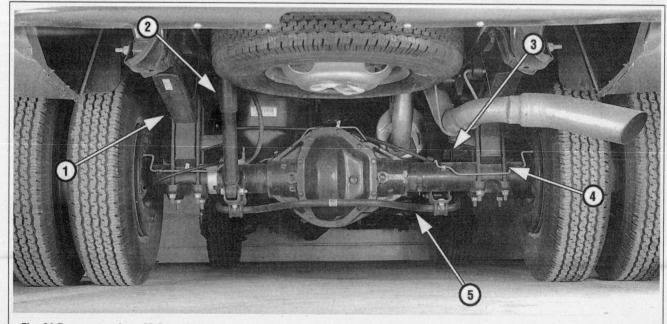

Fig. 64 Rear suspension - 99 Ram shown

1 Leaf spring
2 Rear shock absorber
3 Rear axle
4 U-bolts
5 Stabilizer bar

SUSPENSION AND STEERING 8-17

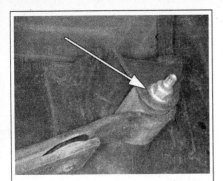

Fig. 65 Leaf spring eye bolt - typical

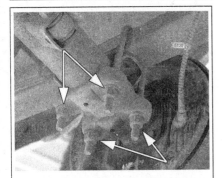

Fig. 66 The leaf spring attaches to the axle with two U-bolts on each side. Be sure that the spring is properly centered before tightening the nuts

Fig. 68 Removing lower shock mount hardware

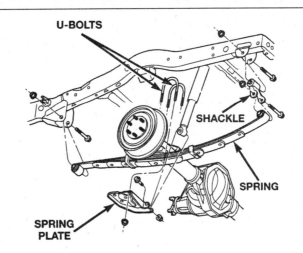

Fig. 67 Leaf spring mounting components - typical

Shock Absorbers

REMOVAL & INSTALLATION

See Figures 68 and 69

1. Raise and safely support the vehicle securely on jackstands.
2. Support the rear axle with a floor jack placed under the axle tube closest to the shock absorber being removed.
3. Remove the upper shock mounting hardware.
4. Remove the lower mounting hardware.
5. Remove the shock absorber.

To install:

6. When installing the shock absorber, make sure the upper bushings are in the correct position.
7. Some shocks should be purged of air before fitting: extend the unit, then turn the unit upside down and compress it several times. There may be more resistance to extension than compression.
8. Replace any worn or cracked bushings.
9. Tighten the shock absorber mounting hardware:
   - Ram: 100 ft. lbs. (136 Nm)
   - Dakotas and Durangos: 70 ft. lbs. (95 Nm)
10. Lower the vehicle.

TESTING

Refer to the "Front Suspension" for shock absorber testing information.

Stabilizer Bar

REMOVAL & INSTALLATION

See Figure 70

1. Raise and support the vehicle.
2. Remove the hardware bolting the links to the stabilizer bar.
3. Remove the hardware bolting the stabilizer bar retainers to the axle.
4. Remove the stabilizer bar.
5. Unbolt and remove the links from the frame, if desired.
6. Replace worn, cracked or distorted bushings.

To install:

7. Install the stabilizer bar and center it with equal spacing on both sides. Tighten retainer hardware to 40 ft. lbs. (54 Nm).
8. Install the links and tighten the hardware on frame and at the bar hand tight.
9. Lower the vehicle onto the suspension.

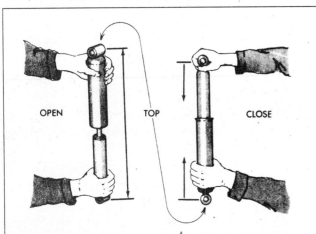

Fig. 69 Some new shocks should have the air purged before fitting

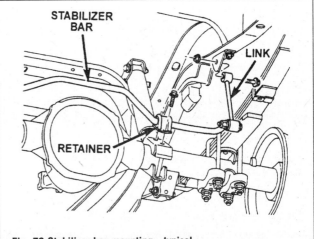

Fig. 70 Stabilizer bar mounting - typical

8-18 SUSPENSION AND STEERING

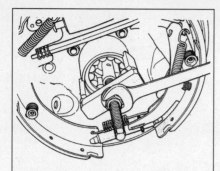

Fig. 71 This is the factory tool in use for wheel bearing removal. But a slide hammer with internal fittings can be used

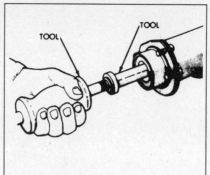

Fig. 72 Using a slide hammer to remove the axle bearing

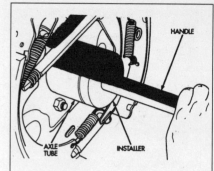

Fig. 73 Bearings must be driven in to the correct depth. Unless the factory tool is available, measure the depth before removal

10. Tighten the links as follows:
• Durango: link upper, 70 ft. lbs. (95 Nm); link lower, 50 ft. lbs. (68 Nm)
• Dakota: 40 ft. lbs. (54 Nm)
• 97 Ram: link upper, 70 ft. lbs. (95 Nm); link lower, 50 ft. lbs. (68 Nm)
• 98 - 00 Ram: link upper, 50 ft. lbs. (68 Nm); link lower, 50 ft. lbs. (68 Nm)

Rear Wheel Bearings

REMOVAL & INSTALLATION

See Figures 71, 72 and 73

Note: Damaged wheel bearings produce noise and vibration. The noise often changes when the load on the bearing changes. A road test (in an empty parking lot, for example) would involve veering sharply left and right. If the noise level changes, suspect the wheel bearings.

Note: The removal procedure often damages rear wheel bearings. Removal is not recommended unless the bearings are going to be replaced. New seals must always be used.

1. Loosen the wheel lug nuts. Raise and support the rear of the vehicle in a safe manner.
2. Remove the rear wheel.
3. Remove the brake drum.
4. Remove the differential cover.
5. Remove the axle (see Chapter 7).
6. Pry out the wheel bearing seal with a suitable tool.
7. A clue to bearing condition can be obtained without removal. Check for obvious signs of damage, scoring, cracked rollers, blueing, etc. Rotation must be a smooth and without binding. Damaged bearings must of course be replaced.

Note: Wheel bearings are lubricated by the oil in the differential. If damage is noted, insufficient or inadequate lubricant may be the cause. The differential should be thoroughly cleaned and filled to the proper level with fresh lubricant. In this case, it would be best to replace the wheel bearings on both sides.

8. The factory removal/installation tool set is designed to install the bearing in the tube to the proper depth. If this is not being used, measure the distance from the end of the axle tube to the bearing so that the new bearing can be installed to the proper depth.
9. Use a slide hammer and suitable internal fittings to pull out the wheel bearing, or the factory tool, if available.
10. Clean the inside of the axle tube before installing the new bearing.
11. Drive the new bearing straight in until it reaches the depth measured in the removal steps. In the absence of the factory tool, a large socket can be used. Apply force only to the outer race.

Note: Install bearings with the part number facing outwards.

12. Press in a new seal. Apply some oil to the lips of the seal before inserting the axle shaft.
13. Don't forget to refill the differential with the right grade and quantity of oil.

STEERING

Steering Wheel

REMOVAL & INSTALLATION

See Figures 74, 75 and 76

CAUTION:

The Supplemental Restraint System (SRS) uses an air bag. Whenever working near any of the SRS components, such as the impact sensors, the air bag module, steering column and instrument panel, disable the SRS. Refer to "Chassis Electrical" for air bag removal, handling and installation procedures.

Note: Do not move the front wheels or steering shaft once the steering wheel is removed. Moving the front wheels or steering shaft will disturb the relationship between the clock spring and the shaft. If the relationship has been disturbed, the clock spring must be centered prior to reinstalling the steering wheel.

1. Disconnect the negative battery cable.
2. Wait two minutes. This time is required to discharge the capacitor and prevent unintended air bag deployment. The system is now disarmed.
3. Ensure the steering wheel and the vehicle's front wheels are in the straight ahead position.
4. As required, remove the speed control switches from the horn pad by unscrewing the attaching screws.
5. Remove the four air bag module retaining fasteners at the back of the steering wheel.
6. Pull the module far enough away from the steering wheel to disconnect the clockspring wire harness.
7. Label and disconnect the clockspring wire harness.
8. Remove the air bag module from the steering wheel.
9. Matchmark the steering wheel and shaft.
10. Remove the steering wheel retaining nut.
11. Using a steering wheel puller, remove the steering wheel from the shaft.

To install:

12. Align and install the steering wheel on the column shaft.
13. Install a new retaining nut and tighten the nut to 45 ft. lbs. (61 Nm) on Rams or 35 ft. lbs. (47 Nm) on Dakotas and Durangos.
14. Connect the clockspring wire harness.
15. Install the four air bag module retaining nuts and tighten to 80 - 100 inch lbs. (9 - 11 Nm).
16. As required, install the speed control switches on the horn pad.
17. Connect the negative battery cable.
18. Check the air bag system for proper function prior to driving the vehicle.

Fig. 74 Removing the air bag module hardware

SUSPENSION AND STEERING   8-19

Fig. 75 Removing the steering wheel nut

Fig. 76 Using a puller to remove the steering wheel

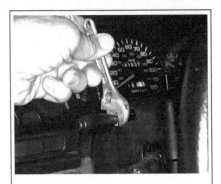

Fig. 77 Unscrew and remove tilt lever

Turn Signal (Combination) Switch

REMOVAL & INSTALLATION

See Figures 77 and 78

CAUTION:

These models are equipped with a Supplemental Restraint System (SRS) which uses an air bag. Whenever working near any of the SRS components, such as the impact sensors, air bag module, steering column and instrument panel, disable the SRS by disconnecting the battery negative cable(s) and allowing at least two minutes for the air bag capacitor to discharge.

1. Disconnect the battery negative cable(s).
2. If equipped with a tilt column, remove the tilt lever (turn CCW).
3. Remove upper and lower steering column shrouds.
4. Remove the mounting screws which require a tamper-proof Torx bit.
5. Gently pull the switch away from the steering column.
6. To disconnect the wiring, remove the screw which secures the connector to the switch.
7. Installation is the reverse of removal.

Ignition Switch & Lock Cylinder

REMOVAL & INSTALLATION

See Figures 79, 80, 81, 82, 83, 84, 85 and 86

CAUTION:

These models are equipped with a Supplemental Restraint System (SRS) which uses an air bag. Whenever working near any of the SRS components, such as the impact sensors, air bag module, steering column and instrument panel, disable the SRS by disconnecting the battery negative cable(s) and allowing at least two minutes for the air bag capacitor to discharge.

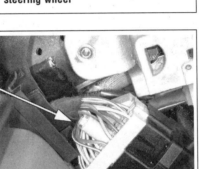

Fig. 78 Remove this screw to disconnect the combination switch wiring

1. Disconnect the battery negative cable(s).
2. If equipped with a tilt column, remove the tilt lever (turn CCW).
3. Remove upper and lower steering column shrouds.
4. Place automatic transmissions in PARK.
5. Turn the key to the RUN position.
6. Press in on the retaining pin while pulling the key from the ignition switch.
7. Remove the key lock cylinder.
8. Remove the three ignition switch mounting screws which require a tamper-proof Torx bit.
9. Gently pull the switch away from the steering column. Release the connector locks on the seven-terminal wiring connector.
10. Release the connector lock on the four-termi-

Fig. 79 Remove the two tamper-proof Torx screws to remove the combination switch

nal halo lamp wiring connector and remove connector.

To install:

11. Rotate the flag on the rear of the ignition switch until it is in the RUN position. This is to allow the tang on the key cylinder to fit into the slots within the ignition switch.
12. With the key in the ignition key cylinder, rotate the key CW until the retaining pin can be depressed.
13. Install the key cylinder into the ignition switch by aligning the retaining pin into the retaining pin slot. Push the key cylinder into the switch until the retaining pin engages. After the pin engages, rotate the key to OFF or LOCK.

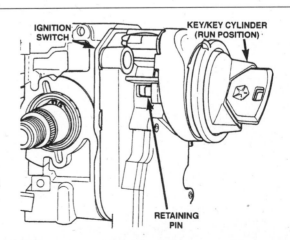

Fig. 80 Press in on the retaining pin while pulling the key cylinder from the ignition switch

14. Check for proper retention of the key cylinder by attempting to pull the cylinder from the switch.

15. On automatic transmission models, before attaching the ignition switch to the steering column, ensure that the shifter is in PARK. The park lock dowel pin on the rear of the ignition switch must also be properly indexed into the park lock linkage before installing the switch.

16. The flag at the rear of the ignition switch must be properly indexed into the steering column before installing the switch. This flag is used to operate the steering wheel lock lever in the steering column. This lever allows the steering wheel position to be locked when the key switch is in the LOCK position.

17. Place the ignition switch in the LOCK position. The switch is in the LOCK position when the column lock flag is parallel to the ignition switch terminals.

18. On automatic transmission models, apply a light coating of grease to the park lock dowel pin and park lock slider linkage. Before installing the switch, push the park lock slider linkage forward until it bottoms. Do a final positioning by pulling it rearward about 1/4 in.

19. Apply a light coating of grease to both column lock flag and shaft at end of flag.

20. Place the ignition switch into openings on steering column.

21. On automatic transmission models, be sure

Fig. 81 Three tamper-proof Torx screws secure the ignition switch

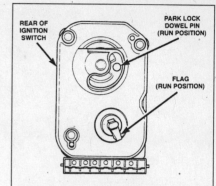

Fig. 82 Flag in RUN position for key cylinder installation

the park lock dowel pin on the rear of the ignition switch enters the slot in the park lock slider linkage.

22. Be sure the flag on the rear of the switch is positioned above the steering wheel lock lever.

23. Align the dowel pins on the rear of the switch into the holes on the side of the steering column.

24. Install the mounting screws and tighten to 17 inch lbs. (2 Nm).

25. After installing the switch, rotate the key from LOCK to ON. Verify that the park lock slider moves in the slider slot, allowing the gearshift lever to be moved out of PARK on automatic transmission mod-

els. If the slider does not move, and the gearshift lever is locked in PARK, the ignition switch park lock dowel pin, on the rear of the ignition switch, is not properly installed in the slot of the park lock slider linkage. Remove the ignition switch and reinstall.

26. Connect the electrical connectors to the halo lamp. Make sure that the switch locking tabs are fully seated in the wiring connectors.

27. Install the steering column covers.

28. The remainder of the procedure is the reverse of removal.

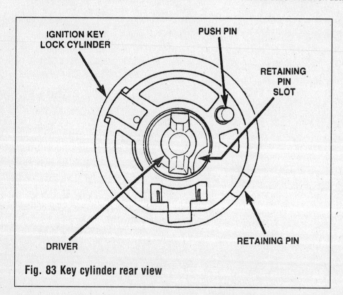

Fig. 83 Key cylinder rear view

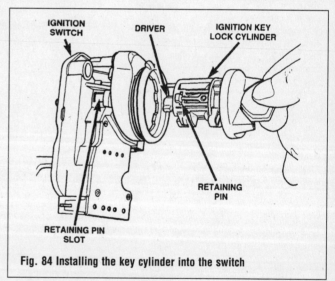

Fig. 84 Installing the key cylinder into the switch

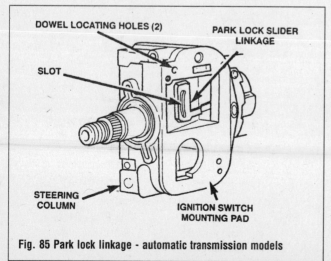

Fig. 85 Park lock linkage - automatic transmission models

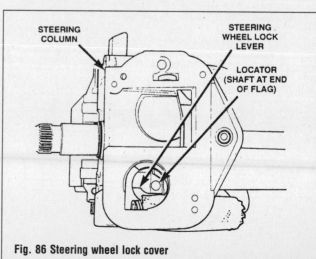

Fig. 86 Steering wheel lock cover

SUSPENSION AND STEERING 8-21

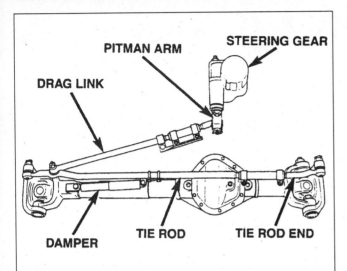

Fig. 87 Major components of the steering linkage are similar for all models - HD link/coil shown

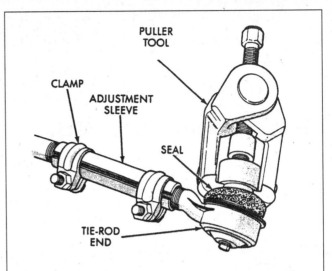

Fig. 88 Steering linkage components must be separated with pullers or pickle forks

Ram Steering Linkage

REMOVAL & INSTALLATION

See Figures 87, 88 and 89

The Ram steering linkage consists of a Pitman arm, drag link or center link, tie rod, and on independent front suspension models, idler arms. Steering linkages vary in detail depending on front suspension (link/coil or independent), and service rating (LD or HD). Service procedures apply to all types, however. Note the following points:

- Pullers or pickle forks are required to separate the components.
- If any components are replaced or serviced, an alignment must be performed.
- Be sure to matchmark all components so that they can be reinstalled in their original positions.
- Components attached with a nut and cotter pin must be torqued to specification. Then, if the slot in the nut does not line up with the cotter pin hole, tighten the nut until alignment is achieved. Never loosen the nut to insert the cotter pin.
- Replace any torn seals. When assembling the steering linkage, be sure to obey all torque specifications for your model. Use new cotter pins.
- Lubricate all components with a good grade of chassis grease.

Independent Front Suspension

See Figure 90

1. Raise and safely support the vehicle.
2. Remove the tie rod cotter pins and nuts.
3. remove the tie rod end ball studs from the steering knuckles with a suitable puller.
4. Remove the tie rod ends from the center link.
5. Remove the idler arm from the center link. Remove the idler arm bolt from the frame bracket.
6. Remove the pitman arm ball stud from the center link.
7. Mark the pitman arm and shaft positions for assembly reference. Remove the pitman arm with a suitable puller.

To install:

8. Check the Torque Specifications chart for proper torque values.
9. Position the idler arm on the frame bracket and tighten the bolt to specification.
10. Center the steering gear to the matchmarks and install the pitman arm.
11. Install the lockwasher and retaining nut on the pitman shaft. Tighten the nut to the proper torque.
12. Install the center link to the ball studs and tighten the nuts to specification.
13. Install the tie rod ends into the center link and tighten hardware to specification. Use new cotter pins.
14. Install the tie rods into the steering knuckles and tighten hardware to spec.
15. Lower the vehicle, center the steering wheel and adjust toe.

Note: Position the clamp on the sleeve so that the retaining bolt is on the bottom side. Tighten to specification.

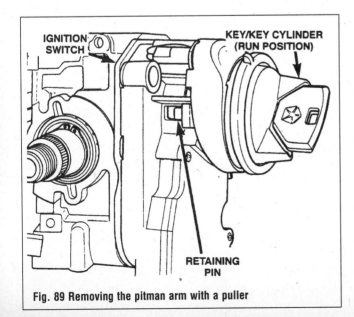

Fig. 89 Removing the pitman arm with a puller

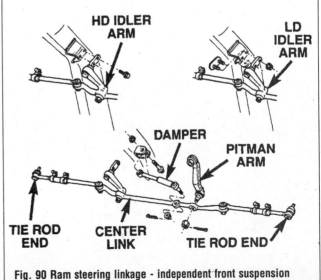

Fig. 90 Ram steering linkage - independent front suspension

8-22 SUSPENSION AND STEERING

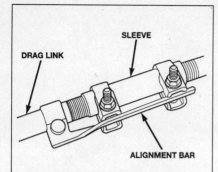

Fig. 91 Do not loosen or move the alignment bar or clamp. The bar is used as a locator for the adjuster clamps

Fig. 92 Removing the tie rod end cotter pin

Fig. 93 Removing the tie rod end nut

Link/Coil Suspension

See Figure 91

WARNING:

Do not loosen or move the alignment bar or clamp on the drag link. The bar is used as a locator for the adjuster clamps.

1. Raise and safely support the vehicle.
2. Remove the steering damper.
3. Remove the tie rod nuts.
4. Remove the tie rod from the drag link and left knuckle with a suitable puller.
5. Remove the drag link and nuts.
6. Remove the drag link from the right knuckle and pitman arm with a suitable puller.
7. Mark the pitman arm and shaft positions for assembly reference. Remove the pitman arm with a suitable puller.

To install:

8. Align the reference marks and install the pitman arm. Install the washer and nut and torque to the proper specification.
9. Install the drag link and properly tighten the hardware.
10. Install the drag link to the right steering knuckle. Torque hardware to the given specification.
11. Install the tie rod to the left steering knuckle and drag link. Tighten the hardware to the correct specification.
12. Install the steering damper. Tighten the fastening hardware to the proper specification.
13. Lower the vehicle, center the steering wheel and adjust toe.

Note: Position the clamp on the sleeve so that the retaining bolt is on the bottom side. Tighten to specification.

Tie Rod Ends

See Figures 92, 93, 94, 95 and 96

1. Remove the cotter pin and the castellated nut from the outer tie rod end. Discard the cotter pin.
2. Separate the outer tie rod end from the steering knuckle using an appropriate tie rod end remover.
3. Mark the outer tie rod jam nut on one side with a reference line for installation.
4. Hold the outer tie rod end with a wrench and loosen the tie rod end jam nut.
5. Back the tie rod end jam nut off ONE FULL TURN ONLY.
6. Remove the outer tie rod end from the inner tie rod spindle.

To install:

7. Clean the threads on the inner tie rod spindle (front wheel spindle connecting rod).
8. Thread the new outer tie rod end onto the inner tie rod until it bottoms on the jam nut.
9. Back the tie rod and jam nut out one full turn until the reference line is in the same position as before.
10. Place the outer tie rod end stud into the steering knuckle. Set the front wheels in a straight ahead position.
11. Install a new castellated nut onto the outer tie rod end stud.
12. Torque the nut to the proper specification.
13. Continue to tighten the castellated nut until a new cotter pin can be inserted through the hole in the stud. Install a new cotter pin.
14. If required, repeat the procedure for the opposite side.
15. Check the alignment and set the toe adjustment to specification.
16. Torque the outer tie rod end jam nut to the proper torque.

Fig. 94 Pressing out the tie rod end with a puller

Fig. 95 Separating the tie rod from the steering knuckle

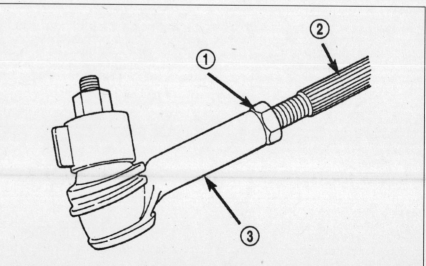

Fig. 96 Tie rod end components: jam nut (1), tie rod (2), tie rod end (3)

SUSPENSION AND STEERING 8-23

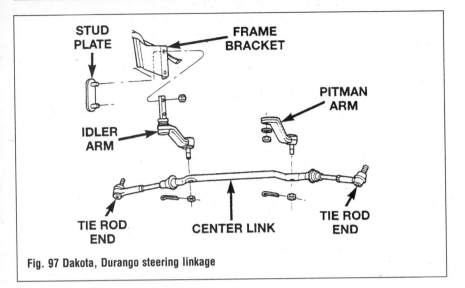

Fig. 97 Dakota, Durango steering linkage

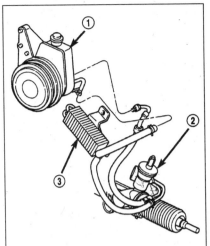

Fig. 98 Power steering pump (1), gear (2) and oil cooler (3) - a typical rack and pinion steering system

Dakota, Durango Steering Linkage

See Figure 97

1999 and earlier Dakota, Durango (non-rack and pinion) steering linkages consists of a pitman arm, center link, tie rod ends, and an idler arm. Note the following points:

• Pullers or pickle forks are required to separate the components.
• If any components are replaced or serviced, an alignment must be performed.
• Be sure to matchmark all components so that they can be reinstalled in their original positions.
• Components attached with a nut and cotter pin must be torqued to specification. Then, if the slot in the nut does not line up with the cotter pin hole, tighten the nut until alignment is achieved. Never loosen the nut to insert the cotter pin.
• Replace any torn seals. When assembling the steering linkage, be sure to obey all torque specifications for your model. Use new cotter pins.
• Lubricate all components with a good grade of chassis grease.

1. Raise and safely support the vehicle.
2. Remove the tie rod cotter pins and nuts.
3. Remove the tie rod end ball studs from the steering knuckles with a suitable puller.
4. Remove the tie rod ends from the center link.
5. Remove the idler arm from the center link. Remove the idler arm bolt from the frame bracket.
6. Remove the pitman arm ball stud from the center link.
7. Mark the pitman arm and shaft positions for assembly reference. Remove the pitman arm with a suitable puller.

To install:

8. Check the Torque Specifications chart for proper torque values.
9. Position the idler arm on the frame bracket and tighten the bolt to specification.
10. Center the steering gear to the matchmarks and install the pitman arm.
11. Install the lockwasher and retaining nut on the pitman shaft. Tighten the nut to the proper torque.
12. Install the center link to the ball studs and tighten the nuts to specification.
13. Install the tie rod ends into the center link and tighten hardware to specification. Use new cotter pins.
14. Install the tie rods into the steering knuckles and tighten hardware to spec.
15. Lower the vehicle, center the steering wheel and adjust toe.

Note: Position the clamp on the sleeve so that the retaining bolt is on the bottom side. Tighten to specification.

Tie Rod Ends

See Figure 96

1. Remove the cotter pin and the castellated nut from the outer tie rod end. Discard the cotter pin.
2. Separate the outer tie rod end from the steering knuckle using an appropriate tie rod end remover.
3. Mark the outer tie rod jam nut on one side with a reference line for installation.
4. Hold the outer tie rod end with a wrench and loosen the tie rod end jam nut.
5. Back the tie rod end jam nut off ONE FULL TURN ONLY.
6. Remove the outer tie rod end from the inner tie rod spindle.

To install:

7. Clean the threads on the inner tie rod spindle (front wheel spindle connecting rod).
8. Thread the new outer tie rod end onto the inner tie rod until it bottoms on the jam nut.
9. Back the tie rod and jam nut out one full turn until the reference line is in the same position as before.
10. Place the outer tie rod end stud into the steering knuckle. Set the front wheels in a straight ahead position.
11. Install a new castellated nut onto the outer tie rod end stud.
12. Torque the nut to the proper specification.
13. Continue to tighten the castellated nut until a new cotter pin can be inserted through the hole in the stud. Install a new cotter pin.
14. If required, repeat the procedure for the opposite side.
15. Check the alignment and set the toe adjustment to specification.
16. Torque the outer tie rod end jam nut to the proper torque.

Rack & Pinion Steering

See Figures 98 and 99

1999 and earlier 2wd Dakotas and Durangos, and all 2000 and later Dakota and Durangos are equipped with rack and pinion steering.

REMOVAL & INSTALLATION

See Figures 99 and 100

1. Raise and support the vehicle.
2. 4WD: remove the splash shield.
3. Remove the nuts from the tie rod ends.
4. Separate the tie rod ends from the knuckles with a suitable puller.

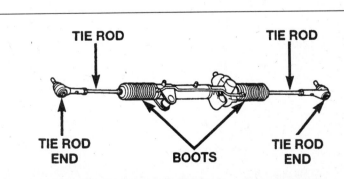

Fig. 99 Rack and pinion steering gear components - Dakota, Durango

8-24 SUSPENSION AND STEERING

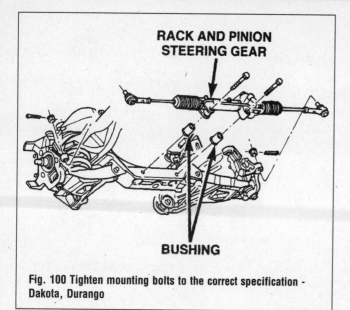

Fig. 100 Tighten mounting bolts to the correct specification - Dakota, Durango

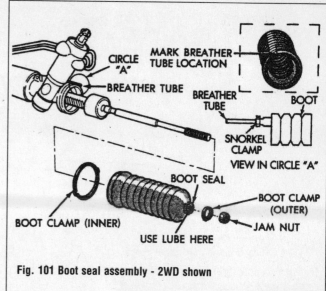

Fig. 101 Boot seal assembly - 2WD shown

5. Remove the power steering lines from the gear.
6. Remove the lower coupler bolt and slide the coupler off the gear.
7. Remove the mounting bolts from the gear to front crossmember and remove the gear.

To install:

8. Reverse the removal procedure. Note the torque specifications for the fastening hardware.
9. Fill the system with fluid and perform the power steering pump initial operation procedure.
10. Realign the front end.

Tie Rod Ends

See Figure 96

1. Remove the cotter pin and the castellated nut from the outer tie rod end. Discard the cotter pin.
2. Separate the outer tie rod end from the steering knuckle using an appropriate tie rod end remover.
3. Mark the outer tie rod jam nut on one side with a reference line for installation.
4. Hold the outer tie rod end with a wrench and loosen the tie rod end jam nut.
5. Back the tie rod end jam nut off ONE FULL TURN ONLY.
6. Remove the outer tie rod end from the inner tie rod spindle.

To install:

7. Clean the threads on the inner tie rod spindle (front wheel spindle connecting rod).
8. Thread the new outer tie rod end onto the inner tie rod until it bottoms on the jam nut.
9. Back the tie rod and jam nut out one full turn until the reference line is in the same position as before.
10. Place the outer tie rod end stud into the steering knuckle. Set the front wheels in a straight ahead position.
11. Install a new castellated nut onto the outer tie rod end stud.
12. Torque the nut to the proper specification.
13. Continue to tighten the castellated nut until a new cotter pin can be inserted through the hole in the stud. Install a new cotter pin.
14. If required, repeat the procedure for the opposite side.

15. Check the alignment and set the toe adjustment to specification.
16. Torque the outer tie rod end jam nut to the proper torque.

Boot Seal

See Figure 101

1. Remove the steering gear.
2. Loosen the jam nut and remove the tie rod end and jam nut.
3. Remove the outer clamp from the rubber boot.
4. Remove the boot inner clamp.
5. On 2WD, mark the breather tube location on steering gear before removing the rubber boot.

To install:

6. Lubricate the boot outer groove (tie rod) with silicone type lubricant. Ensure that the boot is not twisted.
7. On 2WD, align the breather tube with the reference mark on the steering gear.
8. Position and align the new boot over the housing.
9. Install the inner clamp on the rubber boot.
10. Install the snorkel clamp on 2WD vehicles.
11. Install the outer clamp on the inner tie rod.
12. Install the jam nut and the tie rod end.
13. Install the steering gear.

Power Steering Gear

REMOVAL & INSTALLATION

Ram

See Figure 102

1. Place the wheels in the straight ahead position.
2. Disconnect the fluid hoses from the steering gear and cap the lines.
3. Remove the coupler pinch bolt and slide the shaft off the gear.
4. Matchmark the pitman shaft and arm and remove the arm.
5. Remove the steering gear fasteners and remove the gear.
6. Installation is the reverse of removal.

Note: The use of new mounting hardware is recommended. Alternately, use a medium strength thread-locking compound on the mounting hardware.

7. Be sure to tighten all hardware to the proper specification.
8. After adding fluid, carry out the power steering pump "Initial Operation" procedure.

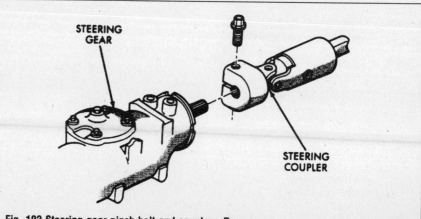

Fig. 102 Steering gear pinch bolt and coupler - Ram

SUSPENSION AND STEERING 8-25

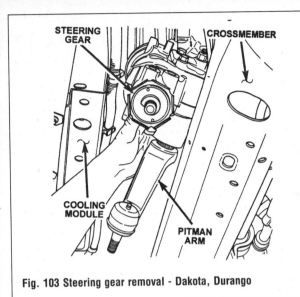

Fig. 103 Steering gear removal - Dakota, Durango

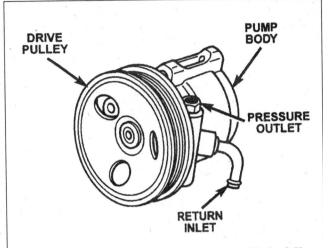

Fig. 104 TC series power steering pump as used with the 2.5L engine

Dakota, Durango

See Figure 103

This procedure applies to 1999 and earlier 4x4 models.

1. Place the wheels in the straight ahead position.
2. Disconnect the fluid hoses from the steering gear and cap the lines.
3. Remove the lower coupler pinch bolt and slide the shaft off the gear.
4. Raise and support the vehicle.
5. Matchmark the pitman shaft and arm and remove the arm.
6. Remove the splash shield under the cooling module.
7. Remove the steering gear fasteners and remove the gear by lowering it through the opening between the cooling module and the frame crossmember.
8. Installation is the reverse of removal.

Note: The use of new mounting hardware is recommended. Alternately, use a medium strength thread-locking compound on the mounting hardware.

9. Be sure to tighten all hardware to the proper specification.
10. After adding fluid, carry out the power steering pump "Initial Operation" procedure.

Power Steering Pump

REMOVAL & INSTALLATION

See Figures 104 and 105

No unusual procedures are required. Note any sensor wiring and vacuum lines at the pump. Bracket design varies depending on model and equipment fitted and may require removal of adjacent components for access.

1. Remove the pump drive belt.
2. Place a drip pan beneath the power steering pump and disconnect the hoses. Cap the hose ends.
3. Remove the power steering pressure switch connector and remove the pressure line from the bottom of the pump, if fitted. Remove the oil cooler mounting bolt from the pump bracket, if fitted.
4. Remove the pump and/or bracket mounting bolts.

To install:

5. Reverse the removal procedure. Observe torque specifications of all mounting hardware.
6. Refill the reservoir with power steering fluid and carry out the "Initial Operation" procedure.

INITIAL OPERATION

This procedure should be carried out whenever the power steering pump lines have been disconnected.

1. Turn the steering wheel all the way to the left.
2. Add fluid to bring the level up to the proper level and let the fluid settle for at least 2 minutes.
3. Raise the front wheels off the ground.
4. Slowly turn the steering wheel lock - to - lock 20 times with the engine OFF.

Note: On vehicles with long return lines or oil coolers, turn the wheel 40 times.

5. Start the engine. Check level and add fluid, if necessary.
6. Lower the front wheels and let the engine idle for two minutes.
7. Turn the steering wheel in both directions and verify power assist and quiet operation of the pump.
8. If the fluid is extremely foamy or milky looking, allow the vehicle to stand for a few minutes and repeat the procedure.

WARNING:

Do not run a vehicle with a foamy fluid for an extended period. This may cause pump damage.

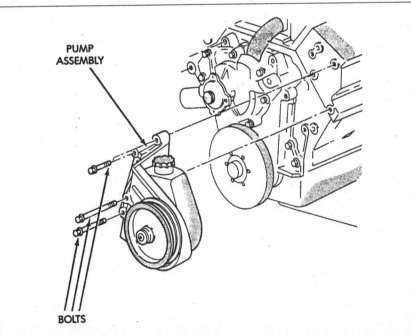

Fig. 105 Power steering pump assembly - typical

SUSPENSION AND STEERING

Torque Specifications (Ram)

Components	English	Metric
Lug Nuts		
BR1500 (5-stud wheel)	95 ft. lbs.	130 Nm
BR2500 (8-stud wheel)	135 ft. lbs.	180 Nm
BR3500 (8-stud dual wheel)	145 ft. lbs.	195 Nm
Front Suspension		
97 Ram, Link/Coil Front Suspension		
Lower suspension arm		
Axle nut	95 ft. lbs.	129 Nm
Frame nut	125 ft. lbs.	170 Nm
Upper suspension arm		
Axle nut	89 ft. lbs.	121 Nm
Frame nut	80 ft. lbs.	108 Nm
Stabilizer bar		
Clamp bolt	40 ft. lbs.	54 Nm
Link upper nut	27 ft. lbs.	37 Nm
Link lower nut	87 ft. lbs.	118 Nm
Track bar		
Ball stud nut	70 ft. lbs.	95 Nm
Axle bracket bolt	130 ft. lbs.	176 Nm
5-Stud hub bearing bolts	125 ft. lbs.	170 Nm
5-Stud hub axle nut	175 ft. lbs.	237 Nm
98, 99 Ram, Link/Coil Front Suspension		
Lower suspension arm		
Axle nut	95 ft. lbs.	129 Nm
Frame nut	130 ft. lbs.	176 Nm
Upper suspension arm		
Axle nut	89 ft. lbs.	121 Nm
Frame nut	85 ft. lbs.	115 Nm
Stabilizer bar		
Clamp bolt	40 ft. lbs.	54 Nm
Link upper nut	27 ft. lbs.	37 Nm
Link lower nut	50 ft. lbs.	68 Nm
Track bar		
Ball stud nut	70 ft. lbs.	95 Nm
Axle bracket bolt	130 ft. lbs.	176 Nm
Hub bearing bolts	125 ft. lbs.	170 Nm
Hub axle nut	175 ft. lbs.	237 Nm
00 Ram, Link/Coil Front Suspension		
Lower suspension arm	140 ft. lbs.	190 Nm
Upper suspension arm	120 ft. lbs.	163 Nm
Stabilizer bar		
Clamp bolt	40 ft. lbs.	54 Nm
Link upper nut	27 ft. lbs.	37 Nm
Link lower nut	35 ft. lbs.	47 Nm
Track bar		
Ball stud nut	70 ft. lbs.	95 Nm
Axle bracket bolt	130 ft. lbs.	176 Nm
Hub bearing bolts	122 ft. lbs.	166 Nm
Hub axle nut	180 ft. lbs.	245 Nm
Independent Front Suspension		
Frame nuts (97-99)	145 ft. lbs.	197 Nm
Frame nuts (00)	125 ft. lbs.	169 Nm
LD ball stud nut	95 ft. lbs.	129 Nm
HD ball stud nut	110 ft. lbs.	149 Nm
Upper suspension arm		
Pivot bar nuts (97-98)	150 ft. lbs.	203 Nm
Pivot bar nuts (99)	155 ft. lbs.	210 Nm

SUSPENSION AND STEERING 8-27

Pivot bar nuts (00)	125 ft. lbs.	169 Nm
Ball stud nut	60 ft. lbs.	81 Nm
Stabilizer bar		
Clamp bolt	40 ft. lbs.	54 Nm
Link nuts	27 ft. lbs.	37 Nm
Hub/bearing spindle nut (00)		
LD 1500 nut	185 ft. lbs.	251 Nm
HD 2500/3500 nut	280 ft. lbs.	380 Nm
Front Shock Absorber		
Independent Front Suspension		
Upper	40 ft. lbs	54 Nm
Lower	105 ft. lbs.	142 Nm
Coil/Link Front Suspension		
Upper	35 ft. lbs.	47 Nm
Lower	100 ft. lbs.	135 Nm
Upper Shock Bracket	55 ft. lbs.	75 Nm
Wheel Bearings (97-99 IFS)		
Step 1:	30-40 ft. lbs.	41-54 Nm
Step 2:	Back off 2 turns	
Step 3:	Finger-tighten	
Rear Suspension		
Leaf spring U-bolts & spring mounts	120 ft. lbs.	163 Nm
Stabilizer bar	40 ft. lbs.	54 Nm
Stabilizer bar links		
97 Ram		
Upper	70 ft. lbs.	95 Nm
Lower	50 ft. lbs.	68 Nm
98-00 Ram (upper and lower)	50 ft. lbs.	68 Nm
Shock Absorbers	100 ft. lbs.	136 Nm
Steering		
Steering wheel	45 ft. lbs.	61 Nm
Air bag module	80–100 inch lbs.	9–11 Nm
Pitman arm gear shaft	185 ft. lbs.	251 Nm
Steering Linkage, Link/Coil Suspension (97,98)		
Drag link ball stud	65 ft. lbs.	88 Nm
Tie rod end	65 ft. lbs.	88 Nm
Ball stud	65 ft. lbs.	88 Nm
Adjuster clamp	40 ft. lbs.	54 Nm
Tie rod ball stud	65 ft. lbs.	88 Nm
Steering damper		
Frame	65 ft. lbs.	88 Nm
Drag link	50 ft. lbs.	68 Nm
Steering Linkage, Link/Coil Suspension (99)		
Drag link		
Pitman arm	75 ft. lbs.	101 Nm
Tie rod	65 ft. lbs.	88 Nm
Adjuster clamp	45 ft. lbs.	61 Nm
Alignment bar	45 ft. lbs.	61 Nm
Tie rod end		
Knuckle (LD)	65 ft. lbs.	88 Nm
Knuckle (HD)	75 ft. lbs.	101 Nm
Adjuster clamp (LD)	45 ft. lbs.	61 Nm
Adjuster clamp (HD)	60 ft. lbs.	81 Nm
Steering damper		
Axle and tie rod	70 ft. lbs.	95 Nm
Bracket	40 ft. lbs.	55 Nm
Steering Linkage, Link/Coil Suspension (00)		
Drag link		
Pitman arm	80 ft. lbs.	108 Nm

SUSPENSION AND STEERING

Tie rod	65 ft. lbs.	88 Nm
Adjuster clamp	45 ft. lbs.	61 Nm
Tie rod end		
Knuckle	80 ft. lbs.	108 Nm
Adjuster clamp	45 ft. lbs.	61 Nm
Steering damper		
Axle	70 ft. lbs.	95 Nm
Tie rod	60 ft. lbs.	81 Nm
Steering Linkage, IFS Suspension (97, 98)		
Pitman arm		
Shaft nut	185 ft. lbs.	250 Nm
Ball stud nut	65 ft. lbs.	88 Nm
Idler Arm		
Ball stud nut	65 ft. lbs.	88 Nm
Mounting nuts (LD)	50 ft. lbs.	68 Nm
Mounting bolts (HD)	195 ft. lbs.	264 Nm
Tie rod		
Ball stud nut	65 ft. lbs.	88 Nm
Adjuster clamp	45 ft. lbs.	61 Nm
Steering damper	50 ft. lbs.	68 Nm
Steering Linkage, IFS Suspension (99)		
Pitman arm		
Gear nut	185 ft. lbs.	250 Nm
Center link nut	65 ft. lbs.	88 Nm
Idler Arm		
Center link nut	65 ft. lbs.	88 Nm
Mounting nuts (LD)	70 ft. lbs.	95 Nm
Mounting bolts (HD)	200 ft. lbs.	271 Nm
Tie rod		
Knuckle nut, center link nut	65 ft. lbs.	88 Nm
Adjuster clamp	45 ft. lbs.	61 Nm
Steering damper		
Frame nut	70 ft. Lbs.	95 Nm
Center link nut	50 ft. lbs.	68 Nm
Steering Linkage, IFS Suspension (00)		
Pitman arm		
Gear nut	185 ft. lbs.	250 Nm
Center link nut	85 ft. lbs.	115 Nm
Idler Arm		
Mounting bolts	200 ft. lbs.	271 Nm
Center link nut	65 ft. lbs.	88 Nm
Tie rod		
Knuckle nut	80 ft. lbs.	108 Nm
Center link nut	65 ft. lbs.	88 Nm
Adjuster clamp	45 ft. lbs.	61 Nm
Power Steering Gear		
Gear-to-frame bolts	100 ft. lbs.	136 Nm
Coupler bolts	36 ft. lbs.	49 Nm
Power Steering Pump (Gasoline Engines)		
Bracket-to-pump	35 ft. lbs.	47 Nm
Bracket-to-engine	30 ft. lbs.	41 Nm
Rear bracket-to-front bracket (8.0L)	18 ft. lbs.	24 Nm
Lines	25 ft. lbs.	35 Nm
Power Steering Pump (Diesel Engine)		
Pump assembly-to-engine	57 ft. lbs.	77 Nm
Pump-to-support bracket	18 ft. lbs.	24 Nm
Steering gear		
Mounting bolts	130 ft. lbs.	176 Nm
Coupler pinch bolt	36 ft. lbs.	49 Nm

SUSPENSION AND STEERING 8-29

Torque Specifications (Dakota, Durango)

Components	English	Metric
Lug Nuts	85–115 ft. lbs.	115–155 Nm
Front Suspension		
Dakota 2WD		
Lower suspension arm		
Front nut		
97	140 ft. lbs.	190 Nm
98, 99, 00	130 ft. lbs.	175 Nm
Rear nut	80 ft. lbs.	108 Nm
Ball joint nut	94 ft. lbs.	127 Nm
Upper suspension arm		
Pivot shaft nuts	130 ft. lbs.	167 Nm
Pivot shaft to frame nuts	155 ft. lbs.	210 Nm
Ball joint nut	60 ft. lbs.	81 Nm
Stabilizer bar		
Link upper nut	27 ft. lbs.	37 Nm
Link ball stud nut	35 ft. lbs.	47 Nm
Retainer bolts		
97	35 ft. lbs.	47 Nm
98, 99, 00	45 ft. lbs.	60 Nm
Hub/bearing spindle nut	185 ft. lbs.	251 Nm
Dakota (4WD), Durango		
Lower suspension arm		
Front bolt	80 ft. lbs.	108 Nm
Rear bolt	140 ft. lbs.	190 Nm
Ball joint nut	135 ft. lbs.	183 Nm
Upper suspension arm		
Pivot shaft nuts	95 ft. lbs.	129 Nm
Pivot shaft to frame bolts	165 ft. lbs.	224 Nm
Ball joint nut	60 ft. lbs.	81 Nm
Stabilizer bar		
Frame retainer bolt	80 ft. lbs.	108 Nm
Frame retainer nut	140 ft. lbs.	190 Nm
Control arm retainer bolts	25 ft. lbs.	34 Nm
Hub/bearing spindle nut	123 ft. lbs.	166 Nm
Front Shock Absorber		
Dakota, Durango 2WD	20 ft. lbs.	27 Nm
Dakota, Durango 4WD		
Upper	19 ft. lbs.	26 Nm
Lower	80 ft. lbs.	108 Nm
Wheel Bearings		
Step 1:	240-300 inch lbs.	27-34 Nm
Step 2:	Back off 1/4 turn (90 degrees)	
Step 3:	Finger-tighten	
Rear Suspension		
Leaf spring U-bolt nuts & spring mount bolts		
Dakota	120 ft. lbs.	163 Nm
Durango	90 ft. lbs.	122 Nm
Stabilizer bar	40 ft. lbs.	54 Nm
Stabilizer bar links		
Dakota	40 ft. lbs.	54 Nm
Durango		
Upper	70 ft. lbs.	95 Nm
Lower	50 ft. lbs.	68 Nm
Shock Absorbers	70 ft. lbs.	95 Nm
Steering		
Steering wheel	35 ft. lbs.	47 Nm
Air bag module	80–100 inch lbs.	9–11 Nm

SUSPENSION AND STEERING

Rack & Pinion Steering Gear		
Gear to frame bolts	190 ft. lbs.	258 Nm
Intermediate shaft bolt	36 ft. lbs.	49 Nm
Tie rod end		
Knuckle nut	80 ft. lbs.	108 Nm
Jam nut	55 ft. lbs.	75 Nm
PS lines	25 ft. lbs.	35 Nm
Steering Linkage		
Centerlink	65 ft. lbs.	88 Nm
Tie rod		
Knuckle nut	65 ft. lbs.	88 Nm
Jam nut	55 ft. lbs.	75 Nm
Inner tie rod	50 ft. lbs.	68 Nm
Idler arm frame nuts	110 ft. lbs.	136 Nm
Power Steering Pump		
Bracket bolts, 3.9L, 5.2L, 5.9L	30 ft. lbs.	41 Nm
Bracket bolts, 2.5L, 4.7L	21 ft. lbs.	28 Nm
Lines	25 ft. lbs.	35 Nm
Steering gear		
Mounting bolts	65 ft. lbs.	88 Nm
Coupler pinch bolt	36 ft. lbs.	49 Nm

BRAKE OPERATING SYSTEM
BASIC OPERATING PRINCIPLES 9-2
BLEEDING THE BRAKE SYSTEM 9-10
BRAKE HOSES AND LINES 9-10
BRAKE LIGHT SWITCH 9-6
COMBINATION VALVE 9-9
HYDRAULIC BRAKE BOOSTER 9-8
MASTER CYLINDER 9-6
VACUUM BRAKE BOOSTER 9-8

DISC BRAKES
BRAKE CALIPER 9-15
BRAKE DISC (ROTOR) 9-16
BRAKE PADS 9-12

DRUM BRAKES 9-17
BRAKE DRUMS 9-18
BRAKE SHOES 9-19
WHEEL CYLINDERS 9-21

FOUR WHEEL ANTILOCK BRAKE SYSTEM 9-29
BLEEDING THE ABS SYSTEM 9-32
DIAGNOSIS AND TESTING 9-30
EXCITER RING 9-32
SPEED SENSORS 9-30
TONE WHEEL 9-31

PARKING BRAKE
FRONT PARKING BRAKE CABLE 9-23
REAR PARKING BRAKE CABLES 9-24

REAR WHEEL ANTILOCK BRAKE SYSTEM
BLEEDING THE RWAL SYSTEM 9-29
DIAGNOSIS AND TESTING 9-28
EXCITER RING 9-29
FAULT CODES 9-27
GENERAL INFORMATION 9-27
RWAL VALVE 9-29
SPEED SENSOR 9-28

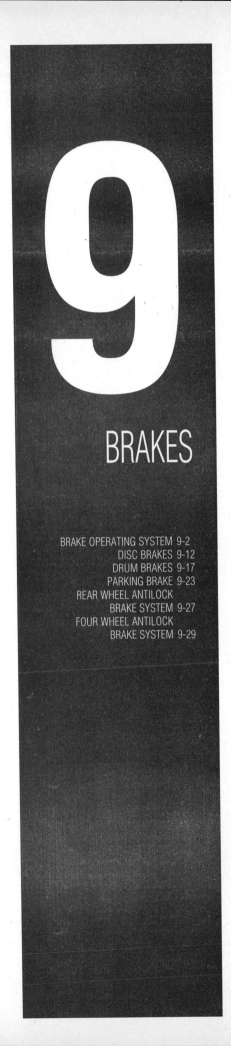

9
BRAKES

BRAKE OPERATING SYSTEM 9-2
DISC BRAKES 9-12
DRUM BRAKES 9-17
PARKING BRAKE 9-23
REAR WHEEL ANTILOCK BRAKE SYSTEM 9-27
FOUR WHEEL ANTILOCK BRAKE SYSTEM 9-29

9-2 BRAKES

BRAKE OPERATING SYSTEM

Basic Operating Principles

Hydraulic systems are used to actuate the brakes of all modern motor vehicles. The system transports the power required to force the frictional surfaces of the braking system together from the pedal to the individual brake units at each wheel. A hydraulic system is used for two reasons.

First, fluid under pressure can be carried to all parts of the vehicle by small pipes and flexible hoses without taking up a significant amount of room or posing routing problems.

Second, a great mechanical advantage can be given to the brake pedal end of the system, and the foot pressure required to actuate the brakes can be reduced by making the surface area of the master cylinder pistons smaller than that of any of the pistons in the wheel cylinders or calipers.

The master cylinder consists of a fluid reservoir along with a double cylinder and piston assembly. Double type master cylinders are designed to separate the front and rear braking systems hydraulically in case of a leak. The master cylinder converts mechanical motion from the pedal into hydraulic pressure within the lines. This pressure is translated back into mechanical motion at the wheels by either the wheel cylinder (drum brakes) or the caliper (disc brakes).

Steel lines carry the brake fluid to a point on the vehicle's frame near each of the vehicle's wheels. The fluid is then carried to the calipers and wheel cylinders by flexible tubes in order to allow for suspension and steering movements.

In drum brake systems, each wheel cylinder contains two pistons, one at either end, which push outward in opposite directions and force the brake shoe into contact with the drum.

In disc brake systems, the cylinders are part of the calipers. At least one cylinder in each caliper is used to force the brake pads against the disc.

All pistons employ some type of seal, usually made of rubber, to contain pressure within its cylinder. A rubber dust boot seals the outer end of the cylinder against dust and dirt. The boot fits around the outer end of the piston on disc brake calipers, and around the brake actuating rod on wheel cylinders.

The hydraulic system operates as follows: When at rest, the entire system, from the piston(s) in the master cylinder to those in the wheel cylinders or calipers, is full of brake fluid. Upon application of the brake pedal, fluid trapped in front of the master cylinder piston(s) is forced through the lines to the wheel cylinders. Here, it forces the pistons outward, in the case of drum brakes, and inward toward the disc, in the case of disc brakes.

Dual circuit master cylinders employ two pistons, located one behind the other, in the same cylinder. The primary piston is actuated directly by mechanical linkage from the brake pedal through the power booster. The secondary piston is actuated by fluid trapped between the two pistons. If a leak develops in front of the secondary piston, it moves forward until it bottoms against the front of the master cylinder, and the fluid trapped between the pistons will operate the rear brakes. If the rear brakes develop a leak, the primary piston will move forward until direct contact with the secondary piston takes place, and it will force the secondary piston to actuate the front brakes. In either case, the brake pedal moves farther when the brakes are applied, and less braking power is available.

All dual circuit systems use a switch to warn the driver when only half of the brake system is operational. This switch is usually located in a valve body which is mounted on the firewall or the frame below the master cylinder. A hydraulic piston receives pressure from both circuits, each circuit's pressure being applied to one end of the piston. When the pressures are in balance, the piston remains stationary. When one circuit has a leak, however, the greater pressure in that circuit during application of the brakes will push the piston to one side, closing the switch and activating the brake warning light.

In disc brake systems, this valve body also contains a metering valve and, in some cases, a proportioning valve. The metering valve keeps pressure from traveling to the disc brakes on the front wheels until the brake shoes on the rear wheels have contacted the drums, ensuring that the front brakes will never be used alone. The proportioning valve controls the pressure to the rear brakes to lessen the chance of rear wheel lock-up during very hard braking.

Warning lights may be tested by depressing the brake pedal and holding it while opening one of the wheel cylinder bleeder screws. If this does not cause the light to go on, substitute a new lamp, make continuity checks, and, finally, replace the switch as necessary.

The hydraulic system may be checked for leaks by applying pressure to the pedal gradually and steadily. If the pedal sinks very slowly to the floor, the system has a leak. This is not to be confused with a springy or spongy feel due to the compression of air within the lines. If the system leaks, there will be a gradual change in the position of the pedal with a constant pressure.

Check for leaks along all lines and at wheel cylinders. If no external leaks are apparent, the problem is inside the master cylinder.

DISC BRAKES

Instead of the traditional expanding brakes that press outward against a circular drum, disc brake systems utilize a disc (rotor) with brake pads positioned on either side of it. An easily-seen analogy is the hand brake arrangement on a bicycle. The pads squeeze the rim of the bike wheel, slowing its motion. Automobile disc brakes use the identical principle but apply the braking effort to a separate disc instead of the wheel.

The disc (rotor) is a casting, usually equipped with cooling fins between the two braking surfaces. This enables air to circulate between the braking surfaces making them less sensitive to heat buildup and more resistant to fade. Dirt and water do not drastically affect braking action since contaminants are thrown off by the centrifugal action of the rotor or scraped off the by the pads. Also, the equal clamping action of the two brake pads tends to ensure uniform, straight line stops. Disc brakes are inherently self-adjusting. There are three general types of disc brake:

- A fixed caliper.
- A floating caliper.
- A sliding caliper.

The fixed caliper design uses pistons mounted on either side of the rotor (in each side of the caliper). The caliper is mounted rigidly and does not move.

The sliding and floating designs are quite similar. In fact, these two types are often lumped together. In both designs, the pad on the inside of the rotor is moved into contact with the rotor by hydraulic force. The caliper, which is not held in a fixed position, moves slightly, bringing the outside pad into contact with the rotor.

DRUM BRAKES

Drum brakes employ two brake shoes mounted on a stationary backing plate. These shoes are positioned inside a circular drum which rotates with the wheel assembly. The shoes are held in place by springs. This allows them to slide toward the drums (when they are applied) while keeping the linings and drums in alignment. The shoes are actuated by a wheel cylinder which is mounted at the top of the backing plate. When the brakes are applied, hydraulic pressure forces the wheel cylinder's pistons (and in some designs, the actuating links) outward. Since the pistons or links bear directly against the top of the brake shoes, the tops of the shoes are then forced against the inner side of the drum. This action forces the bottoms of the two shoes to contact the brake drum by rotating the entire assembly slightly (known as servo action). When pressure within the wheel cylinder is relaxed, return springs pull the shoes back away from the drum.

Most modern drum brakes are designed to self-adjust during application when the vehicle is moving in reverse. This motion causes both shoes to rotate very slightly with the drum, rocking an adjusting lever, thereby causing rotation of the adjusting screw. Some drum brake systems are designed to self-adjust during application whenever the brakes are applied. This on-board adjustment system reduces the need for maintenance adjustments and keeps both the brake function and pedal feel satisfactory.

POWER BOOSTERS

Virtually all modern vehicles use a vacuum assisted power brake system to multiply the braking force and reduce pedal effort. Since vacuum is always available when the engine is operating, the system is simple and efficient. A vacuum diaphragm is located on the front of the master cylinder and assists the driver in applying the brakes, reducing both the effort and travel he must put into moving the brake pedal.

The vacuum diaphragm housing is normally connected to the intake manifold by a vacuum hose. A check valve is placed at the point where the hose enters the diaphragm housing, so that during periods of low manifold vacuum, brake assist will not be lost.

Depressing the brake pedal closes off the vacuum source and allows atmospheric pressure to enter on one side of the diaphragm. This causes the master cylinder pistons to move and apply the brakes. When the brake pedal is released, vacuum is applied to both sides of the diaphragm and springs return the diaphragm and master cylinder pistons to the released position.

If the vacuum supply fails, the brake pedal rod will contact the end of the master cylinder actuator rod and the system will apply the brakes without any power assistance. The driver will notice that much higher pedal effort is needed to stop the car and that the pedal feels harder than usual.

BRAKES 9-3

Vacuum Leak Test

1. Operate the engine at idle without touching the brake pedal for at least one minute.
2. Turn off the engine and wait one minute.
3. Test for the presence of assist vacuum by depressing the brake pedal and releasing it several times. If vacuum is present in the system, light application will produce less and less pedal travel. If there is no vacuum, air is leaking into the system.

System Operation Test

1. With the engine OFF, pump the brake pedal until the supply vacuum is entirely gone.
2. Put light, steady pressure on the brake pedal.
3. Start the engine and let it idle. If the system is operating correctly, the brake pedal should fall toward the floor if the constant pressure is maintained.

Power brake systems may be tested for hydraulic leaks just as ordinary systems are tested.

WARNING:

Clean, high quality brake fluid is essential to the safe and proper operation of the brake system. You should always buy the highest quality brake fluid that is available. If the brake fluid becomes contaminated, drain and flush the system, then refill the master cylinder with new fluid. Never reuse any brake fluid. Any brake fluid that is removed from the system should be discarded.

Troubleshooting the Brake System

Problem	Cause	Solution
Low brake pedal (excessive pedal travel required for braking action.)	Excessive clearance between rear linings and drums caused by inoperative automatic adjusters	Make 10 to 15 alternate forward and reverse brake stops to adjust brakes. If brake pedal does not come up, repair or replace adjuster parts as necessary.
	Worn rear brakelining	Inspect and replace lining if worn beyond minimum thickness specification
	Bent, distorted brakeshoes, front or rear	Replace brakeshoes in axle sets
	Air in hydraulic system	Remove air from system. Refer to Brake Bleeding.
Low brake pedal (pedal may go to floor with steady pressure applied.)	Fluid leak in hydraulic system	Fill master cylinder to fill line; have helper apply brakes and check calipers, wheel cylinders, differential valve tubes, hoses and fittings for leaks. Repair or replace as necessary.
	Air in hydraulic system	Remove air from system. Refer to Brake Bleeding.
	Incorrect or non-recommended brake fluid (fluid evaporates at below normal temp).	Flush hydraulic system with clean brake fluid. Refill with correct-type fluid.
	Master cylinder piston seals worn, or master cylinder bore is scored, worn or corroded	Repair or replace master cylinder
Low brake pedal (pedal goes to floor on first application—o.k. on subsequent applications.)	Disc brake pads sticking on abutment surfaces of anchor plate. Caused by a build-up of dirt, rust, or corrosion on abutment surfaces	Clean abutment surfaces
Fading brake pedal (pedal height decreases with steady pressure applied.)	Fluid leak in hydraulic system	Fill master cylinder reservoirs to fill mark, have helper apply brakes, check calipers, wheel cylinders, differential valve, tubes, hoses, and fittings for fluid leaks. Repair or replace parts as necessary.
	Master cylinder piston seals worn, or master cylinder bore is scored, worn or corroded	Repair or replace master cylinder
Decreasing brake pedal travel (pedal travel required for braking action decreases and may be accompanied by a hard pedal.)	Caliper or wheel cylinder pistons sticking or seized	Repair or replace the calipers, or wheel cylinders
	Master cylinder compensator ports blocked (preventing fluid return to reservoirs) or pistons sticking or seized in master cylinder bore	Repair or replace the master cylinder
	Power brake unit binding internally	Test unit according to the following procedure: (a) Shift transmission into neutral and start engine (b) Increase engine speed to 1500 rpm, close throttle and fully depress brake pedal (c) Slow release brake pedal and stop engine (d) Have helper remove vacuum check valve and hose from power unit. Observe for backward movement of brake pedal. (e) If the pedal moves backward, the power unit has an internal bind—replace power unit

9-4 BRAKES

Troubleshooting the Brake System (cont.)

Problem	Cause	Solution
Spongy brake pedal (pedal has abnormally soft, springy, spongy feel when depressed.)	• Air in hydraulic system • Brakeshoes bent or distorted • Brakelining not yet seated with drums and rotors • Rear drum brakes not properly adjusted	• Remove air from system. Refer to Brake Bleeding. • Replace brakeshoes • Burnish brakes • Adjust brakes
Hard brake pedal (excessive pedal pressure required to stop vehicle. May be accompanied by brake fade.)	• Loose or leaking power brake unit vacuum hose • Incorrect or poor quality brakelining • Bent, broken, distorted brakeshoes • Calipers binding or dragging on mounting pins. Rear brakeshoes dragging on support plate.	• Tighten connections or replace leaking hose • Replace with lining in axle sets • Replace brakeshoes • Replace mounting pins and bushings. Clean rust or burrs from rear brake support plate ledges and lubricate ledges with molydisulfide grease. NOTE: If ledges are deeply grooved or scored, do not attempt to sand or grind them smooth—replace support plate.
	• Caliper, wheel cylinder, or master cylinder pistons sticking or seized • Power brake unit vacuum check valve malfunction	• Repair or replace parts as necessary • Test valve according to the following procedure: (a) Start engine, increase engine speed to 1500 rpm, close throttle and immediately stop engine (b) Wait at least 90 seconds then depress brake pedal (c) If brakes are not vacuum assisted for 2 or more applications, check valve is faulty
	• Power brake unit has internal bind	• Test unit according to the following procedure: (a) With engine stopped, apply brakes several times to exhaust all vacuum in system (b) Shift transmission into neutral, depress brake pedal and start engine (c) If pedal height decreases with foot pressure and less pressure is required to hold pedal in applied position, power unit vacuum system is operating normally. Test power unit. If power unit exhibits a bind condition, replace the power unit.
	• Master cylinder compensator ports (at bottom of reservoirs) blocked by dirt, scale, rust, or have small burrs (blocked ports prevent fluid return to reservoirs). • Brake hoses, tubes, fittings clogged or restricted • Brake fluid contaminated with improper fluids (motor oil, transmission fluid, causing rubber components to swell and stick in bores • Low engine vacuum	• Repair or replace master cylinder CAUTION: Do not attempt to clean blocked ports with wire, pencils, or similar implements. Use compressed air only. • Use compressed air to check or unclog parts. Replace any damaged parts. • Replace all rubber components, combination valve and hoses. Flush entire brake system with DOT 3 brake fluid or equivalent. • Adjust or repair engine

Troubleshooting the Brake System (cont.)

Problem	Cause	Solution
Grabbing brakes (severe reaction to brake pedal pressure.)	• Brakelining(s) contaminated by grease or brake fluid	• Determine and correct cause of contamination and replace brakeshoes in axle sets
	• Parking brake cables incorrectly adjusted or seized	• Adjust cables. Replace seized cables.
	• Incorrect brakelining or lining loose on brakeshoes	• Replace brakeshoes in axle sets
	• Caliper anchor plate bolts loose	• Tighten bolts
	• Rear brakeshoes binding on support plate ledges	• Clean and lubricate ledges. Replace support plate(s) if ledges are deeply grooved. Do not attempt to smooth ledges by grinding.
	• Incorrect or missing power brake reaction disc	• Install correct disc
	• Rear brake support plates loose	• Tighten mounting bolts
Dragging brakes (slow or incomplete release of brakes)	• Brake pedal binding at pivot	• Loosen and lubricate
	• Power brake unit has internal bind	• Inspect for internal bind. Replace unit if internal bind exists.
	• Parking brake cables incorrrectly adjusted or seized	• Adjust cables. Replace seized cables.
	• Rear brakeshoe return springs weak or broken	• Replace return springs. Replace brakeshoe if necessary in axle sets.
	• Automatic adjusters malfunctioning	• Repair or replace adjuster parts as required
	• Caliper, wheel cylinder or master cylinder pistons sticking or seized	• Repair or replace parts as necessary
	• Master cylinder compensating ports blocked (fluid does not return to reservoirs).	• Use compressed air to clear ports. Do not use wire, pencils, or similar objects to open blocked ports.
Vehicle moves to one side when brakes are applied	• Incorrect front tire pressure	• Inflate to recommended cold (reduced load) inflation pressure
	• Worn or damaged wheel bearings	• Replace worn or damaged bearings
	• Brakelining on one side contaminated	• Determine and correct cause of contamination and replace brakelining in axle sets
	• Brakeshoes on one side bent, distorted, or lining loose on shoe	• Replace brakeshoes in axle sets
	• Support plate bent or loose on one side	• Tighten or replace support plate
	• Brakelining not yet seated with drums or rotors	• Burnish brakelining
	• Caliper anchor plate loose on one side	• Tighten anchor plate bolts
	• Caliper piston sticking or seized	• Repair or replace caliper
	• Brakelinings water soaked	• Drive vehicle with brakes lightly applied to dry linings
	• Loose suspension component attaching or mounting bolts	• Tighten suspension bolts. Replace worn suspension components.
	• Brake combination valve failure	• Replace combination valve
Chatter or shudder when brakes are applied (pedal pulsation and roughness may also occur.)	• Brakeshoes distorted, bent, contaminated, or worn	• Replace brakeshoes in axle sets
	• Caliper anchor plate or support plate loose	• Tighten mounting bolts
	• Excessive thickness variation of rotor(s)	• Refinish or replace rotors in axle sets

9-6 BRAKES

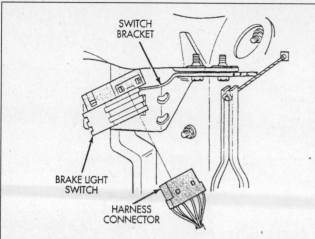

Fig. 1 Disconnecting the harness connector from the brake light switch - typical

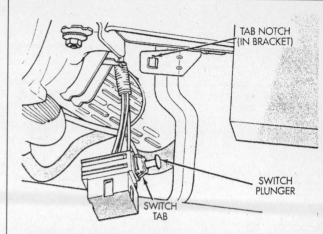

Fig. 2 Align switch tab with notch in bracket, then turn switch 30-degrees to lock - adjustable type

Brake Light Switch

REMOVAL & INSTALLATION

1997 - 99 Ram, 1997 - 98 Dakota, Durango

See Figures 1 and 2

The brake light switch is located on a bracket beneath the dashboard.
1. Remove the knee bolster, if necessary for access.
2. Disconnect the wiring harness at the switch.
3. Press and hold the brake pedal in the applied position.
4. Rotate the switch CCW about 30-degrees to align the lock tabs with the bracket notch and pull the switch out of the bracket.

To install:

5. Pull the switch plunger all the way out to the fully extended position.
6. Push the plunger switch inward 4 detent positions (or clicks). Plunger will extend about half an inch out of the housing.
7. Connect the wiring.
8. Install the switch, first aligning the lock tabs with the bracket notch, then turning it 30-degrees CW after installation to lock it in place.
9. Release the brake pedal.
10. Lightly pull the pedal fully rearward. Do not use excessive force. Plunger adjustment will be made automatically.

All 2000 Models, 1999 Dakota, Durango

See Figure 3

The brake light switch is located on a bracket beneath the dashboard. The switch used on these vehicles is adjusted once on installation. Routine adjustment is neither necessary nor possible.
1. Remove the knee bolster, if necessary for access.
2. Disconnect the wiring harness at the switch.
3. Press and hold the brake pedal in the applied position.
4. Rotate the switch CCW about 30-degrees to align the lock tabs with the bracket notch and pull the switch out of the bracket.

To install a new switch:

5. Connect the wiring to the new switch.
6. Press the brake pedal and hold it down.
7. Install the switch, first aligning the lock tabs with the bracket notch, then turning it 30-degrees CW after installation to lock it in place.
8. Release the brake pedal.
9. Move the release lever on the switch towards the brake pedal to engage the switch plunger. The switch is now adjusted and cannot be adjusted again.

ADJUSTMENT

1997 - 99 Ram, 1997 - 98 Dakota, Durango

1. Remove the knee bolster, if necessary for access.
2. Push and hold the brake pedal down.
3. Pull the switch plunger all the way out to the fully extended position.
4. Push the plunger switch inward 4 detent positions (or clicks). Plunger will extend about half an inch out of the housing.
5. Release the brake pedal. Lightly pull the pedal fully rearward. Do not use excessive force. Plunger adjustment will be made automatically.

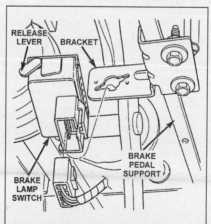

Fig. 3 After the switch is installed, move the release lever towards the brake pedal to engage the switch plunger

All 2000 Models, 1999 Dakota, Durango

The unit is adjusted on installation. Routine adjustment is neither necessary nor possible.

Master Cylinder

See Figure 4

The master cylinder is a two-piece component with an aluminum cylinder body and a nylon reservoir. The reservoir can be removed and replaced, if need be. The cylinder, however, is not repairable and must be replaced as an assembly if it fails.

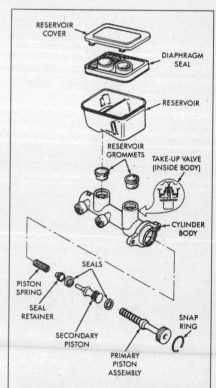

Fig. 4 Exploded view of typical master cylinder. Piston assembly can be removed after prying out the snap ring, but the unit is not rebuildable

BRAKES

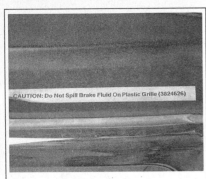

Fig. 5 You have been warned

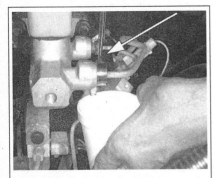

Fig. 6 Place a container beneath the fitting and use a flare nut wrench (arrow) to loosen it

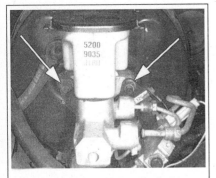

Fig. 7 After disconnecting the brake lines, remove the nuts (arrows) to remove the master cylinder

CAUTION:

Bleeding of four-wheel antilock brake systems require the use a dedicated scan tool. Lines should not be disconnected unless one is available.

REMOVAL & INSTALLATION

See Figures 5, 6, 7 and 8

WARNING:

Exercise care when removing and installing the master cylinder, combination valve and antilock valve connecting lines. The threads in the cylinder and valve fluid ports can be damaged easily. Use a flare nut wrench to loosen and tighten fittings and start all brake system fittings by hand to avoid crossthreading.

Note: Read the "Brake Hoses And Lines" section for useful hints and tips on brake line work.

1. Place a container beneath the master cylinder brake line fittings to catch the fluid which will drip out when the fittings are removed.

Note: Protect painted surfaces from contact with brake fluid which will remove many types of paint, and damage certain types of plastics, in short order.

2. Using a flare nut wrench, unscrew the fittings on the master cylinder.
3. Remove the nuts securing the master cylinder to the power brake booster. On some models, the combination valve bracket may have to be removed as well.
4. Remove the master cylinder.
5. Drain off the brake fluid.

To install:

6. If a new master cylinder is being installed, carry out the "Bench Bleeding" procedure first.
7. Install the master cylinder on the booster studs.
8. Fit the combination valve bracket on the booster stud, if so equipped.
9. Tighten the mounting nuts to 17 ft. lbs. (23 Nm).
10. Start the brake line fittings by hand. Move the lines as necessary to make sure the fittings engage the threads correctly. If resistance is felt, back them off and try again. Do not put a wrench on the fittings until you are sure they are correctly threaded.
11. Tighten the line fittings to 14 ft. lbs. (19 Nm). Do not overtighten. Do not attempt to stop leaks by overtightening the fittings. This is unlikely to help.
12. Add new fluid and bleed the brake system.

BENCH BLEEDING

See Figure 9

WARNING:

All new master cylinders should be bench bled prior to installation. Bleeding a new master cylinder on the vehicle is not a good idea. With air trapped inside, the master cylinder piston may bottom in the bore and possibly cause internal damage.

1. Secure the master cylinder in a bench vise using soft jaws.
2. Remove the master cylinder reservoir cap.
3. Manufacture or purchase bleeding tubes and install them on the master cylinder as illustrated.
4. Fill the master cylinder reservoir with clean, fresh brake fluid until the level is within 0.25 in. of the reservoir top.

Note: Ensure the bleeding tubes are below the level of the brake fluid, otherwise air may get into the system making your bleeding efforts ineffective.

5. Use a blunt tipped rod (a long socket extension works well) to slowly depress the master cylinder piston. Make sure the piston travels its full stroke.
6. As the piston is depressed, bubbles will come out of the bleeding tubes. Continue depressing and releasing the piston slowly until all bubbles cease.
7. Refill the master cylinder with fluid.
8. Remove the bleeding tubes.
9. Install the master cylinder reservoir cap.
10. Install the master cylinder on the vehicle and bleed the system.

Fig. 8 Removing the master cylinder

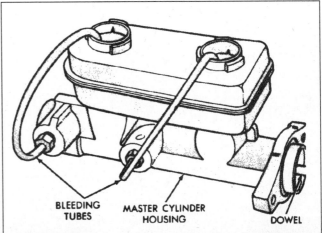

Fig. 9 Bench bleeding the master cylinder using bleeding tubes

9-8 BRAKES

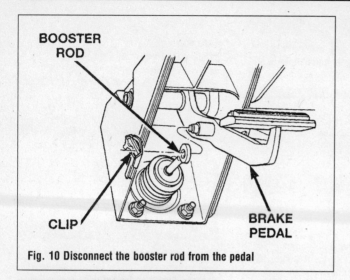

Fig. 10 Disconnect the booster rod from the pedal

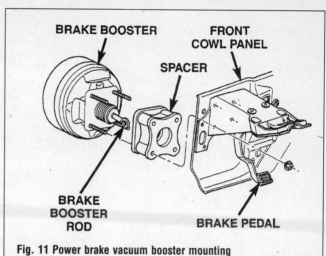

Fig. 11 Power brake vacuum booster mounting

Vacuum Brake Booster

Vacuum brake boosters are used on all models except those with diesel engines.

REMOVAL & INSTALLATION

See Figures 10 and 11

1. Remove the master cylinder.
2. Disconnect the vacuum lines at the booster.
3. Remove the steering column knee bolster for access to the booster mounting nuts, if necessary.
4. Remove the clip securing the booster push rod to the brake pedal.
5. Remove the nuts from the booster mounting studs.
6. Remove the booster, spacer and gaskets from the front cowl panel.

To install:

Note: If the booster is being replaced, ensure that the replacement unit is identical to the original. Refer to identification code letters stamped on the forward face.

7. Reverse the removal procedure.
8. Tighten the mounting nuts to 21 ft. lbs. (28 Nm).

9. The remainder of the procedure is the reverse of removal.

TESTING

See Figures 12 and 13

Vacuum boosters can be checked with a vacuum gauge.

1. Connect the vacuum gauge to the booster check valve with a short length of hose and a T-fitting (see illustration).
2. Start the engine and allow it to idle for one minute.
3. Check for vacuum at the gauge. If there is no vacuum or the reading is very low, the problem is either the booster or the check valve.
4. Assuming vacuum reading is normal, clamp the hose shut between the vacuum source and the check valve.
5. Stop the engine and observe the gauge. If the vacuum drops more than 1 in. Hg within 15 seconds, the booster or check valve is at fault.
6. To test the check valve, proceed as follows:
   a. Disconnect the vacuum hose from the check valve.
   b. Remove the check valve and valve seal from the booster.
   c. With a hand-operated vacuum pump, apply 15 - 20 in. Hg at the large end of the check valve.
   d. Vacuum should hold steady. If the gauge indicates vacuum loss, the check valve is faulty and should be replaced.

Hydraulic Brake Booster

Hydraulic brake boosters are used on trucks with diesel engines.

REMOVAL & INSTALLATION

See Figures 14 and 15

CAUTION:

The accumulator contains high pressure gas. Do not carry the booster by the accumulator or drop the unit on the accumulator.

1. With the engine OFF, depress the brake pedal several times to discharge the accumulator.
2. Remove the master cylinder.
3. Remove the return hose and the two pressure lines from the booster.

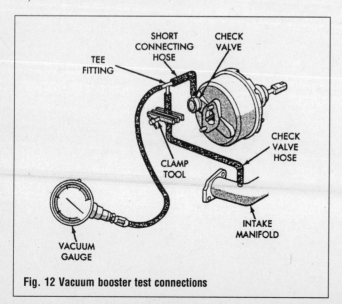

Fig. 12 Vacuum booster test connections

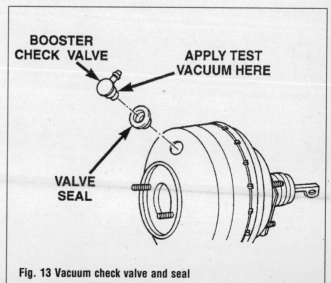

Fig. 13 Vacuum check valve and seal

BRAKES 9-9

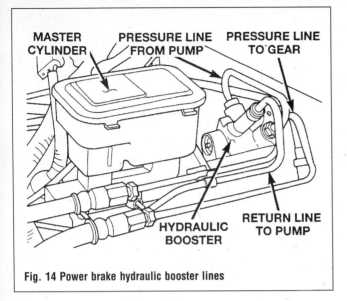

Fig. 14 Power brake hydraulic booster lines

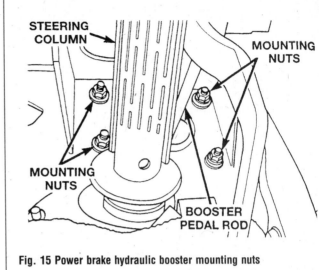

Fig. 15 Power brake hydraulic booster mounting nuts

4. Remove the steering column knee bolster for access to the booster mounting nuts, if necessary.
5. Remove the clip securing the booster push rod to the brake pedal and disconnect the push rod from the pedal.
6. Remove the nuts from the booster mounting studs.
7. Remove the booster.

To install:

Note: If the booster is being replaced, ensure that the replacement unit is identical to the original. Refer to identification code letters stamped on the forward face.

8. Reverse the removal procedure.
9. Tighten the mounting nuts to 21 ft. lbs. (28 Nm).
10. Inspect the O-rings on the pressure line fittings to ensure they are in good condition. Replace them if questionable.
11. Tighten the pressure lines to 21 ft. lbs. (28 Nm).
12. The remainder of the procedure is the reverse of removal.
13. Bleed the booster after installation is completed:
  a. Fill the power steering pump reservoir.
  b. Disconnect the fuel shutdown relay and crank the engine for several seconds.
  c. Check the fluid level and add if necessary.
  d. Reconnect the fuel shutdown relay and start the engine.
  e. Turn the steering wheel slowly from lock to lock.
  f. Stop the engine and discharge the accumulator by depressing the brake pedal five times.
  g. Start the engine and turn the steering wheel slowly from lock-to-lock twice.
  h. Shut off the engine and make a final check of fluid level.

Combination Valve

The combination valve contains a pressure differential valve and switch and a proportioning valve. The pressure differential switch is connected to the brake warning lamp and monitors fluid pressure in the separate front/rear hydraulic circuits. The proportioning valve is used to balance front-rear brake action at high decelerations when a percentage of the rear weight is transferred to the front wheels. The valve allows normal fluid pressure during moderate braking.

REMOVAL & INSTALLATION

See Figure 16

CAUTION:

Bleeding of four-wheel antilock brake systems require the use a dedicated scan tool. Lines should not be disconnected unless one is available.

Note: Read the "Brake Hoses And Lines" section for useful hints and tips on brake line work.

The unit is not serviceable. If it fails, it must be replaced.
1. Disconnect the wiring at the valve.
2. Place a container beneath the valve before loosening the fittings.
3. Disconnect the brake lines using a flare nut wrench on the fittings.
4. Remove the valve mounting hardware and take off the valve.

To install:

5. Reverse the removal procedure.
6. Start the brake line fittings by hand. Move the lines as necessary to make sure the fittings engage the threads correctly. If resistance is felt, back them off and try again. Do not put a wrench on the fittings until you are sure they are correctly threaded.
7. Tighten the line fittings to 14 ft. lbs. (19 Nm). Do not overtighten. Do not attempt to stop leaks by overtightening the fittings. This is unlikely to help.
8. Add new fluid and bleed the brake system.

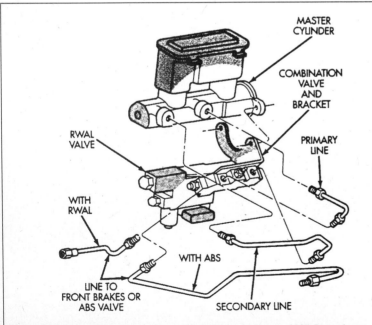

Fig. 16 Typical combination valve and related brake lines - 1997 Ram shown

9-10 BRAKES

Brake Hoses and Lines

Metal lines and rubber brake hoses should be checked frequently for leaks and external damage. Metal lines are particularly prone to crushing and kinking under the vehicle. Any such deformation can restrict the proper flow of fluid and therefore impair braking at the wheels. Rubber hoses should be checked for cracking or scraping; such damage can create a weak spot in the hose and it could fail under pressure.

Any time the lines are removed or disconnected, extreme cleanliness must be observed. Clean all joints and connections before disassembly (use a stiff bristle brush and clean brake fluid); be sure to plug the lines and ports as soon as they are opened. New lines and hoses should be flushed clean with brake fluid before installation to remove any contamination.

CAUTION:

Bleeding of four-wheel antilock brake systems requires the use of a dedicated scan tool. Lines should not be disconnected unless one is available.

Note: When adding brake fluid to the system, never use fluid from an unsealed container. Brake fluid absorbs moisture when exposed to the atmosphere.

REMOVAL & INSTALLATION

See Figures 17, 18, 19, 20 and 21

1. Disconnect the negative battery cable(s).
2. Raise and safely support the vehicle on jackstands.
3. Remove any wheel and tire assemblies necessary for access to the particular line you are removing.
4. Thoroughly clean the surrounding area at the joints to be disconnected.
5. Place a suitable catch pan under the joint to be disconnected.
6. Using two wrenches (one to hold the joint and one to turn the fitting), disconnect the hose or line to be replaced.
7. Disconnect the other end of the line or hose, moving the drain pan if necessary. Always use a back-up wrench to avoid damaging the fitting.
8. Disconnect any retaining clips or brackets holding the line and remove the line from the vehicle.

Note: If the brake system is to remain open for more time than it takes to swap lines, tape or plug each remaining clip and port to keep contaminants out and fluid in.

To install:

9. Install the new line or hose, starting with the end farthest from the master cylinder. Connect the other end, then confirm that both fittings are correctly threaded and turn smoothly using finger pressure. Make sure the new line will not rub against any other part. Brake lines must be at least 1/2 in. (13mm) from the steering column and other moving parts. Any protective shielding or insulators must be reinstalled in the original location.

Fig. 17 Always use a flare nut ("line") wrench for brake work

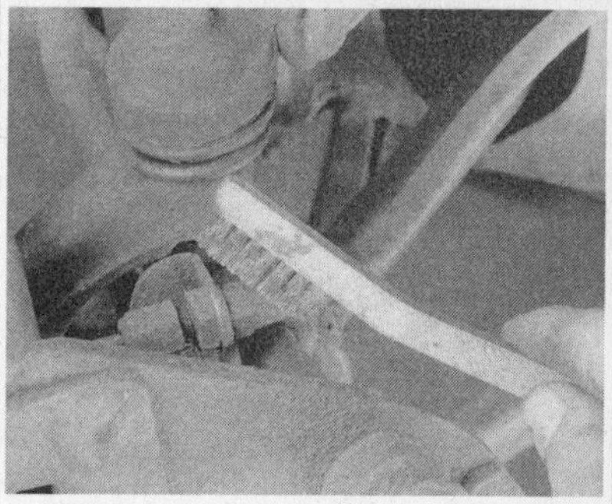

WARNING:

Make sure the hose is NOT kinked or touching any part of the frame or suspension after installation. These conditions may cause the hose to fail prematurely.

10. Using two wrenches as before, tighten each fitting.
11. Install any retaining clips or brackets on the lines.
12. If removed, install the wheel and tire assemblies, then carefully lower the vehicle to the ground.
13. Refill the brake master cylinder reservoir with clean, fresh brake fluid, meeting DOT 3 specifications. Properly bleed the brake system.
14. Connect the negative battery cable.

Bleeding The Brake System

CAUTION:

Bleeding of four-wheel antilock brake systems requires the use of a dedicated scan tool. Lines should not be disconnected unless one is available. The recommended procedure is to carry out a standard bleeding procedure, purge the HCU with the scan tool, then carry out another standard bleeding procedure.

Note: Read the "Brake Hoses And Lines" section for useful hints and tips on brake line work.

Fig. 20 Any gaskets/sealing washers should be replaced with new ones during installation

Fig. 18 Use a brush to clean the fittings of any debris

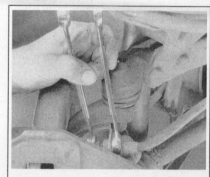
Fig. 19 Use two wrenches to loosen the fitting. If available, use flare nut type wrenches

Note: Add only fresh, clean brake fluid from a sealed container when bleeding the brakes.

Note: If pressure bleeding equipment is used, the front brake metering valve will have to be held open to bleed the front brakes. The valve stem is located in the forward end or top of the combination valve. The stem must either be pressed inward or held outward slightly. Follow equipment manufacturer's instructions carefully when using pressure equipment. Do not exceed the maker's pressure recommendations. Generally, a tank pressure of 15 - 20 psi is sufficient. Do not pressure bleed without the proper master cylinder adapter.

Fig. 21 Tape or plug the line to prevent contamination

BRAKES 9-11

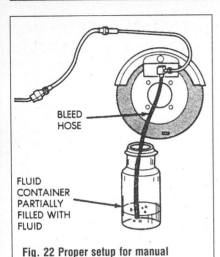

Fig. 22 Proper setup for manual bleeding procedure

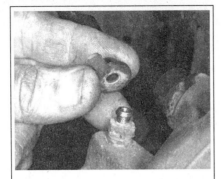

Fig. 23 Bleed screw caps are a must to keep the bleed screw passages clear

When any part of the hydraulic system has been disconnected for repair or replacement, air may get into the lines and cause spongy pedal action (because air can be compressed and brake fluid cannot). To correct this condition, it is necessary to bleed the hydraulic system so to be sure all air is purged.

Bleeding must start where the lines were disconnected. If lines were disconnected at the master cylinder, for example, bleeding must be done at that point before proceeding downstream.

When bleeding the brake system, bleed one brake cylinder at a time, beginning at the cylinder with the longest hydraulic line (farthest from the master cylinder) first. Failure to do so may result in more air being drawn into the lines.

If the existing system fluid seems dirty or if the vehicle has covered considerable mileage, it is recommended that the system be completely purged and refilled with fresh, clean fluid. The best way to start is to siphon the old fluid out of the master cylinder reservoir and fill it completely with fresh fluid.

Brake fluid tends to darken over time. This does not necessarily indicate contamination. Examine fluid closely for foreign matter.

The primary and secondary hydraulic brake systems are separate and are bled independently. During the bleeding operation, do not allow the reservoir to run dry. Keep the master cylinder reservoir filled with brake fluid. Never use brake fluid that has been drained from the hydraulic system, no matter how clean it seems.

1. Clean all dirt from around the master cylinder fill cap, remove the cap and fill the master cylinder with brake fluid until the level is within 1/4 in. (6mm) of the top edge of the reservoir.
2. Clean the bleeder screws at all 4 wheels. The bleeder screws are located on the back of the brake backing plate (drum brakes) and on the top of the brake calipers (disc brakes).
3. Bleeder screws should be protected with rubber caps. If they are missing, the orifice may easily become clogged with road dirt. If the screw refuses to bleed when loosened, remove it and blow clear. Aftermarket caps are readily available.

Manual Bleeding

See Figures 22, 23 and 24

Manual bleeding requires two people and a degree of patience and cooperation.

1. Follow the preparatory steps, above.
2. Attach a length of rubber hose over the bleeder screw and place the other end of the hose in a glass jar, submerged in brake fluid.
3. Have your assistant press down on the brake pedal, then open the bleeder screw 1/2 – 3/4 turn.
4. The brake pedal will go to the floor.
5. Close the bleeder screw - preferably before the pedal reaches the floor. Tell your assistant to allow the brake pedal to return slowly.
6. Repeat these steps to purge all air from the system.
7. When bubbles cease to appear at the end of the bleeder hose, close the bleeder screw and remove the hose. Check that the pedal is firm or at least more firm than it was when you started. If not, continue the procedure.
8. Check the master cylinder fluid level and add fluid accordingly. Do this after bleeding each wheel.
9. Repeat the bleeding operation at the remaining three wheels, ending with the one closest to the master cylinder.
10. Fill the master cylinder reservoir to the proper level.

Note: If there is excessive air in the system, it is possible that the stroke of the brake pedal will be insufficient to purge the lines. In this case a pressure bleeder or vacuum bleeder is the easiest solution.

Vacuum Bleeding

See Figures 23, 24 and 25

Vacuum bleeding can be carried out by one person. Since a good vacuum bleeder will normally move more fluid than a brake pedal stroke, this procedure is preferred. These tools are inexpensive and readily available at auto parts outlets.

1. Follow the preparatory steps, above.
2. Attach the vacuum bleeder according to the manufacturer's recommendations.
3. Pump up the unit until maximum vacuum is reached. Loosen the bleeder screw slightly until bubbles and fluid issue forth. Close the screw before the vacuum is equalized.
4. Repeat the procedure until fluid without bubbles issues from the bleeder screw.
5. Keep a close check on master cylinder fluid level during this procedure as vacuum bleeders move considerable amounts of fluid.

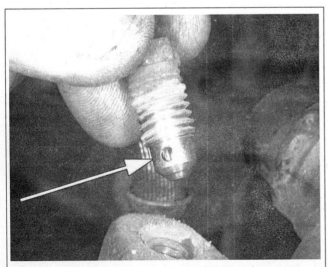

Fig. 24 Lack of cap may cause bleeder screw passages to become clogged

Fig. 25 There are tools, such as this Mighty-Vac, available to assist in vacuum bleeding of the brake system

9-12 BRAKES

DISC BRAKES

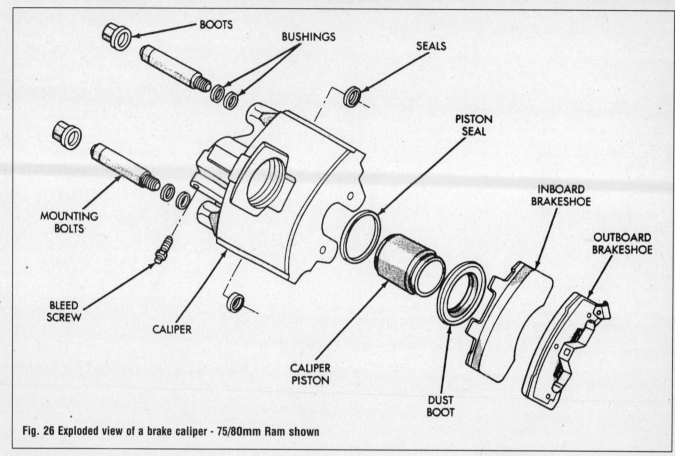

Fig. 26 Exploded view of a brake caliper - 75/80mm Ram shown

See Figure 26

CAUTION:

Brake pads may contain asbestos, which has been determined to be a cancer causing agent. Never clean the brake surfaces with compressed air! Avoid inhaling any dust from any brake surface! When cleaning brake surfaces, use a commercially available brake cleaning fluid.

Brake Pads

REMOVAL & INSTALLATION

Dakota, Durango

See Figures 27, 28, 29, 30, 31 and 32

1. Clean the area around the master cylinder reservoir and filler cap(s).
2. Remove the filler cap(s) and siphon off about 25 percent of the fluid. This is to prevent overflow as the piston is pushed into the caliper. Fitting new pads will cause a rise in the fluid level.
3. Raise and support the vehicle.
4. Remove the wheel.
5. Pry the piston side pad away from the rotor with a tool which won't scratch the rotor surface.
6. On 2000 models, remove the caliper spring by prying up and pulling it out of the caliper holes.
7. Remove the caliper slide pins (mounting bolts).
8. Remove the caliper and brake pads from the rotor.
9. Remove the outboard brake pad: pry one end of the pad retainer spring away from the caliper. Then tilt the pad upward and rotate it out of the caliper.
10. Remove the inboard pad by tilting the pad outward until the retainer spring is clear of the caliper piston.
11. Support the caliper with a wire or twine: do not allow it to hang by the hose.

Note: Do not apply the brake pedal with the caliper off the rotor. Use a soft wedge (such as a piece of wood) to hold the pads apart if the caliper is to be off the rotor for any length of time.

Fig. 27 Removing the caliper spring

To install:

12. Clean the slide surfaces of adapter ledges with a wire brush. Then lubricate the surfaces with a thin coat of high temperature grease.
13. Install new slide pin bushings if required.
14. Install the inboard pad. Be sure the retainer spring is firmly seated in the caliper piston.
15. Insert the outboard pad in the caliper.
16. Degrease the rotor with a safe solvent.
17. Push the piston in far enough to allow the pads to clear the rotor. This can be done by hand or with a special tool: several types are available.
18. Install the caliper on the rotor.
19. Install and tighten the caliper slide pins to 22 ft. lbs. (30 Nm).

Note: Start the slide pins by hand before tightening. Do not crossthread the pins.

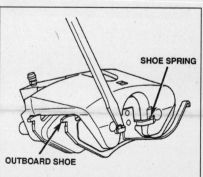

Fig. 28 Removing the outboard brake pad

BRAKES 9-13

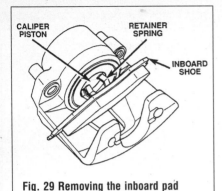

Fig. 29 Removing the inboard pad

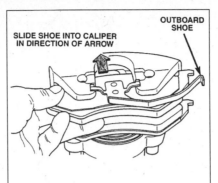

Fig. 30 Inserting the outboard pad into the caliper

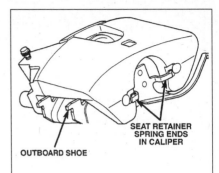

Fig. 31 Be sure the outboard pad retainer spring is seated in the caliper

Fig. 32 A suitable tool, like the one here from Lisle®, can be used to compress the piston into the caliper

20. 2000 models: install the caliper spring into one caliper hole and under the adapter. Pull down on the opposite end of the spring and hold the end under the adapter. With a screwdriver, pry up on the spring to seat it into the other caliper hole.
21. Install the wheel.
22. Apply the brake several times to seat the pads.
23. Check the brake fluid level at the master cylinder reservoir and add if necessary.

Ram

See Figures 33, 34, 35, 36, 37, 38, 39, 40, 41, 42, 43, 44, 45, 46 and 47

Single-piston calipers of various sizes were used from 1997-99. For the 2000 model year a twin-piston caliper was used.

1. Clean the area around the master cylinder reservoir and filler cap(s).

Fig. 34 Removing the brake caliper bolts - Ram shown

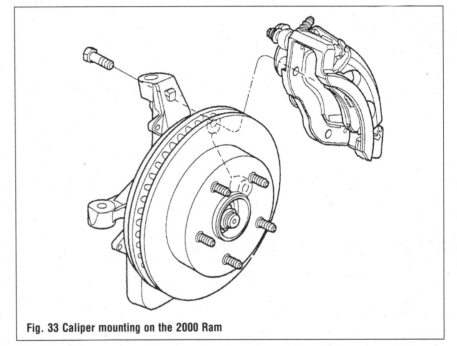

Fig. 33 Caliper mounting on the 2000 Ram

2. Remove the filler cap(s) and siphon off about 25 percent of the fluid. This is to prevent overflow as the piston is pushed into the caliper. Fitting new pads will cause a rise in the fluid level.
3. Raise and support the vehicle.
4. Remove the wheel.
5. Pry the piston side pad away from the rotor with a tool which won't scratch the rotor surface.
6. Remove the caliper mounting bolts.
7. Remove the caliper and brake pads from the

Fig. 35 If necessary, tap a stuck caliper off the rotor with a plastic mallet

rotor. Use a plastic mallet to tap it free, if necessary.
8. Pry out the outboard brake pad.
9. Remove the inboard shoe by tilting the pad outward until the retainer spring is clear of the caliper piston.
10. Remove the anti-rattle springs, if fitted (2000 models).

Note: The anti-rattle springs are not interchangeable.

11. Support the caliper with a wire or twine: do not allow it to hang by the hose.

Note: Do not apply the brake pedal with the caliper off the rotor. Use a soft wedge (such as a piece of wood) to hold the pads apart if the caliper is to be off the rotor for any length of time.

To install:

12. Clean the caliper and steering knuckle slide surfaces with a wire brush. Then lubricate the surfaces with a thin coat of high temperature grease.
13. On 1997 - 99 Rams, lubricate the caliper mounting bolts, collars, bushings and bores with silicone grease.
14. Lubricate and install the anti-rattle springs (2000 models).
15. Install the inboard shoe. Be sure the retainer

9-14 BRAKES

Fig. 36 Support the caliper with wire or twine - do not let it hang by the brake hose

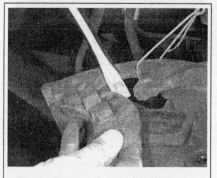

Fig. 37 Prying out the outboard pad

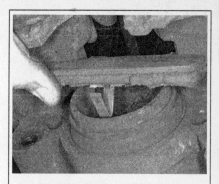

Fig. 38 Removing the inboard pad

spring is firmly seated in the caliper piston.
16. Insert the outboard shoe in the caliper.
17. Degrease the rotor with a safe solvent.
18. Push the piston(s) in far enough to allow the pads to clear the rotor. This can be done by hand or with a special tool: several types are available.
19. Install the caliper and rotor assembly. Be sure the caliper is seated flush on the mounting arm surfaces.
20. Start the mounting bolts by hand to prevent crossthreading. Tighten the bolts to 38 ft. lbs. (51 Nm) on 1997 - 99 models, tighten to 24 ft. lbs. (33 Nm) on 2000 models.

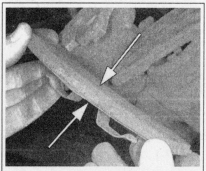

Fig. 39 Measure the thickness of each pad and compare with the specification given

Fig. 40 Using a special tool to press the piston into the caliper

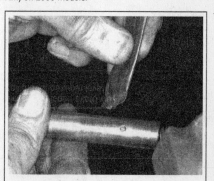

Fig. 41 On 1997 - 99 Rams, lubricate the caliper bushings and bolts with silicone grease

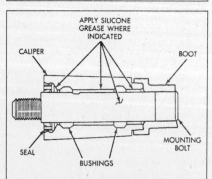

Fig. 42 Mounting bolt lubrication areas on 75mm caliper: 1997 - 99 Ram

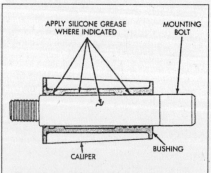

Fig. 43 Mounting bolt lubrication areas on 80/86mm caliper: 1997 - 99 Ram

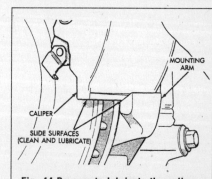

Fig. 44 Be sure to lubricate the caliper and steering knuckle slide surfaces

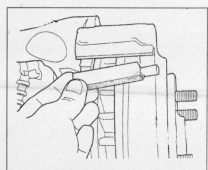

Fig. 45 Installing the top anti-rattle spring - 2000 Ram

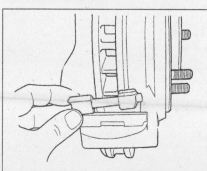

Fig. 46 Installing the bottom anti-rattle spring - 2000 Ram

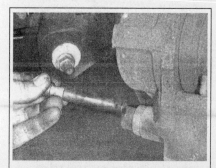

Fig. 47 Start the bolts by hand to avoid crossthreading

BRAKES 9-15

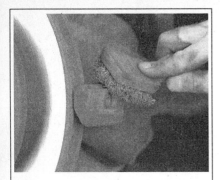

Fig. 48 Clean the caliper contact slides with a wire brush

Fig. 49 Lubricate the caliper slides with high-temperature grease after cleaning

Fig. 50 Anytime the brake line is disconnected, new washers must be used to ensure a proper seal

21. Install the wheel.
22. Apply the brake several times to seat the pads.
23. Check the brake fluid level at the master cylinder reservoir and add if necessary.

INSPECTION

1. Pads must be replaced as a set on both front wheels at the same time.
2. Measure the thickness of both pads and compare to the minimum acceptable limit given in the specifications chart. Replace both pads on both wheels if any one pad is below the limit. State safety inspection limits may differ from these limits and would naturally take precedence if greater (thicker) than the factory values.
3. Replace the pads if they are scored, contaminated with brake fluid or otherwise damaged. Check rotor as well. The perfect brake job would involve new pads and a resurfaced rotor to ensure maximum braking effectiveness and maximum component service life.

Brake Caliper

REMOVAL & INSTALLATION

See Figures 27, 33, 34, 35, 36, 40, 44, 48, 49 and 50

1. Raise and support the vehicle.
2. Remove the wheel.
3. Disconnect the brake line from the caliper if the unit is to be removed from the vehicle. Plug the end of the brake line. Be sure to use new washers on installation. Disconnecting the line will require bleeding of the system.

CAUTION:

Bleeding of four-wheel antilock brake systems requires the use of a dedicated scan tool. Lines should not be disconnected unless one is available.

4. Pry the piston side pad away from the rotor with a tool which won't scratch the rotor surface.
5. Remove the caliper bolts or pins.
6. Remove the caliper from the rotor.

Note: Do not apply the brake pedal with the caliper off the rotor. Use a soft wedge (such as a piece of wood) to hold the pads apart if the caliper is to be off the rotor for any length of time.

To install:

7. Clean the caliper slide surfaces with a wire brush where applicable. Then lubricate the surfaces with a thin coat of high temperature grease.
8. Push the piston(s) in far enough to allow the pads to clear the rotor. This can be done by hand or with a special tool; several types are available.
9. Install the caliper on the rotor.
10. Dakota, Durango: install and tighten the caliper slide pins to 22 ft. lbs. (30 Nm).
11. Ram: tighten the caliper bolts to 38 ft. lbs. (51 Nm) on 1997 - 99 models; tighten to 24 ft. lbs. (33 Nm) on 2000 models.

Note: Start the bolts or pins by hand before tightening to avoid crossthreading.

12. Connect the brake line using new washers.
13. Bleed the caliper.

CAUTION:

Bleeding of four-wheel antilock brake systems requires the use of a dedicated scan tool.

14. Install the wheel.
15. Apply the brake several times to seat the pads before operating the vehicle.
16. Check the brake fluid level at the master cylinder reservoir and add if necessary.

OVERHAUL

See Figures 51, 52, 53, 54, 55, 56 and 57

Note: Read the "Brake Hoses And Lines" section for useful hints and tips on brake line work.

Note: Late model Rams are equipped with dual piston calipers. The procedure to overhaul the caliper is essentially the same with the exception of multiple pistons, O-rings and dust boots.

1. Remove the caliper from the vehicle and place on a clean workbench.

CAUTION:

NEVER place your fingers in front of the pistons in an attempt to catch or protect the pistons when applying compressed air. This will result in personal injury!

2. Stuff a shop towel or a block of wood into the caliper to catch the piston.

Fig. 51 Use compressed air to drive the piston out of the caliper, but be sure to keep your fingers clear

3. Remove the caliper piston using compressed air applied into the caliper inlet hole. Keep pressure low! It doesn't take more than a few psi in a short burst to move the piston! On dual-piston calipers, cover one of the pistons with a piece of wood and use a C-clamp to hold it in place while the other piston is removed. Then place the piece of wood over the empty bore, secure it in place with the C-clamp and remove the other piston.
4. Inspect the piston for scoring, nicks, corrosion and/or worn or damaged chrome plating. The piston must be replaced if any of these conditions are found.
5. If equipped, remove the anti-rattle clip.
6. Use a prytool to remove the caliper boot, being careful not to scratch the housing bore.
7. Remove the piston seals from the groove in the caliper bore.

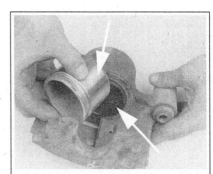

Fig. 52 Withdraw the piston from the caliper bore, then inspect the piston and bore

9-16 BRAKES

Fig. 53 On some vehicles, you must remove the anti-rattle clip

Fig. 54 Use a prytool to carefully pry around the edge of the boot . . .

Fig. 55 . . . then remove the boot from the caliper housing, taking care not to score or damage the bore

8. Carefully loosen the brake bleeder valve cap and valve from the caliper housing.
9. Inspect the caliper bores, pistons and mounting threads for scoring or excessive wear.
10. Use crocus cloth to polish out light corrosion from the piston and bore.
11. Clean all parts with denatured alcohol and dry with compressed air.

To assemble:

12. Lubricate and install the bleeder valve and cap.
13. Install the new seals into the caliper bore grooves, making sure they are not twisted.
14. Lubricate the piston bore.
15. Install the pistons and boots into the bores of the calipers and push to the bottom of the bores.
16. Use a suitable driving tool to seat the boots in the housing.
17. Install the caliper in the vehicle.
18. Properly bleed the brake system.

CAUTION:

Bleeding of four-wheel antilock brake systems requires the use of a dedicated scan tool. Lines should not be disconnected unless one is available.

19. Install the wheel, then carefully lower the vehicle.

Brake Disc (Rotor)

REMOVAL & INSTALLATION

Dakota, Durango, Ram (5-Stud Rotor)

See Figures 58, 59, 60 and 61

1. Raise and support the vehicle.
2. Remove the wheel.

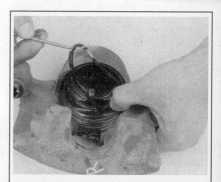

Fig. 56 Use extreme caution when removing the piston seal; DO NOT scratch the caliper bore

Fig. 57 Use the proper size driving tool and a mallet to properly seal the boots in the caliper housing

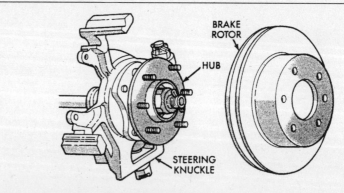

Fig. 58 Brake rotor and hub assembly - Dakota shown

Fig. 59 Removing a rotor retainer

Fig. 60 Balky rotors can be freed with a light-duty puller...

Fig. 61 ...or a plastic mallet wielded with care

BRAKES 9-17

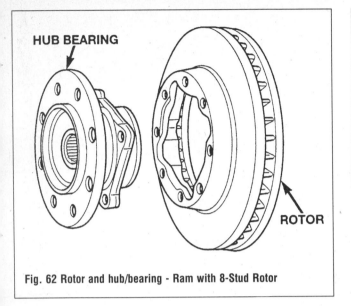

Fig. 62 Rotor and hub/bearing - Ram with 8-Stud Rotor

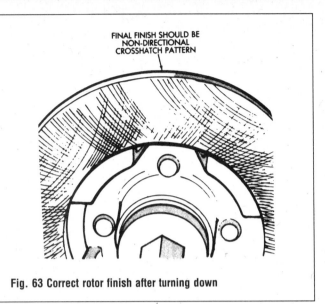

Fig. 63 Correct rotor finish after turning down

3. Remove the brake caliper.
4. Remove the retainers on the wheel studs and remove the rotor. If stuck, the rotor can be freed with a light duty puller or a plastic mallet. Do not use excessive force.
5. Installation is the reverse of removal.

Ram (8-Stud Rotor)

See Figure 62

1. Raise and support the vehicle.
2. Remove the wheel.
3. Remove the hub extension.
4. Remove the brake caliper assembly.
5. Remove the axle nut cotter pin. Remove the axle nut.
6. Disconnect the ABS wheel speed sensor wire from under the hood. Remove the sensor wire from the frame and steering knuckle, if equipped.
7. Back off the hub/bearing mounting bolts 1/4 inch each. Then tap the bolts with a hammer to loosen the hub/bearing mounting bolts.
8. Remove the bolts and remove the hub/bearing assembly.
9. Installation is the reverse of removal. Observe all tightening torques as found in the specifications charts. Use a new cotter pin and apply anti-seize compound to the axle splines.

INSPECTION

See Figure 63

1. Rotor surface should be smooth and featureless. Scoring, ripples, scratches, etc., should be removed.
The final finish should be a non-directional crosshatch pattern.
2. Minimum acceptable thickness is stamped on the rotor hub. Measurement should be taken after refinishing.
3. Check run-out and compare with the specification given.

DRUM BRAKES

See Figures 64 and 65

CAUTION:

Brake shoes may contain asbestos, which has been determined to be a cancer causing agent. Never clean the brake surfaces with compressed air! Avoid inhaling any dust from any brake surface! When cleaning brake surfaces, use a commercially available brake cleaning fluid.

Fig. 64 Typical drum brake assembly - 11 in. version shown

1. Wheel cylinder
2. Brake shoe
3. Hold-down spring and retainer
4. Return spring
5. Adjuster lever spring
6. Adjuster
7. Shoe spring
8. Actuator lever spring
9. Shoe guide plate
10. Parking brake strut and spring
11. Adjuster lever
12. Actuator lever

9-18 BRAKES

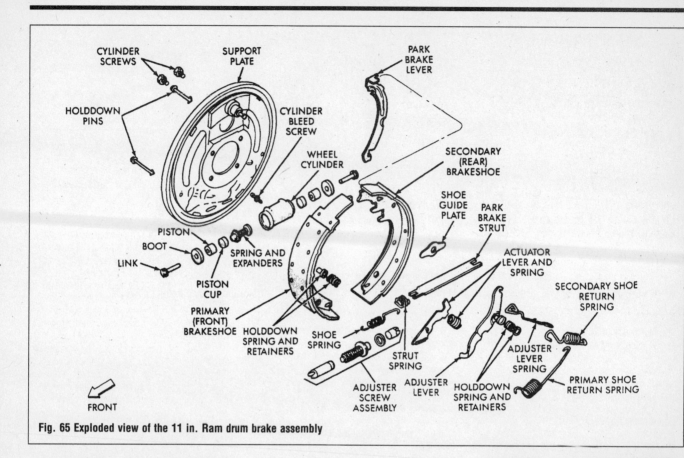

Fig. 65 Exploded view of the 11 in. Ram drum brake assembly

Brake Drums

REMOVAL & INSTALLATION

See Figures 66, 67, 68 and 69

1. Park the vehicle on a level surface. Place the transmission in PARK or in gear. Block the wheels.
2. Release the parking brake fully.
3. Raise the back of the truck and support it securely on jackstands. Remove the wheel.
4. Remove any clip nuts securing the drum to the studs.
5. Remove the brake drum. If the drum is stuck, a light-duty puller can be used, but first check that the parking brake has been fully released.
6. Difficulty may also be caused by brake shoes too tightly adjusted. Some models have a slot in the back of the support plate (protected by a rubber plug) which allows access to the brake adjuster. A thin screwdriver can be used to push the adjuster lever off the star wheel which can then be turned with a brake adjusting tool to release the brake shoes. Do not use excessive force on the drums. If removal is difficult, investigate possible causes before proceeding.
7. Installation is the reverse of removal. Be sure to readjust the brakes if the adjuster was moved, or the shoes were removed or replaced, or if the drum was turned down. Tighten lug nuts to the proper specification.

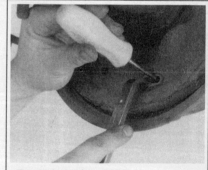

Fig. 66 If there is an adjustment slot in the back of the support plate, brake shoes can be backed off with a small slot head screwdriver and brake adjusting tool

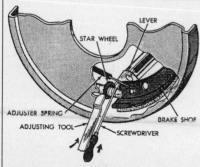

Fig. 67 Push the lever off the star wheel with the screwdriver and rotate the wheel with the adjusting tool

Fig. 68 A light duty puller is often helpful if the brake drum is stuck, but if removal is a problem investigate more closely before using excessive force

Fig. 69 Removing the brake drum - handle with care as you can damage the casting

BRAKES 9-19

Fig. 70 The maximum allowable brake drum diameter is stamped on the casting. This is a Ram 11 inch brake drum

Fig. 71 Clean brake assemblies with brake system cleaner. Do not attempt to blow them clean!

INSPECTION

See Figure 70

1. Remove loose rust and other foreign matter from the outside surfaces of the drum with a wire brush.
2. Check the casting itself for cracks or other obvious damage. Bluing indicates overheating. The cause should be determined.
3. Check the lug stud holes for ovality.
4. Clean the inside of the drum. Check for scoring or grooves. Any imperfections must be removed.
5. Check the inside diameter of the drum. Maximum allowable inside diameter is stamped on the drum. If machining is needed to clean up the contact surface, the measurement must be taken afterwards. Replace the drum if it exceeds the specification.
6. Clean up the brake lining contact area with medium-grit sandpaper. Be sure that you don't leave a directional pattern on the contact surface.
7. Perform a final cleaning of the contact area with a purpose-made commercial brake cleaning solvent to remove fingerprints and oily residue.

Brake Shoes

INSPECTION

1. Brake shoes must be replaced as a set and preferably on both wheels at the same time.
2. Measure the thickness of both linings and compare to the minimum acceptable limit given in the specifications chart. Make your measurement at the thinnest point, if one is noted. Replace both shoes on both wheels if any one lining is below the limit. State safety inspection limits may differ from these limits and would naturally take precedence if greater (thicker) than the factory values.
3. Replace the linings if they are scored, contaminated with brake fluid or oil or otherwise damaged. Check drum as well. The perfect brake job would involve new linings and a resurfaced drum to ensure maximum braking effectiveness and maximum component service life.
4. Signs of fluid leakage must be investigated. Check the wheel bearing seals if oil is noted. Check the wheel cylinder if brake fluid is found on the brake plate or linings.
5. Examine the lining contact pattern to determine if the shoes are bent or the drum tapered. The lining should exhibit contact across its entire width. Shoes exhibiting contact only on one side should be replaced and the drum checked for runout or taper.

REMOVAL & INSTALLATION

See Figures 71, 72, 73, 74, 75, 76, 77, 78, 79 and 80

Due to the obvious safety concerns, brake work must not be undertaken casually.

For the newcomer, the safest approach to disassembling drum brakes is to work on one side at a time. Drum brakes have many small, indistinguishable parts and even assuming you get them back on the right side and the right way around there is the added complication of which part goes on first, second and next. Note that the many springs have unequal arms and must be fitted the right way. Steel stampings (plates, arms, etc.) are often fitted one on top of the other. Proper reassembly is extremely critical. Removing wheels and drums on both sides of the axle and using one side as a "go-by" for reassembly is a safe bet.

Specialized tools are available for drum brake jobs without which things can be very difficult. These include spring removal and installation tools, a brake gauge and adjusting tool.

1. Block the front wheels. Raise the rear wheels and support the vehicle safely.
2. Be sure that the parking brake is fully released. Back off the parking brake adjuster, if necessary.
3. Remove the wheels.
4. Remove the brake drums.
5. Clean the brake assembly using brake system cleaner.

CAUTION:

Do not blow dust from the assembly with compressed air. Air-borne particulates may be a health hazard.

6. With a spring tool, disconnect the return springs and adjuster lever spring.
7. With a suitable tool, compress the hold-down springs and turn the retainers 90-degrees. Release the spring and remove retainers and springs from the pins.
8. Remove the brake shoe assembly from the support plate.

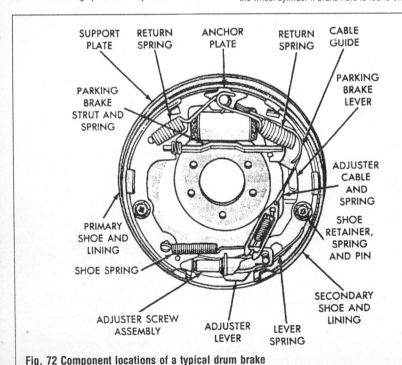

Fig. 72 Component locations of a typical drum brake

Fig. 73 Disconnecting a return spring with a special tool

9-20 BRAKES

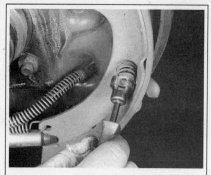

Fig. 74 Depress the hold-down spring and turn it 90-degrees to release. This special tool makes it easier

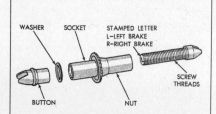

Fig. 77 Adjuster screw assembly: they are not interchangeable. Adjusters are marked "L" and "R" for the wheel they go with

Fig. 78 The adjuster lever spring is installed first since one of the return springs is anchored on it

Fig. 79 Return springs must be hooked into the holes in the shoe (arrow). The spring long arm is attached to the pivot pin or the adjuster lever spring hook

Fig. 75 Removing the brake shoe assembly from the support plate

9. Do not mix up parts from the two wheels. The brake adjusters are not interchangeable. The adjuster screw is stamped "L" or "R" for the wheel in which it is installed.
10. Clean the support plate. Wire brush the lining contact areas on the plate and apply a thin coat of grease to prevent binding.
11. Check that the adjusters turn freely. Brush off corrosion and lubricate the mechanisms so that the threaded rod turns freely and without binding.
12. Installation is the reverse of removal.
13. When installing the brakes, ensure that all components are properly seated. Do not force anything into place. Be sure the wheel cylinder links are seated on the brake shoes.

ADJUSTMENTS

Brakes are fitted with automatic adjusters. If they are functioning properly, brake adjustment should only be required if components are removed or replaced. Check that the adjuster star wheel moves freely, that the adjuster threads are clean and free of rust. A bit of high-temperature grease on the threads will keep adjusters in shape over the long haul.

Adjustment With Brake Gauge

See Figures 81 and 82

1. Ensure that the parking brake is fully released. There should be a bit of slack in the cable. Loosen the adjuster if necessary.
2. Raise the rear of the vehicle and remove the wheels and brake drums.
3. Check that both automatic adjuster cables and levers are properly connected.
4. Measure the drum diameter with the gauge and lock it into position.

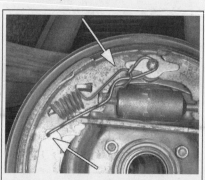

Fig. 80 Installation of the adjuster lever spring and return spring

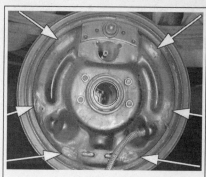

Fig. 76 Wire brush the lining contact areas (arrows) and lubricate them with grease

5. Turn the gauge around and check the shoe diameter at the center of the shoes. The gauge should be a light drag fit over the shoes.
6. If the gauge is not a light drag fit over the shoes, lift the adjuster lever off the star wheel. Turn the star wheel by hand to move the shoes in or out so that the correct clearance can be achieved.
7. Install brake drums and wheels. Lower the vehicle.
8. Check brake operation before driving.
9. In a safe area, drive the vehicle forward a few yards and make one full stop. Then reverse and make one full stop. The vehicle must come to a complete stop each time. Rolling stops will not activate the adjusters.
10. Repeat this procedure 8 - 10 times to activate the automatic adjusters and equalize the brakes.
11. Adjust the parking brake.
12. Make a final check of brake operation before driving in traffic.

Adjustment With Adjusting Tool

See Figure 83

If the brake plate is equipped with adjuster slots, and if a brake adjusting tool is available, shoes can be adjusted without removing the drum.

1. Ensure that the parking brake is fully released. There should be a bit of slack in the cable. Loosen the adjuster if necessary.
2. Raise the rear of the vehicle.
3. Remove the rubber plugs from the adjuster slots.
4. Check that the wheels rotate freely. If not, determine the cause. If the shoes are too tightly adjusted, skip the next step.

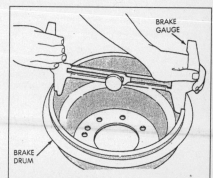

Fig. 81 Setting the brake gauge at the drum

BRAKES 9-21

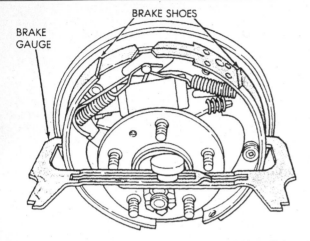

Fig. 82 Checking shoe OD with the gauge: it should be a light drag after the gauge is set at the drum

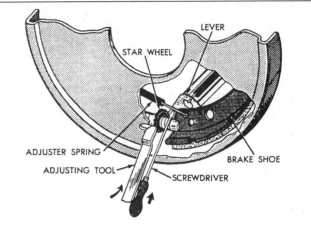

Fig. 83 The adjusting tool will expand the shoes, but to contract them, the lever must be pushed off the star wheel with a screwdriver

5. Insert the adjusting tool into the slot and engage the teeth of the adjusting star wheel. Move the tool handle up which will expand the brake shoes. Rotate the wheel while doing this until a slight drag can be felt.
6. Insert a thin screwdriver into the adjuster slot and push the adjuster lever away from the star wheel. Back off the adjuster star wheel until brake drag is eliminated.
7. Install the adjuster slot plug.
8. Repeat the procedure at other wheel. Be sure adjustment is equal at both wheels.
9. Lower the vehicle.
10. Check brake operation before driving.
11. In a safe area, drive the vehicle forward a few yards and make one full stop. Then reverse and make one full stop. The vehicle must come to a complete stop each time. Rolling stops will not activate the adjusters.
12. Repeat this procedure 8 - 10 times to activate the automatic adjusters and equalize the brakes.
13. Adjust the parking brake.
14. Make a final check of brake operation before driving in traffic.

Wheel Cylinders

REMOVAL & INSTALLATION

See Figures 84 and 85

Note: Read the "Brake Hoses And Lines" section for useful hints and tips on brake line work.

1. Raise and safely support the vehicle securely on jackstands.
2. Remove the wheels.
3. Remove the brake drums and shoes.
4. Loosen the brake fitting with a flare nut wrench. Disconnect and cap the brake hose.
5. Loosen the two wheel cylinder attaching bolts and remove the wheel cylinder from the support plate.

To install:

6. Install the wheel cylinder and tighten the attaching bolts securely.
7. Uncap and connect and the brake hose. Tighten fitting securely with a flare nut wrench.

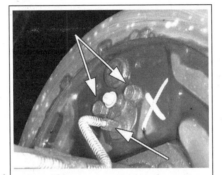

Fig. 84 Disconnect the line and remove the two bolts (arrows) to remove the wheel cylinder

8. Install the brake drums and shoes.
9. Refill and bleed the brake system.
10. Install the wheels.
11. Lower the vehicle.

OVERHAUL

See Figures 86, 87, 88, 89, 90, 91, 92, 93, 94, 95 and 96

Wheel cylinder overhaul kits may be available, but often at little or no savings over a reconditioned wheel cylinder. It often makes sense with these components to substitute a new or reconditioned part instead of attempting an overhaul.

If no replacement is available, or you would prefer to overhaul your wheel cylinders, the following pro-

Fig. 85 Removing the wheel cylinder from the brake plate

cedure may be used. When rebuilding and installing wheel cylinders, avoid getting any contaminants into the system. Always use clean, new, high quality brake fluid. If dirty or improper fluid has been used, it will be necessary to drain the entire system, flush the system with proper brake fluid, replace all rubber components, then refill and bleed the system.

1. Remove the wheel cylinder from the vehicle and place on a clean workbench.
2. Before disassembly, note that most brake cylinder components have a "side" which must be installed facing the right direction.
3. First remove and discard the old rubber boots, then withdraw the pistons. Seals ("cups") and a spring assembly are located behind the pistons in the cylinder bore.
4. Remove the remaining inner components: seals and spring assembly. Compressed air may be

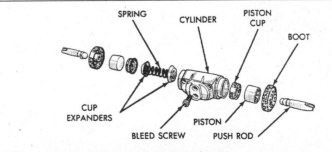

Fig. 86 Exploded view of a typical wheel cylinder

9-22 BRAKES

Fig. 87 Remove the outer boots from the wheel cylinder

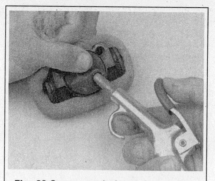

Fig. 88 Compressed air can be used to remove the pistons and seals

Fig. 89 Remove the pistons, cup seals and spring from the cylinder

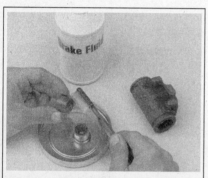

Fig. 90 Use brake fluid and a soft brush to clean the pistons . . .

Fig. 91 . . . and the bore of the wheel cylinder

Fig. 92 Once cleaned and inspected, the wheel cylinder is ready for assembly

useful in removing these components. If no compressed air is available, be VERY careful not to score the wheel cylinder bore when removing parts from it. Discard all components for which replacements were supplied in the rebuild kit.

5. Wash the cylinder and metal parts in denatured alcohol, brake system cleaner or clean brake fluid.

WARNING:
Never use a mineral-based solvent such as gasoline, kerosene or paint thinner for cleaning purposes. These solvents will swell rubber components and quickly deteriorate them.

6. Allow the parts to air dry or use filtered, unlubricated compressed air. Do not use rags for cleaning, since lint will remain in the cylinder bore.

7. Inspect the pistons and replace them if scratched.
8. Check the inside of the cylinder for pitting or scoring. Slight discoloration and stains from the brake fluid is normal.
9. Light scoring can be removed with a fine crocus cloth, but if defects cannot be removed in this manner, the cylinder should be replaced. Honing the cylinder is not recommended.
10. Lubricate the cylinder bore and seals using clean brake fluid.
11. Position the spring assembly.
12. Install the cup expanders, inner seals ("cups"), then the pistons.

WARNING:
Be sure the lips of the piston seals are facing inward towards the spring.

Fig. 93 Lubricate the cup seals with brake fluid

13. Install the boots over the wheel cylinder ends. Do not lubricate the boots.
14. Install the wheel cylinder.

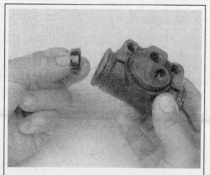

Fig. 94 Install the spring, then the cup seals in the bore: be sure the lips face inward

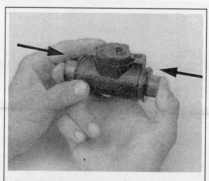

Fig. 95 Lightly lubricate the pistons, then install them

Fig. 96 The boots can now be installed over the wheel cylinder ends

BRAKES 9-23

PARKING BRAKE

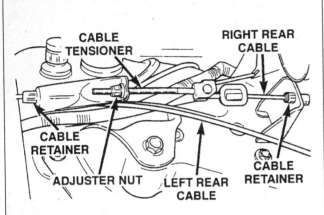

Fig. 97 To remove the cable, first loosen the adjusting nut - Dakota

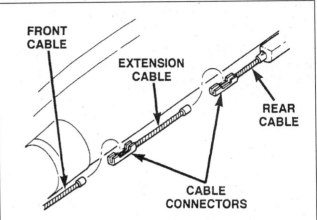

Fig. 98 Then disengage the front cable from the rear or extension cable - Dakota

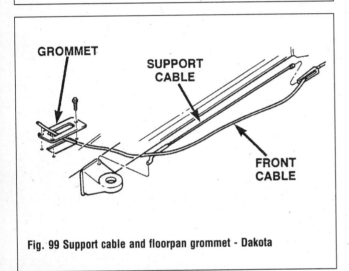

Fig. 99 Support cable and floorpan grommet - Dakota

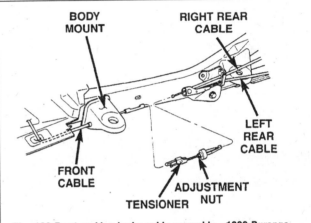

Fig. 100 Front parking brake cable assembly - 1999 Durango shown

Front Parking Brake Cable

REMOVAL & INSTALLATION

Dakota

See Figures 97, 98 and 99

1. Raise and support the vehicle.
2. Loosen the cable adjusting nut.
3. Disengage the front cable from the rear or extension cable.
4. Remove the support cable from the front cable.
5. Lower the vehicle.
6. Remove the left kick panel.
7. Fold back the carpet and remove the cable grommet from the floorpan.
8. Engage the parking brake pedal and disconnect the cable from the pedal assembly.
9. Work the cable and housing assembly through the floor pan.

To install:

10. Reverse the removal procedure.
11. Adjust cable as outlined in that section.

Durango

See Figures 100 and 101

1. Raise and support the vehicle.
2. Remove the cable adjusting nut on 1998 models, loosen it on others.
3. Pull the cable through the body mount.
4. Lower the vehicle.
5. Remove the left kick panel.
6. Fold back the carpet and remove the cable grommet from the floorpan.
7. Engage the parking brake pedal and disconnect the cable from the pedal assembly.
8. Work the cable and housing assembly through the floor pan.

To install:

9. Reverse the removal procedure.
10. Adjust cable as outlined in that section.

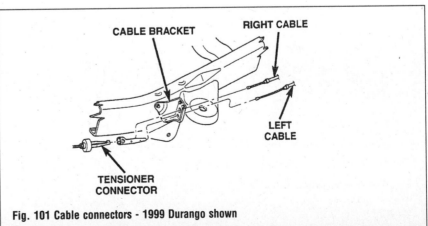

Fig. 101 Cable connectors - 1999 Durango shown

9-24 BRAKES

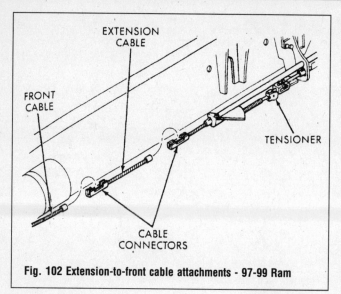

Fig. 102 Extension-to-front cable attachments - 97-99 Ram

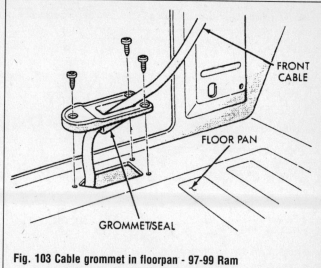

Fig. 103 Cable grommet in floorpan - 97-99 Ram

1997 - 99 Ram

See Figures 102, 103 and 104

1. Remove the knee bolster.
2. Release the parking brake.
3. Raise the vehicle and support it securely on jackstands.
4. Loosen the tensioner nut to create slack in the front and intermediate cables.
5. Disengage the front cable from the extension cable connector.
6. Lower the vehicle.
7. Roll back the carpet and loosen the cable grommet and retainer. Then pull the cable through the floorpan grommet.
8. Disengage the cable from the arm on the foot pedal assembly.

To install:

9. Insert cable through the floorpan grommet and connect to pedal assembly arm.
10. Hook cable T-connector in arm on pedal assembly.
11. Secure floorpan grommet/seal and cable retainer.
12. Realign carpet and install knee bolster.
13. Engage front cable and extension cable in cable connectors. Make sure right rear cable is secured in tensioner connector.
14. Adjust the cable as outlined in that section.

2000 Ram

See Figure 104

1. Raise and support the vehicle safely.
2. Loosen the adjusting nut to provide slack in the cable.
3. Remove the front cable from the connector.
4. Compress the cable end fitting at underbody bracket and remove the cable from the bracket.
5. Lower the vehicle.
6. Push the ball end of the cable out of the pedal clevis with a small screwdriver.
7. Compress the cable end fitting at the pedal bracket and remove the cable.
8. Remove the left cowl trim and sill plate.
9. Pull up the carpet and remove the cable from the body clip.
10. Pull up on the cable and remove it with the body grommet.

To install:

11. From inside the vehicle, insert the cable end fitting into the hole in pedal assembly.
12. Seat the cable retainer in the pedal assembly.
13. Engage the cable ball end in the clevis on pedal assembly.
14. Route the cable along the top of the wheelwell and clip in place.
15. Route cable through the floorpan and install the body grommet.
16. Replace the carpet and left cowl trim and sill plate.
17. Raise the vehicle.
18. Route the cable through the underbody bracket and seat cable housing retainer in bracket.
19. Connect cable to cable connector.
20. Adjust the cable as outlined in that section.

Rear Parking Brake Cables

REMOVAL & INSTALLATION

1997 - 99 Dakota

See Figure 105

1. Raise and support the vehicle.
2. Remove the rear wheels.
3. Loosen the tensioner adjuster nut.
4. Disconnect the right rear cable from the front cable or extension cable, depending on set-up.
5. Remove the right cable from the tensioner.
6. Compress the tabs on each cable retainer with a hose clamp or pliers and release the hose clamp or pliers.
7. Remove the right cable from the frame bracket and left cable from the adjuster nut bracket.
8. Remove the brake drum and shoe assembly.
9. Disconnect the rear cable from the lever on the secondary brake shoe and remove the cable spring.
10. Compress the tabs on each cable retainer with hose clamp or pliers. Release the hose clamp or pliers and remove the cable from the brake support plate.
11. Remove the cable from the vehicle.

To install:

12. Insert the cable through the support plate hole and cable guide until the cable retainer tabs lock into place. Pull on the cable to ensure it is locked into the support plate.
13. Lube the parking brake lever pivot points.
14. Slide the spring over the cable, then connect the spring and cable to the parking brake lever. Ensure that the spring and cable is seated in the lever.
15. Install the parking brake lever to the secondary brake shoe.
16. Install the brake assembly on the support plate. Ensure that the parking brake lever strut and spring are properly positioned before installing the return springs.
17. Adjust the brakes. Refer to that section.
18. Install the cables into the frame brackets until the cable retainer tabs lock in place.
19. Install the right cable to tensioner and left cable to front cable or extension cable.
20. Adjust the parking brake.

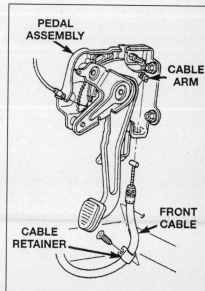

Fig. 104 Cable attachment at foot pedal - Ram

BRAKES 9-25

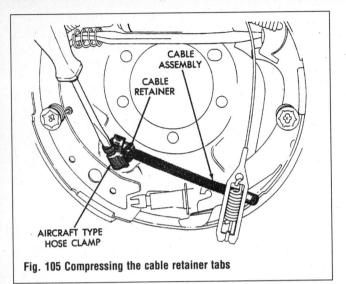

Fig. 105 Compressing the cable retainer tabs

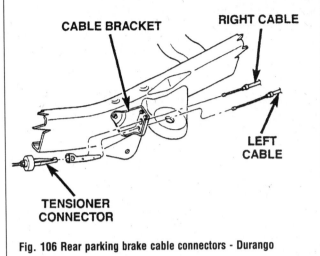

Fig. 106 Rear parking brake cable connectors - Durango

2000 Dakota

1. Raise and support the vehicle.
2. Remove the rear wheels.
3. Loosen the tensioner adjuster nut.
4. Disconnect the right rear cable from the extension cable.
5. Remove the right cable from the frame clip and pull the cable housing through the left cable connector.
6. Remove the left cable connector from the left cable.
7. Pull both cables through the cable bracket.
8. Remove the right cable mounting retainers from the differential housing.
9. Remove the brake drums.
10. Disconnect each cable from the park brake lever.
11. Compress the tabs on the cable housing retainer at the brake support plate.
12. Remove the cables.

To install:

13. Push each cable housing through the brake support plate hole until the cable housing retainer tabs lock into place.
14. Pull back on the end of the cable. Then push the cable in to engage the cable in the park brake lever. Pull on the cable end to ensure it is attached.
15. Install the brake drums.

16. Install the cable mount retainers.
17. Push the cables housing through the cable bracket.
18. Install the left cable onto the cable connector.
19. Push the right cable housing through the left cable connector and connect the cable to the extension cable.
20. Install the wheels.
21. Adjust the parking brake.

Durango

See Figure 106

1. Raise and support the vehicle.
2. Remove the rear wheels.
3. Loosen the tensioner adjuster nut.
4. Remove the right cable from the tensioner and pull the cable housing through the left cable connector.
5. Remove the left cable connector.
6. Pull both cables through the cable bracket.
7. Remove the right cable mounting retainer.
8. Remove the brake drums.
9. Disconnect each cable from the parking brake lever.
10. Compress the tabs on each cable housing retainer at the brake support plate.
11. Remove the cables.

To install:

12. Push each cable housing through the brake support plate hole until the cable housing retainer tabs lock into place. Pull on the cable housing to ensure it is locked.
13. Pull back on the end of the cable. Then push the cable in to engage the cable in the park brake lever. Pull on the cable end to ensure it is attached to the park brake lever.
14. Install the brake drums.
15. Install the right cable mounting retainers.
16. Push the cables housing through the cable bracket.
17. Install the left cable onto the cable connector.
18. Push the right cable housing through the left cable connector and connect the cable to the tension rod.
19. Install the wheels.
20. Adjust the parking brake.

1997 - 99 Ram, 2000 Ram 1500

See Figures 107 and 108

1. Raise the vehicle and remove the rear wheels and brake drums.
2. Remove the secondary brake shoe and disconnect the cable from the parking brake lever.
3. Compress the rear cable retainer with a hose clamp or pliers and pull the cable out of the support plate.

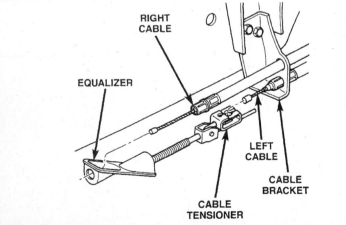

Fig. 107 Cable and tensioner attachment - 1997 - 99 Ram, 2000 Ram 1500

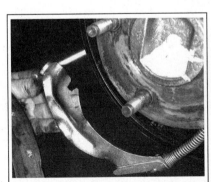

Fig. 108 On most Rams, the parking brake lever is behind the secondary brake shoe

9-26 BRAKES

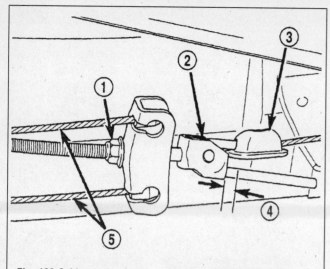

Fig. 109 Cables and tensioner - 2000 Ram 2500/3500: adjuster nut (1), tensioner (2), cable connector (3), adjuster mark (1/4 in.), rear cables (5)

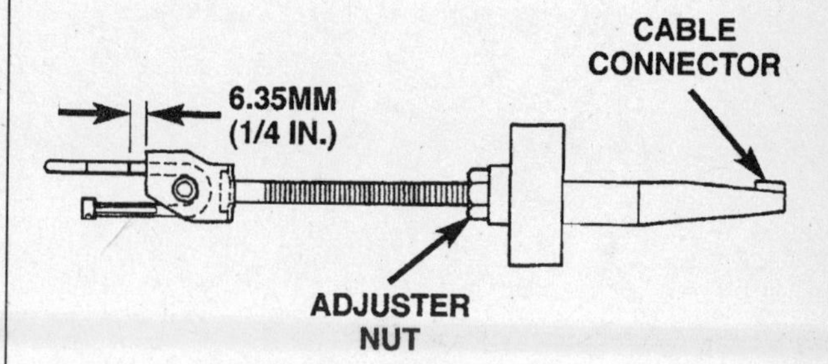

Fig. 110 Parking brake adjustment mark - Dakota, Durango

4. Remove one or both cables from the reaction bracket on left rear frame rail.
5. Disengage the rear cable from the tensioner.
6. Compress the cable retainer with hose clamp or pliers and slide the cable out of the bracket.

To install:
7. Route the new cable to the rear brake support plate.
8. Insert the cable through the support plate, seat cable retainers and attach cable to parking brake lever on the secondary brake shoe.
9. Install the brake shoe assembly.
10. Seat cable in body clips, reaction bracket and frame bracket.
11. Connect cable to tensioner.
12. Adjust cable.
13. The remainder of the procedure is the reverse of removal.

2000 Ram 2500/3500

See Figure 109

1. Raise and support the vehicle.

2. Loosen the cable adjuster nut.
3. Remove the rear cables from the cable tensioner bracket.
4. Remove the right rear cable O-ring. Then pull the cable through the frame bracket.
5. Pull the right rear cable through the brake hose bracket and remove the cable retainers from the axle.
6. Remove the rear wheels and brake drums.
7. Disconnect each cable from the park brake lever.
8. Remove the cable guide spring.
9. Compress the cable tabs on each cable end fitting at the brake support plate.
10. Remove the cables from the brake support plates.

To install:
11. Install the cable guide spring and brake drums.
12. Pull back on the cable then push the cable through the brake support plate hole to engage the cable in the parking brake lever. Pull on the cable end to ensure it is attached to the parking brake lever.

13. Push each cable housing through the brake support plate hole until the cable end fitting tabs lock into place. Pull to make sure.
14. Install the right cable retainers on the axle. Then push the right cable through the hole in the brake hose bracket.
15. Push both cables through the frame bracket. Push the left cable until the cable end fitting tabs lock into place. Install the O-ring on the right cable.

Note: The right cable must be installed in the top hole of the bracket and the left cable in the bottom hole.

16. Install the cables onto the cable tensioner bracket and install the cables into the cable connectors.
17. Adjust the parking brake. The remainder of the procedure is the reverse of removal.

ADJUSTMENT

See Figures 110 and 111

Note: Adjustment is only needed when the tensioner or a cable has been replaced or disconnected for service. To avoid faulty operation, only the procedure below should be carried out.

1. Brakes must be operating correctly and properly adjusted.
2. Check that the parking brake cables operate freely.
3. Raise the vehicle and check that the wheels turn without drag with the parking brake released.
4. Apply the parking brake fully.
5. Mark the tensioner rod 1/4 in. (6.35mm) from the edge of the tensioner (Dakota, Durango or from tensioner bracket (Ram). There may be a factory mark already here.
6. Tighten the adjusting nut on the tensioner (Dakota, Durango) or equalizer (Ram) until the mark is no longer visible.
7. Release the parking brake and ensure that the wheels turn without drag.

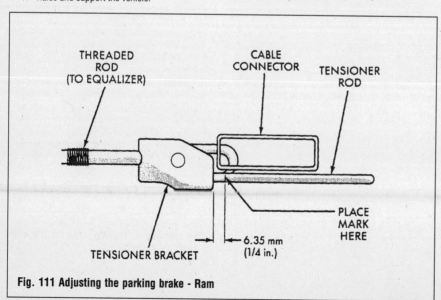

Fig. 111 Adjusting the parking brake - Ram

BRAKES 9-27

REAR WHEEL ANTILOCK BRAKE SYSTEM

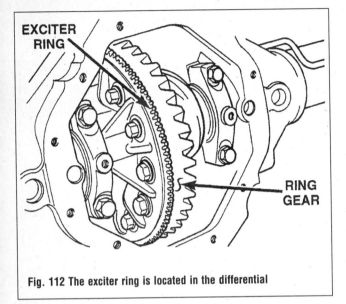

Fig. 112 The exciter ring is located in the differential

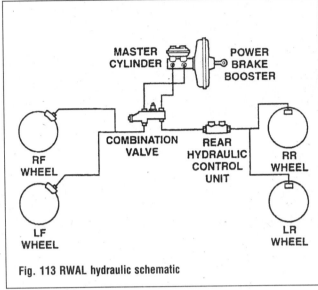

Fig. 113 RWAL hydraulic schematic

General Information

See Figures 112, 113 and 114

The Rear Wheel Antilock Brake System (RWAL) is designed to prevent rear wheel lockup under heavy braking conditions. The main system components are: the RWAL valve, controller (CAB), rear wheel speed sensor mounted on the axle, and an exciter ring on the differential.

When the brakes are applied, hydraulic fluid is routed from the master cylinder's secondary circuit through the combination valve, to the RWAL valve. From there the fluid is routed to the rear brake wheel cylinders. The CAB monitors rear wheel speed through the sensor on the axle. If a wheel is about to lock up, the CAB signals the RWAL valve which then modulates hydraulic pressure to the rear wheels to prevent lock up.

During light brake application, rear wheel deceleration is not sufficient to activate the antilock system. During a normal stop the brake fluid flows without restriction to the rear wheel cylinders to stop the vehicle. The antilock solenoids are inactive. The isolation valve is open and the dump valve is closed allowing normal fluid flow to the rear wheel cylinders.

If the CAB senses impending rear wheel lockup, it will energize the isolation solenoid. This prevents a further increase of driver-induced brake pressure to the rear wheels. If this initial action is not enough to prevent rear wheel lock-up, the CAB will momentarily energize a dump solenoid. This opens the dump valve to vent a small amount of isolated rear brake pressure to an accumulator. The action of fluid moving to the accumulator reduces the isolated brake pressure at the wheel cylinders. The dump (pressure venting) cycle is limited to very short time periods (milliseconds). The CAB will pulse the dump valve until the rear wheel deceleration reaches the desired slip rate programmed into the CAB. The system will switch to normal braking once wheel locking tendencies are no longer present.

Note that the RWAL system controls both rear wheel brakes simultaneously.

Fault Codes

See Figure 115

The red brake warning lamp and amber ABS warning lamp are located in the instrument cluster. The red brake warning lamp is used to alert the driver of a hydraulic fault or that the parking brake is applied. The red brake warning lamp also is used to alert the driver of a problem with the RWAL system.

The red brake warning lamp illuminates when a message is sent over the bus to the cluster to illuminate the bulb. A ground is provided when:
- The parking brake is applied and the parking brake switch is actuated
- A hydraulic fault has occurred and the pressure differential switch is actuated
- A RWAL fault has occurred

The amber ABS warning lamp is used to alert the driver of an RWAL problem and identify DTCs stored in the CAB memory.

If a system fault is detected during the self-test, or at any other time, the control module will illuminate the Antilock indicator lamp and store the fault code in the microprocessor memory. If a fault code is generated, the module will remember the code after the ignition is switched OFF. The microprocessor memory will store and display only one fault code at a time. The stored code can be displayed by grounding the RWAL diagnostic connector and counting the number of flashes on the indicator lamp. To clear the fault code, disconnect the control module connector or disconnect the battery for at least 5 seconds. During system retest, wait 30 seconds to make sure the fault code does not reappear.

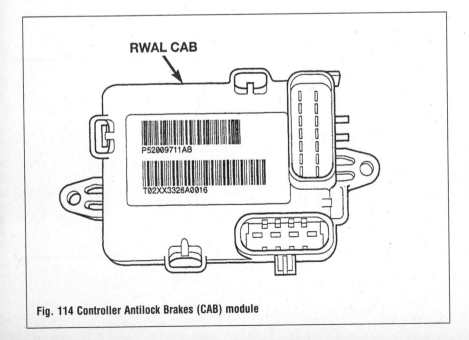

Fig. 114 Controller Antilock Brakes (CAB) module

9-28 BRAKES

FAULT CODE NUMBER	TYPICAL FAILURE DETECTED
1	Not used.
2	Open isolation valve wiring or bad control module.
3	Open dump valve wiring or bad control module.
4	Closed RWAL valve switch.
5	Over 16 dump pulses generated in 2WD vehicles (disabled for 4WD).
6	Erratic speed sensor reading while rolling.
7	Electronic control module fuse pellet open, isolation output missing, or valve wiring shorted to ground.
8	Dump output missing or valve wiring shorted to ground.
9	Speed sensor wiring/resistance (usually high reading).
10	Sensor wiring/resistance (usually low reading).
11	Brake switch always on. RWAL light comes on when speed exceeds 40 mph.
12	Not used.
13	Electronic control module phase lock loop failure.
14	Electronic control module program check failure.
15	Electronic control module RAM failure.

Fig. 115 RWAL diagnostic trouble codes

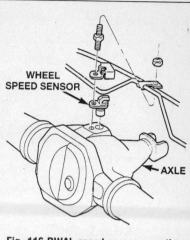

Fig. 116 RWAL speed sensor mounting - typical

Diagnosis And Testing

Diagnosis of base brake conditions which are mechanical in nature should be performed first. This includes brake noise, lack of power assist, parking brake, or vehicle vibration during normal braking.

The RWAL brake system performs several self-tests every time the ignition switch is turned on and the vehicle is driven. The CAB monitors the system inputs and outputs circuits to verify the system is operating properly. If the CAB senses a malfunction in the system it will set a DTC into memory and trigger the warning lamp.

Speed Sensor

REMOVAL & INSTALLATION

See Figure 116

1. Raise the vehicle.
2. Remove the brake line mounting nut, if fitted, and remove the brake line from the sensor stud.
3. Remove the mounting stud from the sensor and shield.
4. Remove the sensor and shield from the differential housing.
5. Disconnect the wire and remove the sensor.

To install:

6. Connect harness to sensor. Be sure the seal is securely in place between the sensor and wiring connector.
7. Install the O-ring on the sensor if removed.
8. Install the sensor on the differential housing, sensor shield, mounting stud. Tighten to 18 ft. lbs. (24 Nm).
9. Install the brake line on the sensor stud and install the nut.
10. Lower the vehicle.

TESTING (1997 MODELS)

Resistance

1. Raise and safely support the vehicle securely on jackstands.
2. Ensure the ignition key is in the OFF position.
3. Disconnect the speed sensor electrical harness.
4. Measure resistance between the terminals on the speed sensor.
5. Resistance should be 1000-2500 ohms.
6. If resistance is not within specification, check the harness for continuity.
7. If continuity exists, the sensor is faulty.

Air Gap

See Figures 117 and 118

1. Raise and safely support the vehicle securely on jackstands.
2. Remove the sensor from the differential.
3. Measure and record the distance from the underside of the sensor flange to the end of the sensor pole piece. This distance represents dimension "B". This dimension should be 1.07-1.08 in. (27.18-27.43mm). If the dimension is not within specification, replace the speed sensor.
4. Measure and record the distance between the sensor mounting surface of the differential case and the teeth at the top of the exciter ring. This distance represents dimension "A". This dimension should be 1.085-1.120 in. (27.56-28.45mm). If dimension is not within specification, replace the exciter ring.
5. Subtract dimension "B" from dimension "A" to determine sensor air gap. The gap should be 0.005-0.050 in. (0.127-1.27mm).

BRAKES 9-29

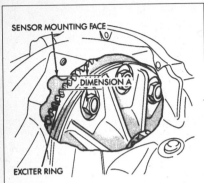

Fig. 117 Measure and record the distance between the sensor mounting surface of the differential case and the teeth at the top of the exciter ring

3. Remove the valve mounting bolt and remove the valve from the bracket.
4. Installation is the reverse of removal. Tighten the mounting bolt to 15 - 20 ft. lbs. (20 - 27 Nm). Tighten the brake line fittings to 14 ft. lbs. (19 Nm).

Exciter Ring

REMOVAL & INSTALLATION

See Figure 120
1. Remove the differential case from the axle housing.
2. Clamp the differential case in a soft-jaw vise.
3. Remove and discard the ring gear bolts.
4. Tap the ring gear off with a rawhide or plastic mallet.
5. Remove the exciter ring with a soft-faced hammer.

To install:
6. Be sure to align the exciter ring tab with the slot in the differential case.
7. Invert the differential case and start two ring gear bolts. This will provide case-to-ring gear bolt hole alignment.
8. Press the exciter ring onto the differential case using the ring gear as a pilot.
9. Install new ring gear bolts and tighten in a cross pattern to 75 ft. lbs. (102 Nm) for 8 1/4 in.

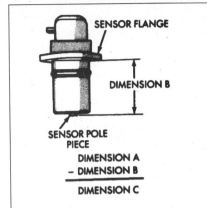

Fig. 118 Measure and record the distance from the underside of the sensor flange to the end of the sensor pole piece

axles, 115 ft. lbs. (157 Nm) for 9 1/4 in. axles, 130 ft. lbs. (180 Nm) for 248 and 267 RBI axles, 220 ft. lbs. (300 Nm) on 286 RBI axles.

Bleeding the RWAL System

This is accomplished using the same methods as for the base brake system. Refer to those procedures.

RWAL Valve

REMOVAL & INSTALLATION

See Figure 119
1. Remove the RWAL valve harness connector from the controller.
2. Remove the brake lines from the valve.

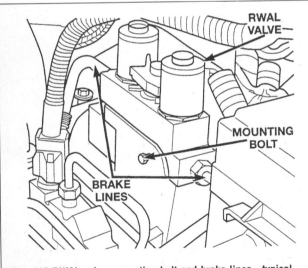

Fig. 119 RWAL valve, mounting bolt and brake lines - typical

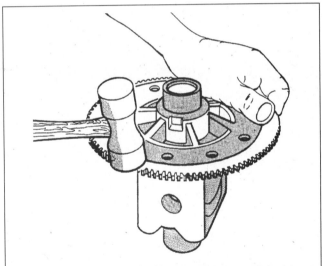

Fig. 120 Removing the exciter ring from the differential case

FOUR WHEEL ANTILOCK BRAKE SYSTEM

General Information

See Figure 121

The Four Wheel Antilock Brake System (ABS) is designed to prevent wheel lock-up during braking under virtually any road surface conditions. This allows the driver to retain greater control of the vehicle during braking.

The major components of the system are:
- Controller Antilock Brakes (CAB)
- Hydraulic Control Unit (HCU)
- Wheel speed sensors
- ABS warning light

The Four Wheel ABS has a three-channel design. The front brake antilock valve provides two channel pressure control of the front brakes. Each front wheel brake unit is controlled separately. Two solenoid valves are used in each control channel.

The rear brake antilock valve controls the rear wheel brakes in tandem. The rear brake valve contains two solenoid valves.

The front and rear antilock valves contain electrically operated solenoid valves. The solenoid valves modulate brake fluid apply pressure during antilock braking. The valves are operated by the antilock electronic module.

The antilock electrical system is separate from other electrical circuits in the vehicle. A specially programmed electronic control module is used to operate the system components.

Some models have Electronic Brake Distribution (EBD) designed into the system which eliminates the combination/proportioning valve.

The ABS electronic control module monitors wheel speed sensor inputs continuously while the vehicle is in motion. The module will not activate the ABS system as long as sensor inputs indicate normal braking.

During normal braking, the master cylinder, power

9-30 BRAKES

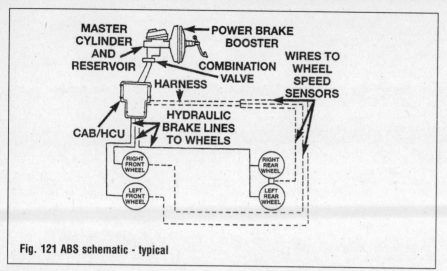

Fig. 121 ABS schematic - typical

booster and wheel brakes units all function as they would in a vehicle without ABS. The solenoid valves are not activated.

The wheel speed sensors converts wheel speed into electrical signals. These signals are transmitted to the module for processing and determine wheel lock-up and deceleration rate. When a wheel speed sensor signal indicates the onset of wheel lock-up the ABS braking is activated.

The antilock system retards the lockup conditions by modulating fluid apply pressure to the wheel brake units. The pressure is modulated according to wheel speed, degree of lock-up and rate of deceleration. The solenoid valves are cycled continuously to modulate pressure. Solenoid cycle time in antilock mode can be measured in milliseconds.

Diagnosis And Testing

Diagnosis of base brake conditions which are mechanical in nature should be performed first. This includes brake noise, lack of power assist, parking brake, or vehicle vibration during normal braking.

The static and dynamic checks occur at ignition start up. During the dynamic check, the CAB briefly cycles the pump and solenoids to verify operation. A noise may be heard during this self-check.

If an ABS component exhibits a fault during initialization, the CAB illuminates the amber warning light and registers a fault code in the memory.

Speed Sensors

REMOVAL & INSTALLATION

Front Wheel Speed Sensor, 2WD

1997 - 1999 RAM

See Figure 122

1. Raise and support the vehicle.
2. Remove the wheel.
3. Remove the brake caliper and rotor.
4. Remove the sensor attaching bolts.
5. Disconnect the wire and remove the sensor from the vehicle.

To install:

6. Tighten the bolts to 18 ft. lbs. (23 Nm).

Note: Use the original or replacement sensor bolts only. The bolt is special and must not be substituted.

FOUR WHEEL ANTI-LOCK BRAKE DIAGNOSTIC CODES

The anti-lock brake system module may report any of the following diagnostic trouble codes:

- (21) Right Front Sensor Open
- (22) No Signal From Right Front Sensor
- (23) Intermittent Signal From Right Front Sensor

- (25) Left Front Sensor Open
- (26) No Signal From Left Front Sensor
- (27) Intermittent Signal From Left Front Sensor

- (35) Rear Sensor Open
- (36) No Signal From Rear Sensor
- (37) Intermittent Signal From Rear Sensor

- (38) Wheel Speed Mismatch

- (41) Right Front Isolation Solenoid Open
- (42) Right Front Dump Solenoid Open
- (43) Right Front Isolation Solenoid Shorted
- (44) Right Front Dump Solenoid Shorted

- (45) Left Front Isolation Solenoid Open
- (46) Left Front Dump Solenoid Open
- (47) Left Front Isolation Solenoid Shorted
- (48) Left Front Dump Solenoid Shorted

- (51) Rear Isolation Solenoid Open
- (52) Rear Dump Solenoid Open
- (53) Rear Isolation Solenoid Open
- (54) Rear Dump Solenoid Shorted

- (61) Right Front Reset Switch Closed
- (62) Left Front Reset Switch Closed
- (63) Rear Reset Switch Closed

- (65) Main Relay Open
- (66) Main Relay Shorted

- (67) Pump Motor Circuit Open
- (68) Pump Motor Stalled

- (70) Controller/Vehicle Mismatch
- (71) RAM Read/Write
- (72) ROM Checksum
- (73) Watchdog
- (78) Foundation Brake

- (81) Brake Switch Circuit
- (88) Brake Warning Lamp (Red) Circuit
- (89) ABS Warning Lamp Circuit

BRAKES 9-31

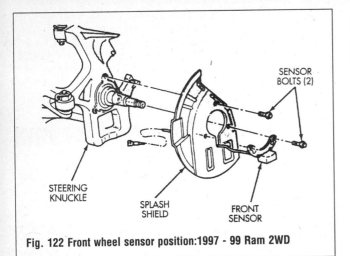

Fig. 122 Front wheel sensor position: 1997 - 99 Ram 2WD

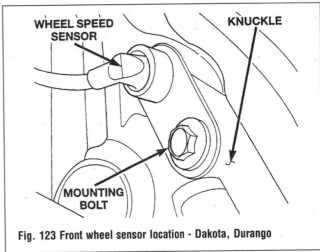

Fig. 123 Front wheel sensor location - Dakota, Durango

DAKOTA, DURANGO AND 2000 RAM

See Figure 123

1. Raise and support the vehicle.
2. Remove the sensor bolt from the steering knuckle and remove the sensor.
3. Disconnect the ABS wheel speed sensor wire and detach the wire from securing clips.

To install:

4. Tighten the sensor bolt to 17 ft. lbs. (23 Nm).

Note: Use the original or replacement sensor bolt only. The bolt is special and must not be substituted.

5. The remainder of the procedure is the reverse of removal.

Front Wheel Speed Sensor, 4WD

1997 RAM

1. Remove the bolt attaching the sensor to the inside of the steering knuckle.
2. Disconnect the sensor wire at the connector.
3. Remove the sensor.

To install:

4. Tighten the sensor bolt to 11 ft. lbs. (14 Nm).

Note: Use the original or replacement sensor bolt only. The bolt is special and must not be substituted.

5. The remainder of the procedure is the reverse of removal.

OTHER MODELS

See Figure 124

1. Raise and support the vehicle.
2. Remove the wheel.
3. Remove the brake caliper.
4. On 8-stud wheels, remove the rotor hub bearing assembly and separate the rotor from the hub bearing.
5. On other wheels, remove the rotor.
6. Remove the sensor attaching bolts.
7. Disconnect the wire and remove the sensor from the vehicle.

To install:

8. Tighten the bolts to 13 ft. lbs. (18 Nm).

Note: Use the original or replacement sensor bolts only. The bolts are special and must not be substituted.

9. The remainder of the procedure is the reverse of removal.

Rear Wheel Sensor

See Figure 116

1. Raise the vehicle and support it securely on jackstands.
2. Remove the brake line mounting nut, if fitted, and remove the brake line from the sensor stud.
3. Remove the mounting stud from the sensor and shield.
4. Remove the sensor and shield from the differential housing.
5. Disconnect the wire and remove the sensor.

To install:

6. Connect harness to sensor. Be sure the seal is securely in place between the sensor and wiring connector.
7. Install O-ring on sensor if removed.
8. Install sensor on differential; housing, sensor shield, mounting stud. Tighten to 18 ft. lbs. (24 Nm).
9. Install the brake line on the sensor stud and install the nut.

Note: Use the original or replacement sensor bolt only. The bolt is special and must not be substituted.

10. Lower the vehicle.

Tone Wheel

REMOVAL & INSTALLATION

See Figure 125

The "tone wheel" works in conjunction with the front wheel speed sensors as part of the ABS. There is one on each wheel.

The tone wheel is located on the hub/bearing, on the axle outer stub shaft or inside the hub/bearing housing, depending on model and equipment.

This component is not serviceable or removable and if damaged, the assembly of which it is a part must be replaced.

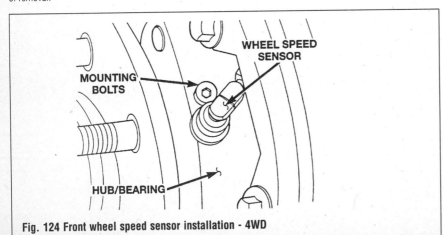

Fig. 124 Front wheel speed sensor installation - 4WD

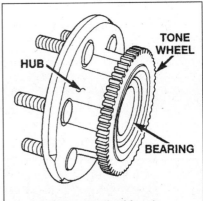

Fig. 125 The tone wheel is not replaceable by itself - 2WD shown

9-32 BRAKES

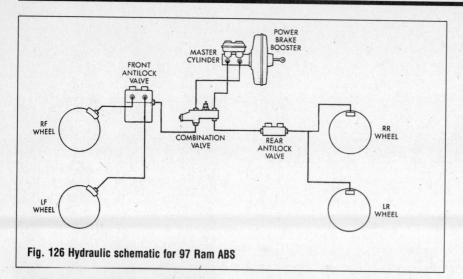

Fig. 126 Hydraulic schematic for 97 Ram ABS

lbs. (180 Nm) for 248 and 267 RBI axles, 220 ft. lbs. (300 Nm) on 286 RBI axles.

Bleeding the ABS System

1997 RAM

See Figure 126

This is accomplished using the same methods as for the base brake system. Refer to those procedures.

OTHER MODELS

ABS system bleeding requires conventional bleeding methods plus use of the DRB scan tool. The procedure involves performing a base brake bleeding, followed by use of the scan tool to cycle and bleed the HCU pump and solenoids. A second base brake bleeding procedure is then required to remove any air remaining in the system.

1. Perform base brake bleeding. Refer to the appropriate section.
2. Connect the scan tool to the data link connector beneath the dashboard.
3. Select "Antilock Brakes" followed by "Miscellaneous", then "Bleed Brakes". Follow the instructions displayed until the unit displays "Test Complete", then disconnect the scan tool and proceed.
4. Perform a base brake bleeding a second time.
5. Top up the master cylinder.

Exciter Ring

REMOVAL & INSTALLATION

See Figure 120

1. Remove the differential case from the axle housing.
2. Clamp the differential case in a soft-jaw vise.
3. Remove and discard the ring gear bolts.
4. Tap the ring gear off with a rawhide or plastic mallet.
5. Remove the exciter ring with a soft-faced hammer.

To install:

6. Be sure to align the exciter ring tab with the slot in the differential case.
7. Invert the differential case and start two ring gear bolts. This will provide case to ring gear bolt hole alignment.
8. Press the exciter ring onto the differential case using the ring gear as a pilot.
9. Install new ring gear bolts and tighten in a cross pattern to 75 ft. lbs. (102 Nm) for 8 1/4 in. axles, 115 ft. lbs. (157 Nm) for 9 1/4 in. axles, 130 ft.

BRAKE SPECIFICATIONS
All measurements in inches unless noted

Year	Model	Master Cylinder Bore	Brake Disc Original Thickness	Brake Disc Minimum Thickness	Brake Disc Maximum Run-out	Brake Drum Original Inside Diameter	Brake Drum Max. Wear Limit	Brake Drum Maximum Machine Diameter	Minimum Lining Thickness Front	Minimum Lining Thickness Rear	Brake Caliper Bracket Bolts (ft. lbs.)	Brake Caliper Mounting Bolts (ft. lbs.)
1997	Dakota ①	NA	0.861	0.810	0.004	9.00	9.09	9.06	0.060	②	47	22
	Dakota ③	NA	0.861	0.810	0.004	10.00	10.09	10.06	0.060	②	47	22
	Ram 1500 Pick-up	1.125	1.260	④	0.004	11.00	11.09	11.06	0.062	②	—	38
	Ram 2500 Pick-up	1.250	1.500	④	0.005	13.00	13.09	13.06	0.062	②	—	38
	Ram 3500 Pick-up	1.250	1.500	④	0.005	13.00	13.09	13.06	0.062	②	—	38
1998	Dakota ①	NA	0.944	0.890	0.004	9.00	⑤	⑤	⑥	⑦	47	22
	Dakota ③	NA	0.944	0.890	0.004	10.00	⑤	⑤	⑥	⑦	47	22
	Durango	1.06	0.900	0.890	0.004	11.00	⑤	⑤	⑥	⑦	47	22
	Ram 1500 Pick-up	1.25	⑧	⑨	0.005	11.00	11.09	11.06	⑥	⑦	—	38
	Ram 2500 Pick-up	1.25	⑩	⑪	0.005	11.00	11.09	11.06	⑥	⑦	—	38
	Ram 3500 Pick-up	1.25	⑫	⑬	0.005	12.00	12.09	12.06	⑥	⑦	—	38
1999-00	Dakota ①	NA	0.944	0.890	0.004	9.00	⑤	⑤	⑥	⑦	47	22
	Dakota ③	NA	0.944	0.890	0.004	10.00	⑤	⑤	⑥	⑦	47	22
	Durango	1.06	0.900	0.890	0.004	11.00	⑤	⑤	⑥	⑦	47	22
	Ram 1500 Pick-up	1.25	⑧	⑨	0.005	11.00	11.09	11.06	⑥	⑦	—	38
	Ram 2500 Pick-up	1.25	⑩	⑪	0.005	11.00	11.09	11.06	⑥	⑦	—	38
	Ram 3500 Pick-up	1.25	⑫	⑬	0.005	12.00	12.09	12.06	⑥	⑦	—	38

NA - Not Available
① With 9 inch rear brakes
② Riveted brake shoes: 0.031 in.
   Bonded brake shoes: 0.0620 in.
③ With 10 inch rear brakes
④ Minimum thickness indicated on rotor hub
⑤ Maximum allowable drum diameter, either from wear or machining, is stamped on the drum.
⑥ Riveted brake pads: 0.0625 in.
   Bonded brake pads: 0.1875 in.
⑦ Riveted brake shoes: 0.031 in.
   Bonded brake shoes: 0.0625 in.
⑧ 2WD: 1.20 in.
   4WD: 1.5 in.
⑨ 2WD: 1.215 in.
   4WD: 1.269 in.
⑩ 2WD: 1.5 in.
   4WD LD: 1.5 in.
   4WD HD: 1.75 in.
⑪ 2WD: 1.269 in.
   4WD LD: 1.269 in.
   4WD HD: 1.521 in.
⑫ 2WD: 1.75 in.
   4WD: 1.75 in.
⑬ 2WD: 1.518 in.
   4WD: 1.521 in.

EXTERIOR
ANTENNA 10-4
DOORS 10-2
FRONT FENDERS 10-6
GRILLE 10-3
HOOD 10-3
LIFTGATE 10-3
OUTSIDE MIRRORS 10-4
REAR FENDERS (DUAL WHEELS) 10-7
TAILGATE 10-3

INTERIOR
DOOR GLASS AND REGULATOR 10-15
DOOR LATCHES 10-12
DOOR LOCKS 10-14
DOOR PANELS 10-10
ELECTRIC WINDOW MOTOR 10-16
FLOOR CONSOLE 10-9
INSIDE REAR VIEW MIRROR 10-17
INSTRUMENT PANEL 10-7
LIFTGATE LATCH 10-14
LIFTGATE LOCK 10-15
POWER SEAT MOTORS 10-19
SEATS 10-19
TAILGATE LATCH 10-13
WINDSHIELD AND FIXED GLASS 10-16
WINDSHIELD CHIP REPAIR 10-17

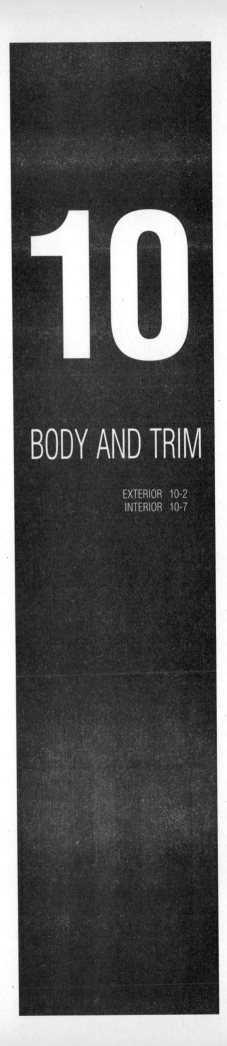

10

BODY AND TRIM

EXTERIOR 10-2
INTERIOR 10-7

10-2 BODY AND TRIM

EXTERIOR

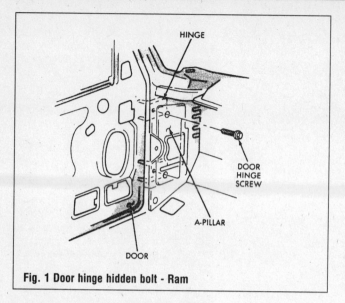

Fig. 1 Door hinge hidden bolt - Ram

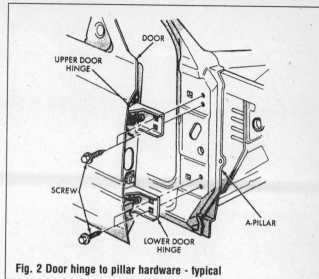

Fig. 2 Door hinge to pillar hardware - typical

Doors

REMOVAL & INSTALLATION

See Figures 1, 2 and 3

1. Open the door.
2. Ram: Remove the cowl trim panel.
3. Disengage the door wire harness connector.
4. Ram: Remove the hidden bolts attaching the door hinge to the hinge pillar behind the cowl panel.
5. Using a suitable marker, mark the outline of the door hinges on the hinge pillar to aid installation.
6. Support the door with a suitable lifting device.
7. Remove the hardware attaching the lower door hinge to the hinge pillar.
8. While holding the door steady on a lift, remove the bolts attaching the upper door hinge to the hinge pillar.
9. Remove the door from the vehicle.
10. Installation is the reverse of removal. Tighten the hinge hardware to 21 ft. lbs. (28 Nm).

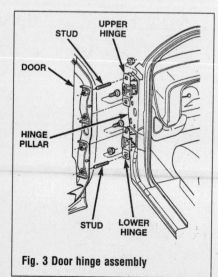

Fig. 3 Door hinge assembly

ADJUSTMENT

Durango

Minor adjustment for alignment of the door is made by moving the latch striker.

IN & OUT ADJUSTMENT

1. Loosen the latch striker.
2. Tap the striker inward if the door character line is outboard of the body character line. Tap the striker outward if the door character line is inboard of the body character line.
3. When position is correct, tighten the striker to 21 ft. lbs. (28 Nm).

UP & DOWN ADJUSTMENT

1. Loosen the latch striker.
2. Tap the striker downward if the door character line is higher than the body character line. Tap the striker upward if the door character line is lower than the body character line.
3. When position is correct, tighten the striker to 21 ft. lbs. (28 Nm).

Dakota Rear Door

Minor adjustment for alignment of the door is made by moving the latch striker.

IN & OUT ADJUSTMENT

1. Loosen the latch striker.
2. Tap the striker inward if the door character line is outboard of the body character line. Tap the striker outward if the door character line is inboard of the body character line.
3. When position is correct, tighten the striker to 21 ft. lbs. (28 Nm).

UP & DOWN ADJUSTMENT

1. Loosen the latch striker.
2. Tap the striker downward if the door character line is higher than the body character line. Tap the striker upward if the door character line is lower than the body character line.
3. When position is correct, tighten the striker to 21 ft. lbs. (28 Nm).

Dakota, Ram

RAM FORE/AFT ADJUSTMENT

Fore/aft (lateral) door adjustment is done by loosening the hinge to cowl screws one hinge at a time. The move the door to the correct position.

1. Support the door with a padded floor jack.
2. Loosen the applicable hinge to cowl screws. Move the door to the correct position.
3. Tighten the screws.

IN & OUT ADJUSTMENT

In/out door adjustment is done by loosening the hinge to door fasteners. Then move the door to the correct position.

1. Support the door with a padded floor jack.
2. Loosen the applicable hinge to door fasteners. Move the door to the correct position.
3. Tighten the fasteners.

UP & DOWN ADJUSTMENT

In/out door adjustment is done by loosening the hinge to cowl fasteners at both hinges. Then move the door to the correct position.

1. Support the door with a padded floor jack.
2. Loosen the applicable hinge to cowl fasteners at both hinges. Move the door to the correct position.
3. Tighten the fasteners.

Ram Cargo Door

1. Remove the C-pillar trim to access the bolts attaching the cargo door.
2. Support the door with a padded floor jack.
3. For up/down or fore/aft adjustment, loosen the applicable C-pillar to hinge bolts and move the door to the correct position.
4. For in/out adjustment, loosen the applicable hinge to door fasteners and move the door to the cor-

BODY AND TRIM 10-3

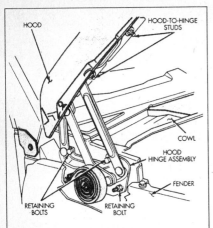

Fig. 4 Hood and hinge mounting hardware - typical

rect position.
5. Proper torque for this hardware is 21 ft. lbs. (28 Nm).

Hood

REMOVAL & INSTALLATION

See Figures 4 and 5

1. Open the hood.
2. Disconnect the underhood lamp wire connector, if fitted.
3. Disconnect the air temperature sensor wire connector, if fitted.
4. Mark all nut and hinge attachment locations with a grease pencil or other suitable marker to provide reference marks for installation.
5. Remove the nuts attaching the hood to the hinge and loosen the bottom nuts until they can be removed by hand.
6. With the assistance of a helper at the other side of the vehicle to support the hood, remove the bottom nuts.
7. Remove the hood from the vehicle.

To install:

8. Position hood on vehicle. Loosely install bottom nuts.
9. Align all marks, install top nuts and tighten bottom nuts.
10. Connect applicable wiring.
11. Check alignment as outlined below.

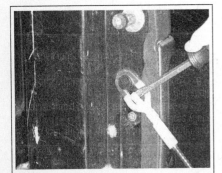

Fig. 6 Disconnecting the tailgate check cable

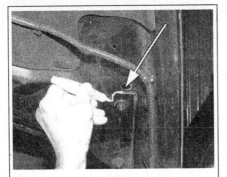

Fig. 5 Mark hinge position with an indelible marker before removal so that the hood can be properly aligned when it is refitted

ALIGNMENT

1. The hood should be aligned to 0.2 in. (5 mm) gap to the front fenders and flush across the top surfaces of the fenders.
2. Loosen mounting hardware and position hood as needed to achieve the alignment.

Tailgate

REMOVAL & INSTALLATION

See Figure 6

1. Open the tailgate.
2. Disconnect the marker light harness, if equipped.
3. Disconnect the tailgate check cables.
4. Close the tailgate until the notch in the right hand collar aligns with the pivot pin.
5. Slip the tailgate hinge collar from the pivot pins.
6. Slide the tailgate to the right and separate the left hand collar from the pivot pin.
7. Remove the tailgate.
8. Installation is the reverse of removal.

Liftgate

REMOVAL & INSTALLATION

See Figures 7 and 8

1. Remove the liftgate upper trim.

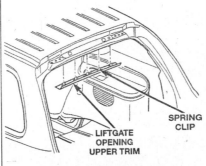

Fig. 7 Liftgate opening upper trim installation

2. Remove the D-pillar trim.
3. Disconnect the liftgate wire harness from the body wire harness.
4. Disconnect the rear window washer hose from the spray nozzle.
5. Support the liftgate with a suitable lift.
6. Remove the screws attaching the support cylinders to the liftgate.
7. Using a suitable marker, mark the position of the hinges on the roof panel.
8. Carefully pull the headliner down and remove the bolts attaching liftgate hinge to roof panel.
9. Remove the liftgate from the vehicle.
10. Installation is the reverse of removal.

ALIGNMENT

The position of the liftgate can be adjusted upward or downward and inward or outward by the use of hinge shims. The liftgate slam bumpers must also be adjusted if liftgate hinges are adjusted. The inward/outward position of each slam bumper is adjusted by the use of shims.

1. To move the position of the liftgate inward or outward, remove or add shims between the hinge-halves and liftgate.
2. To move the position of the liftgate upward or downward, remove or add shims between the hinge-halves and roof panel.
3. To move the position of the liftgate slam bumpers inward or outward, remove or add shims between the slam bumper screws and anchors.

Grille

REMOVAL & INSTALLATION

See Figures 9 and 10

1. Open the hood.
2. Remove the fasteners securing the grille to the hood and the grille bracket.
3. Remove the grille.
4. Installation is the reverse of removal. Install all the hardware loosely before tightening. Check that the grille is centered before final tightening.

Note: Handle the grille carefully as it is fragile. Some grille materials can be damaged by brake fluid and other chemicals.

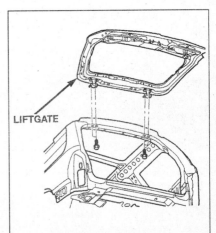

Fig. 8 Liftgate hinge installation

10-4 BODY AND TRIM

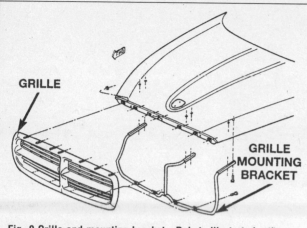

Fig. 9 Grille and mounting bracket - Dakota illustrated, others similar

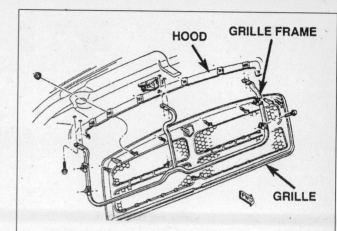

Fig. 10 Another grille mounting style - this time a RAM SLT

Outside Mirrors

REMOVAL & INSTALLATION

Sideview Mirror

See Figures 11, 12, 13 and 14

1. Remove the door trim panel.
2. Remove the mirror flag door seal.
3. Disengage the wiring, if so equipped.
4. Remove the nuts attaching the mirror to the door frame.
5. Separate the harness grommet from the door frame, if so equipped.
6. Remove the mirror.
7. Installation is the reverse of removal.

Low Mounted Sideview Mirror

See Figure 15

1. Remove the bolts attaching the mirror bracket to the door.
2. Installation is the reverse of removal.

Antenna

REPLACEMENT

Durango, Dakota

See Figures 16, 17, 18 and 19

1. Disconnect the battery negative cable.
2. Remove the trim cover from the right cowl side inner panel.

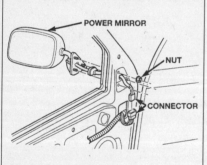

Fig. 11 Power mirror wiring connector - typical

Fig. 12 Removing the mirror flag door seal

Fig. 13 Remove the three nuts to remove the mirror

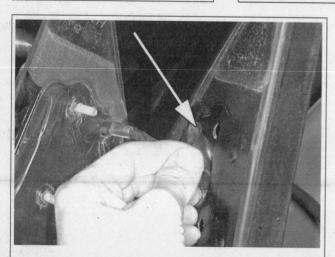

Fig. 14 Separate the wiring harness grommet

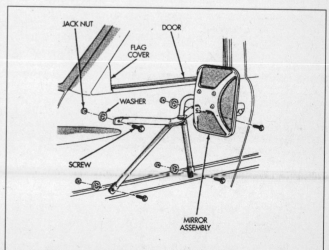

Fig. 15 Low mounted sideview mirror assembly

BODY AND TRIM 10-5

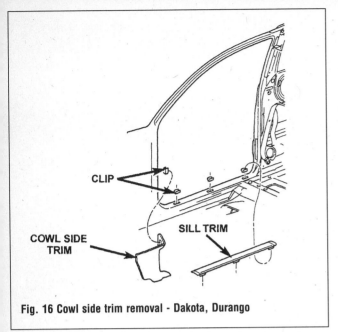

Fig. 16 Cowl side trim removal - Dakota, Durango

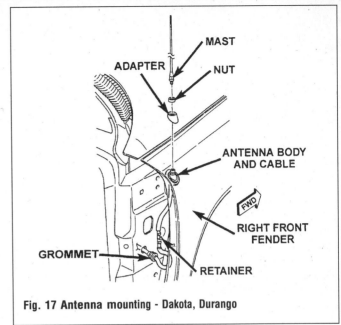

Fig. 17 Antenna mounting - Dakota, Durango

3. Reach under the instrument panel below the glove box to access and disconnect the antenna coax cable connector.
4. Disconnect the connector by pulling it apart while twisting the metal connector halves. Do not pull on the cable.
5. Disengage the antenna coax cable from the retainer clips on the lower instrument panel reinforcement and the blower housing.

6. Disengage the antenna coax cable retainers at the right side inner panel and inside the right front fender.
7. Unscrew the mast from the antenna body.
8. Remove the antenna cap nut.
9. Remove the antenna adapter from the top of the fender.
10. Lower the antenna body through the mounting hole in the top of the fender.
11. Pull the antenna body and cable out through the opening between the right cowl side outer panel and the fender through the front door opening.

12. Disengage the antenna coax cable grommet from the hole in the right cowl side outer panel.
13. Pull the antenna coax cable out of the passenger compartment through the hole in the right cowl side outer panel.
14. Installation is the reverse of removal.

Ram

See Figures 20 and 21

1. Disconnect the battery negative cable.
2. On 1997 models, remove the five screws that secure the right cowl side kick panel/sill trim and remove the sill trim.
3. On other models, reach under the passenger side of the instrument panel near the right cowl side inner panel to disengage the coax cable connector from the retainer clip located on the bottom of the blower housing. Remove the foam tape to access the coax cable connector.
4. Disconnect the connector by pulling it apart while twisting the metal connector halves. Do not pull on the cable.
5. Securely tie a suitable length of twine to the antenna half of the coax cable connector.

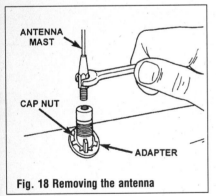

Fig. 18 Removing the antenna

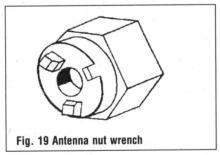

Fig. 19 Antenna nut wrench

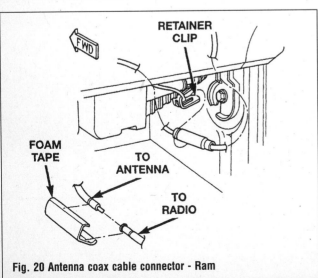

Fig. 20 Antenna coax cable connector - Ram

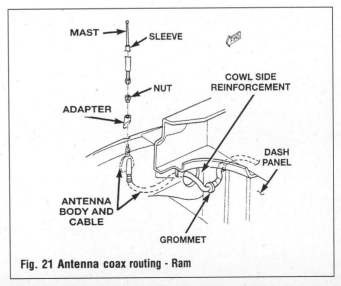

Fig. 21 Antenna coax routing - Ram

10-6 BODY AND TRIM

6. Reach above the PCM on the right side of the dash panel in the engine compartment to disengage the antenna coax cable grommet from the hole in the dash panel.
7. Pull the cable out of the passenger compartment and into the engine compartment through the hole in the dash panel.
8. Raise the sleeve on the antenna mast far enough to access and unscrew the mast from the body.
9. Remove the antenna cap nut.
10. Remove the antenna adapter from the top of the fender.
11. Lower the antenna body and cable through the mounting hole in the top of the fender.
12. Pull the antenna body and cable out through the opening between the right cowl side outer panel and the top of the fender, while feeding the antenna coax cable out of the engine compartment through the hole in the right cowl side reinforcement.
13. Untie the twine from the antenna body and coax connector leaving the twine in the place of the cable through the vehicle.
14. Remove the antenna from the vehicle.
15. Installation is the reverse of removal.

Front Fenders

REMOVAL & INSTALLATION

Dakota, Durango
See Figure 22

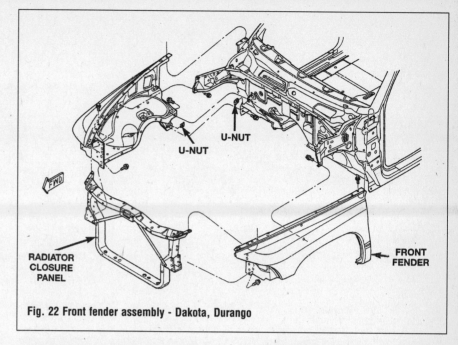

Fig. 22 Front fender assembly - Dakota, Durango

LEFT FRONT FENDER

1. Remove the battery.
2. Raise and support the vehicle.
3. Remove the left front wheel.
4. Remove the wheel opening molding.
5. Remove the wheelwell.
6. Remove the left headlamp module.
7. Remove the Power Distribution Center (PDC).
8. Remove the battery tray and support bracket.
9. Remove the ABS HCU, if fitted.
10. Disengage the clips attaching the hood release cable to the inner fender.
11. Disengage the clips attaching the wire harness to inner fender and wheelwell.
12. Remove the bolt attaching the fender to the lower rocker panel.
13. Remove the bolts attaching the fender to lower radiator closure panel.
14. Remove the bolts attaching the fender to the hood hinge support bracket.
15. Remove the bolts attaching the fender to upper cowl.
16. Remove the bolts attaching the fender to upper radiator closure panel.
17. Remove the fender.
18. Installation is the reverse of removal.

RIGHT FRONT FENDER

1. Remove the battery.
2. Raise and support the vehicle.
3. Remove the right front wheel.
4. Remove the wheel opening molding.
5. Remove the wheelwell.
6. Remove the right headlamp module.
7. Remove the air cleaner element housing.
8. Remove the PDC.
9. Disengage the clips attaching the wire harnesses to the inner fender and wheelwell.
10. Remove the bolt attaching the fender to the lower rocker panel.
11. Remove the bolts attaching the fender to the lower radiator closure panel.
12. Remove the bolts attaching the fender to hood hinge support bracket.
13. Remove the bolts attaching the fender to the upper cowl.
14. Remove the bolts attaching the fender to upper radiator closure panel.
15. Remove the fender.
16. Installation is the reverse of removal.

Ram

LEFT FRONT FENDER

See Figures 23, 24 and 25

1. Open hood.
2. Remove the front bumper.
3. Remove the air cleaner from the wheelwell (Diesel).
4. Remove the coolant overflow bottle (V-10).
5. Remove the battery and tray.
6. Remove the PDC from the left wheelwell.
7. Disengage the wire harness tie-downs from the wheelwell.

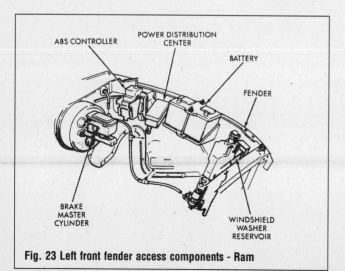

Fig. 23 Left front fender access components - Ram

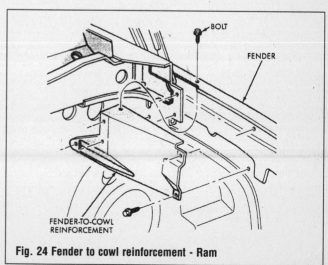

Fig. 24 Fender to cowl reinforcement - Ram

BODY AND TRIM 10-7

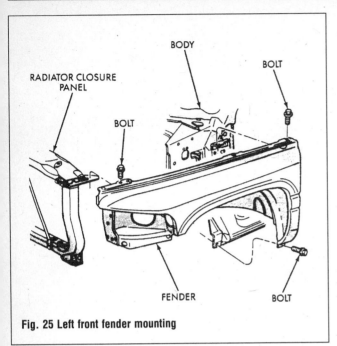

Fig. 25 Left front fender mounting

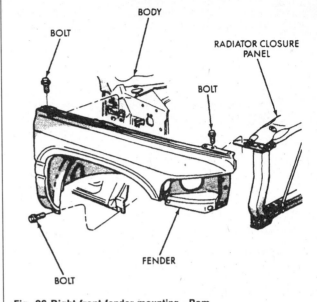

Fig. 26 Right front fender mounting - Ram

8. Disconnect the wiring harness to headlamp connector.
9. Disconnect the wiring harness to airbag sensor and remove the airbag sensor from the wheelwell.
10. Remove the anti-lock brake controller from the wheelwell, if so equipped.
11. Disengage the windshield washer tubing tie-downs from wheelwell.
12. Remove the bolts attaching the front fender to cowl reinforcement.
13. Remove the bolts attaching the front fender to radiator closure panel.
14. Remove the bolts attaching the bottom of the front fender to rocker panel lower flange.
15. Open the left door.
16. Remove the bolt attaching the front fender to hinge pillar mounting bracket.
17. Remove the bolts attaching the top of the fender to the radiator closure panel.
18. Remove the fender.
19. To install, reverse the removal procedure.

RIGHT FRONT FENDER

See Figure 26

1. Remove the front bumper.
2. Disconnect the battery negative cable(s).
3. Remove the auxiliary battery and tray on the right side, if fitted.
4. Disengage the wire harness tie - downs from the wheelwell.
5. Disconnect the wiring harness to headlamp connector.
6. Disconnect the wiring harness to airbag sensor and remove airbag sensor from the wheelwell.
7. Remove the front wheelwell liner.
8. Disengage the A/C tubing from the inner fender clips.
9. Remove the bolts attaching front fender to cowl reinforcement.
10. Remove the bolts attaching front fender to radiator closure panel.
11. Remove the bolts attaching the bottom of the front fender to rocker panel lower flange.
12. Open the right door.
13. Remove the bolt attaching the front fender to hinge pillar mounting bracket.
14. Remove the bolts attaching the top of the fender to radiator closure panel.
15. Separate the right front fender from the vehicle.
16. To install, reverse the removal procedure.

Rear Fenders (Dual Wheels)

1. For left side removal, remove the screws attaching the fuel fill neck to rear fender opening.
2. Remove the tail lamp.
3. Remove the nuts attaching the rear fender to the cargo box side panel through the tail lamp opening.
4. Remove the clearance lamps.
5. Remove the sockets from the clearance lamps.
6. Remove the bolts attaching the bottom of the fender to cargo box forward of the rear wheel.
7. Remove the bolts attaching the bottom of the fender to cargo box rearward of rear wheel.
8. Remove the rear wheelwell splash shields and filler.
9. Remove the nuts attaching the front of rear fender to the cargo box from behind the side panel forward of wheelwell.
10. Remove the screws attaching the access panel to the top of the wheelwell.
11. Remove the nuts attaching rear fender to cargo box through the access hole in top of wheelwell.
12. Separate the rear fender from the cargo box side panel.
13. To install, reverse the removal procedure.

INTERIOR

Instrument Panel

REMOVAL & INSTALLATION

See Figure 27

Dakota, Durango

Note: Before beginning, be sure to turn the steering wheel so that the wheels are in the straight-ahead position.

1. Disconnect and isolate the battery negative cable.

CAUTION:

On air bag-equipped vehicles, wait at least two minutes for the system to discharge before proceeding. Failure to take proper precautions could result in accidental deployment and possible personal injury.

2. Remove the trim from the right and left door sills.
3. Remove the trim from the left and right cowl side inner panels.
4. Remove the steering column opening cover from the instrument panel.
5. Remove the screws that secure the inside hood latch release handle to the instrument panel lower reinforcement and lower the release handle to the floor.
6. Disconnect the driver side airbag module wire harness connector from the instrument panel wire harness at the instrument panel lower reinforcement.
7. Disconnect the overdrive lockout switch wire harness connector, if equipped, from the instrument panel wire harness near the panel lower reinforcement.
8. Remove the steering column from the vehicle complete with airbag, steering wheel and switches. Be sure that the steering wheel is locked and secured from rotation to prevent the loss of clockspring centering.

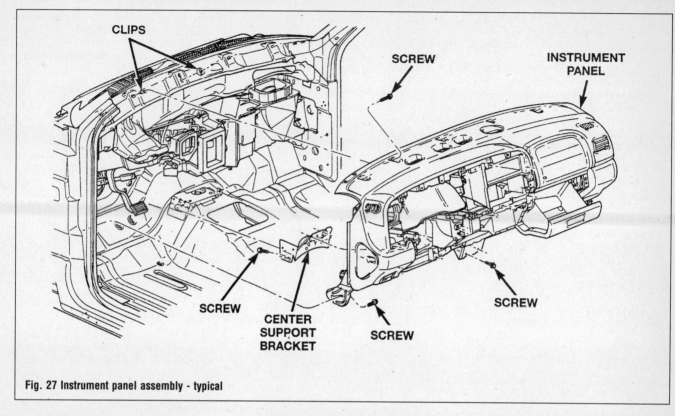

Fig. 27 Instrument panel assembly - typical

9. Remove the screw from the center of the headlamp and dash to instrument panel bulkhead wire harness connector and disconnect the connector.

10. Disconnect the two body wire harness connectors from the two instrument panel wire harness connectors that are secured to the outboard side of the instrument panel bulkhead connector.

11. Disconnect the three wire harness connectors from the three junction block connector receptacles located closest to the dash panel.

12. Unsnap the plastic retainer clip that secures the park brake release linkage rod to the lever on the back side of the park brake release handle and disengage the linkage rod from the lever on the handle.

13. Disconnect the instrument panel wire harness connector from the stop lamp switch connector receptacle.

14. Disconnect the vacuum harness connector located near the left end of the heater-A/C housing.

15. Remove the center support bracket from the instrument panel.

16. Remove the screw that secures the instrument panel wire harness ground eyelets to the left side of the ACM mount on the floor panel transmission tunnel.

17. Disconnect the instrument panel wire harness connector from the ACM connector receptacle.

18. Remove the glove box from the instrument panel.

19. Disconnect the radio antenna coax. Disconnect the cable from the clips.

20. Disconnect the blower motor wire harness near the support brace on the inboard side of the instrument panel glove box opening.

21. Disconnect the sound system wiring, if fitted.

22. Remove the nut that secures the radio ground to the stud.

23. Loosen the right and left instrument panel cowl side roll-down bracket screws about 1/4 in.

24. Remove the five screws that secure the top of the instrument panel to the top of the dash panel, removing the center screw last.

25. Pull the lower instrument panel rearward until the right and left cowl side roll-down bracket screws are in the roll-down slot position of both brackets.

26. Roll down the instrument panel and install a temporary hook in the center hole on top of the instrument panel. Secure the other end of the hook to the center hole in the top of dash panel. The hook should support the instrument panel in its rolled down position about 18 in. from the dash panel.

27. With the instrument panel supported in the roll-down position, disconnect the two instrument panel wire harness connectors from the door jumper wire harness connectors located on a bracket near the right end of the instrument panel.

28. Disconnect the instrument panel wire harness connector from the blower motor resistor connector receptacle on the dash panel.

29. Disconnect the temperature control cable flag retainer from the top of the heater-A/C housing and pull the cable core adjuster clip of the blend-air door lever.

30. Disconnect the demister duct flexible hose from the adapter on the top of the heater-A/C housing.

31. Remove the instrument panel from the vehicle.

To install:

32. Install the instrument panel assembly onto the roll-down bracket screws in the vehicle. Install a temporary hook in the center hole on top of the instrument panel. The hook should support the instrument panel in its rolled down position about 18 in. from the dash panel.

33. With the instrument panel supported in the roll-down position, reconnect the two instrument panel wire harness connectors to the door jumper wire harness connectors located on a bracket near the right end of the instrument panel.

34. Reconnect the instrument panel wire harness connector to the blower motor resistor connector receptacle on the dash panel.

35. Reconnect the temperature control cable flag retainer onto the top of the heater-A/C housing and push the cable core adjuster clip onto the blend-air door lever.

36. Reconnect the demister duct flexible hose to the adapter on the top of the heater-A/C housing.

37. Push the lower instrument panel forward until the right and left cowl side roll-down bracket screws are in the installed slot position of both brackets.

38. Remove the temporary hook from the instrument panel and roll the panel up to the installed position against the dash panel.

39. Install and tighten the screws that secure the top of the instrument panel to the top of the dash panel. Tighten to 28 inch lbs. (3.2 Nm).

40. Tighten the right and left instrument panel cowl side roll-down bracket screws. Tighten the screws to 105 inch lbs. (12 Nm).

41. If fitted, reconnect the two instrument panel wire harness connectors to the speaker amplifier connector receptacles on the right cowl side inner panel.

42. Install and tighten the nut that secures the instrument panel wire harness radio ground eyelet to the stud.

43. Reconnect the two halves of the radio antenna coax.

44. Engage the antenna half of the radio cable into the retainer clip near the outboard side of the lower instrument panel glove box opening.

45. Reconnect the blower motor wire harness connector located near the heater-A/C housing support brace on the inboard side of the instrument panel glove box opening.

46. Install the glove box onto the instrument panel.

47. Install and tighten the screw that secures the

BODY AND TRIM 10-9

instrument panel wire harness ground eyelets to the left side of the ACM mount on the floor panel transmission tunnel. Tighten the screw to 30 inch lbs. (3.4 Nm).

48. Reconnect the instrument panel wire harness connector to the ACM connector receptacle.
49. Install the center support bracket onto the instrument panel.
50. Reconnect the headlamp and dash to the instrument panel bulkhead wire harness connector and tighten the screw in the center of the connector. Tighten to 31 inch lbs. (3.5 Nm).
51. Reconnect the two body wire harness connectors to the two instrument panel wire harness connectors that are secured to the outboard side of the instrument panel bulkhead connector.
52. Reconnect the three wire harness connectors to the junction block connector receptacles located closest to the dash panel.
53. Engage the linkage rod into the lever on the back side of the park brake release handle and snap the plastic retainer clip over the linkage rod that secures it to the lever.
54. Reconnect the instrument panel wire harness connector to the stop lamp switch connector receptacle.
55. Reconnect the vacuum harness connector located near the left end of the heater-A/C housing.
56. Install the steering column into the vehicle. Be sure that the steering wheel was locked and secured from rotation to prevent the loss of clockspring centering.
57. Reconnect the overdrive lockout switch wire harness connector, if fitted, to the instrument panel wire harness near the instrument panel lower reinforcement.
58. Reconnect the driver side airbag module wire harness connector to the instrument panel wire harness at the panel lower reinforcement.
59. Position the inside hood latch release handle to the instrument panel lower reinforcement.
60. Install and tighten the two screws that secure the inside hood latch release handle to the instrument panel lower reinforcement.
61. Install the steering column cover onto the instrument panel.
62. Install the trim onto the left and right cowl side inner panels.
63. Install the trim onto the right and left door sills.
64. Reconnect the battery negative cable.

Ram

Note: Before beginning, be sure to turn the steering wheel so that the wheels are in the straight-ahead position.

1. Disconnect and isolate the battery negative cable(s).

CAUTION:

On air bag-equipped vehicles, wait at least two minutes for the system to discharge before proceeding. Failure to take proper precautions could result in accidental deployment and possible personal injury.

2. Remove the ACM and bracket from the floor panel transmission tunnel.
3. Remove the trim from the left and right cowl side inner panels.
4. Remove the steering column opening cover from the instrument panel.
5. Remove the screws that secure the inside hood latch release handle to the instrument panel lower reinforcement and lower the release handle to the floor.
6. Disconnect the driver side airbag module wire harness connector from the instrument panel wire harness at the instrument panel lower reinforcement.
7. If the vehicle is so equipped, disconnect the overdrive lockout switch wire harness connector from the instrument panel wire harness near the instrument panel lower reinforcement.
8. Remove the steering column from the vehicle, but do not remove the airbag module, steering wheel or switches. Be sure the wheel is locked and secured from rotation to prevent the loss of clockspring centering.
9. Disengage the parking brake release handle linkage rod from the park brake mechanism on the left cowl side inner panel.
10. Disconnect the instrument panel wire harness connector from the park brake switch on the park brake mechanism.
11. Disconnect the wire harness connectors from the junction block connector receptacles located closest to the dash panel.
12. Remove the screw from the center of the headlamp and dash to instrument panel bulkhead wire harness connector and disconnect the connector.
13. Disconnect the instrument panel to the door wire harness connector located directly below the bulkhead wire harness connector.
14. If the vehicle is equipped with the premium sound system, disconnect the wire harness connector from the instrument panel wire harness connector that is secured to the outboard side of the instrument panel bulkhead connector.
15. Disconnect the instrument panel wire harness connector from the stop lamp switch.
16. Disconnect the vacuum harness connector located near the left end of the heater-A/C housing.
17. Disconnect the radio antenna coax.
18. Loosen the right and left instrument panel cowl side roll-down bracket screws about 1/2 in.
19. Remove the screws that secure the top of the instrument panel to the top of the dash panel, removing the center screw last.
20. Roll down the instrument panel and install a temporary hook in the center hole on top of the instrument panel. Secure the other end of the hook to the center hole in the top of the dash panel. The hook should support the instrument panel in its rolled down position about 18 in. from the dash panel.
21. With the instrument panel supported in the roll-down position, disconnect the instrument panel wire harness connectors from the connectors located on the heater-A/C housing.
22. Remove the assembly from the vehicle.

To install:

23. Install the instrument panel assembly onto the roll-down bracket screws in the vehicle. Install a temporary hook in the center hole on top of the instrument panel. Secure the other end of the hook to the center hole in the top of the dash panel. The hook should support the instrument panel in its rolled down position about 18 in. from the dash panel.
24. With the instrument panel supported in the roll-down position, reconnect the instrument panel wire harness connectors to the connectors located on the heater-A/C housing.
25. Remove the temporary hook from the instrument panel and roll the instrument panel up to the installed position against the dash panel.
26. Install and tighten the screws that secure the top of the instrument panel to the top of the dash panel. Tighten to 28 inch lbs. (3.2 Nm).
27. Tighten the right and left instrument panel cowl side roll-down bracket screws. Tighten the screws to 105 inch lbs. (12 Nm).
28. From under the passenger side of the instrument panel, reconnect the radio antenna coax.
29. Engage the park brake release handle linkage rod with the park brake mechanism on the left cowl side inner panel.
30. Reconnect the instrument panel wire harness connector to the park brake switch on the park brake mechanism.
31. Reconnect the wire harness connectors to the junction block connector receptacles located closest to the dash panel.
32. Reconnect the headlamp and dash to instrument panel bulkhead wire harness connector and tighten the screw in the center of the connector to 31 inch lbs. (3.5 Nm).
33. Reconnect the instrument panel to the door wire harness connector located directly below the bulkhead wire harness connector.
34. If the vehicle is equipped with the premium sound system, reconnect the wire harness connector to the instrument panel wire harness connector that is secured to the outboard side of the instrument panel bulkhead connector.
35. Reconnect the instrument panel wire harness connector to the stop lamp switch.
36. Reconnect the vacuum harness connector located near the left end of the heater-A/C housing.
37. Install the steering column. Be sure that the steering wheel was locked and secured from rotation to prevent the loss of clockspring centering.
38. If equipped, reconnect the overdrive lockout switch wire harness connector to the instrument panel wire harness near the instrument panel lower reinforcement.
39. Reconnect the driver side airbag module wire harness connector to the instrument panel wire harness at the instrument panel lower reinforcement.
40. Install and tighten the two screw that secure the inside hood release handle to the instrument panel lower reinforcement.
41. Install the steering column opening cover onto the instrument panel.
42. Install the trim onto the left and right cowl side inner panels.
43. Install the ACM and bracket.
44. Reconnect the battery negative cable(s).

Floor Console

REMOVAL & INSTALLATION

Dakota, Durango
See Figure 28

1. Open the console lid and remove the bolts attaching the console to the floor pan.
2. Lift the cup holder bin mat and remove the bolt attaching the console to the floor pan.
3. Lift the rear of the console and pull the console rearward to separate it from the shift bezel.
4. Remove the console from the vehicle.
5. Installation is the reverse of removal.

10-10 BODY AND TRIM

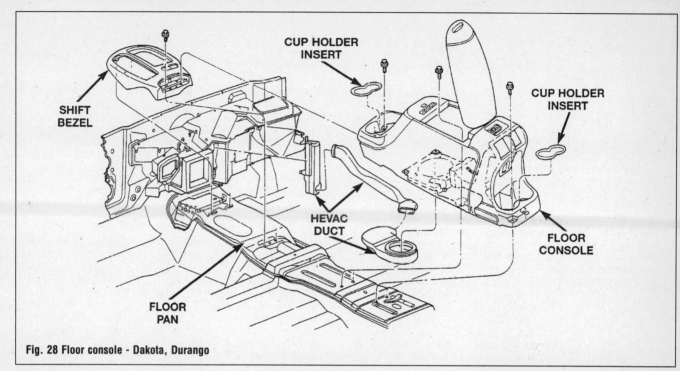

Fig. 28 Floor console - Dakota, Durango

Ram

See Figure 29

1. Using a trim stick, pry the corner of the shift boot up and expose the fasteners.

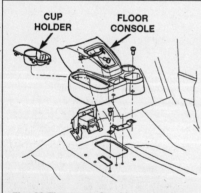

Fig. 29 Floor console with cup holder - Ram

2. Remove the screws attaching the shift boot to the console.
3. Remove the shift knob.
4. Remove the shifter boot.
5. Remove the screws attaching the console to the mounting brackets.
6. Lift the console and disengage the wire harness connector, if fitted.
7. Installation is the reverse of removal.

Door Panels

REMOVAL & INSTALLATION

Dakota, Durango

See Figures 30, 31, 32, 33, 34 and 35

1. Open the door.
2. Roll the window down.
3. Remove the window crank, if fitted.
4. Remove the screws attaching the trim panel to the door.

WARNING:

Do not forcibly pull the trim panel from the door as damage may occur.

5. Simultaneously lift upward and outward to release the retainer steps from the inner door panel.
6. Disengage the inside handle linkage rod from the inside handle.
7. Disconnect the speaker harness wire, if fitted.
8. Disengage the power mirror wire connector, if fitted.
9. Disengage the clips attaching the power window/lock switch panel to the trim panel, if fitted. Disconnect the wiring.
10. Separate the trim panel from the vehicle.
11. Installation is the reverse of removal.

Ram (Front)

See Figures 36, 37, 38, 39, 40, 41 and 42

1. Open the door.
2. Roll down the window.

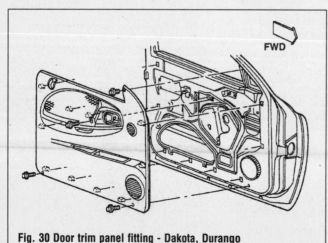

Fig. 30 Door trim panel fitting - Dakota, Durango

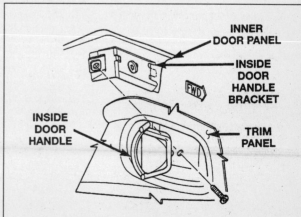

Fig. 31 Trim panel screw adjacent to door handle - Dakota, Durango

BODY AND TRIM 10-11

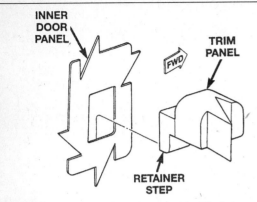

Fig. 32 Trim panel retainer steps must be lifted clear of the cutouts in the inner door panel - Dakota, Durango

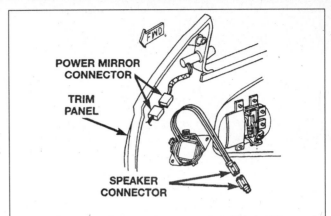

Fig. 33 Speaker and power mirror connector - Dakota, Durango

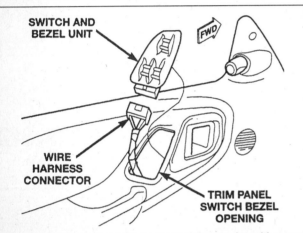

Fig. 34 Switch panel can be pried out of the opening with a small slot head screwdriver

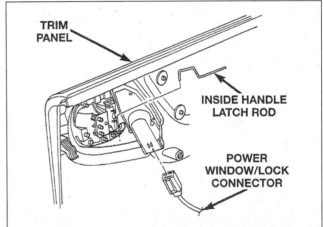

Fig. 35 Latch rod and power system connector - Durango rear door

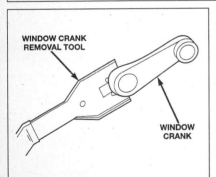

Fig. 36 Removing the window crank with the special tool

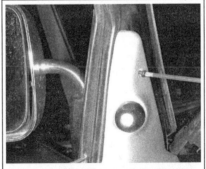

Fig. 37 Remove the screw attaching the trim panel to the outside mirror frame

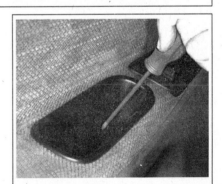

Fig. 38 Removing the pull cup screws

Fig. 39 Pry out the power window/lock switch panel

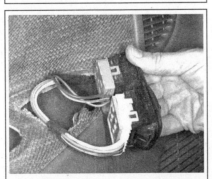

Fig. 40 Disconnect the wiring from the switch panel

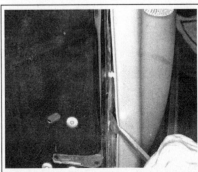

Fig. 41 Prying off the trim panel with a removal tool

10-12 BODY AND TRIM

Fig. 42 Removing the trim panel. Arrow indicates the water dam

clips around the perimeter of the trim panel attaching the panel to the door.

8. While holding the bottom of the trim panel away from the door, lift upward and inboard.
9. Disengage the power mirror switch connector, if fitted.
10. Remove the trim panel from the door.
11. Installation is the reverse of removal.

Ram (Cargo)
See Figure 43

1. Open the cargo door.
2. Remove the screws securing the pull cup to the door.
3. Remove the screw attaching the inside release handle to the door.
4. The trim panel is secured with spring clips and push-in fasteners. Using a trim panel removal tool, remove the push-in fasteners.
5. Pull the trim panel outward to disengage the spring clips.
6. Remove the trim panel.
7. Disengage the cargo door release cable from the inside release handle.
8. Installation is the reverse of removal.

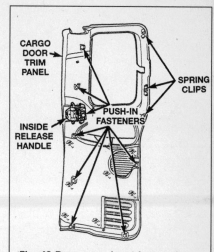

Fig. 43 Ram cargo door trim panel fasteners

3. Remove the window crank, if fitted.
4. Pry out the power window/lock switch panel from the trim panel, if so equipped. Disconnect the wires from the switch panel.
5. Remove the screw attaching the trim panel to the outside mirror frame.
6. Remove the screws attaching the pull cup to the door.
7. With a trim panel removal tool, disengage the

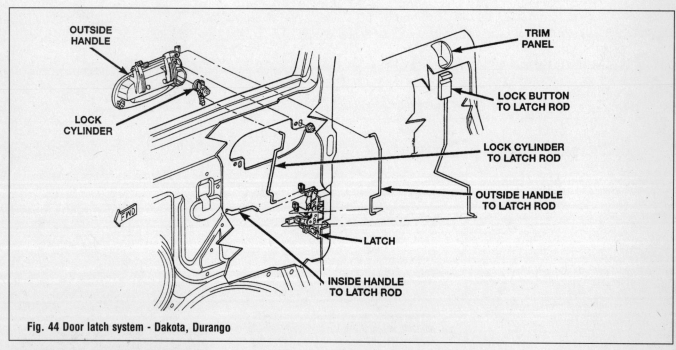

Fig. 44 Door latch system - Dakota, Durango

Door Latches

REMOVAL & INSTALLATION

Dakota, Durango (Front)
See Figures 44 and 45

1. Roll the window up.
2. Remove the door trim panel.
3. Peel back the water dam.
4. Remove the bolts attaching rearward glass run channel to door. Move and secure glass run channel.
5. Remove screws attaching latch to door shut face.
6. Disengage wire harness connector for power door locks, if fitted.
7. Disengage the lock button-to-latch rod from the latch.

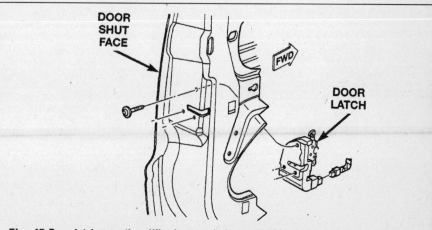

Fig. 45 Door latch mounting. Wire harness is for power door locks

BODY AND TRIM 10-13

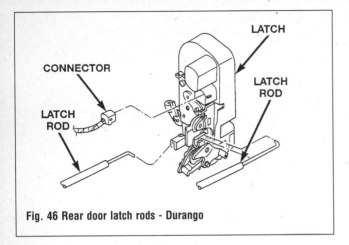

Fig. 46 Rear door latch rods - Durango

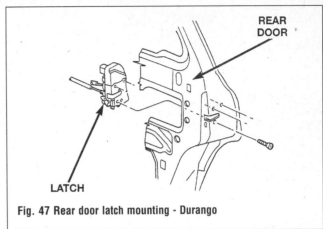

Fig. 47 Rear door latch mounting - Durango

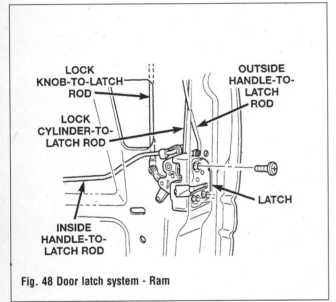

Fig. 48 Door latch system - Ram

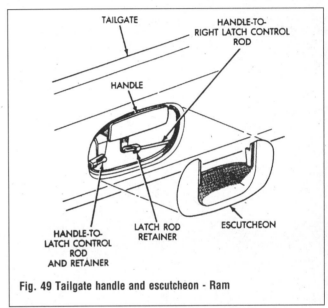

Fig. 49 Tailgate handle and escutcheon - Ram

8. Disengage the lock cylinder-to-latch rod from the latch.
9. Disengage the inside handle-to-latch rod from the latch.
10. Disengage the outside handle-to-latch rod from the latch.
11. Separate the latch from the door.
12. Installation is the reverse of removal.

Durango (Rear)

See Figures 46 and 47

1. Roll up the window.
2. Remove the trim panel.
3. Peel back the water dam.
4. Disconnect the latch rods from the latch.
5. Disconnect the latch harness connector.
6. Remove the screws attaching the latch to the door.
7. Installation is the reverse of removal.

Ram

See Figure 48

1. Roll the window up.
2. Remove the door trim panel.
3. Peel back the water dam.
4. Disengage the clips attaching lock and latch rods to the door latch.

5. Disconnect the power door lock/latch connector, if fitted.
6. Remove the screws attaching the door latch-to-door end panel.
7. Remove the latch from the door.
8. Installation is the reverse of removal.

Tailgate Latch

REMOVAL & INSTALLATION

Ram

See Figures 49, 50, 51 and 52

1. Lift and hold the tailgate latch release handle.
2. Using a trim stick, pry the bottom of the escutcheon outward to disengage the clips.
3. Rotate the escutcheon upward to disengage the clip above the release handle.
4. Push the escutcheon downward from behind to clear handle.
5. Open the tailgate.
6. Disengage the linkage rod from latch handle.
7. Remove the screws attaching the latch to the tailgate.
8. Separate latch from tailgate.
9. Pull the latch and linkage rod from the tailgate.

To install:

10. Position latch and linkage rod in tailgate.
11. Install upper screw attaching latch to tailgate.
12. Install lower screw attaching check cable and latch to tailgate.
13. Engage linkage rod to latch handle.
14. Install escutcheon.

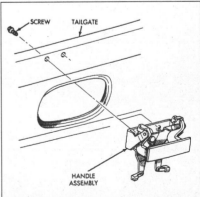

Fig. 50 Tailgate latch handle mounting - Ram

10-14 BODY AND TRIM

Liftgate Latch

REMOVAL & INSTALLATION

See Figure 53

1. Remove the liftgate trim panels.
2. Disconnect the liftgate lamp harness connector.
3. Peel back the liftgate latch watershield.
4. Disconnect the liftgate to outside handle latch rod.
5. Remove the screws attaching the latch to the liftgate and remove the latch.

To install:

6. Reverse the removal procedure. Tighten the screws to 21 ft. lbs. (28 Nm).

Door Locks

REMOVAL & INSTALLATION

Dakota, Durango

See Figure 54

1. Remove the door trim panel.
2. Remove the outside handle.
3. Disengage the lock cylinder-to-latch rod from the lock cylinder.
4. Using a small flat blade, pry the lock cylinder retaining clip from a lock cylinder housing/outside handle.
5. Push the lock cylinder out of the housing/outside handle.

To install:

6. Push the lock cylinder into lock cylinder housing/outside handle. Ensure that the lock cylinder is fully seated in the handle.
7. Install lock cylinder retaining clip. Ensure the clip is fully seated.
8. Engage the lock cylinder to latch rod to the lock cylinder.
9. Install the outside handle and trim panel.

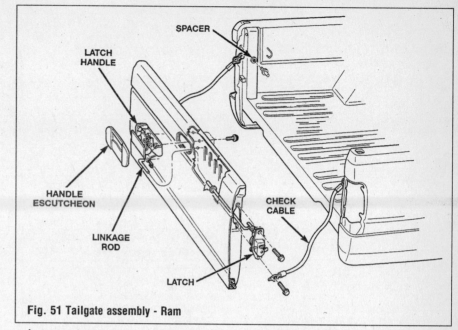

Fig. 51 Tailgate assembly - Ram

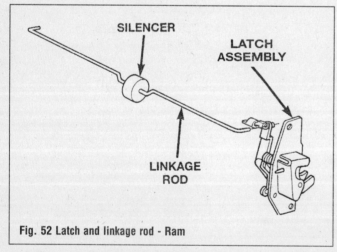

Fig. 52 Latch and linkage rod - Ram

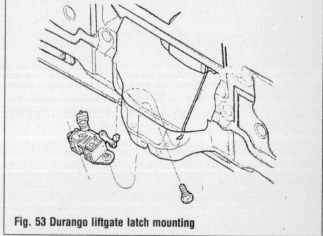

Fig. 53 Durango liftgate latch mounting

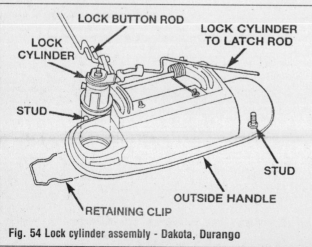

Fig. 54 Lock cylinder assembly - Dakota, Durango

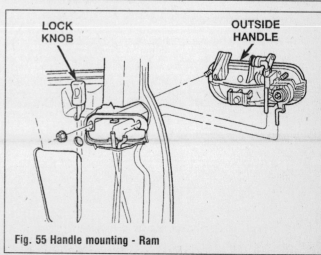

Fig. 55 Handle mounting - Ram

BODY AND TRIM 10-15

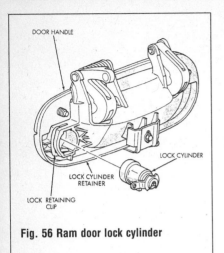

Fig. 56 Ram door lock cylinder

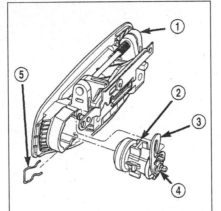

Fig. 57 Liftgate lock cylinder assembly: handle (1), lock cylinder (2), actuator lever (3), C - clip (4), retaining clip (5)

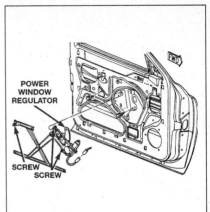

Fig. 58 Power window regulator mounting - Dakota, Durango

Ram

See Figures 55 and 56

1. Remove the outside door handle:
    a. Remove the door trim panel.
    b. Remove the water dam.
    c. Roll up the glass.
    d. Remove the fastener access plug from the door end panel.
    e. Disengage the clips holding the latch and lock rods to the door latch.
    f. Separate the latch and lock rods from the door latch.
    g. Remove the nuts attaching outside door handle to door.
2. Remove the clip securing the lock cylinder to outside door handle.
3. Pull the lock cylinder from the door handle.
4. Installation is the reverse of removal.

Liftgate Lock

REMOVAL & INSTALLATION

See Figure 57

1. Remove the liftgate trim panels.
2. Remove the liftgate handle.
3. Remove the clip retaining lock cylinder in outside handle.
4. Remove the clip retaining actuator link to lock cylinder. Remove the actuator link.
5. Separate the lock cylinder from the handle.
6. Installation is the reverse of removal.

Door Glass and Regulator

REMOVAL & INSTALLATION

Dakota, Durango (Front)

See Figures 58 and 59

1. Remove the door trim panel.
2. Remove the water dam.
3. Remove the inner and outer door belt weatherstrips.
4. Lower the window to the full down position and align the glass regulator arm with the access holes in the inner door panel.
5. Remove the front glass run channel.
6. Remove the nuts attaching the glass channel to regulator arm.
7. Separate the glass and lift it up and out.
8. Disengage the power window motor wire connector from the door harness, if equipped.
9. Loosen the bolts in slotted holes attaching regulator to door inner panel.
10. Remove the bolts attaching window regulator to inner door panel.
11. Remove regulator through access hole in inner door panel.
12. Installation is the reverse of removal. Tighten the nuts attaching the glass to the lift plate to 7 ft. lbs. (9 Nm).

WARNING:

Do not overtighten the nuts. This risks breaking the glass.

Durango (Rear)

See Figures 60 and 61

1. Remove the upper door trim extension panel.
2. Remove the door trim panel.
3. Position the glass to access the fasteners. Remove the fasteners.
4. Disengage the glass from the regulator lift channel.
5. Position the glass up into the door frame. Rest glass on a block of wood on door reinforcement.
6. Lower the water dam.
7. Disconnect the regulator wire harness, if fitted.
8. Remove the screws attaching the window regulator to door inner panel.
9. Remove the regulator.
10. Installation is the reverse of removal.

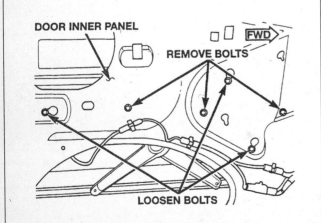

Fig. 59 Window regulator removal - Dakota, Durango

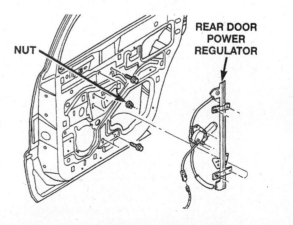

Fig. 60 Rear door power window regulator - Durango

10-16 BODY AND TRIM

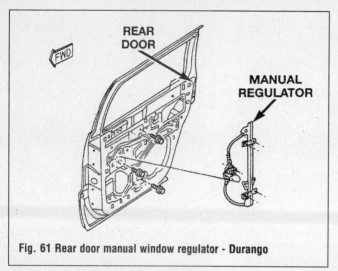

Fig. 61 Rear door manual window regulator - Durango

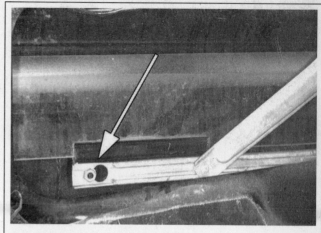

Fig. 62 Two nuts secure the door glass to the lift plate...

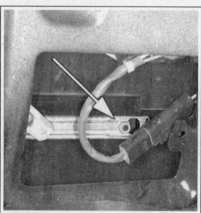

Fig. 63 ...and are accessible through cutouts in the inner door panel

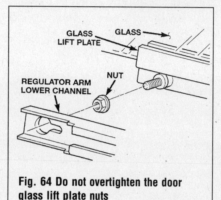

Fig. 64 Do not overtighten the door glass lift plate nuts

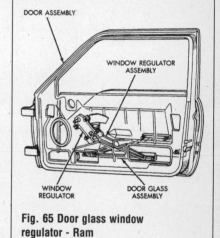

Fig. 65 Door glass window regulator - Ram

Ram
See Figures 62, 63, 64 and 65

1. Remove the door trim panel.
2. Remove the water dam.
3. Remove the inner door belt weatherstrip.
4. Raise or lower the window to access holes in the inner door panel.
5. Loosen the bolts attaching the front lower run channel to the inner door panel.
6. Remove the nuts attaching the door glass to the lift plate.
7. Separate the glass from the lift plate.
8. Lift the glass upward and out of the opening at the top of the door.
9. Disengage the power window motor wire connector from the door harness, if equipped.
10. Remove the bolts attaching the window regulator to the inner door panel.
11. Remove the window regulator through the hole in the inner door panel.
12. Installation is the reverse of removal. Tighten the nuts attaching the glass to the lift plate to 7 ft. lbs. (9 Nm).

WARNING:
Do not overtighten the nuts. This risks breaking the glass.

Electric Window Motor

REMOVAL & INSTALLATION

Power window motors are integrated with the window regulators. A motor failure involves replacing the entire assembly.

TESTING

Motors can be tested by disconnecting the wiring at the connector and applying 12 VDC across the motor connector. The window should move up and down smoothly and powerfully. Replace the motor/regulator assembly if operation is faulty.

Windshield and Fixed Glass

REMOVAL & INSTALLATION

If your windshield, or other fixed window, is cracked or chipped, you may decide to replace it with a new one yourself. However, there are two main reasons why replacement windshields and other window glass should be installed only by a professional automotive glass technician: safety and cost.

The most important reason a professional should install automotive glass is for safety. The glass in the vehicle, especially the windshield, is designed with safety in mind in case of a collision. The windshield is specially manufactured from two panes of specially-tempered glass with a thin layer of transparent plastic between them. This construction allows the glass to "give" in the event that a part of your body hits the windshield during the collision, and prevents the glass from shattering, which could cause lacerations, blinding and other harm to passengers of the vehicle. The other fixed windows are designed to be tempered so that if they break during a collision, they shatter in such a way that there are no large pointed glass pieces. The professional automotive glass technician knows how to install the glass in a vehicle so that it will function optimally during a collision. Without the proper experience, knowledge and tools, installing a piece of automotive glass yourself could lead to additional harm if an accident should ever occur.

Cost is also a factor when deciding to install automotive glass yourself. Performing this could cost you much more than a professional may charge for the same job. Since the windshield is designed to break under stress, an often life saving characteristic, windshields tend to break VERY easily when an inexperienced person attempts to install one. Do-it-yourselfers buying two, three or even four windshields from a salvage yard because they have broken them during installation are common stories. Also, since the automotive glass is designed to prevent the outside elements from entering your vehicle, improper installation can lead to water and air leaks. Annoying whining noises at highway speeds from air leaks or inside body panel rusting from water leaks can add to your stress level and subtract from your wallet. After

BODY AND TRIM 10-17

Fig. 66 Small chips on your windshield can be fixed with an aftermarket repair kit, such as the one from Loctite®

Fig. 67 To repair a chip, clean the windshield with glass cleaner and dry it completely

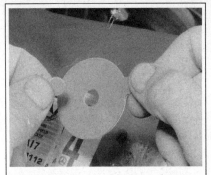

Fig. 68 Remove the center from the adhesive disc and peel off the backing from one side of the disc . . .

Fig. 69 . . . then press it on the windshield so that the chip is centered in the hole

Fig. 70 Be sure that the tab points upward on the windshield

Fig. 71 Peel the backing off the exposed side of the adhesive disc . . .

buying two or three windshields, installing them and ending up with a leak that produces a noise while driving and water damage during rainstorms, the cost of having a professional do it correctly the first time may be much more alluring. We here at Chilton, therefore, advise that you have a professional automotive glass technician service any broken glass on your vehicle.

WINDSHIELD CHIP REPAIR

See Figures 66 through 80

Note: Check with your state and local authorities on the laws for state safety inspection.

Some states or municipalities may not allow chip repair as a viable option for correcting stone damage to your windshield.

Although severely cracked or damaged windshields must be replaced, there is something that you can do to prolong or even prevent the need for replacement of a chipped windshield. There are many companies which offer windshield chip repair products, such as Loctite's® Bullseye™ windshield repair kit. These kits usually consist of a syringe, pedestal and a sealing adhesive. The syringe is mounted on the pedestal and is used to create a vacuum which pulls the plastic layer against the glass. This helps make the chip transparent. The adhesive is then injected which seals the chip and helps to prevent further stress cracks from developing. Refer to the sequence of photos to get a general idea of what windshield chip repair involves.

Note: Always follow the specific manufacturer's instructions.

Inside Rear View Mirror

Note: Breakaway mounts are used with the inside rear view mirror. The breakaway mount is designed to detach from the mirror bracket in the event of an air bag deployment during a collision. Excessive force, up-and-down, or side-to-side movement can cause the mirror to detach from the windshield glass.

Fig. 72 . . . then position the plastic pedestal on the adhesive disc, ensuring that the tabs are aligned

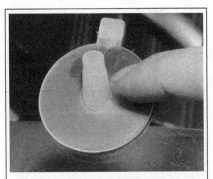

Fig. 73 Press the pedestal firmly on the adhesive disc to create an adequate seal . . .

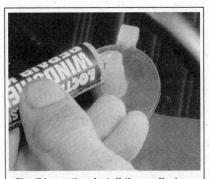

Fig. 74 . . . then install the applicator syringe nipple in the pedestal's hole

10-18 BODY AND TRIM

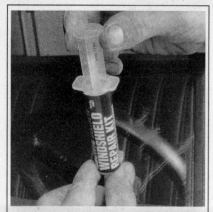
Fig. 75 Hold the syringe with one hand while pulling the plunger back with the other hand

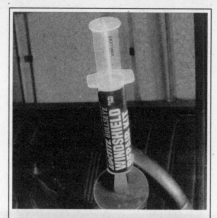
Fig. 76 After applying the solution, allow the entire assembly to sit until it has set completely

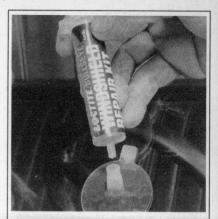
Fig. 77 After the solution has set, remove the syringe from the pedestal . . .

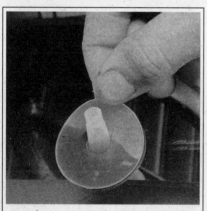
Fig. 78 . . . then peel the pedestal off of the adhesive disc . . .

Fig. 79 . . . and peel the adhesive disc off of the windshield

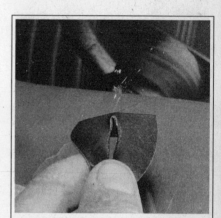
Fig. 80 The chip will still be slightly visible, but it should be filled with the hardened solution

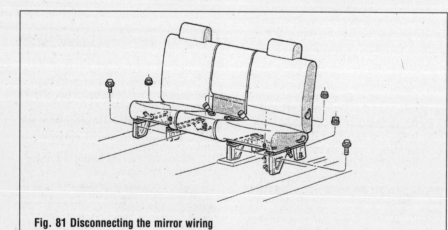
Fig. 81 Disconnecting the mirror wiring

Fig. 82 Remove the setscrew (arrow) to remove the mirror

REPLACEMENT

See Figures 81 and 82

1. Disconnect the mirror harness wiring, if so equipped.
2. Loosen the mirror base setscrew. Slide the mirror up and off the bracket.
3. Installation is the reverse of removal. Tighten the setscrew to 9 inch lbs. (1 Nm).

REAR VIEW MIRROR SUPPORT BRACKET

1. Disconnect the mirror wiring, if so equipped.
2. Loosen the mirror assembly setscrew.
3. Remove the mirror assembly by sliding it upward and away from the mounting bracket.
4. If the bracket mounting pad remains on the windshield, apply low heat from an electric heat gun until the glue softens. Peel the mounting pad off the windshield and discard.
5. Mark the mirror mounting bracket location on the outside surface of the windshield with a wax pencil.
6. Clean the bracket contact area on the glass. Use a mild powdered cleanser on a cloth saturated with rubbing alcohol. Finally, clean the glass with a paper towel dampened with the alcohol.
7. Sand the surface on a support bracket with fine grit sandpaper. Wipe the bracket surface clean with a paper towel.
8. Crush the accelerator vial to saturate the felt applicator.
9. Remove the paper sleeve.

BODY AND TRIM 10-19

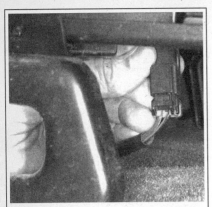

Fig. 83 Disconnect the wiring from under the seat

Fig. 84 Seat track mounting hardware locations - typical

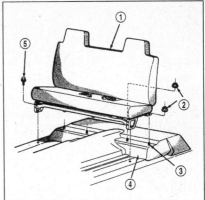

Fig. 85 Ram bench seat components: seat (1), nut (2), stud (3), floor pan (4), bolt (5)

Fig. 86 Split bench seat assembly - Ram

10. Apply accelerator to the contact surface on the bracket.

11. Allow the accelerator to dry for five minutes. Do not touch the bracket contact surface after the accelerator has been applied.

12. Apply adhesive accelerator to the bracket contact surface on the windshield glass. Allow it to dry for one minute. Do not touch the glass contact surface after the accelerator has been applied.

13. Apply one drop of adhesive at the center of the bracket contact surface on the windshield glass.

14. Apply an even coat of adhesive to the contact surface on the bracket.

15. Align the bracket with the marked position on the windshield glass.

16. Press and hold the bracket in place for at least one minute.

Note: Verify that the mirror support bracket is correctly aligned because the adhesive will cure rapidly.

17. Allow the adhesive to cure for about ten minutes. Remove any excess adhesive with an alcohol-dampened cloth.

18. Allow the adhesive to cure for an additional ten minutes before installing the mirror.

Seats

REMOVAL & INSTALLATION

See Figures 83, 84, 85, 86, 87 and 88

1. Clamp the seat belts to keep them from retracting.

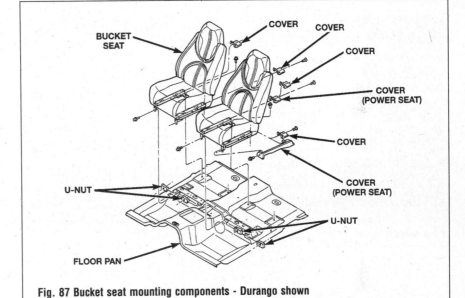

Fig. 87 Bucket seat mounting components - Durango shown

2. Locate and disconnect seat wiring at the connector(s).
3. Move front seats forward or back for access to the hold-down hardware.
4. Put rear seats in the stowed position.
5. Durango: on 3rd seats, lift cushion upward and forward to floor cargo position.
6. Remove the hardware securing the seat.
7. On split bench models, lift the center seat upward to clear attachment stud
8. Remove the seat from the vehicle.

Note: Do not activate the track adjusters once the fastening hardware is removed.

9. Installation is the reverse of removal.

Power Seat Motors

REMOVAL & INSTALLATION

1. Disconnect the battery negative cable(s).
2. Remove the seat assembly from the vehicle.
3. Unplug the power seat wire harness connectors at each of the motors.
4. Release the wire harness retainers from the seat adjuster and motors assembly.
5. Remove the fasteners that secure the center seat cushion section to the brackets on the power seat adjuster, if so equipped.
6. Remove the screws that secure the power seat adjuster and motors assembly to the seat cushion frame.
7. Remove the power seat adjuster and motors assembly from the cushion frame.
8. Installation is the reverse of removal.

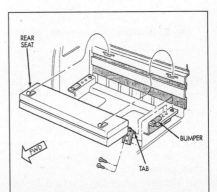

Fig. 88 Rear seat removal/installation

Torque Specifications (Ram)

Components	English	Metric
Bench seat front anchor bolts	40 ft. lbs.	54 Nm
Bench seat rear inboard anchor nuts	30 ft. lbs.	40 Nm
Bench seat rear outboard anchor nuts	40 ft. lbs.	54 Nm
Bench seat, seat track to frame bolts	18 ft. lbs.	25 Nm
Cargo door to hinge bolts	21 ft. lbs.	28 Nm
Cargo door hinge to C-pillar bolt	21 ft. lbs.	28 Nm
Front door hinge to A-pillar bolts	21 ft. lbs.	28 Nm
Front door striker	21 ft. lbs.	28 Nm
Glass to lift plate nuts	7 ft. lbs.	9 Nm
Rear seat to floorpan	21 ft. lbs.	28 Nm
Rear seatbelt anchor nut	30 ft. lbs.	40 Nm
Split bench seat front anchor bolt	40 ft. lbs.	54 Nm
Split bench seat rear anchor nuts	40 ft. lbs.	54 Nm
Split bench seat track to frame bolt	18 ft. lbs.	25 Nm
Cargo box bolts	40 ft. lbs.	54 Nm
Front shoulder belt anchor bolts (conv cab)	28 ft. lbs.	39 Nm
Front belt buckle inboard anchor nut (conv cab)	33 ft. lbs.	45 Nm

Torque Specifications (Dakota)

Components	English	Metric
Bench seat track to seat frame bolts	17 ft. lbs.	24 Nm
Bench seat outer track to seat frame bolts	17 ft. lbs.	24 Nm
Bench seat to floor pan front bolts	30 ft. lbs.	40 Nm
Bench seat to floor pan rear bolts	20 ft. lbs.	28 Nm
Bucket seat track to seat frame bolts	17 ft. lbs.	24 Nm
Bucket seat track to floor pan front bolts	20 ft. lbs.	28 Nm
Bucket seat track to floor pan rear inboard bolt	30 ft. lbs.	40 Nm
Bucket seat track to floor pan rear outboard bolt	20 ft. lbs.	28 Nm
Cab mounting bolts	60 ft. lbs.	81 Nm
Cargo box bolts	20 ft. lbs.	27 Nm
Center seat to bucket seat inboard track bolts	17 ft. lbs.	24 Nm
Front bucket seat belt buckle anchor bolts	29 ft. lbs.	40 Nm
Front door hinge to pillar bolts	21 ft. lbs.	28 Nm
Front door hinge to door nuts and bolts	21 ft. lbs.	28 Nm
Front door striker	21 ft. lbs.	28 Nm
Front seat belt retractor bolt	32 ft. lbs.	44 Nm
Front lower belt anchor bolt	32 ft. lbs.	44 Nm
Front seat rear inboard seat track to floor pan bolts	30 ft. lbs.	40 Nm
Front seat rear outboard seat track to floor pan bolts	11 ft. lbs.	16 Nm
Front seat front seat track to floor pan bolts	11 ft. lbs.	16 Nm
Rear seat belt retractor bolt	32 ft. lbs.	44 Nm
Rear lower belt anchor bolts	32 ft. lbs.	44 Nm
Rear seat belt/buckle anchor bolts	32 ft. lbs.	44 Nm
Glass to lift plate nuts	7 ft. lbs.	9 Nm

Torque Specifications (Durango)

Components	English	Metric
Center seat to seat track bolt	17 ft. lbs.	24 Nm
Front seat belt anchor nut	29 ft. lbs.	40 Nm
Front seat belt retractor bolt	28 ft. lbs.	38 Nm
Front seat belt buckle anchor bolt	30 ft. lbs.	40 Nm
Front seat rear inboard seat track to floor pan	40 ft. lbs.	54 Nm
Front seat rear outboard seat track to floor pan	20 ft. lbs	28 Nm
Front seat front seat track to floor pan	20 ft. lbs	28 Nm
2nd row seat back hinge bolts	19 ft. lbs.	27 Nm
2nd row center seat back to seat cushion	19 ft. lbs.	27 Nm
2nd row seat belt/buckle anchor bolt	70 ft. lbs.	95 Nm
2nd row inboard seat anchor bolt	70 ft. lbs.	95 Nm
2nd row outboard seat anchor bolt	20 ft. lbs	28 Nm
3rd row seat belt anchor bolt	28 ft. lbs.	38 Nm
3rd row seat belt/buckle anchor bolt	70 ft. lbs.	95 Nm
Liftgate latch striker	16 ft. lbs.	22 Nm
Front door hinge to pillar bolts	20 ft. lbs.	28 Nm
Front door hinge to door hardware	20 ft. lbs.	28 Nm
Front door striker	20 ft. lbs.	28 Nm
Rear door hinge to pillar bolt	20 ft. lbs.	28 Nm
rear door hinge to door bolt	20 ft. lbs.	28 Nm
Glass to lift plate nuts	7 ft. lbs.	9 Nm

GLOSSARY

AIR/FUEL RATIO: The ratio of air-to-gasoline by weight in the fuel mixture drawn into the engine.

AIR INJECTION: One method of reducing harmful exhaust emissions by injecting air into each of the exhaust ports of an engine. The fresh air entering the hot exhaust manifold causes any remaining fuel to be burned before it can exit the tailpipe.

ALTERNATOR: A device used for converting mechanical energy into electrical energy.

AMMETER: An instrument, calibrated in amperes, used to measure the flow of an electrical current in a circuit. Ammeters are always connected in series with the circuit being tested.

AMPERE: The rate of flow of electrical current present when one volt of electrical pressure is applied against one ohm of electrical resistance.

ANALOG COMPUTER: Any microprocessor that uses similar (analogous) electrical signals to make its calculations.

ARMATURE: A laminated, soft iron core wrapped by a wire that converts electrical energy to mechanical energy as in a motor or relay. When rotated in a magnetic field, it changes mechanical energy into electrical energy as in a generator.

ATMOSPHERIC PRESSURE: The pressure on the Earth's surface caused by the weight of the air in the atmosphere. At sea level, this pressure is 14.7 psi at 32°F (101 kPa at 0°C).

ATOMIZATION: The breaking down of a liquid into a fine mist that can be suspended in air.

AXIAL PLAY: Movement parallel to a shaft or bearing bore.

BACKFIRE: The sudden combustion of gases in the intake or exhaust system that results in a loud explosion.

BACKLASH: The clearance or play between two parts, such as meshed gears.

BACKPRESSURE: Restrictions in the exhaust system that slow the exit of exhaust gases from the combustion chamber.

BAKELITE: A heat resistant, plastic insulator material commonly used in printed circuit boards and transistorized components.

BALL BEARING: A bearing made up of hardened inner and outer races between which hardened steel balls roll.

BALLAST RESISTOR: A resistor in the primary ignition circuit that lowers voltage after the engine is started to reduce wear on ignition components.

BEARING: A friction reducing, supportive device usually located between a stationary part and a moving part.

BIMETAL TEMPERATURE SENSOR: Any sensor or switch made of two dissimilar types of metal that bend when heated or cooled due to the different expansion rates of the alloys. These types of sensors usually function as an on/off switch.

BLOWBY: Combustion gases, composed of water vapor and unburned fuel, that leak past the piston rings into the crankcase during normal engine operation. These gases are removed by the PCV system to prevent the buildup of harmful acids in the crankcase.

BRAKE PAD: A brake shoe and lining assembly used with disc brakes.

BRAKE SHOE: The backing for the brake lining. The term is, however, usually applied to the assembly of the brake backing and lining.

BUSHING: A liner, usually removable, for a bearing; an anti-friction liner used in place of a bearing.

CALIPER: A hydraulically activated device in a disc brake system, which is mounted straddling the brake rotor (disc). The caliper contains at least one piston and two brake pads. Hydraulic pressure on the piston(s) forces the pads against the rotor.

CAMSHAFT: A shaft in the engine on which are the lobes (cams) which operate the valves. The camshaft is driven by the crankshaft, via a belt, chain or gears, at one half the crankshaft speed.

CAPACITOR: A device which stores an electrical charge.

CARBON MONOXIDE (CO): A colorless, odorless gas given off as a normal byproduct of combustion. It is poisonous and extremely dangerous in confined areas, building up slowly to toxic levels without warning if adequate ventilation is not available.

CARBURETOR: A device, usually mounted on the intake manifold of an engine, which mixes the air and fuel in the proper proportion to allow even combustion.

CATALYTIC CONVERTER: A device installed in the exhaust system, like a muffler, that converts harmful byproducts of combustion into carbon dioxide and water vapor by means of a heat-producing chemical reaction.

CENTRIFUGAL ADVANCE: A mechanical method of advancing the spark timing by using flyweights in the distributor that react to centrifugal force generated by the distributor shaft rotation.

CHECK VALVE: Any one-way valve installed to permit the flow of air, fuel or vacuum in one direction only.

CHOKE: A device, usually a moveable valve, placed in the intake path of a carburetor to restrict the flow of air.

CIRCUIT: Any unbroken path through which an electrical current can flow. Also used to describe fuel flow in some instances.

CIRCUIT BREAKER: A switch which protects an electrical circuit from overload by opening the circuit when the current flow exceeds a predetermined level. Some circuit breakers must be reset manually, while most reset automatically.

COIL (IGNITION): A transformer in the ignition circuit which steps up the voltage provided to the spark plugs.

COMBINATION MANIFOLD: An assembly which includes both the intake and exhaust manifolds in one casting.

GL-2 GLOSSARY

COMBINATION VALVE: A device used in some fuel systems that routes fuel vapors to a charcoal storage canister instead of venting them into the atmosphere. The valve relieves fuel tank pressure and allows fresh air into the tank as the fuel level drops to prevent a vapor lock situation.

COMPRESSION RATIO: The comparison of the total volume of the cylinder and combustion chamber with the piston at BDC and the piston at TDC.

CONDENSER: 1. An electrical device which acts to store an electrical charge, preventing voltage surges. 2. A radiator-like device in the air conditioning system in which refrigerant gas condenses into a liquid, giving off heat.

CONDUCTOR: Any material through which an electrical current can be transmitted easily.

CONTINUITY: Continuous or complete circuit. Can be checked with an ohmmeter.

COUNTERSHAFT: An intermediate shaft which is rotated by a mainshaft and transmits, in turn, that rotation to a working part.

CRANKCASE: The lower part of an engine in which the crankshaft and related parts operate.

CRANKSHAFT: The main driving shaft of an engine which receives reciprocating motion from the pistons and converts it to rotary motion.

CYLINDER: In an engine, the round hole in the engine block in which the piston(s) ride.

CYLINDER BLOCK: The main structural member of an engine in which is found the cylinders, crankshaft and other principal parts.

CYLINDER HEAD: The detachable portion of the engine, usually fastened to the top of the cylinder block and containing all or most of the combustion chambers. On overhead valve engines, it contains the valves and their operating parts. On overhead cam engines, it contains the camshaft as well.

DEAD CENTER: The extreme top or bottom of the piston stroke.

DETONATION: An unwanted explosion of the air/fuel mixture in the combustion chamber caused by excess heat and compression, advanced timing, or an overly lean mixture. Also referred to as "ping".

DIAPHRAGM: A thin, flexible wall separating two cavities, such as in a vacuum advance unit.

DIESELING: A condition in which hot spots in the combustion chamber cause the engine to run on after the key is turned off.

DIFFERENTIAL: A geared assembly which allows the transmission of motion between drive axles, giving one axle the ability to turn faster than the other.

DIODE: An electrical device that will allow current to flow in one direction only.

DISC BRAKE: A hydraulic braking assembly consisting of a brake disc, or rotor, mounted on an axle, and a caliper assembly containing, usually two brake pads which are activated by hydraulic pressure. The pads are forced against the sides of the disc, creating friction which slows the vehicle.

DISTRIBUTOR: A mechanically driven device on an engine which is responsible for electrically firing the spark plug at a predetermined point of the piston stroke.

DOWEL PIN: A pin, inserted in mating holes in two different parts allowing those parts to maintain a fixed relationship.

DRUM BRAKE: A braking system which consists of two brake shoes and one or two wheel cylinders, mounted on a fixed backing plate, and a brake drum, mounted on an axle, which revolves around the assembly.

DWELL: The rate, measured in degrees of shaft rotation, at which an electrical circuit cycles on and off.

ELECTRONIC CONTROL UNIT (ECU): Ignition module, module, amplifier or igniter. See Module for definition.

ELECTRONIC IGNITION: A system in which the timing and firing of the spark plugs is controlled by an electronic control unit, usually called a module. These systems have no points or condenser.

END-PLAY: The measured amount of axial movement in a shaft.

ENGINE: A device that converts heat into mechanical energy.

EXHAUST MANIFOLD: A set of cast passages or pipes which conduct exhaust gases from the engine.

FEELER GAUGE: A blade, usually metal, or precisely predetermined thickness, used to measure the clearance between two parts.

FIRING ORDER: The order in which combustion occurs in the cylinders of an engine. Also the order in which spark is distributed to the plugs by the distributor.

FLOODING: The presence of too much fuel in the intake manifold and combustion chamber which prevents the air/fuel mixture from firing, thereby causing a no-start situation.

FLYWHEEL: A disc shaped part bolted to the rear end of the crankshaft. Around the outer perimeter is affixed the ring gear. The starter drive engages the ring gear, turning the flywheel, which rotates the crankshaft, imparting the initial starting motion to the engine.

FOOT POUND (ft. lbs. or sometimes, ft.lb.): The amount of energy or work needed to raise an item weighing one pound, a distance of one foot.

FUSE: A protective device in a circuit which prevents circuit overload by breaking the circuit when a specific amperage is present. The device is constructed around a strip or wire of a lower amperage rating than the circuit it is designed to protect. When an amperage higher than that stamped on the fuse is present in the circuit, the strip or wire melts, opening the circuit.

GEAR RATIO: The ratio between the number of teeth on meshing gears.

GENERATOR: A device which converts mechanical energy into electrical energy.

HEAT RANGE: The measure of a spark plug's ability to dissipate heat from its firing end. The higher the heat range, the hotter the plug fires.

GLOSSARY GL-3

HUB: The center part of a wheel or gear.

HYDROCARBON (HC): Any chemical compound made up of hydrogen and carbon. A major pollutant formed by the engine as a byproduct of combustion.

HYDROMETER: An instrument used to measure the specific gravity of a solution.

INCH POUND (inch lbs.; sometimes in.lb. or in. lbs.): One twelfth of a foot pound.

INDUCTION: A means of transferring electrical energy in the form of a magnetic field. Principle used in the ignition coil to increase voltage.

INJECTOR: A device which receives metered fuel under relatively low pressure and is activated to inject the fuel into the engine under relatively high pressure at a predetermined time.

INPUT SHAFT: The shaft to which torque is applied, usually carrying the driving gear or gears.

INTAKE MANIFOLD: A casting of passages or pipes used to conduct air or a fuel/air mixture to the cylinders.

JOURNAL: The bearing surface within which a shaft operates.

KEY: A small block usually fitted in a notch between a shaft and a hub to prevent slippage of the two parts.

MANIFOLD: A casting of passages or set of pipes which connect the cylinders to an inlet or outlet source.

MANIFOLD VACUUM: Low pressure in an engine intake manifold formed just below the throttle plates. Manifold vacuum is highest at idle and drops under acceleration.

MASTER CYLINDER: The primary fluid pressurizing device in a hydraulic system. In automotive use, it is found in brake and hydraulic clutch systems and is pedal activated, either directly or, in a power brake system, through the power booster.

MODULE: Electronic control unit, amplifier or igniter of solid state or integrated design which controls the current flow in the ignition primary circuit based on input from the pick-up coil. When the module opens the primary circuit, high secondary voltage is induced in the coil.

NEEDLE BEARING: A bearing which consists of a number (usually a large number) of long, thin rollers.

OHM: (Ω) The unit used to measure the resistance of conductor-to-electrical flow. One ohm is the amount of resistance that limits current flow to one ampere in a circuit with one volt of pressure.

OHMMETER: An instrument used for measuring the resistance, in ohms, in an electrical circuit.

OUTPUT SHAFT: The shaft which transmits torque from a device, such as a transmission.

OVERDRIVE: A gear assembly which produces more shaft revolutions than that transmitted to it.

OVERHEAD CAMSHAFT (OHC): An engine configuration in which the camshaft is mounted on top of the cylinder head and operates the valve either directly or by means of rocker arms.

OVERHEAD VALVE (OHV): An engine configuration in which all of the valves are located in the cylinder head and the camshaft is located in the cylinder block. The camshaft operates the valves via lifters and pushrods.

OXIDES OF NITROGEN (NOx): Chemical compounds of nitrogen produced as a byproduct of combustion. They combine with hydrocarbons to produce smog.

OXYGEN SENSOR: Use with the feedback system to sense the presence of oxygen in the exhaust gas and signal the computer which can reference the voltage signal to an air/fuel ratio.

PINION: The smaller of two meshing gears.

PISTON RING: An open-ended ring with fits into a groove on the outer diameter of the piston. Its chief function is to form a seal between the piston and cylinder wall. Most automotive pistons have three rings: two for compression sealing; one for oil sealing.

PRELOAD: A predetermined load placed on a bearing during assembly or by adjustment.

PRIMARY CIRCUIT: the low voltage side of the ignition system which consists of the ignition switch, ballast resistor or resistance wire, bypass, coil, electronic control unit and pick-up coil as well as the connecting wires and harnesses.

PRESS FIT: The mating of two parts under pressure, due to the inner diameter of one being smaller than the outer diameter of the other, or vice versa; an interference fit.

RACE: The surface on the inner or outer ring of a bearing on which the balls, needles or rollers move.

REGULATOR: A device which maintains the amperage and/or voltage levels of a circuit at predetermined values.

RELAY: A switch which automatically opens and/or closes a circuit.

RESISTANCE: The opposition to the flow of current through a circuit or electrical device, and is measured in ohms. Resistance is equal to the voltage divided by the amperage.

RESISTOR: A device, usually made of wire, which offers a preset amount of resistance in an electrical circuit.

RING GEAR: The name given to a ring-shaped gear attached to a differential case, or affixed to a flywheel or as part of a planetary gear set.

ROLLER BEARING: A bearing made up of hardened inner and outer races between which hardened steel rollers move.

ROTOR: 1. The disc-shaped part of a disc brake assembly, upon which the brake pads bear; also called, brake disc. 2. The device mounted atop the distributor shaft, which passes current to the distributor cap tower contacts.

GL-4 GLOSSARY

SECONDARY CIRCUIT: The high voltage side of the ignition system, usually above 20,000 volts. The secondary includes the ignition coil, coil wire, distributor cap and rotor, spark plug wires and spark plugs.

SENDING UNIT: A mechanical, electrical, hydraulic or electro-magnetic device which transmits information to a gauge.

SENSOR: Any device designed to measure engine operating conditions or ambient pressures and temperatures. Usually electronic in nature and designed to send a voltage signal to an on-board computer, some sensors may operate as a simple on/off switch or they may provide a variable voltage signal (like a potentiometer) as conditions or measured parameters change.

SHIM: Spacers of precise, predetermined thickness used between parts to establish a proper working relationship.

SLAVE CYLINDER: In automotive use, a device in the hydraulic clutch system which is activated by hydraulic force, disengaging the clutch.

SOLENOID: A coil used to produce a magnetic field, the effect of which is to produce work.

SPARK PLUG: A device screwed into the combustion chamber of a spark ignition engine. The basic construction is a conductive core inside of a ceramic insulator, mounted in an outer conductive base. An electrical charge from the spark plug wire travels along the conductive core and jumps a preset air gap to a grounding point or points at the end of the conductive base. The resultant spark ignites the fuel/air mixture in the combustion chamber.

SPLINES: Ridges machined or cast onto the outer diameter of a shaft or inner diameter of a bore to enable parts to mate without rotation.

TACHOMETER: A device used to measure the rotary speed of an engine, shaft, gear, etc., usually in rotations per minute.

THERMOSTAT: A valve, located in the cooling system of an engine, which is closed when cold and opens gradually in response to engine heating, controlling the temperature of the coolant and rate of coolant flow.

TOP DEAD CENTER (TDC): The point at which the piston reaches the top of its travel on the compression stroke.

TORQUE: The twisting force applied to an object.

TORQUE CONVERTER: A turbine used to transmit power from a driving member to a driven member via hydraulic action, providing changes in drive ratio and torque. In automotive use, it links the driveplate at the rear of the engine to the automatic transmission.

TRANSDUCER: A device used to change a force into an electrical signal.

TRANSISTOR: A semi-conductor component which can be actuated by a small voltage to perform an electrical switching function.

TUNE-UP: A regular maintenance function, usually associated with the replacement and adjustment of parts and components in the electrical and fuel systems of a vehicle for the purpose of attaining optimum performance.

TURBOCHARGER: An exhaust driven pump which compresses intake air and forces it into the combustion chambers at higher than atmospheric pressures. The increased air pressure allows more fuel to be burned and results in increased horsepower being produced.

VACUUM ADVANCE: A device which advances the ignition timing in response to increased engine vacuum.

VACUUM GAUGE: An instrument used to measure the presence of vacuum in a chamber.

VALVE: A device which control the pressure, direction of flow or rate of flow of a liquid or gas.

VALVE CLEARANCE: The measured gap between the end of the valve stem and the rocker arm, cam lobe or follower that activates the valve.

VISCOSITY: The rating of a liquid's internal resistance to flow.

VOLTMETER: An instrument used for measuring electrical force in units called volts. Voltmeters are always connected parallel with the circuit being tested.

WHEEL CYLINDER: Found in the automotive drum brake assembly, it is a device, actuated by hydraulic pressure, which, through internal pistons, pushes the brake shoes outward against the drums.

AIR BAG (SUPPLEMENTAL RESTRAINT SYSTEM)
GENERAL INFORMATION 6-7
   ARMING THE SYSTEM 6-8
   DEPLOYED MODULE 6-8
   DISARMING THE SYSTEM 6-8
   SERVICE PRECAUTIONS 6-8
   SYSTEM COMPONENTS 6-7
   SYSTEM OPERATION 6-7
GENERAL INFORMATION HANDLING A LIVE MODULE 6-8

AIR POLLUTION
AUTOMOTIVE POLLUTANTS
   HEAT TRANSFER 4-2
   TEMPERATURE INVERSION 4-2
INDUSTRIAL POLLUTANTS 4-2
NATURAL POLLUTANTS 4-2

AUTOMOTIVE EMISSIONS
CRANKCASE EMISSIONS 4-4
EVAPORATIVE EMISSIONS 4-4
EXHAUST GASES
   CARBON MONOXIDE 4-3
   HYDROCARBONS 4-3
   NITROGEN 4-3
   OXIDES OF SULFUR 4-3
   PARTICULATE MATTER 4-3

AUTOMATIC TRANSMISSION 7-8
AUTOMATIC TRANSMISSION ASSEMBLY
   ADJUSTMENTS 7-11
   REMOVAL & INSTALLATION 7-10
EXTENSION HOUSING SEAL
   REMOVAL & INSTALLATION 7-10
NEUTRAL SAFETY SWITCH
   REMOVAL & INSTALLATION 7-9
   TESTING 7-9
UNDERSTANDING THE AUTOMATIC TRANSMISSION
   HYDRAULIC CONTROL SYSTEM 7-9
   PLANETARY GEARBOX 7-8
   SERVOS AND ACCUMULATORS 7-9
   TORQUE CONVERTER 7-8

BATTERY CABLES
DISCONNECTING THE CABLES 6-7

BASIC FUEL SYSTEM DIAGNOSIS
PRECAUTIONS 5-2

BRAKE OPERATING SYSTEM
BASIC OPERATING PRINCIPLES 9-2
   DISC BRAKES 9-2
   DRUM BRAKES 9-2
   POWER BOOSTERS 9-2
BLEEDING THE BRAKE SYSTEM 9-10
BRAKE HOSES AND LINES 9-10
   REMOVAL & INSTALLATION 9-10
BRAKE LIGHT SWITCH 9-6
   ADJUSTMENT 9-6
   REMOVAL & INSTALLATION 9-6

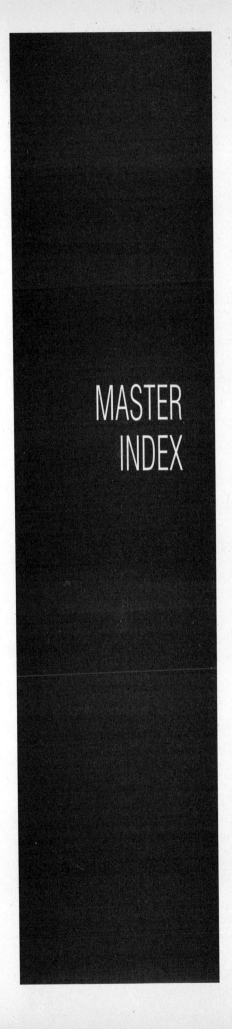

MASTER INDEX

IND-2 MASTER INDEX

COMBINATION VALVE 9-9
   REMOVAL & INSTALLATION 9-9
HYDRAULIC BRAKE BOOSTER 9-8
   REMOVAL & INSTALLATION 9-8
MASTER CYLINDER 9-6
   BENCH BLEEDING 9-7
   REMOVAL & INSTALLATION 9-7
VACUUM BRAKE BOOSTER 9-8
   REMOVAL & INSTALLATION 9-8
   TESTING 9-8
CAPACITIES 1-54
CHARGING SYSTEM
ALTERNATOR 2-10
ALTERNATOR SYSTEM
   DIAGNOSTIC TROUBLE CODES 2-10
   REMOVAL & INSTALLATION 2-11
   TESTING 2-10
ELECTRONIC VOLTAGE REGULATOR 2-12
GENERAL INFORMATION 2-9
PRECAUTIONS 2-9
CLEARING CODES
CONTINUOUS MEMORY CODES 4-26
CIRCUIT PROTECTION
FUSES 6-26
   JUNCTION BLOCK 6-26
   POWER DISTRIBUTION CENTER 6-26
   REPLACEMENT 6-26
CLUTCH
CLUTCH HYDRAULICS
   HYDRAULIC SYSTEM SERVICE 7-8
   REMOVAL & INSTALLATION 7-7
DRIVEN DISC AND PRESSURE PLATE
   REMOVAL & INSTALLATION 7-4
DRIVEN DISC AND PRESSURE PLATE ADJUSTMENTS 7-7
PILOT BEARING
   REMOVAL & INSTALLATION 7-7
RELEASE BEARING
   REMOVAL & INSTALLATION 7-7
UNDERSTANDING THE CLUTCH 7-4
CRUISE CONTROL
GENERAL INFORMATION 6-11
DIESEL FUEL SYSTEM
DIESEL INJECTION PUMP 5-14
DIESEL INJECTION PUMP INJECTION PUMP TIMING 5-16
DIESEL INJECTION PUMP REMOVAL & INSTALLATION 5-14
FUEL TRANSFER PUMP
   LOW PRESSURE BLEEDING 5-13
   REMOVAL & INSTALLATION 5-12
FUEL/WATER SEPARATOR FILTER
   REMOVAL & INSTALLATION 5-16
FUEL/WATER SEPARATOR FILTER DRAINING WATER FROM
   THE SYSTEM 5-17

IDLE SPEED ADJUSTMENT (1997 - 98 MODELS) 5-17
INJECTION LINES 5-10
INJECTORS
   GENERAL INFORMATION 5-10
   REMOVAL & INSTALLATION 5-11
   TESTING (ALL MODELS) 5-12
DISC BRAKES
BRAKE CALIPER
   OVERHAUL 9-15
   REMOVAL & INSTALLATION 9-15
BRAKE DISC (ROTOR)
   INSPECTION 9-17
   REMOVAL & INSTALLATION 9-16
BRAKE PADS
   INSPECTION 9-15
   REMOVAL & INSTALLATION 9-12
DISTRIBUTOR IGNITION SYSTEM
CRANKSHAFT AND CAMSHAFT POSITION SENSORS 2-6
DIAGNOSIS AND TESTING
   HELPFUL TOOLS 2-2
   NO-SPARK CONDITION 2-3
   SECONDARY SPARK TEST 2-3
   SPARK DROP TEST 2-3
   SYSTEM INSPECTION 2-2
DISTRIBUTOR
   INSTALLATION 2-5
   REMOVAL 2-4
DISTRIBUTOR CAP/ROTOR
   REMOVAL & INSTALLATION 2-3
GENERAL INFORMATION 2-2
IGNITION COIL
   LOCATION 2-4
   TESTING 2-4
   TESTING IN PLACE 2-3
IGNITION COIL REMOVAL & INSTALLATION 2-4
DISTRIBUTORLESS IGNITION, 8.0L ENGINE
CRANKSHAFT AND CAMSHAFT POSITION SENSORS 2-8
GENERAL INFORMATION 2-7
IGNITION COIL PACK
   REMOVAL & INSTALLATION 2-8
   TESTING 2-7
DISTRIBUTORLESS IGNITION SYSTEM, 4.7L ENGINE
CRANKSHAFT AND CAMSHAFT POSITION SENSORS 2-7
DIAGNOSIS AND TESTING 2-7
GENERAL INFORMATION 2-6
IGNITION COIL 2-7
   REMOVAL & INSTALLATION 2-7
   TESTING 2-7
DRIVELINE
CENTER BEARING
   ADJUSTMENT 7-18
   REMOVAL & INSTALLATION 7-19

MASTER INDEX   IND-3

FRONT DRIVESHAFT AND U-JOINTS
   REMOVAL & INSTALLATION 7-14
   U-JOINT REPLACEMENT 7-15
REAR DRIVESHAFT AND U-JOINTS
   DRIVESHAFT BALANCING 7-17
   REMOVAL & INSTALLATION 7-16
   U-JOINT REPLACEMENT 7-17
TROUBLESHOOTING 7-14

DRUM BRAKES 9-17
BRAKE DRUMS
   INSPECTION 9-19
   REMOVAL & INSTALLATION 9-18
BRAKE SHOES
   ADJUSTMENTS 9-20
   INSPECTION 9-19
   REMOVAL & INSTALLATION 9-19
WHEEL CYLINDERS
   OVERHAUL 9-21
   REMOVAL & INSTALLATION 9-21

ELECTRONIC ENGINE CONTROLS
CAMSHAFT POSITION (CMP) SENSOR
   OPERATION 4-19
   REMOVAL & INSTALLATION 4-20
   TESTING 4-19
CRANKSHAFT POSITION (CKP) SENSOR
   OPERATION 4-21
   REMOVAL & INSTALLATION 4-21
   TESTING 4-21
ENGINE COOLANT TEMPERATURE (ECT) SENSOR
   OPERATION 4-15
   REMOVAL & INSTALLATION 4-15
   TESTING 4-15
IDLE AIR CONTROL MOTOR (IAC)
   OPERATION 4-13
   REMOVAL & INSTALLATION 4-14
   TESTING 4-14
INTAKE AIR TEMPERATURE (IAT) SENSOR
   OPERATION 4-16
   REMOVAL & INSTALLATION 4-16
   TESTING 4-16
MANIFOLD ABSOLUTE PRESSURE (MAP) SENSOR
   OPERATION 4-16
   REMOVAL & INSTALLATION 4-17
   TESTING 4-17
OXYGEN (O2) SENSOR
   OPERATION 4-12
   REMOVAL & INSTALLATION 4-13
   TESTING 4-13
POWERTRAIN CONTROL MODULE (PCM)
   OPERATION 4-12
   REMOVAL & INSTALLATION 4-12
THROTTLE POSITION SENSOR (TPS)
   OPERATION 4-18
   REMOVAL & INSTALLATION 4-18
   TESTING 4-18

EMISSION CONTROLS
AIR INJECTION SYSTEM
   COMPONENT TESTING 4-11
   OPERATION 4-11
   REMOVAL & INSTALLATION 4-11
CATALYTIC CONVERTER
   OPERATION 4-10
CRANKCASE VENTILATION SYSTEM
   COMPONENT TESTING 4-6
   OPERATION 4-6
EVAPORATIVE EMISSION CONTROLS
   COMPONENT TESTING 4-8
   OPERATION 4-6
   REMOVAL & INSTALLATION 4-9
EXHAUST GAS RECIRCULATION SYSTEM
   COMPONENT TESTING 4-10
   OPERATION 4-9
   REMOVAL & INSTALLATION 4-10
MALFUNCTION INDICATOR LAMP
   RESETTING 4-11
POSITIVE CRANKCASE
   COMPONENT TESTING 4-5
   OPERATION 4-4
   REMOVAL & INSTALLATION 4-6

ENGINE MECHANICAL 3-2
BELT-DRIVEN ENGINE FAN
   REMOVAL & INSTALLATION 3-29
CAMSHAFT BEARINGS AND LIFTERS
   INSPECTION 3-45
   REMOVAL & INSTALLATION 3-43
CRANKSHAFT DAMPER
   REMOVAL & INSTALLATION 3-38
CYLINDER HEAD
   REMOVAL & INSTALLATION 3-32
ELECTRIC ENGINE FAN (2.5L)
   REMOVAL & INSTALLATION 3-29
ENGINE, REMOVAL & INSTALLATION
   DIESEL ENGINE 3-19
   GASOLINE ENGINES (EXCEPT 4.7L) 3-18
EXHAUST MANIFOLD
   REMOVAL & INSTALLATION 3-27
FLYWHEEL/DRIVEPLATE
   REMOVAL & INSTALLATION 3-47
INTAKE MANIFOLD
   REMOVAL AND INSTALLATION 3-23
OIL PAN
   REMOVAL & INSTALLATION 3-35
OIL PUMP
   REMOVAL & INSTALLATION 3-36
RADIATOR
   REMOVAL & INSTALLATION 3-28
REAR MAIN SEAL
   REMOVAL & INSTALLATION 3-45
ROCKER ARMS
   REMOVAL & INSTALLATION 3-20

IND-4 MASTER INDEX

THERMOSTAT
    REMOVAL & INSTALLATION 3-22
TIMING CHAIN AND SPROCKETS
    INSPECTION 3-42
    REMOVAL & INSTALLATION 3-40
TIMING COVER AND SEAL (GASOLINE ENGINES)
    REMOVAL & INSTALLATION 3-39
TIMING GEAR COVER AND SEAL (DIESELS)
    REMOVAL & INSTALLATION 3-40
TURBOCHARGER
    REMOVAL & INSTALLATION 3-28
VALVE COVER
    REMOVAL & INSTALLATION 3-19
WATER PUMP
    REMOVAL & INSTALLATION 3-30

ENGINE RECONDITIONING
BUY OR REBUILD? 3-50
CYLINDER HEAD 3-52
    ASSEMBLY 3-58
    DISASSEMBLY 3-53
    INSPECTION 3-54
    REFINISHING & REPAIRING 3-57
DETERMINING ENGINE CONDITION 3-49
    COMPRESSION TEST 3-50
    OIL PRESSURE TEST 3-50
ENGINE BLOCK
    ASSEMBLY 3-61
    DISASSEMBLY 3-58
    GENERAL INFORMATION 3-58
    INSPECTION 3-59
    REFINISHING 3-61
ENGINE OVERHAUL TIPS
    CLEANING 3-51
    OVERHAUL TIPS 3-51
    REPAIRING DAMAGED THREADS 3-52
    TOOLS 3-51
ENGINE PREPARATION 3-52
ENGINE START-UP AND BREAK-IN
    BREAKING IT IN 3-64
    KEEP IT MAINTAINED 3-64
    STARTING THE ENGINE 3-63

ENTERTAINMENT SYSTEMS
RADIO RECEIVER/AMPLIFIER/TAPE PLAYER/CD PLAYER 6-12
    REMOVAL & INSTALLATION 6-12
SPEAKERS 6-13
    REMOVAL & INSTALLATION 6-13

EXHAUST SYSTEM 3-47
COMPONENT REPLACEMENT 3-48
SAFETY PRECAUTIONS 3-48

EXTERIOR
ANTENNA 10-4
    REPLACEMENT 10-4
DOORS
    ADJUSTMENT 10-2
    REMOVAL & INSTALLATION 10-2
FRONT FENDERS
    REMOVAL & INSTALLATION 10-6
GRILLE
    REMOVAL & INSTALLATION 10-3
HOOD
    ALIGNMENT 10-3
    REMOVAL & INSTALLATION 10-3
LIFTGATE
    ALIGNMENT 10-3
    REMOVAL & INSTALLATION 10-3
OUTSIDE MIRRORS
    REMOVAL & INSTALLATION 10-4
REAR FENDERS (DUAL WHEELS) 10-7
TAILGATE
    REMOVAL & INSTALLATION 10-3

FASTENERS MEASUREMENTS AND CONVERSIONS 1-5
BOLTS NUTS AND OTHER THREADED RETAINERS 1-5
STANDARD AND METRIC MEASUREMENTS 1-9
TORQUE 1-6
    TORQUE ANGLE METERS 1-9
    TORQUE WRENCHES 1-8

FIRING ORDERS 2-8

FLUIDS AND LUBRICANTS 1-36
AUTOMATIC TRANSMISSION 1-40
    DRAIN AND REFILL 1-40
    FLUID RECOMMENDATIONS 1-40
    LEVEL CHECK 1-40
BODY LUBRICATION AND MAINTENANCE 1-47
    CARE OF YOUR TRUCK 1-47
    HOOD LATCH AND HINGES 1-47
    TAILGATE HINGES 1-48
BRAKE MASTER CYLINDER 1-45
    FLUID RECOMMENDATIONS 1-45
    LEVEL CHECK 1-45
CHASSIS GREASING 1-46
    AUTOMATIC TRANSMISSION LINKAGE 1-47
    PARKING BRAKE LINKAGE 1-47
    STEERING LINKAGE 1-46
CLUTCH MASTER CYLINDER 1-45
COOLING SYSTEM 1-42
    FLUID RECOMMENDATIONS 1-43
    FLUSHING & CLEANING THE SYSTEM 1-43
    INSPECTION 1-43
    LEVEL CHECK 1-42
DRIVE AXLE (REAR AND/OR FRONT) 1-41
    DRAIN AND REFILL 1-42
    FLUID RECOMMENDATIONS 1-41
    LEVEL CHECK 1-42
ENGINE 1-37
    OIL & FILTER CHANGE 1-37
    OIL LEVEL CHECK 1-37

MASTER INDEX

FLUID DISPOSAL 1-36
FUEL AND ENGINE OIL RECOMMENDATIONS 1-36
    DIESEL ENGINES 1-36
    GASOLINE ENGINES 1-36
MANUAL TRANSMISSION 1-38
    DRAIN AND REFILL 1-39
    FLUID RECOMMENDATIONS 1-38
    LEVEL CHECK 1-39
POWER STEERING PUMP 1-46
    FLUID RECOMMENDATIONS 1-46
    LEVEL CHECK 1-46
TRANSFER CASE 1-41
    DRAIN AND REFILL 1-41
    FLUID RECOMMENDATIONS 1-41
    LEVEL CHECK 1-41
WHEEL BEARINGS 1-48
    REMOVAL REPACKING & INSTALLATION 1-48
WINDSHIELD WASHER RESERVOIR 1-44

FOUR WHEEL ANTILOCK BRAKE SYSTEM 9-29
BLEEDING THE ABS SYSTEM 9-32
    1997 RAM 9-32
    OTHER MODELS 9-32
DIAGNOSIS AND TESTING 9-30
EXCITER RING 9-32
    REMOVAL & INSTALLATION 9-32
SPEED SENSORS 9-30
    REMOVAL & INSTALLATION 9-30
TONE WHEEL 9-31
    REMOVAL & INSTALLATION 9-31

FRONT DRIVE AXLE 7-19
AXLE SHAFT BEARING AND SEAL
    REMOVAL & INSTALLATION 7-22
CV JOINTS
    CV JOINT BOOTS 7-20
    DISASSEMBLY & ASSEMBLY 7-20
FRONT AXLE HOUSING ASSEMBLY
    REMOVAL & INSTALLATION 7-24
FRONT DRIVESHAFT
    REMOVAL & INSTALLATION 7-19
PINION SEAL
    REMOVAL & INSTALLATION 7-23
TROUBLESHOOTING 7-19

FRONT SUSPENSION 8-4
COIL SPRINGS 8-5
    REMOVAL & INSTALLATION 8-5
FRONT HUB AND BEARING 8-11
    REMOVAL & INSTALLATION 8-11
LOWER BALL JOINT 8-8
    INSPECTION 8-8
    REMOVAL & INSTALLATION 8-9
LOWER SUSPENSION ARM 8-10
    REMOVAL & INSTALLATION 8-10
SHOCK ABSORBERS 8-6
    REMOVAL & INSTALLATION 8-6
    TESTING 8-8
STABILIZER BAR 8-9
    REMOVAL & INSTALLATION 8-9
STEERING KNUCKLE 8-11
    REMOVAL & INSTALLATION 8-11
TORSION BARS 8-6
    REMOVAL & INSTALLATION 8-6
    SUSPENSION HEIGHT ADJUSTMENT 8-6
UPPER BALL JOINT 8-8
    INSPECTION 8-8
    REMOVAL & INSTALLATION 8-8
UPPER SUSPENSION ARM 8-9
    REMOVAL & INSTALLATION 8-9
WHEEL ALIGNMENT 8-14
    INDEPENDENT FRONT SUSPENSION 8-14
    SOLID FRONT AXLE FRONT SUSPENSION 8-15

FUEL LINES AND FITTINGS
CONVENTIONAL TYPE FUEL FITTINGS
    REMOVAL & INSTALLATION 5-2
FUEL LINES AND HOSES 5-2
QUICK-CONNECT FUEL FITTINGS 5-2
    REMOVAL & INSTALLATION 5-2

FUEL TANK
TANK ASSEMBLY 5-17
    REMOVAL & INSTALLATION 5-17

FUSES
CARTRIDGE FUSE 6-27
    REMOVAL & INSTALLATION 6-27
CIRCUIT BREAKERS 6-27
    RESETTING AND/OR REPLACEMENT 6-27
FLASHERS 6-28
    REPLACEMENT 6-28

GASOLINE FUEL INJECTION SYSTEM
FUEL FILTER/PRESSURE REGULATOR ASSEMBLY 5-6
    REMOVAL & INSTALLATION 5-6
FUEL GAUGE SENDING UNIT 5-6
    REMOVAL & INSTALLATION 5-6
    TESTING 5-6
FUEL INJECTORS 5-10
    REMOVAL & INSTALLATION 5-10
    TESTING 5-10
FUEL PUMP 5-5
    REMOVAL & INSTALLATION 5-5
    TESTING 5-5
FUEL RAIL ASSEMBLY
    REMOVAL & INSTALLATION 5-7
GENERAL INFORMATION 5-4
RELIEVING FUEL SYSTEM PRESSURE
    BYPASSING THE PRESSURE TEST PORT 5-4
    USING THE PRESSURE TEST PORT 5-4
ROLLOVER VALVE
    REMOVAL & INSTALLATION 5-6

IND-6 MASTER INDEX

THROTTLE BODY
  REMOVAL & INSTALLATION 5-7
HEATING AND AIR CONDITIONING
AIR CONDITIONING COMPONENTS 6-10
BLOWER MOTOR
  REMOVAL & INSTALLATION 6-10
CONTROL PANEL
  REMOVAL & INSTALLATION 6-11
HEATER - A/C HOUSING
  REMOVAL & INSTALLATION 6-9
HEATER CORE 6-10
TEMPERATURE CONTROL CABLE 6-10
  REMOVAL & INSTALLATION 6-10
HOW TO USE THIS BOOK 1-2
AVOIDING THE MOST COMMON MISTAKES 1-2
AVOIDING TROUBLE 1-2
MAINTENANCE OR REPAIR? 1-2
SPECIAL TOOLS 1-4
TOOLS AND EQUIPMENT 1-2
WHERE TO BEGIN 1-2
JUMP STARTING A DEAD BATTERY
BATTERY 1-50
JACKING 1-52
JACKING PRECAUTIONS 1-52
JUMP STARTING PRECAUTIONS 1-50
JUMP STARTING PROCEDURE 1-50
  DUAL BATTERY DIESEL MODELS 1-51
  SINGLE BATTERY VEHICLES 1-50
INSTRUMENTS AND SWITCHES
FOG LAMP SWITCH
  REMOVAL & INSTALLATION 6-18
GAUGES & BULBS
  REMOVAL & INSTALLATION 6-17
HEADLIGHT SWITCH
  REMOVAL & INSTALLATION 6-18
INSTRUMENT CLUSTER
  REMOVAL & INSTALLATION 6-17
WINDSHIELD WIPER SWITCH
  IGNITION SWITCH 6-17
INTERIOR
DOOR GLASS AND REGULATOR
  REMOVAL & INSTALLATION 10-15
DOOR LATCHES
  REMOVAL & INSTALLATION 10-12
DOOR LOCKS
  REMOVAL & INSTALLATION 10-14
DOOR PANELS
  REMOVAL & INSTALLATION 10-10
ELECTRIC WINDOW MOTOR
  REMOVAL & INSTALLATION 10-16
FLOOR CONSOLE
  REMOVAL & INSTALLATION 10-9
INSIDE REAR VIEW MIRROR 10-18

REAR VIEW MIRROR SUPPORT BRACKET 10-18
  REPLACEMENT 10-18
INSTRUMENT PANEL
  REMOVAL & INSTALLATION 10-7
LIFTGATE LATCH
  REMOVAL & INSTALLATION 10-14
LIFTGATE LOCK
  REMOVAL & INSTALLATION 10-15
POWER SEAT MOTORS
  REMOVAL & INSTALLATION 10-19
SEATS
  REMOVAL & INSTALLATION 10-19
TAILGATE LATCH
  REMOVAL & INSTALLATION 10-13
WINDSHIELD AND FIXED GLASS
  REMOVAL & INSTALLATION 10-16
WINDSHIELD CHIP REPAIR 10-17
LIGHTING
FOG/DRIVING LIGHTS
  AIMING 6-25
  AIMING FACTORY FOG LIGHTS 6-23
  BULB REMOVAL & INSTALLATION 6-23
  FOG LAMP MODULE REMOVAL & INSTALLATION 6-24
  INSTALLING AFTERMARKET AUXILIARY LIGHTS 6-24
HEADLIGHTS
  AIMING THE HEADLIGHTS 6-19
  HEADLAMP MODULE REMOVAL & INSTALLATION 6-19
  REMOVAL & INSTALLATION 6-18
SIGNAL AND MARKER LIGHTS
  REMOVAL & INSTALLATION 6-20
MAINTENANCE INTERVAL CHARTS 1-52
MANUAL TRANSMISSION
BACK-UP LIGHT SWITCH
  REMOVAL & INSTALLATION 7-2
  TESTING 7-2
EXTENSION HOUSING SEAL
  REMOVAL & INSTALLATION 7-3
MANUAL TRANSMISSION ASSEMBLY
  REMOVAL & INSTALLATION 7-3
SHIFT HANDLE
  REMOVAL & INSTALLATION 7-2
UNDERSTANDING THE MANUAL TRANSMISSION 7-2
PARKING BRAKE
FRONT PARKING BRAKE CABLE
  REMOVAL & INSTALLATION 9-23
REAR PARKING BRAKE CABLES
  ADJUSTMENT 9-26
  REMOVAL & INSTALLATION 9-24
PDC & PCM LOCATIONS 2-2
REAR AXLE 7-25
AXLE SHAFT BEARING AND SEAL
  REMOVAL & INSTALLATION 7-26

MASTER INDEX IND-7

PINION SEAL
   REMOVAL & INSTALLATION 7-28
REAR AXLE HOUSING ASSEMBLY
   REMOVAL & INSTALLATION 7-29
TROUBLESHOOTING 7-25

REAR SUSPENSION 8-16
LEAF SPRINGS
   REMOVAL & INSTALLATION 8-16
REAR WHEEL BEARINGS
   REMOVAL & INSTALLATION 8-18
SHOCK ABSORBERS
   REMOVAL & INSTALLATION 8-17
   TESTING 8-17
STABILIZER BAR
   REMOVAL & INSTALLATION 8-17

REAR WHEEL ANTILOCK BRAKE SYSTEM
BLEEDING THE RWAL SYSTEM 9-29
DIAGNOSIS AND TESTING 9-28
EXCITER RING
   REMOVAL & INSTALLATION 9-29
FAULT CODES 9-27
GENERAL INFORMATION 9-27
RWAL VALVE
   REMOVAL & INSTALLATION 9-29
SPEED SENSOR
   TESTING (1997 MODELS) 9-28
   SPEED SENSOR REMOVAL & INSTALLATION 9-28

ROUTINE MAINTENANCE AND TUNE-UP 1-12
AIR CLEANER (ELEMENT)
   REMOVAL & INSTALLATION 1-13
AIR CONDITIONING SYSTEM
   PREVENTIVE MAINTENANCE 1-31
   SYSTEM SERVICE & REPAIR 1-31
BATTERY
   BATTERY FLUID 1-18
   CABLES 1-19
   CHARGING 1-19
   GENERAL MAINTENANCE 1-18
   PRECAUTIONS 1-18
BELTS
   ADJUSTMENT 1-21
   INSPECTION 1-20
   REMOVAL & INSTALLATION 1-21
CV-BOOTS
   INSPECTION 1-23
DISTRIBUTOR CAP AND ROTOR
   INSPECTION 1-27
   REMOVAL & INSTALLATION 1-26
EVAPORATIVE CANISTER 1-17
   SERVICING 1-18
FUEL FILTER (GASOLINE ENGINES) 1-15
FUEL FILTER/WATER SEPARATOR (DIESEL ENGINES)
   REMOVAL & INSTALLATION 1-15
HOSES
   INSPECTION 1-23
   REMOVAL & INSTALLATION 1-23
IDLE SPEED AND MIXTURE ADJUSTMENTS
   DIESEL ENGINE 1-30
   GASOLINE ENGINES 1-30
IGNITION TIMING (GASOLINE ENGINES)
   GENERAL INFORMATION 1-27
   INSPECTION & ADJUSTMENT 1-28
PCV VALVE 1-16
   REMOVAL & INSTALLATION 1-17
REPLACEMENT 1-20
SPARK PLUG WIRES
   REMOVAL & INSTALLATION 1-26
   TESTING 1-26
SPARK PLUGS
   INSPECTION & GAPPING 1-25
   RECOMMENDED SPARK PLUGS 1-24
   REMOVAL & INSTALLATION 1-24
   SPARK PLUG HEAT RANGE 1-24
SYSTEM INSPECTION 1-31
TIRES AND WHEELS 1-33
TIRES AND WHEELS CARE OF SPECIAL WHEELS 1-35
TIRES AND WHEELS INFLATION & INSPECTION 1-35
TIRES AND WHEELS TIRE DESIGN 1-34
TIRES AND WHEELS TIRE ROTATION 1-33
TIRES AND WHEELS TIRE STORAGE 1-35
VALVE LASH 1-29
WINDSHIELD WIPER (ELEMENTS) 1-32
   ELEMENT (REFILL) CARE & REPLACEMENT 1-32

SENDING UNITS AND SENSORS
BATTERY TEMPERATURE SENSOR 2-17
   REMOVAL AND INSTALLATION 2-17
COOLANT TEMPERATURE GAUGE SENDING UNIT
   REMOVAL & INSTALLATION 2-15
   TESTING 2-16
ELECTRIC FAN
   TESTING 2-17
OIL PRESSURE SENSOR
   REMOVAL & INSTALLATION 2-16
   TESTING 2-16

SERIAL NUMBER IDENTIFICATION 1-10
DRIVE AXLE 1-11
ENGINE 1-10
EQUIPMENT IDENTIFICATION 1-10
TRANSFER CASE 1-12
TRANSMISSION 1-10
VEHICLE 1-10
VEHICLE SAFETY CERTIFICATION LABEL 1-10

SERVICING YOUR VEHICLE 1-4
DO'S 1-4
DON'TS 1-5
SAFELY 1-4

IND-8 MASTER INDEX

STARTING SYSTEM
GENERAL INFORMATION
   PRECAUTIONS 2-12
   TESTING 2-12
STARTER
   PRECAUTIONS 2-13
   RELAY 2-14
   SOLENOID
      TESTING 2-13
   STARTER MOTOR REMOVAL & INSTALLATION 2-14

STEERING 8-18
DAKOTA/DURANGO STEERING LINKAGE 8-23
IGNITION SWITCH & LOCK CYLINDER
   REMOVAL & INSTALLATION 8-19
POWER STEERING GEAR
   REMOVAL & INSTALLATION 8-24
POWER STEERING PUMP
   INITIAL OPERATION 8-25
   REMOVAL & INSTALLATION 8-25
RACK & PINION STEERING
   REMOVAL & INSTALLATION 8-23
RAM STEERING LINKAGE
   REMOVAL & INSTALLATION 8-21
STEERING WHEEL
   REMOVAL & INSTALLATION 8-18
TURN SIGNAL (COMBINATION) SWITCH
   REMOVAL & INSTALLATION 8-19

TRAILER TOWING
COOLING
   ENGINE 1-49
   TRANSMISSION 1-49
GENERAL RECOMMENDATIONS 1-49
HANDLING A TRAILER 1-50
HITCH (TONGUE) WEIGHT 1-49
MANUFACTURER'S RECOMMENDATIONS
   SAFETY PRECAUTIONS 1-50
   TOWING WITH FRONT END LIFTED 1-50
   TOWING WITH REAR END LIFTED 1-50
TOWING THE VEHICLE 1-50
TRAILER WEIGHT 1-49

TRAILER WIRING 6-25

TRANSFER CASE
FRONT OUTPUT SHAFT SEAL
   REMOVAL & INSTALLATION 7-13
REAR OUTPUT SHAFT SEAL
   REMOVAL & INSTALLATION 7-12
SHIFT LINKAGE
   REMOVAL & INSTALLATION 7-12
TRANSFER CASE ASSEMBLY
   REMOVAL & INSTALLATION 7-13

TROUBLE CODES
DIAGNOSTIC CONNECTOR 4-22
GENERAL INFORMATION 4-21
   MALFUNCTION INDICATOR LAMP (MIL) 4-22
READING CODES 4-22

UNDERSTANDING AND TROUBLESHOOTING ELECTRICAL SYSTEMS
BASIC ELECTRICAL THEORY 6-2
   HOW DOES ELECTRICITY WORK: THE WATER ANALOGY 6-2
   OHM'S LAW 6-2
ELECTRICAL COMPONENTS 6-2
   CONNECTORS 6-4
   GROUND 6-3
   LOAD 6-3
   POWER SOURCE 6-2
   PROTECTIVE DEVICES 6-3
   SWITCHES & RELAYS 6-3
   WIRING & HARNESSES 6-4
TEST EQUIPMENT
   JUMPER WIRES 6-4
   MULTIMETERS 6-5
   TEST LIGHTS 6-5
TESTING
   OPEN CIRCUITS 6-6
   RESISTANCE 6-6
   SHORT CIRCUITS 6-6
   VOLTAGE 6-6
   VOLTAGE DROP 6-6
TROUBLESHOOTING ELECTRICAL SYSTEMS 6-5
WIRE AND CONNECTOR REPAIR 6-7

VACUUM DIAGRAMS 4-27

WHEELS
INSPECTION 8-3
REMOVAL & INSTALLATION 8-2
SPECIAL NOTES 8-2
WHEEL LUG STUDS
   REMOVAL & INSTALLATION 8-3
WHEELS 8-2

WINDSHIELD WIPER SYSTEM
WASHER FLUID LEVEL SENSOR
   REMOVAL & INSTALLATION 6-15
WINDSHIELD WASHER PUMP
   REMOVAL & INSTALLATION 6-15
WINDSHIELD WASHER RESERVOIR
   REMOVAL & INSTALLATION 6-16
WINDSHIELD WIPER MOTOR
   REMOVAL & INSTALLATION 6-14
WIPER ARMS
   REMOVAL & INSTALLATION 6-14
WIPER BLADES
   REMOVAL & INSTALLATION 6-14

WIRING DIAGRAMS 6-29